ARCHITECTURAL ACOUSTICS HANDBOOK

Ning Xiang
Editor

ISBN-13: 978-1-60427-004-4

Printed and bound in the U.S.A. Printed on acid-free paper.

10 9 8 7 6 5 4 3 2 1

Library of Congress Cataloging-in-Publication Data

Xiang, Ning, editor.
Architectural acoustics handbook / edited by Ning Xiang.
Plantation, FL : J. Ross Publishing, 2017. | Includes
bibliographical references.
LCCN 2016052914 | ISBN 9781604270044 (hardcover : alk. paper)
LCSH: Architectural acoustics—Handbooks, manuals, etc.
LCC NA2800 .A684 2017
DDC 690/.2—dc23
LC record available at https://lccn.loc.gov/2016052914

Direct all inquiries to J. Ross Publishing, Inc., 300 S. Pine Island Rd., Suite 305, Plantation, FL 33324.

Phone: (954) 727-9333
Fax: (561) 892-0700
Web: www.jrosspub.com

Cover Photographs (top down):
Wild Beast Music Pavilion, CalArts, Valencia, CA
Photographer: Tom Bonner Photography

Granada Theatre Restoration, Santa Barbara, CA
Photographer: Lawrence Anderson/ESTO

Zipper Concert Hall, Colburn School for Performing Arts, Los Angeles, CA
Photographer: Foaad Farah Images

ACOUSTICS: INFORMATION AND COMMUNICATION SERIES

Ning Xiang, Editor-in-Chief

Worship Space Acoustics
by Mendel Kleiner, David Lloyd Klepper, Rendell R. Torres

Acoustics and Audio Technology, Third Edition
by Mendel Kleiner

Head-Related Transfer Function and Virtual Auditory Display, Second Edition
by Bosun Xie

Architectural Acoustics Handbook
by Ning Xiang

Contents

PART II Architectural Acoustics Practice

Preface

The *Architectural Acoustics Handbook* attempts to summarize the present state of knowledge evolved from both the research and consulting communities in this important field. To this end, the handbook contains two Parts; Part I: *Architectural Acoustics Essentials* and Part II: *Architectural Acoustics Practice*—contributed by authorities in various subfields, though it is not always possible to establish a clean division between the two. It is meant to serve as a handy reference and a useful resource for research scientists, undergraduate and graduate students studying architectural acoustics, and for acoustic consultants and engineers who are professionally engaged in architectural acoustics practice.

As such, this volume aims to provide for audiences who are interested and engaged in frontier research with the latest progress and findings in vibrant research fields that were otherwise treated largely in specific acoustical journals. The topics and subfields covered include geometrical and wave-based room-acoustic modeling methods (Chapters 1 and 2), acoustics in long and coupled spaces (Chapters 3 and 4), measurement methods for architectural acoustics (Chapter 5), advanced room-acoustic energy decay analysis (Chapter 6), sound insulation in buildings (Chapter 7), auditory perception and auralization in rooms (Chapters 8 and 9), room-related sound representations using loudspeakers (Chapter 10) and environmental acoustics around the built environment (Chapter 11). To also serve architectural acoustics design practice, Part II of this volume provides guidance for the practical design of sound systems (Chapter 12), and heating, ventilating, and air conditioning systems in buildings (Chapter 13), as well as the acoustical design and renovations of various types of venues, including worship spaces (Chapter 14), and music performance halls, dramatic arts, and music instruction spaces (Chapter 15). To keep the book to an appropriate size, the authors were given a page limit. Most of the chapters in Part I were kept within this limit, while some chapters covering design practice in Part II were allotted more pages.

Recognizing that no single individual possesses all the expert knowledge on such a diverse field as architectural acoustics, the editor of this book wishes to extend his sincere appreciation to all the chapter authors, who alongside their professional work load, have dedicated themselves to the laborious task of presenting their respective fields of expertise in an extensive, yet compact form. We are particularly indebted to Tim Pletscher and Stephen Buda at J. Ross Publishing for their effective help and guidance in the production of this book.

Needless to say, this effort spans many years. Two esteemed chapter authors—who were delightful colleagues and highly respected acoustical consultants—passed before seeing this work published. Ronald McKay, right after delivering his chapter on *Music Performance*

Spaces, passed in December 2011.[1] Ewart (Red) Wetherill, who submitted his entirely completed chapter on *Acoustics in Worship Spaces* on September 1, 2013, after two rounds of thorough revisions, passed in November 2015. This book is dedicated to the memory of our esteemed colleagues, Ronald L. McKay (1932–2011) and Ewart A. Wetherill (1928–2015).

Ning Xiang, Troy, July 2016

[1]After his passing, a number of partially completed illustrations were finished with the help of Yiqiao Hou.

About the Editor

Ning Xiang, Professor of acoustics, director of the Graduate Program in Architectural Acoustics at Rensselaer Polytechnic Institute, is a Fellow of the Acoustical Society of America (ASA) and a Fellow of the Institute of Acoustics, United Kingdom. He has over 300 publications including peer-reviewed journal papers, books and book chapters, and conference proceeding papers. In 2014, he received the Wallace Clement Sabine Medal from the ASA. He served the Chair of the Technical Committee on Signal Processing in Acoustics of the ASA from 2012 to 2015, and he has been serving as an Associate Editor of the *Journal of the Acoustical Society of America* (JASA) for over 10 years. He is also an Editorial Board member of ASA-Press (Springer books).

List of Contributors

Wolfgang Ahnert, Acoustic Design Ahnert and Ahnert Feistel Media Group, Berlin, Germany
Jens P. Blauert, Institute of Communication Acoustics, Ruhr-University, Bochum, Germany
Jonathan Botts, Department of Media Technology, Aalto University School of Science and Technology, Espoo, Finland
Jonas Braasch, Graduate Program in Architectural Acoustics, Center of Cognition, Communication and Culture, Rensselaer Polytechnic Institute, Troy, New York, USA
David A. Conant, McKay Conant Hoover, Inc., Westlake Village, CA, USA
Stefan Feistel, Ahnert Feistel Media Group, Berlin, Germany
K. Anthony Hoover, McKay Conant Hoover, Inc., Westlake Village, CA, USA
Carl Hopkins, School of Architecture, University of Liverpool, United Kingdom
Jian Kang, School of Architecture, University of Sheffield, United Kingdom
Ronald L. McKay, McKay Conant Hoover, Inc., Westlake Village, CA, USA
Rudolf Rabenstein, Multimedia Communications and Signal Processing, Friedrich-Alexander-University Erlangen-Nürnberg, Germany
Lauri Savioja, Department of Media Technology, Aalto University School of Science and Technology, Espoo, Finland
Douglas H. Sturz, AcenTech Inc. Cambridge, MA, USA
U. Peter Svensson, Acoustics Research Centre, Department of Electronics and Telecommunications, Norwegian University of Science and Technology, Trondheim, Norway
Samuel Siltanen, Department of Media Technology, Aalto University School of Science and Technology, Espoo, Finland
Michael Vorländer, Institute of Technical Acoustics, RWTH Aachen University, Germany
Ewart A. Wetherill, Acoustical Consultant, Alameda, CA, USA
Ning Xiang, Graduate Program in Architectural Acoustics, Rensselaer Polytechnic Institute, Troy, New York, USA

BIOGRAPHY OF CONTRIBUTORS

Wolfgang Ahnert

CEO of the Acoustic Design Company ADA, Berlin, Germany, Dr. Ahnert has been a Professor at the Film University Babelsberg FUB in Potsdam-Babelsberg, Germany and has been doing research in electroacoustics, sound reinforcement systems, and architectural acoustics since 1970. He has over 100 publications in national and international journals and proceedings, and has published 8 books; some of them translated into Russian, English, and Chinese. Some of his academic distinctions include: Fellow of the Acoustical Society of America, Fellow of the Audio Engineering Society, Fellow of the Institute of Acoustics (United Kingdom) and recipient of the Peter Barnett Award by the Institute, and the honorary title of "Foreign Professor" by the Lomonossov University, Moscow.

Jens Blauert

Jen Blauert, Dr.-Ing., Dr. Tech. h. c., AES and ASA fellow and medalist of AES (gold) and ASA (silver), is emeritus professor at the Institute of Communication Acoustics of Ruhr-University, Bochum, Germany. He is also a distinguished visiting professor at the Rensselaer Polytechnic Institute, Troy, NY—an adjunct to their architectural-acoustics program. Jens Blauert is cofounder and was chairman of the board of the European Acoustics Association, EAA, and president of the German Acoustical Society, DEGA. His career spans 35 years as a chartered acoustical consultant in architectural acoustics, electroacoustics, binaural technology, speech technology, and sound-quality assessment.

Jonathan Botts

Jonathan Botts received a B.S. degree in physics and mathematics from Drake University in 2008. He received M.S. and Ph.D. degrees from Rensselaer Polytechnic Institute in Architectural Acoustics in 2009 and 2012, respectively. From 2012–2014, he was a postdoctoral researcher with Aalto University and from 2014–2015 with Rensselaer Polytechnic Institute. His research interests include numerical vibro-acoustic modeling and data analysis.

Jonas Braasch

Jonas Braasch is a psychoacoustician, aural architect, and experimental musician. His research work focuses on functional models of the auditory system, large-scale immersive and interactive virtual reality systems, and intelligent music systems. Dr. Braasch received a Master's Degree in Physics from the Technical University of Dortmund in 1998, and doctoral degrees from the University of Bochum in Electrical Engineering and Information Technology in 2001 and Musicology in 2004. Afterward, he worked as an Assistant Professor in McGill University's Sound Recording Program before joining Rensselaer Polytechnic Institute in 2006, where he is now Associate Professor in the School of Architecture and Director of the Center for Cognition, Communication, and Culture.

David A. Conant

David Conant, FASA, is a generalist in architectural acoustics with nearly 40 years' experience across virtually all building types. His concentration has been in fine and performing arts and higher education, with such projects extending from Western Europe across the USA to the Far East. He earned his B.S. in Physics from Union College, M.A. and in Geology from Columbia University, and his B. Arch. and M. Arch. from Rensselaer Polytechnic Institute. Notable projects include the Guggenheim Museum in Bilbao, Spain, MIT's Stata Center,

Los Angeles' Valley Performing Arts Center, the Mesa Arts Center (AZ), and multiple renovations of historic theaters.

Stefan Feistel

Stefan Feistel studied physics at the Humboldt University, Berlin, Germany, and received a Master's degree in theoretical physics in 2004. He received his Ph.D. on computational modeling of sound systems from the department of technical acoustics at the RWTH Aachen University in 2014. Dr. Feistel authored or coauthored more than 70 papers focusing on software projects and the related mathematical, numerical, and experimental background studies. The JAES article on Methods and Limitations of Line Source Simulation was distinguished with the AES Publications Award 2010. Dr. Stefan Feistel is the author of the book *Modeling the Radiation of Modern Sound Systems in High Resolution*, and a coauthor of the books *Messtechnik der Akustik*, edited by M. Möser, and *Handbook for Sound Engineers*, edited by G. Ballou.

K. Anthony Hoover

Tony Hoover has consulted on over 2,000 architectural acoustical projects throughout the U.S. and abroad. He earned his B.A. from Notre Dame and his M.S. in Acoustics from Penn State. He has served in numerous leadership positions, including President—National Council of Acoustical Consultants, and Chair—ASA Technical Committee on Architectural Acoustics. He has lectured widely, chaired numerous technical sessions, and was Assistant Professor at Berklee College of Music and Boston Architectural Center. Music education facility projects include various Berklee renovations; Bose World Headquarters; Universidad Americas, Puebla, Mexico; Tufts Granoff, Boston, MA; and Pepperdine Ahmanson, Malibu, CA.

Carl Hopkins

Carl Hopkins is a Professor in Acoustics at the University of Liverpool in the United Kingdom where he is Head of the Acoustics Research Unit within the School of Architecture. He is also a Fellow of the Institute of Acoustics. His research is primarily concerned with the prediction and measurement of sound and structure-borne sound in the built environment. He is involved in European and International Standardization groups on building acoustics as a convenor or member of working groups that draw up and revise standards, and is chairman of the British Standards committee on building acoustics.

Jian Kang

Jian Kang obtained his first degree and MSc from Tsinghua University and his Ph.D. from the University of Cambridge. He has been Professor of Acoustics at the School of Architecture, University of Sheffield, since 2003. Before joining Sheffield, he worked as a senior research associate at the Martin Centre, University of Cambridge, and as an A. v. Humboldt Fellow at the Fraunhofer Institute of Building Physics in Germany. His main research area is architectural and environmental acoustics. He has published three books, more than 200 refereed journal papers and book chapters, and more than 400 conference papers.

Ronald McKay

Ronald McKay graduated from the Massachusetts Institute of Technology with a B.A. (1954), and an M.A. (1958). Mr. McKay had nearly fifty years of consulting experience covering the entire field of architectural acoustics and noise control and was responsible for 1,000 acoustics projects, from Chicago's 100-story John Hancock Center to the highly regarded Ambassador and Royce Hall auditoria in Los Angeles. He was renowned for his work in performance, rehearsal, recording, and teaching facilities for music and drama, and was awarded the AIA's coveted Institute Honor for Collaborative Achievement. He taught architectural acoustics at several universities, created a television series and textbook on advanced architectural acoustics, and was a popular lecturer at scientific and engineering societies and institutions. He was a Fellow of the Acoustical Society of America. Ronald McKay passed away in 2011.

Rudolf Rabenstein

Rudolf Rabenstein studied Electrical Engineering at the University of Erlangen-Nuremberg, Germany, and at the University of Colorado at Boulder, USA. He received the degrees "Doktor-Ingenieur" in electrical engineering and "Habilitation" in signal processing from the University of Erlangen-Nuremberg, Germany in 1991 and 1996. He worked with the Physics Department of the University of Siegen, Germany, and as a Professor at the University of Erlangen-Nuremberg. His research interests are in the fields of multidimensional systems theory and multimedia signal processing.

Lauri Savioja

Lauri Savioja is a professor at the Department of Computer Science, Aalto University, Finland. He received the degree of Doctor of Science in Technology from the Helsinki University of Technology, Espoo, Finland, in 1999. His research interests include room acoustics, virtual reality, and parallel computation. Prof. Savioja is a fellow of the Audio Engineering Society (AES), a senior member of the IEEE, and a life member of the Acoustical Society of Finland.

Douglas H. Sturz

Douglas H. Sturz is the principal consultant at Acentech, Cambridge, MA. He obtained his Bachelor of Architectural Engineering from Pennsylvania State University. During his professional career, Doug has been engaged in a variety of projects involving mechanical system noise control, vibration isolation, noise control, sound isolation, and room acoustics. Having been principally involved with the design of over a hundred science/laboratory facilities, the majority of which include noise and/or vibration sensitive equipment, this type of building is one of Doug's specialties. He also consults on projects where community noise due to mechanical systems is a concern and works to reduce noise in order to meet community standards. Doug has taught Architectural Acoustics at the Boston Architectural Center. He also lectures on acoustics, noise control, and vibration control to professional organizations. He has co-authored two chapters in the *Handbook of Acoustical Measurements and Noise Control—Third Edition*, edited by Cyril Harris.

U. Peter Svensson

U. Peter Svensson has been a Professor of Electroacoustics at the Norwegian University of Science and Technology, Trondheim, Norway, since 1999. He obtained a Ph.D. from Chalmers University of Technology, Gothenburg, Sweden in 1994. His research has dealt with auralization, especially computational methods involving diffraction modeling and loudspeaker reproduction techniques. He has also worked on beamforming techniques for microphone arrays, measurement techniques, reverberation enhancement, and interaction over Internet/video conferencing. He has been on the boards of the acoustical societies of Sweden and Norway and the European Acoustics Association.

Samuel Siltanen

Dr. Samuel Siltanen has worked as a researcher in Aalto University, Espoo, Finland, since 2005. He has a background in computer science, and his interests include efficient computational methods for room acoustics modeling. The title of his doctoral thesis was "Efficient Physically-Based Room-Acoustics Modeling and Auralization". More recently, he led an Academy of Finland-funded project with the goal of finding algorithms for automatic optimization of room acoustics.

Michael Vorländer

Michael Vorländer is a professor at RWTH Aachen University. After a university education in physics and a doctoral degree with a thesis in room acoustical computer simulation, he worked in various fields of acoustics. His first research activities were focused on psychoacoustics, electroacoustics, and on room and building acoustics. Since 1996 he has been the Director of the Institute of Technical Acoustics, ITA, at RWTH Aachen University. He was President of the European Acoustics Association, EAA, and of the International Commission for Acoustics, ICA. The research focus of ITA is auralization and acoustic virtual reality in its various applications in psychoacoustics, architectural acoustics, automotive, and noise control.

Ewart A. Wetherill

Red Wetherill had been a licensed architect and professor of architecture in both the U.S. and Canada, and since 1960 worked as an acoustical consultant with firms in Massachusetts and California. For the past decade, he had worked as an independent consultant, specializing in spaces for worship and music performance. In retirement, he wrote on ways to enhance hearing conditions in buildings by raising the level of understanding between designers and builders. Ewart A. Wetherill passed away in 2015.

Ning Xiang

Ning Xiang is a professor of acoustics and director of the Graduate Program in Architectural Acoustics at Rensselaer Polytechnic Institute. He is a Fellow of the Acoustical Society of America (ASA) and a Fellow of the Institute of Acoustics, United Kingdom. He has over 300 publications including peer-reviewed journal papers, books and book chapters, along with conference proceeding papers. In 2014, he received the Wallace Clement Sabine Medal from the ASA. He served as the Chair of the Technical Committee on Signal Processing in Acoustics of the ASA from 2012 to 2015, and he has been serving as an Associate Editor of the *Journal of the Acoustical Society of America* (JASA) for over 10 years. He is also an Editorial Board member of ASA-Press (Springer books).

This book has free material available for download from the Web Added Value™ resource center at *www.jrosspub.com*

At J. Ross Publishing we are committed to providing today's professional with practical, hands-on tools that enhance the learning experience and give readers an opportunity to apply what they have learned. That is why we offer free ancillary materials available for download on this book and all participating Web Added Value™ publications. These online resources may include interactive versions of material that appears in the book or supplemental templates, worksheets, models, plans, case studies, proposals, spreadsheets and assessment tools, among other things. Whenever you see the WAV™ symbol in any of our publications, it means bonus materials accompany the book and are available from the Web Added Value Download Resource Center at www.jrosspub.com.

Downloads for *Architectural Acoustics Handbook* include various animations and Powerpoint presentations to reinforce material found in the book.

1

Computational Modeling of Room Acoustics I: Wave-Based Modeling

U. Peter Svensson, Acoustics Research Centre, Department of Electronics and Telecommunications, Norwegian University of Science and Technology, Trondheim, Norway

Jonathan Botts and **Lauri Savioja**, Department of Media Technology, Aalto University School of Science, Espoo, Finland

1.1 ROOM ACOUSTIC MODELING

Room acoustics offer challenging problems for computational and numerical modeling. The geometry of the problem can be very large relative to wavelengths that span many orders of magnitude. At the same time, requirements for precision and accuracy might be high if the computed results are to be used for auralization or other evaluation of perceived quality.

The result is a situation where both more physically accurate wave-based methods and faster, but more approximate, geometrical methods might be necessary to cover the wide frequency range of interest with adequate accuracy. Physically motivated wave-based methods like the *boundary element method* (BEM), the *finite element method* (FEM), the *finite difference method* (FDM), and many other related variants are both computationally feasible and relevant at low frequencies or in small geometries. Analytical solutions are also available for simplified room geometries, for rooms composed of simple subdomains, and to augment partial solutions from other numerical methods.

The limited resolution of our hearing makes modeling fine details at higher frequencies less important, which also implies that these accurate but computationally costly methods might give unnecessarily precise results at high frequencies. (*Note: they are potentially precise but the input data is not available with the required precision.*) Other chapters demonstrate that different computational methods have been developed for different problems and scenarios,[1, 2] and the importance of auralization[3] and the auditory system must also be kept in mind. This chapter outlines the methods that are relevant and well-suited to modeling physical wave mechanics for low frequencies and small geometries, where effects of wave physics are important in order to get accurate results.

1.2 ANALYTICAL SOLUTIONS

Explicit analytical solutions to the wave equation without medium losses are available for a few geometries and types of boundary conditions, and in room acoustics where the canonical room shape is parallelepiped, as illustrated in Figure 1.1. Other potentially useful geometries often correspond to common, orthogonal coordinate systems, like the cylinder (including wedges) and the sphere (including hemispheres). These are not presented here but are available.[4] Section 2.3 also demonstrates how simplified geometries might be combined to represent more general structures. The differential equation governing linear acoustics is the second-order wave equation:

$$\nabla^2 p(\mathbf{x},t) - \frac{1}{c^2}\frac{\partial^2 p(\mathbf{x},t)}{\partial t^2} = q_S(\mathbf{x},t), \tag{1.1}$$

where c is the speed of sound, and $p(x, t)$ is the sound pressure field. The quantity $q_S(\mathbf{x},t)$ on the right-hand side is a source term, which might, e.g., be a Dirac function of space to indicate a point source. If we consider single-frequency sources, that is, sources with a time dependence of the form, $q(\mathbf{x},t) = q_S(\mathbf{x})e^{j\omega t}$, then the partial differential equation (1.1) reduces to the Helmholtz equation:

$$\nabla^2 p(\mathbf{x}) + \left(\frac{\omega}{c}\right)^2 p(\mathbf{x}) = q_S(\mathbf{x}), \tag{1.2}$$

where the function $p(\mathbf{x})$ and the source function $q_S(\mathbf{x})$ will depend on (angular) frequency ω. If the geometry is one of the few canonical shapes,[5] separation of variables can be applied, and explicit solutions can be written as a classical modal summation. Furthermore, if the study is restricted to point sources (located in $\mathbf{x}_S$), then the solution can be written in the general form:

$$p(\mathbf{x}) = \frac{j\omega U_0 \rho_0 c^2}{V} \sum_{m,n,q\in[0,\infty]} \Lambda_{mnq} \frac{\Phi_{mnq}(\mathbf{x}_S)\Phi_{mnq}(\mathbf{x})}{\omega^2 - \omega_{mnq}^2}, \tag{1.3}$$

where the summation is over all combinations of integer values m, n, q; $\Phi(\mathbf{x})$ are the so-called mode functions which depend on geometry and boundary conditions (BC); U_0 is the volume velocity amplitude of the point source in $\mathbf{x}_S$; ρ_0 is the density of air; V is the room volume;

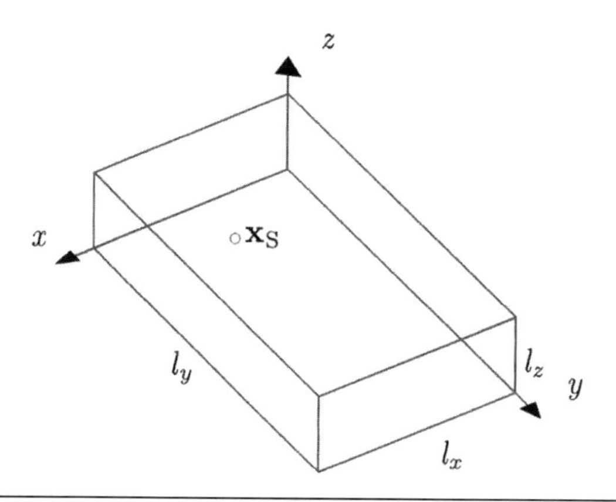

Figure 1.1 The parallelepipedic room for which an analytical solution is available

ω_{mnq} are the so-called eigenvalues; and Λ_{mnq} is a mode number normalization factor: 2 if two of m, n, q are zero, 4 if one of m, n, q is zero, and 8 if none of m, n, q is zero. If $U_0 = 1$, then this solution corresponds to a transfer function in space—a Green's function—and the solution for an extended source in space can be produced by convolving the Green's function from Eq. (1.3) with the source distribution function $q_S(\mathbf{x})$.[5] The harmonic time-dependence, $e^{j\omega t}$, is left out here and in all subsequent constant-frequency expressions. It could be noted that this solution corresponds to an ideal, lossless situation, which would reach infinite amplitude if the source frequency was chosen as one of the eigenvalues ω_{mnq}. The common solution to model small amounts of losses which are evenly distributed across the walls, is to introduce complex eigenvalues:

$$\underline{\omega}_{mnq} = \omega_{mnq} + j\delta_{mnq}, \tag{1.4}$$

where the underline indicates a complex value, and δ_{mnq} is a loss factor, which is related to the reverberation time T_{60} via $\delta = 3\ln 10 / T_{60}$.

By assuming a distributed loss, the mode functions, $\Phi(\mathbf{x})$, will be very similar to those for a lossless case. This assumption requires that losses are small, that is $\delta_{mnq} << \omega_{mnq}$, which is usually fulfilled in room acoustical cases. This small-loss assumption leads to the modal sum with losses:[6]

$$p(\mathbf{x}) = \frac{j\omega U_0 \rho_0 c^2}{V} \sum_{m,n,q} \Lambda_{mnq} \frac{\Phi_{mnq}(\mathbf{x}_S)\Phi_{mnq}(\mathbf{x})}{\omega^2 - \omega_{mnq}^2 - j2\delta_{mnq}\omega_{mnq}}. \tag{1.5}$$

If the eigenfunctions and eigenvalues can be computed for the geometry and boundary conditions at hand, this spectral solution can be used to compute the sound pressure amplitude for a given source frequency ω.

The expression in Eq. (1.5) involves an infinite summation over three indices. At low frequencies, where the modal density is low, a single term might dominate the sum, particularly near the eigenvalues ω_{mnq}. Above the so-called Schröder frequency, $f_{Sch.}$, however, the modal density is so high that there are large numbers of terms of similar amplitudes at any given source frequency ω. The value of this important frequency is:[7]

$$f_{Sch.} = 2000\sqrt{\frac{T_{60}}{V}}, \tag{1.6}$$

where the numerical constant obviously has the unit $(m/s)^{3/2}$. Also below the Schröder frequency, for frequencies between eigenvalues, a large number of terms might have significant amplitudes and consequently, the sum might converge very slowly. The amplitude of higher-order terms falls off as $1/\omega_{mnq}^2$, but the number of higher-order terms is large thanks to the triple summation. A numerical example in Section 3.1 illustrates this effect. One demonstration of the slow convergence is the case when the receiver position is placed exactly at the position of the point source. This case is expected to give an infinite amplitude as a result, since the free-field (direct sound) singularity at the point source location should become imminent. But, the form in Eq. (1.5) does not seem to indicate that $\mathbf{x} = \mathbf{x}_S$ leads to any singularity. The explanation is that when $\mathbf{x} = \mathbf{x}_S$, then $\Phi_{mnq}(\mathbf{x}_S)\Phi_{mnq}(\mathbf{x})$ is always positive and consequently, the summation will diverge—for other cases, that mode function product will have alternating signs, rendering the sum convergent, albeit slowly.

Time-domain expressions, that is, impulse responses, can be found via an inverse Fourier transform of the result of Eq. (1.5), or via explicit time-domain modal summation forms.[8]

1.2.1 Parallelepipedic (Shoebox) Room

As previously stated, a small number of canonical shapes are *analytically* solvable by separation of variables. In practice, many rooms and buildings are essentially rectilinear in shape, so the parallelepipedic *shoebox* room is an important representative example in room acoustics. The most important boundary condition (BC) to study is the Neumann BC, formulated as:

$$\left.\frac{\partial p}{\partial n}\right|_{\text{at surface}} = 0 \Rightarrow v_n = 0, \tag{1.7}$$

which corresponds to a perfectly rigid wall with an absorption coefficient of zero. More realistic cases are discussed below, but as mentioned above, a common technique for introducing (small) losses is to maintain a lossless BC, while introducing a modal loss factor δ_{mnq}. For the parallelepipedical room in Figure 1.1, with a Neumann BC on all six walls, the modal function set (the so-called eigenfunctions) has the form:[6]

$$\Phi_{mnq}(x) = \cos\frac{m\pi x}{l_x}\cos\frac{n\pi y}{l_y}\cos\frac{q\pi z}{l_z}, \tag{1.8}$$

to be used in Eqs. (1.3) and (1.5). The modal resonance frequencies (the so-called eigenfrequencies) are given by:

$$\omega_{mnq} = \pi c\sqrt{\left(\frac{m}{l_x}\right)^2 + \left(\frac{n}{l_y}\right)^2 + \left(\frac{q}{l_z}\right)^2}, \tag{1.9}$$

where l_x, l_y and l_z are the side lengths of the room as indicated in Figure 1.1.

1.2.2 Modal Solution + Propagating Waves

The 3-D eigenfunction form given in the previous section, for the case of a Neumann BC on all surfaces, might be practical to write in a form with propagating waves in one of the dimensions. As one example, we might have a locally reacting material described by an impedance Z_{wall} on a single wall, e.g., at $y = l_y$, or a source distribution in the form of a vibrating wall at $y = 0$. Then the solution could be written:

$$p(x,y,z) = \sum_{m,n} \frac{\Phi_{mn}(x,z)}{\omega^2 - \omega_{mn}^2 - 2j\delta_{mn}\omega_{mn}}\left[A_{mn}e^{-jk_y y} + B_{mn}e^{jk_y y}\right], \tag{1.10}$$

$$k_y = \sqrt{k^2 - \frac{\omega_{mn}^2}{c^2}}, \quad \omega_{mn} = \frac{c}{2}\sqrt{\left(\frac{m}{l_x}\right)^2 + \left(\frac{n}{l_z}\right)^2}, \tag{1.11}$$

and the coefficients A_q, B_q are derived to fulfill the boundary conditions at $y = l_y$ and at the source. Using this form, a number of cases can be handled, e.g., a source distribution in the wall of $y = 0$, as illustrated in Figure 1.2 can be studied, where the shaded area of the wall at $y = 0$ is vibrating as a piston. The boundary condition to fulfill at $y = 0$ is then:

$$v_z(x,0,z) = v_{z0}\sum_{m,n} V_{mn}\Phi_{mn}(x,z), \tag{1.12}$$

where the mode amplitudes V_{mn} are found by expanding the source distribution function in the modal functions:

$$V_{mn} = \iint_{[0,l_x],[0,l_z]} v_z(x,0,z)\Phi_{mn}(x,z)\,dx\,dz, \tag{1.13}$$

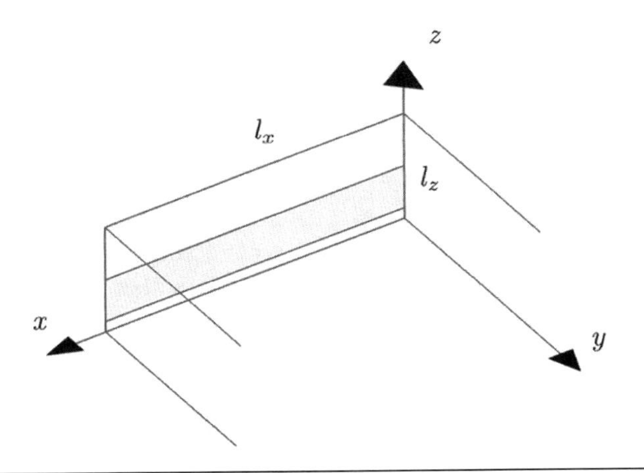

Figure 1.2 Example with a section of a wall as a vibrating surface (shaded area)

$$\Phi_{mn}(x,z) = \cos\frac{m\pi x}{l_x}\cos\frac{n\pi z}{l_z}. \tag{1.14}$$

1.2.3 Domain Matching

Many complex geometries can be constructed of simpler, connected subdomains, for which we have individual analytical solutions. This can be as illustrated in Figure 1.3 where two parallelepipedic domains are connected.

The formulation in Subsection 2.2, with propagating waves in one of the three dimensions, can then be used to give us one description in each subdomain:

$$p^I(x,y,z) = \sum_{m,n}\frac{\Phi^I_{mn}(x,y)}{\omega^2 - \left(\omega^I_{mn}\right)^2 - 2j\delta^I_{mn}\omega^I_{mn}}\left[A^I_{mn}e^{-jk^I_z z} + B^I_{mn}e^{jk^I_z z}\right],\ z \in [0, l_z], \tag{1.15}$$

$$p^{II}(x,y,z) = \sum_{r,s}\frac{\Phi^{II}_{rs}(x,y)}{\omega^2 - \left(\omega^{II}_{rs}\right)^2 - 2j\delta^{II}_{rs}\omega^{II}_{rs}}\left[A^{II}_{rs}e^{-jk^{II}_z z} + B^{II}_{rs}e^{jk^{II}_z z}\right],\ z \in [l_z, l_{z2}]. \tag{1.16}$$

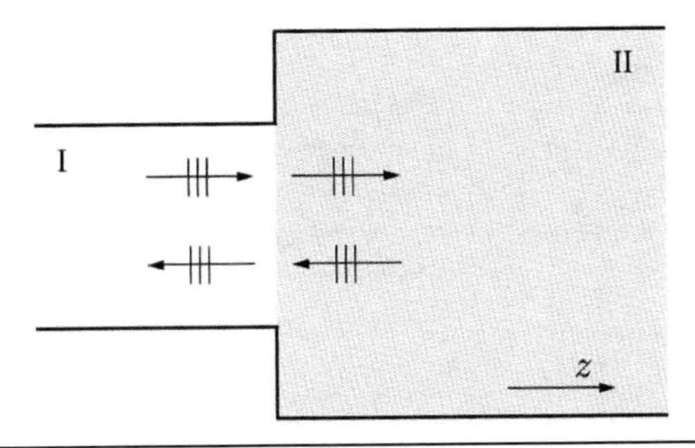

Figure 1.3 An example geometry which can be described as one rectangular domain connected to another rectangular domain

Across the interface between the two domains, the sound pressure and its gradient must be continuous, and on parts of the interface surface, a Neumann BC might apply. The fulfillment of such matching on the interface leads to a set of equations.

To solve for the unknowns, A_q^I, B_q^I, A_q^{II}, B_q^{II}, one must decide on a maximum number of modes that will be taken into account, n_{max}, m_{max}, r_{max}, s_{max}. That choice specifies the number of unknowns and one consequently has to distribute at least that many surface sample points in which the equality should be fulfilled. Finally, a direct inversion or a solution of the overdetermined equation system via regularization must be employed.

A special case of domain matching is a decomposition of the domain under study into parallelepipedical blocks.[9] Yet another example of this approach is used in the study of ducts.[10] An analogous approach has also been used with discrete, spectral approximations of the solution in the subdomains instead of analytical solutions.[11, 12]

1.3 NUMERICAL SOLUTIONS

Frequently in room acoustics modeling, the geometry and boundary conditions render an analytical solution intractable, and numerical methods must be used to generate an approximate solution. Common numerical methods in room acoustics include FEMs, BEMs, and FDMs. Each can be adapted to produce solutions to the stationary problem in the frequency domain or the transient problem in the time-domain. Each method relies on discretization of the operator or solution to make the problem manageable. We present only a brief overview of each method with references to the truly massive body of literature on the subject.

1.3.1 Finite Difference Methods

Historically, the first methods used to generate approximate solutions to partial differential equations (PDEs) were FDMs.[13] The first applications of these methods to 3-D acoustic simulation in rooms date back to 1994.[14, 15] These early approaches have distinct origins; one grew out of methods developed for electromagnetic propagation,[14, 16] and the other comes from the approximation of wave propagation by a delay network[17] or by a transmission-line matrix.[18] All of these approaches and their variants, including the earliest ones,[13] are related, but the problem is perhaps most generally posed as a numerical solution to Eq. (1.1).

FDMs are the simplest and most accessible method described in this section, so we provide a minimal example. Typically, the problem is evaluated on a regular discrete grid in space and time, and the approximate form of the three-dimensional wave equation on the grid is given by:

$$\begin{aligned} p_{i,j,k}^{n+1} = \lambda^2 (p_{i+1,j,k}^n + p_{i-1,j,k}^n + p_{i,j+1,k}^n + p_{i,j-1,k}^n + p_{i,j,k+1}^n + p_{i,j,k-1}^n) \\ + \ 2(1-3\lambda^2) p_{i,j,k}^n - p_{i,j,k}^{n-1}. \end{aligned} \tag{1.17}$$

Superscripts indicate temporal indices, and subscripts indicate spatial indices of grid nodes. The grid is defined by spatial and temporal steps, Δx, Δt, respectively. The time step should be chosen by fixing the spatial step so that it resolves all wavelengths of interest and setting the constant, $\lambda^2 @ c^2 \Delta t^2 / \Delta x^2 \le 1/3$, where c is the speed of sound. The Courant factor, λ, governs stability and the speed of numerical wave propagation.[13] Using the *updated Eq.* (1.17), if the pressure is known everywhere on the grid at times n and $n - 1$, the pressure at time $n + 1$ may be computed from its nearest-neighbor pressure values (six for a 3-D model, four for a 2-D model). In higher-order[19, 20] or interpolated schemes,[21, 22] larger numbers of neighbor values will be involved, resulting in more accurate approximations at slightly higher computational cost. Spectral methods (Section 3.3) are in some sense a limiting case, using *all* field values to compute the derivatives.

Figure 1.4 shows how the values used in the updated equation appear on a typical grid. It is only shown in two dimensions for visual clarity. The open circle is the unknown value being computed or updated. One advantage of the FDM, illustrated in the figure, is sparsity or a locally dependent update.

Although it is relatively old, this simple update continues to be a useful tool for numerical simulation. The properties of this and its more sophisticated variants can be found, for example, in References 19 and 21–24. The practical applicability of finite-difference time-domain (FDTD) techniques is limited by high computational costs at higher frequencies such that doubling the frequency band induced 8-fold memory consumption and 16-fold computational load. In addition, the valid frequency band is limited by inherent dispersion. The actual valid band is different for each scheme, but for the basic scheme of Eq. (1.17), it gives results that are reliable up to approximately one fifth of the sampling frequency.

The previous description is derived for the scalar wave equation, whereas another popular approach solves the coupled first-order equations for pressure and particle velocity. To do so, pressure and three components of velocity are staggered in both space and time to maintain explicit time-stepping. This approach originates from the electro-magnetics literature and is often referred to as the Yee algorithm.[16, 25] However, in linear acoustics this approach is equivalent to the scalar formulation but imposes heavier computational load and memory requirements.

The overall advantage of FDMs, over others like the FEM or BEM, is realized on regular grids when the update may be applied uniformly across the domain. This also makes FDMs extremely well suited to parallelization.[26, 27]

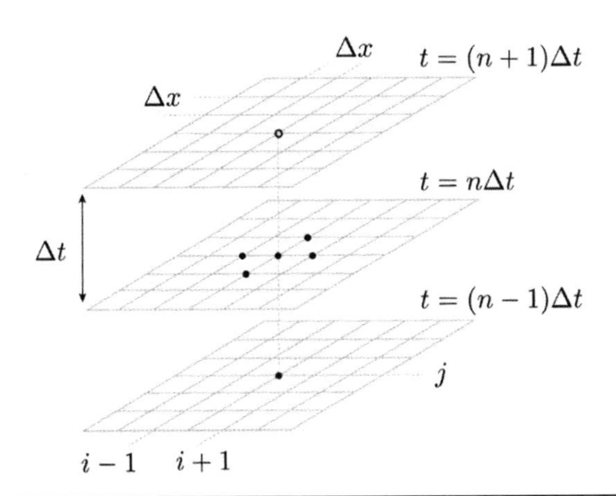

Figure 1.4 Graphical depiction of a local, explicit, two-dimensional finite difference update for the discrete wave equation

1.3.2 Finite Element and Boundary Element Methods

FEMs are also based on a volume discretization of the room, but instead of discretizing the operator, the FEM uses a discrete basis set for representing the solution, often piecewise linear or piecewise polynomial. One of its distinctions from the FDM is that the mesh is often unstructured, which can increase the *geometrical* accuracy of the model. For this reason, the FEM is very popular for solving structural and vibrational problems with highly irregular geometries. It is similarly useful for acoustic problems with complex geometries, and it is most often applied to the Helmholtz equation, i.e., Eq. (1.2).[28, 29] The result in the time-harmonic case is a piecewise linear or piecewise polynomial approximation to the eigenfunctions of Eq. (1.3); however, time-domain FEM solutions are also possible.[30]

BEMs typically approximate solutions to the Helmholtz-Kirchoff integral, which involves the pressure gradient on the boundary and the Green's function. Discretizing boundary surfaces leads to matrices of Green's functions that relate a source to boundary elements. The matrix only scales with surface area instead of volume, so it may be smaller than a finite difference or a finite element matrix for the same problem. However, the matrix is dense in contrast with the sparse matrices of the FEM and FDM, so even with fewer elements, it may be more expensive. One approach that is applicable in some cases is to use fast multipole methods, which can exploit the regularity of dense BEM matrices.[31–33]

1.3.3 Spectral Methods

Spectral methods, whether associated with discretized solutions or operators, are characterized by exponential convergence rates.[24] FDMs and FEMs converge at polynomial rates, but methods using suitable spectral differentiation and spectral elements can converge much faster. As with higher-order or interpolated difference methods, greater accuracy allows coarser discretization, which then leads to less computation. The trade-off essentially reduces to smaller, but denser matrix operators.

The advantage of spectral methods is achieved by expanding the solution or operator into an orthogonal basis, such as a trigonometric series or the Chebyshev polynomials, and the approximation is done in this spectral domain. Using Fourier or Chebyshev bases, spectral methods have been adapted to irregular geometries through coordinate transformations and domain matching,[9, 11, 12] analogous to Section 2.3. Especially when problems are limited by memory storage, spectral methods are potentially a good alternative to finite difference or finite element methods. However, the problems for which they are applicable typically coincide with domains where an analytical solution is available.

REFERENCES

1. J. Kang, "Acoustics in Long Rooms," in *Architectural Acoustics Handbook*, N. Xiang, Ed. J. Ross Publishing, 2017.
2. U. P. Svensson, S. Siltanen, L. Savioja, and N. Xiang, "Computational Modeling of Room Acoustics II: Geometrical Acoustics," in *Architectural Acoustics Handbook*, N. Xiang, Ed. ASA Press, 2013.
3. M. Vorländer, *Auralization: Fundamentals of Acoustics, Modelling, Simulation, Algorithms, and Acoustic Virtual Reality*. Berlin: Springer-Verlag, 2007.
4. E. G. Williams, *Fourier Acoustics*. London, UK: Academic Press, 1999.

5. P. M. Morse and H. Feschbach, *Methods of Theoretical Physics. Part I.* New York: McGraw-Hill, 1953.
6. H. Kuttruff, *Room Acoustics*, 4th ed. London, UK: Spon Press, 2000.
7. M. R. Schroeder and K. H. Kuttruff, "On frequency response curves in rooms. comparison of experimental, theoretical, and monte carlo results for the average frequency spacing between maxima," *J. Acoust. Soc. Am.*, vol. 34, no. 1, pp. 76–80, 1962.
8. M. M. Boone, "Modal superposition in the time domain: Theory and experimental results," *J. Acoust. Soc. Am.*, vol. 97, no. 1, p. 92, 1995.
9. N. Raghuvanshi, B. Lloyd, N. Govindaraju, and M. Lin, "Efficient numerical acoustic simulation on graphics processors using adaptive rectangular decomposition," *In Proc. EAA Symp. Auralization*, Espoo, Finland, 2009.
10. R. J. Alfredson, "The propagation of sound in a circular duct of continuously varying cross-sectional area," *J. Sound Vib.*, vol. 23, no. 4, pp. 433–442, 1972.
11. J. S. Hesthaven, "A stable penalty method for the compressible Navier-Stokes equations: III. multidimensional domain decomposition schemes," *SIAM Journal on Scientific Computing*, vol. 20, no. 1, pp. 62–93, 1998.
12. Y. Q. Zeng, Q. H. Liu, and G. Zhao, "Multidomain pseudospectral time-domain (PSTD) method for acoustic waves in lossy media," *Journal of Computational Acoustics*, vol. 12, no. 03, pp. 277–299, 2004.
13. R. Courant, K. Friedrichs, and H. Lewy, "On the partial difference equations of mathematical physics," *IBM Journal of Research and Development*, vol. 11, no. 2, pp. 215–234, 1967.
14. D. Botteldooren, "Acoustical finite-difference time-domain simulation in a quasi-cartesian grid," *J. Acoust. Soc. Am.*, vol. 95, no. 5, pp. 2313–2319, 1994.
15. L. Savioja, T. Rinne, and T. Takala, "Simulation of room acoustics with a 3-D finite difference mesh," in *Proc. Int. Computer Music Conf.*, 1994, pp. 463–466.
16. K. S. Yee, "Numerical solution of initial boundary value problems involving Maxwell's equations in isotropic media," *IEEE Trans. Antennas and Propagation*, vol. 14, no. 3, pp. 302–307, 1966.
17. J. O. Smith, "Physical modeling using digital waveguides," *Computer Music J.*, vol. 16, no. 4, pp. 74–87, 1992.
18. Y. Kagawa, T. Tsuchiya, B. Fujii, and K. Fujioka, "Discrete Huygens' model approach to sound wave propagation," *J. Sound Vib.*, vol. 218, no. 3, pp. 419–444, 1998.
19. B. Gustafsson, *High order difference methods for time dependent PDE*, Springer Series in Computational Mathematics, vol. 38. Springer Verlag, 2008.
20. S. Sakamoto, H. Nagatomo, A. Ushiyama, and H. Tachibana, "Calculation of impulse responses and acoustic parameters in a hall by the finite-difference time-domain method," *Acoust. Sci. & Tech.*, vol. 29, no. 4, pp. 256–265, 2008.
21. L. Savioja and V. Välimäki, "Interpolated rectangular 3-D digital waveguide mesh algorithms with frequency warping," *IEEE Trans. on Speech and Audio Processing*, vol. 11, no. 6, pp. 783–790, 2003.
22. K. Kowalczyk and M. van Walstijn, "Room acoustics simulation using 3-D compact explicit FDTD schemes," *IEEE Trans. Audio, Speech, Language Process.*, vol. 19, no. 1, pp. 34–46, 2011.
23. J. Strikwerda, *Finite Difference Schemes and Partial Differential Equations.* New York, NY: Chapman & Hall, 1989.
24. L. N. Trefethen, *Spectral methods in MATLAB*, vol. 10. Philadelpia: Soc. Indust. Appl. Math., 2000.
25. P. H. Aoyagi and R. Mittra, "A hybrid Yee algorithm/scalar-wave equation approach," *IEEE Trans. Microwave Theory and Techniques*, vol. 41, no. 9, pp. 1593–1600, 1993.
26. L. Savioja, "Real-time 3D finite-difference time-domain simulation of low- and mid-frequency room acoustics," in *Proc. 13th Int. Conf. Digital Audio Effects (DAFx-10)*, Graz, Austria, September 6–10, 2010.
27. A. Southern, D. Murphy, G. Campos, and P. Dias, "Finite difference room acoustic modelling on a general purpose graphics processing unit," in *Proc. 128th Audio Eng. Soc. Conv., preprint no. 8028*, London, UK, 22–25 May, 2010.
28. O. C. Zienkiewicz and R. L. Taylor, *The finite element method*, vol. 3. London, UK: McGraw-Hill, 1977.

29. L. L. Thompson, “A review of finite-element methods for time-harmonic acoustics,” *J. Acoust. Soc. Am.*, vol. 119, no. 3, pp. 1315–1330, 2006.
30. J.-F. Lee, R. Lee, and A. Cangellaris, “Time-domain finite-element methods,” *IEEE Trans. Antennas and Propagation*, vol. 45, no. 3, pp. 430–442, 1997.
31. L. Greengard and V. Rokhlin, “A fast algorithm for particle simulations,” *Journal of computational physics*, vol. 73, no. 2, pp. 325–348, 1987.
32. Y. J. Liu and N. Nishimura, “The fast multipole boundary element method for potential problems: A tutorial,” *Engineering Analysis with Boundary Elements*, vol. 30, no. 5, pp. 371–381, 2006.
33. N. A. Gumerov and R. Duraiswami, *Fast multipole methods for the Helmholtz equation in three dimensions*. Amsterdam: Elsevier Science, 2005.

2

Computational Modeling of Room Acoustics II: Geometrical Acoustics

U. Peter Svensson, Acoustics Research Centre, Department of Electronics and Telecommunications, Norwegian University of Science and Technology, Trondheim, Norway

Samuel Siltanen and **Lauri Savioja**, Department of Media Technology, Aalto University School of Science, Espoo, Finland

Ning Xiang, School of Architecture, Rensselaer Polytechnic Institute

2.1 UNIFYING FRAMEWORKS OF GEOMETRICAL ACOUSTICS

Geometrical room acoustics (GA) is based on the assumption that sound can be modeled with rays traveling along straight paths. It is a clear simplification as it neglects all the wave phenomena of sound, such as diffraction. However, at higher frequencies, GA is a viable approach since the role of the wave phenomena decreases remarkably when the wavelength gets shorter than the dimensions of the objects in the space under study. In addition, the wave-based modeling techniques become computationally very expensive at this frequency range, thus the GA methods are typically applied in practice for mid- and high-frequency modeling of room acoustics. There are two general frameworks that are able to cover all the GA methods as described in the following text. In a separate section, the superposition of specular reflections and diffraction is described.

2.1.1 Acoustic Radiative Transfer Model

The acoustic radiative transfer model (ARTM),[1] also known as the 3-dimensional transport equation,[2] is based on the transport theory. In that model, sound particles are considered to carry energy and propagate along straight lines at speed *c*. In enclosures, the sound particles strike walls or scattering objects which absorb and scatter the sound particles while the interactions of sound particles are negligible.[1] We use sound radiance $L(\mathbf{r},\hat{\mathbf{s}},t)$ at location **r** and time *t*, flowing along direction $\hat{\mathbf{s}}$ to denote the energy flow density of sound particles which is the energy flow per unit normal area per unit solid angle per unit time, also termed sound energy angular flux,[2] where the normal area is perpendicular to the flow direction, its units are $Wm^{-2}sr^{-1}$. The radiative transfer equation can be written as:

$$\frac{1}{c}\frac{\partial L(\mathbf{r},\hat{\mathbf{s}},t)}{\partial t}+\hat{\mathbf{s}}\cdot\nabla L(\mathbf{r},\hat{\mathbf{s}},t)+(\mu_\alpha+\mu_s)L(\mathbf{r},\hat{\mathbf{s}},t)-\mu_s\int_{\Omega'}P(\hat{\mathbf{s}}',\hat{\mathbf{s}})\,L(\mathbf{r},\hat{\mathbf{s}}',t)d\Omega'=q(\mathbf{r},\hat{\mathbf{s}},t)\,, \quad (2.1)$$

where $\mu_s=(1-\bar{\alpha})/\lambda$ and $\mu_a=m+\bar{\alpha}/\lambda$ which take account of scattering and absorption of radiation over the unit of mean free path λ, $\bar{\alpha}$ is averaged absorption coefficient, and m is air attenuation coefficient. $P(\hat{\mathbf{s}}',\hat{\mathbf{s}})$ represents the probability of particles with a propagation direction $\hat{\mathbf{s}}'$ being scattered by scattering objects into the solid angle $d\Omega$ around direction $\hat{\mathbf{s}}$, $q(\mathbf{r},\hat{\mathbf{s}},t)$ is an omnidirectional volume source term.

Equation (2.1) provides a general framework for modeling propagation of sound energy with a capability to incorporate complex sound sources. Furthermore, it explicitly handles partial specular reflections and partial scattering both on room boundaries and in the medium which is of general significance when the fitted enclosures containing objects need to be considered. In case of enclosures without scattering objects, the radiative transfer equation reduces to:

$$\frac{1}{c}\frac{\partial L(\mathbf{r},\hat{\mathbf{s}},t)}{\partial t}+\hat{\mathbf{s}}\cdot\nabla L(\mathbf{r},\hat{\mathbf{s}},t)+(\mu_\alpha+\mu_s)L(\mathbf{r},\hat{\mathbf{s}},t)=q(\mathbf{r},\hat{\mathbf{s}},t)\,. \quad (2.2)$$

This is the so-called interior equation governing the sound particle transport (propagation), which is subject to boundary condition:

$$L(\mathbf{r}_b,\hat{\mathbf{s}},t)=\int_{\Omega^-}R_F(\mathbf{r}_b,\hat{\mathbf{s}}',\hat{\mathbf{s}})L(\mathbf{r}_b,\hat{\mathbf{s}}',t)(\hat{\mathbf{s}}'\cdot-\hat{\mathbf{n}})\,d\Omega' \quad (2.3)$$

in absence of sources at the boundaries. Jing et. al., have applied the radiative transfer equation of Eq. (2.2) along with Eq. (2.3) to room-acoustic modeling of elongated spaces.[2] To model the sound energy distributions in such spaces, they numerically solved this system of equations with specified boundary conditions for the long side walls and two ends of elongated space.[3]

The ARTM forms the base for the diffusion equation model described in subsection 2.3.3.

2.1.2 Room Acoustic Rendering Equation

The room acoustic rendering equation (RARE) offers a framework for the acoustic energy propagation in an enclosure[4] under the assumption of GA. GA modeling algorithms can be derived as special cases of the RARE.

To derive an energy propagation model, an appropriate reflection model is required. Assuming local reaction, the reflection properties of any material can be expressed as a ratio of the outgoing differential radiance, dL, to the incoming differential irradiance, dE. Irradiance refers to the incident power on a point. Radiance is radiant power per projected unit area per unit solid angle. The ratio is:

$$\rho(\hat{\mathbf{s}}',\hat{\mathbf{s}};\mathbf{x}')=\frac{dL(\hat{\mathbf{s}})}{dE(\hat{\mathbf{s}}')}, \quad (2.4)$$

where all variables are angle dependent. Here $\hat{\mathbf{s}}'$ and $\hat{\mathbf{s}}$ are the incoming and outgoing directions expressed as unit vectors. The ratio ρ is called the bidirectional reflectance distribution function (BRDF). It is a four-dimensional function and it is specific to the material of the reflecting surface.

Sound propagates with a finite speed, c, which causes time delay. This can be expressed with an operator S as follows:

$$S_r I(t) = l\left(t - \frac{r}{c}\right). \tag{2.5}$$

For simplicity, time dependence is implicitly assumed in radiance and irradiance written in lower case, l and e, respectively.

Next, the effects of geometry on sound propagation are examined. A differential solid angle covered by differential area dA can be written as:

$$d\omega = \max\{\hat{\mathbf{n}} \cdot \hat{\mathbf{s}}', 0\} \frac{dA}{r^2}, \tag{2.6}$$

where $\hat{\mathbf{n}}$ is the normal vector of the surface in the differential area, $\hat{\mathbf{s}}'$ is the unit vector pointing from the observation point towards the differential area, and r is the corresponding distance. Negative values are clamped to zero because negative solid angles are not meaningful in this context.

It is possible to write the differential incident radiance by using the expressions in Eqs. (2.5) and (2.6) as:

$$de(\mathbf{x}', \hat{\mathbf{s}}') = S_r \max\{\hat{\mathbf{n}}' \cdot \hat{\mathbf{s}}', 0\} l(\mathbf{x}, -\hat{\mathbf{s}}') d\omega, \tag{2.7}$$

where $\hat{\mathbf{n}}'$ is the normal of the surface on which the radiance arrives, while $\mathbf{x}'$ and $\mathbf{x}$ represent points on the room surfaces. Negative values are again clamped to zero. By using the expression for the BRDF in Eq. (2.4), the differential radiance at the observation point becomes:

$$dl(\mathbf{x}', \hat{\mathbf{s}}) = \rho(\hat{\mathbf{s}}', \hat{\mathbf{s}}; \mathbf{x}') S_r l(\mathbf{x}, -\hat{\mathbf{s}}') \max\{\hat{\mathbf{n}}' \cdot \hat{\mathbf{s}}', 0\} \max\{\hat{\mathbf{n}} \cdot \hat{\mathbf{s}}', 0\} \frac{dA}{r^2}. \tag{2.8}$$

A geometry term is defined as:

$$g(\mathbf{x}, \mathbf{x}') = S_r \max\{\hat{\mathbf{n}}' \cdot \hat{\mathbf{s}}', 0\} \max\{\hat{\mathbf{n}} \cdot \hat{\mathbf{s}}', 0\} \frac{1}{r^2}, \tag{2.9}$$

which simplifies the expression for the differential reflected radiance:

$$dl(\mathbf{x}', \hat{\mathbf{s}}) = \rho(\hat{\mathbf{s}}' \cdot \hat{\mathbf{s}}; \mathbf{x}') g(\mathbf{x}, \mathbf{x}') l(\mathbf{x}, -\hat{\mathbf{s}}') dA. \tag{2.10}$$

A visibility term $v(\mathbf{x}, \mathbf{x}')$ equals zero if the line between $\mathbf{x}$ and $\mathbf{x}'$ is obstructed and one otherwise. A reflection kernel is:

$$R(\mathbf{x}, \mathbf{x}', \hat{\mathbf{s}}) = v(\mathbf{x}, \mathbf{x}') \rho(\hat{\mathbf{s}}' \cdot \hat{\mathbf{s}}; \mathbf{x}') g(\mathbf{x}, \mathbf{x}'). \tag{2.11}$$

Then the room acoustic rendering equation can be written as:

$$l(\mathbf{x}', \hat{\mathbf{s}}) = l_0(\mathbf{x}', \hat{\mathbf{s}}) + \iint_G R(\mathbf{x}, \mathbf{x}', \hat{\mathbf{s}}) l(\mathbf{x}, -\hat{\mathbf{s}}') d\mathbf{x}. \tag{2.12}$$

The integral is over all the surfaces, and l_0 describes the emittance by the surfaces or the primary reflected radiance if the sources are not described as part of the surface model. This integral equation describes the propagation of acoustic energy in a general case.

Different GA modeling algorithms can be seen as specializations of the RARE where the BRDF is different for each algorithm. For the image source method (ISM) and specular ray tracing, the reflectance function is:

$$\rho_{spec}(\hat{\mathbf{s}}', \hat{\mathbf{s}}) = \frac{\beta}{\hat{\mathbf{n}} \cdot \hat{\mathbf{s}}'} \delta(\hat{\mathbf{s}}' - M(\hat{\mathbf{s}})), \tag{2.13}$$

where δ is the Dirac delta function, M is the mirror reflection transformation, and β is the (intensity) reflection coefficient. For stochastic ray tracing, the BRDF corresponds to the probability distribution of the outgoing rays, or:

$$\rho_{prob}(\hat{\mathbf{s}}', \hat{\mathbf{s}}) = P(\text{"ray from } \hat{\mathbf{s}}' \text{ reflected to } \hat{\mathbf{s}}\text{"}). \tag{2.14}$$

For acoustic radiosity, the BRDF is constant:

$$\rho_{diff}(\hat{\mathbf{s}}', \hat{\mathbf{s}}) = \frac{\beta}{\pi}. \tag{2.15}$$

Because the RARE follows from the GA principles, it is expected that any new GA algorithm can be expressed by it with a suitable BRDF.

Navarro et al.,[1] have shown that the RARE can be considered as a special case of the radiative transfer equation. The main difference between the two is the ability of the ARTM to model non-ideal media. This means that the ARTM can handle scattering and absorbing media which escape the capabilities of the RARE. Thus the ARTM actually expands the definition of GA to cover even non-linear propagation paths.

In addition, the formulations of the ARTM and RARE are quite different from each other. Where the RARE represents the sound energy at surface elements, the ARTM uses a volumetric integral thus enabling incorporation of effects of the media. In principle, both of them are able to model similarly complex boundary conditions. The ARTM represents the boundary conditions as probabilities of a sound particle to be reflected to a certain direction when coming in another given angle and is thus similar to the BRDF formulation of the RARE for stochastic ray tracing.

2.2 DETERMINISTIC MODELING OF SPECULAR REFLECTIONS AND DIFFRACTIONS

In GA, reflections can be specular or diffuse, as described by the reflectance functions in Eqs. (2.13) and (2.15). For specular reflections, exact analytical solutions exist for a few cases. However, most often, approximate methods are used that are asymptotically correct for high frequencies.

2.2.1 Image Source Method for a Single Surface

The ISM is based on the idea that a boundary condition can be fulfilled by two free-field radiating sources—the original source and a fictive *image source* (IS). For an infinite plane surface of locally reacting impedance Z_n, the IS is given the amplitude:

$$A_{IS} = A_S \frac{Z_n \cos\theta - \rho c}{Z_n \cos\theta + \rho c} = A_S R, \tag{2.16}$$

that is, a plane-wave (sound pressure) reflection coefficient, R, is assigned to the IS. The presence of the angle θ in the expression implies that the IS is directional. See Figure 2.1 for a definition of the angle θ. For a Neumann or Dirichlet boundary condition (BC), that is, $Z_n = \infty$ or $Z_n = 0$, respectively, the directivity terms vanish and the IS solution will give the exact solution. For general locally-reacting materials, the error with this approach is small if the sum of the perpendicular distances of the source and the receiver from the surface is large relative to the wavelength.[5] For the special case of grazing incidence, more accurate expressions are available.[6] It should be noted that the complex impedance is required for the application of this formula. The plane-wave approximation is avoided using the spherical reflection coefficient approach, which requires the computation of an integral expression rather than the explicit formula in Eq. (2.16).[7, 8]

2.2.2 Exact Image Source Solution for Shoebox-Shaped Rooms

When there is more than one surface, some simple geometries can be handled exactly by a set of IS. In addition, perfectly rigid walls, i.e., ideal Neumann BC, are assumed as the Dirichlet BC is usually not relevant in room acoustics problems. The most relevant case in room acoustics is the parallelepipedical room, as shown in Figure 2.2 in the form of a 2-D view. The IS impulse response solution presented by Allen & Berkley[9] is:

$$h(t)=\sum_{p=1}^{8}\sum_{q=-\infty}^{\infty} A_{pr}\frac{\delta\left(t-\frac{|\mathbf{R}_p+\mathbf{R}_q|}{c}\right)}{|\mathbf{R}_p+\mathbf{R}_q|}, \tag{2.17}$$

where $\mathbf{R}_p$ represents the 8 coordinate sets created by:

$$\mathbf{R}_p=(x_S \pm x_R, y_S \pm y_R, z_S \pm z_R), \tag{2.18}$$

while (x_S, y_S, z_S) and (x_R, y_R, z_R) are the source and receiver coordinates respectively, and in the sums, q represents an integer triplet (n, l, m), and:

$$\mathbf{R}_q = 2\left(nL_x, lL_y, mL_z\right), \tag{2.19}$$

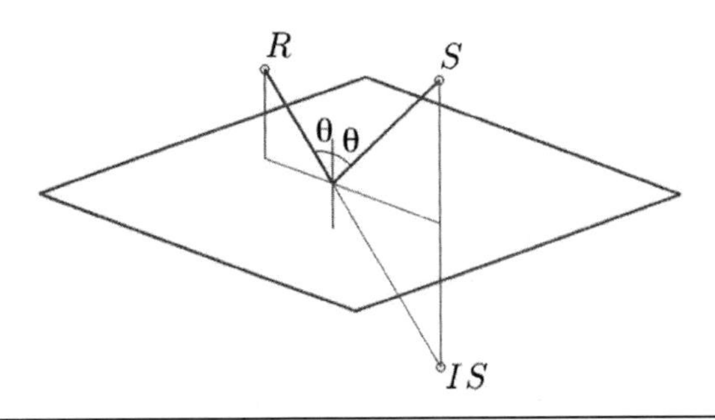

Figure 2.1 Source, S, above a reflecting surface, the image source, IS, and a receiver position, R

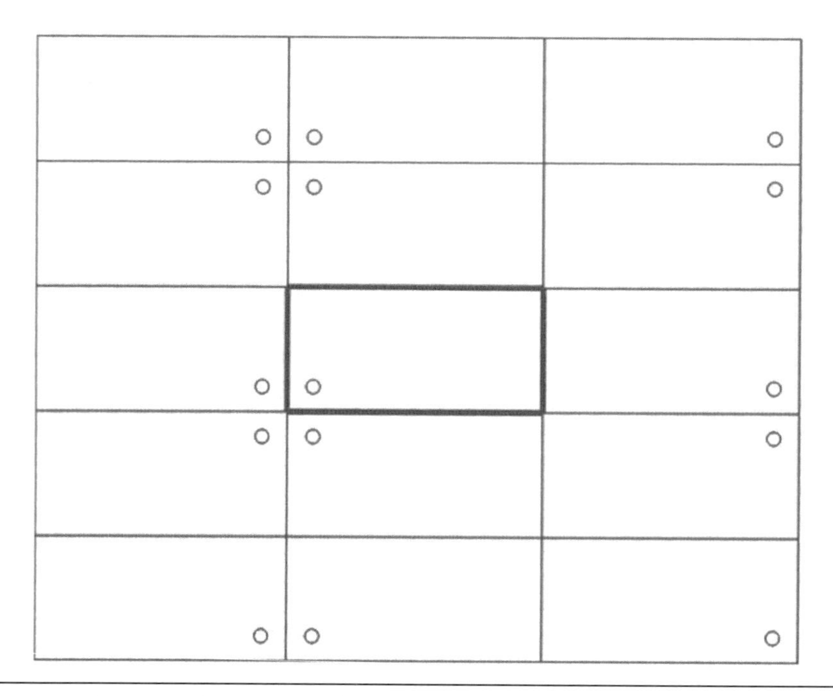

Figure 2.2 Two-dimensional view of a parallelepipedical room, for which a set of ISs can fulfill the rigid-wall boundary conditions exactly by assigning all the ISs the same source strength as the original source

where L_x, L_y, L_z are the lengths of the room in the x-, y-, and z-direction, respectively. The corresponding frequency-domain solution is:

$$H(\omega)=\sum_{p=1}^{8}\sum_{q=-\infty}^{\infty} A_{pr}\frac{e^{-jk\left|\mathbf{R}_p+\mathbf{R}_q\right|}}{\left|\mathbf{R}_p+\mathbf{R}_q\right|}, \tag{2.20}$$

where it should be noted that a free-field impulse response/transfer function of:

$$h_{FF}(t)=\frac{\delta\left(t-\frac{r}{c}\right)}{r},\quad H_{FF}(\omega)=\frac{e^{-jkr}}{r}, \tag{2.21}$$

for distance r was assumed. Furthermore, the amplitude A_{pq} is given by the involved wall reflection factors, R:

$$A_{pq}=\prod_{p,q} R_{pq}, \tag{2.22}$$

where a very simple wall reflection model is assumed:[9] the reflection factor for each wall, $\boldsymbol{R}_{pq}$, is a real-valued and angle-independent constant, which typically would be calculated from the diffuse-field absorption coefficient $\alpha_{\text{diffuse},pq}$:

$$R_{pq}=\sqrt{1-\alpha_{\text{diffuse},pq}}. \tag{2.23}$$

More refined reflection expressions can be introduced. Utilizing the plane-wave reflection coefficient in Eq. (2.16) is straightforward if the complex wall impedances Z_{n} are known. In a comparison between the use of plane-wave reflection coefficients and the much more involved spherical reflection coefficients, very small differences were found between the two.[10]

The formulations in Eqs. (2.17) or (2.20) are useful but suffer from being slowly converging. A more efficient hierarchical modeling approach has been suggested, where image sources in close proximity to each other are replaced by multipoles.[11]

Other geometries exist for which the IS solution is complete and exact. Any room which has only interior angles that are 180 degrees divided by an integer number, that is, 90 degrees, 60 degrees, 45 degrees, etc., can be represented completely by a set of ISs. However, interior angles which are 60 degrees or smaller are obviously rare in rooms. This IS solution is exactly equivalent to the modal solution when $A_{pr} = 1$.[9]

2.2.3 Image Source Method for Arbitrarily Shaped Rooms

For most practical rooms of a more complex geometry, an approximate approach can be used based on the GA principle: a specular reflection occurs if there is an unobstructed path from the receiver to the mirror image of the source through the reflecting plane. This is illustrated in Figure 2.3, where receivers inside the indicated beam can see the IS through the reflection plane. Thus, the IS is active, or valid, only for those receivers.

The ISM, and several other GA implementations, share the same two-stage approach for computing the response at a receiver: (1) identification of valid sound paths, in the form of combinations of reflections and, possibly, diffractions, and (2) generation of an echogram or an impulse response.

2.2.3.1 General Algorithm

The classical recursive ISM algorithm, as detailed by Borish,[12] is an extension of the illustration in Figure 2.3: the primary source can generate *potentially valid ISs, of first-order* by reflecting the primary source position in every single reflecting polygon that the source can see. The next step is that *potentially valid second-order ISs* are generated by reflecting all the potentially valid first-order ISs in all the reflecting polygons that can be seen through the last reflection polygon. Third-order potentially valid ISs are then generated from the second-order ISs, etc. This should make clear both the recursive nature of the algorithm, and the computational problem—the number of potentially valid ISs can grow exponentially. However, the final number of ISs that are finally valid for one specific receiver position will be orders of magnitude smaller than this hypothetical high number, as discussed e.g., by Mechel.[5]

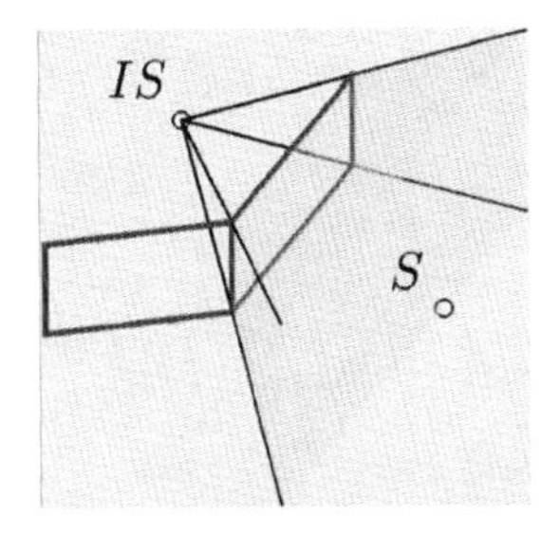

Figure 2.3 Illustration of an IS's visibility for a first-order reflection. Only receivers within the beam indicated by the four beam faces can be reached by a specular reflection according to the ISM solution.

The ISM algorithm can be split up into two consecutive steps. Step I is independent of the receiver position and involves a validity test. Step II depends on the receiver position and involves three tests (visibility, proximity, obstruction), as briefly described in the following text.

The list of all the *potentially valid ISs* forms a so-called IS tree, which is independent of the receiver position. The list is generated by a recursive generation as described previously.

Once a receiver location is specified, the IS tree is traversed to check the visibility, proximity, and obstruction, of every single IS candidate in the tree, for that particular receiver position.

In the visibility check, all the reflection points of the $n + 1$ specular reflections are tested to make sure that they are inside their respective reflecting polygons, as illustrated in Figure 2.4. The problem to solve is a sequence of the point-in-polygon (PIP) problem, which is a classical problem in computer graphics, and the literature gives several algorithms and efficient implementation.

After the visibility of an IS has been evaluated, then the sequence of $n + 1$ reflection path segments is tested for *obstruction*, that is, the $n + 1$ path segments on the way from source to receiver must not hit any obstructions, or protruding planes. For the case that a room forms a completely convex polyhedron, then the obstruction test is not needed.

The truncation criterion for the whole IS finding process could be a proximity criterion, that is, if the impulse response is computed up to a finite time t_{end}, then the distance from the IS candidate to the receiver is tested.

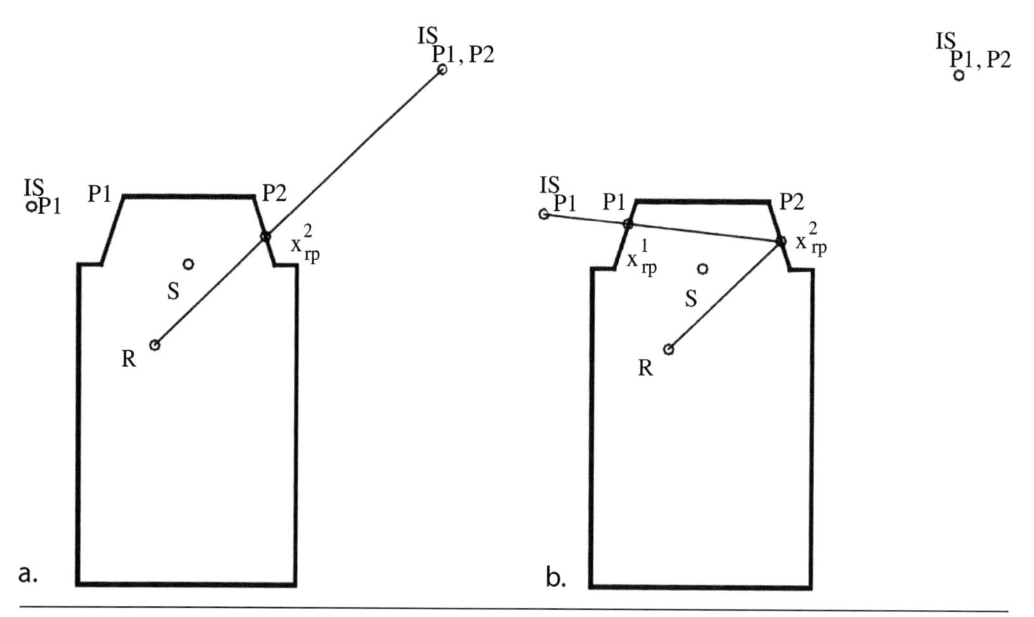

Figure 2.4 Illustration of the visibility check of a second-order reflection in a simple hall model. (a) The second reflection point, $\mathbf{x}_{rp}^{2}$, is computed, and is found to be inside reflection plane P2. (b) The first reflection point, $\mathbf{x}_{rp}^{1}$, is computed and is found to be inside reflection plane P1.

2.2.3.1.1 The Contribution by an Image Source

Once valid ISs have been found, their contributions are added into the total signal of sound pressure, or sound pressure squared. Thus, the complex sound pressure amplitude for a single IS could be calculated in a more refined way than by Eqs. (2.20) and (2.22),

$$p_i(\omega)=p_{ref}r_{ref}\sqrt{\frac{\Gamma(\theta_S)\Gamma(\theta_R)}{\Gamma(\theta_S,ref)}}\frac{1}{r}\prod_{j=1}^{n_{refl,i}}\frac{Z_{n,i,j}\cos\theta_{i,j}-\rho_0 c}{Z_{n,i,j}\cos\theta_{i,j}+\rho_0 c}e^{-mr/2}e^{-j\omega r/c}, \tag{2.24}$$

where a number of factors have been introduced. First, the plane-wave reflection model of Eq. (2.16) is used. Losses in the medium are represented by the factor $e^{-mr/2}$, where m is the air attenuation coefficient, and the directivity of the sound source and the receiver is represented by directivity factors $\gamma(\theta)$, which are assumed to be rotationally symmetrical for brevity. Any phase distortion of the source or receiver directivity is ignored by this expression. The scaling of the sound pressure is handled by the free-field sound pressure, p_{ref}, at distance r_{ref}, e.g., 1 m. Commonly, the intensity/pressure squared version is employed, rather than the complex sound pressure version:

$$p_i^2=\frac{p_{ref}^2 r_{ref}^2}{\Gamma(\theta_{S,ref})}\frac{\Gamma(\theta_S)\Gamma(\theta_R)}{r^2}\prod_{j=1}^{n_{refl,i}}\beta_{i,j}e^{-mr}, \tag{2.25}$$

where β is the intensity reflection coefficient used in Section 1.

2.2.3.2 Algorithm with Diffraction

For some cases where the pure ISM cannot generate the exact solution, adding diffraction wave components gives an improved solution. Figure 2.5 illustrates an external corner constructed by two half planes. The geometrical acoustics solution will give rise to three different zones, in which a receiver will receive different GA components: (I) direct sound and a specular reflection, (II) direct sound only, and (III) no direct sound or specular reflection.

These zones give rise to zone boundaries, drawn as the two thin lines in Figure 2.5, at which a sound field discontinuity will appear. Diffraction waves are such that they can be added to

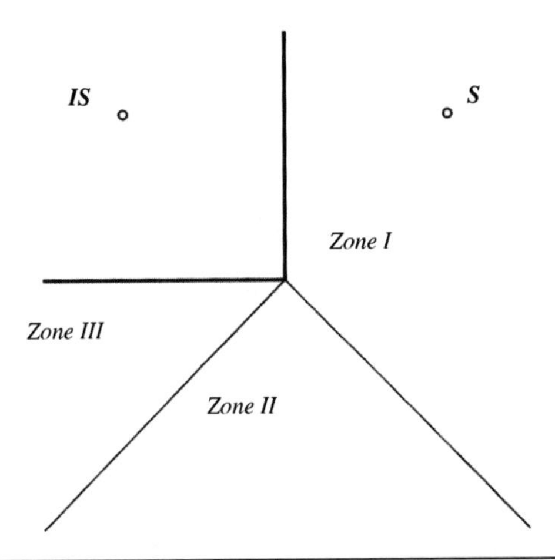

Figure 2.5 Illustration of the zone boundaries which cause sound field discontinuities for GA modeling

the GA solution and give a total field that is continuous across zone boundaries. Diffraction waves will necessarily be discontinuous as well across the zone boundaries.

Because diffraction waves are simply added to the specular reflections, the ISM algorithm can quite easily be extended with the concept of edge diffraction, as briefly outlined in the following.

For the combined ISM+diffraction algorithm, one can use the term edge source (ES) for apparent sources that are placed along each edge. The ESs are reflected in the same way as the original source and the ISs, since these behave as ISs with the exception that they are distributed along an edge, rather than localized to a point.

The validity, visibility, proximity, and obstruction tests for ESs are very similar to their IS counterpart.[14] One aspect that severely increases the computational complexity is the fact that ESs can see a large part of the room model: an ES on a protruding 90 degree corner can see 3/4 of the entire space. This is in sharp contrast to the ISs of higher order that can see a small portion of the rest of the room through the last reflection plane/polygon.

2.2.3.2.1 Contribution by a First-Order Diffraction ES

When diffraction components are introduced, sound-pressure expressions must be employed, because energy summation would often lead to huge errors in the frequency range of interest. This is caused by the fact that the diffraction contributions often arrive shortly after some specular reflection, and with opposite polarity. This leads to significant interference effects that are missed if pure energy-summation is used.

The geometry of an edge is illustrated in Figure 2.6. The source, S, and receiver, R, can be replaced by an IS and/or an image receiver and then handle preceding and subsequent specular reflections. Different formulations for the diffraction contribution are available. Below, a

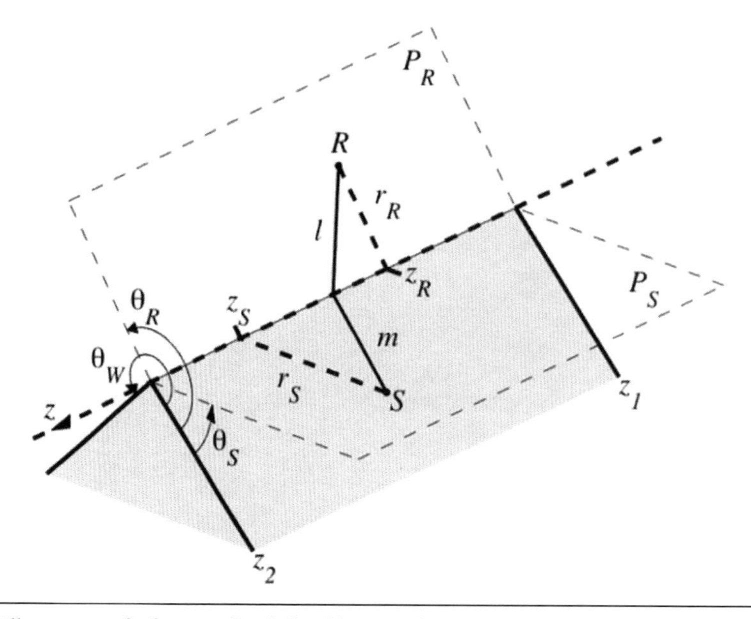

Figure 2.6 Illustration of a finite wedge defined by two planes. A source, *S*, and a receiver, *R*, are indicated, and the cylindrical coordinates (r, θ, z) for the source, receiver, and wedge are given, too.

formulation is presented that is based on the Biot-Tolstoy solution, which in turn is exact for an infinite rigid wedge.

The contribution to the sound pressure at the receiver for a finite edge is, in the time domain and in the frequency domain, given by an integral over the contributions of secondary ESs.[15, 16, 17]

$$h(t) = -\frac{\nu}{4\pi}\sum_{i=1}^{4}\int_{z_1}^{z_2}\delta\left(t-\frac{m+l}{c}\right)\frac{\beta_i}{ml}dz,\quad H(\omega) = -\frac{\nu}{4\pi}\sum_{i=1}^{4}\int_{z_1}^{z_2}e^{-jk(m+l)}\frac{\beta_i}{ml}dz, \tag{2.26}$$

where z is the position along the edge, z_1 and z_2 are the end points of the edge, ν is the so-called wedge index $= \pi / \theta_W$, , and the edge source directivity functions are:

$$\beta_i = \frac{\sin(\nu_i)}{\cosh(\nu\eta) - \cos(\nu_i)}, \tag{2.27}$$

where $\varphi_1 = \pi + \theta_S + \theta_R$, $\varphi_2 = \pi + \theta_S - \theta_R$, $\varphi_3 = \pi - \theta_S + \theta_R$, $\varphi_4 = \pi - \theta_S - \theta_R$. The auxiliary function η can be written as:

$$\eta = \cosh^{-1}\frac{(z - z_S)(z - z_R) + ml}{r_S r_R}. \tag{2.28}$$

It could be noted that in the line integrals, the factors, m, l, β, and η are all functions of the integration variable z.

The edge diffraction contribution will display a discontinuity in space for receivers on a zone boundary, i.e., angles where $\cos(\nu\varphi_i) = 0$ in Eq. (2.27). The discontinuity leads to the conclusion that the diffraction component, as computed by Eq. (2.26), will be numerically hard to compute because the β-factor of the integrand will have an increasing singularity-like behavior the closer the receiver position comes to a zone boundary. An expansion of the integrand can then be employed as in Reference 18. The diffraction solution is exact for first-order diffraction of rigid boundaries. It is crucial that an IS solution has a visibility testing algorithm that matches the diffraction algorithm, since the discontinuities must match perfectly. Also, the diffraction contribution should be 0 on the zone boundary, whereas the specular reflection is given half the amplitude of its *full-visibility value*.

The integration along each edge in Eq. (2.26) is quite costly, and other popular approaches use the geometrical theory of diffraction (GTD)[19] or the uniform geometrical theory of diffraction (UTD)[20] formulation. These approaches can be used in applications where computational performance is more important than the accuracy, such as in some virtual reality systems.[21]

Higher-order diffraction will be generated by any finite edge, but those higher-order components are typically of smaller amplitude. Efficient methods for handling them have been presented, e.g., by Antani et al.[22]

Non-rigid surfaces have no exact diffraction solutions but for the UTD, the high-frequency asymptotic approach is to scale the diffraction contribution by the reflection factor.[21]

2.2.4 Beam Tracing

There are several variations on how the search of valid ISs can be speeded up, but the most typical approach is *beam tracing*.[5, 23, 24, 25, 26, 27] In beam tracing, each IS, generated by a certain last reflection plane, constructs a beam/frustum through its last reflection plane and thus clips out a small subset of the full set of planes. The actual beam tracing is done by emitting beams, that is, a group of rays that form a pyramidal beam which emanates from the source and extends radially outwards. Once the beam hits a wall surface, it is reflected specularly.

Moreover, the beam can be split up into sub-beams by the edges of the reflecting planes. A beam is a set of three rays, forming a triangular beam, see Figure 2.7(a), or higher numbers of rays. The beams for larger numbers of rays could be convex or concave (with indents). In the same way as for the polygons that build up a room model, beams that have a convex cross section will be easier to process, in particular triangular beams.

The method requires these consecutive steps:

1. Emit beams from a source position.
2. Follow all these beams through the model and let them be reflected at all room boundaries that are visible inside the beam, after the last reflection event. Repeat this for one reflection order at a time until some interrupt criterion is reached, either given by maximum reflection order, or maximum total traveling time. Beams that hit edges of room boundary polygons can be split up into smaller beams, see Figure 2.7(b). The resulting beam tree is independent of the receiver position and thus corresponds to the IS tree for the ISM algorithm.
3. Check which beams contain the receiver position inside the beam. Each such beam will either generate a valid reflection path automatically, if beams have been split up exactly, or will generate a potential reflection path, which must be checked for validity using the same methods as for the ISM algorithm.

One efficient implementation is presented in Reference 27, where the involved geometrical tests are detailed. One example of a beam which hits the edge of a polygon, and then gets split, is illustrated in Figure 2.7(b). Such a split up will lead to an exponentially increasing number of beams, as discussed by Funkhouser et al.[24]

Since beam tracing is an efficient method for finding the validity regions of ISs, and since the beam splitting also identifies diffraction paths, such paths can be identified in the process.[21]

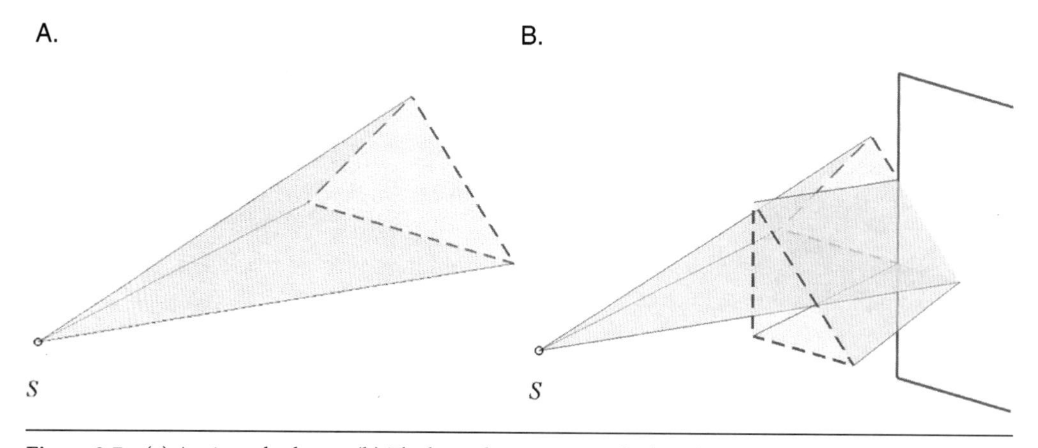

Figure 2.7 (a) A triangular beam. (b) The beam hits an exposed edge of a polygon and gets split up.

The beam-tracing technique is just an alternative method for finding, geometrically, the valid paths in a room model, and thus gives as an output a list of such paths for a specific source-receiver combination. If diffraction is implemented, then the contributions from all such paths are added, following the expressions in Eq. (2.24) and (2.27). For beam tracing without diffraction, contributions could either be computed by Eq. (2.25), or by Eq. (2.24). The latter is often referred to as *phased beam tracing.*

2.3 STATISTICAL MODELING OF GEOMETRICAL ACOUSTICS

Finding all the possible reflection paths as obtained with the deterministic methods is not always necessary. Instead, it is often sufficient to find a representative set of the reflection paths that can be used to reliably predict the response. This section presents some approximate and efficient modeling techniques to obtain an echogram, from which a room impulse response can be constructed.

2.3.1 Ray Tracing

Ray tracing is the most often used stochastic modeling technique in room acoustics. It is based on Monte Carlo sampling of the possible reflection paths to find the ones that actually contribute to the echogram. Ray tracing is unique in that it can easily handle specular reflections as well as diffuse reflections. The ray-tracing method does the following steps, as illustrated in Figure 2.8.

1. A large number of rays are sent out from the source, in directions that are either randomly or systematically distributed over the entire sphere. For each ray, either a history will be kept of which surfaces have been hit, or the current ray intensity.
2. For each ray, one or more truncation criteria are tested:
   a. Has a maximum propagation time been exceeded? This is similar to the proximity test for the ISM.

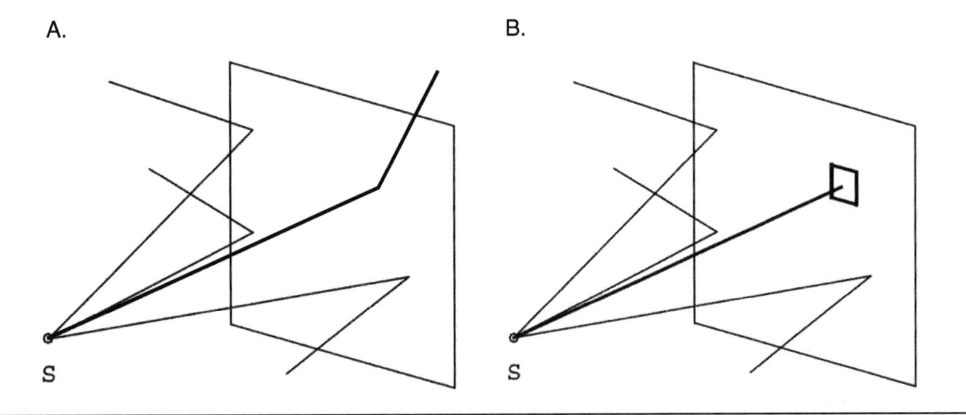

Figure 2.8 Illustration of the ray-tracing principle. (a) Reflection at a surface. One ray, in bold, is reflected specularly and three rays are reflected diffusely. (b) A square receiver area, with bold lines, is hit by one of the rays, drawn in bold.

   b. Has a maximum reflection order been reached? This is similar to maximum IS tree depth for the ISM.
   c. Is the remaining ray intensity high enough to be relevant?
3. If Test 2 doesn't lead to the truncation of a ray, then two tests are performed:
   a. Intersection test with one or more receiver volumes/areas. If so, then a hit is registered with the reflection history/intensity stored for that ray, see Figure 2.8(b).
   b. Intersection with all visible wall polygons is tested. If there is an intersection, the ray is reflected back into the room and the new reflected ray is fed back to Step 2, see Figure 2.8(a).

Originally, ray tracing was introduced to model specular reflections,[28] but it is easy to model non-specular reflections as well by performing the reflection according to the reflection distribution function as described in subsection 2.1. One implementation would be to generate several outgoing rays at reflections randomly from an angle distribution, instead of using only one outgoing ray. However, this approach leads to an exponential growth of rays with propagation time. This growth can be avoided by summing the diffuse contributions across fixed-size wall elements.[29] Classical ray tracing uses a randomized approach such that the reflected ray is either specular or diffuse, based on the surface's scattering coefficient,[30] see Figure 2.8(a). An important consequence of this approach is that frequency-dependent scattering coefficients will require a separate ray-tracing process for each frequency band. Another common version of ray tracing is to construct a separate path from each diffuse reflection point to the receiver—so-called *diffuse rain*.[30] The approaches by Dalenbäck[29] and Schröder[30] both borrow elements from radiance exchange methods, see section 2.3.2. Finally, an approach based on ray divergence near edges has been suggested for the incorporation of diffraction effects into ray tracing.[31]

The choice of detector type will differentiate between some versions of ray tracing. Classical ray tracing[28] uses a *fixed detector size*, and then the spherical spreading/divergence of the rays will lead to the deduction that a large number of rays may follow the same reflection path and reach the detector. Consequently, the $1/r^2$-factor for the intensity/pressure squared in Eq. (2.25) is not included in the contribution by each ray—its effect will be handled by the number of rays hitting the detector. As discussed by Stephenson,[26] some further details apply to the choice of area versus volume detectors.

Another version is approximate beam tracing, or cone tracing, where each ray carries a *circular detector disc*, and a receiver hit is registered if the receiver point is hit by this ray cross-section disc. This corresponds to representing circular beams by its central ray. For this version, a single ray hit, or possibly a few, will be registered for each reflection sequence so, as a consequence, the $1/r^2$-factor is included in the contribution by each ray. The possibility to register more than one hit is because rays with circular cross sections will necessarily overlap a little.

As mentioned for the ISM and beam tracing, the outcome of this method could be a list of reflection paths, and contributions for each path will be computed by Eq. (2.24) (without the $1/r^2$-factor for classical ray tracing). The use of a receiver detector volume or area indicates a certain degree of approximation—sampling errors will occur because of this. If the reflection history has been saved for each ray, then the so-called hybrid method[32] can be used for replacing each registered hit (with an all-specular reflection history) by its exact IS. A separate visibility test has to be carried out for each such potentially visible IS, but the number of such candidates is greatly reduced compared to classical ISM approaches.

Adaptive frustum tracing is another version of the approximate beam tracing.[33] In this technique, square detectors are carried by each ray, forming so-called frusta, and in addition, these detectors are subdivided when needed upon reflections so that the frustum size will adapt to the size of the reflecting polygons. Still, each frustum is represented by its central ray so there will be a certain amount of sampling error, but very efficient implementations are possible.[33]

2.3.2 Radiance Exchange Methods

While specular reflections are physically correct representations of reflections from rigid flat surfaces with infinite extent, in practice some surfaces produce more diffuse-like reflections. Diffusors in concert halls are an example of such surfaces. They can be modeled either with the ray-tracing algorithm as described in the previous subsection, or with an algorithm specialized in diffuse reflection modeling. They can be used in combination with the other algorithms or as such for room acoustics modeling.

The fundamental idea behind this class of modeling algorithms is that the surface of the room is divided into small elements and the acoustic radiance exchange between them is modeled.[25, 34, 35, 36, 37] Initially, the sound source radiates onto the surface elements; then each of these elements reflects the incoming radiance in all directions according to the diffuse reflection model. This radiance arrives at other elements and is reflected further. The radiance exchange between the elements can be explicitly calculated by utilizing their geometric properties. Part of the radiant energy is absorbed at each reflection and the total energy in the system decays as the radiance is iteratively propagated between the elements. When the total energy is below a threshold, the propagation can be terminated. In the end, each element has stored an echogram based on the radiant energy that it has received over time. Then, a final gathering step is performed, where the stored echograms contribute to the echogram at the receiver position according to their geometric relationships.

The results of the radiance exchange methods resemble those of ray tracing because the final result is an echogram. On the other hand, because the number of elements is fixed in the beginning, these methods do not suffer from the exponential explosion in the number of ISs. Thus, they are well fit for modeling the late reverberation. The main limitation is the reflection model, which is either ideal diffuse or directional diffuse. Early reflections from rigid flat surfaces are better modeled with the other methods, even though attempts have been made to include specular reflections in a radiance exchange formulation.[38]

2.3.3 Diffusion Equation Method

The diffusion equation was first developed by Fourier for heat conduction in solid media, while Ollendorff was the first to propose the diffusion model to room-acoustic predictions.[39] The diffusion equation can be considered an asymptotic approximation of the ARTM described in Section 1.1; however, for the purpose of computationally efficient modeling, the transport-equation model[2,3] is less compelling than the diffusion equation models. Picaut and his colleagues have demonstrated the practical usability of the diffusion model in a variety of enclosure types.[40,41]

Allowing the assumption of ergodicity, energy density $w(\mathbf{r}, t)$ as a function of position $\mathbf{r}$ and time t is given to first order by the diffusion equation:

$$\frac{\partial w(\mathbf{r},t)}{\partial t} - D\nabla^2 w(\mathbf{r},t) + c\,m\,w(\mathbf{r},t) = F(\mathbf{r},t) \in V, \tag{2.29}$$

subject to the boundary condition on the interior surface S:

$$D\frac{\partial w(\mathbf{r},t)}{\partial n} + c\,A\,w(\mathbf{r},t) = 0, \tag{2.30}$$

where Eq. (2.29) is the interior equation for the room denoted by domain V. The domain contains a source term, $F(\mathbf{r},t)$, which is zero for any room positions where no source is present. The diffusion coefficient D is given by $D = \lambda c/3$, where λ is the mean free path. The term $c\ m\ w(\mathbf{r},t)$ accounts for air dissipation in the room(s).[42] The absorption term, A can take the following forms:

$$A_S = \frac{\alpha}{4};\quad A_E = \frac{-\ln(1-\alpha)}{4};\text{or}\quad A_M = \frac{\alpha}{2(2-\alpha)}. \tag{2.31}$$

The term A_S has been used in room-acoustics predictions since 1969.[39] The other absorption term, A_E in Eq. (2.31), is suitable for modeling cases where surfaces of a room feature higher overall absorption than that when using A_S.[43] Diffusion equations using the absorption terms A_S or A_E in Eq. (2.31) are designated as the diffusion-Sabine model or diffusion-Eyring model, respectively. The term A_M in Eq. (2.31) is theoretically grounded and can model both low and high absorption surfaces.[44] This boundary condition is especially important when modeling configurations in spaces, which allows the absorption coefficient to be 1.0 for boundary surfaces within the domain of interest.

Xiang et al., have shown that the diffusion equation model is valid after the space under investigation has already undergone sufficient mixing of sound energy density, on order of 2–3 mean-free-path time.[45, 46] For this reason, valid modeling of a sound field using the diffusion equation model does not include the direct sound and early reflections.

The diffusion equation method can use any spatial grid spacing such that a denser grid gives more reliable results in the whole frequency band. This is in contrast to finite-difference time-domain (FDTD) techniques in which the valid frequency band is strictly limited by the grid spacing. In practice, it has been shown that a spatial grid spacing of less than 0.5m is sufficient when the update rate is at least 8kHz (i.e., the temporal step is less than 1/8 ms).[47]

REFERENCES

1. J. M. Navarro, F. Jacobsen, J. Escolano, and J. J. López, "A theoretical approach to room acoustic simulations based on a radiative transfer model," *Acta Acust. Acust.*, vol. 96, no. 6, p. 12, 2010.
2. Y. Jing, E. W. Larsen, and N. Xiang, "One-dimensional transport equation models for sound energy propagation in long spaces: theory," *J. Acoust. Soc. Am.*, vol. 127, no. 4, pp. 2312–2322, Apr. 2010.
3. Y. Jing and N. Xiang, "One-dimensional transport equation models for sound energy propagation in long spaces: simulations and experiments," *J. Acoust. Soc. Am.*, vol. 127, no. 4, pp. 2323–2331, Apr. 2010.
4. S. Siltanen, T. Lokki, S. Kiminki, and L. Savioja, "The room acoustic rendering equation," *J. Acoust. Soc. Am.*, vol. 122, no. 3, pp. 1624–1635, 2007.

5. F. P. Mechel, "Improved mirror source method in room acoustics," *J. Sound Vib.*, vol. 256, no. 5, pp. 873–940, Oct. 2002.
6. C. F. Chien and W. W. Soroka, "Sound propagation along an impedance plane," *J. Sound Vib.*, vol. 43, no. 1, pp. 9–20, Nov. 1975.
7. M. Ochmann, "The complex equivalent source method for sound propagation over an impedance plane," *J. Acoust. Soc. Am.*, vol. 116, no. 6, pp. 3304–3311, Dec. 2004.
8. G. Taraldsen, "A note on reflection of spherical waves," *J. Acoust. Soc. Am.*, vol. 117, no. 6, pp. 3389–3392, Jun. 2005.
9. J. Allen and D. Berkley, "Image method for efficiently simulating small-room acoustics," *J. Acoust. Soc. Am.*, vol. 65, no. 4, pp. 943–950, 1979.
10. Y. W. Lam, "Issues of computer modelling of room acoustics in non-concert hall settings," *Acoust. Sci. Tech.*, vol. 26, no. 2, pp. 145–155, 2005.
11. R. Duraiswami, D. Zotkin, and N. Gumerov, "Fast evaluation of the room transfer function using multipole expansion," *IEEE Trans. on Acoustics, Speech and Language Proc.*, vol. 15, no. 2, 2007.
12. J. Borish, "Extension of the image model to arbitrary polyhedra," *J. Acoust. Soc. Am.*, vol. 75, no. 6, pp. 1827–1836, 1984.
13. J. Foley, A. van Dam, S. Feiner, and John F. Hughes, *Computer Graphics: Principles and Practice, Second Edition in C.* Addison-Wesley Professional, 1996, p. 1175.
14. R. R. Torres, U. P. Svensson, and M. Kleiner, "Computation of edge diffraction for more accurate room acoustics auralization," *J. Acoust. Soc. Am.*, vol. 109, no. 2, pp. 600–610, Feb. 2001.
15. U. P. Svensson, R. I. Fred, and J. Vanderkooy, "An analytic secondary source model of edge diffraction impulse responses," *J. Acoust. Soc. Am.*, vol. 106, no. 5, pp. 2331–2344, 1999.
16. U. P. Svensson and U. R. Kristiansen, "Computational modelling and simulation of acoustic spaces," in *Proc. AES 22nd Conf. Virtual, Synthetic and Entertainment Audio*, 2002, pp. 11–30.
17. U. P. Svensson, P. T. Calamia, and S. Nakanishi, "Frequency-domain edge diffraction for finite and infinite edges," *Acta Acust. Acust.*, vol. 95, no. 3, pp. 568–572, 2009.
18. U. Svensson and P. Calamia, "Edge-diffraction impulse responses near specular-zone and shadow-zone boundaries," *Acta Acustica united with Acustica*, vol. 92, no. 4, pp. 501–512, 2006.
19. J. Keller, "Geometrical theory of diffraction," *J. Optical Soc. Am.*, vol. 52, no. 2, pp. 116–130, 1962.
20. R. D. Kouyoumjian and P. H. Pathak, "A uniform geometrical theory of diffraction for an edge in a perfectly conducting surface," *Proc. IEEE*, vol. 62, no. 11, pp. 1448–1461, 1974.
21. N. Tsingos, T. Funkhouser, A. Ngan, and I. Carlbom, "Modeling acoustics in virtual environments using the uniform theory of diffraction," in *SIGGRAPH '01: Proc. 28th Conf. on Computer Graphics and Interactive Techniques*, 2001.
22. L. Antani, A. Chandak, M. Taylor, and D. Manocha, "Efficient finite-edge diffraction using conservative from-region visibility," *Appl. Acoust.*, vol. 73, no. 3, pp. 218–233, Mar. 2012.
23. N. Dadoun, D. G. Kirkpatrick, and J. P. Walsh, "The geometry of beam tracing," in *Proc. First Annual Symposium on Computational Geometry—SCG '85*, 1985, pp. 55–61.
24. T. Funkhouser, I. Carlbom, G. Elko, G. Pingali, M. Sondhi, and J. West, "A beam-tracing approach to acoustic modeling for interactive virtual environments," in *SIGGRAPH '98: Proc. 25th Conf. on Computer Graphics and Interactive Techniques*, 1998, pp. 21–32.
25. T. Lewers, "A combined beam tracing and radiant exchange computer model of room acoustics," *Appl. Acoust.*, vol. 38, no. 2–4, pp. 161–178, 1993.
26. U. Stephenson, "Quantized pyramidal beam tracing—a new algorithm for room acoustics and noise immission prognosis," *Acta Acust. Acust.*, vol. 82, pp. 517–525, 1996.
27. S. Laine, S. Siltanen, T. Lokki, and L. Savioja, "Accelerated beam-tracing algorithm," *Appl. Acoust.*, vol. 70, no. 1, pp. 172–181, 2009.
28. A. Krokstad, S. Strøm, and S. Sørsdal, "Calculating the acoustical room response by the use of a ray tracing technique," *J. Sound Vib.*, vol. 8, no. 1, pp. 118–125, 1968.
29. B.-I. Dalenbäck, "Room acoustic prediction based on a unified treatment of diffuse and specular reflection," *J. Acoust. Soc. Am.*, vol. 100, no. 2, pp. 899–909, 1996.
30. D. Schröder, *Physically based real-time auralization of interactive virtual environments.* Berlin: Logos Verlag, 2011, p. 59.

31. U. Stephenson, "An Energetic Approach for the Simulation of Diffraction within Ray Tracing Based on the Uncertainty Relation," *Acta Acust. Acust.*, vol. 96, no. 3, pp. 516–535, 2010.
32. M. Vorländer, "Simulation of the transient and steady-state sound propagation in rooms using a new combined ray-tracing/image-source algorithm," *J. Acoust. Soc. Am.*, vol. 86, no. 1, pp. 172–178, 1989.
33. C. Lauterbach, A. Chandak, and D. Manocha, "Interactive sound rendering in complex and dynamic scenes using frustum tracing," *IEEE Trans. on Visualization and Computer Graphics*, vol. 13, no. 6, pp. 1672–1679, 2007.
34. D. Immel, M. Cohen, and D. Greenberg, "A radiosity method for non-diffuse environments," in *SIGGRAPH '86: Proc. 13th Conf. on Computer Graphics and Interactive Techniques*, 1986.
35. N. Tsingos and J.-D. Gascuel, "A general model for the simulation of room acoustics based on hierachical radiosity," in *SIGGRAPH '97: SIGGRAPH 97 Visual Proceedings: The art and interdisciplinary programs of SIGGRAPH '97*, 1997.
36. E.-M. Nosal, M. Hodgson, and I. Ashdown, "Improved algorithms and methods for room sound-field prediction by acoustical radiosity in arbitrary polyhedral rooms," *J. Acoust. Soc. Am.*, vol. 116, no. 2, pp. 970–980, 2004.
37. S. Siltanen, T. Lokki, and L. Savioja, "Frequency domain acoustic radiance transfer for real-time auralization," *Acta Acust. Acust.*, vol. 95, no. 1, pp. 106–117, 2009.
38. N. Korany, J. Blauert, and O. Abdel Alim, "Acoustic simulation of rooms with boundaries of partially specular reflectivity," *Appl. Acoust.*, vol. 62, no. 7, pp. 875–887, Jul. 2001.
39. F. Ollendorff, "Statistical room-acoustics as a problem of diffusion—A proposal," *Acustica*, vol. 21, pp. 236–245, 1969.
40. J. Picaut, L. Simon, and J. D. Polack, "A mathematical model of diffuse sound field based on a diffusion equation," *Acta Acust. Acust.*, vol. 83, no. 4, pp. 614–621, 1997.
41. A. Billon, C. Foy, J. Picaut, and V. Valeau, "Modeling the sound transmission between rooms coupled through partition walls by using a diffusion model," *J. Acoust. Soc. Am.*, vol. 123, no. 6, pp. 4261–4271, 2008.
42. A. Billon, J. Picaut, C. Foy, V. Valeau, and A. Sakout, "Introducing atmospheric attenuation within a diffusion model for room-acoustic predictions," *J. Acoust. Soc. Am.*, vol. 123, no. 6, pp. 4040–4043, Jun. 2008.
43. Y. Jing and N. Xiang, "A modified diffusion equation for room-acoustic predication," *J. Acoust. Soc. Am.*, vol. 121, no. 6, pp. 3284–3287, 2007.
44. Y. Jing and N. Xiang, "On boundary conditions for the diffusion equation in room-acoustic prediction: Theory, simulations, and experiments," *J. Acoust. Soc. Am.*, vol. 123, no. 1, pp. 145–153, Jan. 2008.
45. J. Escolano, J. M. Navarro, and J. J. López, "On the limitation of a diffusion equation model for acoustic predictions of rooms with homogeneous dimensions," *J. Acoust. Soc. Am.*, vol. 128, no. 4, pp. 1586–1589, Oct. 2010.
46. N. Xiang, Y. Jing, and A. C. Bockman, "Investigation of acoustically coupled enclosures using a diffusion-equation model," *J. Acoust. Soc. Am.*, vol. 126, no. 3, pp. 1187–1198, Sep. 2009.
47. J. M. Navarro, J. Escolano, and J. J. López, "Implementation and evaluation of a diffusion equation model based on finite difference schemes for sound field prediction in rooms," *Appl. Acoust.*, vol. 73, no. 6–7, pp. 659–665, Jun. 2012.

3

Acoustics in Long Rooms

Jian Kang, School of Architecture, University of Sheffield, United Kingdom

INTRODUCTION

This chapter examines the acoustics of long rooms, where one dimension is much greater than the other two, although the other two are still relatively large compared to the acoustic wavelength. Examples of long rooms include underground/railway stations, corridors, concourses, as well as road or railway tunnels. Acoustics is often a major concern in many long rooms. For example, in underground/railway stations, poor speech intelligibility of public address (PA) systems can cause loss of important travel information and misunderstanding of vital instructions during an emergency; in public buildings, the noise disturbance between rooms through a corridor is often significant; and in road tunnels, noise pollution is a serious problem.

This chapter starts with a discussion of fundamental acoustic characteristics in long rooms and inapplicability of classic room acoustic theories, given that the assumption of a diffuse field does not hold with the extreme dimensional condition. It then presents a number of computer models for long rooms based on the *image source method, radiosity method*, and *ray tracing*. A number of statistical, empirical, and analytical formulae are then reviewed and summarized. This is followed by a series of parametric studies using computer models to examine the effects of designable factors in long rooms. Based on measurements in two 1:16 scale models of underground stations, the effectiveness of strategic architectural acoustic treatments are then discussed. Finally, the chapter presents some measurement results in actual long rooms, including a corridor, an underground shopping street, and several underground stations.

3.1 FUNDAMENTALS OF ACOUSTIC CHARACTERISTICS IN LONG ROOMS

In a diffuse field, sound rays are uniformly distributed over all directions of propagation, and there is no difference between time average and directional average. By taking the direct sound into account, the sound pressure level (SPL) becomes approximately constant beyond the reverberation radius, a parameter that is strongly dependent on the room absorption conditions. The reverberation time (T_{30}) in a diffuse field is a single value across the room and the logarithmic decay curves are linear. In long rooms, the sound field is fundamentally different from that in a diffuse field, as analyzed below in terms of sound distribution and reverberation, using both ray theory and wave theory.

3.1.1 Ray Theory

Figure 3.1 compares a rectangular long room and a cube in terms of the reflection pattern and the SPL directional distribution at a typical receiver, where the boundaries are considered as geometrically reflective, with an absorption coefficient of $\alpha = 0.05$ [Kang, 1996f]. The image source method (ISM) is used in the analysis. For the long room, the first order image source (IS) plane plays a more important role than the others, especially in the areas that are relatively further from the end walls. As a result, the SPL decreases continuously along the length due to the increased length of reflected sound paths, although less so, or even increases slightly, in the area near the end walls. Moreover, from Figure 3.1(a) it can be seen that in the long room the rays are not uniformly distributed in all directions. In the cube, conversely, the rays are more uniformly distributed in all directions, and the difference of sound energy from various IS planes is much less, so that the sound distribution is more even.

Diffusers on boundaries usually make the sound field more diffuse in regularly shaped rooms. In long rooms, conversely, diffusers could increase the relative SPL attenuation along the length, making the sound field even more nondiffuse. This can be explained with a simplified case, as illustrated in Figure 3.2 [Kang, 1996f]. It can be seen that, compared to the

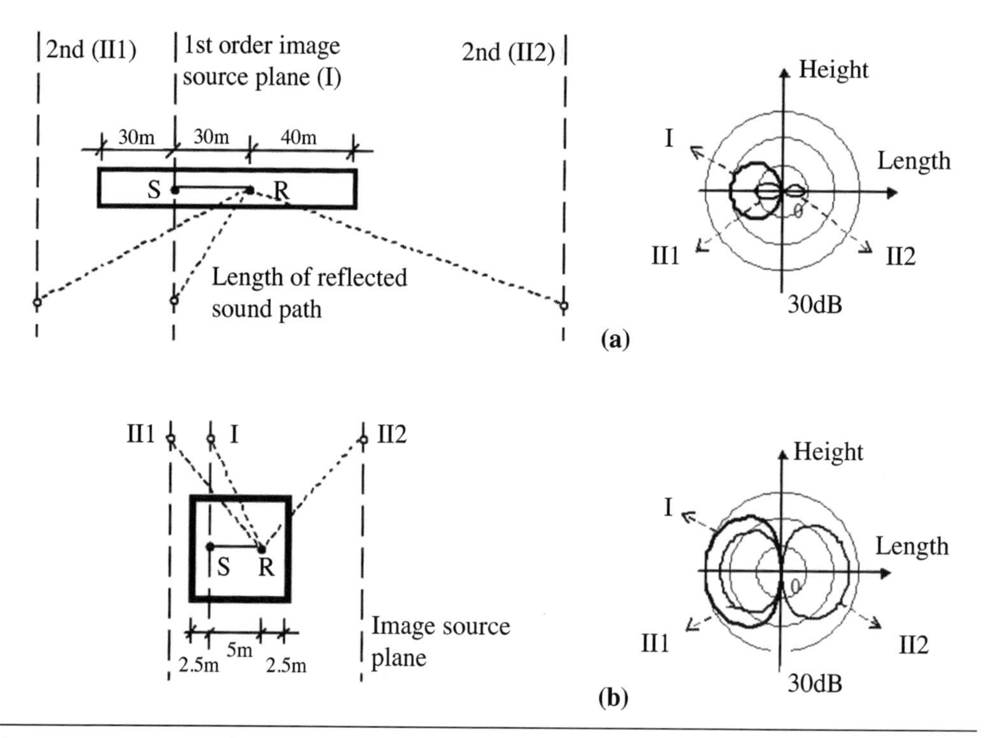

Figure 3.1 Comparison between a rectangular long room of 100m × 3m × 3m (a) and a cube of 10m × 10m × 10m (b) in terms of the reflection pattern (left) and the directional SPL distribution (right) at a typical receiver, where the boundaries are geometrically reflective, with an absorption coefficient of 0.05.

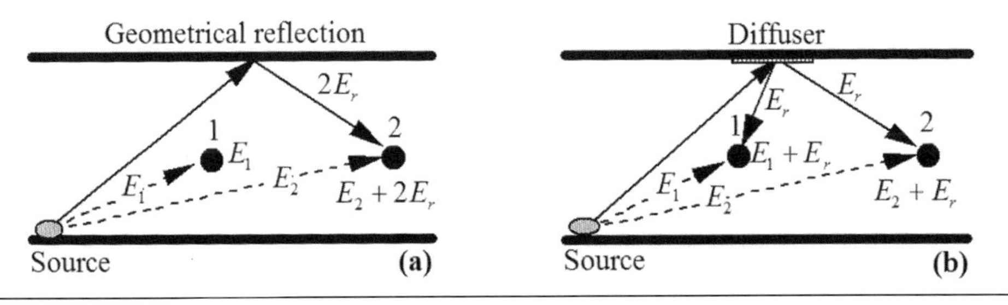

Figure 3.2 Comparison in sound attenuation along the length in long rooms with geometrically (a) and diffusely reflecting boundaries (b).

situation with geometrically reflecting boundaries, with diffusely reflecting boundaries the sound energy is increased at Receiver 1 and decreased at Receiver 2. Moreover, with diffusely reflecting boundaries the total energy in a long room is lower than that with geometrically reflecting boundaries. This is because with diffusely reflecting boundaries the average length of the reflected sound path between a source and receiver pair is generally longer and thus, more energy is absorbed by the medium, and the sound rays have more chances of impinging upon the boundaries, causing more energy loss.

In terms of the reverberation in long rooms, consider the first-order IS array in Figure 3.1. It can be seen that the difference in sound-path length between various image sources is greater with a shorter source-receiver distance and consequently, the reverberation in long rooms will vary systematically along the length, rather than a single value like in a diffuse field. With diffusely reflecting boundaries, it has been demonstrated that the variation in free path length is greater when the length of a rectangular room becomes longer [Kuttruff, 2000] and thus, in long rooms the use of mean free path, as in classic reverberation theories based on the assumption of diffuse field, becomes unreasonable.

3.1.2 Wave Theory

In Figure 3.3 a comparison is made between a regularly shaped room of 5m × 7m × 9m and a long room of 5m × 7m × 120m in terms of the *eigenfrequency* distribution. It can be seen that compared to the regularly shaped room, in the long room the eigenfrequency distribution is rather uneven, and the unevenness could be even greater if the width and height of the long room becomes the same [Kang, 2002c].

Since in long rooms, for a given frequency, the resonant forms of various modes could be significantly different and the difference in damping constants among various modes is likely to be great, the logarithmic decay curves in long rooms tend to be nonlinear, given that linear decay only occurs when the damping constants are equal for all resonant modes.

3.1.3 Definition of Long Rooms

In long rooms, the SPL decreases along the length continuously, instead of becoming stable beyond the reverberation radius; the reverberation time varies along the length, instead of a single value; and the decay curves tend to be nonlinear. As a result, for predicting the sound field in long rooms, classic theories such as the Sabine and Eyring formulae, which are based on the assumption of a diffuse field, must not be used.

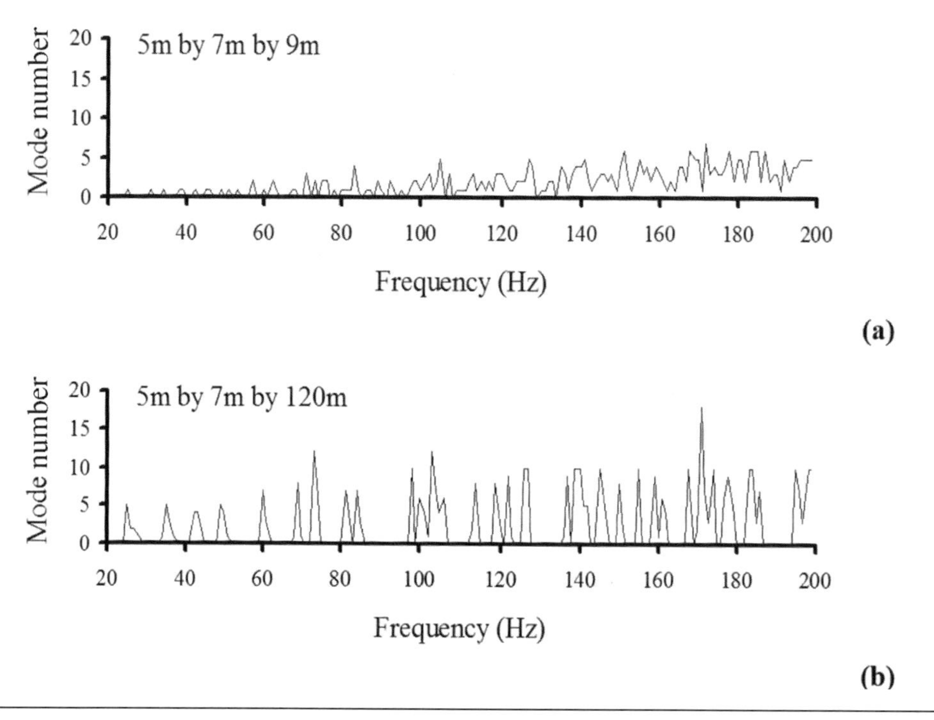

Figure 3.3 Comparison between a regularly shaped room of 5m × 7m × 9m (a) and a long room of 5m × 7m × 120m (b) in terms of the eigenfrequency distribution, where the vertical axis is the number of modes in a bandwidth of 1Hz.

A room is usually defined as a long room when the length is greater than six times the width and the height [Kang, 2002c]. However, the inapplicability of classic theories also depends on other factors:

1. The absorption condition of end walls is important. The variation in T_{30} along the length can be diminished by highly reflective end walls. Conversely, with totally absorbent end walls the reverberation also varies in regularly shaped rooms.
2. Boundary conditions along the length, such as diffusion, could affect the sound field significantly.
3. The length/cross-section ratio characterizing the long room could be different for reverberation and for sound distribution.
4. The width/height ratio should be considered. If the width is much greater than the height, the theories for flat rooms should be applied [Kang, 2002a], even if the length/cross-section ratio is still great.

3.2 ACOUSTIC SIMULATION OF LONG ROOMS

Given the special characteristics of long rooms, a number of computer models have been developed to predict the acoustic indices including T_{30}, steady-state SPL, and speech transmission index (STI). For reverberation, both T_{30} and early decay time (EDT) are considered. In a diffuse sound field a logarithmic decay curve is perfectly linear and thus, the T_{30} and EDT should have the same value, whereas in a nondiffuse field, like in a long room, the T_{30}/EDT ratio is an indication of nondiffuseness. The STI is a commonly used index for intelligibility, which is based on the modulation transfer function (MTF) between source and receiver. An attractive feature of the MTF is that the effects of reverberation, ambient noise, and the contribution of direct field are combined in a natural way in the single function [Steeneken and Houtgast, 1980].

3.2.1 Image Source Model

Figure 3.4 shows the distribution of IS in an infinitely long room with geometrically reflecting boundaries, where a and b are the cross-sectional height and width [Kang, 1996c]. Consider an impulse from the source, if the arrival time of direct sound is defined as $t = 0$, then the reflection of the image source (m, n) arrives at a receiver at time $t_{z,m,n}$:

$$t_{z,m,n} = \frac{D_{z,m,n}}{c} - \frac{z}{c}, \tag{3.1}$$

where $D_{z,m,n}$ is the distance between an IS (m, n) and the receiver, c is the speed of sound, and z is the source-receiver distance along the length. Between time t and $t + \Delta t$, a short time L_{eq} (equivalent continuous sound level) at a receiver can be calculated by:

$$L(t)_z = 10\log\left[\sum_{m=-\infty}^{\infty}\sum_{n=-\infty}^{\infty} e^{-MD_{z,m,n}} E_{z,m,n}\right], \tag{3.2}$$

where M is the intensity related attenuation constant in air, and $\boldsymbol{E}_{z,m,n}$ is the sound energy contributed by the IS (m, n) at the receiver:

$$E_{z,m,n} = \frac{K_W}{D_{z,m,n}}(1-\alpha)^{|m|+|n|} \qquad t \le t_{d,m,n} < t + \Delta t \tag{3.3}$$

$$= 0 \qquad \text{otherwise},$$

where K_W is a constant relating to the sound power of the source, and α is the absorption coefficient of the boundaries, by assuming that all the boundaries have the same absorption coefficient and the absorption is angle independent.

With Eq. (3.1) acoustic indices including the T_{30}, EDT, and the steady-state SPL can be derived using the reverse-time integration.

The above calculation can be simplified by using a statistical method, as illustrated in Figure 3.5 [Kang, 1996c]. This is to calculate the average reflection distance D_0, the approximate number of image sources N, and the average order of image sources R, all between time t and $t + \Delta t$, providing $D_p >> a,b$:

$$D_0 = c\left(\frac{z}{c} + t + \frac{1}{2}\Delta t\right), \tag{3.4}$$

$$N = \frac{\pi c^2}{S}\left[2\left(\frac{z}{c} + t\right) + \Delta t\right]\Delta t, \tag{3.5}$$

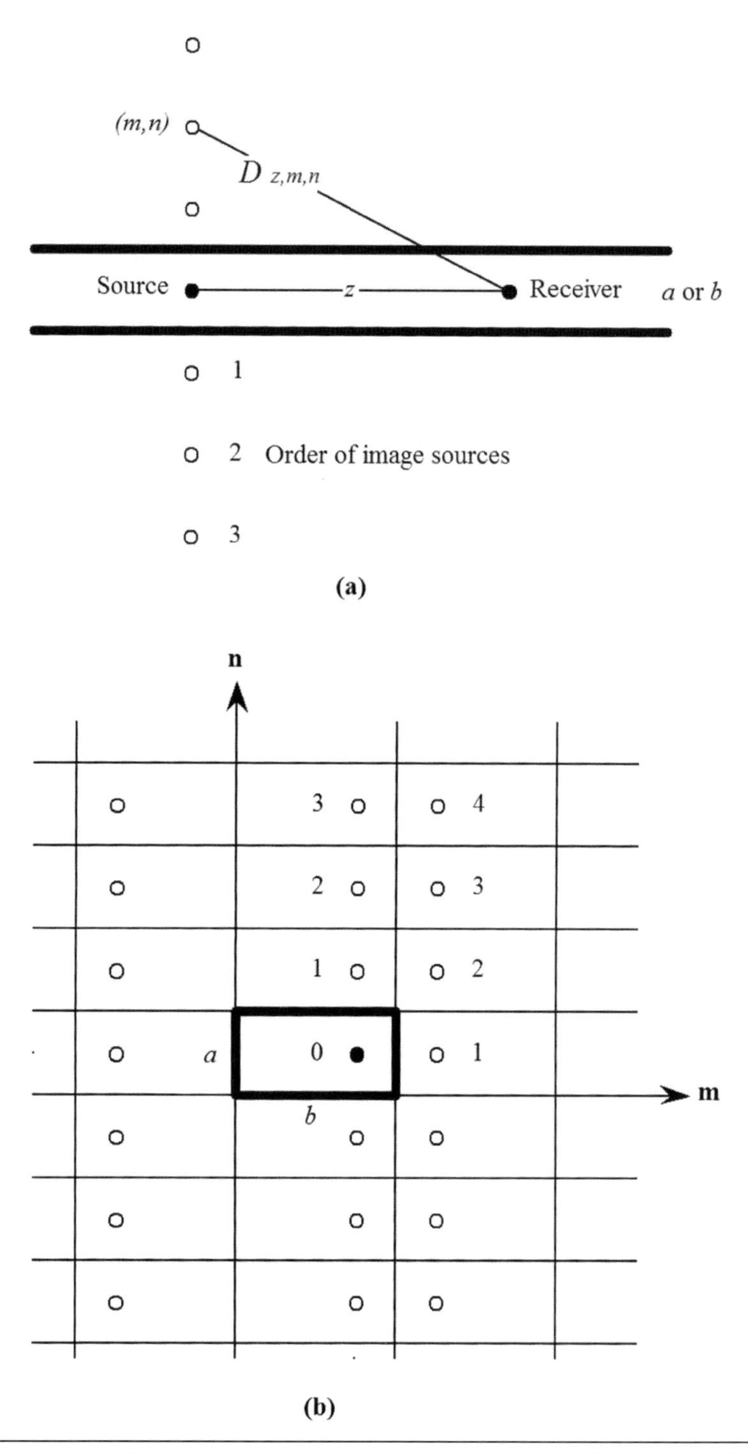

Figure 3.4 Distribution of ISs in length-wise section (a) and cross section (b) in an infinitely long room.

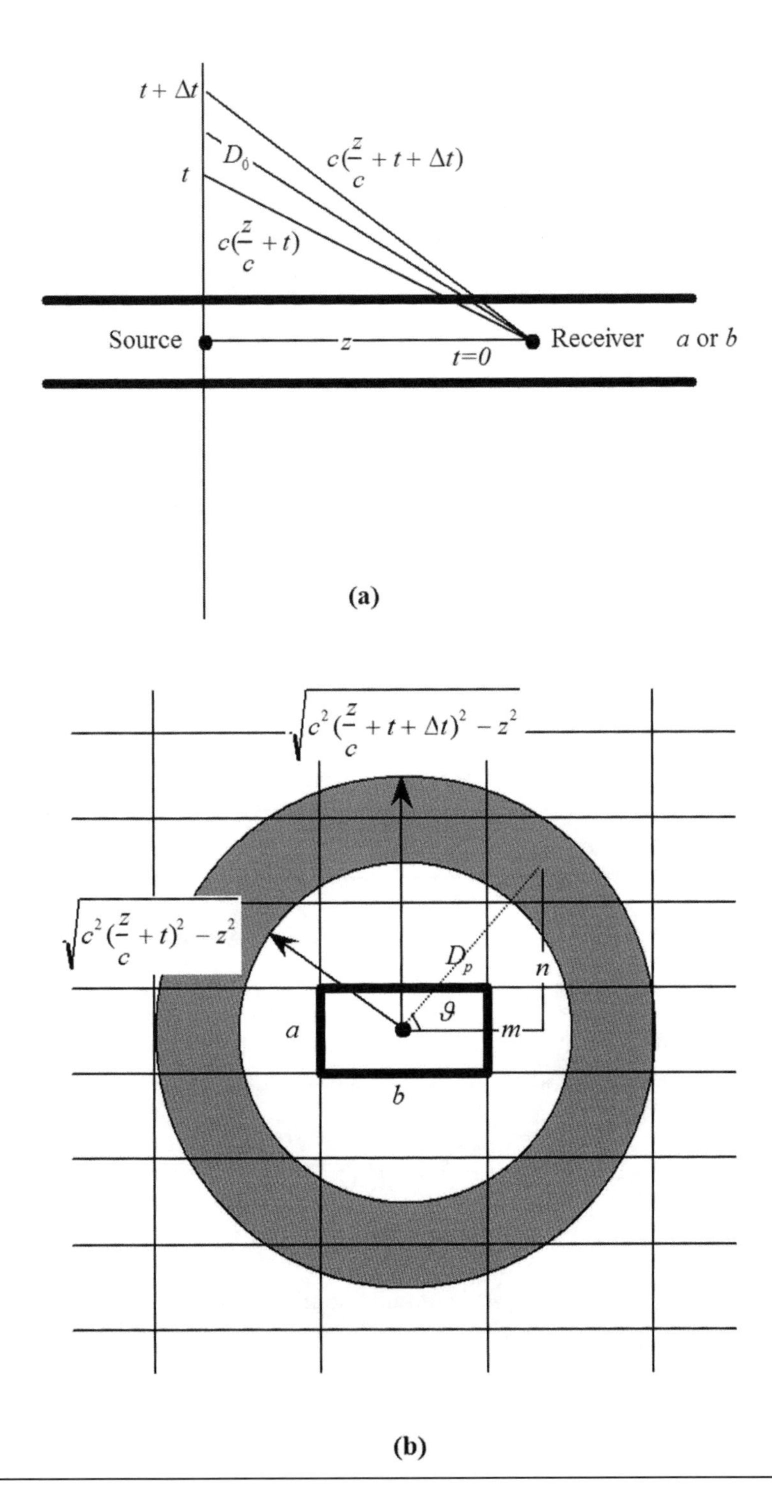

Figure 3.5 Statistical distribution of ISs in length-wise section (a) and cross section (b) in an infinitely long room, for the calculation of the average reflection distance D_0, the approximate number of ISs N, and the average order of ISs R.

$$R = \frac{2\Delta\vartheta}{\pi} \sum_{\vartheta=0, step:\Delta\vartheta}^{\frac{\pi}{2}} \left(\frac{D_p \sin\vartheta}{a} + \frac{D_p \cos\vartheta}{b} \right). \tag{3.6}$$

In Eqs. (3.4) and (3.5) the term z/c is introduced to scale the arrival time of direct sound to zero. S is the cross-sectional area. In Eq. (3.6) ϑ is an angle for determining the position of ISs, and D_p is the projection of D_0 into the IS plane. Consequently, $L(t)_z$ can be calculated with:

$$L(t)_z = 10\log\left[N \frac{K_W}{D_0^2}(1-\alpha)^R \right] - MD_0, \tag{3.7}$$

where M is the air absorption coefficient in dB/m.

Eq. (3.7) again shows that the reverberation in long rooms depends on the source-receiver distance, which is fundamentally different from that of the diffuse field.

If the source is directional, Eqs. (3.1) and (3.7) can be easily modified by multiplying an additional term with the directionality characteristics. For long rooms with a finite length, namely, if the end walls are reflective, more IS planes, as shown in Figure 3.1, should be integrated by modifying the above equations with different source-receiver distances [Kang, 1996c].

The ISM has been validated against measurements in a corridor of 42.5m × 1.56m × 2.83m and in several underground stations in London, where the boundaries could be approximately considered as geometrically reflective [Kang, 2002c].

It is noted that when relatively small cross sections and low frequencies are considered, the interference between reflections should be taken into account, for which coherent models have been developed [Lam and Li, 2007].

3.2.2 Radiosity Model

While in the ISM the boundaries are assumed to be acoustically smooth and geometrically reflective, the consideration of diffusely reflecting boundaries is important and several methods have been explored [Picaut et al., 1999; Ollendorff, 1976]. A radiosity model has also been developed for long rooms, and the main simulation steps are [Kang, 2002b]:

1. Divide each boundary into a number of patches, in rectangular or other forms such as triangles. While smaller patches will generally lead to a better accuracy, there is a square-law increase of calculation time in the patch number. Since for a given patch size the calculation of form factor becomes less accurate the closer the patch is to an edge, the boundaries are so divided that a patch is smaller when it is closer to an edge.
2. Distribute the sound energy of an impulse source to the patches. The fraction of the energy received by each patch is the same as the ratio of the solid angle subtended by the patch at the source to the total solid angle. The mean beam length between the source and patch, an important factor for the calculation accuracy, can be determined by subdividing a patch into a number of equal elements and then calculating their average distance to the source. It can also be approximated by the distance from the source to the center of the patch if the patch size is sufficiently small. The patches can then be regarded as sound sources, which are called first-order patch sources. Directional sources can also be modeled at this stage.

3. Determine the form factors between pairs of patches, namely the fraction of the sound energy diffusely emitted from one patch which arrives at the other by direct energy transport. In the model it is assumed that the sound energy reflected from a boundary is dispersed over all directions according to the Lambert cosine law.
4. Redistribute the sound energy of each first-order patch source to every other patch and thus generate the second-order patch sources. Continue this process and the k th order patch sources can be obtained ($k = 1 \ldots \infty$). This process is *memory-less*, namely, the energy exchange between patches depends only on the form factors and the patch sources of preceding order. Due to this feature, the computation speed can be greatly increased.
5. Calculate the energy response at each receiver by considering all orders of patch sources from which the acoustic indices can be derived.

In rectangular long rooms, the relative location between any two patches is either parallel or orthogonal. For the former, the form factor can be calculated by projecting the receiving patch onto the upper half of a cube centered about the radiation patch. For the latter, computing form factor is equivalent to projecting the receiving patch onto a unit hemisphere centered about the radiation patch, projecting this projected area orthographically down onto the hemisphere's unit circle base, and dividing by the area of the circle. Figure 3.6, as an example, illustrates the calculation from emitter $A_{l',n'}$ to receiver $C_{l,m}$. By considering the absorption of patch $A_{l',n'}$ and air absorption, the energy from $A_{l',n'}$ to $C_{l,m}$, $AC_{(l',n'),(l,m)}$, can be calculated by:

$$AC_{(l',n'),(l,m)} = (1-\alpha_{A_{l',n'}})e^{-Md_{(l',n'),(l,m)}}\frac{1}{2\pi}\cdot \left|\cos^2\gamma_{(l',n'),(l,m)} - \cos^2(\gamma_{(l',n'),(l,m)} + \Delta\gamma_{(l',n'),(l,m)})\right|\vartheta_{(l',n'),(l,m)}, \tag{3.8}$$

where $l,l' = 1 \ldots N_X$, $m = 1 \ldots N_Y$, and $n' = 1 \ldots N_Z$ are the number of patches along the three dimensions of the room; $d_{(l',n'),(l,m)}$ is the mean beam length between patches $A_{l',n'}$ and $C_{l,m}$; $\gamma_{(l',n'),(l,m)}$, $\Delta\gamma_{(l',n'),(l,m)}$, and $\vartheta_{(l',n'),(l,m)}$ are the angles for determining the relative location of the two patches; and $\alpha_{A_{l',n'}}$ is the angle-independent absorption coefficient of patch $A_{l',n'}$.

At time t the energy at a receiver R from the k th order patch sources on a boundary, for example, C, with patch energy as $C_k(t-R_{l,m}/c)_{l,m}$, , can be calculated by:

$$E_k(t)_C = \sum_{l=1}^{N_X}\sum_{m=1}^{N_Y}\left[\frac{C_k\left(t-\frac{R_{l,m}}{c}\right)_{l,m}}{\pi R_{l,m}^2}\cos(\xi_{l,m})\right]e^{-MR_{l,m}}\quad\left(t-\frac{R_{l,m}}{c}\geq 0\right), \tag{3.9}$$

where $R_{l,m}$ is the mean beam length between the receiver and patch $C_{l,m}$, and $\xi_{l,m}$ is the angle between the normal of patch $C_{l,m}$ and the line joining the receiver and the patch.

By considering all orders of patch sources as well as the direct energy transport from source to receiver, the energy response at receiver R can be calculated from which the acoustic indices including T_{30}, EDT, and the steady-state SPL can be determined. In the simulation, the time interval for calculating energy response is typically 3–5ms. Calculation stops when the total energy reduces to a certain amount, typically 10^{-6} of the source energy.

The above model can be modified to simulate the sound field in other spaces, including regularly shaped rooms and street canyons [Kang, 2006]. The models have been validated in

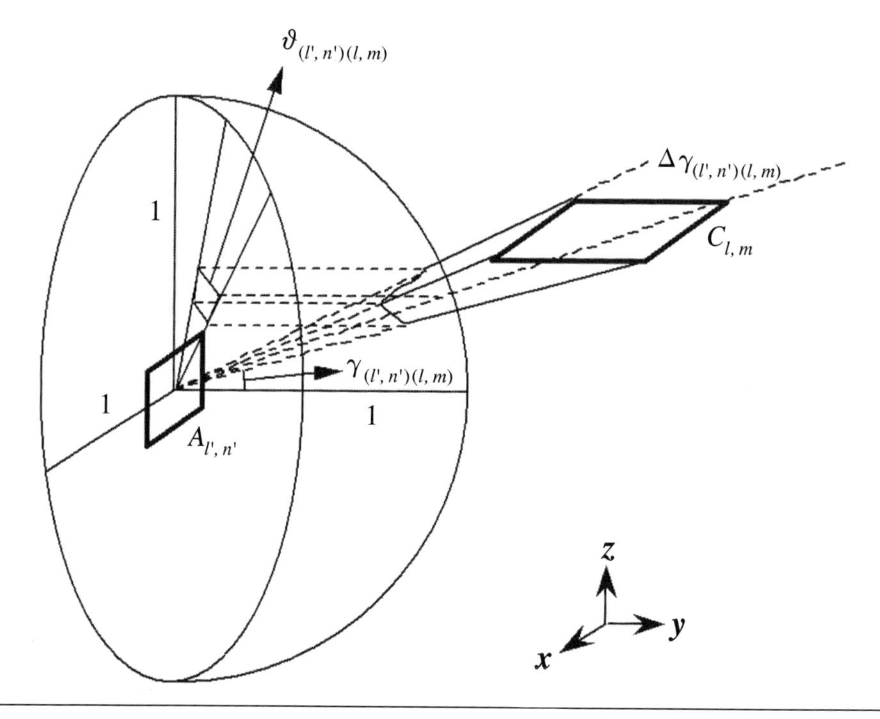

Figure 3.6 An example showing the determination of the form factor between two orthogonal patches, from emitter patch $A_{l',n'}$ to receiver patch $C_{l,m}$.

a number of spaces, compared to measurements, as well as to theoretical values, for example, the sum of the form factors from any patch to all the other patches should be unit.

3.2.3 Ray Tracing and Combined Ray Tracing and Radiosity

Ray tracing models have been developed for long rooms [Yang and Shield, 2001]. In comparison with the IS model, ray tracing models can consider more complex configurations. In ray tracing or beam tracing models the simulation of diffuse reflections is usually approximated by generating random ray directions, which is a possible source of inaccuracy. A combined ray tracing and radiosity model has thus been developed [Meng and Kang, 2008].

3.3 ACOUSTIC FORMULAE FOR LONG ROOMS

Given the inapplicability of classic theories and formulae for long rooms, a number of statistical, empirical, and analytical formulae have been developed, as summarized in this section. It is important to note that these formulae are all based on certain assumptions and it has been shown that the application outside their valid range could cause significant errors [Kang, 1996d].

3.3.1 SPL with Geometrically Reflecting Boundaries

Consider an infinitely long room with a rectangular cross section. Assume that all the boundaries have the same absorption coefficient, and a point source and a receiver are positioned at the center of the cross section. The sound energy density u can be determined using the ISM, as expressed in Eqs. (3.1) to (3.3) and Figure 3.4, and an approximation was proposed by Kuttruff [1989] as:

$$u=\frac{P}{4\pi c}\sum_{m=-\infty}^{\infty}\sum_{n=-\infty}^{\infty}\frac{\rho^{|m|+|n|}}{(ma)^2+(nb)^2+z^2}\approx\frac{P}{4\pi cz^2}[1+\frac{4\rho}{(1-\rho)^2}],\quad (z>>a,b), \tag{3.10}$$

where P is the sound power of the source, m and n are the IS orders, and ρ is the mean reflection coefficient of the boundaries.

Another approximation of the IS summation is [Kuno et al., 1989]:

$$u=-\frac{P}{2S}[\cos a_2 z ci(a_2 z)+\sin a_2 z si(a_2 z)], \tag{3.11}$$

where $a_2 = (U/\pi S)\log\rho$, U is the cross-sectional perimeter. Eq. (3.11) was validated by the measurements in a corridor of 2m × 2.6m × 57m with hard boundaries, and the accuracy was about ± 1.5dB.

Corresponding to Eq. (3.10), an approximation for a line source across the width and at half the height of the cross section was proposed by Kuttruff [1989], with the receiver at the center of the cross section:

$$u=\frac{P}{4c}\sum_{n=-\infty}^{\infty}\frac{\rho^{|n|}}{\sqrt{(na)^2+z^2}}\approx\frac{P}{4cz}(1+\frac{2\rho}{1-\rho}),\quad (z>>a). \tag{3.12}$$

For rectangular long rooms with different absorption coefficients of the four boundaries in a cross section, a formula was proposed by Yamamoto [1961], again based on the ISM:

$$\begin{aligned}L_z = L_W - 11 - 10\log[&\frac{1}{z^2}\\&+\sum_{m=0}\sum_{n=0}\frac{(\rho_f\rho_c)^n(\rho_{w1}\rho_{w2})^m}{z^2+(2n+1)^2a^2+(2m+1)^2b^2}(\rho_c\rho_{w1}+\rho_f\rho_{w1}+\rho_c\rho_{w2}+\rho_f\rho_{w2})\\&+\sum_{m=0}\sum_{n=1}2\frac{(\rho_f\rho_c)^n(\rho_{w1}\rho_{w2})^m}{z^2+(2n)^2a^2+(2m+1)^2b^2}(\rho_{w1}+\rho_{w2})\\&+\sum_{m=1}\sum_{n=0}2\frac{(\rho_f\rho_c)^n(\rho_{w1}\rho_{w2})^m}{z^2+(2n+1)^2a^2+(2m)^2b^2}(\rho_f+\rho_c)\\&+\sum_{m=1}\sum_{n=1}4\frac{(\rho_f\rho_c)^n(\rho_{w1}\rho_{w2})^m}{z^2+(2n)^2a^2+(2m)^2b^2}\\&+\sum_{n=1}2\frac{(\rho_f\rho_c)^n}{z^2+(2n)^2a^2}+\sum_{n=0}\frac{(\rho_f\rho_c)^n}{z^2+(2n+1)^2a^2}(\rho_f+\rho_c)\\&+\sum_{m=1}2\frac{(\rho_{w1}\rho_{w2})^m}{z^2+(2m)^2b^2}+\sum_{m=0}\frac{(\rho_{w1}\rho_{w2})^m}{z^2+(2m+1)^2b^2}(\rho_{w1}+\rho_{w2})],\end{aligned} \tag{3.13}$$

where ρ_c, ρ_f, ρ_{w1}, ρ_{w2} are the reflection coefficients of the ceiling, floor, and two side walls, and the source and receiver are still at the center of the cross section.

Eq. (3.13) was validated in a corridor of cross section 1.76m × 2.74m, and the measurement was made with a maximum source-receiver distance of 18m. The absorption coefficients of the ceiling, floor, and side walls were $\alpha_c = 0.34\text{-}0.63$, $\alpha_f = 0.03$, and $\alpha_w = 0.31\text{-}0.74$, respectively. The accuracy was within about ± 2dB from 150Hz to 4.8kHz.

To consider the different absorption coefficients for the floor, ceiling, and side walls in rectangular road tunnels, with a line source along the center of the cross section and of the same length as the tunnel, a formula and its approximation were derived by Said [1981; 1982]:

$$E_R = \frac{P'}{4a_1 c}\sum_{g,h,q}\frac{(1-\alpha_c)^g(1-\alpha_f)^h(1-\alpha_w)^q a_1}{\pi a_{ghq}}\left(\arctan\frac{x_0}{a_{ghq}} + \arctan\frac{L-x_0}{a_{ghq}}\right) \tag{3.14}$$

$$\approx \frac{-P'\pi}{2cU\ln(1-\bar{\alpha})}\left[2 - e^{\frac{\ln(1-\bar{\alpha})\pi S x_0}{U}} - e^{\frac{\ln(1-\bar{\alpha})\pi S(L-x_0)}{U}}\right],$$

where E_R is the sound energy density from ISs, P' is the sound power per unit length of the line source, g, h and q are the orders of ISs on the ceiling, floor and side walls, a_1 = 1m is the reference distance, a_{ghq} is the distance between the receiver and the IS (g,h,q), x_0 is the horizontal distance between the receiver and a tunnel end, L is the tunnel length, and $\bar{\alpha}$ is the average absorption coefficient of all the boundaries.

It is noted that with highly absorbent boundaries the exclusion of angle-dependent absorption in Eqs. (3.10) to (3.13) might be unreasonable, as suggested through some comparisons between calculation and measurement [Kang, 2002c]. The angle dependence of the absorption coefficient was considered by Sergeev [1979a; 1979b] when he derived a series of formulae for long spaces in a similar manner as the aforementioned, considering relatively hard boundaries.

3.3.2 T_{30} with Geometrically Reflecting Boundaries

Assume $\Delta t << t$, D_p can be approximated as $\sqrt{ct(2z+ct)}$ and Eqs. (3.4), (3.5), and (3.6) can be further simplified to:

$$D_0 \approx z + ct\,, \tag{3.15}$$

$$N \approx \frac{2\pi c}{S}(z+ct)\Delta t\,, \tag{3.16}$$

$$R \approx 0.62 D_p\left(\frac{1}{a}+\frac{1}{b}\right). \tag{3.17}$$

A formula can then be derived to calculate the T_{30} in long rooms:

$$T_{30}(z) = \frac{60t_0}{ct_0 M - 10\log\left[\frac{z}{z+ct_0}(1-\alpha)^{0.62\left(\frac{1}{a}+\frac{1}{b}\right)\sqrt{ct_0(2z+ct_0)}}\right]}, \tag{3.18}$$

where t_0 is a coefficient introduced to determine the decay slope. For a linear decay curve $T_{30}(z)$ is independent of t_0. Since the decay curves are normally not linear, the calculation by Eq. (3.18) would be better if t_0 is closer to the actual T_{30}. As an example, with t_0 = 2.5s, which is a typical T_{30} of many underground stations, Eq. (3.18) becomes:

$$T_{30}(z) = \frac{150}{850M - 10\log\left[\frac{z}{z+850}(1-\alpha)^{25.6\left(\frac{1}{a}+\frac{1}{b}\right)\sqrt{z+425}}\right]}. \quad (3.19)$$

3.3.3 SPL with Diffusely Reflecting Boundaries

With diffusely reflecting boundaries in long rooms, analytic solutions are only available for limited cases. For a point source at the center of a circle cross section, the sound energy density at receivers along the center of the cross section can be calculated by [Kuttruff, 1989]:

$$u = \frac{P}{4\pi cz^2} + \frac{2\rho P}{\pi^2 a_0^2 c}\int_0^\infty \frac{[\chi(\xi)]^2 \cos(\xi \frac{z}{a_0})}{1-\rho\lambda(\xi)} d\xi, \quad (3.20)$$

where $\chi(\xi) = \xi K_1(\xi)$, K_1 is the modified Hankel function, a_0 is the radius, and:

$$\lambda(\xi) \approx \frac{1}{1+\frac{4}{3}\xi^2}. \quad (3.21)$$

For a line source along the width and in the middle of a rectangular cross section with geometrically reflecting side walls and diffusely reflecting ceiling and floor, the sound energy density can be calculated by [Kuttruff, 1989]:

$$u = \frac{P}{4cz} + \frac{2\rho P}{\pi ac}\int_0^\infty \frac{e^{-\xi}\cos(\xi \frac{z}{a})}{1-\rho\chi(\xi)} d\xi. \quad (3.22)$$

3.3.4 SPL Based on Wave Theory

In a manner of plane wave decomposition, Davies [1973] derived a series of formulae to estimate the sound attenuation along a rectangular corridor, following the propagation of plane waves on a form of geometrical acoustics. The analyses were at high frequency, at which the modal summations could be replaced by suitable integrals. The diffraction was ignored. Attention was limited to the total acoustic power flow, and not to the details of the cross-sectional variations of the sound pressure field. Only two cases were considered: either highly absorbent materials, where all the energy of a plane wave was absorbed effectively after two reflections, or relatively hard boundary materials. Two idealized sound sources were assumed: the equal energy source with which the propagating modes had equal energy, and a simple source with which there was relatively more energy in the higher order modes. With the preceding assumptions, a formula for calculating the receiver/input energy ratio was given:

$$\frac{P_{SO}}{P_{IN}} = \left(1 + \sum_{i=1}^{4} \frac{P_{ABS,i}}{P_{IN} - P_{ABS,i}}\right)^{-1}, \quad (3.23)$$

where $P_{ABS,i} = P_{IN} - P_{SO,i}$. P_{SO} is the power flow at the receiver, P_{IN} is the input power, i = 1 – 4 is the boundary index, and $P_{SO,i}$ is the power flow at the receiver caused by boundary i. The calculation of $P_{SO,i}$ should be made in four categories, corresponding to the above conditions. A comparison with measurement shows that although the agreement was satisfactory, the calculation tended to underestimate the actual attenuation—probably because the assumed conditions of the theory were too strict to be practically achieved.

3.3.5 An Empirical SPL Formula

Based on a series of measurements in a 1:8 scale model of 1.6–3.2m high, 1.28–2.48m wide, and 18.4–36.8m long (full-scale dimension), Redmore [1982a; 1982b] developed an empirical formula to calculate the sound distribution in long rooms:

$$L_z = 10\log\rho_0 c + 10\log P + 10\log\left(\frac{1}{2\pi z^2} + \frac{B}{U\bar{\alpha}}10^{-\nabla z/10}\right) - L_{ref}, \tag{3.24}$$

$$L_z = 10\log\left\{\frac{1}{2\pi z^2} + \frac{B}{U\bar{\alpha}}10^{-\nabla z/10} + \rho_E\left[\frac{1}{2\pi(2L-z)^2} + \frac{B}{U\bar{\alpha}}10^{-\nabla(2L-z)/10}\right]\right\}, \tag{3.25}$$

where ρ_0 is the density of air, L_{ref} is the reference sound level, P is the sound power of the source, $B = 0.14$ is an empirical attenuation coefficient, $\nabla = 1.4U\bar{\alpha}/S$ is an empirical rate of the sound attenuation along the length outside the direct field, and ρ_E is the reflection coefficient of the end wall opposite the sound source. In Eq. (3.25) the effect of end walls is taken into account. Eq. (3.24) is used for the absolute SPL and Eq. (3.25) is applied to at least two receivers to predict comparative levels outside the direct field.

3.4 EFFECTS OF DESIGNABLE FACTORS

To examine the effects of designable factors in long rooms, including cross-sectional area and aspect ratio, amount and distribution of absorption, reflection characteristics of boundaries, number and position of sources, source directionality, and air absorption, a series of calculations are made using the ISM for geometrically reflecting boundaries, and the radiosity method for diffusely reflecting boundaries. The calculation is carried out by assuming a rectangular cross section with a point source and receivers along the center of the cross section, and air absorption is not considered except where indicated.

3.4.1 Sound Distribution

A comparison in the relative sound attenuation along the length, with reference to the sound power level L_W, between geometrically and diffusely reflecting boundaries is shown in Figure 3.7 [Kang 1997b; 1998a], where the length of the long room is L = 60m, the cross-section is S = 6m × 4m, the boundary absorption coefficient is α = 0.2, and the two end walls are totally absorbent. A point source is at one end of the long room. It is interesting to note that with geometrically reflecting boundaries, the SPL attenuation is considerably less than that with diffusely reflecting boundaries.

To examine the effects of cross-sectional area and aspect ratio, six long rooms are considered—including 3m × 3m, 6m × 1.5m, 5m × 5m, 10m × 2.5m, 10m × 10m, and 20m × 5m. In other words, there are three cross-sectional areas, namely 9m^2, 25m^2, and 100m^2, and two aspect ratios, namely 1:1 and 4:1. The receiver is at 120m and the boundary absorption

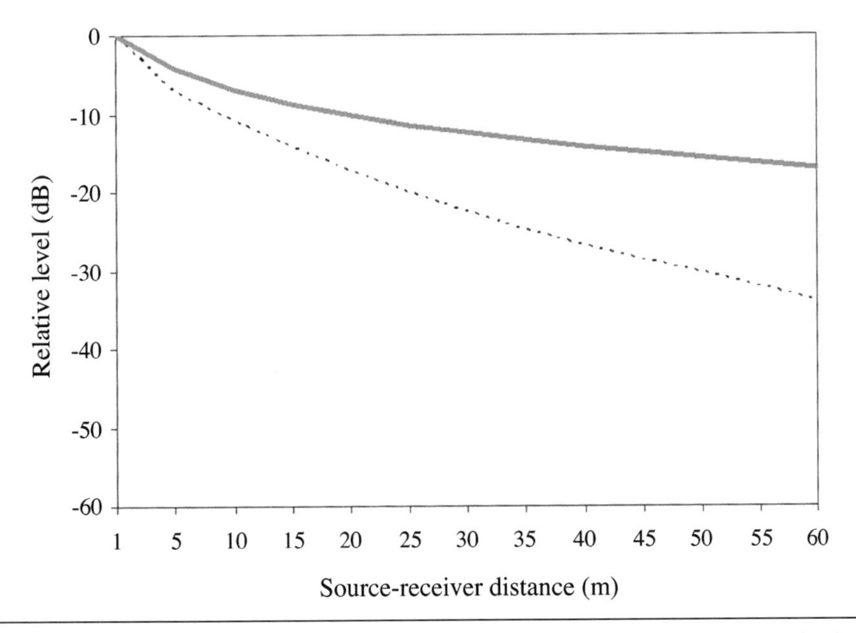

Figure 3.7 Comparison of the relative SPL attenuation with reference to the sound power level L_W along the length between geometrically (——) and diffusely (······) reflecting boundaries, calculated using the ISM and radiosity model, respectively.

coefficient is 0.05. Two kinds of sound attenuation are considered. One is the relative attenuation with reference to a given distance, 20m from the source, and the other is the absolute attenuation with reference to the source power level, L_W. The calculation based on the ISM show that with increasing cross-sectional area the relative SPL attenuation from 20m to 120m becomes less, but the SPL attenuation at 120m with reference to L_W becomes greater. With Eq. (3.23), namely using the wave theory, similar results can be obtained [Kang, 2002c]. With diffusely reflecting boundaries, in terms of the relative attenuation with reference to 20m, the result is similar to that with geometrically reflecting boundaries, whereas in terms of the absolute attenuation with reference to L_W the result is opposite to that with geometrically reflecting boundaries. For both types of boundary, with a constant cross-sectional area, when the width/height ratio is 4:1, the sound attenuation tends to be slightly greater than that of the square section, namely a width/height ratio of 1:1.

The changes in SPL attenuation with increasing boundary absorption from 0.05 to 0.5 is shown in Figure 3.8, where S = 3m × 3m, the source-receiver distance is 50m, and the SPL is with reference to L_W. It is seen that for both kinds of boundary, as the absorption coefficient increases linearly, the attenuation increases with a decreasing gradient. This means that the efficiency of absorbers per unit area is greater when there are fewer absorbers.

A series of calculations with various distributions of absorption in a cross section shows that with geometrically reflecting boundaries, the SPL attenuation is significantly higher when the absorbers are on three or four rather than on one or two boundaries. Perhaps this is because with more than one hard boundary, especially when two of them are parallel, it is possible for some reflections to reach the receiver without impinging upon any absorbent boundaries. With diffusely reflecting boundaries, conversely, the differences in sound

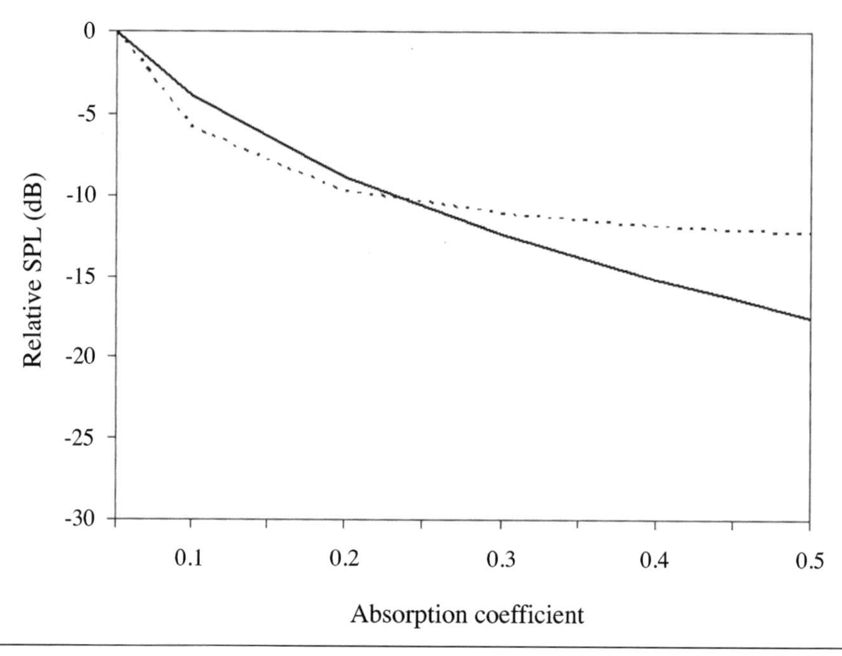

Figure 3.8 Changes in SPL attenuation with increasing boundary absorption from 0.05 to 0.5, with geometrically (—) and diffusely (·····) reflecting boundaries.

attenuation between various absorption distributions are much less, given that in this case a sound ray has more chances to hit all the boundaries.

3.4.2 Reverberation with a Single Source

The variation in T_{30} and EDT along the length in a long room of L = 60m and S = 6m × 4m is shown in Figure 3.9, where both geometrically and diffusely reflecting boundaries are considered, the sound source is at one end of the long room and at the center of the cross section, the two ends are open, and the absorption coefficient of all the boundaries is $\alpha = 0.2$. It can be seen that when the boundaries are geometrically reflective, with the increase of source-receiver distance, the T_{30} and EDT increase rapidly until maximized and then decrease slowly. This variation is mainly caused by two opposing factors, namely the relative change in reflection path length and the number of reflections at a given time after the direct sound [Kang, 2002c]. In the case of diffusely reflecting boundaries, with the increase of source-receiver distance, the T_{30} increases continuously, and the EDT increases rapidly until it reaches a maximum and then decreases slowly.

In comparison with geometrically reflecting boundaries, the T_{30} and EDT with diffusely reflecting boundaries are considerably longer, typically by 30–60%. A possible reason for the difference is that with diffusely reflecting boundaries the sound path is generally longer. It is noted that if one boundary is strongly absorbent, the reverberation with geometrically reflecting boundaries could be longer than that with diffusely reflecting boundaries [Kang, 2002c].

With both boundary types the decay curves are not linear as the T_{30}/EDT ratio is not close to 1. With geometrically reflecting boundaries the T_{30}/EDT ratio is greater than 1, whereas

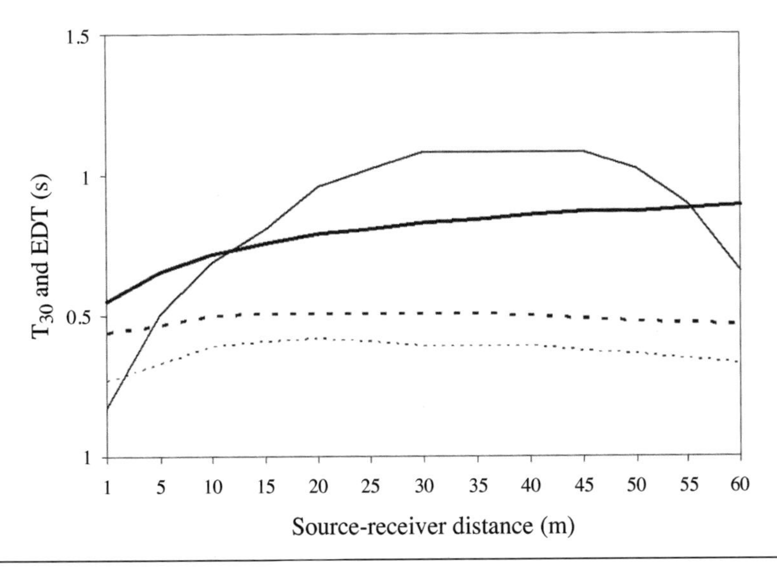

Figure 3.9 Comparison of T_{30} between geometrically (······· , T_{30}; ············ , EDT) and diffusely (▬▬ , T_{30}; —— , EDT) reflecting boundaries in a long room of L = 60m and S = 6m × 4m.

with diffusely reflecting boundaries this ratio is generally less than 1 beyond a certain source-receiver distance. This is probably because the number of reflections with diffusely reflecting boundaries becomes greater with increasing source-receiver distance than with geometrically reflecting boundaries and, consequently, the decay becomes faster with a larger source-receiver distance.

The effects of end walls on the reverberation could be rather significant in long rooms, especially with geometrically reflecting boundaries. An example is shown in Figure 3.10 in

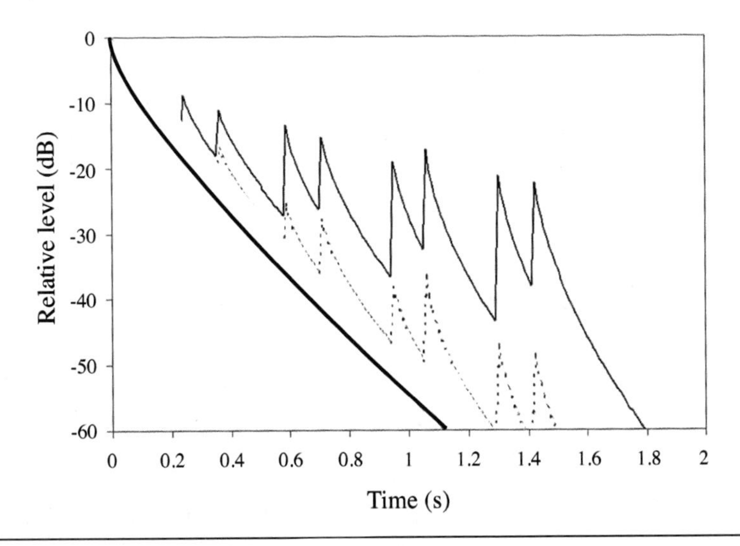

Figure 3.10 Energy response curve $L(t)$ in a finite long room of L = 120m and S = 6m × 4m with three absorption coefficients of the end walls: α_e = 0.1 (——), 0.8 (······) and 1 (▬▬).

terms of the energy response curve $L(t)$, where L = 120m, S = 6m × 4m, α = 0.1; the distance between the source and an end wall is 60m; the source-receiver distance is z = 20m; and three absorption coefficients, α_e = 0.1, 0.8, and 1 are considered for end walls. In the calculation, eight IS planes are taken into account. It can be seen that the end walls can bring a more than 20dB increase on the curves when α_e is 0.1, causing significant difference in reverberation. Conversely, the effect of end walls on the sound attenuation along the length is much less significant, with a difference of about 1dB between α_e = 0.1 and 1, given that the SPL mainly depends on early reflections, whereas end walls are normally effective for later reflections. With diffusely reflecting boundaries, the effects of end walls on both reverberation and sound attenuation are insignificant. By opening the ends, the T_{30} and EDT are only slightly reduced [Kang, 2002c].

To examine the effects of a cross-sectional area and aspect ratio, five infinitely long rooms with geometrically reflecting boundaries are considered, with different cross sections: 12m × 8m, 16m × 6m, 24m × 4m, 6m × 4m, and 8m × 8m, where α = 0.1. The calculated EDT is shown in Figure 3.11(a) [Kang, 1996c]. It can be seen that with a constant aspect ratio, the reverberation is longer for a larger cross section (compare S = 6m × 4m and 12m × 8m). Unlike the results obtained by classic formulae, the increase of reverberation with an increasing cross-sectional area is more significant for a longer source-receiver distance. For a given cross-sectional area, the EDT reaches a maximum as the aspect ratio tends to 1, which can be seen by comparing S =12m × 8m, 16m × 6m, and 24m × 4m. This is perhaps because the average order of ISs decreases as the cross section tends toward square. By comparing S = 24m × 4m and 8m × 8, it can be seen that the reverberation may be longer with a smaller cross-sectional area but a square cross section.

With diffusely reflecting boundaries, similar calculations are carried out, as shown in Figure 3.11(b), where L = 120m, α = 0.2, and the source (30m from an end wall) and receivers are along the center of the cross section. Again, it can be seen that with a constant aspect ratio, the reverberation is longer for a larger cross section, and for a given cross-sectional area, the reverberation could vary significantly with the aspect ratio and become greater as the cross section tends toward square. In Figure 3.11(b) the T_{30} is also given, and it is seen that with a greater cross section, the T_{30}/EDT ratio becomes greater in the near field, and the point where T_{30} = EDT shifts further from the source.

With a given amount of absorption, it is interesting to study the effectiveness of the absorption distribution in a cross section. With diffusely reflecting boundaries, it has been shown that the T_{30} and EDT are the longest with absorption that is evenly distributed in a cross section, and the shortest when one boundary is strongly absorbent. The variation is about 10–30% [Kang, 2002b].

The effect of air absorption is different with the two kinds of boundaries. For typical configurations such as those described previously, with geometrically reflecting boundaries the decrease in reverberation caused by air absorption could be about 20–40% less than that with diffusely reflecting boundaries, where the sound path is generally longer and thus air absorption is more effective. With diffusely reflecting boundaries, the extra SPL attenuation along the length caused by a given air absorption is also greater. Moreover, since the effect of air absorption is greater with a longer reflection distance, the reduction of sound energy caused by air absorption increases with time. This would consequently change the form of decay curves. For example, for geometrically reflecting boundaries, with air absorption the decay curves are more linear [Kang, 1996c; 2002b].

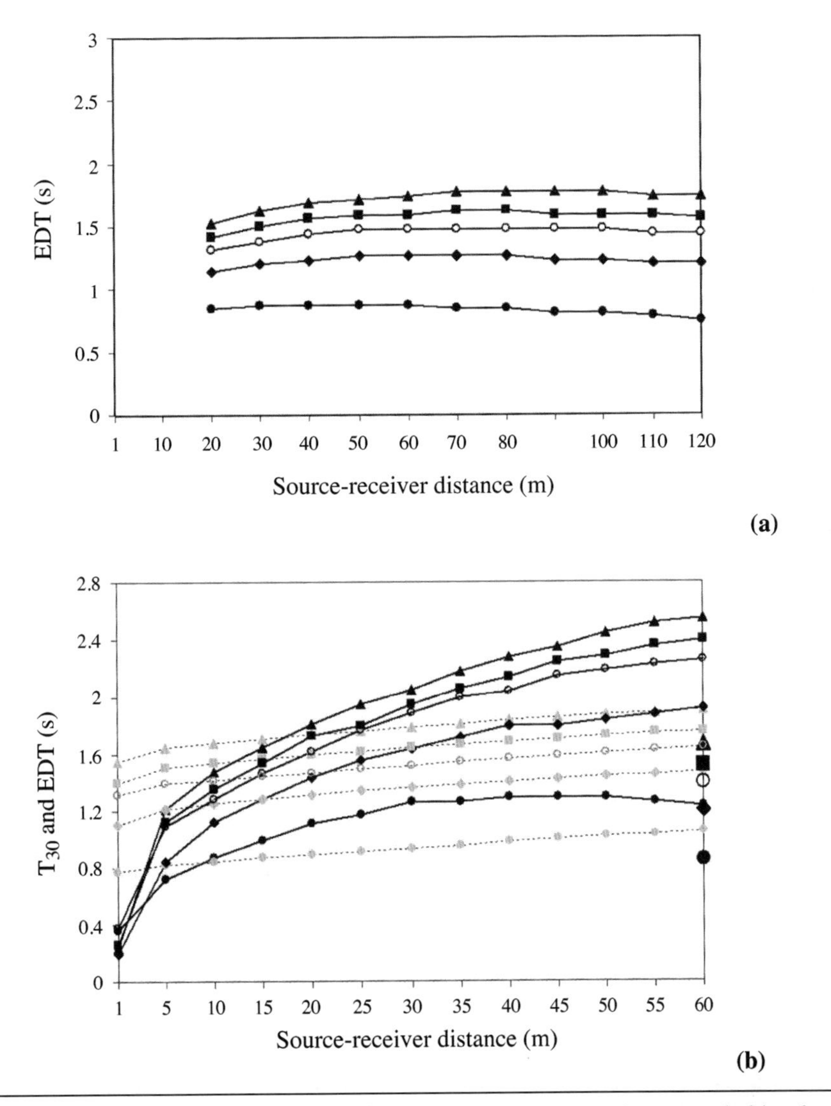

Figure 3.11 T_{30} (solid lines with solid symbols) and EDT (dotted lines with grey symbols) in long rooms with different cross sections with geometrically (a) and diffusely (b) reflecting boundaries: ▲, 12m × 8m; ■, 16m × 6m; ◆, 24m × 4m; ●, 6m × 4m; ○, 8m × 8m.

While the above results are all based on omnidirectional sources, calculations with directional sources show that both reverberation and SPL attenuation along the length can be changed by the source directionality [Kang, 2002c].

3.4.3 Reverberation with Multiple Sources

While multiple loudspeaker PA systems are often used in long rooms, the effects of multiple sources are analyzed below using a multiple-source in underground long spaces (MUL) program [Kang, 1996a] in a rectangular long room with geometrically reflecting boundaries, where the length is L = 120m, the cross section is S = 6m × 4m, the absorption in the cross section is uniform, and the end walls are totally absorbent. A number of point sources are evenly distributed along the whole length with a source spacing σ, and there is neither a time delay nor a sound power difference between the sources. In Figure 3.12 the configurations used in the analysis are shown, where two source spacings, $\sigma = 5$m and 15m, are considered, corresponding to the source number of 25 and 9, respectively. Two typical receiver positions are used. R_0 is in the same cross section with the central source (60m to both end walls), and R_m is with a distance of $\sigma/2$ from the cross-section with the central source.

The comparisons in energy response $L(t)$ between a single source and multiple sources at R_0 and R_m are shown in Figures 3.13(a) and 3.13(b), respectively [Kang, 1996a]. It is interesting to note in Figure 3.13(a) that the reverberation from multiple sources is significantly longer than that caused by the central source alone, and this is also the case in Figure 3.13(b), where two boundary absorption coefficients are considered. This means that the T_{30} can be increased by adding sources beyond a single source. This is mainly because the later energy of the decay process is increased by the added sources. In contrast, from Figure 3.13(a) it can be seen that the reverberation from multiple sources is shorter than that caused by the source at 60m. In other words, the T_{30} can be decreased by adding sources between a single source and the receiver. The decrease is due to the reduction of the minimum source-receiver distance. The two opposite effects occur simultaneously when the number of evenly distributed sources is changed in a given space, and it appears that there is no simple relationship between the T_{30} from multiple sources and the source spacing. However, in any case, it is important that the reverberation in long rooms is dependent on the source position and the number of sources, which is fundamentally different from that of the diffuse field.

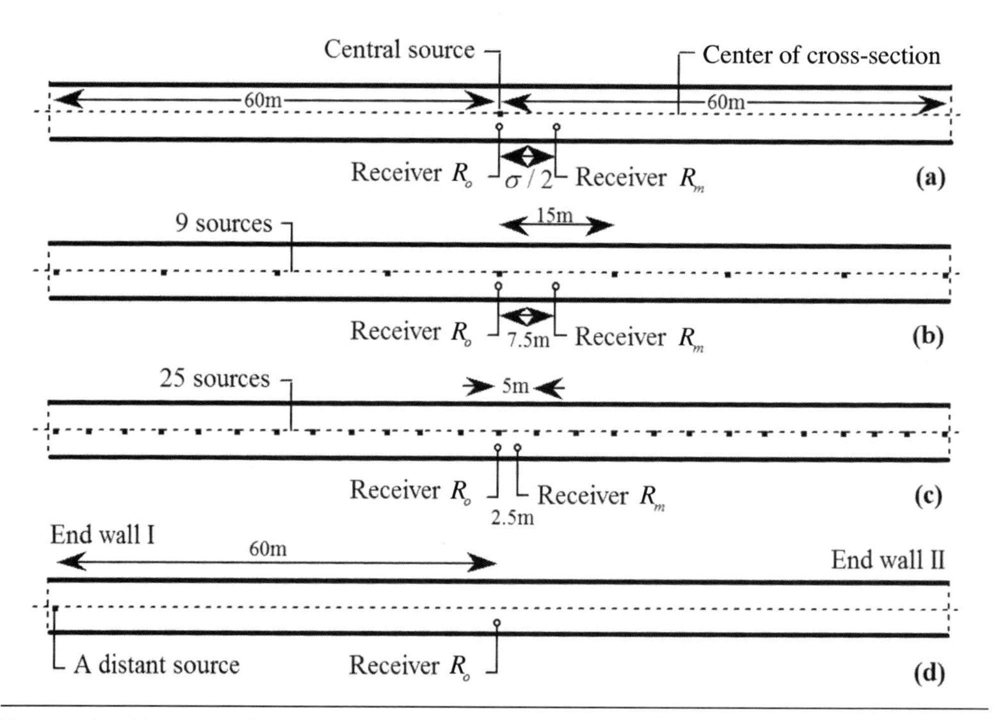

Figure 3.12 Plan or length-wise section of the long rooms showing the source and receiver locations: (a) the central source alone at 2m from R_0 along the width/height; (b) 9 sources; (c) 25 sources; (d) a single source with a distance of 60m from R_0.

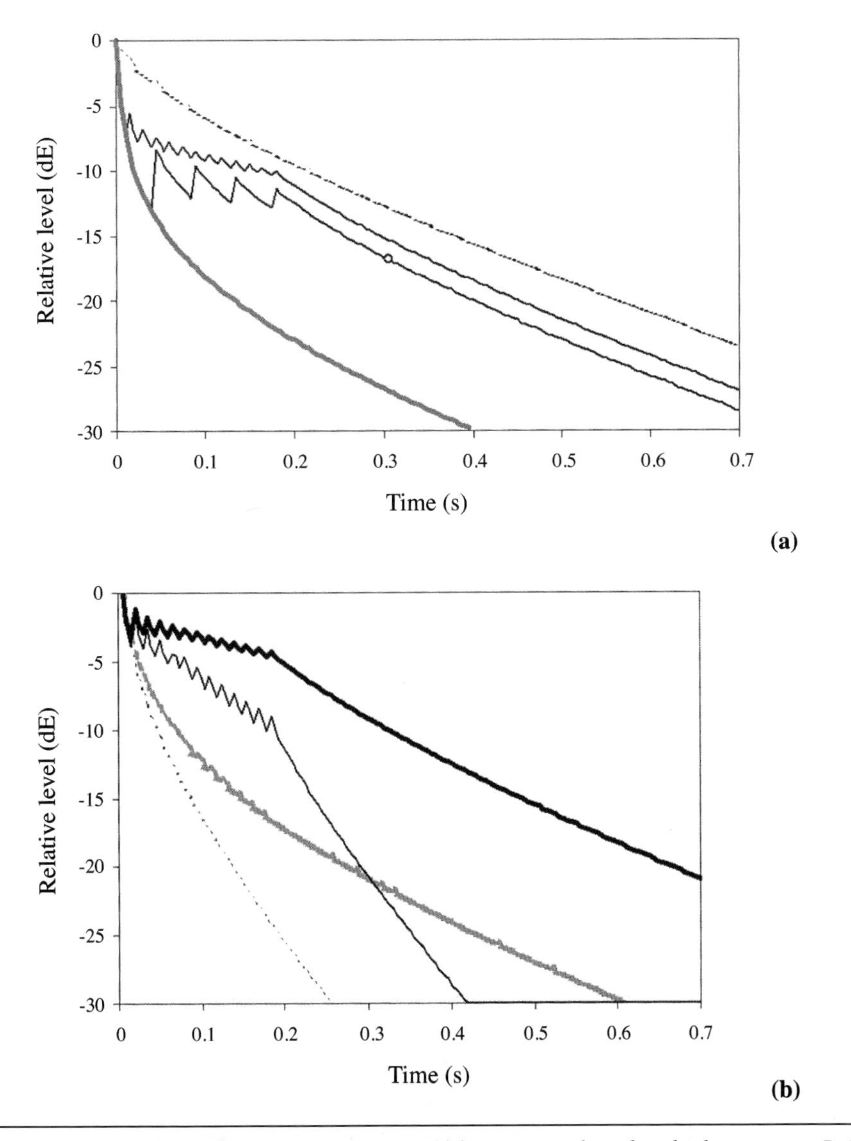

Figure 3.13 Comparison in energy response curve $L(t)$ between single and multiple sources at R_0 (a) (▬, central source; ⋯⋯, a point source with a distance of 60m from R_0; —, 25 sources, σ = 5m; -○-, 9 sources, σ = 15m); and at R_m (b) (▬, central source with α = 0.05; ⋯⋯, central source with α = 0.15; —, multiple sources of σ = 5m with α = 0.05; —, multiple sources of σ = 5m with α = 0.15).

To increase the quality of multiple loudspeaker PA systems, as measured by the STI, it is important to reduce the reverberation. An effective way is to increase the absorption, but this could also decrease the sound level and the spatial evenness of the loudspeaker signals and thus, the STI is not necessarily better with increased absorption. The reverberation from multiple sources can also be adjusted by using directional sources, which should be so designed that the radiation intensity is the same, in the range of $\pm\ \sigma/2$, and decreases significantly beyond this range. Consequently, the reverberation from multiple sources is close to that of the nearest source alone, which is much shorter. A systematic analysis in an underground station shows that with the provision of room and loudspeaker conditions, an optimal loudspeaker spacing can be determined by considering the minimum requirement of the STI, the distribution of the STI along the length, and the cost of the whole PA system [Kang, 1996g].

3.5 CASE STUDIES BASED ON SCALE MODELING

Based on a series of measurements in two 1:16 scale models of underground stations [Kang, 1996e; 1997c; 1998b], of rectangular and circular cross-sectional shapes, respectively, this section further examines the effectiveness of strategic architectural acoustic treatments, including absorbers, diffusers, reflectors, obstructions, and their combinations. Compared to theoretical/computer models, scale models are more suitable for investigating relatively complicated configurations, and for considering certain physical phenomena such as diffraction. The rectangular scale model was 4,880mm long (78.08m full-scale), representing approximately 1/3 to 1/2 of modern underground stations. The cross section of the model was 300mm (4.8m full-scale) high and 937.5mm (15m full-scale) wide. Acoustically hard boundaries were modeled with well-varnished timber. The circular scale model was a plastic pipe with a length of 8,000mm (128m full-scale) and a diameter of 405mm (6.48m full-scale), simulating typical underground stations in London. Note in this section the dimensions and frequencies relate to full-scale, except where indicated.

3.5.1 Diffusers

A ribbed structural element, which can often be found in underground tunnels, has been simulated as a diffuser [Kang, 1995]. In the rectangular underground station, the effectiveness of the ribbed diffusers on the EDT and SPL attenuation along the length is shown in Figures 3.14(a) and 3.14(b), respectively, where the end walls are strongly absorbent using a 10mm thick (16cm full-scale) plastic foam. It can be seen that with diffusers the EDT is systematically lower, especially beyond a certain source-receiver distance (say 10m) and the SPL attenuation along the length is systematically greater. Further experiments have shown that when the long room is relatively absorbent, the effectiveness of the diffusers is less than that under the low absorption condition, but is still noticeable. An important reason is that with highly absorbent boundaries the number of reflections becomes fewer and, thus, the efficiency of diffusers is diminished.

In Figure 3.14 a comparison is also made between absorbent and reflective end walls. It can be seen that the variation in EDT and SPL along the length can be diminished by highly reflective end walls, which corresponds to the previous theoretical analysis.

In the circular cross-sectional underground station, with ribbed diffusers, the SPL increases slightly in the near field, corresponding to the theoretical results, and at greater distances the extra attenuation is significant. Further experiments have been carried out using Schroeder diffusers, and the effectiveness is even greater. Again, under highly absorbent conditions, the extra attenuation by diffusers becomes slightly less compared with empty conditions. With different diffuser distributions, there is generally no significant difference in terms of extra SPL attenuation.

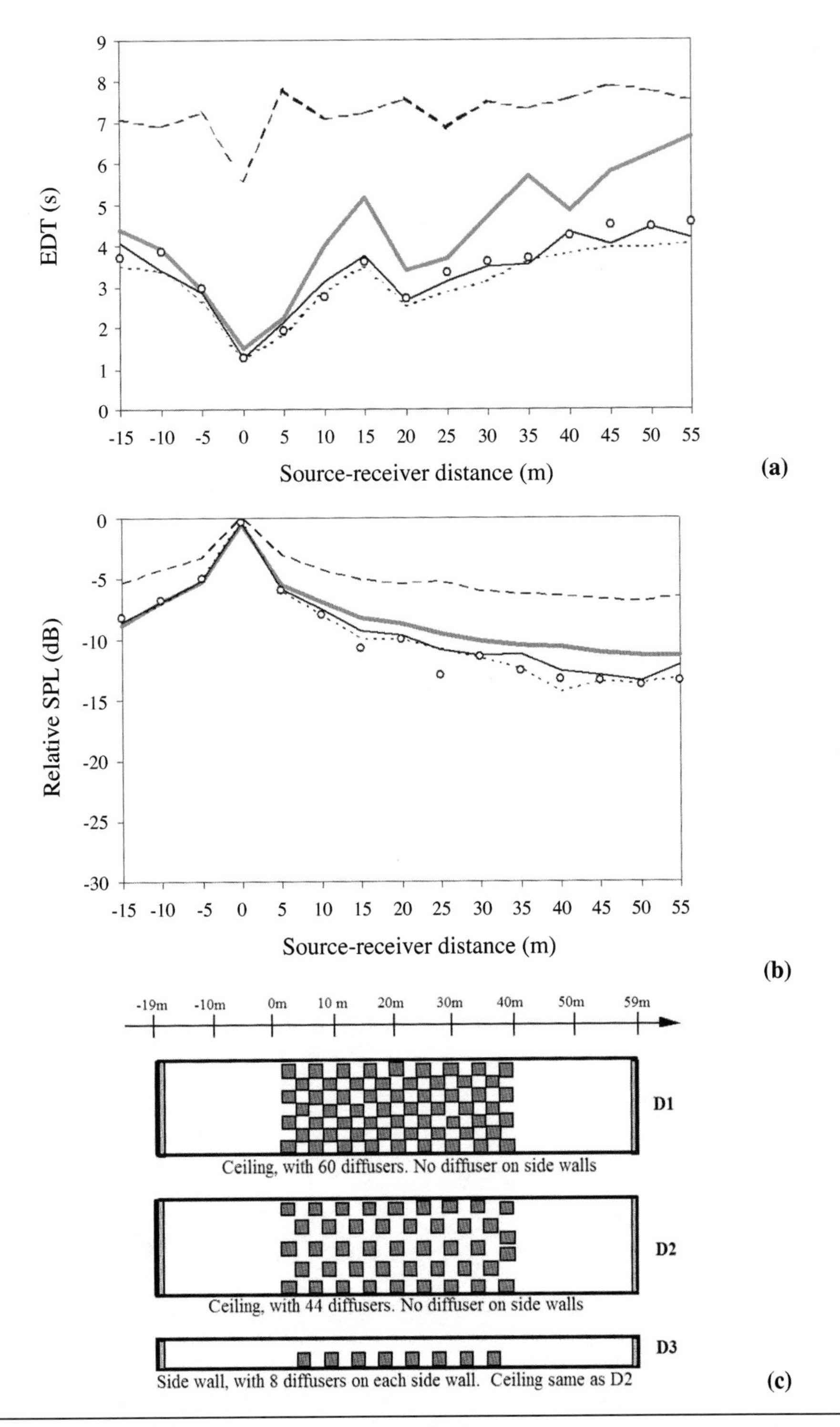

Figure 3.14 EDT (a) and SPL (b) along the length with various arrangements (c) of ribbed diffusers (○, Arrangement D1; —, Arrangement D2; ......, Arrangement D3), based on the average of 500Hz and 1kHz, compared with no-diffuser situation with (▬) and without (• • • •) absorbent end walls. The sound source is at 0m.

3.5.2 Absorbers

Two typical absorbers have been considered in the rectangular underground station, including a porous absorber, namely a 10mm thick (16cm full-scale) plastic foam, and a resonant membrane absorber, namely an airtight plastic film with an airspace from a hard boundary. In Figure 3.15 the EDT and SPL with various arrangements of the porous absorber are shown, where both the absorption amount and distribution are considered. It can be seen that the efficiency of absorbers can be significantly affected by their arrangements. Moreover, some absorber arrangements, which are very effective for increasing the sound attenuation along the length, are not necessarily effective for decreasing reverberation. With a strategically located absorber arrangement, the acoustic performance could be better with fewer absorbers, especially in terms of the STI from multiple loudspeakers, which integrates the effects of both reverberation and the SPL.

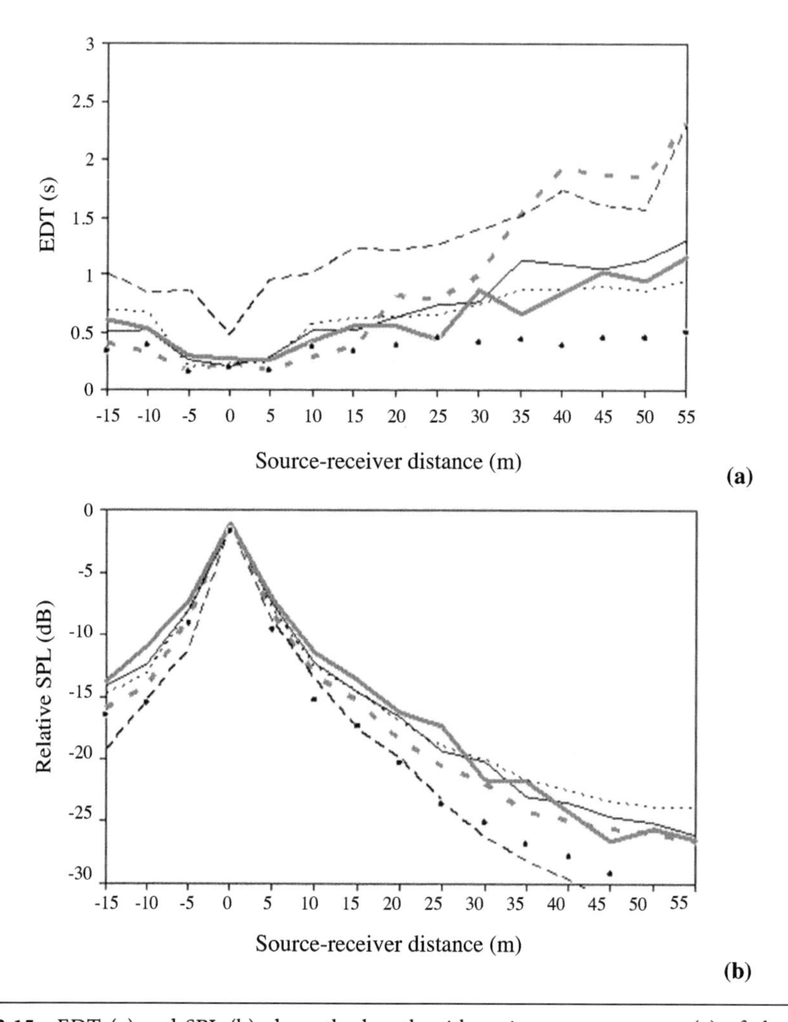

Figure 3.15 EDT (a) and SPL (b) along the length with various arrangements (c) of absorbers (——, Arrangements A1; ······· , A2; ▬, A3; • • • , A4; - - - - , A5; •, A6), based on the average of 500Hz and 1kHz, where the absorption coefficient of the porous absorber is over 0.9.

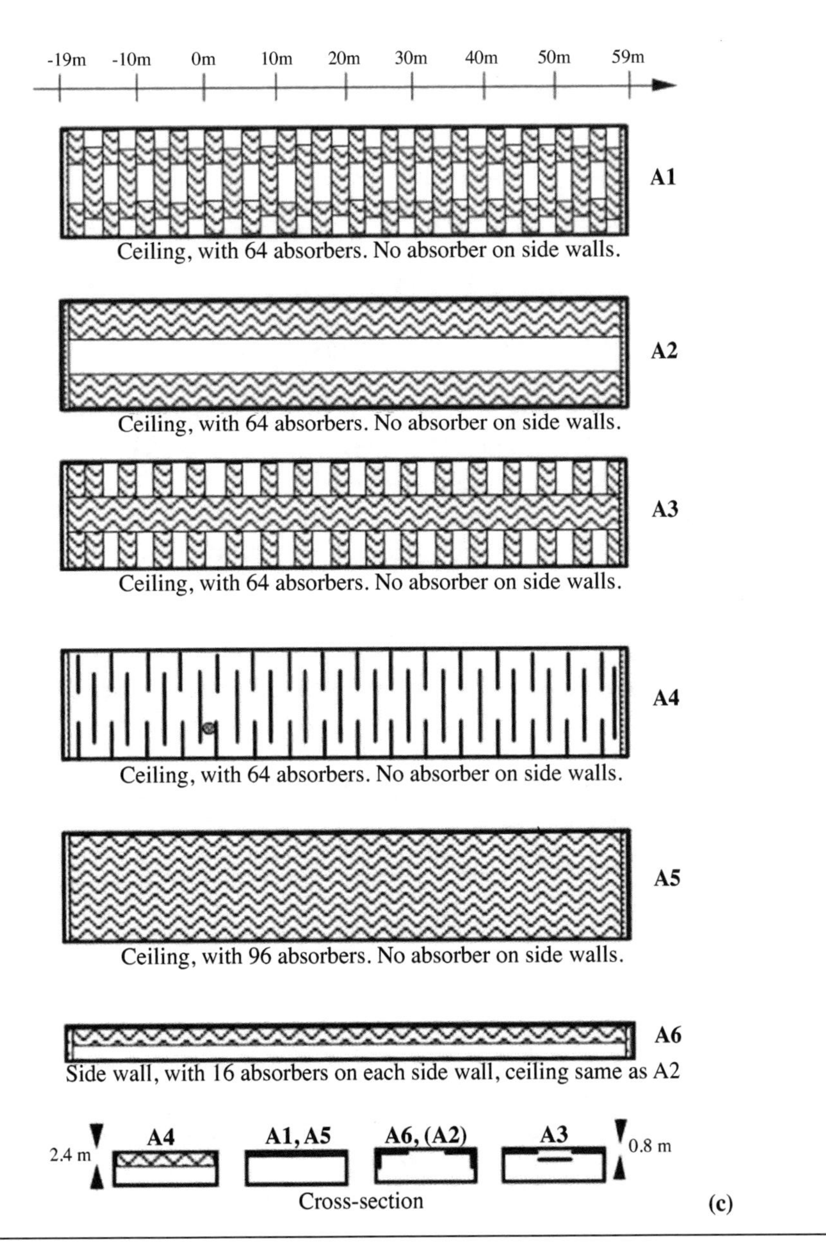

Figure 3.15c Continued.

Despite the relatively large angle of sound incidence in long rooms, resonant absorbers like membranes can still be effective for increasing the sound attenuation along the length and decreasing reverberation, and thus increasing the speech intelligibility of multiple loudspeaker PA systems [Kang, 2002c].

3.5.3 Reflectors and Obstructions

The effects of obstructions and reflectors have also been investigated in scale models. For the former, a number of barriers along the length, and two blocks in the middle of the rectangular scale model have been considered in the rectangular model. For the latter, reflectors regularly arranged corresponding to the loudspeaker spacing have been tested in the circular scale model. It has been shown that strategically designed obstructions and reflectors can increase the speech intelligibility from multiple loudspeakers [Kang, 2002c].

3.5.4 Train Noise and the STI in Underground Stations

In underground stations, train noise is a major consideration in terms of acoustic comfort as well as the intelligibility of PA systems. In the circular scale model, a model train was made, and a tweeter was put underneath a model train section to simulate the wheel/rail noise. With the scale model some basic characteristics of train noise have been examined [1997d].

The scale modeling results have also been used as the input/database of a train noise in stations (TNS) computer model, which can predict the temporal and spatial distribution of train noise in stations and tunnels [Kang, 1996b]. The logic of the modeling process is to consider a train as a series of sections and to calculate the train noise distribution in an underground station by inputting the sound attenuation along the length from a train section source. The calculation sums the sound energy from all the train sections by considering train movement and the position of the train section in the train.

It has been shown using TNS that the overall level of train noise in the area near the end walls is slightly less than that in the other areas. Some conventional architectural acoustic treatments in a station are effective when a train is still in the tunnel but not as helpful when the train is already in the station. Train noise typically decreases the STI of PA systems from 0.5 to below 0.2 on entering the station. With train noise, instead of a single value, the intelligibility of a PA system becomes a dynamic process.

To calculate the speech intelligibility in long rooms, a comprehensive acoustic calculation in long spaces (ACL) model that combines the advantages of both the computer simulation and scale modeling has been developed [Kang, 1997a]. ACL first treats an actual long room as having a relatively simple geometry and calculates the acoustic indices of a single source using theoretical formulae or computer simulation models. The possible errors caused by the simplifications are then corrected using a database based on scale model measurements or site measurements.

3.6 CASE STUDIES BASED ON SITE MEASUREMENTS

In this section, the measurement data in a number of actual long rooms are provided. Figure 3.16 shows the measured T_{30} in a corridor of 42.5m × 1.56m × 2.83m in a university building in Cambridge, UK [Kang, 2002c]. The boundaries of this corridor were acoustically hard and could be considered as geometrically reflective. The increase in reverberation along the length can be clearly seen. Similar results have also been obtained in other measurements [Clausen and Rekkebo, 1976].

Figure 3.17 shows the measured T_{30} and SPL attenuation in an underground shopping center 160m long in Harbin, China [Zhang et al., 2008]. A row of cubicles/shops were located at both sides along the length, and the tests were carried out with the cubicles enclosed with vertical folding doors that were semi-transparent acoustically. It can be seen that the reverberation increases and the SPL decreases continuously along the length.

Figure 3.18 shows the variation in T_{30} and EDT in three deep tube stations of circular cross-sectional shape in London, UK [Orlowski, 1994], which represent both high and low absorbent conditions. It can be seen that with the increase of source-receiver distance, the T_{30} and EDT increase along the length until a certain point, and then become approximately

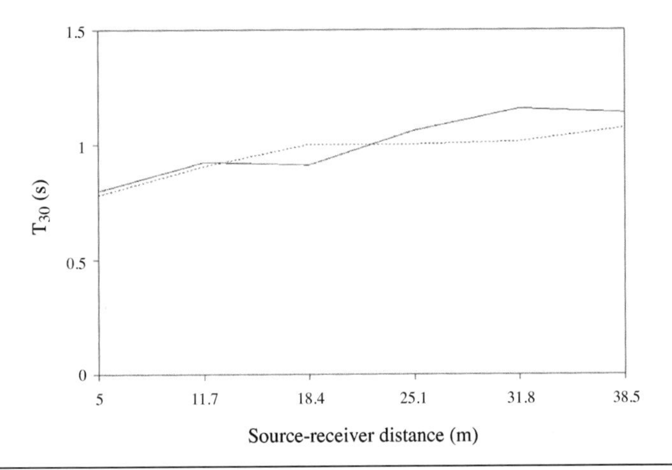

Figure 3.16 Measured T_{30} at 500Hz (——) and 1kHz (······) in a corridor in Cambridge, UK.

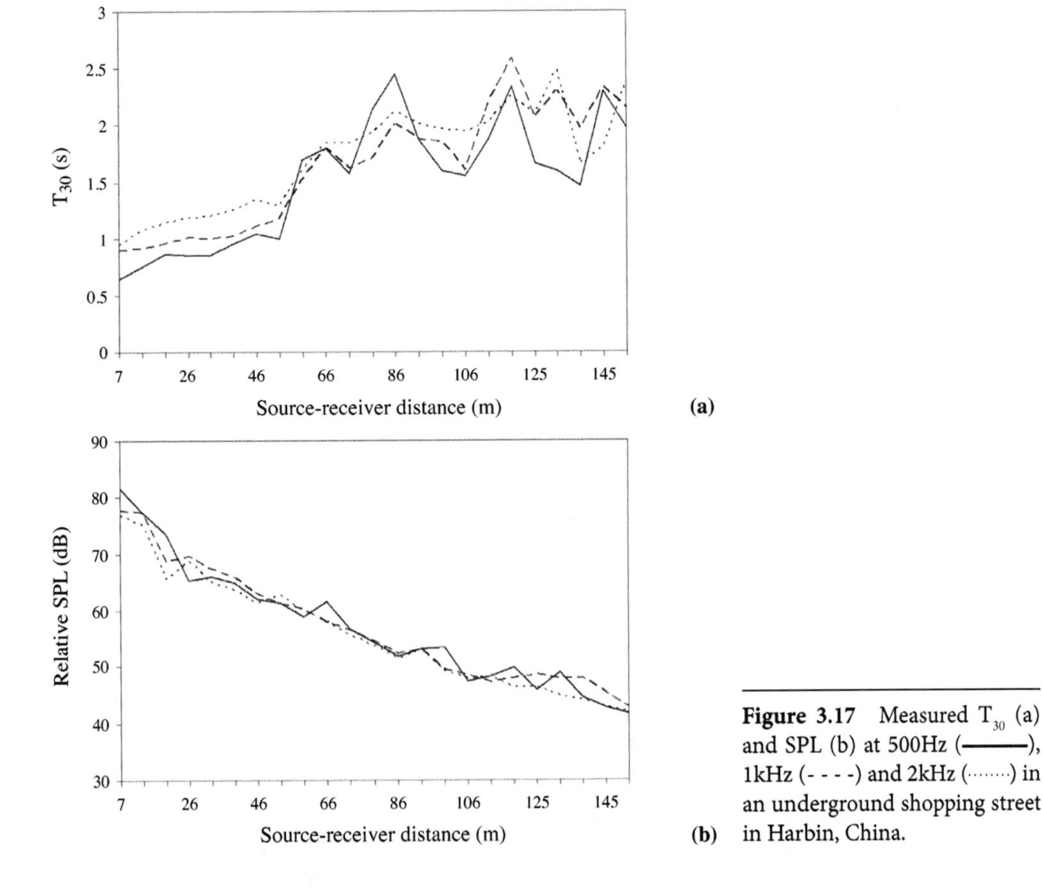

Figure 3.17 Measured T_{30} (a) and SPL (b) at 500Hz (———), 1kHz (- - - -) and 2kHz (········) in an underground shopping street in Harbin, China.

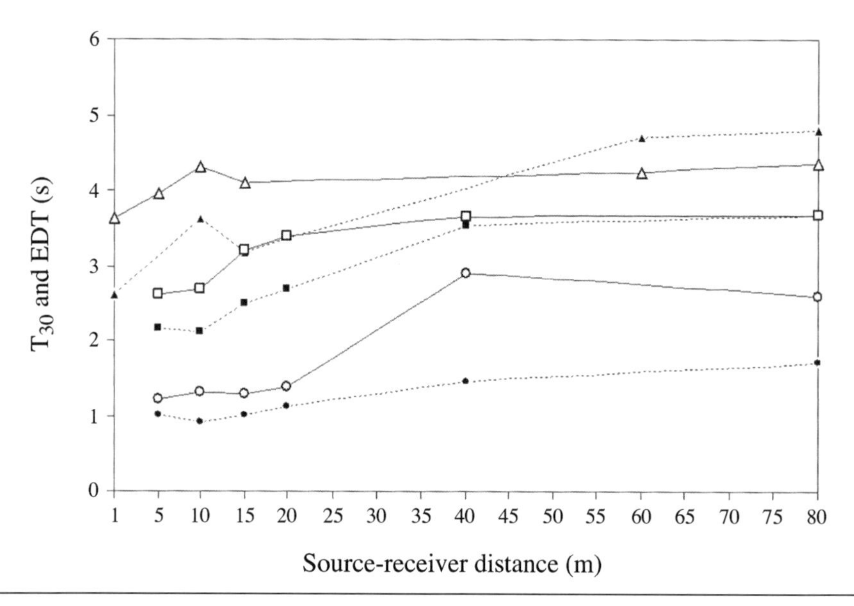

Figure 3.18 Measured T_{30} (open symbols △, □, ○) and EDT (solid symbols ▲, ■, ●) in three deep tube stations in London, UK. Average of 500Hz and 1kHz.

stable or decrease slightly. T_{30} is generally greater than the EDT, indicating that the decay curves are not linear.

A number of measurements have also been made in long rooms with branches [Tang et al., 2008; Liu, 2006], where the sound fields are more complicated. For example, the increase in reverberation could be diminished by side corridors.

DEFINING TERMS

Image source model: This method treats a flat surface as a mirror and creates an IS. The reflected sound is modeled with a sound path directly from the IS to a receiver. Multiple reflections are achieved by considering further images of the IS. At each reflection the strength of the IS is reduced due to surface absorption. With the ISM, the situation of a source in an enclosure is replaced by a set of mirror sources in a free field, visible from the receiver considered. The acoustic indices at the receiver are determined by summing the contribution from all the image sources.

Radiosity model: This method—also called radiation balance, radiation exchange, or radiant interchange—was first developed in the 19th century for the study of radiant heat transfer in simple configurations. Basically, the radiosity method divides boundaries in a room into a number of patches (i.e., elements) and replaces the patches and receivers with nodes in a network. The sound propagation in the room can then be simulated by energy exchange between the nodes.

Ray tracing model: A sound ray can be regarded as a small portion of a spherical wave with a vanishing aperture, which originates from a certain point. Ray tracing creates a dense spread of rays that are subsequently reflected around a room and tested for intersection with a detector (receiver) such as a sphere or a cube. An echogram can be constructed using the energy attenuation of the intersecting rays and distances traveled. Beams are rays with a non-vanishing cross section. The beams may be cones with a circular cross section or pyramids with a polygonal cross-section. By using beams, a point detector can be used. Beams are reflected around a room and tested for illumination of the detector.

Eigenfrequencies: In a room of given dimensions, standing waves could form at certain frequencies, traveling not only between two opposite, parallel boundaries, but also around the room involving the boundaries at various angles of incidence. The frequencies are called resonant frequencies, natural frequencies, normal frequencies, or eigenfrequencies.

REFERENCES

1. Clausen, J. E. and Rekkebo, J. A., 1976, *Acoustical Properties of Corridors*, Siv. ing. Thesis. The Norwegian Institute of Technology (in Norwegian).
2. Davies, H. G., 1973, "Noise Propagation in Corridors," *Journal of the Acoustical Society of America*, vol. 53, pp. 1253–1262.
3. Kang, J., 1995, "Experimental Approach to the Effect of Diffusers on the Sound Attenuation in Long Enclosures," *Building Acoustics*, vol. 2, pp. 391–402.
4. Kang, J., 1996a, "Acoustics in Long Enclosures with Multiple Sources," *Journal of the Acoustical Society of America*, vol. 99, pp. 985–989.
5. Kang, J., 1996b, "Modelling of Train Noise in Underground Stations," *Journal of Sound and Vibration*, vol. 195, pp. 241–255.
6. Kang, J., 1996c, "Reverberation in Rectangular Long Enclosures with Geometrically Reflecting Boundaries," *Acustica united with acta acustica*, vol. 82, pp. 509–516.
7. Kang, J., 1996d, "Sound Attenuation in Long Enclosures." *Building and Environment*, vol. 31, pp. 245–253.
8. Kang, J., 1996e, "Speech Intelligibility Improvement for Multiple Loudspeakers by Increasing Loudspeaker Directionality Architecturally," *Building Services Engineering Research and Technology*, vol. 17, pp. 203–208.
9. Kang, J., 1996f, "The Unsuitability of the Classic Room Acoustical Theory in Long Enclosures," *Architectural Science Review*, vol. 39, pp. 89–94.
10. Kang, J., 1996g, *Station Acoustics Study*. Hong Kong Mass Transit Railway Corporation Report, Consultancy No. 92110800-94E.
11. Kang, J., 1997a, "A Method for Predicting Acoustic Indices in Long Enclosures," *Applied Acoustics*, vol. 51, pp. 169–180.
12. Kang, J., 1997b, "Acoustics in Long Underground Spaces," *Tunneling and Underground Space Technology*, vol. 12, pp. 15–21.
13. Kang, J., 1997c, "Improvement of the STI of Multiple Loudspeakers in Long Enclosures by Architectural Treatments," *Applied Acoustics*, vol. 51, pp. 169–180.
14. Kang, J., 1997d, "Scale Modelling of Train Noise Propagation in an Underground Station," *Journal of Sound and Vibration*, vol. 202, pp. 298–302.
15. Kang, J., 1998a, "Acoustic Theory of Long Enclosures and Its Application to Underground Stations," *Underground Engineering and Tunnels*, no. 4, pp. 37–40 (in Chinese).
16. Kang, J., 1998b, "Scale Modelling for Improving the Speech Intelligibility of Multiple Loudspeakers in Long Enclosures by Architectural Acoustic Treatments," *Acustica united with acta acustica*, vol. 84, pp. 689–700.
17. Kang, J., 2002a, "Comparison of Sound Fields in Regularly-shaped, Long and Flat Enclosures with Diffusely Reflecting Boundaries," *International Journal of Acoustics and Vibration*, vol. 7, pp. 165–171.
18. Kang, J., 2002b, "Reverberation in Rectangular Long Enclosures with Diffusely Reflecting Boundaries," *Acustica united with acta acustica*, vol. 88, pp. 77–87.

19. Kang, J., 2002c, *Acoustics of Long Spaces: Theory and Design Guidance*, Thomas Telford, London.
20. Kang, J., 2006, *Urban Sound Environment*, Taylor & Francis incorporating Spon, London.
21. Kuno, K., Kurata, T., Noro, Y. and Inomoto, K., 1989, "Propagation of Noise along a Corridor—Theory and Experiment," *Proceedings of the Acoustical Society of Japan*, pp. 801–802 (in Japanese).
22. Kuttruff, H., 1989, "Schallausbreitung in Langräumen," *Acustica*, vol. 69, pp. 53–62.
23. Kuttruff, H., 2000. *Room Acoustics*, 4th ed., Spon Press, London.
24. Lam, P. M. and Li, K. M., 2007, "A Coherent Model for Predicting Noise Reduction in Long Enclosures with Impedance Discontinuities," *Journal of Sound and Vibration*, vol. 299, pp. 559–574.
25. Liu, J. C. C., 2006, "Sound Propagation in Long Enclosures with a Vertical or Inclined Branch," *Applied Acoustics*, vol. 67, pp. 1022–1030.
26. Meng, Y. and Kang, J., 2008, "Combined Ray Tracing and Radiosity Modeling for Micro-scale Urban Sound Environment," *Journal of the Acoustical Society of America*, submitted for publication.
27. Ollendorff, F., 1976, "Diffusionstheorie des Schallfeldes im Strassentunnel," *Acustica*, vol. 34, pp. 311–315.
28. Orlowski, R. J., 1994, "Underground Station Scale Modelling for Speech Intelligibility Prediction," *Proceedings of the Institute of Acoustics (UK)*, vol. 16 (4), pp. 167–172.
29. Picaut, J., Simon, L. and Polack, J.D., 1999, "Sound Field in Long Rooms with Diffusely Reflecting Boundaries," *Applied Acoustics*, vol. 56, pp. 217–240.
30. Redmore, T. L., 1982a, "A Method to Predict the Transmission of Sound through Corridors," *Applied Acoustics*, vol. 15, pp. 133–146.
31. Redmore, T. L., 1982b, "A Theoretical Analysis and Experimental Study of the Behaviour of Sound in Corridors," *Applied Acoustics*, vol. 15, pp. 161–170.
32. Said, A., 1981, "Schalltechnische Untersuchungen im Strassentunnel," *Zeitschrift für Lärmbekämpfung*, vol. 28, pp. 141–146.
33. Said, A., 1982, "Zur Wirkung der schallabsorbierenden Verkleidung eines Strassentunnels auf die Spitzenpegel," *Zeitschrift für Lärmbekämpfung*, vol. 28, pp. 74–78.
34. Sergeev, M. V., 1979a, "Acoustical Properties of Rectangular Rooms of Various Proportions," *Soviet Physics—Acoustics*, vol. 25, pp. 335–338.
35. Sergeev, M. V., 1979b, "Scattered Sound and Reverberation on City Streets and in Tunnels," *Soviet Physics—Acoustics*, vol. 25, pp. 248–252.
36. Steeneken, H. J. M. and Houtgast, T., 1980, "A physical method for measuring speech-transmission quality," *Journal of the Acoustical Society of America*, vol. 67, pp. 318–326.
37. Tang, Z. Z., Jin, H. and Kang, J., 2007, "Sound Field in Underground Long Spaces—A Case Study in a Corridor," *Journal of Harbin Institute of Technology*, vol. 40, pp. 205–210 (in Chinese).
38. Yamamoto, T., 1961, "On the Distribution of Sound in a Corridor," *Journal of the Acoustical Society of Japan*, vol. 17, pp. 286–292 (in Japanese).
39. Yang, L. N. and Shield, B. M., 2001, "The Prediction of Speech Intelligibility in Underground Stations of Rectangular Cross Section," *Journal of the Acoustical Society of America*, vol. 109, pp. 266–273.
40. Zhang, J., Jin, H. and Kang, J., 2008, "Sound Fields in Underground Spaces of Different Spatial Forms," *Proceedings of inter-noise*, Shanghai, China.

FOR FURTHER INFORMATION

More details on the results in this chapter can be found in the author's monograph *Acoustics of Long Spaces: Theory and Design Guidance*, published in 2002 by Thomas Telford, London.

Jing et al. (2010) have been able to solve the transport equation without reducing to the diffusion equation for simulating long rooms (see Sec. 2.1.1 and Ref. 2 and Ref. 3 in Chapter 2 of this book).

4

Acoustics in Coupled Volume Systems

Ning Xiang, School of Architecture, Rensselaer Polytechnic Institute

ABSTRACT

Recent applications of coupled reverberation chambers in performing arts venues have prompted active research on sound fields in coupled rooms. This chapter reviews advances in the room acoustic research of coupled rooms, including wave-theory-based modal analysis and modal expansion methods, statistical room acoustics, geometrical room acoustic approaches and, with particular emphasis, a diffusion equation-based approach. Acoustical scale modeling-based experimental tools developed with high spatial resolution, automatic scanning capabilities, and advanced Bayesian energy decay analysis applied in room acoustic research for coupled rooms are also briefly discussed.

4.1 INTRODUCTION

In building acoustics, early studies on sound transmission loss of wall partitions, doors and windows, as well as coupled chambers for experimental measurements of sound transmission loss stimulated interest in acoustics in coupled spaces.[1, 2] In recent years coupled reverberation chambers purposely applied in performance venues have prompted increasing research interest. One such application is the design and adaptation of theater stage shells to couple with reverberant stage houses[3, 4] as illustrated in Figure 4.1. For symphonic purposes, the stage shell features acoustic coupling between the reverberant stage house behind and above the shell and the main audience floor. Another application purposely implements secondary reverberation chambers around the concert hall with coupling doors coupled to the main floor.[5–7] Figure 4.2 illustrates a concert hall cross section in which the main floor is surrounded by the reverberation chambers. These reverberation chambers, partially coupled to the primary space, can generate nontraditional, so-called double-rate or multiple-rate sound energy decays—simultaneously achieving two desirable yet competing auditory attributes: clarity and reverberance. The opening and closing of combinations of these secondary chambers has become an important tool for generating the range of acoustic conditions needed for the widely varying music performed today.[8]

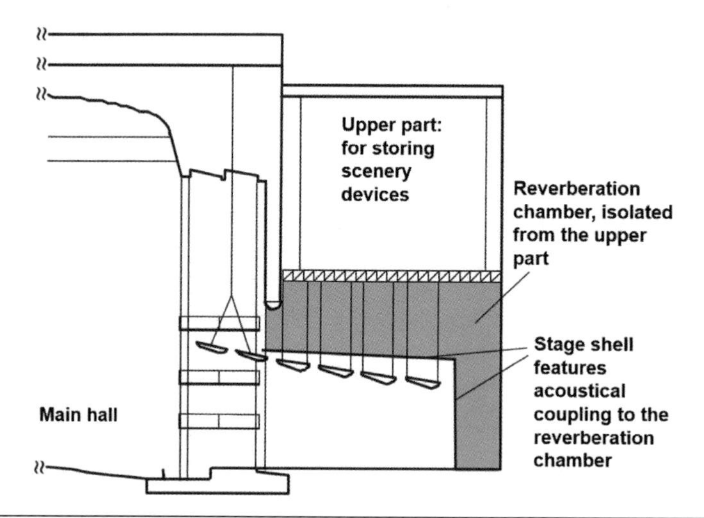

Figure 4.1 Stage houses of opera/theater venues feature a stage shell (concert hall shaper) that is readily deployable to convert from an opera/ballet setting to a symphonic performance setting (by Jaffe, see Reference 4). For symphonic purposes, the stage shell features acoustic coupling between the reverberant stage house behind and above the shell and the main audience floor.

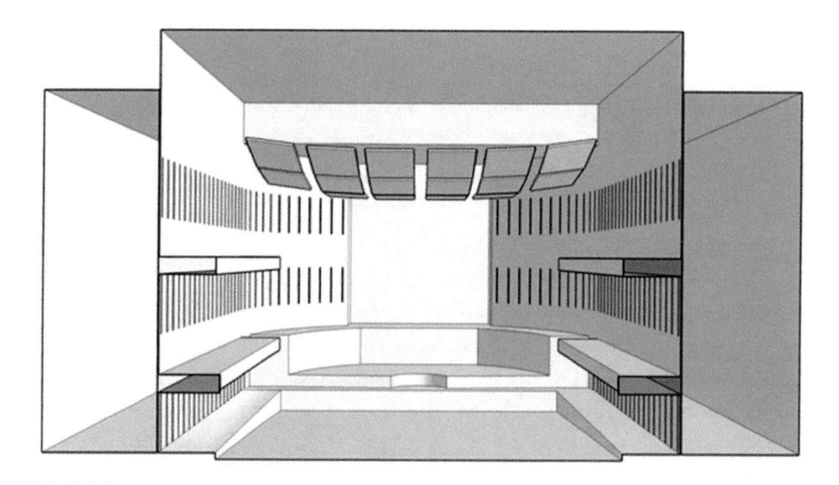

Figure 4.2 Reverberation chambers around the main audience floor of a concert hall with distributed coupling apertures.[6, 7] The acoustic coupling when the apertures are properly adjusted may result in double-slope, even multiple-sloped sound energy decays, which can simultaneously meet two desirable, yet competing, auditory features: clarity and reverberance.

4.2 STATISTICAL ACOUSTICS MODELS

Recent textbooks on room acoustics[9, 10] often follow the statistical theory of room acoustics first laid out by Davis,[1] providing brief discussion on sound energy in diffuse sound fields of coupled spaces. Early experimental investigations conducted by Eyring[11] on coupled spaces revealed that steady-state sound energy decays are of a non-single-slope (rather than double-slope) nature. Cremer and Müller's work[9] provides a thorough discussion of sound energy decays in two coupled rooms within the scope of the statistical theory, in which the sound energy and surface absorption are assumed to be uniformly distributed within each single room because the sound fields are considered diffuse. In two coupled rooms, two coupled partial differential equations provide two solutions of sound energy decays when the sound source is turned off:[9]

$$\begin{aligned} E_I(t) &= E_{I1}e^{-\delta_1 t} + E_{I2}e^{-\delta_2 t}, \\ E_{II}(t) &= E_{II1}e^{-\delta_1 t} + E_{II2}e^{-\delta_2 t}, \end{aligned} \tag{4.1}$$

where $E_I(t)$, $E_{II}(t)$ are the energy decay functions of Rooms I and II, respectively. E_{I1}, E_{I2} and E_{II1}, E_{II2} are the initial values of each decay mode. δ_1, δ_2 are decay constants with $\delta_i = 13.8/T_i$, T_i being ith decay time (in seconds). Note that in the case of a single-slope decay often occurring in single-space enclosures, T is termed *reverberation time*. If multiple slopes are apparent in the energy decay, a single reverberation time is said not to exist.[12] In this case, multiple decay times T_i represent pertinent quantities to characterize the steady-state energy decay profiles.

In architectural acoustics practice we are primarily concerned with the conditions where a sound source and receivers are within the same room—say Room 1 (also called the primary room) and another Room 2 (secondary room)—with $T_1 < T_2$ and the decay process associated with decay time T_2 starts at a lower level than the first decay process associated with T_1. Otherwise, the decay processes described by Eq. (4.1), even with two distinct decay constants, will result in single-slope energy decay functions.[9, 13]

The previous discussion, simplified by using the statistical theory, can be generalized to N coupled rooms to establish a group of N partial differential equations,[10, 14] their corresponding solutions describe steady-state sound energy decays, generally with N exponential decay modes.

Recently Summers et al.,[15] according to Lyle's suggestion,[16] have incorporated a delay time $\bar{\tau}$ into a group of N partial differential equations:

$$\frac{d\varepsilon_i}{dt} = -\eta_i\left(2\zeta_i + \frac{c}{4V_i}\sum_{\substack{j=1\\ j\neq 1}}^{N} S_{ij}\right)\varepsilon_i(t) + \frac{c}{4V_i}\sum_{\substack{j=1\\ j\neq 1}}^{N}\eta_j S_{ij}\varepsilon_j(t-\bar{\tau}_{ij}),\ i=1,\ldots,N, \tag{4.2}$$

in order to include the physical nature of sound propagation, that sound energies have to travel across spaces from one room to the other. In the partial differential Eq. (4.2), $\varepsilon_i(t)$ represents sound energy as a function of time, η_i is a coefficient ($\eta_i = 1$ when using the Sabine model it will be different using the Eyring or Kuttruff models[11]), ζ_i is a decay constant in the decoupled condition determined from the averaged absorption coefficient $\overline{\alpha_i}$; $(cS_i\overline{\alpha_i}/8V_i)$ in the ith room, with S_i being the total interior surface area of the ith room. $\bar{\tau}_{ij}$ is the averaged delay time from the ith room to the jth room, S_{ij} is the coupling area from the ith room to the jth room, c is the speed of sound, and V_i is the ith room volume. This partial differential equation has the general solution:

$$\varepsilon_i(t) = \sum_{j=1}^{N}\varepsilon_{j0}^{(i)}\exp(-2\delta_j t)\ \text{with}\ i=1,\ldots,N. \tag{4.3}$$

Taking two coupled rooms as an example, when $\zeta_1 > \zeta_2$, we can obtain $2\delta_1 > \eta_1(2\delta_1 + cS_{12}/4V_1) > \eta_2(2\delta_2 + cS_{12}/4V_2) > 2\delta_2$.

Summers[14, 15] experimentally validated the statistical-acoustics model expressed in Eq. (4.2) using acoustical scale modeling. The statistical-acoustic model continues to find application in design due to its computational efficiency and the ease of use it offers. While it is well-known that statistical acoustics models of this type must make multiple assumptions and approximations, they can offer reasonable accuracy when certain conditions are met. Of particular relevance are the approximations that each subsystem of the coupled rooms has to be considered to be at *one point* in the spatial and time-domain—at this point the diffuse sound energy and overall absorption take on a single value, no spatial variations are modeled. Accuracy can often be further improved through semi-empirical corrections, e.g., the above assumed averaged delay time can be further approximated through integrating over the subspace if the delay time is known as a distribution density over the subspace.[15]

4.3 WAVE-ACOUSTICAL METHODS

Classical wave-acoustic investigations in enclosures evolved into wave-theoretical room acoustics.[9] Harris and Feshbach[2] already studied coupled rooms in the late 1940s and early 1950s. Their work was based on wave-acoustics theory using room-acoustic modal analysis—approaching the problem of coupled rooms as a boundary value problem, yet limited to two extreme coupling cases: weak and strong coupling. The insight gained from their wave-theoretical investigation, validated by experimental effort, was limited to the discussion of modal frequency shifts in terms of coupling.

In the 1980s, Thompson[17] applied a wave-based method of matched asymptotic expansions, approximating sound fields in coupled rooms, both near the coupling window and far away from it. In the regions far from the coupling aperture and the discontinuity caused by the dividing wall with the aperture in it, the field quantities (pressure and particle velocity) are expanded in terms of an asymptotic sequence. The coefficients of the asymptotic sequence for the sound pressure still satisfy the homogeneous Helmholtz equation. In the regions near the aperture and the discontinuity, the coefficients for the asymptotic sequence for sound pressure are governed by the Laplace equation.[16] Thompson solved the Laplace equation for the field quantities in the coupled enclosures using the Schwartz Christoffel transformation. Lagrange's expansion theorem was applied. In the end, Thompson matched solutions of both regions in terms of zeroth, first, and second order approximations for sound pressure. Thompson's work demonstrated that the coupling aperture and discontinuity of the dividing wall will cause the modal distributions to skew. Experimental results support his wave-based asymptotic approximation solutions at a number of modal frequencies.

In more recent years, wave-based modal expansion methods have also been investigated by Meissner,[18] who simulated sound pressure distributions for different room geometries, source and receiver locations, and impedance boundaries, and also obtained sound pressure level decay functions with double-slope characteristics. Most recently, Meissner[19] has conducted further studies of coupled rooms. Still based on the modal wave theory, solutions of the wave equation along with the boundary conditions are derived for a room's sound pressure response when subjected to a sound source where the sound source is expressed as a series of eigenfunctions and eigenvalues. Meissner then analyzed the time-averaged potential, kinetic sound energy densities, and the active and reactive components of sound intensities in coupled rooms. The active component of the complex sound intensity describes propagation of the acoustic energy flux, while the reactive sound intensity represents nonpropagating oscillatory sound energy flux. Meissner's study reveals there are circulating energy flows in the steady-state sound fields when the coupled rooms are excited by a harmonic sound source as illustrated in Figure 4.3(a). In numerical, simulated active intensity distributions over two coupled rooms, Meissner is able to demonstrate in Figure 4.3 that discrete vortices follow the modal patterns of the enclosures under investigation. The vortices clearly indicate acoustic energy flowing continually around the closed paths around the vortices' center as shown in Figure 4.3(a), with the centers of vortices corresponding exactly to local nodes of the modal pressure distributions as shown in Figure 4.3(b).

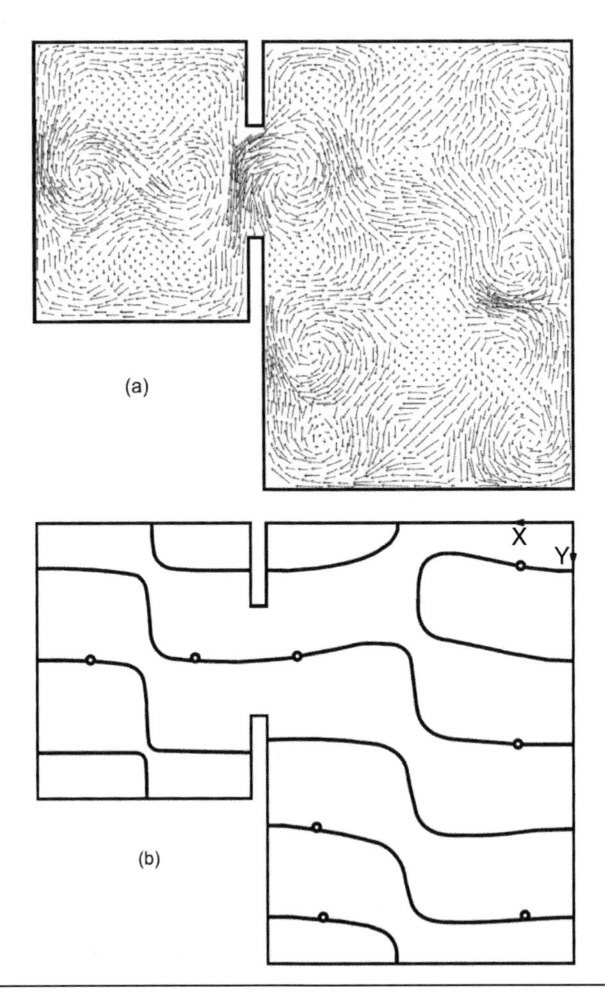

Figure 4.3 Nodal sound pressure and intensity distributions at a specific modal frequency (110.7Hz) in two coupled rooms [with room dimensions (H × W × L) 3 × 4 × 5m and 3 × 5.7 × 8m] by Messner.[19] The sound source was positioned in the larger room at x = 2m, y = 5m and z = 1m. (a) Modal distributions of active sound intensities on the observation plane at height 1.8m (results overtaken from Meissner for illustrative purposes. See more accurate distributions in original work by Meissner[19]). (b) Nodal lines indicating zeros of sound pressure via the modal eigenfunctions for the specific mode number (m = 48). Circles denote vortex centers.

Although wave-theoretical modal expansion methods can help shed light into the sound pressure distributions, potential and kinetic sound energy densities, and the propagating energy flows in terms of active and reactive sound intensity distributions in coupled spaces, the frequency range is limited to low, modal frequencies.

4.4 GEOMETRICAL ACOUSTICS METHODS

A historic review of how ray tracing entered into the geometrical acoustics (GA) room simulations by the end of the 1960s and early '70s can be found in a recent book.[20] In the 1990s, active research produced rapid development with GA methods.[21] Image source (IS)[22] and ray-tracing[23, 24] methods—two traditional methods used for computational modeling of sound fields in rooms according to the laws of GA—display no fundamental limitations unique to coupled-room configurations. Rather, they only require careful implementation, allowing for more complex visibility checks and the possible requirement of more rays[25–28] to provide for the more stringent requirements of spatial sampling. However, IS and ray-tracing methods have been largely superseded commercially by more modern variations on these algorithms (cone- or beam-tracing) and hybrid approaches[28–31] (see also Chapter 2 by Svensson et al. in this book).

With the increasing need of commercial software to simulate coupled rooms, the validity of these modeling tools has been reexamined.[28] In commercial beam-tracing software, beam tracing is most often applied approximately by tracing only the central axes of the beams—so-called beam-axis tracing. This approach offers the advantage of decreased computing time by covering the 4π *sr* of solid angle outwardly visible from the source more efficiently than ray tracing (i.e., fewer beam axes are needed than rays due to the increased detection probability of a beam face relative to a ray). However, it can result in incorrect detections of energy arrivals at receivers in the late part of the decay, which must be addressed by a tail correction procedure. Likewise, hybrid methods often require some form of tail correction or must make assumptions regarding the nature of the late decay. Conventionally, correction algorithms assume that the reflection density $n(t)$ is quadratic in time following the expression:

$$n(t)=\frac{4\pi\, c^3 t^2}{V}, \tag{4.4}$$

or, more generally, assume that $n(t) \propto t^2$ and determine the constant of proportionality by curve fitting. However, these assumptions do not hold in coupled rooms, as the reflection density associated with each subroom may differ, such that the function describing $n(t)$ can vary with both time and position.

In the work by Summers et al.,[28] the beam-axis tracing algorithm, the so-called *randomized tail-corrected cone-tracing*, used by the commercial software CATT-Acoustic was modified to eliminate both of these sources of errors when modeling coupled rooms. The new algorithm behaves as the randomized tail-corrected cone-tracing until any one of the expanding detector spheres contacts a surface boundary. From that point on, rather than applying tail correction, the radius of each detection sphere is held fixed and the detection procedure at each receiver is altered to that of ray tracing (i.e., the propagation law of r^{-2} is addressed by changes in detection probability due to ray divergence, rather than explicit reduction of the energy associated with each ray). In addition, the effects of diffraction are often significant in coupled rooms,[27] a recent effort in improving/including diffraction phenomena in GA modeling[32] is expected to contribute to the effective modeling of coupled rooms. Recent work has seen the application of improved GA approaches in investigations of acoustics in coupled spaces.[7, 33]

The classical phonon picture of GA is the basis for the most commonly used modeling techniques in room acoustics. Within this framework, the time-dependent energy (phonon) density $w(\mathbf{r},t)$ at any point in an enclosure having ideal diffuse reflection (Lambert's law) on the boundaries is given by Kuttruff's integral equation:[36]

$$w(\mathbf{r}_r,t)=\frac{W(t-R_{sr}/c)}{4\pi cR_{sr}^2}e^{-mR_{sr}}+\frac{1}{\pi c}\int_S\frac{B(\mathbf{r},t-R_r/c)[1-\alpha(\mathbf{r})]\cos\theta_r}{R_r^2}e^{-mR_r}dS, \tag{4.5}$$

where $\mathbf{r}_r$ is the position of the receiver, $W(t)$ is the time-dependent power of the omnidirectional source, R_{sr} is the distance between the source and the receiver, c is the speed of sound, m is the energy dissipation coefficient of air, $\mathbf{r}$ is the position of an infinitesimal surface element dS, $\alpha(\mathbf{r})$ is the angle-independent absorption coefficient at the surface element, and θ_r is the angle between the normal to the surface element dS and the line joining dS and the receiver position. The integration in the second term on the right-hand side in Eq. (4.5) is carried out over the entire room surface S. The expression can be modified to account for a general reflection law,[36–38] but yields analytic solutions only in the case of spherical enclosures[36] and infinite parallel plates.[38]

Computational GA approaches comprise direct or indirect methods of solving this integral. Acoustical radiosity attempts to solve some form of Eq. (4.5) through a direct computational approach of discretization.[39] Typically, it begins with the diffuse reflection form given above. More conventional methods of computational GA, such as ray tracing, are (for Lambert's diffuse reflection on the boundaries) formally equivalent to Monte Carlo methods of solving the integral.[37] While Kang,[40] Nosal et al.,[39] and Zhang[41] have detailed using the acoustic radiosity method in room-acoustics applications, work has only recently been directed toward using it in coupled rooms.[42] As with other computational GA methods, implementation may require some specific adaptations for coupled rooms.

4.5 DIFFUSION EQUATION METHODS

Diffusion equations have already been widely applied in physical, biological, geological, and even social sciences[43] (see also Chapter 2 by Svensson et al.). In 1969 Ollendorf[44] first applied the diffusion equation in room acoustic simulations, yet this did not draw much attention from the room-acoustics community. The late 1990s[45] and early 2000s[46] have seen active research in the diffusion equation modeling. Valeau et al.[46] detailed boundary conditions associated with the interior diffusion equation in enclosed spaces. Jing and Xiang[47] expanded the boundary conditions within the framework of the diffusion equation models based on rigorous physical-mathematical theory, so as to significantly extend the valid ranges in which the diffusion equation models can be applied. Allowing the assumption of ergodicity, and considering sound particles (phonons) as traveling along straight lines at the speed of sound, energy density $w(\mathbf{r},t)$ as a function of position ($\mathbf{r}$) and time (t) is given to first-order by the diffusion equation:[28, 46]

$$\frac{\partial w(\mathbf{r},t)}{\partial t}-D\nabla^2 w(\mathbf{r},t)+cmw(\mathbf{r},t)=F(\mathbf{r},t),\ \in V, \tag{4.6}$$

subject to the boundary condition on the interior surface S:

$$-D\frac{\partial w(\mathbf{r},t)}{\partial \mathbf{n}} = A(\mathbf{r},\alpha)cw(\mathbf{r},t), \in S, \tag{4.7}$$

with $A(\mathbf{r}, \alpha)$ in three alternative forms:[47–50]

$$A_S(\mathbf{r},\alpha)=\frac{\alpha(\mathbf{r})}{2}, \quad A_E(\mathbf{r},\alpha)=\frac{-\ln[1-\alpha(\mathbf{r})]}{4}, \quad A_M(\mathbf{r},\alpha)=\frac{\alpha(\mathbf{r})}{2[2-\alpha(\mathbf{r})]}, \tag{4.8}$$

with $A_S(\mathbf{r}, \alpha)$ termed Sabine,[46] $A_E(\mathbf{r}, \alpha)$ Eyring,[49, 50] and $A_M(\mathbf{r}, \alpha)$[47] modified boundary condition, $\alpha(\mathbf{r})$ represents the energy-based sound absorption coefficient at surface location $\mathbf{r}$. Eq. (4.6) represents the interior equation for the subroom denoted by domain V having a source term $F(\mathbf{r}, t)$, which is zero for any subroom where no source is present. The diffusion coefficient D is given by $D = \lambda c/3$, with λ being the mean free path. ∇^2 is the Laplace operator. The term $cmw(\mathbf{r},t)$ accounts for air dissipation in the room(s), and can be extended to account for absorption due to scattering objects inside the room(s).

When applied in two coupled rooms (see Figure 4.4), another boundary condition is the continuous boundary of the coupling aperture and it has to fulfill the following condition:

$$\hat{\mathbf{n}}\cdot\left[D_1\nabla w(\mathbf{r}_b,t)-D_2\nabla w(\mathbf{r}_b,t)\right]=0, \tag{4.9}$$

which represents a continuity boundary condition on interior boundaries at the aperture position $\mathbf{r}_b$, where D_1 is the diffusion coefficient in the primary room and D_2 is the diffusion coefficient for the secondary room.

In essence, the diffusion equation modeling can be considered either as GA, or statistical acoustics. The finite element meshing condition of the solutions often is dictated by the mean-free path given that the coupling aperture dimensions are not extremely small.[51] This will bring advantages in terms of computational load over those of the classical GA modeling on one hand, and the diffusion equation modeling in the form of parabolic partial differential equations containing both spatial ($\mathbf{r}$) and temporal (t) variables on the other hand. The solution of the diffusion equation reflects the detailed, gradual changes/distributions of reverberation processes over the entire enclosed volumes under investigation, which go beyond the inherent limits of classical statistical acoustics theory. Figure 4.4 illustrates the sound energy level distributions calculated using the diffusion equation over two coupled rooms. More importantly, the sound energy flux vector, $\vec{\mathbf{J}}$, according to Fick's law[46, 52] is determined by the gradient of sound energy density $w(\mathbf{r},t)$ by:

$$\vec{\mathbf{J}}=-D\,\text{grad}\,w(\mathbf{r},t), \tag{4.10}$$

which provides an efficient way to study the sound energy flux across the coupled rooms via the aperture.

Billon et al.,[53] first applied diffusion equations in the acoustics research of coupled rooms. Xiang et al.,[51] have conducted a large-scale experimental campaign, systematically comparing diffusion equation modeled results with those obtained from scaled-down models. The work systematically validated the correctness of the diffusion equation modeling, particularly after the diffuse sound fields have formed in the spaces under test.

The validated diffusion equation models become some of the efficient prediction tools available to obtain the sound energy flux distributions—as determined by Eq. (4.10)—in addition to the energy density distributions, and steady-state energy decays. The sound energy flux corresponds to the sound intensity vector. Xiang et al.,[51] relying on advanced sound

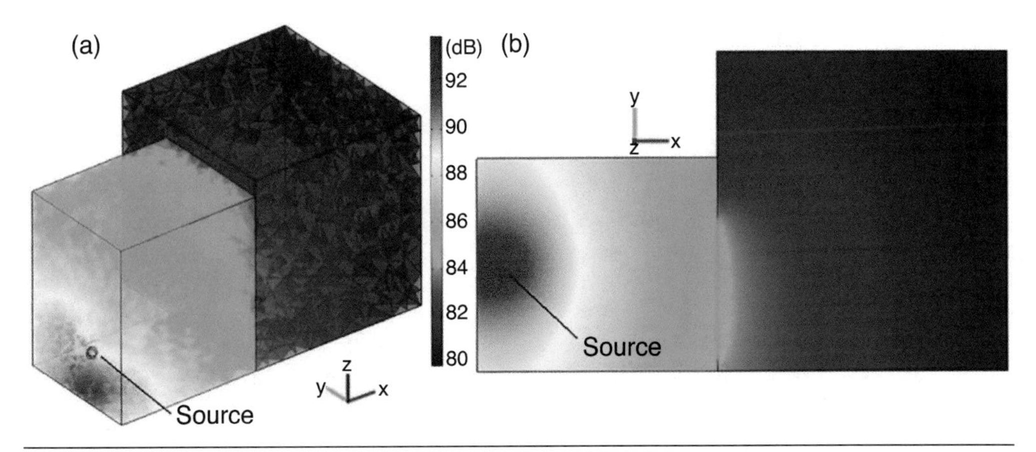

Figure 4.4 Sound energy level distributions by the diffusion model in the scale model coupled rooms (room dimensions for the source room: H = 6.4m, W = 5.6m, L = 4.8m and the secondary room H = 6.4m, W = 6.8m, L = 7.2m).[48] (a) Three-dimensional presentation with a partial transparent pseudo-colored scale showing finite element meshing. (b) Two-dimensional presentation in the *X*-*Y* plane.

energy decay analysis (see following section), revealed the mechanism and necessary conditions of sound energy exchanges from one room to the other, finding that the *turning point* of the double-slope energy decay corresponds to the moment of reversing energy flux directions. Jing and Xiang[52] visualized the sound energy fluxes in two coupled rooms; a number of animated energy flux visualizations can be viewed from the aforementioned publication in its online version.[52] A number of stationary frames the energy fluxes modeled using the diffusion equation are illustrated in Figure 4.5.

Recently, room-acoustic studies of coupled volume systems have been reported.[53–55] Pu et al.[54] applied diffusion equation modeling and scaled-down modeling to investigations of concave and convex energy decay profiles near the coupling aperture. More recently, Luizard et al.[55] have further analyzed the diffusion equation solutions. When removing the term containing the Laplace operator in Eq. (4.6), containing the spatial second-order partial differentiations, the solution of the so-simplified diffusion equation for steady-state switch-off excitations reduces to a simple exponential decay function in time removing the spatial variations that have long been familiar to the statistical theory in classical room acoustics. This analysis efficiently demonstrates why the diffusion equation can be considered as a higher order statistical theory being able to model the spatially varied sound energy distributions. Luizard et al.[55] further proposed analytical solutions of the diffusion equation models. In the case of two coupled rooms, the solution in the primary room can be expressed as:

$$E_I(\mathbf{r},t) = E_{I1}\left(\frac{a}{r_{\mathrm{SR}}}e^{-\varepsilon r_{\mathrm{SR}}} + b\right)e^{-\delta_1 t} + E_{I2}\left(\frac{a}{r_{\mathrm{AR}}}e^{-\varepsilon r_{\mathrm{AR}}} + b\right)e^{-\delta_2 t}, \tag{4.11}$$

where E_{Ii} are the initial sound energies in each room—see also Eq. (4.1). r_{SR} and r_{AR} are the source-receiver and aperture-receiver distances, respectively. $\varepsilon = \sqrt{\sigma_1 / D_1}$ is the spatial decay constant in the primary room. In Eq. (4.11), coefficients a and b can be used to specify the receiver region (distance) from the sound source or from the aperture, respectively. For example if $b >> ae^{-\varepsilon r_{SR}}/r_{SR}$ and $b >> ae^{-\varepsilon r_{AR}}/r_{AR}$, then the receiver position is far away from both

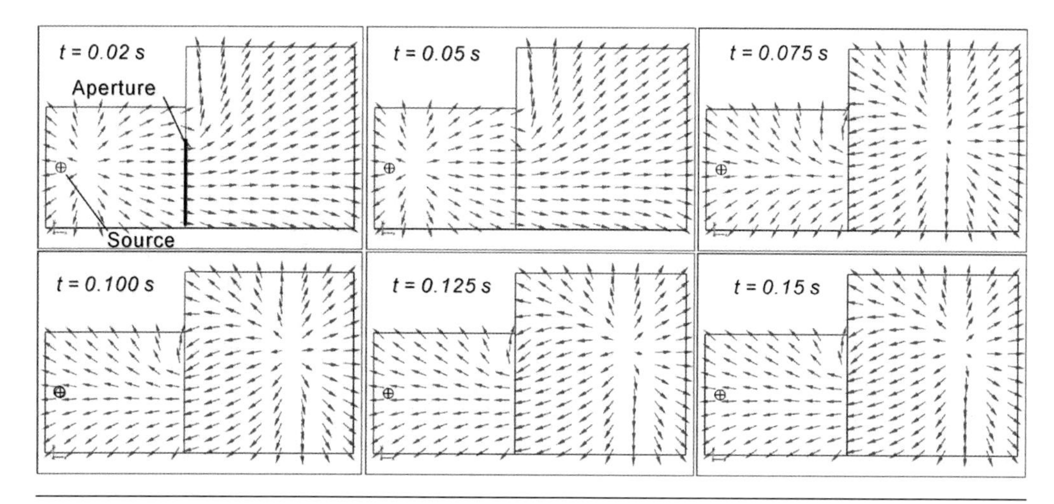

Figure 4.5 Two-dimensional mapping of steady-state-derived sound energy flow vectors using Eq. (4.10) for six different snapshots on a plane at z = 3m in the coupled rooms (room dimensions for the source room: H = 6.4m, W = 5.6m, L = 4.8m and the secondary room H = 6.4m, W =6.8m, L = 7.2m).[48] Overall acoustic conditions are characterized by the natural reverberation times (T_1', T_2') in two rooms: $T_1' < T_2'$, where T_1' is the natural reverberation time in the source room, and T_2' in the secondary room.

the sound source and the aperture so that the solution is reduced to the solution as shown in Eq. (4.1), asymptotically approximating the solution of classical statistical theory as given by Cremer and Müller.[8] When $b << ae^{-\varepsilon r_{SR}}/r_{SR}$, the solution of Eq. (4.11) predicts the results of the receiver being close to the sound source; the sound energy contains the location-dependence with respect to the sound source.

4.6 EXPERIMENTAL INVESTIGATIONS AND ANALYSIS TOOLS

Investigations of acoustics in coupled volume systems can also be carried out via experimental approaches. Many authors have to rely on experimental investigations to validate their findings/theories or observations,[2, 7, 14, 15, 17, 28, 33, 48, 53–55] particularly using acoustical scale modeling. When combined with advanced room acoustic measurement techniques,[56–58] it exhibits technological advantages in experimental investigations of coupled volume systems. Scale modeling inherently contains acoustical wave phenomena, and enables convenient maneuvering of space configurations for systematic investigations, so as to shorten the time required for experimental investigations. Many numerical studies and theoretical discussions have to be experimentally validated. In order to increase the efficiency of the experimental investigation of coupled rooms, Xiang et al. recently developed a scale-modeling platform particularly suitable for research of coupled rooms.[51] Figure 4.6 illustrates the high spatial resolution room scanning system in the form of an eighth-scale model of two coupled rooms. The sound fields inside the two coupled rooms can be measured from an automatic scanning mechanism installed outside the rooms under investigation, sound sources and receivers inside the rooms can be moved via a magnetic coupling, e.g., when moving a microphone inside the rooms in a stop-stare manner, only the microphone holder is moving from one position to the other. On this platform, a two-dimensional plane (X-Y) at the microphone

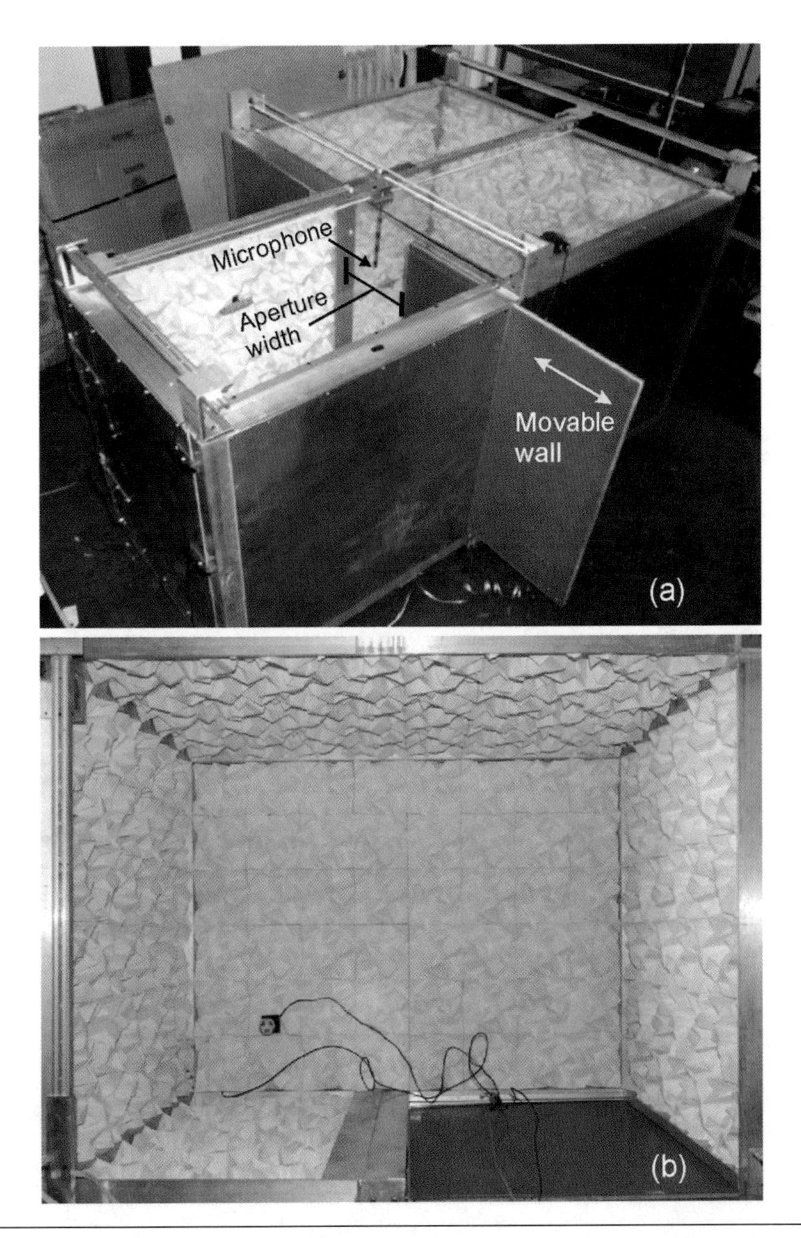

Figure 4.6 A high spatial resolution room scanning system in the form of an eighth-scale model of two coupled rooms.[51] (a) Photo of the model rooms with a movable wall between the two rooms for adjusting the coupling aperture sizes. A step-motor-driven scanning mechanism is installed on top of and outside the models. It moves a microphone holder inside via magnetic coupling. (b) Top view into the larger model room to show the diffusely reflecting surfaces of interior walls. A scale model sound source in dodecahedron form, covering a frequency range between 1kHz and 45kHz, is positioned in one corner.

(receiver) height can be sampled at a high spatial resolution. The spatial resolution (Δx, Δy) is dictated by the spatial Nyquist theorem:

$$\Delta x, \Delta y \le \frac{c}{2 f_u}, \tag{4.12}$$

where c is the speed of sound and f_u is the upper limit frequency. When defining the finest scanning grid sizes over the X-Y plane, reposition uncertainties must be taken into account to obtain reliable experimental data up to the upper limit frequency, f_u. In the experimental platform, the coupling aperture is designed to be easily adjustable. Interior surfaces of the two scaled-down coupled rooms are treated by pseudo-random numbers to obtain highly diffuse reflections. The lower degrees of diffuse reflections of the interior surfaces can be adjusted by gradually covering the rough surfaces by specular reflecting surfaces. Using this automatic scanning platform, experimental data of two-dimensional sound fields can be efficiently obtained, so as to create animations of sound pressure distributions and sound energy coupling/feedback completely based on acoustical experiments. The high spatial resolution automatic scanning system can obtain experimental data valid for animating wave propagations between tens of Hertz up to 1kHz. Furthermore, all measurements at strategic (yet fixed) measurement positions can also provide experimental data valid even to broad frequency ranges extending to the upper audio frequencies,[51] which can effectively support experimental investigations/validations of wave-based acoustics modeling, modeling using statistical acoustics, GA, and using diffusion equations.

As discussed previously in many studies on acoustics in coupled rooms, one often encounters double-slope or even multi-slope sound energy decays.[64, 66] Already in the 1930s, Eyring[10] pointed out that double-slope sound energy decays occur not only in two coupled rooms, it can also be observed in single-volume spaces when sound absorption and scattering are not evenly distributed. With the increased interest in studying acoustics in coupled volume systems, architectural acousticians are challenged by the highly demanding tasks of energy decay analysis. Quantitative analysis and characterization of multi-slope sound energy decays become crucial throughout both theoretical and experimental investigations in coupled rooms (also see Chapter 6 by Xiang). Using nonlinear regression to tackle the problems of single- and double-slope decay analysis[59, 60] in the late 1990s, the focus was transitioned immediately to the application of model-based Bayesian analysis.[7, 12, 61] Room impulse responses, either from experimental measurements or numerical simulations, are used to derive the steady-state energy decay as a function of time using Schroeder's backward integration[62] and can be approximated by a parametric mode:[63–66]

$$\mathbf{H}_M(\boldsymbol{\Theta}) = \theta_0 (t_K - t_k) + \sum_{m=1}^{M} \theta_{2m-1} \left[\exp\left(-\frac{13.8}{T_m} t_k \right) - \exp\left(-\frac{13.8}{T_m} t_K \right) \right]; \quad 0 \leq k < K, \qquad (4.13)$$

with $T_m = \theta_{2m}$ being the mth decay time (parameter), one can take $\boldsymbol{\Theta}$ to include all $(2M + 1)$ parameters with M being the number of slopes. t_k represents a discrete time variable, t_K is the upper limit of Schroeder's integration.

According to Eyring's early investigations[10] and a number of recent experimental investigations,[65–69] experimenters often are not able to determine the number of slopes present in the sound energy decay functions prior to data analysis. Experimenters should not even confine themselves to only single-slope or double-slope energy decays.[66] It often occurs that the sound energy decay data across different (octave) frequency bands possibly exhibit decay characteristics from single slope to multiple slope in nature. It is therefore not helpful to apply an incorrect decay model (with an incorrect number of slopes) to analyze the parameters appearing in that model.[67] This model-based sound energy decay analysis requires two levels of inference, both the decay model selection and decay parameter estimations.[65] At the level of the model selection, giving some energy decay data in the form of the Schroeder decay function, a finite set of decay models $\mathbf{H}_1(\mathbf{\Theta})$, $\mathbf{H}_2(\mathbf{\Theta})$, ... may potentially be competing with each other to describe the data. Solving the model selection problem is of primary concern to narrow down the options to one specific decay model. At the parameter estimation level, once the decay model is selected by the data (say, $\mathbf{H}_2(\mathbf{\Theta})$), the task is to estimate the parameter vector $\mathbf{\Theta}$ appearing in that model. Bayesian probability inference logically provides a unified framework capable of handling both the model selection and the parameter estimation.[65]

SUMMARY

A number of recent performance venues intentionally incorporating coupled volumes have prompted active research interest in the acoustics of coupled volume systems. In studying acoustics in coupled rooms, statistical acoustics approaches still remain effective tools for design due to their ease of use and the prediction speed often required in practice. It is well known that statistical acoustics approaches must rely on multiple assumptions and approximations; they can offer reasonable accuracy under certain conditions. For the acoustical modeling of coupled rooms, GA approaches have also been further developed and applied. Wave-based acoustics approaches have also been significantly developed, particularly for the modal analysis of kinetic and potential sound energy, and the real and imaginary parts of sound intensity distributions in coupled rooms. Due to the existence of the coupled apertures, systematic analysis and a deeper understanding of sound energy flux (corresponding to the sound intensity vector field) become feasible. Diffusion equation models are considered across statistical and GA theory; they are useful tools for predicting/studying sound energy distributions and sound energy fluxes across coupling apertures in the frequency range valid for GA. The diffusion equation model has been experimentally validated via an extensive systematic experimental campaign with large amounts of experimental data supporting the validation effort. Data collection has been possible thanks to an automatic room scanning platform based on scale modeling, which is particularly suitable for investigations of acoustics in coupled rooms. This automatic experimental platform will potentially become an effective tool for future efforts in modal analysis of sound energy fluxes, and sound intensity vector field analysis. In quantifying/characterizing the double-slope and multi-slope energy decays, Bayesian probability inference as a unified framework is available for room acoustic energy decay analysis. Nevertheless, room-acousticians will be facing even more challenges; at the same time, there will be a potential opportunity for them to make new contributions in the field of acoustics of coupled volume systems.

ACKNOWLEDGMENTS

The author is grateful to Dr. Jason Summers and Dr. Yun Jing for many years' collaborations, and stimulating discussions of various aspects related to this topic. Thanks also go to Douglas Beaton for insightful discussions and proofreading of early drafts of this chapter.

REFERENCES

1. Davis, A. H., "Reverberation equations for two adjacent rooms connected by an incompletely sound proof partitions," *Phil. Mag.*, **50**, pp. 75–81 (1925).

2. Harris, C. M. and Feshbach, H., "On the acoustics in coupled rooms," *J. Acoust. Soc. Am.*, **22**, pp. 572–578 (1950).
3. Jaffe, J. C., "Design Considerations for a Demountable Concert Enclosures (Symphonic Shell)," *J. Audio Eng. Soc.*, **22**, pp. 163–170 (1974).
4. Jaffe, J. C., "Innovative approaches to the design of symphony halls," *Acoust. Sci. & Tech.*, **26**, pp. 240–243 (2005).
5. Johnson, R., "The source of the McDermott Concert Hall Design," *Proc. W.C. Sabine Centennial Symposium*, Cambridge, MA, USA, 5–7 June, 1994, pp. 295–298.
6. Johnson, R., Kahle, E. and Essert, R., "Variable coupled cubage for music performance," *Proc. Music and Concert Hall Acoustics (MCHA 1995)*, Kirishima, Japan, 15–18 May, 1995, —
7. Xiang, N., Goggans, P., Jasa, T. and Robinson, P., "Bayesian characterization of multiple-slope sound energy decays in coupled-volume systems," *J. Acoust. Soc. Am.*, **129**, pp. 741–752 (2011).
8. Xiang, N. and Goggans, P. M., "Evaluation of decay times in coupled spaces: Bayesian parameter estimation," *J. Acoust. Soc. Am.*, **110**, pp. 1415–1424 (2001).
9. Cremer, L. and Müller, H. A., "*Die wissenschaftlichen Grundlagen der Raum-akustik,*" (S. Hirzel-Verlag, Stuttgart, 1978), [translation in English: L. Cremer, H. A. Müller, and T. J. Schultz: *Principles and Applications of Room Acoustics*, (Applied Science Publishers, London, UK, 1982)]
10. Kuttruff, H., "*Room Acoustics,*" John Wiley & Sons, 1973.
11. Eyring, C. F. "Reverberation time measurements in coupled rooms," *J. Acoust. Soc. Am.*, **3**, 181–206 (1931).
12. ISO 3382-1:2009, "Acoustics—Measurement of room acoustic parameters—Part 1: Performance spaces," International organization for standardization, 2009.
13. Xiang, N. and Goggans, P. M., "Evaluation of decay times in coupled spaces: Bayesian decay model selection," *J. Acoust. Soc. Am.*, **113**, pp. 2685–2697 (2003).
14. Summers, J. E., Torres, R. R. and Shimizu, Y., "Statistical acoustics models of energy decay in systems of large, coupled rooms and their relation to geometrical acoustics," *J. Acoust. Soc. Am.*, **116**, pp. 958–969 (2004).
15. Summers, J. E., "Accounting for delay of energy transfer between coupled rooms in statistical-acoustics models of reverberant-energy decay," *J. Acoust. Soc. Am. Express Let.*, **132**, pp. EL129–134 (2012).
16. Lyle, C. D., "An improved theory for transient sound behavior in coupled diffuse spaces," *Acoust. Lett.* **4**, pp. 248–252 (1981).
17. Thompson, C., "On the acoustics of a coupled space," *J. Acoust. Soc. Am.*, **75**, pp. 707–714 (1984).
18. Meissner, M., "Computational studies of steady-state sound field and reverberant sound decay in a system of two coupled rooms," *CEJP*, **4**, pp. 293–312 (2007).
19. Meissner, M., "Acoustic energy density distribution and sound intensity vector field inside coupled spaces," *J. Acoust. Soc. Am.*, 132, pp. 228–238 (2012).
20. Krokstad, A., Svensson, U. P. and Strom, S., "The early history of ray tracing" in "*Acoustics, Information, and Communication,*" Eds. Xiang, N. and Sessler, G., Chapter 2, Springer 2015, Cham, Heidelberg, New York, Dordrecht, London.
21. Xiang, N. and Summers, J., "Acoustics in Coupled Rooms: Modelling and Data Analysis," *Proc. Intl. Symp. Room Acoust.*, (2007).
22. Gibbs, B. M. and Jones, D. B., "A simple image method for calculating the distribution of sound pressure levels within an enclosure," *ACUSTICA*, **26**, pp. 24–32 (1972).
23. Schroeder, M. R., Atal, B. S. and Bird, C., "Digital computers in room acoustics," *Proc. 4th Intl. Congress Acoust.*
24. Krokstad, A., Strøm, S. and Sørsdal, S., "Calculating the acoustical room response by the use of a ray tracing technique," *J.Sound & Vib.*, **8**, pp. 118–125 (1968).
25. Ayr, U., Cirillo, E. and Martellotta, F., "Theoretical and experimental analysis for coupled rooms transient behaviour and relevant acoustical parameters," Proc. 17th Intl. Congress Acoust., Roma 2001.
26. Nijs, L., Jansens, G., Vermeir, G. and van der Voorden, M., "Absorbing surfaces in ray-tracing programs for coupled spaces," *Appl. Acoust.*, **63**, pp. 611–626 (2002).

27. Summers, J. E., "Comments on 'Absorbing surfaces in ray-tracing programs for coupled spaces'" *Applied Acoustics*, **64**, pp. 825–831 (2003).
28. Summers, J. E., Torres, R. R., Shimizu, Y. and Dalenbäck, B.-I. L., "Adapting a computational geometrical-acoustics algorithm based on randomized beam-axis tracing to modeling of coupled rooms via late-part ray tracing," *J. Acoust. Soc. Am.*, **118**, pp. 1491–1502 (2005).
29. Svensson, U. P., Siltanen, S., Savioja, L. and Xiang, N., "Computational Modeling of Room Acoustics II: Geometrical Acoustics," *Architectural Acoustics Handbook*, J. Ross Publishing, Chapter 2 (2017).
30. Vorländer, M., "Simulation of the transient and steady-state sound propagation in rooms using a new combined ray-tracing/image source algorithm," *J. Acoust. Soc. Am.*, **86** pp. 172–178 (1989).
31. Naylor, G. M., "ODEON—Another hybrid room acoustical model," *Applied Acoust.*, **38**, pp. 131–143 (1993).
32. Svensson, U. P. and Calamia, P., "Edge-diffraction impulse responses near specular-zone and shadow-zone boundaries," *Acta Acustica united with Acustica,* **92**, pp. 501–512 (2006).
33. Ermann, M., "Coupled volumes: Aperture size and the double-sloped decay of concert halls," *Build. Acoust.*, **12**, pp. 114 (2005).
34. Bradley, D. and Wang, L. M., "Optimum absorption and aperture parameters for realistic coupled volume spaces determined from analysis and subjective testing results," *J. Acoust. Soc. Am.*, **127**, pp. 223–232 (2010).
35. Kuttruff, H., "Simulierte Nachhallkurven in rechteckräumen mit diffuseen schallfeld, (Simulated decay curves in rectangular rooms with diffuse sound fields.) *ACUSTICA*, **25**, pp. 333–342 (1971).
36. Joyce, W. B., "The exact effect of surface roughness on the reverberation time of a uniformly absorbing spherical enclosure," *J. Acoust. Soc. Am.*, **64**, pp. 1429–1436 (1978).
37. Le Bot, A., "A functional equation for the specular reflection of rays," *J. Acoust. Soc. Am.*, **112 pp.** 1276–1287 (2002).
38. Kuttruff, H., "Stationary propagation of sound energy in flat enclosures with partially diffuse surface reflection," *ACUSTICA*, **86**, pp. 1028–1033 (2000). (See also references therein.)
39. Nosal, E.-M., Hodgson, M. and Ashdown, I., "Improved algorithms and methods for room sound-field prediction by acoustical radiosity in arbitrary polyhedral rooms," *J. Acoust. Soc. Am.*, **116**, pp. 970–980 (2004).
40. Kang, J., "Reverberation in Rectangular Long Enclosures with Diffusely Reflecting Boundaries," *Acustica united with Acta Acustica*, **88**, pp. 77–87 (2002).
41. Zhang, H., "Relaxation of sound fields in rooms of diffusely reflecting boundaries and its application in acoustical radiosity simulation," *J. Acoust. Soc. Am.*, **119**, pp. 2189–2200 (2006).
42. Jiang, G.-R. and Zhang, X.-L., "A radiosity model for sound decay in coupled rooms," J. South China Univ. of Tech. (Natural Science Edition), **35**, pp. 28–31 (2007).
43. Narasimhan, T.N., "The dichotomous history of diffusion," *Physics Today*, July 2009, pp. 48–53.
44. Ollendorff, F., "Statistische raumakustik als diffusions problem," *Acustica*, **21**, pp. 236–245 (1969).
45. Picaut, J., Simon, L. and Polack, J.-D., "A mathematical model of diffuse sound field based on a diffusion equation," *Acust. Acta Acust.*, **83**, pp. 614–621 (1997).
46. Valeau, V., Picaut, J. and Hodgson, M., "On the use of a diffusion equation for room acoustic predictions," *J. Acoust. Soc. Am.*, **119**, pp. 1504–1513 (2006).
47. Jing, Y. and Xiang, N., "On boundary conditions for the diffusion equation in room acoustic prediction: Theory, simulations, and experiments," *J. Acoust. Soc. Am.*, **123**, pp. 145–153 (2008).
48. Xiang, N., Jing, Y. and Bockman, A., "Investigation of acoustically coupled enclosures using a diffusion equation model," *J. Acoust. Soc. Am.*, **126**, pp. 1187–1198 (2009).
49. Jing, Y. and Xiang, N., "A modified diffusion equation for room-acoustic predication (L)," *J. Acoust. Soc. Am.*, **121**, pp. 3284–3287 (2007).
50. Billon, A., Picaut, J. and Sakout, A., "Prediction of the reverberation time in high absorption room using a modified-diffusion model," *Appl. Acoust.*, **69**, pp. 68–74 (2008).
51. Xiang, N., Escolano, J., Navarro J. M. and Jing, Y., "Investigation on the effect of aperture sizes and receiver positions in coupled rooms," *J. Acoust. Soc. Am.*, **135**, pp. 3975–3985 (2013).

52. Jing, Y. and Xiang, N., "Visualization of sound energy across coupled rooms using a diffusion equation model," *J. Acoust. Soc. Am. Express Letters*, **124**, pp. EL360–EL365 (2008).
53. Billon, A., Valeau, V., Sakout, A. and Picaut, J., "On the use of a diffusion model for acoustically coupled rooms," *J. Acoust. Soc. Am.*, **120**, pp. 2043–2054 (2006).
54. Pu, H. J., Qiu, X.-J. and Wang, J.-Q., Different sound decay patterns and energy feedback in coupled volumes, *J. Acoust. Soc. Am.*, **129, pp.** 1972–1980 (2011).
55. Luizard, P., Polack, J.-D. and Katz, B. F. G., "Sound energy decay in coupled spaces using a parametric analytical solution of a diffusion equation," *J. Acoust. Soc. Am.*, **135, pp.** 2765–2776 (2014).
56. Xiang, N. and Schroeder, "Reciprocal maximum-length sequence pairs for acoustical dual source measurements," *J. Acoust. Soc. Am.*, **113**, pp. 2754–2761 (2003).
57. Müller, S., "Measuring Transfer-Functions and Impulse Responses," in *Handbook of Signal Processing in Acoustics*, eds. Havelock, et al., Springer Science + Business Media LLC., **1**, pp. 67–85 (2008).
58. Xiang, N., "Digital Sequences," in *Handbook of Signal Processing in Acoustics*, eds. Havelock, et al. Springer Science + Business Media LLC., **1**, pp. 87–106 (2008).
59. Xiang, N., Evaluation of reverberation times using a nonlinear regression approach, *J. Acoust. Soc. Am.*, **98**, pp. 2112–2121 (1995).
60. Xiang, N. and Vorländer, M., "Using iterative regression for estimating reverberation times in two coupled spaces (A)," *J. Acoust. Soc. Am.*, **104**, p. 1763 (1998).
61. Xiang, N., "Advanced Room-Acoustic Decay Analysis," in "*Acoustics, Information, and Communication,*" eds. Xiang, N. and Sessler, G., Chapter 3, Springer 2015, Cham, Heidelberg, New York, Dordrecht, London.
62. Schroeder, M. R., "New method of measuring reverberation time," *J. Acoust. Soc. Am.*, **37**, pp. 409–412 (1965).
63. Faiget, L., Legros, C. and Ruiz, R., "Optimization of the Impulse Response Length: Application to Noisy and Highly Reverberant Rooms," *J. Audio Eng. Soc.*, **46**, pp.741–750 (1998).
64. Xiang, N., Goggans, P., Jasa, T. and Robinson, P., "Bayesian characterization of multiple-slope sound energy decays in coupled-volume systems," *J. Acoust. Soc. Am.*, **129**, pp. 741–752 (2011).
65. Xiang, N., Room-acoustics energy decay analysis, *Architectural Acoustics Handbook*, Chapter 6, ed. N. Xiang, J. Ross Publisher (2017).
66. Xiang, N., Robinson, P. and Botts, J., "Comments on 'Optimum absorption and aperture parameters for realistic coupled volume spaces determined from computational analysis and subjective testing results' (L)," J. Acoust. Soc. Am., **128**, pp. 2539–2542 (2010).
67. Jasa, T. and Xiang, N., "Efficient estimation of decay parameters in acoustically coupled spaces using slice sampling," *J. Acoust. Soc. Am.*, **126**, pp. 1269–1279 (2009).
68. Jasa, T. and Xiang, N., "Nested sampling applied in the Bayesian room-acoustics decay analysis," *J. Acoust. Soc. Am.*, **132**, pp. 3251–3262 (2012).
69. Xiang, N., Goggans, P. M., Jasa, T. and Kleiner, M., "Evaluation of decay times in coupled spaces: Reliability analysis of Bayesian decay time estimation," *J. Acoust. Soc. Am.*, **117**, pp. 3707–3715 (2005).

5

Advanced Measurements Techniques: Methods in Architectural Acoustics

Wolfgang Ahnert, Acoustic Design Ahnert and Ahnert Feistel Media Group, Berlin, Germany

Stefan Feistel, Ahnert Feistel Media Group, Berlin, Germany

5.1 OVERVIEW

In this chapter we discuss advanced measurement methods and measurement technologies for applications in room acoustics and sound reinforcement.

We start with a short review of the fundamentals and then introduce traditional as well as modern measurement techniques. These range from the conventional impulse tests (pistol shot) or excitation with pink noise over the computer-based time-delay spectrometry (TDS) and maximum-length sequence (MLS) methods to current, commonly used evaluation practices based on Fourier analysis. Remarks about the calibration of measurements and typical errors as well as measurement uncertainties conclude the first section.

In the second part we describe room acoustic applications, including all relevant room acoustic parameters according to ISO standard 3382. A full reference is provided in Chapter 12, Section 12.1.2. Additionally, special types of measurement, such as with applied filtering or in situ, are discussed as well.

The third section represents an overview of commonly used methods for the measurement, evaluation, and tuning of sound reinforcement systems, where we focus particularly on the desired frequency response and level. But also, speech intelligibility measurements—speech transmission index (STI) and the STI developed for fast measurements of public address systems (STIPa) procedure—are explained in detail. The section concludes with typical tests made during the commissioning of a sound system.

5.2 MEASUREMENT METHODS

5.2.1 Traditional Sound Level Measurements and Assessment

The loudness of sound events is closely related to sound pressure level; it has been measured since the existence of the first microphones. To quantify it, until the 1920s the level was usually defined in Nepers, which is a quantity based on the natural logarithm. Nowadays, the common logarithm of the sound pressure/power is normally used and given in the unit of Bels or Decibels. By 1926 Barkhausen proposed the subjective scale Phon for loudness, which is identical to the sound level scale in decibel (dB) at 1000 Hz but accounts for the frequency-dependent sensitivity of the human hearing system (Weber-Fechner law). The weighting curves A, B, and C are derived from that. In particular, the A-weighted level as displayed by a broadband measurement device corresponds to the subjective loudness of sound at normal levels as it is perceived by a listener.

We refer the reader to standard literature for further details about conventional sound level meters which are most commonly used in building acoustics.

To the contrary, in this chapter we would like to discuss primarily computer-based measurements in the field of room acoustics and electro-acoustics. Until the 1950s and '60s, the principal means for the measurement of room acoustic characteristics was the impulse test. These tests have not only been applied to rooms of regular size and shape, but also to scale models beginning in the 1930s. A spark generator was mostly used as the source of excitation, and an oscilloscope with memory was employed to record the sound returned by the room. Back at that time, the experience and knowledge of acousticians determined whether or not useful conclusions could be drawn from the display of the oscilloscope. Except for the reverberation time, there were no objective measures to describe the properties of the room under test, thus, misinterpretation was common and disputes arose often. It was not until the 1970s that an increasing amount of investigation was made regarding the relationship of subjective perception and objective conditions. Objective measurements could now benefit from the first advanced acoustic measurement devices, as well as from limited computer-based processing.

In the 1970s, the so-called time/energy/frequency (TEF) analyzer was developed in the United States. Using TDS, these devices could measure energy-time curves by means of a swept-sine wave excitation signal. We will discuss this method a little later in Section 5.2.2.6.

Also in the 1970s, Schroeder had introduced another idea, namely the notion of using pseudo-random noise based on the MLS to determine the impulse response of the system under test (SUT). But the MLS technique was widely accepted only in 1988 when the first device called a maximum-length sequence system analyzer (MLSSA) became commercially available worldwide (see Section 5.2.2.3). Since then, it has become a common standard for room acoustic measurements. This was also facilitated by the fact that various room acoustic parameters had been established in the meantime and an extensive set of post-processing functions available in the MLSSA allowed the display of these parameters immediately after the measurement. We will come back to this topic in Section 5.2.2.5.

After the introduction of modern personal computers and especially due to the broad availability of laptop computers, several software-based measurement packages (Dirac, WinMLS, Smaart, and recently EASERA and SysTune) have been developed which all allow for a variety of excitation signals to determine impulse responses or transfer functions. They distinguish themselves primarily with respect to the implemented post-processing.

5.2.2 Measurement Techniques Based on Fourier Analysis

5.2.2.1 Fundamentals

Acoustic measurement methods are generally based on the recording of sound signals and their evaluation. In the very beginning, frequency analysis was found to be an important tool for the assessment of recorded data, since it allowed investigations in a way similar to the human perception of sound.

In general, Fourier analysis (Fourier, French mathematician, 1768–1830) is understood as the spectral decomposition of a time signal with respect to the harmonic frequencies. Put in a simple form, the signal amplitude at a given frequency is determined by the scalar product of the time signal, $a(t)$, and the harmonic function at frequency ω:

$$\tilde{A}(\omega)=\frac{1}{2\pi}\int_{-\infty}^{\infty} a(t)\exp(-j\omega t)\,dt\,. \tag{5.1}$$

The resulting complex frequency spectrum, $\tilde{A}(\omega)$, provides insight into the contributions of individual frequencies or tones, respectively, to the sum signal. Therefore, Fourier analysis is principally useful for comparing subjectively perceived spectra of tones with objectively measured ones, but also to draw conclusions from the objective measurement regarding the subjective impression. Furthermore, this method facilitates, for example, the identification of resonances in more complex processes that cannot be recognized by human hearing when masked by the overall signal. Therefore, evaluation in the frequency domain has become a common measurement method and it normally complements investigations in the time domain.

The development of digital signal processing (DSP) and computer-based measurement systems and the use of analog/digital (AD) converters made it necessary to manage signals that were discrete in time and treat them properly (Oppenheim & Schafer).[1] The sampling of the intruding time signal in fixed intervals and the evaluation of data frames of finite length limit the temporal and spectral resolution of digital recordings compared to the real signal. This is well defined by the sampling theorem, which states that the sampling rate, f_s, determines the highest frequency resolved, the so-called Nyquist frequency (Nyquist, Swedish mathematician, 1889–1976):

$$f_{\max}=\frac{1}{2}f_s\,. \tag{5.2}$$

The sampling interval T determines the density Δf of the discrete frequency spectrum:

$$\Delta f\approx\frac{1}{T}\,. \tag{5.3}$$

Accordingly, all digital measurement systems are subject to these principal constraints.

In practice a variety of such measurement systems are being utilized. They can be distinguished by two fundamental principles. Either they only support the analysis of incoming sound in time and frequency, or they also provide for the more complicated determination of the transfer function of the SUT. In the first case, the time signal at the input is simply transformed into the frequency domain for further analysis. In the second case, the so-called reference signal is required to compute the impulse response or complex frequency response by means of deconvolution. For reasons of computational efficiency, this happens mostly in the frequency domain. This second type of measurement is often understood as a measurement based on Fourier analysis in the narrow sense.

Typical representatives of simple measurement systems that only display and analyze the frequency spectrum at the input are hand-held sound level meters (e.g., B&K 2240, Norsonic Nor140) and mobile analyzers (Ivie IE-45, Terrasonde Audio Toolbox, etc.). They provide broadband figures such as the overall sound level as well as results based on a 1/3 octave or octave band resolution. This method also allows for the determination of the frequency response of the system under test, but only with respect to its modulus. For this purpose, a broadband noise signal is employed with a spectral shape that is pink or white. In a display of sum levels based on fractional octave frequency bands, the pink noise turns into a constant function over frequency. As a consequence, the frequency response of the system under test can be assessed immediately by the change of the curve when the signal is fed into the system under test.

Advanced measurement systems can measure the complex transfer function or impulse response of the system of interest. For this purpose the system is excited with a known test signal and its response is recorded:

$$e(t) \longrightarrow \boxed{\text{SUT}} \longrightarrow a(t)$$

Assuming that the system is a linear, time-invariant (LTI) system, the transfer behavior can be obtained from the deconvolution of the two data sets. That is because the response function $a(t)$ is the convolution product of the excitation signal $e(t)$ and the transfer function $h(t)$:

$$a(t) = h(t) * e(t), \tag{5.4}$$

where * stands for linear convolution.

This becomes a simple product in the frequency domain so that one can solve for the transfer function $H(\omega)$:

$$H(\omega) = \frac{A(\omega)}{E(\omega)}, \quad E(\omega) \neq 0 \text{ for all } \omega \text{ of interest.} \tag{5.5}$$

In practice, this technique, also known as inverse filtering, is sensitive to a low signal-to-noise ratio and excitation signals $e(t)$ of insufficient density or limited spectral coverage. Improvements can be implemented, for example, by applying the so-called Wiener method (Wiener, American mathematician, 1894–1964) and similar procedures.[2] In the simplest form, a threshold for the minimum amplitude of the signal is required:

$$H(\omega) = \begin{cases} \dfrac{A(\omega)}{E(\omega)}, & \text{for } \dfrac{1}{|E(\omega)|} < \upsilon \\ A(\omega) \dfrac{\upsilon |E(\omega)|}{E(\omega)}, & \text{otherwise.} \end{cases} \tag{5.6}$$

Better approximations can be accomplished by assuming an additive noise term such as:

$$A(\omega) = H(\omega) E(\omega) + N(\omega), \tag{5.7}$$

and by means of an estimate or a separate measurement of the power spectrum of the noise.

For measurements utilizing the deconvolution method, pseudo-random noise, swept-sine, and other well-defined excitation signals are commonly used. Also, an impulse-like test signal is principally possible. However, this does not find much use in practice since the short duration of the signal requires a high amplitude in order to excite the system sufficiently. This is particularly difficult in the case of room acoustic measurements such as of a concert hall. In addition to the noise floor, limiting factors are also the load capacity of the individual elements of the measurement chain, like the microphone and the loudspeaker.

Real-world systems are only approximately linear. Therefore, several aspects have to be considered. On the one hand, the background noise which always exists and reappears in the impulse response has to be accounted for. On the other hand, higher nonlinearities in the system's response contaminate the measurement results; they may be caused by the loudspeaker system, other parts of the measurement chain, or by inhomogeneities of air as the transfer medium (such as air movements and temperature gradients). Often, time variance of the system under test is a cause for measurement errors, as well.

Therefore the measurement signal should be chosen depending on the nature of the perturbations (following sections). Most swept-sine signals are advantageous because the deconvolution places nonlinearities (harmonics of higher order) at the end of the impulse response, which allows for simple removal by windowing. The duration of the measurement period and the number of time averages determines the reduction of the random part of the background noise. If only short measurements are possible, excitation signals with a high power density should be used in order to maximize the signal-to-noise ratio. Under certain circumstances, it may also be possible to use only excitation signals which are insensitive to small time-variances in the system under test.

In principle, signals from external sources which are unknown *a-priori* can also be used for measurements as long as the signal is known at the point of time when the deconvolution takes place. This method is primarily used for concerts and rehearsals when measurements with dedicated but disturbing test signals must be shortened or even completely avoided. Naturally, data obtained in this way depends significantly on the frequency range of the external source's signal as well as on its level relative to the noise floor. In most cases, longer measurement times and filtering are necessary to achieve results qualitatively comparable to the measurement procedures described before.

The measurement process is often complemented by using multiple evaluation locations. Being not only a necessity for the tuning of modern sound systems, this also allows for spatial averaging and thus are another means to suppress the noise floor. For this purpose, measurements are made at several points of the room and the gathered frequency responses are averaged. From the results, one can derive which correction or equalization must be applied, for instance, to the sound reinforcement system.

5.2.2.2 Conventional Excitation Signals

One of the most common excitation methods is feeding the system under test with pink noise. Similar to white noise, it can be easily generated and has several advantages. At first, it sounds subjectively more pleasant, especially over a longer period of time. Secondly, loudspeaker systems usually have a higher sensitivity and thus lower load capacity in the high frequency range, so that pink noise, which provides a reduced signal level at high frequencies, has a lower probability of damaging the sound reproduction system (e.g., the high-frequency transducer) or of causing nonlinear distortions in the measurement chain.

Historically, pink noise was mainly used for the simple frequency analysis based on a single input signal. Accordingly, this required an average spectral distribution of the exciting random noise over time. However, to determine the complex transfer function of the system under test by means of deconvolution, the exact knowledge about the time function of the excitation signal is needed. Therefore, many computer-based measurement systems actually use pseudo-random noise, which is exactly determined in advance with regard to its amplitude function over time but enjoys the properties of true random noise.

The energy content of the pink noise signal decreases by 3 dB per octave, the power density spectrum can be defined as:

$$S(\omega) \propto \frac{1}{\omega}. \tag{5.8}$$

This relationship is also shown in Figure 5.1 in comparison with white noise.

But pink noise is not completely determined. Both the exact sequence of time samples as well as the crest factor can vary from measurement to measurement or from application to application. When comparing or reproducing measurement results, these limitations have to be taken into account.

Another classical excitation technique is the impulse test. For this purpose, pistols or balloons are typically used. The pistol shot or the sound of the exploding balloon is recorded and analyzed afterwards in a post-processing step. If the signal spectrum is sufficiently known or reproducible (usually approximately white), one can also draw conclusions about the frequency response of the system under test. Similarly, signals with the behavior of the

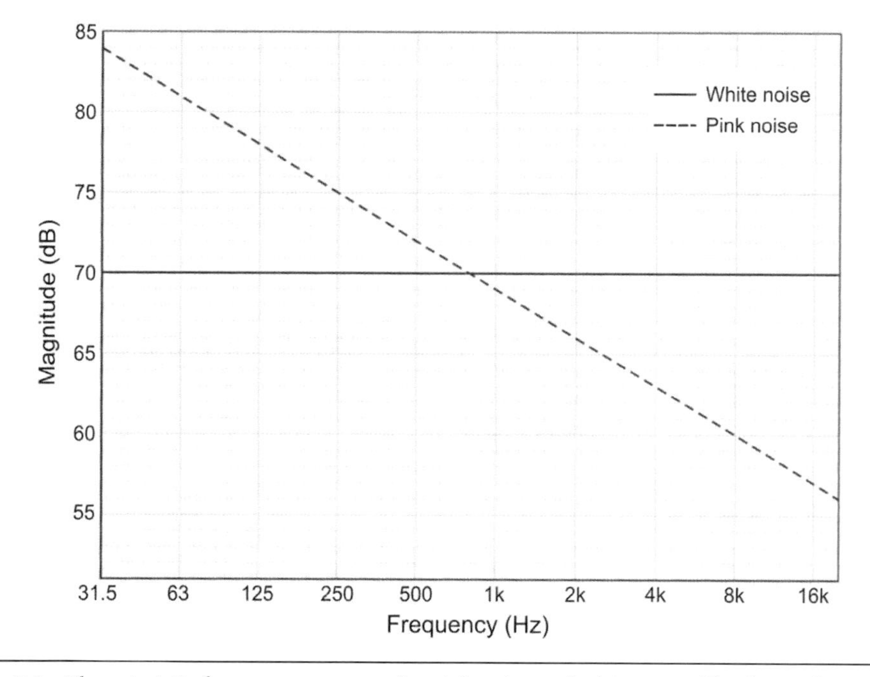

Figure 5.1 Characteristic frequency spectrum for pink noise and white noise. The figure illustrates the power-density spectrum level in dB using an arbitrary normalization.

step function (Heaviside function) can be utilized, such as the fast switch-off of a piece of music. In this way it is possible to use, for example, the recordings of organ concerts for the determination of the reverberation time. However, the applicability of these methods is naturally limited to certain investigations in room acoustics, but not for the tuning and alignment of sound reinforcement systems.

5.2.2.3 Sweep-Based Measurements

Over recent years, the swept-sine signal (also known as sweep or chirp), combined with the deconvolution in the frequency domain has become the most important measurement method. This is primarily due to the fact that this technique benefits from a set of measurement advantages and that computer-based measurement systems are broadly available nowadays along with sufficient processor performance and memory. Only 20 years ago, limited system resources were required to employ specific technologies, such as MLSSA[3] or TEF,[4] in order to comply with the computer hardware available. The downside of this compromise was the limited flexibility regarding the choice of the excitation signals.

Expressed simply, the sweep is a continuous sinusoidal signal $s(t)$ whose frequency is changing over time:

$$s(t) \propto \sin[\varphi(t)], \tag{5.9}$$

where the phase $\varphi(t)$ can depend on time t in different ways. Usually this relationship is given by the instantaneous frequency:

$$\Omega(t) = \frac{d}{dt}\varphi(t) \,. \tag{5.10}$$

If the instantaneous frequency is increasing linearly with time, the sweep is a linear sweep or so-called white sweep:

$$\Omega(t) = \alpha \times t + \omega_0 \,. \tag{5.11}$$

In this case, the sweep rate α in Hz/s is constant since the signal covers the same frequency range in the same period of time. If the dependency is exponential, it is then termed pink sweep or log sweep which has a pink frequency response:

$$\varphi(t) = \exp(\beta_1 \times t + \beta_0) + \varphi_0 \,. \tag{5.12}$$

The pink sweep has a constant sweep rate of $\beta = \beta_1 / \ln(2)$ in octaves/s, because the same number of fractional octave bands is covered in the same period of time. In addition to the sweep rate, the start and stop frequency are also important parameters. These should be defined so that they include the entire frequency range of interest.[5]

Comparing the swept sine with other signal shapes, such as pink noise or maximum length sequences, it represents a continuous function over frequency. This is advantageous, for example, when looking at digital-to-analog converters (DACs), where compared to step-like or discontinuous signals, the probability is lower that the anti-aliasing filters of the DAC overshoot. Depending on the exact type and length of the sweep, nonlinearities caused by distortions in the measurement chain can also be removed fairly easily from the impulse response. In particular, the log sweep allows the precise identification, isolation, and analysis of all higher harmonics. Furthermore, the sweep is also less vulnerable to small time variances of the system under test.[6] Finally, another significant advantage is that sweep measurements allow the engineer to perceive distortions and perturbations subjectively during the measurement, which is usually more difficult with noise signals.

Figure 5.2 shows three fundamental sweep signals—a white sweep, a log sweep, and a weighted sweep. The latter provides an adapted spectral shape which is especially useful for loudspeaker measurements. In contrast to the log sweep, its level reduction in the high frequency range is smaller and therefore the signal-to-noise ratio is usually higher.

5.2.2.4 Noise Applications

Besides the pink noise signal that we already discussed, in practice several other noise signals are employed for acoustic measurements. Worth mentioning are, for example, white noise which has a constant frequency response and red noise which decays more steeply than pink noise with 6 dB/octave. Like the sweep, noise signals can be colored, that is, frequency bands can be weighted relative to others in order to achieve an optimal adaption to the measurement conditions and thus to maximize the signal-to-noise (S/N) ratio and the level difference between the used noise excitation signal and the undesired background noise. For example, one can create a noise signal with a frequency response equivalent to the weighted sweep, so that loudspeaker measurements can be made with noise excitation.

The noise signal is not uniquely defined by its frequency response alone. Therefore, it is common to utilize statistical quantities to characterize the distribution of amplitude values over time. Examples include Gaussian white noise or Poisson-distributed white noise. Pseudo-random noise—for example, the maximum length sequences explained in the next section—has characteristics similar to random noise but it is deterministic.

Recently, other types of dedicated noise signals have found common use. These are related to the assessment of the quality of speech transmission through public address (PA) systems.

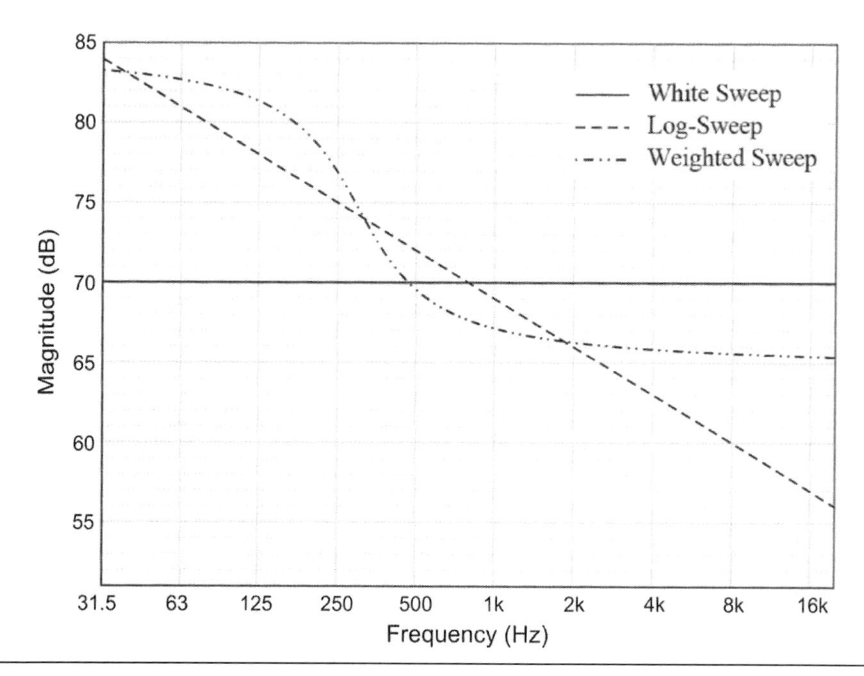

Figure 5.2 Characteristic frequency spectrum for white sweep, log sweep, and weighted sweep. The figure illustrates the power density spectrum with an arbitrary normalization.

Primarily the STIPa measurement method (see Section 5.4.2.5) utilizes a noise-like excitation signal which is subjected to amplitude modulation and a specific octave band filter.

5.2.2.5 Technique Using Maximum-Length Sequences

Acoustic measurements using MLSs are based on the creation of deterministic pseudo-random number sequences and their processing by means of correlation analysis between the sequence itself and the response of the system under test to that excitation. The MLS measurement method is considered as the prototype of data evaluation in the time domain because the computation of the impulse response happens directly in the time domain due to the specific mathematical properties of the MLS.

MLSs are characterized by their order N, a positive integer number, and they have 2^N − 1 samples by definition. The individual samples represent a binary sequence of values which assume either 1 or 0 (measurement systems usually rescale this to the symmetrical range of +1 and −1). The construction of the sequence can be clarified by means of a shift register algorithm (see Figure 5.3 for N = 3). This algorithm combines selected bits (taps) recursively

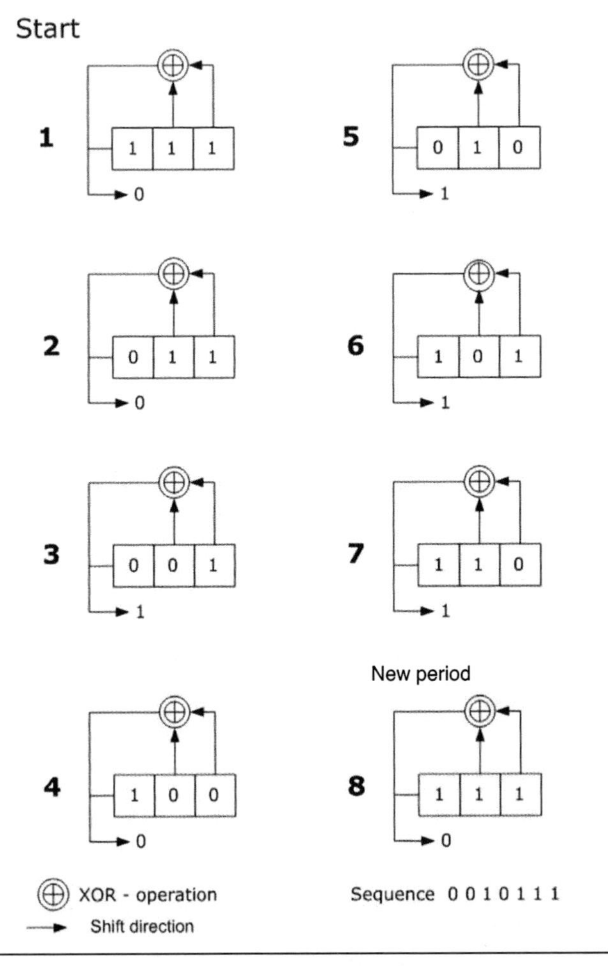

Figure 5.3 Shift register for the construction of the maximal length sequence of order N = 3.

with the register state so that all possible combinations of N bits are assumed except for the zero vector. Creating a sequence of the length of 2^N – 1 samples requires only a minimal amount of memory, namely the number of N register bits.

Depending on the algorithm, there may be several different MLSs for the same order N, see Figure 5.4. In practice, this is advantageous, since one can choose among these different versions, for example, in order to reduce the effect of small nonlinearities.[7]

The MLS measurement itself is a two-step process. First, the MLS is fed into the system under test and the response is recorded. Then, the cross-correlation function between the excitation and the response sequences is computed and hence, the impulse response obtained. Due to the nature of the MLS, the numerically expensive computation of the correlation function can be simplified dramatically. This specific transformation, which exploits the particular properties of the MLS, is also called Hadamard transform. It is most advantageous for impulse response measurements on systems with limited memory and processing capacity.

The spectrum of the MLS signal is constant over frequency (white) except for a DC-component. The crest factor of the MLS is small and allows for a high signal-to-noise ratio. However, the strongly discontinuous course of the time function may cause distortion effects and thus, require applying the MLS test signal at a lower level than, e.g., a sweep signal. For the same reason, the MLS signals are quite sensitive to time variances in the system under test.[7, 8]

The MLS can be understood as a special case of the general noise signal.[3] As a type of pseudo-random noise, its time function is subject to well-defined conditions. Of course, an MLS signal can also be weighted,[9] for example, to achieve a level reduction of the high-frequency component. However, naturally the time function of the MLS adapted in such a way does not comply anymore with the requirements of an MLS, so that the Hadamard transform

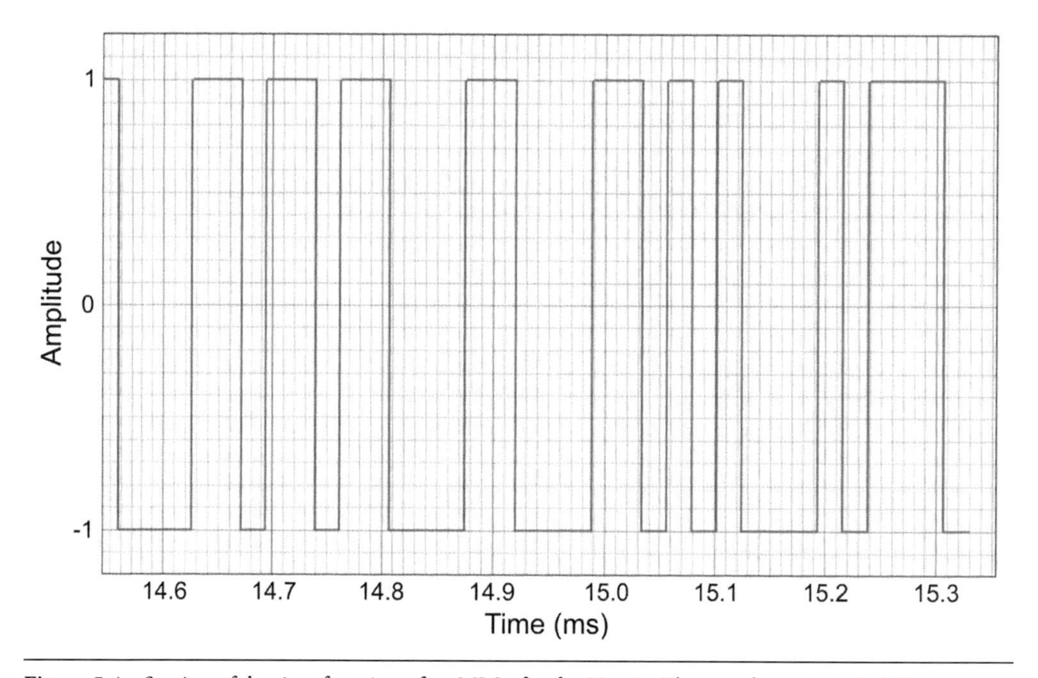

Figure 5.4 Section of the time function of an MLS of order N = 16. The sampling rate is 24 kHz.

cannot be applied. For such a case, a dedicated algorithm has been developed based on the fast Fourier transform.[10]

An overview of MLS fundamentals and diverse applications in system identification (measurement) was recently given by Xiang.[11]

5.2.2.6 Time-Delay Spectrometry Method

At the end of the 1960s, Richard Heyser developed the TDS measurement method for applications in the field of audio engineering.[4] The original hardware implementation consisted of a significant number of standard test devices, such as oscilloscopes and signal generators. This development eventually led to the TEF analyzer (TEF-10, TEF-12), which was based on Pulse-code modulation (PCM) technology and utilized a tracking filter.

The final TEF-20 analyzer consists of a generator that simultaneously creates a swept sine and swept cosine signal with a fixed-phase relationship. The sinusoidal signal is fed into the system under test. The recorded signal is then divided by the original sinusoidal signal to obtain the real part of the transfer function and by the cosine signal to obtain its imaginary part. After that a low-pass filter with a fixed cut-off frequency is applied to the resulting data sets.

TDS can thus be understood as a two-port measurement where the swept sine signal must be created by the measurement system. After feeding it into the system under test, the response is passed through a band-pass filter in which the center frequency increases simultaneously with the frequency of the excitation signal. By changing the delay offset between the excitation signal and tracking filter as well as the bandwidth of the band-pass, selected sections of time response can be evaluated. This can also be understood as a frequency-dependent windowing. Therefore, the signal-to-noise ratio of such measurements can be high. The tracking filter also facilitates the exclusion of reflections, which allows for quasi-anechoic measurements in environments that are not reflection-free. Figure 5.5 demonstrates the measurement principle of this method. The software package EASERA implements the TDS measurement as well, as shown in Figure 5.6.

5.2.2.7 Measurements Using Arbitrary Excitation Signals

The use of excitation signals like sweeps or noise is usually not disturbing when performing laboratory or electronic measurements. However, acoustic measurements in rooms or outdoors often have to be scheduled because they hinder other activities, such as rehearsals. In many cases, the tuning of the sound reinforcement system requires that the rehearsal is paused for some time or shortened. The use of acoustic measurement signals is even more critical when measurements have to be performed in the presence of an audience. Naturally, feeding loud noise or sweep signals into the sound system is a disturbing process and therefore, usually only possible under special conditions—such as at night or in an empty venue. Resulting figures, such as the speech intelligibility, then become less reliable since, for example, the presence of the audience changes the acoustic properties of a room significantly. Such effects have to be accounted for by correcting the original measurements using predictions of the influence of the additional sound-absorbing surface. However, these methods are still very error prone today.

Therefore it would be better in any case if the spectators would not even recognize that acoustic measurements are being made. For more than 20 years, Meyer Sound Lab., Inc., has worked on developing such a system—called *source-independent measurements* (SIM).[12] It

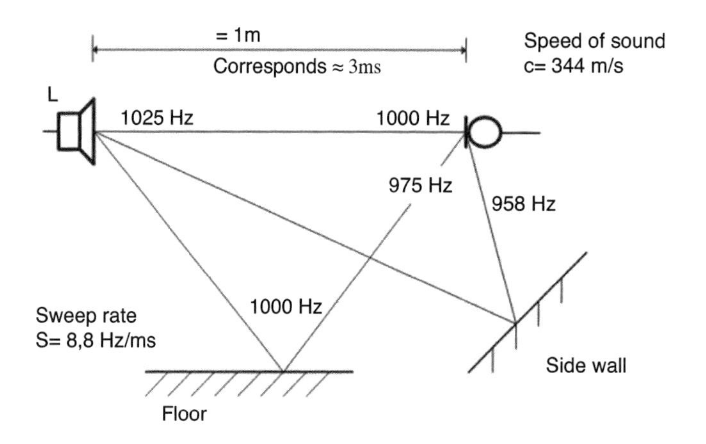

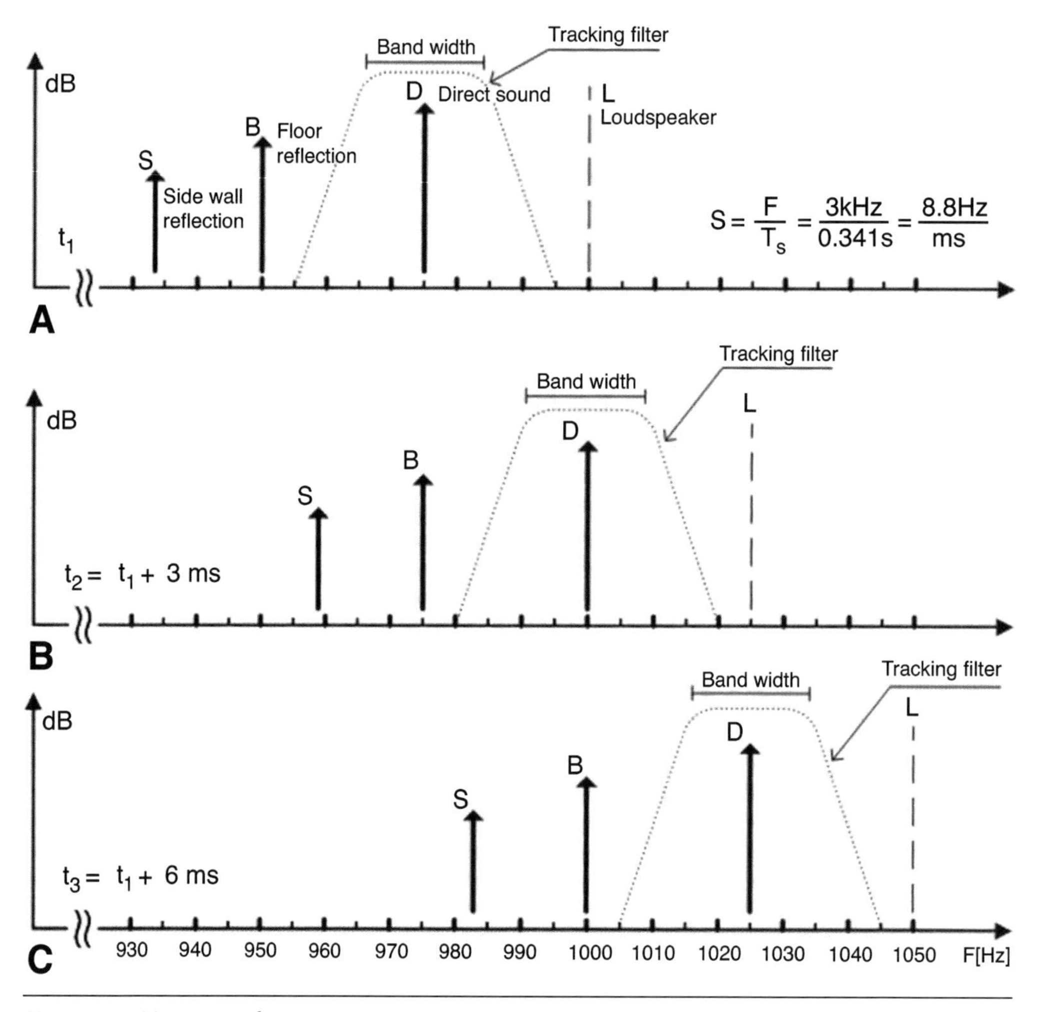

Figure 5.5 TDS principle

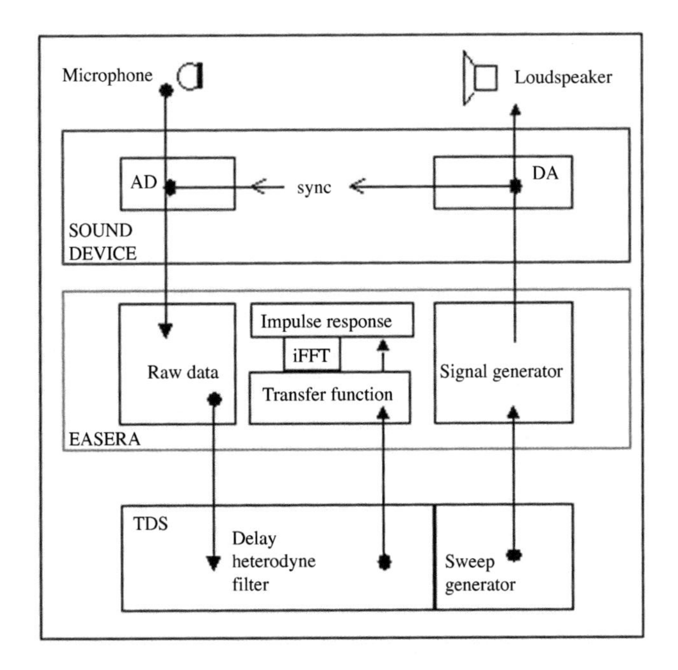

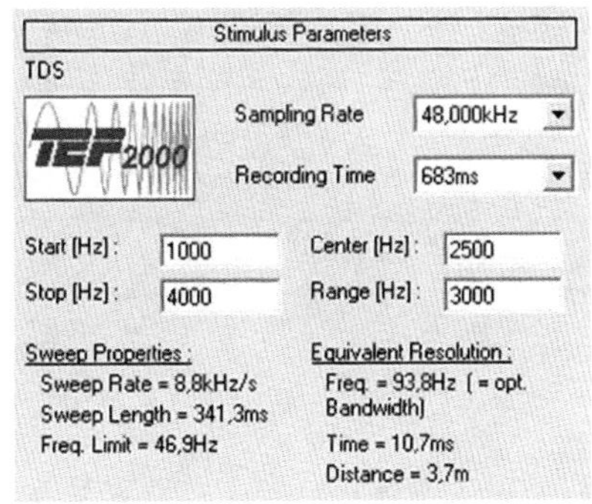

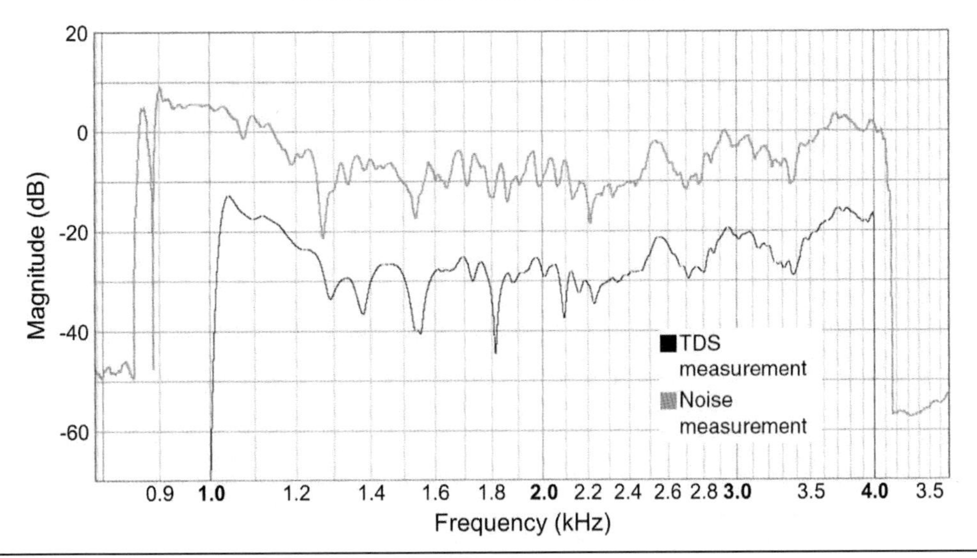

Figure 5.6 Block diagram of TDS measurements with EASERA

utilizes fairly complex measurement hardware and an algorithm based on Fourier analysis. Its main goal is the acquisition of the complex transfer function of loudspeaker systems in real time, but it is limited to the direct field and early reflections. It does not derive the impulse response of the room, in room acoustic length that corresponds approximately to the reverberation time of the venue. In a similar manner, but on a standard computer platform, the software SMAART by Rational Acoustics[13] facilitates a two-channel measurement providing an impulse response by means of calculating the complex transfer function of the two channels and applying the inverse Fourier transform.

The method commonly used until today to acquire room acoustic impulse responses (RIRs) is based on two distinct steps, namely the recording of the raw response and its later post-processing and evaluation. In 2008, the authors introduced a new measurement method which combines these two steps and performs them simultaneously and continuously. In this manner, the full length RIR can be determined in real time by means of so-called real-time deconvolution (see Figure 5.7).

The RIR derived in this dynamic way is equivalent to the static computation in a post-processing step discussed previously, but requires a number of optimization steps. It can assume typical lengths of the order of, e.g., four to eight seconds. The transform between time and frequency domain is performed linearly and at full length—that is, without windowing or any loss of data—and thus provides results identical to the static setup. Electro-acoustic and room acoustic quantities can be derived from the RIR in the same way. Furthermore, averaging and advanced filter techniques can be applied to suppress the background noise as well.

The real-time capability of the measurement system SysTune by AFMG[14] is accomplished by providing high refresh rates for the computation of results as well as for the display and

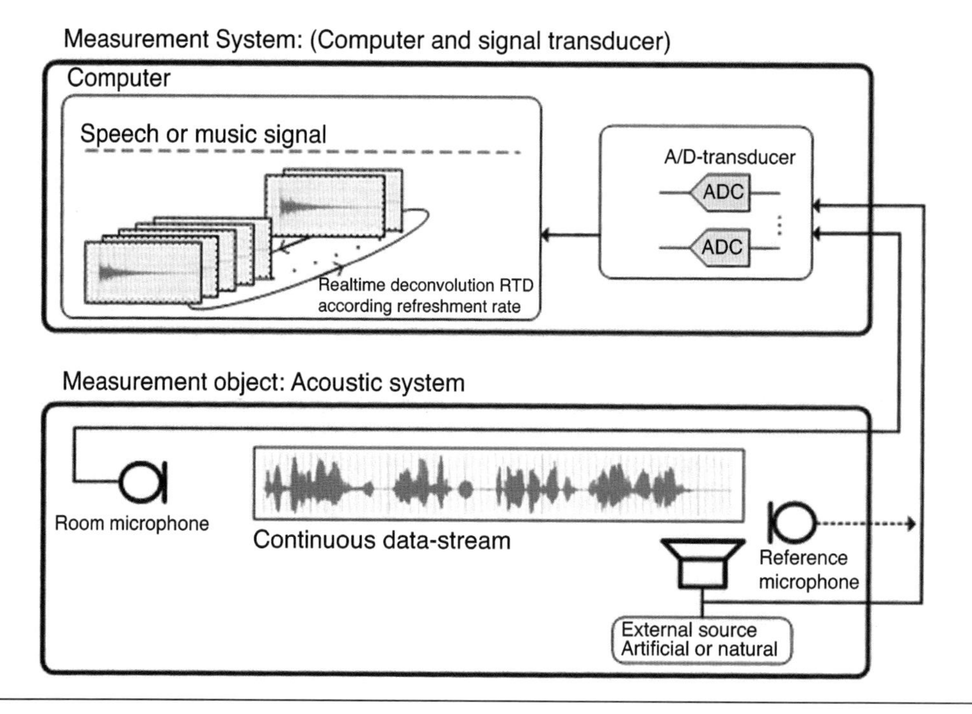

Figure 5.7 Measurement system SysTune

analysis of these results. Simply put, this measurement system can be understood as *oscilloscope for room impulse responses*, allowing for a completely new way of room acoustic evaluation.

5.2.3 Absolute and Relative Measurements, Calibration

5.2.3.1 Measurement Parameters

When performing measurements using computer-based or other digital measurement systems, the analysis and post-processing of the data is performed in the software domain. The analog amplitude of a signal at the input or at the output of the measurement system is related to its digital counterparts by a sensitivity or conversion factor. In practice, it is often the case that these sensitivities consist of a larger number of individual factors contributed by the individual parts of the measurement chain. This includes, but is not limited to, amplification and attenuation units, delay elements, A/D and D/A converters, protection filters, and limiters. As a result, these sensitivities may be known to the software only roughly or not at all. In such cases the calibration of the measurement system is conducted whereby a well-defined calibration signal is passed into the input or read from the output. By relating a value for the digital amplitude (typically given in FS = Full Scale) on the software side, to the physical signal on the analog side, the overall sensitivity is determined accurately and can be employed by the software (see Figure 5.8).

The calibration procedure itself is mostly performed before any actual measurement takes place, and thus, serves not only for the calibration itself but also as a simple test of the proper functioning of the system. In practice, the need for absolute measurements, that

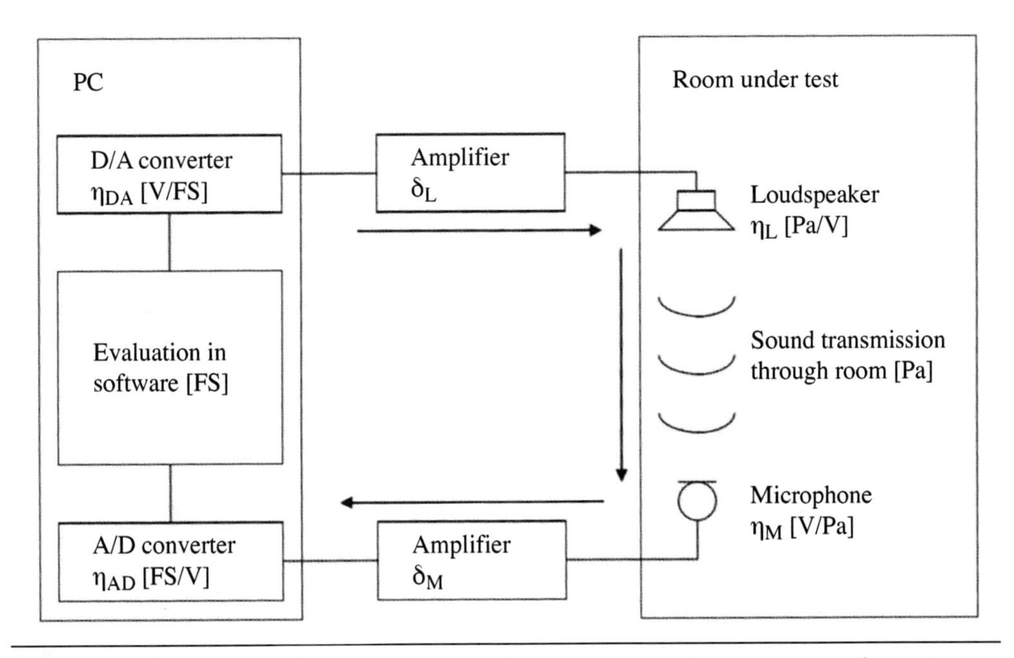

Figure 5.8 Scheme of a typical measurement chain for a computer-based measurement with sensitivities η and amplification/attenuation elements δ.

is, measurements with an exact calibration, depends on the actual goal. For example, most room acoustic measurement actually does not require any knowledge about the signal levels involved, since all room acoustic parameters are defined in a relative manner and therefore do not depend on the absolute levels (ISO 3382).[15] Also, the alignment of delay times of sound reinforcement systems or the relative adjustment of gain structures do not require quantifying the actual sound pressure received.

But calibrating the measurement system is crucial when absolute levels have to be measured; whether at conventions for the purpose of addressing the audience in case of fire alarms, at open-air concerts for the purpose of determining if output sound levels are compliant with sound emission regulations regarding the neighborhood, or for the purpose of measuring the noise impact of airplane traffic in the proximity of an airport. Also, electronic and electro-acoustic measurements often necessitate absolute figures, such as the maximum load capacity of a loudspeaker, its sensitivity or its maximum sound pressure level.

Calibrating measurement devices is also the simplest way when relative quantities are to be compared that are acquired by multiple measurement systems of the same or of different type. Only the process of matching measurement setups allows for certainty since even physically identical units may differ from each other and the whole measurement chain may be complicated as well.

Principally, a digital measurement system is calibrated by defining a sensitivity or conversion factor. This entity relates a digital unit known in the software domain, such as FS, with a physical unit, such as Pa or V. The sensitivity defines how, for example, the signal at the microphone input given in Pa is converted into the digital signal given in FS.

Typical calibration methods of the input of a measurement system utilize a well-defined calibration signal. Often, acoustic measurement devices are calibrated by applying a 1 kHz sinusoidal tone at 94 or 104 dBSPL generated by a pistonphone. Electronic measurement devices are usually calibrated by supplying a pink noise with a given root-mean-square value, like 1 V or 2.83 V.

Calibrating the output of a measurement system is needed less often. In this case, normally the software provides an output signal of known digital amplitude and its physical equivalent is measured using a calibrated device. The resulting absolute value is then entered into the software.

5.2.4 Measurement Errors, Optimization, and Limits of Application

When performing acoustic or electronic measurements, a variety of measurement errors can occur. In the following section, typical sources of error are explained, along with their symptoms and potential solutions. At this point we understand the measurement error very generally as: all undesired components in the measurement result that interfere with the properties of interest of the system under test in a qualitative or quantitative way.

Before discussing the details, let us classify such errors in three categories:

- Errors in the measurement system or measurement chain
- Disturbances during the measurement and undesired components in the response of the SUT
- Errors caused by the post-processing and analysis of the acquired raw data

5.2.4.1 Measurement System and Measurement Chain

With respect to the measurement chain, it has to be ensured first that all parts are driven in the optimal range of function. Most important, the gain or relative magnitude should be checked since the clipping or overshooting of any element will cause distortions and non-linear artifacts, whereas driving components below their optimal signal level will reduce the signal-to-noise ratio. Most often it is just one specific piece of the measurement chain that is adjusted badly and thus dominates the quality of the measurement overall. Identifying and correcting this element accordingly will therefore reduce the measurement error significantly.

Especially, computer-based measurement systems with many components show frequency and latency behavior that is not optimal. As a consequence, most software packages allow compensating for such influences directly—for example, by making a reference measurement using a short-cut input and output of the measurement system. Once such a reference is established, it can be subtracted from all subsequent measurements, therefore removing errors caused by the measurement setup—such as the characteristics of A/D and D/A converters, loudspeaker, microphone, and even the amplifier. This is particularly important for the exact measurement of frequency response and delay times.

Other typical sources of error include asynchronous sampling clocks at the input and output of the digital measurement system or insufficient shielding of measurement channels against each other and against external interference, such as the frequency of the power network. Many measurement errors originate in malfunctioning or broken elements of the measurement chain as well as in defective or lose connectors. It is therefore most advisable that a well-defined test measurement is made before actually investigating the SUT.

We conclude with the remark that a measurement system is always a physical system as well. It undergoes its own transient processes during the measurement. Consequentially, it is necessary to ensure that effects, such as initial excitation, loading, or the decay process are either negligible or can be compensated for.

5.2.4.2 External Influences

In practice, any measurement of any SUT is subject to noise. If the noise is of a random nature, such as most types of background noise, temporal averaging will reduce the noise floor. For each doubling of the measurement time, the signal-to-noise ratio is increased by 3 dB if the noise is not correlated. Also, choosing an appropriate measurement signal (see previous section) can improve the signal-to-noise ratio significantly.

Other common errors are caused by the fact that although a LTI system under test is assumed, the real system actually shows time variances or nonlinearities that are not negligible. For example, acoustic measurements may suffer under strong winds or electronic measurements may be affected by a dynamic equalizer. In such cases, one should first make sure that the SUT can actually be approximated well enough by an LTI system. If this is the case, measurement errors can be removed to some extent by an appropriate choice of the measurement signal and by applying post-processing.

Particularly for room acoustic measurements, it is essential that the transfer behavior of the SUT is measured in steady state, that is, while the room is fully excited by the test signal. Therefore the choice of the duration of excitation and of measurement is critical. To account for this, primarily, measurements using noise excitation feed the system with a signal that includes an initial period (*presend*) whose sole purpose is to initiate the transient shift of the SUT to the steady state. Electrical measurements must often take into account that the measurement system may have an inertia or response time equivalent or slower than the SUT.

Last, it is necessary to ensure that the excitation signal covers the entire frequency range of interest, which may be particularly difficult when using an external signal supply. If it is not possible to excite the full bandwidth of the SUT, appropriate filtering or other post-processing methods must be applied so that frequency bands with insufficient signal or low signal-to-noise ratio are removed and do not affect the rest of the measurement. Also, the sampling rate has to be chosen so that it can cover the frequency range of interest according to the Nyquist frequency (see Section 5.2.2.1).

5.2.4.3 Post-Processing

In most cases, the post-processing of acquired data combines two separate functions in one step. On the one hand, results should be derived from the raw data, and on the other hand, the noise floor as well as systematic errors should be removed from the measurement. Often, artifacts caused by the nature of the measurement process itself have to be reduced or eliminated.

The most elementary form of suppressing random noise is averaging. Averaging in the frequency domain corresponds to smoothing the spectrum over a certain bandwidth. Averaging in the time domain means accumulation and averaging of several measurement periods. But also, spatial averaging is possible by deriving the average result of multiple measurement locations. When averaging data, the same quantities should not be averaged multiple times, otherwise information could be removed that does not originate from disturbances.

Recently the authors also introduced frequency-selective time averaging. This filter called spectrally selective accumulation (SSA) was implemented—for example, in the 1.1 update version of SysTune. It is particularly effective when using excitation signals with discontinuities in time or frequency, such as music and speech.[16]

Windowing measured impulse responses in the time domain is also very common. Noise in time sections before and after the actual response can be removed in this manner. For some measurement techniques (such as using sweeps and the deconvolution method), windowing also allows eliminating nonlinear portions of the impulse response. If one is only interested in a part of the time response, such as the direct sound in acoustics, windowing can be used to isolate this section from others, such as the reflections. However, this procedure should be used with care, because windowing in the time domain always affects the frequency domain as well. At first, the length of the region enclosed by the window defines a lower frequency limit. Secondly, windows that are too steep or discontinuous, e.g., a rectangular window, may cause ripples and artifacts in the frequency domain.

The same is true for the application of filters in the frequency domain. Filter slopes that are too steep should be avoided if a later evaluation will also take place in the time domain. On the other hand, any filters applied have to be steep enough in order to accomplish the isolation of the frequency bandwidth of interest. In the end, it is a matter of fact for both filters as well as windows that depending on the nature of the measurement and of the analysis, there is always a compromise to be found between the effects removed as desired and the undesired effects caused by filtering or windowing.

To overcome some of the effects caused by using a window of fixed length, namely either including too many reflections at the high frequencies or excluding too much information at the low frequencies, the authors[17] introduced a frequency-dependent windowing function, called a time-frequency-constant (TFC) window. It was implemented, for example, in SysTune and its shape is shown in Figure 5.9. It is related to earlier approaches, like the so-called *adaptive window* in MLSSA, but it represents a continuous function over frequency and is not based on a set of fairly broad band-pass filters which can cause discontinuities in the frequency response obtained.

The previous sections already discussed the importance of the transient behavior of the SUT compared to that of the measurement system. This fact must also not be neglected when post-processing and assessing acquired data. For example, a filter has to have a time response that is appropriate for the type of measurement. The impulse response of an electronic system can show a faster decay than the band-pass filter applied to it. In such a case, the filter behavior will mask the properties of the SUT.

In addition, many computer-based software tools allow for compensating random noise.[18] For this purpose, the raw data is analyzed and the background noise is estimated based on statistical assumptions (see also Chapter 6). During the subsequent calculation of results, the software can then subtract this estimated background noise—for example, from the energy sum when calculating figures based on energy ratios.

When comparing measurement data sets and measurement locations (see also next section), it is important to take into account the environmental conditions. Especially acoustic

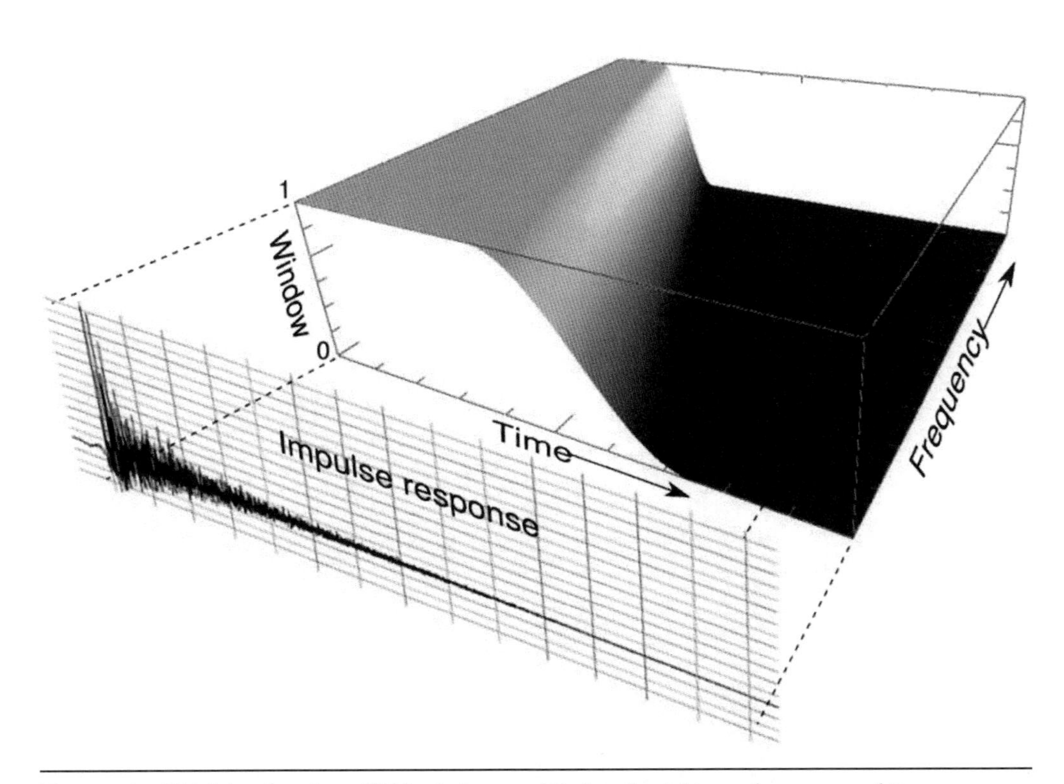

Figure 5.9 Weighting function of TFC window as a 3D plot of weight over frequency and time.

measurements may suffer under significant variations of air, such as air movements, temperature changes, or locally varying temperature gradients. This applies in particular to long measurement series outside and to measurement locations distant from each other. Also electronic measurements or loudspeaker measurements may exhibit temporally slow changes of the object under test. These are mostly depending on the temperature of the object. If these effects cannot be compensated, they should at least be recorded quantitatively to put any comparison or evaluation of the result data in the right context.

5.3 ROOM ACOUSTIC MEASUREMENTS

5.3.1 Introductory Comments

In the past, room acoustic measurements have been performed primarily using stationary, random noise signals. In most cases, pink noise was used, which contains the same energy content in each fractional octave band (see Figure 5.10). The disadvantage of this method is that with a non-deterministic excitation signal, one can only obtain information about average amplitudes while phase relationships between individual sources as well as reflections from the ceiling, floor, or side walls of the room are neglected.

For this reason, present computer-based measurement systems determine full impulse responses, from which one can obtain all relevant energy, time, and frequency information by means of post-processing. However, real-time analyzers may still play an important supportive role.

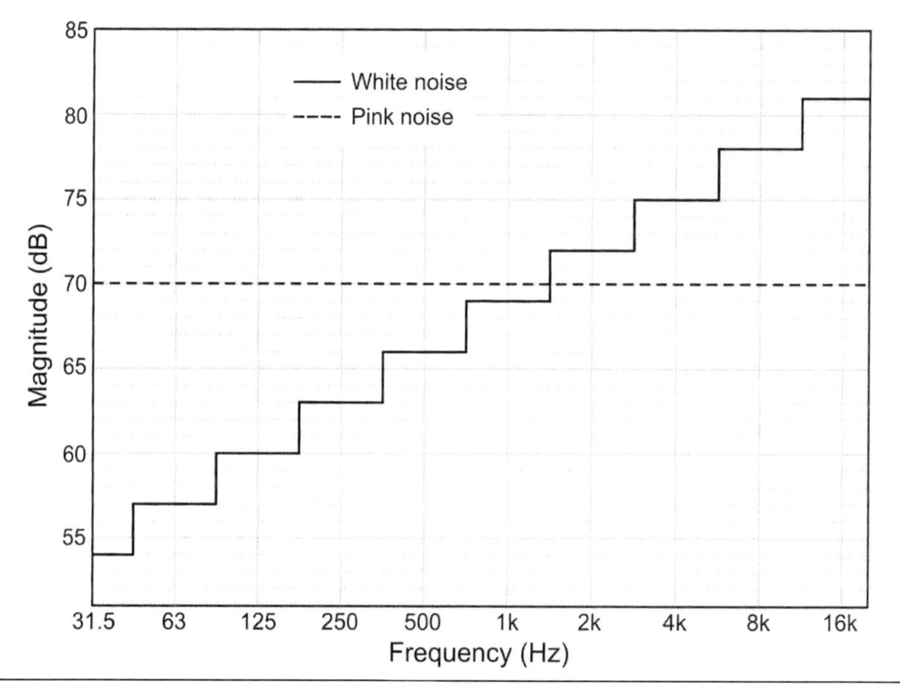

Figure 5.10 Octave-band display of the spectral shape of white noise and pink noise. Graph shows the band-related power sum spectrum with an arbitrary normalization.

5.3.2 Selection of Measurement Locations

To determine the measurement locations, usually a raster of points is defined in the areas of performance and of reception. The granularity and size of the grid depend on the degree of complexity with respect to achieving even signal coverage over the areas of interest. Important and critical locations, such as seats under the balcony or at the edge of a seating area, have to be accounted for (see Figure 5.11).

Our example, the city hall of Hannover, Germany, shows sending locations that are distributed symmetrically over the whole stage as well as receiving locations (R) which are limited to only one half of the venue. This choice is advantageous in symmetrical rooms, as it saves measurement time. However, for asymmetrical rooms, measurement locations should be distributed over the whole auditorium.

5.3.3 Measurement of Room Acoustic Properties

To determine the acoustic properties of a room, measurement routines are utilized that provide the impulse response of the SUT. These methods are usually the same as for the setup of the sound reinforcement system, which we will discuss later. In the past, impulse sound tests have been performed which relied on the excitation of the room by a short and loud

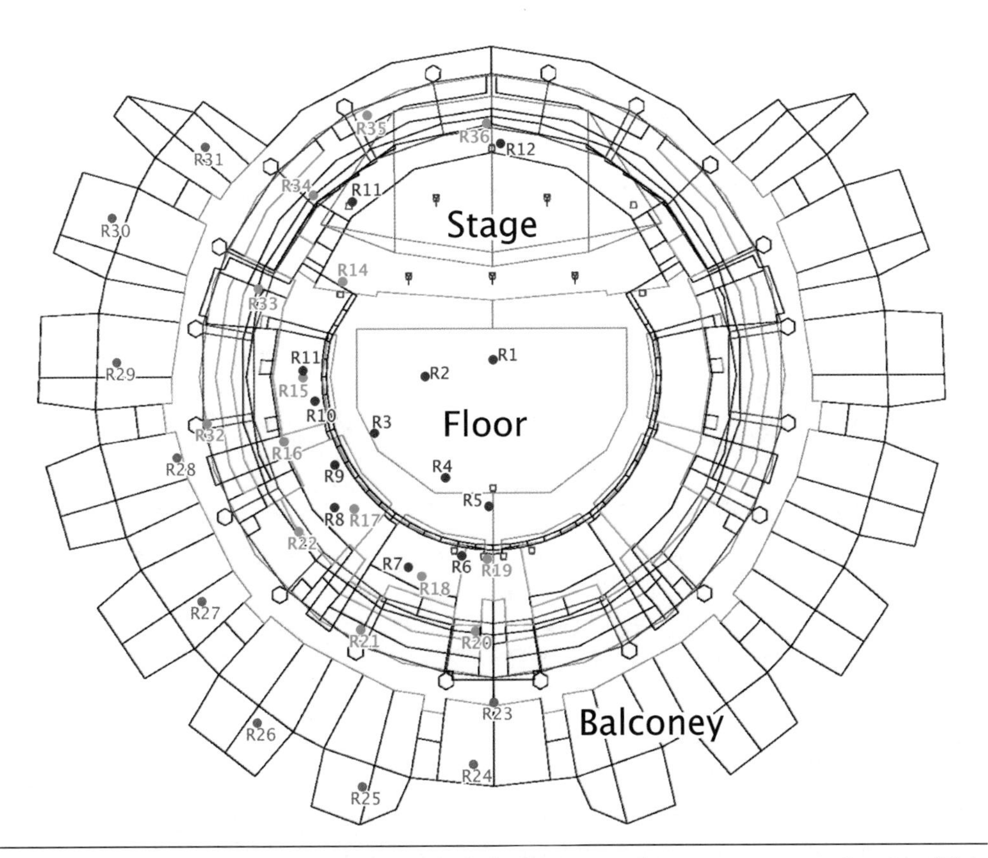

Figure 5.11 Top view drawing of the convention hall of Hannover, showing measurement locations (R) in the auditorium and sending locations on the stage.

burst or impulse, such as from a pistol or a spark generator. For measurements in physical scale models, these methods are still in use due to the limitations of miniaturization of sound transducers.

Room acoustic measurements typically require the use of omnidirectional measurement microphones as well as dummy heads. To determine all common room acoustic figures, such as those defined by ISO 3382, a 4-channel measurement setup is needed. Figure 5.12 shows a typical setup. It allows for obtaining standard measures with an omnidirectional microphone—lateral measures like LE, LF, or LFC using a figure-of-eight microphone, as well as binaural measures based on the left and right ear microphone of the dummy head. It has become quite common to mount all of these on a single stand. Sometimes this configuration is even complemented by some additional microphones or microphone arrays to allow resolving the sound field with respect to the angle of incidence.

To excite the room according to the standards of an omnidirectional source, a dodecahedron loudspeaker is mostly used.

Looking back over the past 50 years, it became possible to determine the acoustic properties of a room with constantly increasing accuracy and detail. Beginning in the 1970s, computer-based measurement systems became available at a price that was considered reasonable for many engineers and consultants. Various measurement platforms have been introduced:

- In 1978, the well-known TEF 10, 12, and 20 system by Crown/USA (later Gold Line) and 10 years later the MLSSA system by DRA Laboratories, as well as Monkey Forest by ITA Aachen/Germany.
- In the 1990s, several measurement systems based on MS Windows followed, such as Smaart by SIA Soft, WinMLS by Morset Sound Development, and Dirac by Brüel & Kjaer.
- Since 2005, the measurement system EASERA has been available, which was developed by the authors and is used for many of the examples.

Compared to the late 1980s, the processing power of modern computers is so great that all measurement steps—that is, processing, signal generation, and data sampling—can be executed in parallel and in real time.

Measurement platforms like EASERA facilitate the use of excitation signals such as sweeps, maximum-length sequences, and noise, but also specific signals such as speech or music samples. EASERA additionally features a TDS module which is based on a white sweep signal and allows performing noise-free measurements in noisy environments. External signals can be recorded too, such as impulse-like excitations generated with pistol shots or other impulse-based methods.

As a result, a single- or a multi-channel impulse response (see Figure 5.13) is acquired and provides the starting point for further post-processing which leads to the room acoustic quantities of interest eventually.

5.3.4 Time Domain Quantities

Each impulse response (monaural, binaural, or bidirectional) is a function of sound pressure over time and can be used to determine time domain quantities and energy ratios in a post-processing step.

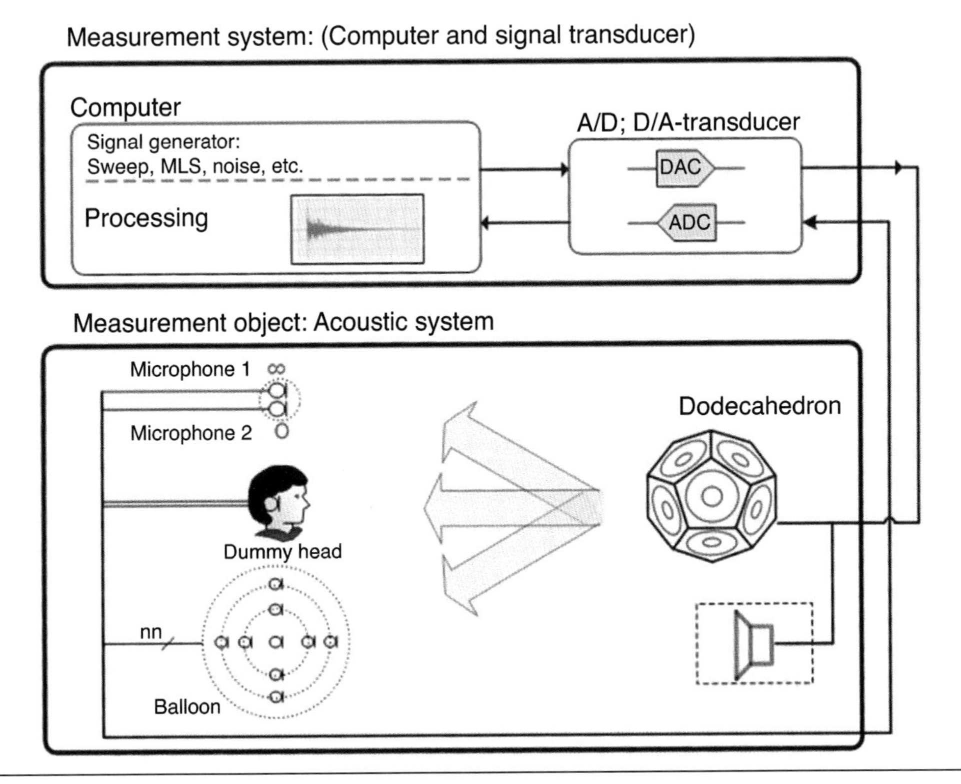

Figure 5.12 Room acoustic measurement setup

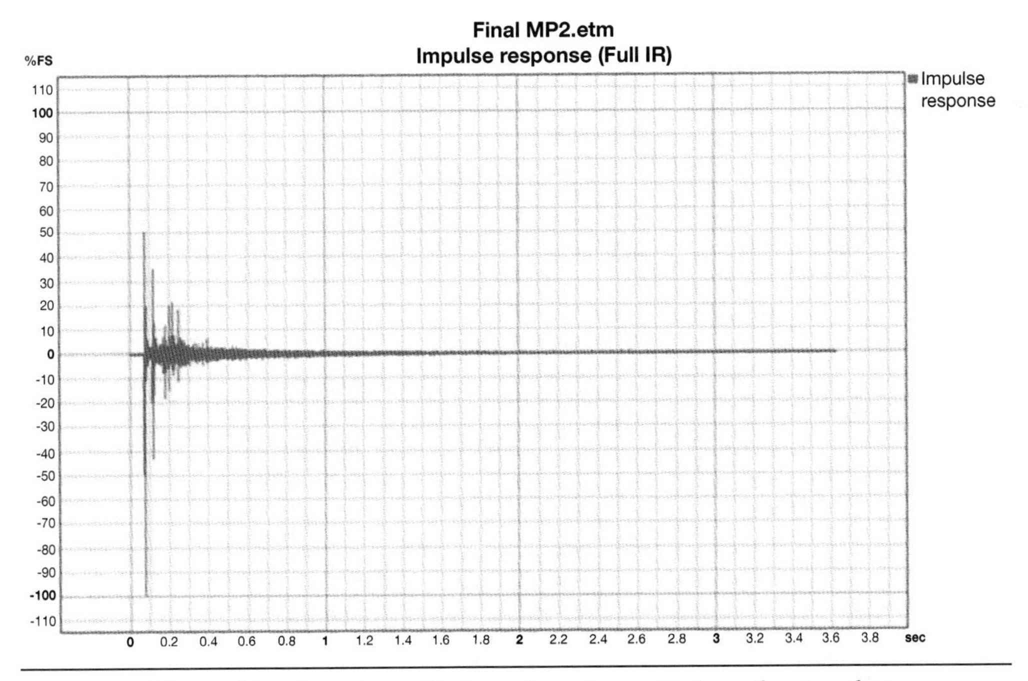

Figure 5.13 Monaural impulse response. The figure shows the amplitude as a function of time.

In this respect, it is usually distinguished between the room acoustic quantities[15] for speech performances (class room, auditorium, convention hall, or church) and for music performances (concert hall or opera). It is also distinguished between general time-based parameters and quality criteria related to the specific location of the listener. The latter are usually described by the combination of a receiver location and the location of one or multiple senders. Specifically for music performances these would be the locations of the conductor and of the musicians.

The most important time-based figures are:

- Reverberation times (T_{30}) as T_{10}, T_{20}, T_{30}
- Early Decay Time (EDT)
- Bass Ratio (BR)

Typically, these general criteria are measured using an undirected, that is, omnidirectional microphone.

The location-related evaluation of the room acoustic quality for speech performances employs the following criteria (also measured using an omnidirectional microphone):

- 50 ms part or definition D_{50}
- Definition C_{50}
- Articulation loss of consonants AL_{cons}
- Speech transmission index STI
- Center time t_s
- Echo criterion EK_{Speech} quantifying the perception of disturbing echoes or reflections

For the location-related room acoustic quality for music performances, the following measures have been developed or proposed, respectively:

- Source-related criteria for measurements with an omnidirectional microphone:
    - Direct sound C_7 for the sensitivity to directness and proximity of the sound source
    - Clarity C_{80} for the transparency of musical structures (time and register clarity)
    - Sound strength G for the perceived volume at the receiver location
- Source-related criteria for measurements with a binaural receiver (artificial head):
    - Interaural cross-correlation coefficient (IACC) for the width or breadth of the sound source as perceived by the listener (Apparent Source Width [ASW])
- Room-related criteria for measurements with an omnidirectional microphone:
    - Reverberation measure R for the enhancement of the dynamics or liveliness of the music presentation
    - Echo criterion EK_{Music} for the perception of disturbing echoes or reflections
- Room-related criteria for measurements with a bidirectional (figure-of-eight) microphone:
    - Lateral measures Lateral Energy (LE), Lateral Energy Fraction (LF) for the ASW, and the envelopment by the reflected sound, Listener Envelopment (LEV)
    - Lateral also measures the Lateral Fraction Coefficient (LFC)

For the evaluation of the room acoustic quality for music performances related to the receiver location, and the location of the conductor or musicians, the following criteria were developed (omnidirectional microphone):

- Early Ensemble Level (EEL) for the room acoustic support of the performance as perceived by musicians on the stage.
- Stage support ST1 and ST2 for the room acoustic support of the performance as well as for the room response as perceived on the stage and in the orchestra pit.

All of the above measures relevant for sound system design will be explained in Chapter 12, Section 12.1.2.

5.3.5 Frequency Domain Quantities

The impulse response can be transformed into a complex transfer function by means of the Fourier transform. This includes the spectral function of both signal amplitude and phase. The frequency-dependent amplitude function is also called frequency response in the narrow sense and serves as the primary criterion of quality when evaluating sound reinforcement systems. In room acoustics, the frequency dependence of T_{30} plays the decisive role. Its quality is determined based on defined tolerance curves.

In electro-acoustic installations, equalizers are often used to smooth the frequency response and thus remove peaks and dips caused by the reproduction system. It is a commonly accepted rule of practice that one should primarily focus on a 1/3 octave smoothed frequency response, because discontinuities of smaller bandwidth, such as 1/24 octave, can normally not be recognized.

The complex transfer function can also be used to investigate the frequency-dependent real and imaginary part, as well as the group delay, which is the negative derivative of phase with respect to frequency. For example, filters with linear phase have a constant group delay over the whole bandwidth. This also includes linear-phase finite-impulse response (FIR) filters which exhibit a constant latency for all frequencies.

5.3.6 Time-Frequency Representation (Waterfall Plots)

When combining both representations of the transfer function in time and frequency, one can obtain a three-dimensional display, a so-called waterfall, that shows the level on the z-axis, and time and frequency on the x- and y-axis, respectively. This method is particularly useful to investigate the transient response of a room or loudspeaker system based on a choice of parameters, like the time and frequency resolution and the type of window. Specific parameter settings also allow for resolving individual reflections with respect to their spectral content. This type of presentation can be enhanced further using wavelet analysis instead of harmonic decomposition (see Figures 5.14 and 5.15).

5.3.7 Special Applications

Among special applications we count such measurements which are not covered by standard measurement or processing techniques. This includes in particular the application of filters, windows, and averaging, as well as arithmetic combination of multiple channels. Also, in situ

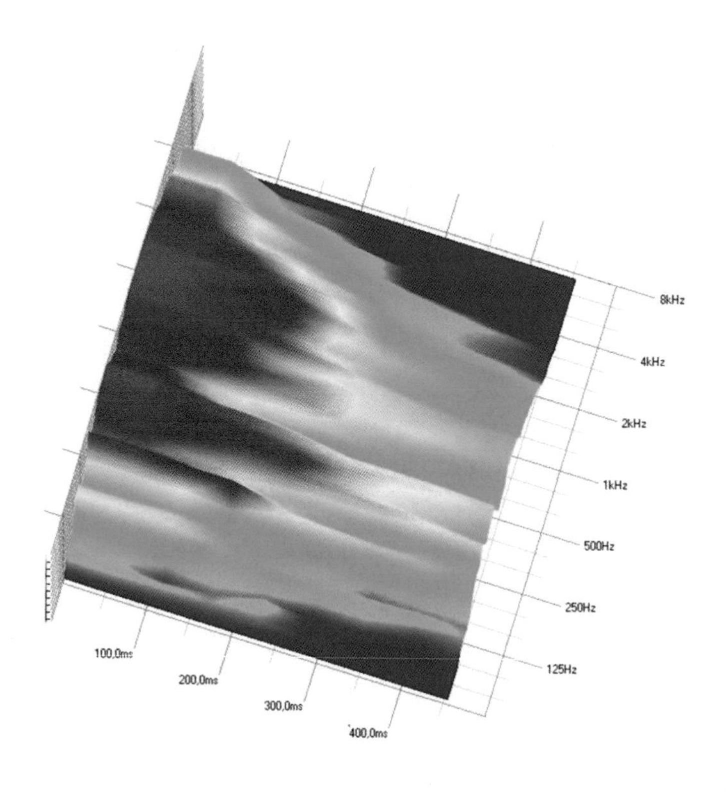

Figure 5.14 Partial spectrogram

measurements[19] and measurements of the scattering coefficient are explained in the following sections. We do not explain the measurement of acoustic parameters that have already been well-known for several years, such as the absorption coefficient or impedance of wall materials. In this case we refer the reader to standard textbooks.[20]

5.3.7.1 Filtering and Averaging

We already introduced filtering and averaging as a means to suppress or remove noise in the measurements (see Section 5.2.2.8). Here we would like to look at other typical areas of application.

Also in room acoustics, the filtering of a signal that was recorded broadband is a good way to suppress noise. A band-pass filter is particularly well-suited to remove perturbations outside of the upper and lower frequencies that limit our perception, without affecting the processing of the purely acoustical part of the measurement. High-frequency noise and low-frequency perturbations can be removed from typical measurement bandwidths like 50 Hz to 18 kHz. Noise inside the frequency range of interest can still be removed if it is narrow-banded. In this case, a steep band-stop filter can be applied to remove distortions like sinusoidal signals. By how much the result can actually be improved depends on the width of the filter.

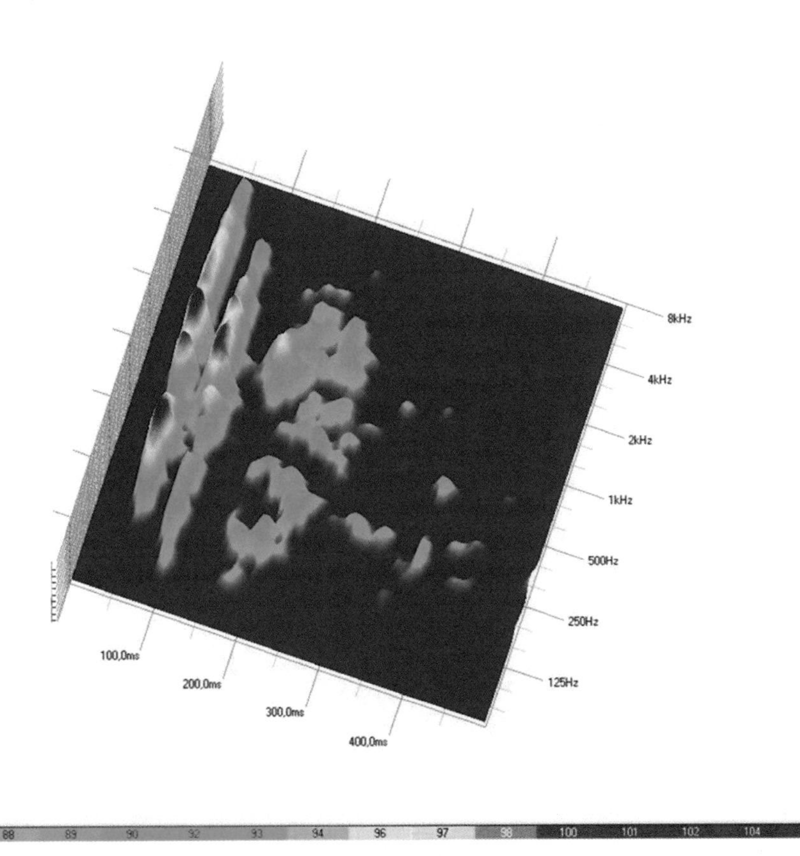

Figure 5.15 Wavelet-like presentation

However, applying a band-pass as described above has no effect on most room acoustic figures since these are only defined on the basis of a 1/1 octave band or 1/3 octave bands. The filters applied to obtain such band-related impulse responses have been standardized in the last years,[21] so that measurement devices provide comparable results.

Computer-based measurement platforms have to implement filters on the software side, whereas in analog measurement, instrument filters are realized directly as part of the hardware. This includes in particular the weighting functions like A-, B-, and C-weighting.[22] Because the stream of audio data available to the software is initially broadband and unweighted, additional processing steps must be applied to achieve the corresponding filtering.

Similarly, computer-based measurement systems have to implement data averaging as a post-processing function. This applies especially if integration times and transient responses of analog systems must be reproduced in order to be able to compare measurements made with a software program with measurements by a hardware platform. A typical example of that is the use of the time constants *Slow* and *Fast* in real-time analyzers.[22]

Spatial averaging is another form of result averaging often employed in room acoustics. In this case location-related quantities (see Section 5.3.4) are combined into a single number that is representative for the entire room or part of it. To be correct, this single figure should always be published with an uncertainty, since information about the spread of result values over the averaged area is lost otherwise.

In contrast, spatial averaging of impulse responses is usually more difficult or not sensible, respectively. Because the impulse response describes the transmission behavior for a specific receiving location, plain averaging will cause problems due to the phase information included in the response and the according rapid change of the amplitude response over space. However, time responses (envelope) and frequency responses (modulus) can usually be averaged energetically, because the energy distribution in the room is mostly a fairly continuous function—that is, it changes only by small amounts throughout the space.

A special form of averaging is applied when determining the scattering coefficient of wall materials by measurements in a reverberation chamber. Here, the averaging is used to remove scattered reflections from the measurement. This method is discussed in greater detail in Section 5.3.7.3.

5.3.7.2 In Situ Measurement of the Absorption Coefficient

Acoustic measurements of T_{30} and of acoustic energy measures in rooms by means of impulse responses or transfer functions have been discussed in Section 5.3.4. Compared to that, it is more difficult in practice to carry out measurements which are normally made in a dedicated acoustic environment, like a reverberation chamber. Such procedures are usually governed by standards. The method of measuring the sound absorption coefficient in a reverberation chamber is described in Cox and D'Antonio.[20]

But what happens if the absorption coefficient, α, of a wall material should be determined which is already built into the room? There are two possibilities to solve this:

- By removing a sample of the material of the size of 5-10 m^2 from the wall and measuring it in a reverberation chamber according to the standard
- By measuring on-site, that is, in situ

The first solution is often impossible because the interior of the room has to stay intact. Therefore, the remaining second method is normally the only way to obtain acoustic data about the materials in the room. For this purpose[23] a measurement setup is used that includes a small loudspeaker as well as a measurement microphone. The loudspeaker and the microphone are mounted rigidly on each end of a beam with the loudspeaker facing toward the microphone. Now, the device is aimed at the wall of interest to measure the impulse response including a reflection from the wall (see Figure 5.16). A second measurement must be made

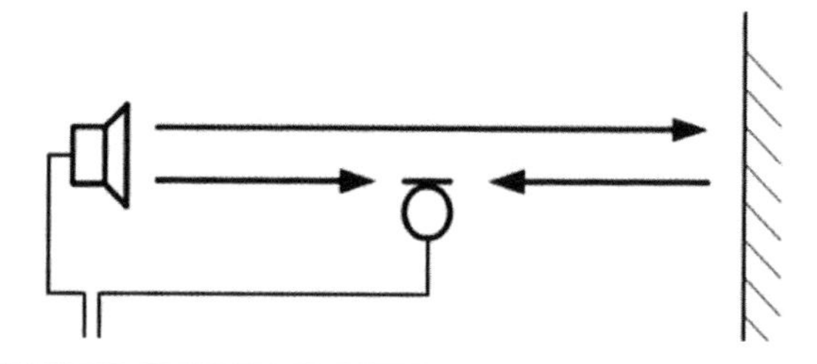

Figure 5.16 Scheme of the in situ measurement setup, showing the loudspeaker, the microphone mounted in front of it, as well as the acoustic surface of interest.

with the device pointing away from any wall in order to capture the direct sound of the loudspeaker only (see Figure 5.17).

Both impulse responses are then used to determine the reflection factor r in a post-processing step, whereby the level difference between direct sound and reflection due to the propagation distance must be accounted for. Using the relationship $\alpha = 1 - r^2$, the sound absorption coefficient is determined. But we denote that this sound absorption coefficient was determined for normal incidence, whereas the measurement in the reverberation chamber yields the random-incidence absorption coefficients.

5.3.7.3 Measurement of Scattering Coefficients

In room acoustics, the scattering of incident sound by the surfaces of a room plays an important role, for example, in the design of concert halls and recording studios. An approximately homogeneous diffuse sound field with the according exponential decay behavior occurs only if the boundaries scatter incident sound waves sufficiently. The quality of this diffuse field determines the acoustic impression of the room significantly, whereby neither total homogeneity of the diffuse field nor the lack of any diffuse sound is desirable. Therefore it is a practically important and often required measurement task to determine the scattering properties of surfaces. Only recently an appropriate measurement method was developed (ISO 17497),[24] and Mommerz.[25, 26] It can be used for scale models but also for real-world measurements.

As described by this standard, the scattering coefficient, s, is understood as a measure for the part of the reflected energy, E_{ref}, that is not reflected geometrically (E_{geo}):

$$s = 1 - \frac{E_{geo}}{E_{ref}}. \tag{5.13}$$

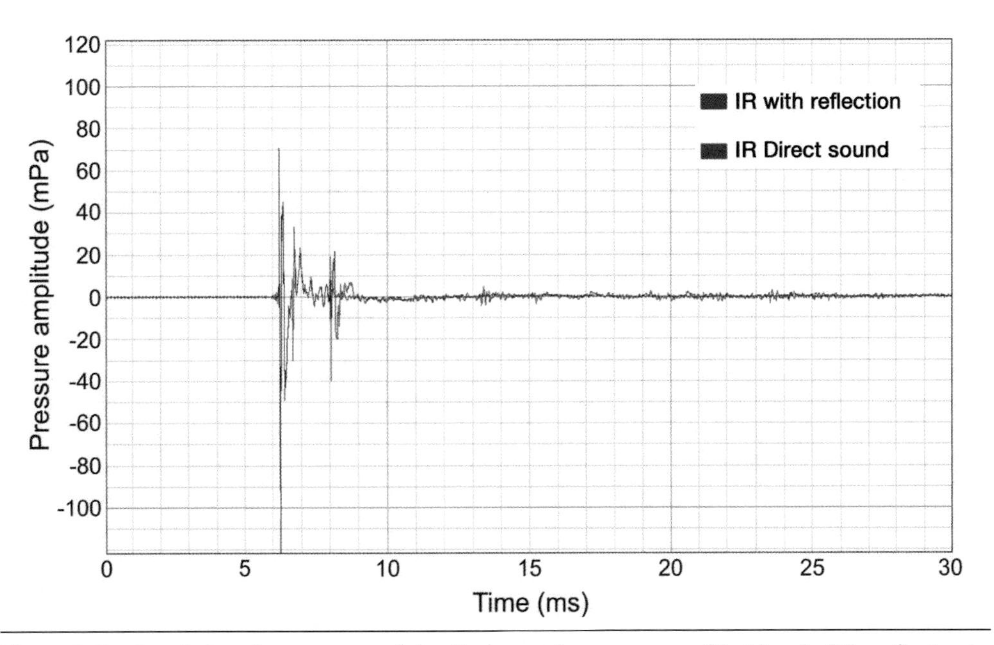

Figure 5.17 Sample impulse responses of the direct sound measurement (black) and of the reflection by the surface (grey).

The entity s is also utilized for the modeling of room transfer functions in various software packages (EASE AURA, CATT, ODEON).[27] But this definition of scattering is not to be confused with the diffusion coefficient of a scattering surface as defined by Cox and D'Antonio,[28] which describes the homogeneity of the reflected energy over angle.

The measurement principle requires that impulse responses of a sample of the material are measured for different rotational angles of the sample and then averaged directly in the time domain in order to separate the consistent geometrical part of the reflection from the varying scattered part of the reflection.

In the simplest form, these impulse responses are measured for a sample of sufficient size in the free field. The measurements are performed at different angles of the sample, which is rotated about the normal of the surface while loudspeaker and microphone location are fixed. In that case, the scattered part of the impulse response will vary randomly between measurements, but the geometrical (coherent) part of the response will remain the same. After that, all impulse responses are averaged. Assuming statistical independence of the scattered components, the result of the averaging can be compared with the measured results with a plane surface of the same size to derive the energy loss, which equates directly to the scattered part of the response (see Figure 5.18).

In practice, such a measurement is usually made in a reverberation chamber. The sample is mounted on a turntable. If the impulse response measurements are performed now, while the table is rotating, the averaging process happens automatically, assuming that the measurement duration is long enough (at least one rotation). Alternatively, the turntable can be rotated in a stepped manner (measurements of real-world samples may require a strong and possibly noisy motor), then the individual measurements are averaged subsequently.

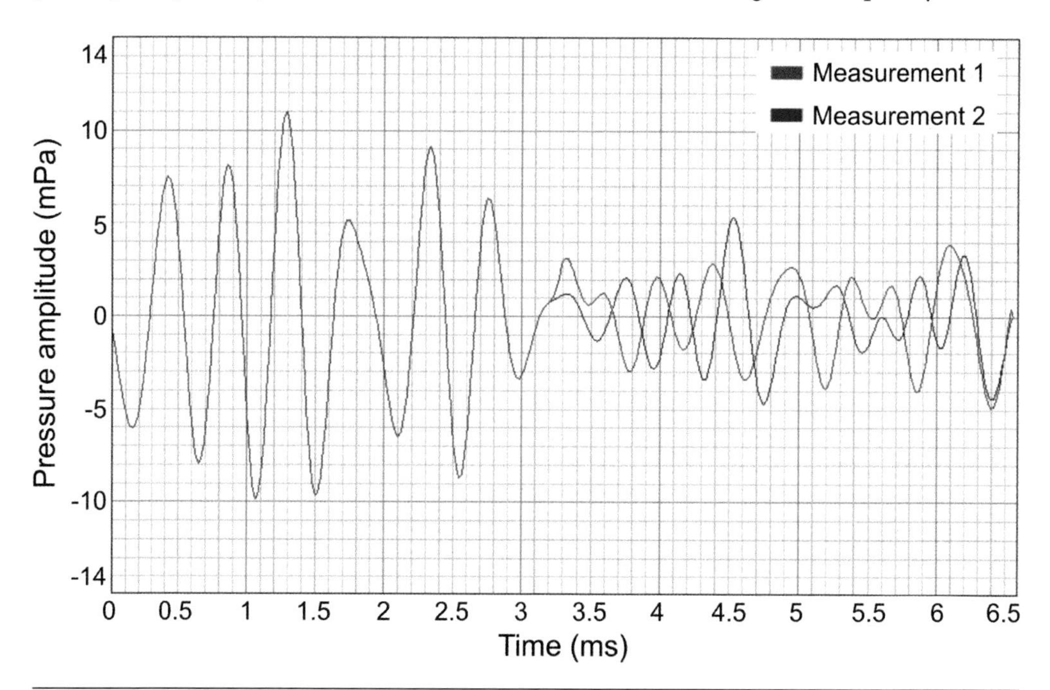

Figure 5.18 Example of two band-passed impulse responses with a coherent geometrical part and an uncorrelated late part.

We note that this method is fairly sensitive to variations in environmental conditions, such as air temperature and humidity, and it also requires great care regarding the setup and the sample properties, like size and edges. More details are given in ISO 17497.

5.3.7.4 Modal Analysis

In room acoustics, the frequency limit below which disturbing modes can be expected, especially in small rooms, is approximated by the Schroeder frequency:

$$f_{Schroeder} = 2000\,\text{Hz} \times \sqrt{\frac{T_{30}}{V}}, \tag{5.14}$$

where T_{30} is the reverberation time in seconds and V the room volume in m^3. For a small studio of 55 m^3 with a reverberation time of 0.23 seconds, we obtain $f_{Schroeder} \approx 130$ Hz, (see Figure 5.19).

In Section 5.3.6 we have already introduced the waterfall analysis as a typical tool for the analysis of spectral changes as a function of time. This form of presentation, also called a spectrogram, is particularly suited for modal analysis and the investigation of room resonances and eigenfrequencies. Besides reflections, which are typically localized in time, this method also allows identifying periodic or constant patterns of the signal over time. This is usually an indicator for the existence of room modes.

As an example, Figure 5.20 shows the waterfall plot (3-D graph of level over time and frequency) of a room impulse response of about 1 second in length measured in a studio. Modal behavior is easily recognized at the frequencies of 200 Hz and 300 Hz; below 100 Hz the low-frequency noise floor becomes visible.

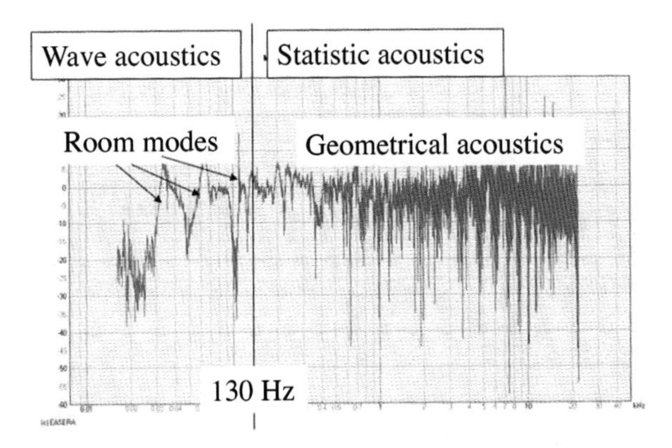

Figure 5.19 Room modes in the low frequency range of a measured spectrum. The graph shows the logarithm of the modulus over frequency.

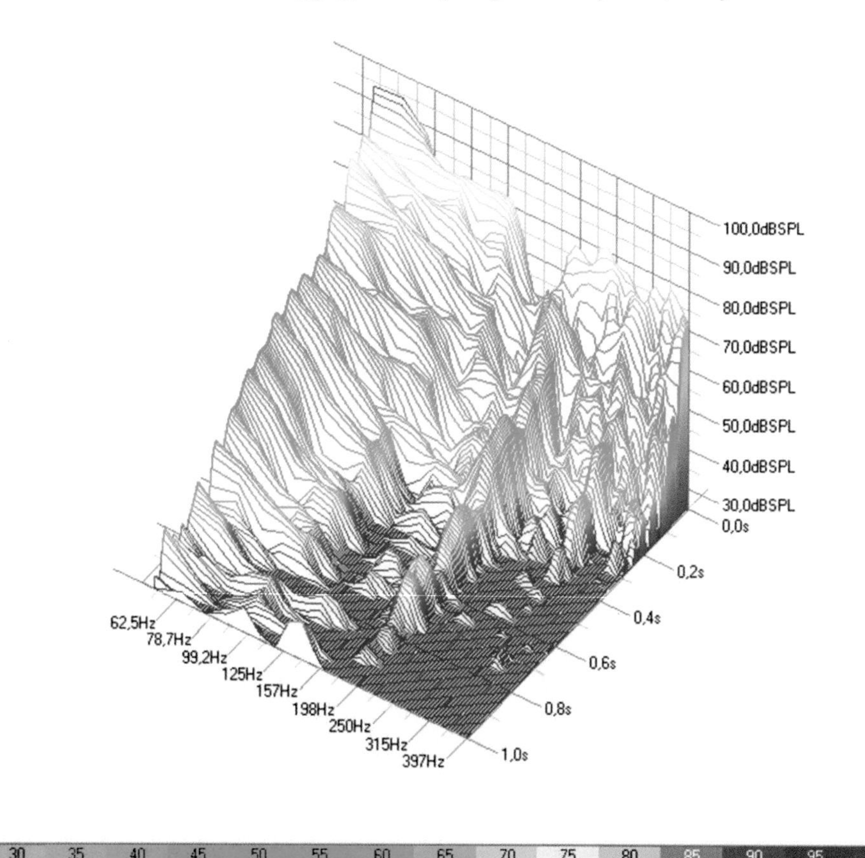

Figure 5.20 Waterfall plot of a room impulse response: level over frequency and time.

5.4 APPLICATIONS IN SOUND REINFORCEMENT

5.4.1 Electrical Verification

5.4.1.1 Subjective Tests

Before turning on a sound reinforcement system, an electrical verification of the signal chain and its quality is necessary. For smaller or portable systems, a subjective test may suffice, but for larger systems this test should be complemented by actual measurements.

After switching on the sound system, distortions of electrical origin sometimes occur. They have the following typical causes:

- Humming
    - Existence of ground loops, that is, the system includes multiple grounding connections which have different electrical potentials. In this case, a small current

flows between the connection points and induces hum in the resistors between these points.
    - Open, high-impedance input with amplification.
    - Different polarity of the net voltage between different parts of the sound system.
- Electrical resonance
    - Loudspeaker and microphone cables in close proximity and therefore high gain between them.
    - Open input into one channel with high amplification on the line.
- Drop-outs at peak levels
    - Clipping of one or multiple power amplifiers.
    - Overloading of a limiter.
    - Overly high impedance termination of a power amplifier.
- Dominant low-frequency part, "boomy" reproduction or periodic drop-out of reproduction:
    - Bad matching of low-impedance output of a power amplifier, e.g., shortcut of a power line or overload due to bad transformer match.
- Dominant high-frequency part, "sharp" reproduction
    - Interruption of the audio connection at the input of the power amplifier (capacitive input of the signal).
    - Single-pin connection of the power amplifier.
- Reduced reproduction level at low frequencies for two loudspeakers when playing a coherent signal:
    - Inverted polarity of the loudspeaker connections.
- Perturbations by radio-frequency interference due to Thyristor light dimmers (maximum effect at half-light intensity):
    - High-power lines of the lighting systems are too close to the audio transmission lines at the input of the sound system (minimum distance is 500 mm).
    - Missing suppression of noise induced by lighting system controls (dimmers), can be improved by inductive blocking.

5.4.1.2 Electrical Measurements

Electrical commissioning and tuning of larger sound reinforcement systems is an important part of the overall tuning process and typically represents a precondition for the subsequent acoustic tuning of the system. Generally it requires a higher quality for concert sound applications than for speech applications and emergency announcements.

The following measurements are especially important for sound reinforcement systems:

- Verification of all input and output channels with respect to:
    - Transmission and attenuation
    - Polarity (A-B reversal)
    - Symmetry of ungrounded connections (symmetrical connections are preferred, especially for long transmission lines, because inductive and capacitive distortions cancel out)
    - Cross-talk
    - Signal-to-noise ratio

- Verification of the electric level at the inputs and outputs of individual elements of the system according to a previously defined gain structure. This level structure takes into account the upper limit of the expected noise level and the maximum dynamic of the system. Only when the configuration accounts for these limitations, the dynamic range of the sound system can be optimized.
- Testing of all switching functions in mixing consoles, distribution and processing units, especially all switches, attenuators, and filters.

5.4.2 Acoustic Measurement and Tuning

5.4.2.1 Introductory Comment

The acoustic measurement and tuning of a sound reinforcement system should always be preceded by the evaluation of the acoustic properties of the room. This is necessary to decide if the quality of the sound transmission by the sound system can be affected negatively by the room acoustic environment. (Of course, any professional design of a sound system should have taken this into account prior to the installation so that there are no surprises.)

Therefore the acoustic tuning (or assessment) of an existing sound system begins with the determination of the measurement locations (see Section 5.3.2).

The acoustic tuning of the system usually starts with a verification of the conditions as they were assumed in the design of the sound reinforcement system, such as:

- Reverberation time
- Intelligibility without sound reinforcement at various, representative locations
- Sound level distribution without the electro-acoustic system
- Background noise level distribution in the venue

Measurements of the reverberation time should be made for all configurations of the room for which the sound system is used. In this respect, it is particularly useful to measure the frequency dependence of the reverberation time for both occupied and unoccupied states and at several measurement locations. For example, in the case of theaters, the configurations of the room may include different setups of the stage, a closed or open stage, or different decorations of the stage. In modern multipurpose halls one can often find a variety of different configurations of the room and podium.

Intelligibility and sound level distribution without the electro-acoustic system can be derived from the room acoustic verification measurements. Parameters and data may also be obtained from the measurement protocols of the acoustical consultant.

Acoustic verification should also include the measurement of the absolute noise level and of the noise power spectrum in the receiving areas. Also in this case, different configurations of the room must be accounted for.

Only after acquiring the knowledge of the room acoustic conditions, the performance of the sound reinforcement system should be evaluated. Again, this consists of computer-based measurements to determine the according impulse response data. But here, due to the lack of binaural quality criteria, usually only an omnidirectional measurement microphone is used.

5.4.2.2 SPL Coverage

Starting from the measured impulse response, the quantity of sound strength G according to Lehman[29] is well-suited to determine the sound level distribution throughout the room or over the audience area outdoors. For this purpose, the G measure which requires the use of an omnidirectional sound source in room acoustic applications is used for the sound system itself. In this case the sound pressure at various measurement locations is compared to the direct field measured at a reference location, such as the front of the house or the mixing desk. It can also be normalized to the total sound pressure level at the reference location. This is allowed because there are no standards or regulations for the tuning of sound systems with respect to the level variation over the audience area.

After measuring the sound strength G at representative locations of the room, these values can be assembled in a tabular format or as mapping graphs similar to the display of modeling results. A variety of software packages are available that facilitate impressive result graphics (see Figure 5.21).

Quite often, random pink noise is used as an excitation signal still. This signal in combination with a processing that uses power-averaging will provide results that are largely insensitive to wave interference phenomena such as comb filtering, as it occurs in closed rooms or in front of large reflecting faces. Also, pink noise can cover the entire bandwidth of operation of the sound source or of sound sources, representatively, for typical audio signals.

If it is to be expected that measurements are distorted by noise in the low frequency range (such as ventilation or traffic noise), it is common practice to apply an A-weighting to the

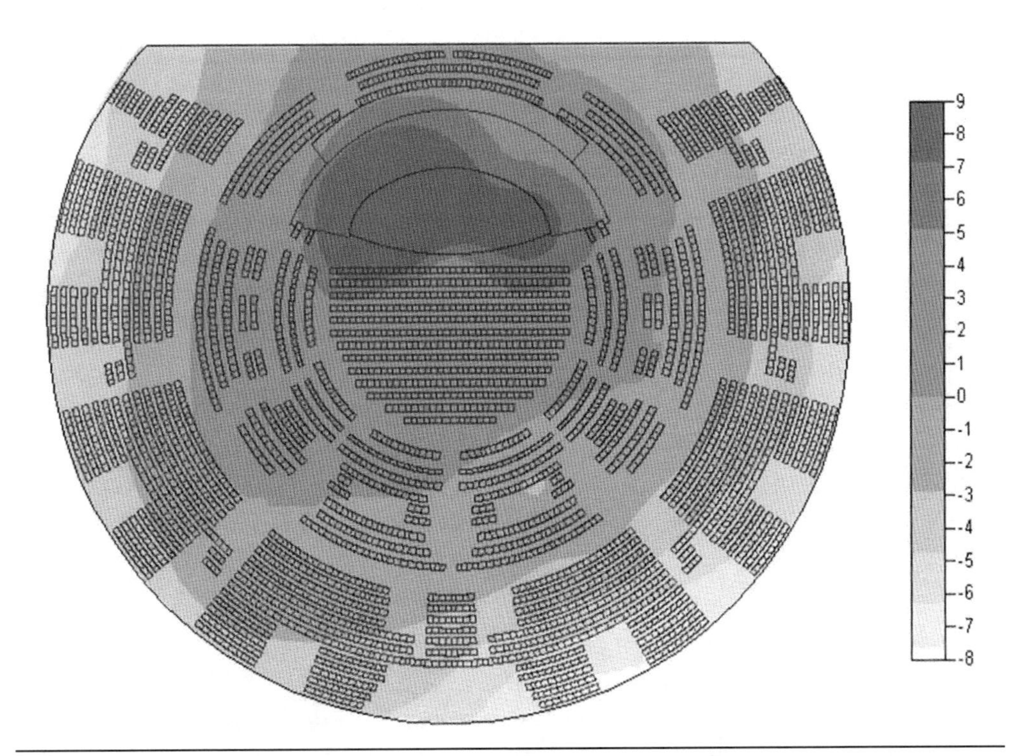

Figure 5.21 Measured distribution of sound strength G in a multipurpose hall.

received frequency response in order to suppress such noise in a way similar to the human hearing system.

Measurements should be performed for individual loudspeaker groups of the sound system, as well as for the sound system as a whole. When measuring subsets of the loudspeaker system, only those audience zones have to be evaluated that are to be covered by that group or can be disturbed by that group. For example, multichannel sound reinforcement systems usually require verification of the neighbor zones to determine the interference between groups and thus potential errors in the intended localization of the reinforced sound source.

The level variation between the highest and the lowest sound level measured should be less than 3 dB, but in any case, less than 5 dB for comparable listening locations.

Additionally, the spectral distribution of the sound level at individual locations is of interest, of course. If the reproduction is too *sharp*, then energy content in the low frequencies is missing. In contrast, if the bass is perceived as strong, then the reproduction of the low-frequency content is dominant. For an objective assessment of these qualities, a measurement of the frequency response is necessary. Frequently, real-time analyzers are used in connection with pink noise. This allows an immediate verification of the spectral composition of the signal received at the measurement location. With the introduction of computer-based measurement and processing, not only can the spectrogram plot, but also three-dimensional graphs of level over time and frequency (so-called waterfall plots) be derived directly from the impulse response (Figures 5.14 and 5.15 are derived from the impulse response in Figure 5.13).

5.4.2.3 Maximum Sound Pressure Level

Several recommendations and regulations for multipurpose halls or stadiums define maximum levels, such as 100 or 105 dB(A). But quite often it is not clear if this is related to the direct sound or the total sound. For a long time the soccer association FIFA did not require a level value except that the sound is loud enough so that the signal level exceeds the noise level caused by the audience.[30] But in an updated document from November 2014, FIFA speaks about a crowd level of (arguably) up to 110 dB(A)! And the signal level must exceed this level by 6 dB.[31] The basketball association FIBA asks for *only* 95 dB(A). Very often such requirements are based on levels as sometimes reached at rock and pop concerts. In this respect, the authors had already been concerned with level requirements of up to 120 dB(A), which is, of course, extremely unhealthy already when exposed to it for only a very short period of time. Some standards—like the German DIN 15905-5[32]—demand a maximum mean level of 99 dB(A) at events of more than two hours length. Since music performances in halls, stadiums, and other venues often endure even longer than two hours, such limits should be taken into account. However, emergency announcements of short duration must and can be louder, since they have to exceed the existing background noise created by the spectators. For that purpose, mean total levels of about 110 dB(A) are possible. But higher levels usually cannot be accomplished in fixed installations and they may cause significant hearing damage in any case. But because the human hearing threshold can increase temporarily, the hearing loss may not be permanent.

Additionally, one should be aware that in the case of high sound levels, masking effects will reduce potential speech intelligibility (see Figure 12.9 in Chapter 12, Section 12.1.2.2.1).

Therefore, the feasible total level that is maximally achievable should be assumed in the order of 105 dB(A). It can be assessed most easily by exciting the system with pink noise and measuring the level with a calibrated, hand-held sound level meter. This procedure can also be used to determine the corresponding settings for the sound system in order to achieve this maximum level. After that, the gain controls can be sealed in order to prevent a further increase of the level by third parties and, thus, to avoid potential hearing damage in the audience.

5.4.2.4 Measurement of the Frequency Response

A set of different effects influences the level and frequency response of the sound system at the receiving location:

- The frequency dependence of the direct sound level of the loudspeaker
- The directivity function of the loudspeaker and its influence on the diffuse sound field
- Differences in the sound absorption of the surfaces of the room
- Different sound radiation conditions due to the configuration of the sound sources
- Interference of the direct sound with reflected sound or the sound of delayed sources

The impulse response as measured at the listening location can be transferred into the frequency domain using the Fourier transform.

By means of equalizing the sound system output, the frequency response of the reproduction system can be smoothed. However, the following aspects should be taken into account:

1. Measurement of the frequency response of reproduction should be performed at multiple locations. That is necessary because electrical corrections of the frequency response of the sound reinforcement system will act differently at different locations, in respect to the sound source or sound sources. The raster of the measurement points can be similarly chosen to the evaluation of the sound pressure level but may be coarser.
2. In small rooms with a relatively strong diffuse sound field, very few representative locations can be chosen. But also in this situation, it has to be ensured that the measurement results do not suffer from random effects. This can be accomplished, for example, by taking additional measurements close to the original measurement location.
3. In order to assess distortions in the entire transmission channel, the test signal is sometimes fed directly into the input of the mixing console.
4. Equalization provides a means to achieve a smooth reproduction frequency response. However, it is usually not possible to obtain a flat frequency response or a frequency response that is independent from the receiving location. A linear, that is, flat frequency response is only important for the listening area close to the stage. In large rooms, the hearing system is trained to expect a level decay at the high frequencies at distances further away from the source. Therefore, a flat frequency response at the back of the hall might be perceived as too *sharp*.
5. It should be clear that interference effects due to sound reflected by room surfaces can usually not be compensated for—or if at all, except in a very limited way.

Tolerances for recommended reproduction curves in different applications are given by Mapp[33] (see Figure 5.22).

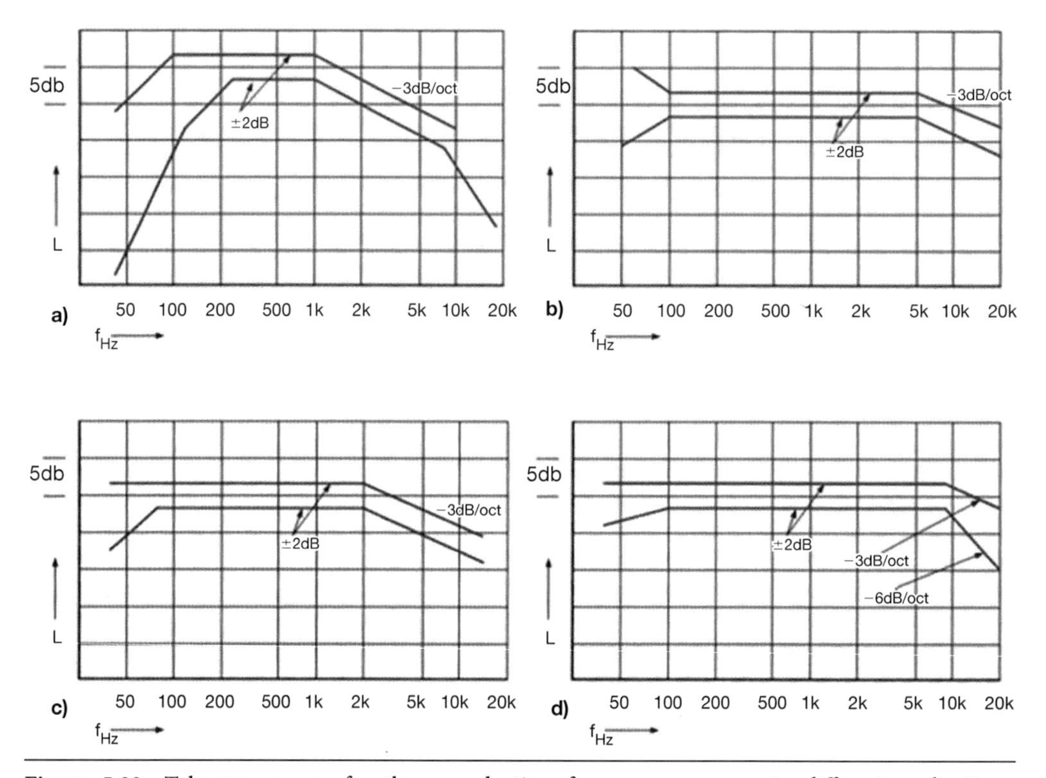

Figure 5.22 Tolerance curves for the reproduction frequency response in different applications: (a) recommended curve for reproduction of speech, (b) recommended curve for studios or monitoring, (c) international standard for cinemas, (d) recommended curve for loud rock and pop music.

One can recognize that the limits defined for the recommended decay of level at higher frequencies are different. The tolerance curve defined for rock and pop music is closest to a flat curve. In this case, it is undesirable that the high frequencies decrease in level at distances further away from the sound source.

Often, the frequency response of the sound system is also corrected in order to suppress or avoid feedback effects. Since feedback always occurs at the frequency that exhibits the highest peak in the transfer function curve at the microphone location, the feedback limit can be expanded by reducing such a peak by means of an appropriate filter. Applying several notch filters at different frequencies can result in a significant increase of the available gain before feedback.

Nowadays, the described steps of verifying and correcting the frequency response of reproduction are usually part of the commissioning of every sound reinforcement system. For sound systems that consist of a variety of loudspeakers on the stage, for special effects and for reinforcement, this may lead to the fact that prior to each performance a so-called sound check is made at representative seats in order to verify that the loudspeaker and filter settings made are correct. Modern measurement methods often utilize the speech and music signals available during the rehearsal to derive corrections with respect to the reproduction frequency response without disturbing the musicians or audience.[12, 13, 14]

5.4.2.5 Measurement of the Speech Intelligibility STI

STI figures for a sound system can be determined by measuring the modulation loss of a signal that propagates from the sound source to the receiver. The frequency range concerned includes the octave bands of 125 Hz to 8000 Hz. For this purpose, Steeneken and Houtgast proposed the use of a special, modulated noise signal to excite the room and measure the reduction of modulation depth.[34] Schroeder could prove that STI values can also be obtained from a measured impulse response,[35, 36] which has become common practice among modern computer-based measurement platforms, like EASERA.

Steeneken and Houtgast assumed that not only reverberation and disturbances, but also signals from other sources and generally changes of the signal itself along the propagation path can reduce speech intelligibility. To determine this influence quantitatively, the modulation transfer function (MTF) is employed for acoustical purposes. In principle, the ratio of the usable signal S and of the noise signal N is calculated. The resulting entity is the modulation reduction factor $m(F)$ which characterizes the effect on speech intelligibility:

$$m(F) = \frac{1}{\sqrt{1+(2\pi F \cdot \mathrm{T}_{30}/13.8)^2}} \cdot \frac{1}{1+10^{-(\frac{S/N}{10dB})}}, \tag{5.15}$$

where F is the modulation frequency in Hz, T_{30} is the reverberation time in s, S/N is the signal-to-noise ratio in dB. More details are given in Chapter 12, Section 12.1.2.2.1.

A related method that was introduced recently is the STIPa measurement. It is defined primarily on the basis of the excitation of the SUT with modulated noise. Deriving the STIPa value from an impulse response is possible, but not preferred.

The frequency spectrum of the noise signal is shown in Figure 5.23. It is typical that the signal is concentrated at the centers of octave bands, but only excites half of each octave. In

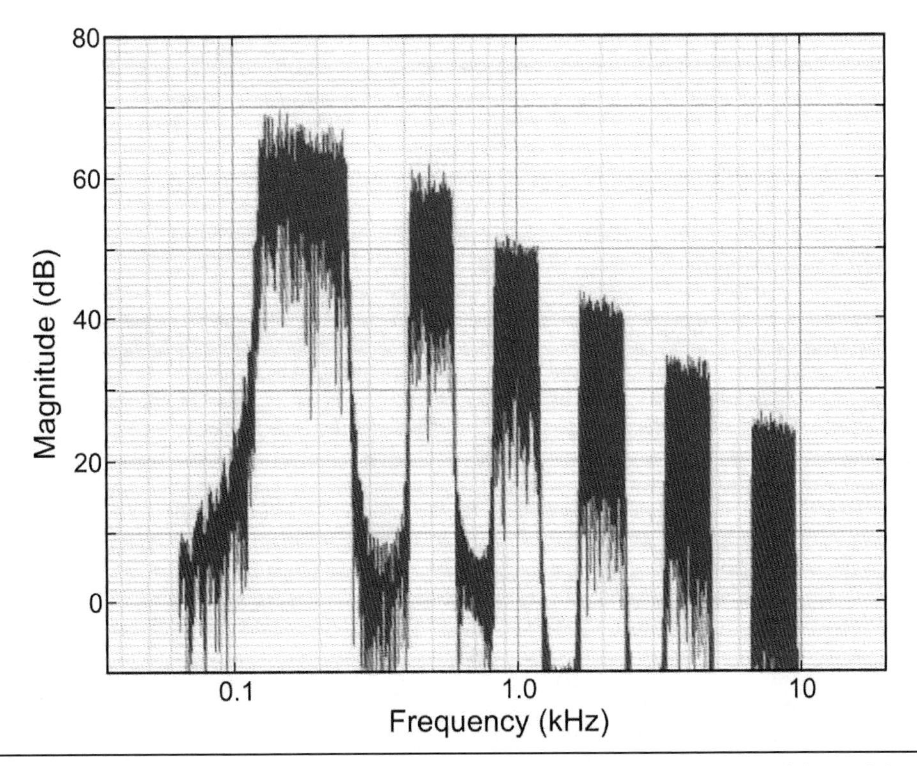

Figure 5.23 Frequency spectrum of a STIPa signal. The graph shows the logarithm of the modulus over frequency.

practice, the excitation signal is radiated into the room using the loudspeaker system. While the signal is being played, a portable receiver can be used to determine the STIPa value at different locations of the room.[37, 38] Because this method does not require any specific knowledge, STIPa measurements can also be made by personnel and technicians without further acoustic or technical education. This method is also increasingly employed for the verification and testing of emergency notification systems.[39]

5.4.2.6 Subjective Assessment of Speech Transmission Index Values

According to the results of subjective listening tests, STI and STIPa values are related to the subjective score for syllable intelligibility[36, 40] as defined in Table 5.1.

5.4.2.7 Signal Roughness and Source Mislocalization

When using a distributed loudspeaker system, with or without delay, there is always the risk that the wave fronts of a radiated signal arrive at the listener at different times. If the time difference between such wave fronts exceeds about 30 ms the definition of the transmission can be reduced. For time differences of more than 50 ms even echoes can be perceived.

Such echoes and audible artifacts can occur for systems which have already been commissioned for the following reasons:

- The mounting point of the sound sources had to be changed, for example, for technical reasons
- Strong reflections occur
- The level relationships among sound sources and groups of sources change
- The delay adjustment was erroneous

In order to verify that the signals from the individual loudspeakers arrive at the desired time, impulse responses can be analyzed. For this purpose, sequential measurements are made; initially just a single loudspeaker close to the stage is active and then additional loudspeaker groups are switched on, step by step. In this way, the influence of each group can be determined individually. Figure 5.24 shows the impulse response of a situation where the direct sound is not correctly localized due to bad signal alignment.

This measurement method is accurate enough to determine differences in propagation times (echoes, comb filtering in general, and reduced definition), but it does not allow for determining sound sources that cause localization errors, such as when the sound system should only reinforce a weak original sound source. In this case, one has to investigate the frequency behavior in addition to the energy-time relationship.

Table 5.1 Subjective assessment of STI values

Subjective Syllable Intelligibility	STI Value
Unsatisfactory	0 ... 0.3
Poor	0.3 ... 0.45
Satisfactory	0.45 ... 0.6
Good	0.6 ... 0.75
Excellent	0.75 ... 1.0

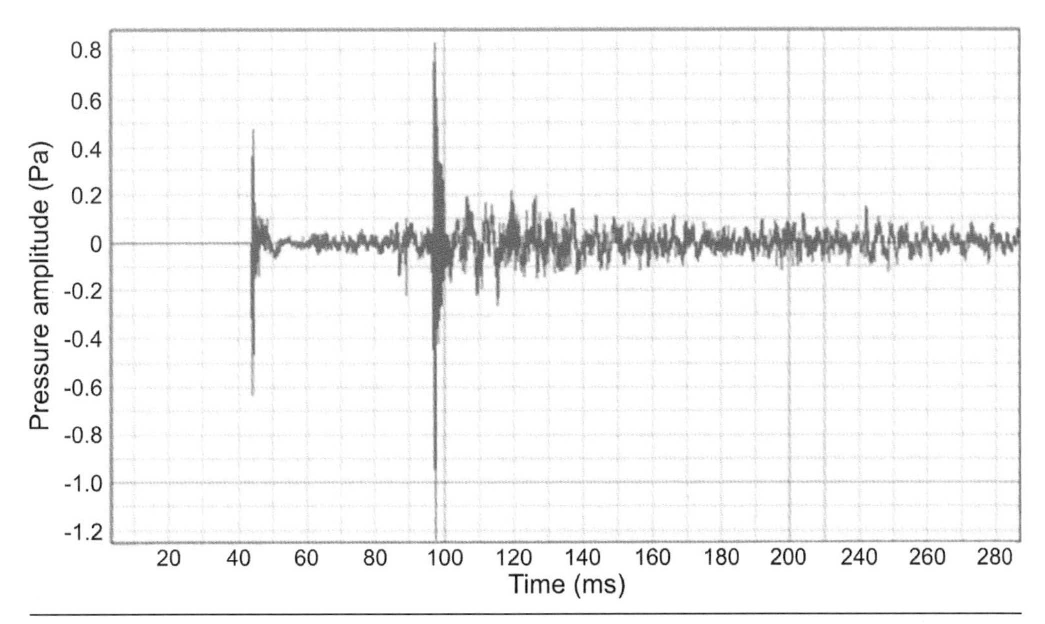

Figure 5.24 Exemplary section of a measured impulse response where the sound reinforcement system (after 90 ms) provides a higher signal level than the original sound source on the stage (at about 44 ms)

Principally, localization errors cannot occur if the initial impulse is not exceeded by more than 10 dB within the first 30 ms and by no more than 6 dB within 30 to 60 ms (rule of the first wave front).[41]

5.4.2.8 *Subjective Assessment*

In addition to objective measurements, the commissioning of a sound reinforcement system may require subjective methods of assessing the quality of the transmission systems. This may happen, for example, if there are no appropriate objective methods available or these methods do not cover the full range of applications. It may also be the case that fairly complex, subjective qualities have to be evaluated such as the operability of the systems or the use of particular devices for special effects.

The assessment of syllable intelligibility using senseless *words* is a well-known method.[42] Only when correctly perceived can the particular syllable be recognized—guessing it is impossible. The ratio of syllables understood compared to syllables read is a measure of the subjective intelligibility, (see also Table 12.4 in Chapter 12, Section 12.1.2.2.1).

It is a more complex task to perform the overall subjective assessment of a sound system by a group of listeners. Such tests are most important when the sound system is also used to influence room acoustic parameters, such as generating artificial reverberation.

The listening group usually consists of 5 to 20 people. Possibly at the same time, different seating areas in the auditory should be checked. To do that, different and typical samples of program material should be used. After each test, the listeners should move to a different seating area. The subjective test results should be entered into questionnaires which have to be evaluated statistically afterwards.

5.4.3 Additional Measurements

5.4.3.1 Signal Alignment

Alignment measurements are acoustic measurements made to achieve a coherent wave front. When a building loudspeaker arrays, it is important that the wave fronts of the individual systems arrive at the audience at approximately the same time. If the cluster is spatially distributed, this is generally not the case without further adjustments. Delays have to be applied in order to compensate for different propagation times among the individual components of the array. When tuning the system, usually the arrival time of the latest signal is used as a reference. Relative to that, all elements of the cluster are delayed electronically, until a coherent wave front is accomplished. These time adjustments are typically made on the basis of impulse response measurements for each system or group. As a result, establishing the alignment and, thus, the coherence of arriving wave fronts will also reduce or eliminate comb filter effects.

There exists a variety of possible measurement methods to determine the alignment settings for a loudspeaker system. It is most simple if the measurement system can store and graphically show the impulse response of the reference system, while the other systems are displayed as overlaid curves. By adding an electronic delay to a particular system according to its overlay measurement, the direct sound of this way or channel can be aligned to the reference channel. After that, the next subsystem can be enabled and the procedure is repeated.

Of course, as for equalizing systems, the alignment of loudspeaker delays is difficult for extended listening zones. A compromise has to be established for all receiving locations because the exact alignment of all sources is only possible for a specific spot. Very often several iterations of the tuning process are necessary to obtain a satisfying result.

5.4.3.2 Feedback Test

It is principally negative when the sound reinforcement system exhibits a degree of feedback, usually indicated by a *ringing* in the reproduced sound. In that case, the loudspeaker radiates too much sound into the microphone feeding it, so that a feedback loop is established. When adjusting the level of the sound system for a performance, it can often be seen that the musicians increase the gain of the system until they recognize the first effects of feedback. In this way, they find out about the feedback threshold and the maximum gain. Naturally this kind of feedback test is disturbing.

It is preferred, for example, to use a real-time analyzer in parallel when increasing the gain in order to determine the frequencies of over-proportional level increase. As a consequence, a narrow-band filter can be applied to attenuate the peaks in the spectrum and, thus, increase the potential gain before feedback. Nowadays, real-time analysis is often replaced by software-based measurement tools which can provide even better indicators for a feedback situation, such as a spectrogram view (similar to Figures 5.15, 5.20). This also simplifies the placement and activation of narrow-band filters in order to suppress feedback probability.

5.4.3.3 Polarity Test

If two loudspeakers are located close to each other and play the same output signal, different polarity will lead to cancellations in a small area in front of the speakers (where the distance to both sources is approximately equal). To determine the polarity, so-called polarity checkers can be used that display the polarity by means of a visual signal after the loudspeaker was excited with an impulse-like signal. Alternatively, the sound pressure distribution in front of the loudspeakers can be used to verify the polarity. In the area of equal distance to both sources, the sound pressure level should increase by 6 dB if the polarity is correct. Otherwise a cancellation or attenuation of the received signal will occur that originates in the misaligned phase of the arriving wave fronts.

5.5 FINAL REMARKS

In this chapter we have tried to introduce the reader to the state of acoustic measurements as they are currently practiced in room acoustics and electro-acoustics. Naturally, the mathematical and physical basis for the determination of impulse response data such as Fourier analysis, convolution, and windowing will hardly change in the future. However, the corresponding measurement platforms based on software and hardware solutions are subject to on-going development and, thus, improvement. Therefore, parts of this section will have only a limited lifetime, which is why we have treated present measurement tools and commercially available solutions just briefly. Nevertheless, the general principles will remain valid and we hope that the reader of this chapter will expand on their knowledge about the computer-based acquisition of measurement results. For more detailed applications of the explained acoustic measures, we refer to Chapter 12, Section 12.1.2 and to the literature referenced.

REFERENCES

1. A. V. Oppenheim and R. W. Schafer, "Zeitdiskrete Signalverarbeitung," (time-discrete signal processing), R. Oldenbourg Verlag, München, 1992.
2. Buttkus, B., "Spectral Analysis and Filter Theory in Applied Geophysics," Springer-Verlag, Berlin, Heidelberg, 2000.
3. Douglas D. Rife and John Vanderkooy, "Transfer Function Measurement with Maximum-Length Sequences," JAES, vol. 37, no 6, June 1989, pp. 419–444.
4. Richard C. Heyser, "Time Delay Spectrometry—An Anthology of the Works of Richard C. Heyser," *AES*, New York, 1988.
5. A. Farina, "Simultaneous Measurement of Impulse Response and Distortion with a Swept-Sine Technique," presented at the AES 108th Convention—Paris, (2000 February 19–22).
6. S. Müller, P. Massarani, "Transfer-Function Measurement with Sweeps," J. Audio Eng. Soc., vol. 49, no 6, pp. 443–471, (2001 June).
7. John Vanderkooy, "Aspects of MLS measuring systems," JAES, vol. 42, April 1994, p. 219.
8. Vorländer, M. and Bietz, H., "Der Einfluss von Zeitvarianzen bei Maximalfolgenmessungen," DAGA, 1995, p. 675.
9. Eckard Mommertz and Swen Müller, "Measuring Impulse Responses with Pre-emphasized Pseudo Random Noise derived from Maximum Length Sequences," *Applied Acoustics*, 1995, vol. 44, p. 195.
10. J. Daigle and N. Xiang, "A specialized cross-correlation algorithm for acoustical measurements using coded sequences," JASA 119, pp. 330–335 (2006).
11. N. Xiang (2008), "Digital sequences," in Handbook of Signal Processing in Acoustics, Vol.1 ed. by D. Havelock, S. Kuwano and M. Vorlaender, Springer, New York, pp. 87–106.
12. SIM Audio Analyzer, http://www.meyersound.com/products/#sim, Meyer Sound, Berkeley, CA, USA; www.meyersound.com.
13. Smaart Software, http://www.rationalacoustics.com, Rational Acoustics, Putnam, CT, USA.
14. EASERA SysTune, http://systune.afmg.eu, AFMG Technologies GmbH, Berlin, Germany, http://afmg.eu.

15. Standard ISO 3382: Measurement of room acoustic parameters.
16. U.S. Patent: 9,060,222 B2.
17. German Patent 10 2007 031 677, U.S. Patent 8,208,647.
18. Lundeby, A., Vorländer, M., Vigran, T. E. and Bietz, H., "Uncertainties of Measurements in Room Acoustics," ACUSTICA, 81 (1995), S. 344–355.
19. Olson, B., Ahnert, W. and Feistel, S., Experience with in situ measurements using Electronic and Acoustic System Evaluation and Response Analysis (EASERA), 153rd Meeting ASA, Salt Lake City, Utah, 4–8 June 2007.
20. DIN 52212 / ISO 354: Measurement of sound absorption in a reverberation room and DIN 52215 / ISO 10534-1: Determination of sound absorption coefficient and impedance in impedance tubes.
21. IEC 61260:1995: Octave-band and fractional-octave-band filters.
22. IEC 61672-1:2013: Sound Level Meters—Part 1: Specifications.
23. ISO 13472-1:2002 Measurement of sound absorption properties of road surfaces in situ—Part 1: Extended surface method.
24. ISO 17497-1:2004 Acoustics—Sound-scattering properties of surfaces—Part 1: Measurement of the random-incidence scattering coefficient in a reverberation room.
25. E. Mommertz, Determination of scattering coefficients from the reflection directivity of architectural surfaces. Applied Acoustics **60** (2000) 201–203.
26. Vorländer, M. and Mommertz, E., Definition and measurement of random-incidence scattering coefficients. Applied Acoustics 60 (2000) 187–199.
27. AURA software module as part of the EASE Software, AFMG Technologies GmbH, Berlin, Germany, http://www.afmg.eu, ODEON software, http://www.dat.dtu.dk/~odeon/, CATT-Acoustic software, http://www.catt.se/.
28. Trevor J. Cox and Peter D'Antonio, Acoustic Absorbers and Diffusers: Theory, Design and Application, Taylor & Francis London and New York, 2nd ed. 2009.
29. P. Lehmann, Über die Ermittlung raumakustischer Kriterien und deren Zusammenhang mit subjektiven Beurteilungen der Hörsamkeit, Dissertation, TU Berlin 1976.
30. "Football Stadiums—Technical recommendations and requirements." Published by Fédération Internationale de Football Association (FIFA), 2007 ed 4, located at: http://www.fifa.com/mm/51/54/02/football_stadiums_technical_recommendations_and_requirements_en_8211.pdf.
31. FIFA World Cup Stadium Requirements Handbook, November, Chapter 50.20.20.50.
32. DIN 15905-5 (Norm 1989-10, Normentwurf 2006-01) "Maßnahmen zum Vermeiden einer Gehörgefährdung des Publikums durch hohe Schallemissionen elektroakustischer Beschallungstechnik."
33. P. Mapp: First published in "Audio System Design & Engineering," Klark Teknik, 1985.
34. T. Houtgast and H. J. M. Steeneken: The modulation Transfer Function in Room Acoustics as a Predictor of Speech Intelligibility, Acustica Band 28, Seiten 66–73, 1973.
35. M. R. Schröeder, Modulation Transfer Functions: Definition and Measurement, Acustica, Vol. 49 (1981), S. 179–182.
36. T. Houtgast and H. J. M. Steeneken: A review of the MTF concept in room acoustics and its use for estimating speech intelligibility in auditoria. J. Acoust. Soc. Amer. 77 (1985), pp. 1060–1077.
37. IEC 60268-16:2011: Sound system equipment—Part 16: Objective rating of speech intelligibility by speech transmission index.
38. K. Jacob, H. Steeneken, J. Verhave and S. McManus, "Development of an Accurate, Handheld Simple-to-Use, Meter for the Prediction of Speech Intelligibility," Proc. IOA Vol 23 Pt 8.
39. ISO 7240-19, Fire detection and alarm systems—Part 19: Design, installation, commissioning and service of sound systems for emergency purposes, International Standard Association, 2007
40. EN ISO 9921:2003: Ergonomics—Assessment of speech communication.
41. H. Haas; Über den Einflußeines Einfachechos auf die Hörsamkeit von Sprache, Acustica, Band 1, Seiten 49–58, 1951.
42. W. Ahnert, W., F. Steffen, Sound Reinforcement Engineering, Fundamentals and Practice, E&FN SPON, London, 2000.

6

Room-Acoustic Energy Decay Analysis

Ning Xiang, School of Architecture, Rensselaer Polytechnic Institute

ABSTRACT

Sound energy decay analysis plays a fundamental role in architectural acoustics practice and research. Schroeder's integration method for sound energy decay analysis broke new ground in classical architectural acoustics. Schroeder's integration method yields sound energy decay functions from room impulse responses. For reverberation time (T_{30}) estimation based on a parametric model, a nonlinear regression method has been proposed. The nonlinear regression method yields T_{30} estimates, insensitive to background noise and the upper limit of integration. Recent interest in acoustically coupled-volume systems has prompted new challenges in analyzing sound energy decay characteristics. This chapter will demonstrate a useful framework for this room-acoustics application using Bayesian inference and Schroeder's integration as the foundation of the advanced model-based energy decay analysis.

6.1 INTRODUCTION

Sound energy decay analysis in enclosures has been one of the main focuses of architectural acoustic measurement techniques since Sabine's time,[1] in which a single-volume system is primarily of interest. Sabine's experimental method for determining T_{30}s, and later, a slightly modified version using band-pass filtered random noise as acoustic excitations had been a primary experimental tool in the architectural acoustics community until Schroeder[2] developed a new method for determining the T_{30} in 1965.

Under certain circumstances, sound energy decays in enclosures may be of a multiple-slope nature. In characterizing the sound energy decays, we are often challenged by the task of identifying multiple decay slopes, as shown by Eyring's experimental work.[3] In this chapter we begin with a discussion of experimental methods for T_{30} determination, followed by the characterization of double-slope or multiple-slope energy decays. Both single and multiple-slope cases are particularly relevant for our discussion on sound fields in both single-volume and coupled-volume venues since coupled-volume enclosures under certain conditions may not present multiple-slope sound decays, while single-volume enclosures with nonuniformly distributed absorption and/or no diffusing scheme may present multiple-slope decays.

Room-acoustic sound energy decays are traditionally determined using random-noise excitations; once this signal has brought the sound field in the enclosure under investigation to its steady-state, the noise excitation is interrupted. Normally, sound pressures at receiver locations are recorded, and the squared pressures over time are termed steady-state *energy-time* functions. Schroeder considered a sound source-receiver arrangement in a room under investigation to be a linear time-invariant system. When the sound source is driven by the interrupted noise, the receiver receives a sound signal represented by convolution of the room impulse response (RIR) with the interrupted noise. The steady-state energy decay functions $D(t)$ can be derived by integration of the energy RIR $h^2(t)$:

$$D(t) = \int_t^{\infty} h^2(\tau)\, d\tau, \tag{6.1}$$

which, since then, has been termed Schroeder's integration. This method, from a single measurement of an RIR, produces an energy decay function which is theoretically equivalent to an ensemble average of a large number of steady-state energy decay functions when using interrupted noise. Following Schroeder, a number of investigations of this method[4–15] have substantiated its advantages. The Schroeder decay functions are not only free of unnecessary random fluctuations, but they contain significant single or multiple exponential processes as well.

6.2 INTEGRATED IMPULSE-RESPONSE METHOD

Schroeder's new approach presents a novel method of obtaining sound energy decays in terms of integration of energy RIRs—*Schroeder integration*. The sound energy decay function obtained by Schroeder integration is termed *Schroeder decay function*, or simply, *Schroeder curve* when presented graphically. This chapter also discusses a parametric model derived from Schroeder integration that has been shown to fully describe Schroeder decay functions. Based on the Schroeder decay model, this chapter outlines some existing approaches to T_{30} estimation—a nonlinear regression approach and a model-based Bayesian inference approach. The Bayesian inference approach represents an advanced room-acoustic energy decay analysis where both decay order (model) selection and decay parameter estimation problems are to be solved.

6.2.1 Schroeder Integration and Energy-Time Function (Curve)

A normalized steady-state sound energy decay function $d(t)$, via Schroeder integration[2, 16] in Eq. (6.1), can be expressed in the digital domain as:

$$d(t_k) = \frac{1}{E}\sum_{\tau=t_k}^{t_K} h^2(\tau) = 1 - \frac{1}{E}\sum_{\tau=0}^{t_k} h^2(\tau), \text{ with } E = \sum_{\tau=0}^{t_K} h^2(\tau),\ 0 \le k < K, \tag{6.2}$$

where variable t_k represents a discrete interval of time and variables k and K are integers with K being the total number of data points contained in the normalized decay function $d(t_k)$, so that t_K represents the total time record length of the *RIR* $h(t)$ as shown in Figure 6.1(a). The *RIR* $h(t)$ is defined between a sound source and a receiver in the room under investigation. In the following t_K is also termed *the upper limit of integration*. E is the total energy contained

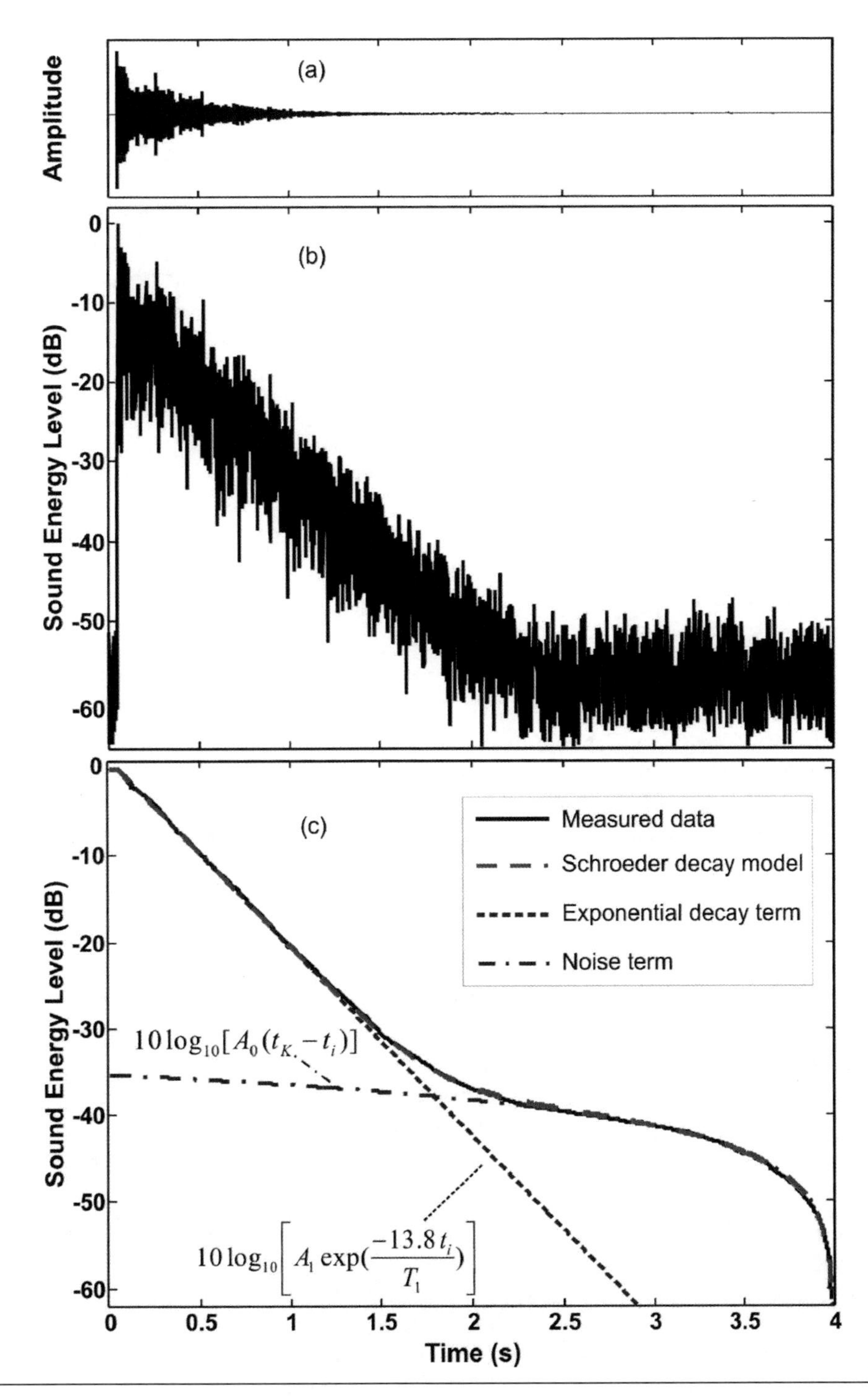

Figure 6.1 Experimentally measured RIR, band-pass filtered at 1 kHz (oct.), its corresponding energy-time curve and Schroeder decay curve (ETC). (a) RIR; (b) Normalized ETC; (c) Normalized Schroeder decay function in comparison with its decay model and two decomposed terms.

in $h(t)$. Schroeder integration essentially establishes the relationship between the *energy RIR* $h^2(t)$ and the *steady-state energy decay function* $d(t)$.[17] The right-hand side of Eq. (6.2) is of both theoretical and practical relevance, as it represents the complementary relationship between the energy decay and build-up;[16] at the same time, it is beneficial for numerical implementation due to a straightforward accumulative summation.

In architectural acoustics practice, logarithmic energy RIRs $10 \log_{10}[h^2(t)]$ are often represented graphically over time—also termed an *energy-time curve* (ETC), as illustrated in Figure 6.1(b). Figure 6.1(c) illustrates the corresponding normalized Schroeder curve.

6.2.2 Schroeder Decay Model

A parametric model (*Schroeder decay model*) with S exponential decay terms:[14–15]

$$\mathbf{H}_S(\mathbf{A}_S, \mathbf{T}_S) = A_0(t_K - t_i) + \sum_{j=1}^{S} A_j \left[\exp\left(\frac{-13.8 \cdot t_i}{T_j} \right) - \exp\left(\frac{-13.8 \cdot t_K}{T_j} \right) \right] \tag{6.3}$$

can fully describe Schroeder decay function $d(t)$. Vector $\mathbf{A}_S = [A_0, A_1, \cdots A_S]$ contains $S + 1$ linear coefficients, while vector $\mathbf{T}_S = [T_1, T_2, \cdots T_S]$ contains S decay time parameters (decay order), denoting that the decay model contains S different exponential decay terms with S different decay times. The models in the present application are nested models—a higher-order model embodies lower-order ones. Particularly, if the second-order model describes the data reasonably well, the first-order or the third-order models often also describe the data well. When $S = 1, 2, 3$, it is said that the sound energy decays are of single-, double-, or triple-slope nature, respectively. For single-slope decays ($S = 1$), the corresponding decay time is termed reverberation time, T_{30}, while in multiple-slope decays ($S \geq 2$), multiple different *decay times* are expected. When t_K is selected large enough, the constant term $\exp(-13.8 \cdot t_K/T_j)$ in Eq. (6.3) is negligible. Figure 6.1(c) also illustrates the decay model for $S = 1$, along with two decomposed model terms.

Schroeder integration yields energy decay functions from experimentally measured RIRs, which sensitively depend on the upper limit of integration[4] and on inevitable background noise.[7] The Schroeder decay model in Eq. (6.3) is helpful to understand the characteristic behavior of the Schroeder decay functions. A full understanding of this characteristic behavior, in turn, facilitates the discussion of the advanced analysis tools later in this chapter.

Figure 6.2 illustrates the corresponding model curves; the parameters of which are derived from the Schroeder decay functions for different upper limits of integration. One experimentally measured RIR is Schroeder integrated over three different upper limits of integration. This shows that the decay model in Eq. (6.3) can cope with different upper limits of integration because the model in Eq. (6.3) includes this limit t_K as one model parameter.

In addition to the effect of the upper limit of integration, the model can also account for the different levels of background noise inherently in RIRs. Figure 6.3 illustrates the Schroeder decay functions derived from RIRs experimentally measured in an enclosure for the same source-receiver arrangement. The measured RIRs are filtered at the 1 kHz octave band. By adjusting the amplifications and number of averages, different peak-to-noise ratios (PNRs) of the measured RIRs result (50 dB, 54 dB, and 58 dB). Figure 6.3 also illustrates the corresponding decay model curves [dash-line curves using Eq. (6.3) with appropriated decay parameters] for comparison. Figure 6.3 demonstrates the characteristic curvature inherent in the Schroeder decay functions that may impede an accurate evaluation of T_{30} (or decay times) when using a linear least-squares fit, even when the PNR of an RIR would be high enough—particularly when the upper limit of integration is selected larger than the T_{30} to be determined.

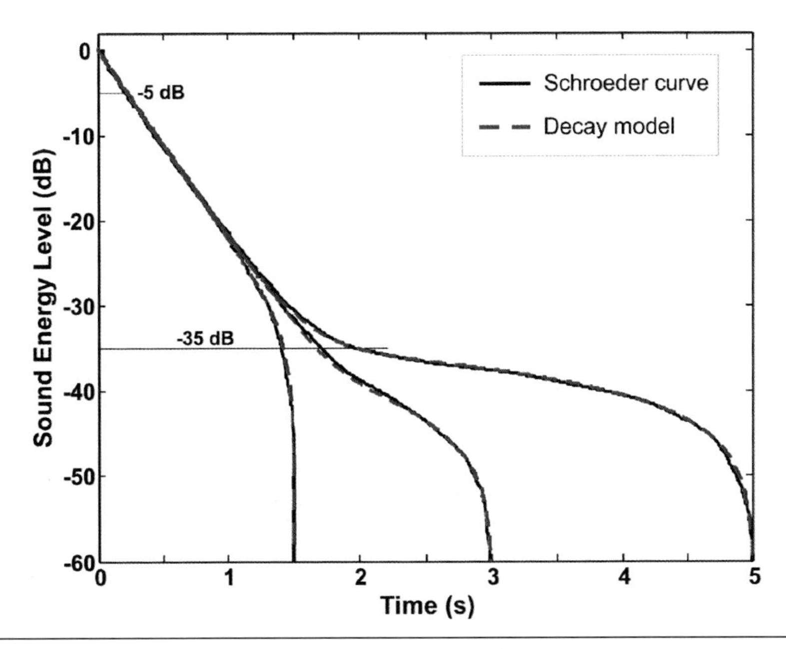

Figure 6.2 Effect of the upper limit of the integration on Schroeder decay curves. Schroeder decay functions evaluated from the same measured RIR, octave band-pass filtered at 1 kHz, but for three different upper limits of integration, 1.5 (s), 3 (s), and 5 (s). The signal-to-noise ratio is about 56 decibel (dB). Each model function is also plotted for ease of comparison.

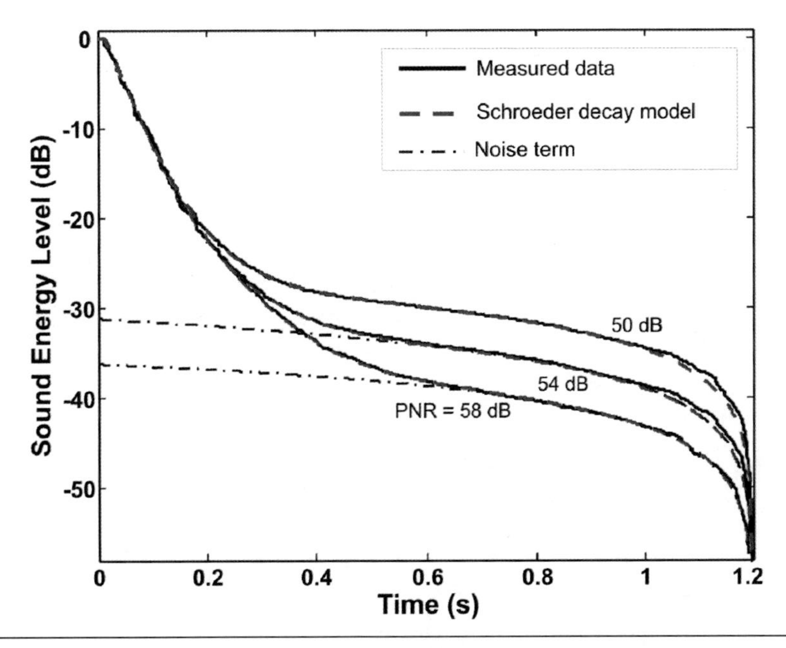

Figure 6.3 Effect of the background noise on Schroeder decay curves. Schroeder decay functions evaluated from the experimentally measured RIR at the same receiver location, octave band-pass filtered at 1 kHz, but for three different PNRs (PNR = 50, 54, and 58 dB). Each model function is also plotted for ease of comparison. The coefficient A_0 in the noise term will obtain different values, while the other parameters—particularly the decay times—are highly similar for these three curves.

Note that the three different curves in both Figures 6.2 and 6.3 are all reflected by the model term $A_0(t_K - t_k)$ in Eq. (6.3). In other words, significant changes of these three decay curves in model parameters is solely registered by either parameter A_0 or t_k. The most relevant decay parameters, particularly the decay times T_j associated in the second term of Eq. (6.3) are insignificantly different when analyzing them with a capable approach elaborated later. The advanced method discussed in this chapter is a model-based approach, as long as we focus on the relevant decay parameters $\{A_i, T_i\}$ while ignoring the irrelevant (so-called nuisance) term $A_0(t_K - t_k)$, which represents no technical issues any more.

6.2.3 *Characteristics of Schroeder Decay Functions*

From Figures 6.1(c), 6.2, and 6.3, the following characteristic features become evident:

1. The Schroeder decay curve will, generally (or very often), not exhibit a linearly decaying curve when graphically presented in a logarithmic scale, even for the single-slope case. It can be characterized by three major portions:
   a. *Beginning portion*: the decay curve decays within this portion with a slope or with slopes predominantly determined by room-acoustic properties, such as T_{30} (decay times).
   b. *Late portion*: the decay curve ends up with a characteristic curvature toward the end of the time record of the RIR (the upper limit of the integration).
   c. *Middle portion*: a curve often appears to decay with a somewhat slower slope in comparison to that in the beginning portion. Only when the upper limit of the integration is long enough, this portion remains observable [see the curves between 1.5 and 2.5 (s) as shown in Figure 6.1(c)], otherwise it can disappear so that the beginning portion transitions directly to the late portion [see an example in Figure 6.2].
2. The linearly decreasing term $A_0(t_K - t_k)$ [see Eq. (6.3)] is predominantly responsible for the characteristic curvature of the late portion in the logarithmic scale, and it accounts for both the effect of the upper limit of the integration through parameter t_K[14] and the effect of background noise through parameter A_0; therefore, $A_0(t_K - t_k)$ is also termed the *noise term* in this chapter. It is actually a nuisance term, irrelevant to the room-acoustics problem at hand.
3. Schroeder decay functions become undefined at the upper limit of integration t_K [see Figure 6.1(c)] when examined logarithmically, since for $t_k \to t_K$, $d(t_k) \to 0$ and the parametric model in Eq. (6.3) accounts for this fact as well.

6.3 TRUNCATION APPROACH

Promptly following Schroeder, Kürer and Kurze[4] reported that Schroeder's integrated impulse response method yields decay functions that are highly dependent on the upper limit of integration as already shown in Figure 6.2.

For evaluation of T_{30}, e.g., the linear least-squares fit would yield reasonable results only from the curves with an upper limit of the integration somewhere between 2 (s) and 3 (s) in the figure. Too small or too large upper limits of integration will lead to unaccepted, erroneous T_{30} estimates when applying the least-squares fit. Due to this sensitivity of the upper limit of integration, Kürer and Kurze suggested to truncate the RIRs by a proper selection of the upper limit of the integration so that the resulting Schroeder decay functions would take the shape of a somewhat linearly decaying function within the required level range. However, their suggested upper limit of integration was also dependent on the T_{30} to be determined. Some more recent investigations[9, 13] reexamined this problem. Vorländer and Bietz[9] suggested, as a rule-of-thumb, to estimate the upper limit of the integration with the help of the energy-time curve [see Figure 6.1(b)].

A proper upper limit of the integration would be taken at a level of −45 dB along the energy-time curve for T_{30}, at −35 dB for T_{20}, respectively. In other words, the rule-of-thumb suggests adding a safety margin of 10 dB to the lower limit of the level range for specific T_{30}s, and estimating the upper limit of the integration first along the energy-time curve. Lundeby et al.,[13] then proposed an iterative scheme to find an optimal upper limit of integration. The reason for this is that the PNR of the RIR also plays a crucial role in the shape of the Schroeder decay functions; this, in turn, is well modeled by the noise term $A_0(t_K - t_k)$. The following section elaborates these phenomena.

6.4 NOISE SUBTRACTION

Different levels of background noise, inevitably, in the experimentally measured RIRs cause a different course of Schroeder curves, as illustrated in Figure 6.3. Previous research work has also suggested ways of mitigating the noise effect when applying the least-squares fit for T_{30} estimations.

6.4.1 Pre-Subtraction

Without revealing the characteristic curvatures inherently in Schroeder curves derived from experimentally measured RIRs (as summarized in Section 6.2.3), some previous reports erroneously considered Schroeder decay functions to provide *insufficient* dynamic ranges even for enough PNR if the upper limit of integration is not carefully selected. Direct application of least-squares fitting may therefore lead to erroneous T_{30} estimates. Chu (1978)[7] proposed subtracting the mean-squared value of noise from RIRs to mitigate *insufficient* dynamic ranges in the Schroeder decay functions. The noise-subtraction proposed by Chu relies on being able to estimate the mean-square value $\overline{n^2}$ of the noise in the RIR. Chu conceptually considered a noisy RIR consisting of an ideal RIR $h_i(t)$ and an additive noise $n(t)$ with a mean-square value $\overline{n^2}$. The subtraction of the estimated $\overline{n^2}$ from the squared RIR prior to performing the Schroeder integration leads to:

$$d(t_k) = \frac{1}{E}\sum_{\tau=t_k}^{t_K}\left\{\left[\, n(\tau)+h_i(\tau)\right]^2 - \overline{n^2}\right\} \approx \frac{1}{E}\sum_{\tau=t_k}^{t_K}\left\{\left[n^2(\tau)-\overline{n^2}\right] + h_i^2(\tau)\right\}, \tag{6.4}$$

where the uncorrelated cross term $2\sum n(\tau)h_i(\tau)$ has been ignored. Chu demonstrated that the apparent dynamic range of the Schroeder decay functions is indeed increased since the residual noise $n^2(\tau)-\overline{n^2}$ becomes much smaller given that $\overline{n^2}$ is properly estimated. Since the noise is subtracted prior to the Schroeder integration, it is therefore termed *pre-subtraction*. One pre-subtraction result is illustrated in Figure 6.4.

The pre-subtraction critically relies on an accurate estimation of the mean-square value of the background noise in the RIR. One simple way to estimate the mean-square noise is to

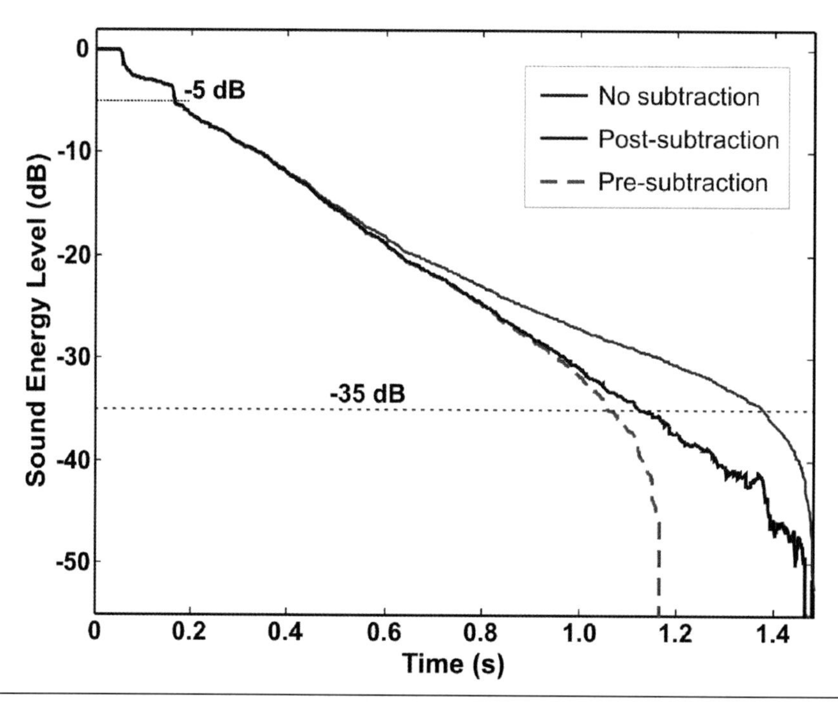

Figure 6.4 Schroeder decay function obtained by post-subtraction, without subtraction, and with pre-subtraction. The Schroeder decay functions are derived from an RIR with a PNR of 48dB, measured in the Berliner Philharmonie, 1 kHz octave band-pass filtered.

take a small segment of length at the end (or in front of the direct sound) in the RIR. Difficulties may occur in estimating the mean-square value optimally, sometimes due to fluctuations in the noise tail of the RIR or the different segment sizes may yield slightly different estimates. This slight variation will more sensitively influence the subtracted results if the PNR of RIRs is low.

6.4.2 Post-Subtraction (Noise Compensation)

Different from the noise subtraction prior to the Schroeder integration, the noise effect leading to the characteristic curvature in Schroeder decay functions can also be compensated after the Schroeder integration.[11] The Schroeder decay model $\mathbf{H}_S(\mathbf{A}_S, \mathbf{T}_S)$ in Eq. (6.3) approximately describes the Schroeder decay functions $d(t_i)$ reasonably well, given that the model parameters $\{A_j, T_j\}$ would be optimally estimated. Subtracting the noise term $A_0(t_K - t_k)$ from the Schroeder decay function $d(t_i)$ in Eq. (6.3) yields:

$$d_a(t_k) \approx d(t_k) - A_0(t_K - t_k) \approx \text{single- or multiple-slope exponential decay.} \tag{6.5}$$

Figure 6.4 also illustrates the post-subtraction, given that A_0 is accurately estimated and the decay function itself contains a single-slope decay process. Figure 6.4 implies that the post-subtraction may outperform the pre-subtraction. Given that the mean-square value of the noise can be optimally estimated, the resulting logarithmic decay function would exhibit a more linear decaying process. Figure 6.5 illustrates post-subtraction results with slightly

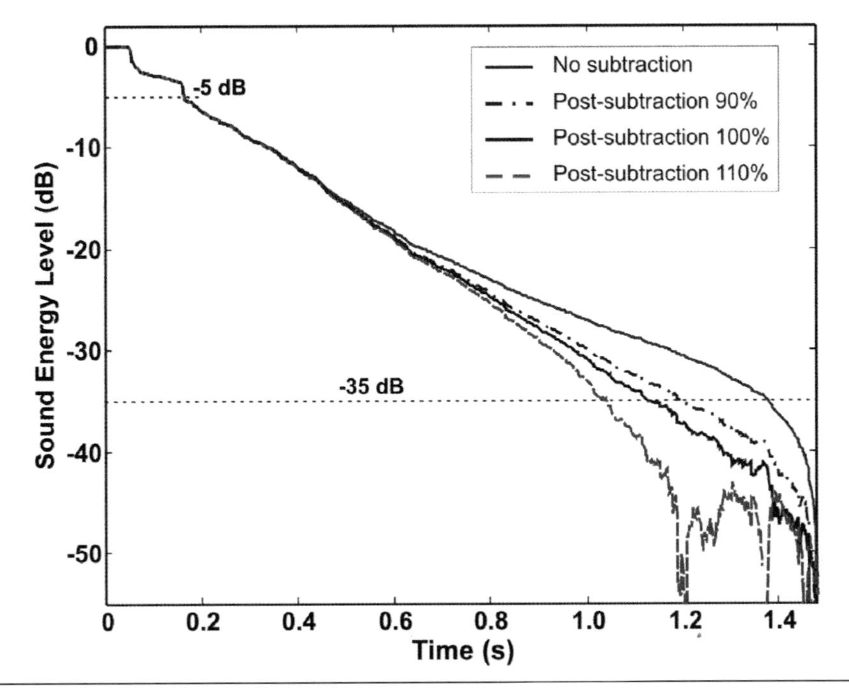

Figure 6.5 Schroeder decay function obtained by the post-subtraction method for different mean-square noise values. Two are ±10% from the fine-tuned nominal value. The Schroeder decay functions are derived from an RIR with a signal-to-noise ratio of 48 dB, measured in the Berliner Philharmonie, 1 kHz octave band-pass filtered.

different noise term estimates. Note that the sensitivity of the noise, however, is highly dependent on the noise level initially contained in the RIR and the upper limit of integration.

6.4.3 Least-Squares Fitting for Noise Estimation

The most sensitive issue with these two subtraction schemes is how the mean-square value of the noise can be *optimally* estimated. Very often, the *optimal* value requires fine-tuning. Due to the fact that the block size cannot be easily selected for an overall optimal estimation, or due to fluctuations in the noise tails, there seems to be no general rules for the optimal estimation. Often, a substantial, individual fine-tuning needs to be undertaken. To mitigate the fluctuations, one may reexamine the parametric model in Eq. (6.3). When the sound energy decays so much until the noise tail becomes dominate, given some length of the noise tail is included in the RIR under investigation, the Schroeder decay function, in its later portion, is approximated by:

$$d(t_k \rightarrow t_K) \approx A_0 t_K - A_0 t_k. \tag{6.6}$$

The right-hand side in Eq. (6.6) represents a linearly decreasing line, with $A_0 t_K$ being a constant offset. $-A_0$ is the slope of the linearly decreasing line. The linear least-square fit can also yield an estimate of A_0. Figure 6.6 illustrates one segment of Schroeder decay function $d(t_k)$ at the last 256 points right before the upper limit of integration, evaluated from an experimentally measured RIR. Note that the curves are presented in their linear scale (without a logarithmic operation).

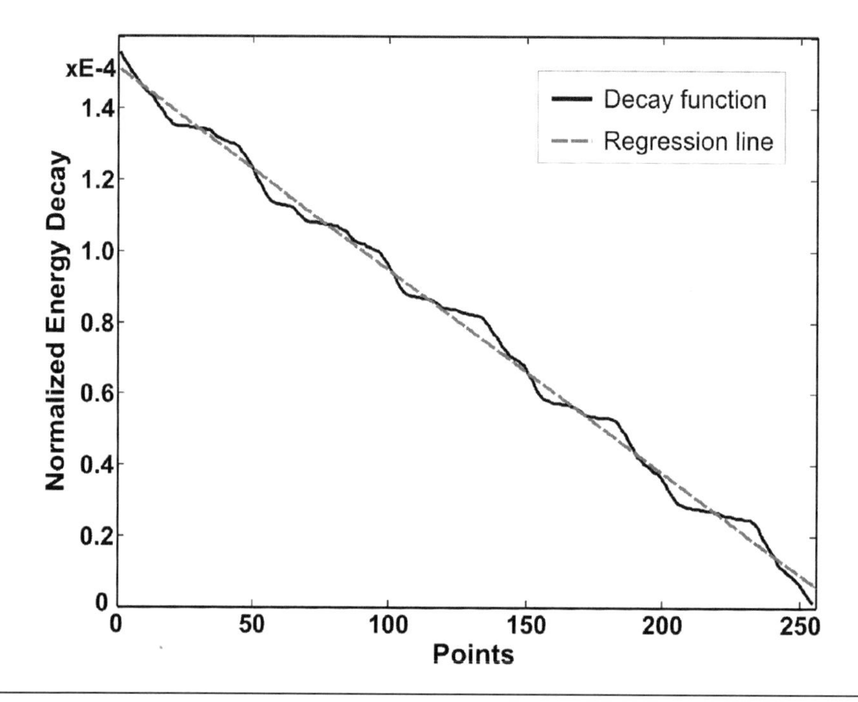

Figure 6.6 Schroeder decay function tail compared with a least-squares linear fit in the linear vs. linear presentation.

The characteristic features of Schroeder decay curves summarized earlier, along with the model in Eq. (6.3), help us understand the fact that the linear least-squares-fit approach to directly fit logarithmic Schroeder decay functions (curves) relies on an inappropriate model. A large number of investigations have been reported in major acoustics journals for decades after Schroeder published this method. Most of the mitigation methods previously outlined, however, employ linear (least-square) fitting, implicitly relying on a linear decay model. The success of T_{30} estimation then depends critically on accurate estimation of the upper limit of integration t_K, accurate estimation of A_0 associated with the background noise[14] or accurate elimination of the background noise. In order to improve T_{30} estimates, iterative processes have also been proposed to accurately estimate the upper limit of the integration.[13] Separate multiple measurements at a single receiver location,[10] or a replacement of the convex curvature at the end of the Schroeder decay curve by a somewhat straight (single-logarithmic) decay have also been proposed to reduce the effect of the background noise. A close look at the model in Eq. (6.3) reveals that these previous methods depend critically on accurate estimations or removal of one or the other of the nuisance parameters (A_0, t_K). In other words, these methods primarily focus on one or two nuisance parameters first, then expecting a single-slope decay, try to fit a linear model to the processed decay functions.

6.5 NONLINEAR REGRESSION

Focusing the effort on the most relevant decay parameters (T_1), a nonlinear regression method[14] was introduced in 1995, based on the nonlinear model in Eq. (6.3). For T_{30} estimates (single-slope case), the nonlinear regression removes the stringent requirement on accurate estimation of nuisance parameters, both the upper limit of integration t_K or the parameter A_0 associated with the background noise. The nonlinear regression elaborated in this section is also a model-based approach; it starts with a rough estimate of initial values of model parameters A_1, T_1, and A_0. It then iteratively improves the model parameters in terms of infinitesimal corrections of each individual parameter by solving a linear system of equations. For a clear description, we adopt $\boldsymbol{\theta}=\{\theta_0,\theta_1,\theta_2,\theta_3,\theta_4,\cdots\}$ to express the model parameters $\{\mathbf{A}, \mathbf{T}\}$ collectively. The Schroeder decay model $H(\boldsymbol{\theta},t_i)=H(\mathbf{A},\mathbf{T},t_i)$ expressed in Eq. (6.3) is then rewritten as:

$$\mathbf{H}_S(\boldsymbol{\theta},t_k)=\theta_0(t_K-t_k)+\sum_{j=1}^{S}\theta_{(2j-1)}\left[\exp\left(-\theta_{(2j)}\cdot t_k\right)-\exp\left(-\theta_{(2j)}\cdot t_K\right)\right], \tag{6.7}$$

with $\theta_{(2j)}=13.8/T_j$, T_j is the decay time of j-th decay slope. This parametric model approximates the Schroeder decay functions $d(t_k)$ in a way of finding $\boldsymbol{\theta}$ such that an error function $e(\boldsymbol{\theta},t_k)=H(\boldsymbol{\theta},t_k)-d(t_k)$ is minimized according to the generalized least-squares principle.[14] The nonlinear regression method is an iterative process. At the n-th iteration a linear equation system:

$$\mathbf{B}[\Delta\boldsymbol{\theta}^{(n)}]=\mathbf{a} \tag{6.8}$$

is solved for infinitesimal corrections $\Delta\boldsymbol{\theta}^{(n)}$ to improve $\boldsymbol{\theta}^{(n)}$ in the parameter space so as to reach the minimum of the least-square error one step further. The superscript n for $\boldsymbol{\theta}^{(n)}$ and $\Delta\boldsymbol{\theta}^{(n)}$ stands for the n-th iteration. $\mathbf{a}=\left[a_0,a_1,\cdots,a_{(2S)}\right]^{Tr}$ is a column vector of $2S + 1$ elements, with S being the number of decay slopes, each of its elements:

$$a_i=\sum_{k=1}^{K}\left[d(t_k)-H(\boldsymbol{\theta}^{(n)},t_k)\right]^{Tr}\frac{\partial H(\boldsymbol{\theta}^{(n)},t_k)}{\partial\theta_i}, \text{ with } i=0,1,\cdots,2S \tag{6.9}$$

is derived from both the decay function data $d(t_k)$ and the parametric Schroeder decay model $H(\boldsymbol{\theta},t_i)$ in Eq. (6.7), Tr stands for matrix transpose. The matrix $\mathbf{B}$ is a square matrix of $2S + 1$ in dimension, each of its elements:

$$B_{ij}=\sum_{k=1}^{K}\left[\frac{\partial H(\boldsymbol{\theta}^{(n)},t_k)}{\partial\theta_i}\frac{\partial H(\boldsymbol{\theta}^{(n)},t_k)}{\partial\theta_j}\right], \text{ with } i,j=0,1,\cdots,2S \tag{6.10}$$

is derived exclusively from the parametric Schroeder decay model. According to Eq. (6.7), the partial derivatives are determined by:

$$\frac{\partial H(\boldsymbol{\theta}^{(n)},t_k)}{\partial\theta_0}=t_K-t_k, \tag{6.11}$$

$$\frac{\partial H(\boldsymbol{\theta}^{(n)},t_k)}{\partial\theta_{(2j-1)}}=\exp\left[-\theta_{(2j)}^{(n)}\cdot t_k\right]-\exp\left[-\theta_{(2j)}^{(n)}\cdot t_K\right], \tag{6.12}$$

$$\frac{\partial H(\boldsymbol{\theta}^{(n)},t_k)}{\partial\theta_{(2j)}}=-\theta_{(2j-1)}^{(n)}\cdot t_k\cdot\exp\left[-\theta_{(2j)}^{(n)}\cdot t_k\right], \tag{6.13}$$

with $j = 1, 2, \cdots, S$, the integration index $0 \le k < K$ is used to specify the data segment used for the evaluation, e.g., Schroeder decay function segment between −5 dB and −45 dB. For solving the linear system of equations in Eq. (6.8), a robust method using the matrix pseudo-inverse based on the singular value decomposition[18] is highly recommended. The iterative regression continues with $\boldsymbol{\theta}^{(n+1)} = \boldsymbol{\theta}^{(n)} + \Delta\boldsymbol{\theta}^{(n)}$ until $\left|\Delta\boldsymbol{\theta}^{(n+1)}\right|$ becomes smaller than a predefined threshold.

At the beginning with $n = 0$ before the iteration starts, $\boldsymbol{\theta}^{(0)}$ is termed as *initial state*. The nonlinear regression should start with a reasonable initial state. For the T_{30} evaluation which is a single-slope case ($S = 1$), Xiang (1995)[14] suggested using the linear least-squares fit of a small segment of the logarithmic decay function $10 \log_{10}[d(t_k)]$ between −5 dB and −10 dB for estimates of $\theta_1^{(0)}$ and $\theta_2^{(0)}$ or simply take two points corresponding to −5 dB and −10 dB while using the linear least-squares fit of the Schroeder decay function tail for $\theta_0^{(0)}$, as described in Section 6.4.3. A breaking mechanism of the iteration can be designed for the T_{30}-related parameter $\theta_2 = 13.8 / T_1$, such as a threshold $\delta\theta_2 \le 10^{-3}\theta_2^{(0)}$, once $\Delta\theta_2^{(n+1)}$ becomes smaller than the threshold, namely $\left|\Delta\theta_2^{(n+1)}\right| \le \delta\theta_2$, the iteration can be stopped. The threshold should actually be determined according to the required precision.

For the single-slope cases, the nonlinear regression easily converges to the global minimum by roughly estimating the initial values of $\theta_1^{(0)}$ and $\theta_2^{(0)}$. Accurate initial estimates of $\theta_0^{(0)}$ and t_K are not critical, and t_K can also be set with ease to be somewhat smaller or larger than T_1, such as one time, or two times of the initial state value of T_1 which has already been estimated,[14] or simply set sufficiently large without any burden of accurate determination.

The nonlinear regression works efficiently for the T_{30} estimation (single-slope case) without the stringent requirement on parameters t_K and A_0 because it involves t_K and A_0 as adjustable model parameters in the regression process. The iterative process of the nonlinear regression will lead to convergence to the global minimum after only a few iterations even with rough estimates of initial state values.[14] In comparison to the linear-squares fit, the iterative nonlinear regression requires some added computational expenses; but these are, on current PC-platforms, barely noticeable, on the order of a few milliseconds.

Following successful application of the nonlinear regression to the T_{30} estimation, it was soon discovered[19] that if the model order is higher than one (double- or multiple-slope cases for $S > 1$), initial value estimation becomes too critical to allow convergence. For this reason, the model-based Bayesian method, elaborated in the following section, has emerged.[20, 21] Before describing Bayesian inference, a close look at experimentally measured data, as illustrated in Figure 6.3, motivates applications of advanced energy decay analysis methods. Figure 6.3 illustrates Schroeder decay curves in a coupled-volume system at one specific receiver location, but with three different resulting PNRs (PNR = 50 dB, 54 dB, and 58 dB, approximately). The decay characteristics represented by these three curves are actually the same. According to Eq. (6.3), a model-based decay parameter estimation with $S = 2$ yields decay parameter values. Using these parameter values, the decay model curves [see Eq. (6.3), plotted by dash lines in Figure 6.3] agree well with the experimental data. In this example, the decay model curves consist of two exponential decay terms and one noise term $[A_0(t_K - t_k)]$. The only parameter differing significantly between the models is $10 \log_{10}(A_0) = -58.5$ dB, −62.2 dB, and −67.2 dB, for the three cases, respectively. Figure 6.3 also illustrates the noise term for the two highest PNRs. In addition, Figure 6.2 also illustrates single-slope decay curves for three different upper limits of integration with a fixed PNR of 56 dB. The Schroeder decay model of Eq. (6.3) accurately describes the course of the experimental decay curves of these examples which vary between single-slope and double-slopes.

Some previous authors reported that Schroeder decay functions present difficulties when fitting the straight lines.[4, 7, 9, 10, 13] One may argue that corrections/compensations can be applied to make Schroeder decay curves look like linear decays by carefully selecting the upper limit of integration or by subtraction/elimination of background noise, even resorting to iterative processes to optimize the corrections. The breaking mechanisms of the iterative process are conventionally set according to linearity of the resulting curves. From the viewpoint of the model-based analysis, these previous approaches have applied an inappropriate (linear) model for analyzing Schroeder decay data. Moreover, if the data contain decay processes more complicated than a single-decay process (as shown in Figure 6.3), these correcting measures, implicitly assuming a straight-line decay function, will significantly destroy inherent decay characteristics in the data. More profound challenges are due to the fact that one cannot know prior to data analysis how many decay processes the data at hand contain, since in single-space rooms, multiple-slope decays may occur, whereas in coupled-volume systems, single-slope decays can often be found. The following section elaborates on an advanced method based on Bayesian inference, which will first determine how many slopes are in the data before using an appropriate model to analyze the data.

6.6 TWO LEVELS OF BAYESIAN DECAY ANALYSIS

Bayesian probabilistic inference applied to solving parameter estimation problems is referred to as the *first level of inference*, while solving model selection problems is referred to as the *second level of inference.*[21–23] Bayesian inference encompasses both the parameter estimation and the model selection problems by extensive use of Bayes' theorem. The following section concisely discusses the application of the two levels of Bayesian decay analysis. The following discussion begins with the second level of inference, decay model selection. This is also the scientifically logical approach; one should determine which of the competing models is appropriate before the relevant decay parameters are inferred.

6.6.1 Model Selection: The Second Level of Inference

Among a set of competing models, increased decay orders will always improve curve fitting but models with increased orders often generalize poorly.[24] To avoid over-parameterization, Bayesian model selection applies Bayes' theorem to one of the competing models $\mathbf{H}_s$ given the background information I and the data $\mathbf{D}$, containing a Schroeder decay function of K elements, as:

$$p(\mathbf{H}_s\,|\,\mathbf{D}) = \frac{p(\mathbf{H}_s)\,p(\mathbf{D}\,|\,\mathbf{H}_s)}{p(\mathbf{D})}, \tag{6.14}$$

while pushing any interest in model parameter values into the background of the current problem. In Eq. (6.14), $p(\mathbf{H}_s)$ is the *prior probability* of the model $\mathbf{H}_s$. Quantity $p(\mathbf{D} \mid \mathbf{H}_s)$ is termed *marginal likelihood*, also termed *evidence*. Quantity $p(\mathbf{D})$ is a constant, and $p(\mathbf{H}_s \mid \mathbf{D})$ is the *posterior probability* of the model $\mathbf{H}_s$, given the data $\mathbf{D}$.

Model comparison between two different models $\{\mathbf{H}_j, \mathbf{H}_k\}$ relies on the posterior ratio:

$$\frac{p(\mathbf{H}_j\,|\,\mathbf{D})}{p(\mathbf{H}_k\,|\,\mathbf{D})} = \frac{p(\mathbf{D}\,|\,\mathbf{H}_j)}{p(\mathbf{D}\,|\,\mathbf{H}_k)}, \tag{6.15}$$

when assigning equal prior probability to each model. This indicates that the marginal likelihood $p(\mathbf{D} \mid \mathbf{H}_s)$ plays a central role in Bayesian model comparison/selection.

6.6.2 Parameter Estimation: The First Level of Inference

The first level of Bayesian inference applies Bayes' theorem:

$$p(\boldsymbol{\theta}_s\,|\,\mathbf{D},\mathbf{H}_s)=\frac{p(\boldsymbol{\theta}_s\,|\,\mathbf{H}_s)p(\mathbf{D}\,|\,\boldsymbol{\theta}_s,\mathbf{H}_s)}{p(\mathbf{D}\,|\,\mathbf{H}_s)} \tag{6.16}$$

to the parameters $\boldsymbol{\theta}_s$ containing $2 \cdot s + 1$ parameters and the data $\mathbf{D}$ once a specific model $\mathbf{H}_s(\boldsymbol{\theta}_s)$ is determined by the model selection, where $p(\boldsymbol{\theta}_s \mid \mathbf{H}_s)$ is the prior probability of parameters $\boldsymbol{\theta}_s$, $p(\mathbf{D} \mid \mathbf{H}_s)$ is exactly the marginal likelihood in Eq. (6.14), and $p(\mathbf{D} \mid \boldsymbol{\theta}_s, \mathbf{H}_s)$ is the likelihood function of the parameters, which is the probability of the residual error $\mathbf{e} = \mathbf{D} - \mathbf{H}_s$. Application of the principle of maximum entropy leads to an assignment of the likelihood function:[23]

$$L(\boldsymbol{\theta}_s)=p(\mathbf{D}\,|\,\boldsymbol{\theta}_s,\mathbf{H}_s)\approx\frac{(2\pi\,\varepsilon)^{-K/2}}{2} \quad \text{with } \varepsilon=\frac{\mathbf{e}^{Tr}\mathbf{e}}{2}, \tag{6.17}$$

where $(\cdot)^{Tr}$ denotes matrix transpose.

The probability $p(\boldsymbol{\theta}_s \mid \mathbf{D}, \mathbf{H}_s)$ on the left-hand side of Eq. (6.16) is the *posterior probability* of the parameters $\boldsymbol{\theta}_s$, representing the updated knowledge about the parameters once the data become available. The integration over the entire parameter space $\int_{\boldsymbol{\theta}_s}$ on both sides of Eq. (6.16), combined with the simplified notations $\pi(\boldsymbol{\theta}_s)=p(\boldsymbol{\theta}_s\,|\,\mathbf{H}_s)$ and $L(\boldsymbol{\theta}_s)=p(\mathbf{D}\,|\,\boldsymbol{\theta}_s,\mathbf{H}_s)$ in Eq. (6.16), yields:

$$p(\mathbf{D}\,|\,\mathbf{H}_s)=Z_s=\int_{\boldsymbol{\theta}_s} L(\boldsymbol{\theta}_s)\pi(\boldsymbol{\theta}_s)\,d\boldsymbol{\theta}_s=\int_{\mu} L(\boldsymbol{\theta}_s)\,d\mu, \tag{6.18}$$

with $\mu=\int_{\boldsymbol{\theta}}\pi(\boldsymbol{\theta}_s)d\boldsymbol{\theta}_s$ and $d\mu=\pi(\boldsymbol{\theta}_s)d\boldsymbol{\theta}_s$. $Z_s=p(\mathbf{D}\,|\,\mathbf{H}_s)$ determined by Eq. (6.18) is exactly the same as the marginal likelihood in Eq. (6.14), often referred to as (Bayesian) *evidence* for model $\mathbf{H}_s$.[24-26] Bayesian evidence encapsulates the principle of parsimony, and quantitatively implements Occam's razor: *when two competing theories explain the data equally, the simpler one is preferred.* Using an asymptotic approximation, the next section elaborates on the quantitative implementation of Occam's razor.

6.6.3 Bayesian Information Criterion

The Bayesian information criterion (BIC) asymptotically approximates Bayesian evidence if a (multidimensional) Gaussian distribution can approximate the posterior probability distribution within a vicinity around the global extreme of the likelihood. Experimental results have shown that this is often the case for energy decay analysis.[17] In this application, the BIC for ranking a set of decay models $\mathbf{H}_1, \mathbf{H}_2, \mathbf{H}_3, \cdots$ is given by:

$$\text{BIC}\approx 2\cdot\ln[\hat{L}(\hat{\boldsymbol{\theta}}_s)]-N_D\ln(K) \;\; [\text{nepers}], \tag{6.19}$$

with $N_D = 2 \cdot s + 1$ being the dimensionality, or the number of parameters involved in model $\mathbf{H}_s$. The quantity $\hat{L}$ is the peak value of the likelihood which location in the parameter space is denoted by $\hat{\boldsymbol{\theta}}_s$. The first term in Eq. (6.19) represents the degree of the model fit to the data, while the second term represents the penalty of over-parameterization. When increasing the decay model order, the first term may also increase, indicating a better fit to the data, but the BIC will dramatically decline because of the second term, which penalizes over-parameterization. It will outweigh the first term.[17] In the scope of energy decay analysis among a set of decay models $\mathbf{H}_1$, $\mathbf{H}_2$, $\mathbf{H}_3$, $\cdots$, the model yielding a significantly larger BIC value in Eq. (6.19) is the most concise model providing the best fit to the decay function data, while at the same time capturing the important exponentially decaying features evident in the data. Following the convention,[24] this chapter expresses the BIC in decibans (db) using $10 \cdot (BIC_{\text{neper}})\log_{10} e$.

Taking the experimentally measured data (PNR = 54 dB, as shown in Figure 6.3), Figure 6.7 illustrates the Bayesian analysis results and shows the decomposed decay terms in the

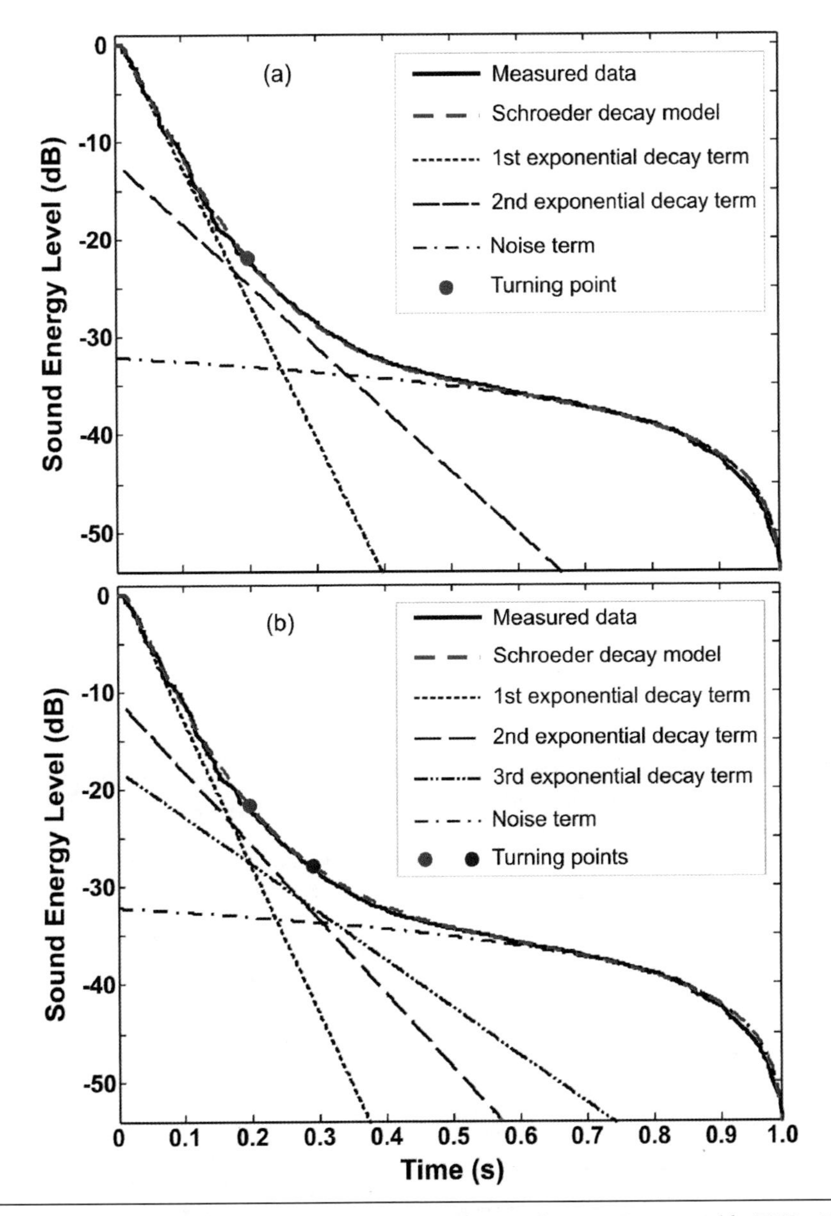

Figure 6.7 Experimentally measured Schroeder decay curves as shown in Figure 6.3 (the PNR = 54 dB). (a) Comparison with Bayesian model curve and decomposition using a double-slope model. (b) Comparison with Bayesian model curve and decomposition using a triple-slope model.

second- and third-order Schroeder decay models. The first-order decay model represents a strong misrepresentation of the data and is not shown here. Figure 6.7(b) illustrates the typical example of an over-parameterized model, which may or may not improve the curve fitting significantly in comparison with that of the second-order model as shown in Figure 6.7(a), but Occam's razor, quantitatively implemented in Bayesian analysis, penalizes the over-parameterized model, since it receives a significantly lower BIC value. Its BIC value is 342 decibans (db) lower than that of the second-order decay model (see Table 6.1, last column).

6.6.4 Advanced Sampling Methods

While accumulating the evidence Z_s in Eq. (6.18), the likelihood function over the prior mass μ over the entire parameter space will be explored. The prior mass is the amount (mass) of the prior density $\pi(\boldsymbol{\theta}_s)$ contained in the parameter space. There are existing advanced Markov chain Monte Carlo sampling methods[25–27] to explore the parameter space. Equation (6.18) implies the posterior values [in Eq. (6.16)] can be estimated through Eqs. (6.17) and (6.18), such that the posterior values are readily determined once the evidence is estimated over the entire parameter space. The maximum *a posteriori* or probabilistic moment calculations (means and covariance) will accomplish the decay parameter estimation task; at the same time, moment calculations also provide estimates of the uncertainty and the parameter interrelationship. If the mean values of decay parameters cannot meet the required accuracy based on the available posterior samples, the mean values and the covariance estimated so far can guide a further effort to refined sampling, using, e.g., importance sampling,[27] where the proposal distribution, which is essential to the importance sampling, can be more accurately assigned.

As an analysis method for quantifying energy decay characteristics, the methods reviewed in this chapter are model-based methods, which critically depend on the model involved. No

Table 6.1

Bayesian analysis of an experimentally measured Schroeder decay function in acoustic scale models as shown in Figures 6.3 and 6.8, derived from a 1 kHz octave-bandpass-filtered room impulse response. Bayesian information criterion (BIC, decibans[25]) along with decay parameters estimated using single-slope, double-slope, and triple-slope models are listed. The double-slope decay obtains the highest BIC values, by which the other BIC values are normalized. The standard derivations dA_1, dT_1, dA_2, and dT_2 are also listed only for the double-slope case. [*: dA_1, dA_2 are listed linearly]

	Single-slope	Double-slope	Triple-slope
BIC (db)	−372	0.0	−342
A_0 (dB)	−62.40	−62.2	−62.26
A_1 (dB)	−5.15	−5.15 (±3.4E-3)*	−5.26
T_1 (s)	0.483	0.411 (±9.3E-3)	0.39
A_2 (dB)	—	−14.91 (±1.3E-3)*	−14.24
T_2 (s)	—	—	0.768
A_3 (dB)	—	—	−20.43
T_3 (s)	—	—	1.22

decay characteristics can be inferred without an appropriate parametric model. Note that linear least-squares fitting is essentially a model-based method as well. The reason that correcting measures must be applied in some previous analyses is that the linear model misrepresents Schroeder decay functions. Bayesian decay analysis can also provide misleading estimates if incorrect models are used. In this specific application, the decay order (number of slopes) must be determined before using the estimated parameters to explain the data. Fortunately, the Bayesian framework also embodies decay model selection, and quantitatively implements Occam's razor.[23] One should apply Bayesian model selection to a proper model set to determine which model among the competing models is appropriate before interpreting the decay parameters (in this case a proper model contains a proper number of decay slopes). The model-based Bayesian decay analysis discussed in this chapter can be applied to both T_{30} estimation and to quantification of multiple-slope decay characteristics. Efficient sampling methods within the framework of Bayesian inference, particularly dedicated to multiple-slope decays, are still on-going research.

SUMMARY

Sound energy decay analysis in room acoustics is of fundamental relevance. This chapter is mainly dedicated to sound energy decay analysis. The parametric model derived from Schroeder integration is used for the task, and at the same time, it is also helpful to understand the behavior of Schroeder decay functions. In room-acoustics practice, Schroeder decay models of different orders are often competing with each other to explain or predict the data. The calculation of Bayesian evidence is therefore of central importance for both decay model selection and decay parameter estimation problems—more specifically in room-acoustic applications where T_{30}s or decay characteristics in a room under investigation need to be quantified. For single-slope decays, the nonlinear regression has been proven to be an efficient method; it requires a computational load on the order of a few milliseconds or less on the current personal-computer platforms. Once the number of slopes is known, some efficient optimization algorithms can also be employed for the energy decay analysis. However, it is the most challenging task in room-acoustic practice to know before getting experimental data how many slopes are to be found in a specific space at specific locations within specific frequency bands, therefore, requiring advanced methods, such as the Bayesian inference method.

ACKNOWLEDGMENT

The author is very grateful to Dr. P. Goggans, Dr. J.-D. Polack, and Dr. T. Jasa for many detailed discussions. He also would like to thank J. Dunham for constructive comments and proofreading of the final text.

REFERENCES

1. W. C. Sabine, *Collected Papers on Acoustics*, 2nd ed. (Peninsula Publishing, Los Altos, California 1992).
2. M. R. Schroeder, New method of measuring reverberation time, *J. Acoust. Soc. Am.*, 37, pp. 409–412 (1965).

3. F. Eyring, Reverberation time measurements in coupled rooms, *J. Acoust. Soc. Am.*, 3, pp. 181–206 (1931).
4. R. Kürer and U. Kurze, Integrationsverfahren zur Nachhallauswertung, *Acustica*, 19, pp. 314–322 (1967/68).
5. H. Kuttruff and M. Jusofie, Messungen des Nachhallverlaufs in meheren Räumen ausgefuehrt nach dem Verfahren der integrieten Impulseantworten (Energy decay measurements in rooms using the integrated impulse-response method), *Acustica*, 11, pp. 1–9 (1969).
6. K. Bodlund, On the use of the integrated impulse response method for laboratory reverberation measurements, *J. Sound & Vib.*, 56, pp. 341–362 (1978).
7. W. T. Chu, Comparison of reverberation measurements using Schroeder's impulse method and decay-curve averaging method, *J. Acoust. Soc. Am.*, 63, 1444–1450 (1978).
8. L. Cremer, H. A. Müller and T.J. Schultz: *Principles and Applications of Room Acoustics*, (Applied Science Publishers, London, UK, 1982).
9. M. Vorländer and H. Bietz, Comparison of Methods for Measuring Reverberation Time, *Acustica*, 80, pp. 205–215 (1994).
10. Y. A. Hirata, A method of eliminating noise in power responses, *J. Sound Vib.*, 82, pp. 593–595 (1982).
11. N. Xiang, and Genuit, Two Approaches to Evaluation of Reverberation Times from Schroeder's Decay Curves, *Proc. 15th ICA, Trondheim*, IV, pp. 175–178 (1995).
12. N. Xiang, W. Ahnert and R. Feistel, Nachhallauswertung mittels nicht-linearer Regression (Reverberation time evaluation using nonlinear regression), *Proc. Fortschritte der Akustik, DAGA93*, pp. 255–258.
13. A. Lundeby, T. E. Vigran, H. Bietz and M. Vorländer, Uncertainties of Measurements in Room Acoustics, *Acustica*, 81, pp. 344–355 (1995).
14. N. Xiang, Evaluation of reverberation times using a nonlinear regression approach, *J. Acoust. Soc. Am.*, 98, pp. 2112-2121 (1995).
15. L. Faiget, C. Legros and R. Ruiz, Optimization of the Impulse Response Length: Application to Noisy and Highly Reverberant Rooms, *J. Audio Eng. Soc.*, 46, pp. 741–750 (1998).
16. M. R. Schroeder, Complementarity of sound buildup and decay, *J. Acoust. Soc. Am.*, 40, pp. 549-551 (1966).
17. N. Xiang, Y. Jing and A. Bockman, Investigation of acoustically coupled enclosures using a diffusion equation model, *J. Acoust. Soc. Am.*, 126, pp. 1187-1198 (2009).
18. G. Golub and W. Kahan, Calculating the singular values and pseudoiverse of a matrix. *J. SIAM Numer. Anal.*, Ser. B 2, pp. 205–224 (1965).
19. N. Xiang and M. Vorländer, Using iterative regression for estimating reverberation times in two coupled spaces (A), *J. Acoust. Soc. Am.*, 104, p. 1763 (1998).
20. N. Xiang and P. M. Goggans, Evaluation of decay times in coupled spaces: Bayesian parameter estimation, *J. Acoust. Soc. Am.*, 110, pp. 1415–1424 (2001).
21. N. Xiang and P. M. Goggans, Evaluation of decay times in coupled spaces: Bayesian decay model selection, *J. Acoust. Soc. Am.*, 113, pp. 2685–2697 (2003).
22. T. Jasa and N. Xiang, Nested sampling applied in the Bayesian room-acoustics decay analysis, *J. Acoust. Soc. Am.*, 132, pp. 3251–3262 (2012).
23. N. Xiang, P. Goggans, T. Jasa, and P. Robinson, Bayesian characterization of multiple-slope sound energy decays in coupled-volume systems, *J. Acoust. Soc. Am.*, 129, pp. 741–752 (2011).
24. H. Jeffreys, *Theory of Probability*, 3rd ed. (Oxford University Press, UK, 1965), pp. 193–244.
25. J. Skilling, Nested Sampling in *Bayesian Inference and Maximum Entropy Methods in Science and Engineering*, 735, ed. R. Fisher, et al., pp. 395–405 (2004).
26. D. Sivia and J. Skilling, *Bayesian Data Analysis—A Bayesian Tutorial 2nd ed.*, Oxford University Press, New York, 2006.
27. T. Jasa and N. Xiang, Efficient estimation of decay parameters in acoustically coupled spaces using slice sampling, *J. Acoust. Soc. Am.*, 126, pp. 1269–1279 (2009).

7

Sound Insulation in Buildings

Carl Hopkins, Acoustics Research Unit, School of Architecture, University of Liverpool, United Kingdom

7.1 INTRODUCTION

Sound insulation in buildings is an essential aspect of building performance. It is needed to ensure that all occupants have a suitable acoustic environment for work, rest, and leisure activities, to allow them to carry out activities that generate sound without annoying other occupants, and to ensure that everyone has sufficient acoustic privacy.

In this chapter, the principles of direct and flanking sound transmission in buildings are introduced alongside prediction models that explain features of airborne and impact sound insulation that are commonly observed in practice. Progress in understanding, quantifying, and predicting flanking transmission has emphasized the fact that in all buildings where sound insulation is important, it is not possible to design a construction by focusing only on the separating element such as the wall or floor—careful design of the flanking construction is equally important. This chapter therefore reviews the state-of-the-art in International Standardization concerning the measurement and prediction of sound insulation in the laboratory and in the field, and provides examples using measured and predicted data to illustrate important features of sound insulation in buildings.

7.2 AIRBORNE SOUND INSULATION—DIRECT TRANSMISSION

Direct sound transmission across a single building element occurs when the element is excited by a sound field on one side and the element radiates or transmits sound from the other side without any other building element taking part in the transmission process (see Figure 7.1). Under laboratory conditions, this can often be achieved in a transmission suite with suppressed flanking transmission.[1]

7.2.1 Descriptors

When all sound is transmitted via direct transmission through the test element, the ratio of the transmitted sound power, W_2, to the sound power that is incident upon the test element, W_1, is defined as the transmission coefficient τ at each frequency:

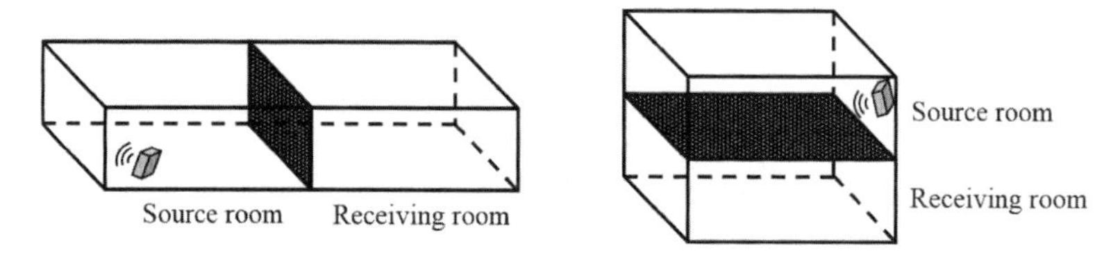

Figure 7.1 Direct sound transmission across a separating wall/floor in a transmission suite.

$$\tau = \frac{W_2}{W_1}. \tag{7.1}$$

This defines the sound reduction index R in decibels (dB), which is equivalent to the transmission loss referred to in the American Society for Testing and Materials (ASTM) Standards:[2]

$$R = -10\lg(\tau). \tag{7.2}$$

In a transmission suite where the sound field in the source and receiving rooms can be assumed to be diffuse, we can infer the incident or radiated sound power from measurement of the sound pressure levels (SPLs) in the source and receiving rooms:

$$R = L_{p1} - L_{p2} + 10\lg\left(\frac{S}{A}\right), \tag{7.3}$$

where L_{p1} and L_{p2} are the time and space average SPLs in the source and receiving rooms respectively (dB re 2×10^{-5} Pa), S is the surface area of the test element (m^2), and A is the absorption area (m^2) of the receiving room that is calculated from the measured reverberation time in the receiving room using the Sabine equation.

For building regulations it is usually more convenient to refer to an integer single-number quantity, rather than several frequency-dependent values that describe the sound insulation. These single-number quantities are the weighted sound reduction index,[3] R_w, and the sound transmission class (STC).[2] Although the weighting processes for these two descriptors use different frequency ranges, for most building elements they tend to give integer values within a few dB of each other.

The sound reduction index is used when the area of the test element is well-defined and the test result is representative of identical elements with different dimensions. While this is often reasonable for large walls and floors, it is not appropriate for small elements, such as ventilators. This requires an alternative descriptor that avoids use of the surface area, called the element-normalized level difference, $D_{n,e}$, which is calculated using:

$$D_{n,e} = L_{p1} - L_{p2} - 10\lg\left(\frac{A}{A_0}\right), \tag{7.4}$$

where the reference absorption area, A_0, is 10 m^2. The single-number quantity is called the weighted element-normalized level difference,[3] $D_{n,e,w}$.

7.2.2 Solid Plates

When a solid plate (e.g., solid concrete wall or floor) is excited by a sound wave in air, structure-borne sound waves are excited on the plate, which results in measurable vibration.

While sound in air is characterized by longitudinal wave motion, structures can support different types of structure-borne sound waves. Bending waves (see Figure 7.2) tend to be the most important type of structure-borne wave for sound insulation because of their large lateral displacements. This allows them to be excited by sound waves in air that impinge upon a plate in the source room, and they are relatively efficient at displacing the air adjacent to the plate surface so that sound is radiated into the receiving room.

A prediction model for sound transmission across a solid plate can be approached by treating the plate as being of either finite or infinite extent. Both approaches lead to useful insights into the sound transmission mechanisms and can be used in combination to solve practical problems.[4] Here we shall consider a model based on statistical energy analysis (SEA) for finite plates, partly for reasons of brevity, but mainly because a basic understanding of SEA will be useful to the reader that moves on to consider sound transmission in the field situation where there is often significant flanking transmission.

SEA is suited to modeling a system which is comprised of coupled subsystems, where each subsystem is defined by its ability to store modal energy. The solution of an SEA model gives the time and space average energy, E, in each subsystem. Energy is simply related to measured parameters such as the mean-square sound pressure for rooms (or cavities) or the mean-square velocity for plates, where:

$$E_{\text{room}} = \frac{\langle p^2 \rangle V}{\rho_0 c_0^2} \qquad E_{\text{plate}} = m\langle v^2 \rangle , \tag{7.5}$$

where $\langle p^2 \rangle$ and $\langle v^2 \rangle$ are the time and space average mean-square pressure and velocity respectively, V is the room volume, (m³), ρ_0 is the density of air, (kg/m³), c_0 is the speed of sound in air, (m/s), and m is the plate mass, (kg).

In buildings, a reverberant room or cavity can be represented by a subsystem because modal energy is stored in the modes of the sound field. Similarly, a solid wall or floor element can be represented by a subsystem because modal energy is stored in the modes of the bending wave field. Figure 7.3 shows the SEA system for airborne sound transmission between two rooms where all sound is transmitted across a solid plate.

A sound source, such as a loudspeaker with a white or pink noise input, provides an input power, $W_{\text{in}(1)}$, into the source room (Subsystem 1). This gives rise to energy in the source room that is transmitted to the receiving room (Subsystem 3) via two different transmission mechanisms: non-resonant and resonant transmission. In each subsystem, there will be power, W_d, that is dissipated into heat energy. In rooms and cavities this is primarily due to sound absorptive surfaces and air absorption, whereas in plates it occurs due to internal damping losses.

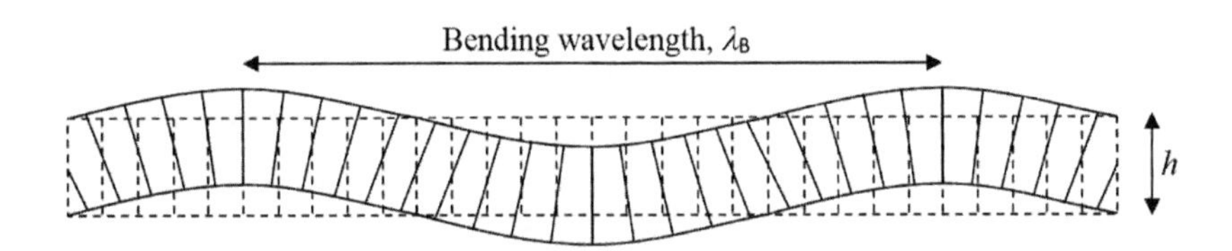

Figure 7.2 Bending wave motion. Cross-sectional view of a plate (thickness, h) with dashed lines showing the undeformed plate.

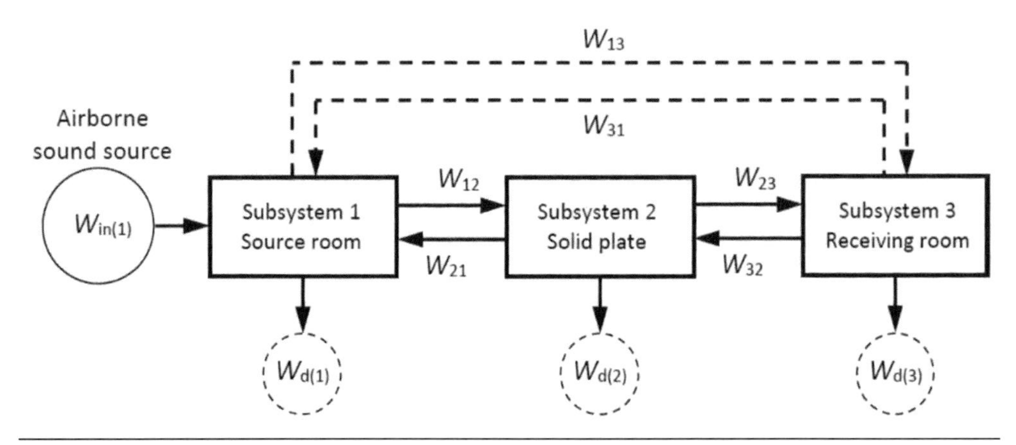

Figure 7.3 A three-subsystem SEA model for airborne sound insulation of a solid plate. Resonant transmission paths are indicated by solid lines and non-resonant transmission paths by dashed lines.

Within the framework of SEA, the net power transferred between two coupled subsystems is proportional to the difference in the modal energies of these subsystems.[5] A thermal analogy can be considered here because for each subsystem, the modal energy can be represented by temperature. Hence, there will be heat (energy) flow from subsystems with a high temperature (high modal energy) to subsystems with a low temperature (low modal energy). For two coupled subsystems, the net power, $W_{\mathrm{net},ij}$, flowing from subsystem i to j is the difference between the power transmitted from subsystem i to j and that from subsystem j to i, as given by:

$$W_{\mathrm{net},ij} = W_{ij} - W_{ji} = \omega\eta_{ij}E_i - \omega\eta_{ji}E_j\,, \tag{7.6}$$

where ω is the angular frequency and the coupling loss factor, η_{ij}, gives the fraction of energy coupled from subsystem i to subsystem j per radian cycle.

Based on Eq. (7.6), it is possible to write down a set of power balance equations for any SEA system that can be generalized into a matrix equation that can be solved using matrix inversion to find the subsystem energies. For a three-subsystem model with power input into Subsystem 1, the matrix equation is given by:

$$\begin{bmatrix} \sum_{n=1}^{3}\eta_{1n} & -\eta_{21} & -\eta_{31} \\ -\eta_{12} & \sum_{n=1}^{3}\eta_{2n} & -\eta_{32} \\ -\eta_{13} & -\eta_{23} & \sum_{n=1}^{3}\eta_{3n} \end{bmatrix} \begin{bmatrix} E_1 \\ E_2 \\ E_3 \end{bmatrix} = \begin{bmatrix} W_{\mathrm{in}(1)}/\omega \\ 0 \\ 0 \end{bmatrix}. \tag{7.7}$$

For small systems such as this three-subsystem model, it is also possible to create simple expressions to calculate the sound reduction index using SEA path analysis for resonant and non-resonant transmission. SEA path analysis can be carried out for power input into Subsystem 1, with transmission along a chain of subsystems. For example, along the path 1→2 →3→...→N, the energy ratio between source Subsystem 1 and receiving Subsystem N is:

$$\frac{E_1}{E_N} = \frac{\eta_2\eta_3\ldots\eta_N}{\eta_{12}\eta_{23}\ldots\eta_{(N-1)N}}. \tag{7.8}$$

For this three-subsystem model, the resonant transmission process involves coupling between the room modes and the plate bending modes. This is characterized by the radiation efficiency, σ, which is defined as the ratio of the radiated power, to the power radiated by a large baffled piston with a uniform mean-square velocity that is equal to the time and space average mean-square velocity of the plate. In contrast to sound in air, bending waves are dispersive, which means that the bending wavespeed is frequency-dependent. At very low frequencies, the bending wavespeed will be lower than the speed of sound in air, and at very high frequencies it will be higher. The frequency at which the bending wavespeed and the speed of sound in air are equal is called the critical frequency. This is the frequency at which the radiation efficiency tends to be highest. At frequencies above the critical frequency, the radiation efficiency can usually be assumed to have a value of unity, whereas below the critical frequency there are different options available to calculate the radiation efficiency depending on the type of plate.[4] The critical frequency, f_c, can be calculated using:

$$f_c = \frac{c_0^2\sqrt{3}}{\pi h c_L}, \tag{7.9}$$

where h is the thickness (m), and c_L is the longitudinal wavespeed (m/s), which is typically 1500–1800 m/s for plasterboard; 2200–2600 m/s for chipboard, medium density fiberboard (MDF), and oriented strand board (OSB); 3800 m/s for concrete; and 5200 m/s for glass.[4] Note that for plates the longitudinal wavespeed is related to the Young's modulus, E, and the bending stiffness, B, according to:

$$B = \frac{Eh^3}{12(1-\nu^2)} = \frac{\rho_s c_L^2 h^2}{12}, \tag{7.10}$$

where ν is the Poisson ratio (typically between 0.2 and 0.3).

The sound reduction index due to resonant transmission, R_R, is determined from Eq. (7.8) as:

$$R_R = 10\lg\left(\frac{\eta_2\eta_3}{\eta_{12}\eta_{23}}\right) + 10\lg\left(\frac{V_3}{V_1}\right) + 10\lg\left(\frac{S}{A}\right) = 10\lg\left(\frac{2\pi^2 h c_L \rho_s^2 f^3 \eta}{\sqrt{3}\rho_0^2 c_0^4 \sigma^2}\right), \tag{7.11}$$

where ρ_s is the mass per unit area, (kg/m^2), f is the frequency, (Hz), η is the total loss factor, (–), and σ is the radiation efficiency, (–). The total loss factor is the sum of all the loss factors for the plate, hence:

$$\eta = \eta_{\text{int}} + \eta_{\text{rad}} + \sum \eta_{\text{str}}, \tag{7.12}$$

where η_{int} is the internal loss factor which accounts for the conversion of vibration energy into heat energy, η_{rad} is the radiation loss factor which accounts for energy losses from the plate due to it radiating sound into air on each of its two sides, and $\Sigma\eta_{\text{str}}$ is the sum of all the structural coupling loss factors which accounts for all energy losses due to structure-borne sound transmission from the plate to other connected parts of the building structure; for heavyweight walls and floors that are connected along all edges, the structural coupling term is often large.

Non-resonant transmission primarily depends on the frequency and the mass per unit area of the plate, and is sometimes referred to as mass-law transmission. For most building elements it can be assumed that it is only applicable below the critical frequency. The main features of the non-resonant sound reduction index can be based on the field incidence mass-law which shows that it increases by 6 dB per octave and 6 dB per doubling of the mass per unit area, although other more accurate theories have slightly different features.[4] The sound reduction index due to non-resonant transmission, R_{NR}, below the critical frequency can also be determined from Eq. (7.8) and is given by:

$$R_{NR} = 10\lg\left(\frac{\eta_3}{\eta_{13}}\right) + 10\lg\left(\frac{V_3}{V_1}\right) + 10\lg\left(\frac{S}{A}\right) = 10\lg\left(1 + \left(\frac{\omega\rho_s}{2\rho_0 c_0}\right)^2\right) - 5\text{ dB}. \quad (7.13)$$

In practice, the sound reduction index that is measured in the laboratory is determined by the combination of the resonant and non-resonant paths; hence the sound reduction index is given by:

$$R = -10\lg\left(10^{-R_R/10} + 10^{-R_{NR}/10}\right). \quad (7.14)$$

For solid plates, the resulting airborne sound insulation is therefore dependent on the mass, stiffness, and damping. Figure 7.4 demonstrates the relative importance of non-resonant and resonant transmission for solid plates with significantly different mass per unit areas. A single sheet of plasterboard has a critical frequency in the high frequency range, and primarily because of its low mass per unit area, non-resonant transmission dominates over the majority of the low- and mid-frequency range. In contrast, sound transmission across a heavyweight masonry wall can be dominated by resonant transmission across the entire building acoustics frequency range.

For solid, nonporous, heavyweight plates such as concrete floors and masonry walls, the airborne sound insulation over a large part of the frequency range is under damping control due to the total loss factor in Eq. (7.11). In the low- and mid-frequency range the total loss factor is dominated by the sum of all structural coupling loss factors in Eq. (7.12). In practice, the structural coupling to the aperture walls varies between different transmission suites; hence a laboratory measurement tends to be of limited use unless it is accompanied by the measured structural reverberation time of the plate, from which the total loss factor can be calculated.[6] For example, assume that a plate has been measured in Laboratory A and has a sound reduction index, R_A, with a total loss factor, η_A. The same plate is then measured in Laboratory B to have a sound reduction index, R_B, and a total loss factor, η_B. Assuming that all measurement errors are negligible, it is possible to convert R_A to the sound reduction index R_B that would be measured in Laboratory B using:

$$R_B = R_A + 10\lg\left(\frac{\eta_B}{\eta_A}\right). \quad (7.15)$$

Note that this can also be used to convert R_A from a laboratory, to the sound reduction index R_B that would be measured *in situ* where η_B is the total loss factor *in situ*.

7.2.3 Cavity Wall and Floor Constructions

With a solid wall or floor there are limitations on the level of airborne sound insulation that can be achieved when using very thick walls or floors to achieve a high mass per unit

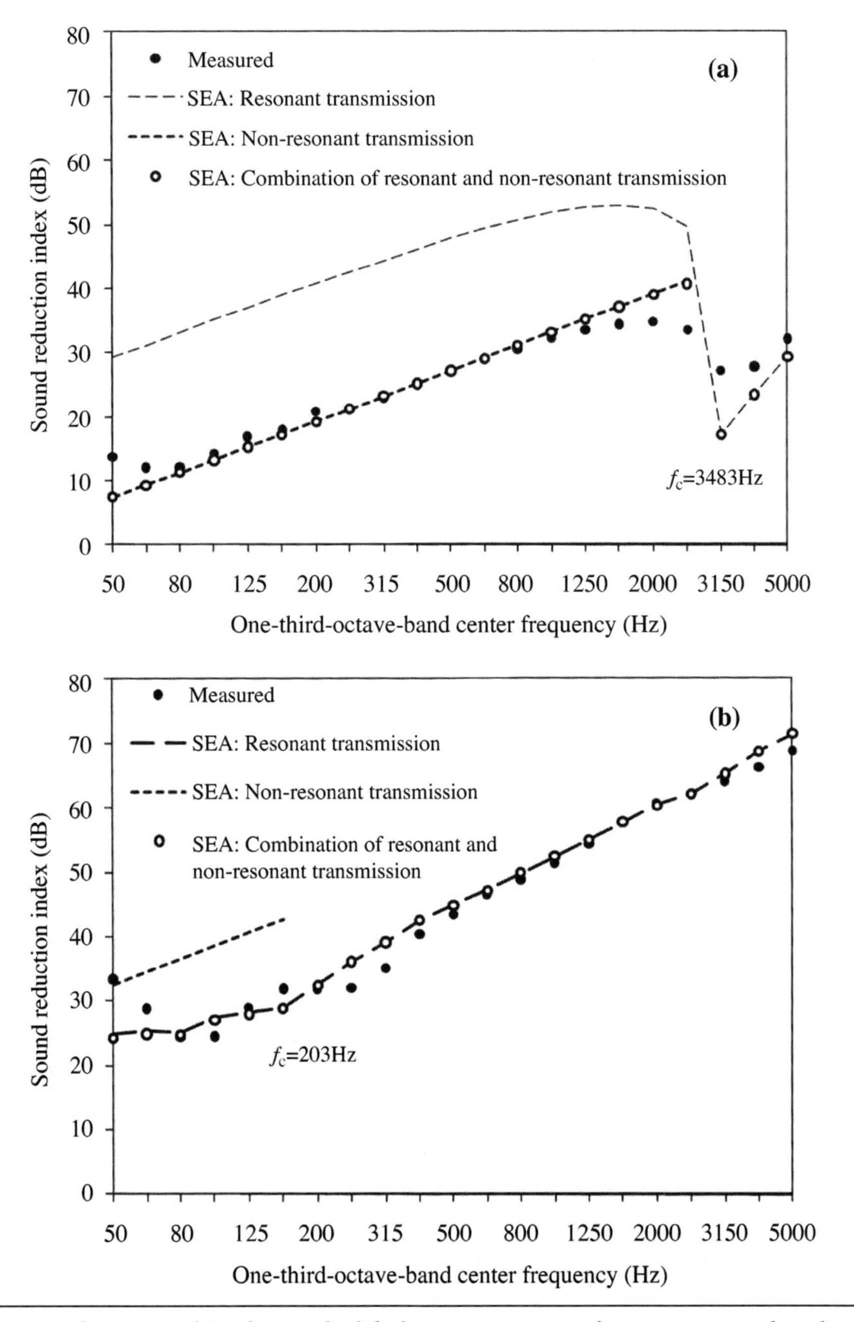

Figure 7.4 Airborne sound insulation of solid plates—comparison of measurement and prediction. (a) 12.5 mm sheet of plasterboard (11 kg/m²). (b) 100 mm masonry wall with plaster finish (200 kg/m²).

area. An efficient alternative is to use a cavity wall or floor that is formed from two plates that are separated by a cavity. Cavity constructions are commonly found in buildings in the form of studwork partitions made with plasterboard, timber floors, masonry cavity walls, or masonry walls with a plasterboard lining. Their ability to provide relatively high levels of sound insulation is due to the fact that the sound insulation not only depends upon the mass, stiffness, and damping of the individual plates, but also on the degree of isolation that exists between them. By reducing vibration transmission between the plates, it is possible to use plates with a significantly lower mass per unit area and achieve higher sound insulation than with an equivalent solid plate. However, this is only possible above a lower limiting frequency called the mass-spring-mass resonance frequency, f_{msm}, where each plate acts as a lump mass and the air in the cavity acts as a spring. At frequencies below f_{msm}, the two plates effectively transmit the same sound as a single solid plate with the combined mass per unit area of the two individual plates. At f_{msm}, there is a resonance effect at which there is significant sound transmission that usually results in a visible dip in the airborne sound insulation. For two plates separated by an air-filled cavity, the mass-spring-mass resonance frequency can be estimated using:

$$f_{msm} = 60\sqrt{\frac{\rho_{s1} + \rho_{s2}}{\rho_{s1}\rho_{s2}d}}, \tag{7.16}$$

where ρ_{s1} and ρ_{s2} are the mass per unit areas of Plates 1 and 2 (kg/m^2), and d is the cavity depth (m).

At frequencies above f_{msm}, it is possible to use a five-subsystem SEA model (see Figure 7.5) to explain typical features of the airborne sound insulation. In this particular model, the structural coupling is simply introduced using a coupling loss factor between the two plates to represent resonant transmission between the plate bending modes on each plate. However, in some cases it will be necessary to expand the model by adding beam subsystems to model the structural frame that couples the plates together.

In order to illustrate typical features of a cavity construction, Figure 7.6 shows idealized sound reduction indices for two plasterboard studwork walls with and without an absorbent material to absorb sound within the cavity. Idealized versions are shown because with lightweight partitions, even seemingly minor differences in material properties, stud profiles, screw fixings, and boundary conditions in the laboratory can significantly change the measured sound reduction index. For this reason there is an abundance of laboratory test data from manufacturers for specific lightweight wall and floor systems.[7]

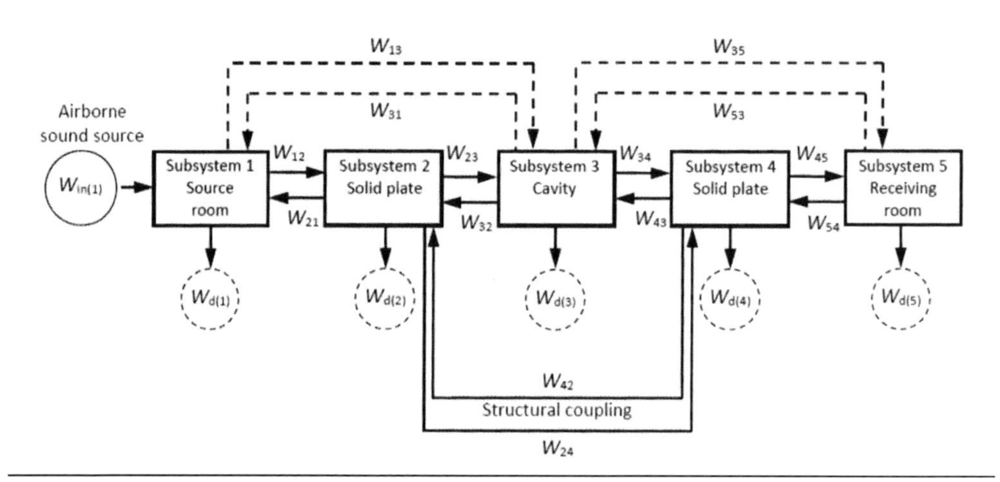

Figure 7.5 A five-subsystem SEA model for airborne sound insulation of a cavity wall.

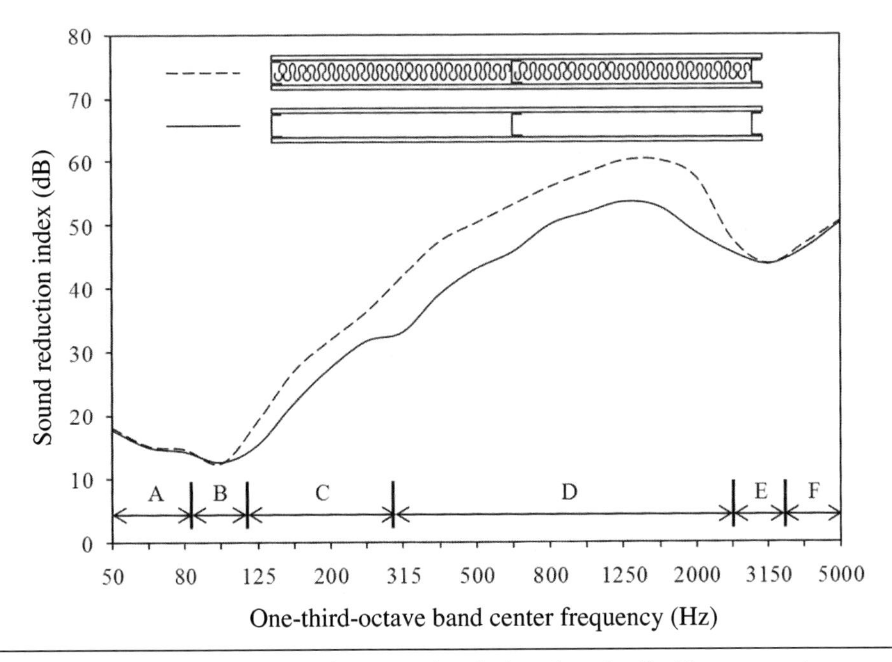

Figure 7.6 Airborne sound insulation for two studwork plasterboard walls. The construction comprises a single sheet of plasterboard (≈ 9 kg/m²) on each side of ≈ 70 mm steel C-studs spaced at ≈ 600 mm centers, with either an empty cavity or a cavity that is half to fully filled with low-density mineral fiber.

Frequency regions are identified in Figure 7.6 for which the transmission occurs via different mechanisms or paths. Region A is defined by the range below the mass-spring-mass resonance where the two plates tend to move together as one; hence they can be represented by a single plate with a mass per unit area and a stiffness that represents the complete construction. Region B contains the mass-spring-mass resonance frequency which tends to manifest itself as a visible dip or plateau in the airborne sound insulation. For a cavity that is fully filled with a porous material such as glass fiber in the cavity, there is isothermal rather than adiabatic compression of the enclosed air within the cavity; hence, f_{msm} can be estimated using Eq. (7.16) with a factor of 51 instead of 60.[4] In Region C the dominant transmission path tends to be 1→3→5, which involves non-resonant transmission between the rooms and the axial modes that form a one-dimensional sound field in each narrow cavity. The strength of this path is primarily determined by the mass per unit area of the plasterboard. Between Regions C and D there can be a change in gradient as the sound field in each cavity changes from a one-dimensional to a two-dimensional field. In Region D the overall sound insulation is determined by a combination of paths involving the sound field in the cavity, 1→3→5, 1→2→3→5, 1→3→4→5, 1→2→3→4→5, and the structural path via the studs, 1→2→4→5, that effectively bypasses the cavity. Across Regions D and E this structural path via the studs becomes increasingly dominant with increasing frequency. The inclusion of sound absorbent material reduces the sound pressure level in each cavity; hence all the paths involving the cavity in Regions C and D become significantly less important than the structural path. This explains the higher airborne sound insulation when absorbent material is used in each cavity. In Region E there is a visible dip in the airborne sound insulation in the vicinity of the critical frequency where the plasterboard radiates sound efficiently. For lightweight plates on a frame, the different stiffness in the two directions can result in an orthotropic plate with two critical frequencies, resulting in an extended frequency region for the dip. The critical frequency dip is usually much deeper for plates made of plasterboard or chipboard compared to plates of concrete or masonry.[4] In Region F the structural path via the studs, 1→2→4→5 dominates; hence the absorbent material in each cavity has negligible effect and both walls have similar sound insulation. To increase the airborne sound insulation across the entire frequency range one would need to significantly reduce transmission by the paths involving the cavities, as well as via the structural path; hence one would consider additional layers of plasterboard on one or both sides, as well as the use of isolated studs to support the plasterboard on each side.

Examples for the airborne sound insulation of plasterboard walls on light steel studwork are given in Table 7.1. The sound insulation increases from left to right across the table to illustrate the beneficial effects of increasing the mass and stiffness (i.e. adding thicker sheets or more sheets of plasterboard), increasing the damping (i.e., adding absorbent material in the cavity) and increasing the isolation (i.e., incorporating wider cavities and using isolated studwork). The same principles determine the performance of timber frame constructions for the examples shown in Table 7.2. However, there are often significant differences between light steel and timber frame because of the way the frame stiffens the entire wall, the connections formed at the interface of the studs and the boarding, and of particular importance, the dynamic properties of the studs which determines the effectiveness with which they transmit vibration between the boards on either side of the frame.

7.2.4 Wall and Floor Linings

A common approach to reduce the sound radiated by walls and floors is to attach a lining such as a suspended ceiling, floating floor, or plasterboard fixed onto the base wall/floor via a frame, resilient material, or other type of connector. Most linings consist of a plate with a low mass per unit area and a critical frequency in the high-frequency range. The latter is used to take advantage of the low radiation efficiency of such a plate well below the critical frequency. However, the air in any cavity between the plate and the base wall/floor can act as a spring element. This results in a mass-spring-mass resonance frequency that can cause the sound insulation to be lower than the base wall/floor without the lining. The addition of a lining tends to increase the sound insulation in the mid- and high-frequency ranges while reducing it in the low-frequency range near the mass-spring-mass resonance frequency; the latter can be estimated using Eq. (7.16).

The improvement due to the lining for direct sound transmission is determined by measuring the sound reduction index of a base wall/floor with and without the lining. The sound reduction improvement index, ΔR, is defined as:

$$\Delta R = R_{\text{with lining}} - R_{\text{base wall/floor}} \,. \tag{7.17}$$

Note that although it is referred to as an improvement index, it will often have negative values near the mass-spring-mass resonance, as indicated by the example in Figure 7.7.

Table 7.1 Laboratory airborne sound insulation for plasterboard walls on light steel studwork. Courtesy of British Gypsum.

	Single walls					Double wall
Plan view						
Plasterboard (Each side)	1 × 12.5 mm sheet (8.6 kg/m²)	2 × 12.5 mm sheets (8.6 kg/m²)		2 × 12.5 mm sheets (10.4 kg/m²)	2 × 12.5 mm sheets (10.5 kg/m²)	3 × 15 mm sheets (12.8 kg/m²)
Mineral fiber	—	—	25 mm glass fiber (16 kg/m³)	3 × 25 mm glass fiber sheets (18 kg/m³)	50 mm glass fiber (13 kg/m³)	2 × 100 mm glass fiber (11 kg/m³)
Studs	48 mm deep steel C-studs at 600 mm centers			92 mm deep steel C-studs at 600 mm centers	146 mm deep steel C-studs at 600 mm centers with resilient bars at 600 mm centers (perpendicular to studs)	92 mm deep steel C-studs at 600 mm centers—studwork on each side is isolated to form a 460 mm deep cavity
	Sound reduction index, *R* (dB)					
100	15.2	18.8	21.4	36.1	43.0	54.3
125	14.0	22.9	29.9	39.0	45.0	54.8
160	16.4	24.0	35.1	42.8	47.9	59.7
200	21.5	29.2	34.5	43.6	53.2	58.7
250	23.9	33.1	38.5	46.1	55.8	63.8
315	27.6	36.3	42.6	48.6	57.2	68.4
400	30.0	40.3	47.5	51.4	58.0	72.9
500	32.4	42.0	51.1	53.6	59.7	75.7
630	35.7	42.9	51.5	56.6	61.4	78.5
800	40.5	47.0	53.8	58.0	65.1	83.5

Table 7.1 (continued)

	Single walls					Double wall
1000	43.1	48.5	54.6	58.9	67.5	88.8
1250	45.5	50.9	56.8	62.0	71.0	93.2
1600	48.8	53.4	59.3	64.1	74.3	96.2
2000	48.0	53.6	61.3	63.7	76.4	96.8
2500	38.8	45.9	55.2	62.8	74.7	95.7
3150	34.7	43.2	48.6	54.1	68.3	92.1
4000	40.5	49.2	52.5	56.1	67.1	90.3
5000	44.4	52.9	57.1	58.5	70.5	87.2
R_w	34	42	49	56	64	76
STC	34	43	50	56	64	76

Table 7.2 Laboratory airborne sound insulation for timber frame walls and floors. Courtesy of British Gypsum.

	Single walls		Double wall	Floors	
Plan (Walls) Section (Floors)					
Plasterboard (Walls—Each side)	1 × 12.5 mm sheet (8.8 kg/m^2)	2 × 12.5 mm sheets (9.1 kg/m^2)	1 × 19 mm sheet (14.9 kg/m^2) 1 × 12.5 mm sheet (8.7 kg/m^2)	Walking surface: 1 × 18 mm tongue and grooved chipboard (≈14 kg/m^2) Ceiling: 1 × 12.5 mm sheets of plasterboard (≈8 kg/m^2)	Walking surface: 1 × 15 mm OSB (9.3 kg/m^2) Ceiling: 2 × 15 mm sheets of plasterboard (12.8 kg/m^2) fixed to the joists with resilient bars at 450 mm centers (perpendicular to the joists)

Mineral fiber	—	—	25 mm glass fiber (17 kg/m^3)	—	100 mm glass fiber (10 kg/m^3)
Studs/Joists	75 mm deep × 38 mm wide timber studs at 600 mm centers		90 mm deep × 38 mm wide timber studs at 600 mm centers with timber noggings—studwork on each side is isolated to form a 230 mm deep cavity	195 mm deep × 45 mm wide timber joists at 450 mm centers	
	Sound reduction index, *R* (dB)				
100	10.9	14.9	31.7	20.3	29.2
125	19.3	29.8	42.5	17.1	31.7
160	24.1	27.3	47.1	15.3	33.5
200	22.6	30.5	51.0	22.9	40.8
250	31.4	35.7	55.6	26.4	46.1
315	28.3	33.3	56.7	29.3	49.5
400	31.2	34.0	61.7	32.6	50.8
500	33.2	37.1	63.6	34.1	51.7
630	32.2	35.7	66.8	34.9	54.8
800	35.6	37.9	68.2	38.1	58.5
1000	37.2	40.1	69.3	40.1	60.5
1250	39.7	42.2	70.7	41.4	60.5
1600	41.2	44.4	70.7	40.5	59.0
2000	39.4	41.8	71.4	38.7	58.1
2500	35.3	38.0	73.4	37.9	57.0
3150	34.0	39.5	76.6	39.5	59.3
4000	36.3	42.4	80.2	44.2	63.3
5000	40.4	45.4	81.9	48.7	69.0
R_w	35	38	63	36	54
STC	35	39	65	35	54

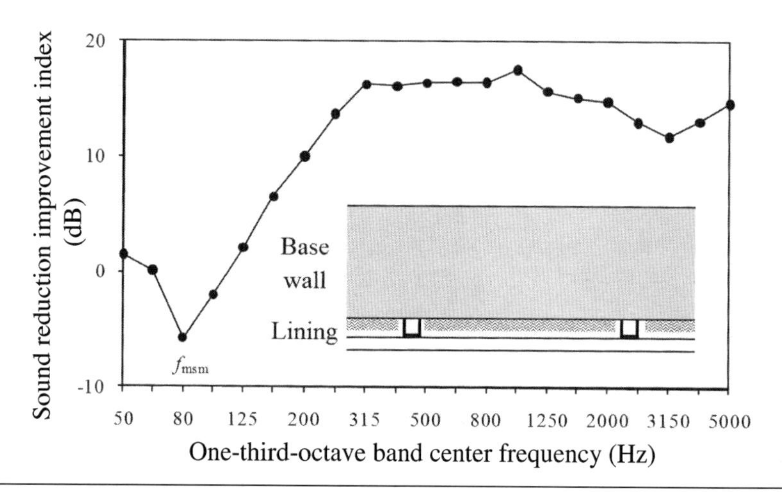

Figure 7.7 Sound reduction improvement index for a plasterboard lining on a masonry wall. Base wall: 100 mm aircrete block wall (51 kg/m^2). Lining: 12.5 mm plasterboard (11 kg/m^2) which is screwed to a light steel frame (37 mm cavity containing 25 mm rock fiber quilt).

7.2.5 Air Paths Due to Holes, Gaps, and Slits

Any unobstructed air path will tend to reduce the airborne sound insulation of a partition. For this reason, it is important that any construction is well-sealed over its surface and around its edges. Poor workmanship on building sites can lead to slits along unsealed edges of walls and floors whereas circular apertures can be found as unsealed holes. It is important to note that the sound reduction index for a narrow slit or small hole cannot simply be assumed to be 0 dB at all frequencies. In practice, the sound reduction index of a slit or hole will depend on its dimensions, internal profile and the position of the slit or hole. The resulting transmission not only varies significantly with frequency but also shows distinct resonance phenomena due to the slit/hole depth at high frequencies. A summary of prediction models for simple slits and circular apertures is given in Hopkins [2007].[4]

The effect of a narrow slit is illustrated using a prediction model in Figure 7.8 for an unsealed edge along the length of a lightweight wall. Such slits may only be 0.5 mm wide and visually difficult to identify but they can significantly reduce the airborne sound insulation in the mid- and high-frequency range. The resonance dips of the slits in the high-frequency range are not usually quite as deep because of damping due to air viscosity.

7.2.6 Glazing and Windows

The airborne sound insulation for single sheets of glass can be estimated by assuming a solid plate as described in Section 7.2.2. Figure 7.9(a) shows typical measured values of airborne sound insulation for 3, 6, and 12 mm float glass and 10 mm laminated glass. For 3, 6, and 12 mm glass, there are dips in the sound insulation at the critical frequencies that fall within the 4000, 2000, and 1000 Hz octave bands respectively. Depending upon the temperature, laminated glass has a much higher internal loss factor than float glass, hence the dip at the critical frequency is not always visible.

The glazing in modern window constructions tends to be an insulating glass unit (IGU). These are also referred to as thermal glazing units (or double glazing) and consist of two

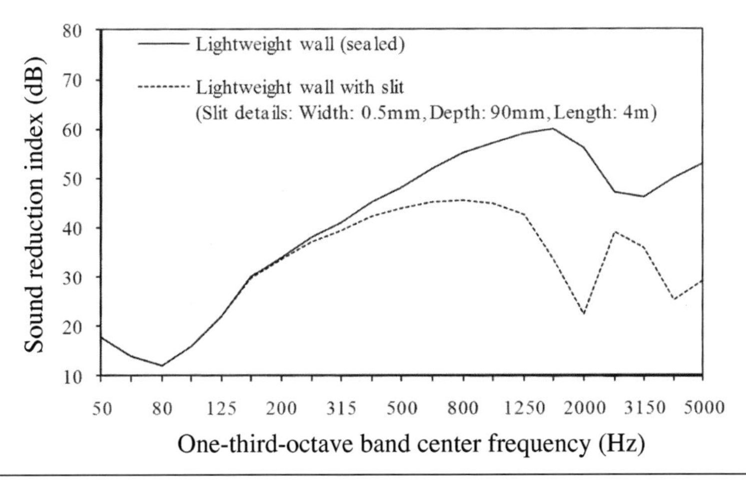

Figure 7.8 Airborne sound insulation of a lightweight wall (90 mm thick) with and without a slit aperture along one edge.

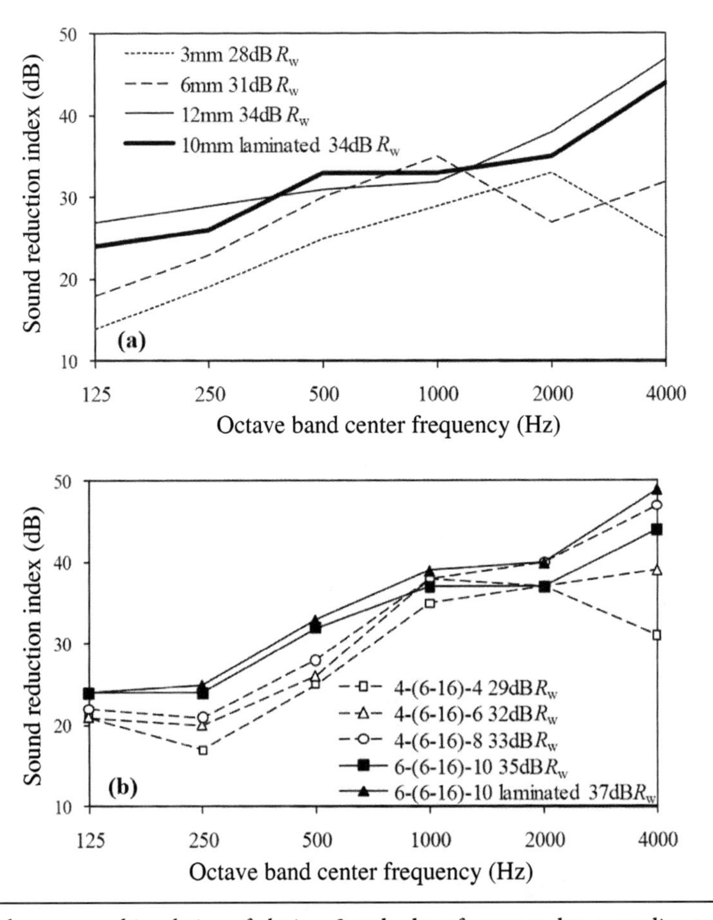

Figure 7.9 Airborne sound insulation of glazing. Standard performance data according to EN 12758. (a) Single panes of glass. (b) IGUs (dimensions in the legend are in millimeters: pane thickness—range of cavity depths—pane thickness).

panes of glass separated by a cavity in a hermetically sealed unit. Predicting the sound insulation of these cavity constructions is often difficult; hence there is a dependence on laboratory measurements for which typical measured values are shown in Figure 7.9(b). For IGUs it can be assumed that there is negligible improvement in increasing the cavity depth from 6 to 16 mm.

Laboratory measurements of glazing are only indicative of what will be achieved with a specific type of window frame; therefore, additional measurements are often required on the complete window. The reasons for this are: first, that the sound insulation of a window depends upon an airtight seal around the perimeter of the glazing to avoid air gaps, and second, the dimensions of glazing used in windows sometimes differs considerably from the single sheet of glazing in the laboratory test.

7.2.7 Doors

The airborne sound insulation provided by doors is primarily limited by the sealing around the door perimeter, followed by the door construction itself. Figure 7.10 shows the typical range of airborne sound insulation for single or double doorsets with no seals and 2–5 mm gaps around the perimeter. The maximum sound insulation performance for a doorset is measured by caulking around the entire perimeter so that it is airtight; this can then be compared with commercially available seals at the threshold, head, jambs, and meeting stile to assess their efficacy. Examples are shown in Figure 7.10. In situations where there are high sound insulation requirements and doors are essential, it is usually necessary to have two

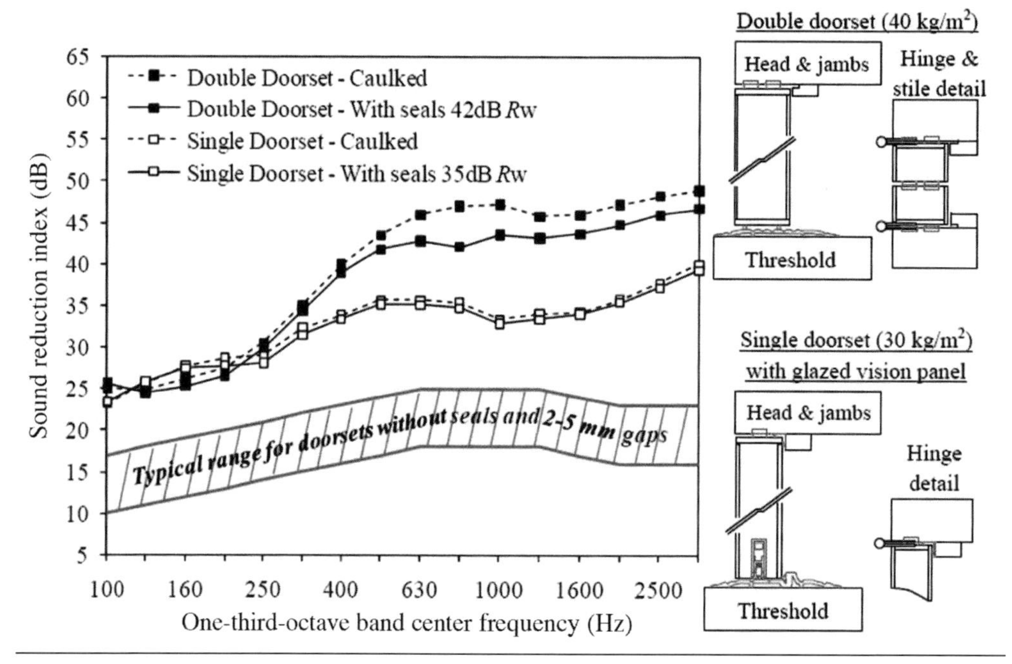

Figure 7.10 Airborne sound insulation of doorsets—effect of perimeter seals. Courtesy of Lorient Polyproducts Ltd.

doors separated by a lobby with absorbent surfaces, and to ensure that the two doors are not in-line with each other.

7.2.8 Combining Sound Reduction Indices for Different Building Elements that Form a Single Surface

In buildings there are often situations where direct transmission occurs across several adjacent building elements on the same surface plane. For example, airborne sound insulation between an office and a corridor across the door, glazed vision panel, and corridor wall, or façade sound insulation for a single façade comprised of windows and an external wall. For building specification purposes it can be useful to create a composite indicator of the airborne sound insulation. Hence for N elements each with a sound reduction index, R_n, and an area, S_n, the combined sound reduction index is:

$$R_{\text{combined}} = -10\lg\left(\frac{1}{\sum_{n=1}^{N} S_n}\sum_{n=1}^{N} S_n 10^{-R_n/10}\right). \tag{7.18}$$

This equation allows a quantitative value to be determined, which explains the qualitative rule-of-thumb which states that the overall sound insulation of several building elements is limited by the element with the lowest sound insulation.

7.3 IMPACT SOUND INSULATION—DIRECT TRANSMISSION

There are many different impact sources in buildings—such as footsteps, children jumping, dropped objects, chopping on kitchen work surfaces, slamming doors, furniture being moved, and machinery. Most, but not all impacts in dwellings are imparted to floors, so building regulations tend to focus on floors with the aim of at least providing impact sound insulation against footsteps. This is done on the basis that footsteps are likely to be the most common form of impact, and because it is difficult or impractical to consider a wide range of different impact sources. While this is the intention, measurement of impact sound insulation in both the laboratory and the field requires a well-defined impact source that provides repeatable measurements as well as a rating system that gives a link to the real impact source of interest. In practice, it has not been possible to find the *perfect* impact source that: (a) is suited to the practicalities of field and laboratory measurements, (b) has an associated rating procedure that can provide a direct link to the subjective evaluation of impact sound insulation in dwellings, and (c) can correctly rank order the impact sound insulation performance of all types of floors against all kinds of human walking, running, and jumping activities on floors. However, the majority of countries make use of the International Standards Organization (ISO) tapping machine as an impact source for their building regulations. This is because pragmatism allows some of its inadequacies to be forgiven due to the complexity of relating objective to subjective ratings of impact sound insulation in dwellings. A discussion of the issues relating to the choice of impact sources and rating systems can be found in Schultz [1981][8] and Hopkins [2007].[4]

7.3.1 Standard Impact Sources

The most widely used impact source is the ISO tapping machine, an example of which is shown in Figure 7.11. Its specifications are described in International Standards [ISO 10140][1] along with the test method. The ISO tapping machine has a line of five equally spaced hammers that are driven in such a way that there are 10 impacts upon the floor every second, with 100 ms between successive impacts. This succession of impacts allows the time and space average SPL to be measured in the receiving room using the equivalent continuous SPL, L_{eq}. The requirement for each tapping machine hammer is that the momentum of each hammer impact should represent a free-falling mass of 0.5 kg with a drop-height of 0.04 m. On thick, heavy, concrete floor slabs it can be assumed that the impacts from the hammers are of short duration; hence the force from the tapping machine will have a flat spectrum up to a maximum frequency of approximately 3 kHz. For such short impacts the mean-square force, F^2_{rms}, is 3.9*B*, where *B* is the bandwidth of the one-third-octave or octave band. This assumption of short impacts is not valid for floors where the walking surface is made of thin wooden boards for which the longer duration impacts lead to lower force levels that rarely have a flat spectrum above 500 Hz.[4] The power, W_{in}, that is input into a plate by the ISO tapping machine depends on the relative size of the plate impedance to the impedance of the tapping machine hammer and can be estimated using:

$$W_{in} = F^2_{rms} \frac{2.3\rho_s c_L h}{(2.3\rho_s c_L h)^2 + (\omega m)^2}, \tag{7.19}$$

where ρ_s, c_L, and h correspond to the plate, and m is the mass of the ISO tapping machine hammer (0.5 kg).

The role of the source and receiver impedances in determining the input power explains why it is often difficult or impractical to consider a wide range of different impact sources

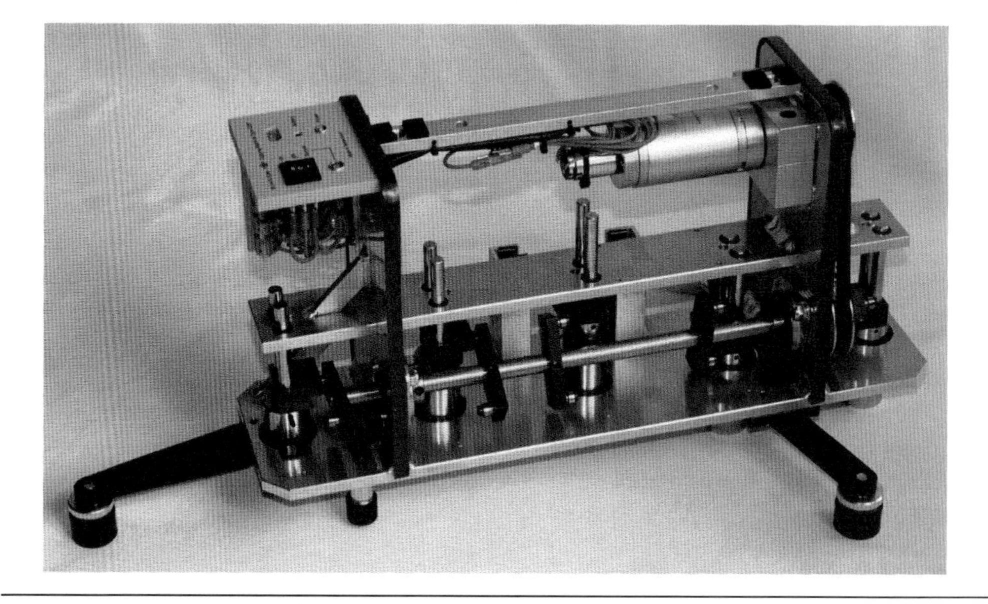

Figure 7.11 ISO tapping machine (external cover removed). Photograph copyright: Brüel and Kjær.

in building regulations using a single impact source. In fact, the impedance of a 0.5 kg mass used in the tapping machine is not a good representation of measured heel impedances of men or women.[9, 10] This is primarily why the ISO tapping machine is not always able to correctly rank order the impact sound insulation of all different types of floors against footsteps, although there are also problems with the non-linear response of some soft floor coverings to the high forces generated by the tapping machine hammers.

A proposal to modify the ISO tapping machine to make the hammer impedance more representative of footsteps has been to insert a resilient material between the face of the metal hammers and the floor.[10, 11, 12] This has been incorporated into the International Standard for laboratory measurement of the improvement of impact sound insulation due to floor coverings on lightweight floors.[1] Springs can either be attached to the ISO tapping machine hammers or a resilient material can be placed on the floor underneath the hammers. The resilient material significantly reduces the impact SPL in the mid- and high-frequency ranges, therefore low background noise levels are needed in the laboratory.

Although the ISO tapping machine does not actually simulate footsteps, the ISO rating system[3 (Part 2)] does provide a link between the subjective and objective rating of impact sound insulation in dwellings. In dwellings it is possible to identify two different types of impact from people on floors: either light, hard impacts such as footsteps in hard-heeled shoes; or heavy, soft impacts such as children running and jumping, adults exercising, and footsteps in bare feet. In previous surveys of impact sound insulation in dwellings[13], it is reasonable to assume that both light and heavy impacts occurred on a daily basis, so the subjective assessments are likely to have considered both types of impact. Hence, by using the ISO tapping machine and a rating system that gives a single-number quantity that correlates well with the subjective assessment, then to some unknown extent, both light and heavy impacts are taken into account.[4] However, Japan and Korea have placed emphasis on providing impact sound insulation against heavy impacts through the use of alternative impact sources such as the rubber ball and the bang machine.

The bang machine is a pneumatic tire dropped by a mechanical device from a height of 0.9 m, as shown in Figure 7.12. Its specifications and the test method are described in Japanese Industrial Standard [JIS A 1418-2]. Compared to the ISO tapping machine, it is awkward to transport, and it produces such a heavy blow that it can cause minor damage to decorated timber-frame dwellings. For these reasons it is mainly suited to laboratory measurements.

The rubber ball specification was developed in Japan and is now included in an International Standard.[1] It is formed from a hollow sphere of 30 mm thick silicone rubber with an outer diameter of 180 mm and a weight of approximately 2.5 kg (see Figure 7.13). A single impact on the floor is produced by manually dropping the ball from a height of 1 m above the floor and catching it. Compared to the bang machine, the rubber ball is easy to transport and because it applies a lower force, it does not cause damage. In contrast to the continuous signal created by the tapping machine, the resulting single impulse is measured using the maximum time-weighted (fast) sound level in the receiving room. This is averaged for a number of excitation positions and stationary microphone positions to give the impact SPL. Compared to the ISO tapping machine, the impedance of the rubber ball is more representative of a human walker in bare feet between 50 and 200 Hz. In addition, the impact sound produced by the rubber ball is more representative of children jumping and running.[14]

Figure 7.12 Bang machine. Photograph copyright: Rion.

Figure 7.13 ISO rubber ball. Photograph copyright: Rion.

7.3.2 Descriptors

When a floor is excited by the ISO tapping machine, the normalized impact SPL, L_n, at individual frequencies is defined as:

$$L_n = L_p + 10\lg\left(\frac{A}{A_0}\right), \tag{7.20}$$

where L_p is the time and space average SPL in the receiving room and the reference absorption area, A_0, is 10 m^2.

The standardized single-number quantities are the weighted, normalized impact SPL,[3] $L_{n,w}$, and the impact insulation class (IIC).[15] However, there is an important difference between them. Lower values of $L_{n,w}$ indicate higher impact sound insulation, whereas higher values of IIC indicate higher impact sound insulation. For many, but not all floor constructions, the relationship between them can be calculated using IIC = 110 – $L_{n,w}$, although when using this conversion there will occasionally be floors with differences up to 7 dB.[16]

7.3.3 Solid Plates

In dwellings, a solid concrete floor will not provide sufficient impact sound insulation without a soft floor covering or floating floor. However, it is useful to be able to estimate its impact sound insulation because it represents a common base floor that is used to assess the efficacy of different floor coverings. For a solid, homogenous plate that is excited by the ISO tapping machine, the normalized impact SPL is:

$$L_n = 10\lg\left(\frac{\rho_0^2 c_0^2 \sigma}{\rho_s^2 h c_L \eta}\right) + X\text{dB}, \tag{7.21}$$

where η is the total loss factor of the plate, X = 78 dB for one-third-octave bands and X = 83 dB for octave bands. Figure 7.14(a) shows a comparison of measured data with predicted values using this equation for a 140 mm solid concrete floor using the measured total loss factor.

From Eq. (7.21), it is seen that the total loss factor directly affects the impact sound insulation; hence, L_n, can differ between laboratories or between the laboratory and the field. In a similar way to airborne sound insulation (Section 7.2.2), it is possible to convert from one situation to another. Assuming that all measurement errors are negligible, it is possible to convert $L_{n,A}$ from Laboratory A, to $L_{n,B}$ that would be measured in Situation B (i.e., a different laboratory or *in situ*) using:

$$L_{n,B} = L_{n,A} + 10\lg\left(\frac{\eta_A}{\eta_B}\right). \tag{7.22}$$

7.3.4 Timber Floor

In dwellings, a basic timber floor will not usually provide sufficient impact sound insulation without a soft floor covering or floating floor, and other modifications to the ceiling. Figure 7.14(b) shows the impact sound insulation of two timber floors. Compared with concrete floors, timber floors tend to have high-impact SPLs in the low-frequency range, and lower levels in the high-frequency range.

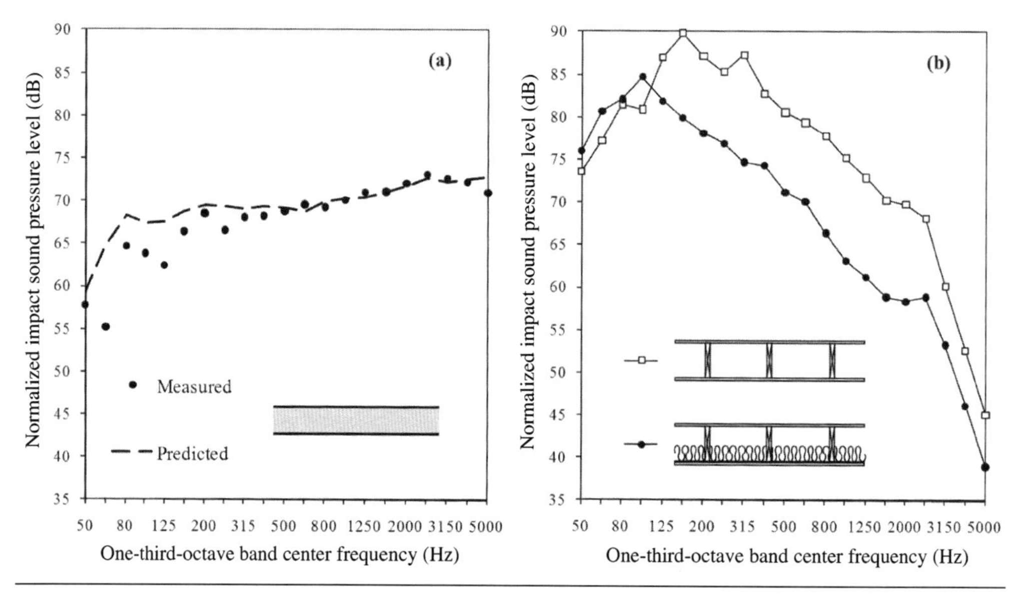

Figure 7.14 Impact sound insulation of floors measured with the ISO tapping machine: (a) Comparison of measured and predicted data for a 140 mm solid concrete floor; (b) Measured data for two timber joist floors: (1) a basic timber floor and (2) an upgraded timber floor with 100 mm mineral wool in the cavities and resilient metal channels supporting the plasterboard ceiling.

7.3.5 Floor Coverings

The most efficient way to improve the impact sound insulation is to tackle the excitation at the point of impact by placing a floor covering between the impact source (e.g., a human walker) and the base floor. Referring back to Figure 7.14(a) and 7.14(b), it is seen that base floors of concrete and timber both need floor coverings to reduce the impact SPL in the room below the floor to an acceptable level. Typical floor coverings include soft coverings such as carpet or linoleum, or those with a rigid walking surface on top of a resilient material, such as timber or screed floating floors.

The effectiveness of a floor covering on a standardized ISO base floor using the ISO tapping machine is quantified in the laboratory using the improvement of impact sound insulation ΔL using:

$$\Delta L = L_{n0} - L_n, \tag{7.23}$$

where L_{n0} is the normalized impact SPL for the ISO base floor without the floor covering and L_n is the normalized impact SPL with the floor covering. Note that the same equation is used for measurements with the ISO rubber ball, except that the SPLs are maximum time-weighted (Fast) levels.

The improvement of impact sound insulation depends upon the type of base floor; this can be seen from the examples in Figure 7.15. For this reason, ISO standards define a heavyweight base floor made of concrete, and three lightweight base floors made of timber and plasterboard. On lightweight base floors, the ISO rubber ball tends to give lower values for the improvement of impact sound insulation than the ISO tapping machine. In the low- and mid-frequency ranges the different impact sources also highlight differences between

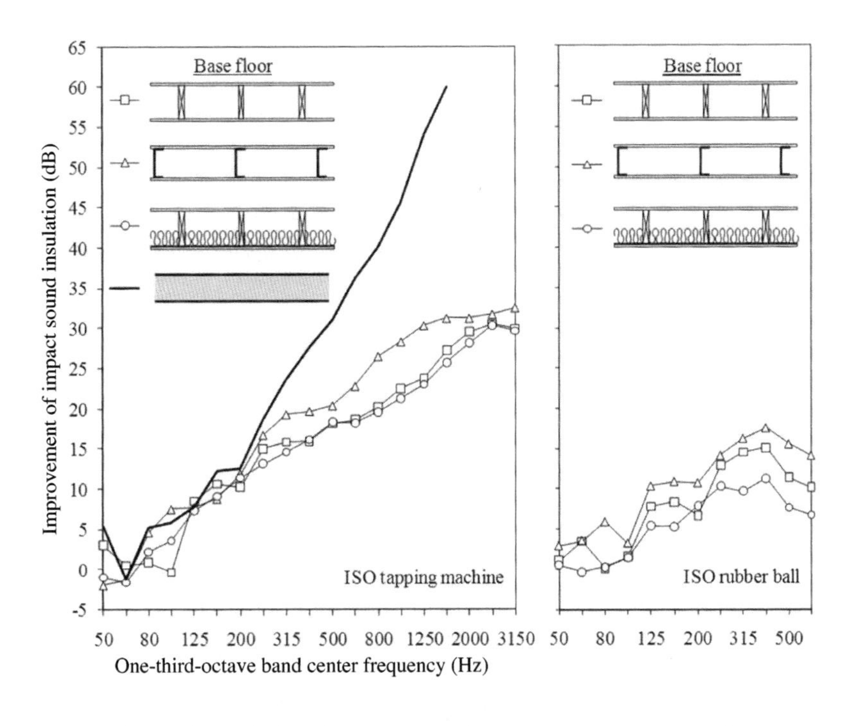

Floating floor: Timber raft (Chipboard on timber battens on mineral wool)

Figure 7.15 Improvement in impact sound insulation for a timber raft floating floor on different base floors (basic timber joist floor, steel C-joist floor, upgraded timber floor with 100 mm mineral wool in the cavities and resilient metal channels supporting the plasterboard ceiling, 150 mm solid concrete floor) using the ISO tapping machine (left) and the ISO rubber ball (right).

lightweight floors where the components have all been rigidly connected to make a stiffer floor, and floors where resilient connectors have been used to isolate components such as the plasterboard ceiling.

For soft floor coverings, the improvement of impact sound insulation on a heavyweight base floor is determined by the contact stiffness of the covering when it can be considered to act as a spring. In the low-frequency range $\Delta L \approx 0$ dB, ΔL typically increases in the mid- and high-frequency ranges at a rate of 12 dB/octave, assuming linear spring stiffness. However, many vinyl coverings and carpets act as nonlinear springs with a frequency-dependent Young's modulus. This results in ΔL increasing with a range of slopes varying from 5 to 22 dB/octave.[4] For this reason it is common to rely on laboratory measurements of ΔL to compare different soft coverings.

For a floating floor comprised of a rigid walking surface on a resilient material that covers the entire surface of a heavyweight base floor, the mass-spring resonance frequency associated with the floating floor is:

$$f_{\text{ms}} = \frac{1}{2\pi}\sqrt{\frac{s'}{\rho_{\text{s}}}}, \tag{7.24}$$

where ρ_s is the mass per unit area of the walking surface (kg/m^2), and s' is the dynamic stiffness per unit area of the resilient material (N/m^3) measured according to ISO 9052-1,[17] which is typically 10–12 MN/m^3 for 25 mm glass fiber and 65–80 MN/m^3 for 50 mm polystyrene. At the mass-spring resonance frequency, there is significant transmission of vibration that usually results in a visible dip in the impact sound insulation.

For floating floors with a walking surface comprised of sand-cement screed, the improvement in impact sound insulation above the mass-spring resonance frequency is typically 30 dB/decade and can be estimated using:

$$\Delta L = 30\lg\left(\frac{f}{f_{\mathrm{ms}}}\right). \tag{7.25}$$

For floating floors with a thin walking surface that has a low mass per unit area such as chipboard, the improvement in impact sound insulation above the mass-spring resonance frequency can be estimated using:

$$\Delta L = 40\lg\left(\frac{f}{f_{\mathrm{ms}}}\right) + 10\lg\left(1 + \left(\frac{\omega m}{2.3\rho_s c_L h}\right)^2\right), \tag{7.26}$$

where ρ_s, c_L, and h correspond to the plate that forms the walking surface and m is the mass of the ISO tapping machine hammer (0.5 kg).

Careful attention to detail is needed during the construction of floating floors because they must be fully isolated from the base floor by a resilient material underneath the walking surface and around the perimeter of the walking surface. Any structural bridge between the walking surface and the base floor or flanking walls will introduce an unwanted transmission path for structure-borne sound that effectively bypasses the resilient material. This can significantly reduce the impact sound insulation. A common cause of such a problem occurs when the resilient material has not been overlapped at its joints, leaving a gap, so that screed pours through the gap to form a structural bridge.

7.4 SOUND INSULATION *IN SITU*

In contrast to the laboratory, sound insulation *in situ* is almost always determined by the combination of direct and flanking transmission; this tends to increase the complexity of prediction. Along with flanking transmission, another complication is the variable quality of workmanship which can cause a wide spread in the field sound insulation performance of constructions that are nominally identical in their design.

7.4.1 Descriptors

This section gives an overview of descriptors commonly used for sound insulation within buildings and for façade sound insulation.

7.4.1.1 Sound Insulation Within Buildings

In the field situation, the field transmission coefficient, τ', accounts for the sound power, W_2, that is radiated by the separating wall or floor and the sound power, W_3, that is radiated by the flanking walls/floors or other elements, such as ducts; hence:

$$\tau' = \frac{W_2 + W_3}{W_1}. \tag{7.27}$$

For field measurements within buildings, the airborne sound insulation is usually quantified using the apparent sound reduction index, R', (equivalent to the apparent transmission loss in ASTM Standards) given as:

$$R' = -10\lg(\tau') = L_{p1} - L_{p2} + 10\lg\left(\frac{S}{A}\right), \tag{7.28}$$

or the standardized level difference, D_{nT}, where:

$$D_{nT} = L_{p1} - L_{p2} + 10\lg\left(\frac{T}{T_0}\right), \tag{7.29}$$

where S is the surface area of the separating wall/floor (m^2), A is the absorption area (m^2) of the receiving room that is calculated from the measured reverberation time, T (s) in the receiving room using the Sabine equation, and the reference reverberation time, T_0, is 0.5 s.

For field measurements within buildings, the impact sound insulation can be quantified using the normalized impact SPL:

$$L'_n = L_p + 10\lg\left(\frac{A}{A_0}\right), \tag{7.30}$$

or the standardized impact SPL:

$$L'_{nT} = L_p - 10\lg\left(\frac{T}{T_0}\right). \tag{7.31}$$

For furnished rooms in dwellings (excluding kitchens and bathrooms) with volumes between 15 and 60 m^3, the reverberation time over the building acoustics frequency range typically lies between 0.4 and 0.6 s.[18] Therefore a reference reverberation time T_0 of 0.5 s is reasonable for furnished dwellings where the reverberation time tends to be independent of the room volume. This is because the absorption area tends to be proportional to the volume, as larger rooms tend to have more absorbent furnishings.[19]

The field descriptors described above are determined for all frequency bands of interest; typically one-third-octave bands between 100 and 3150 Hz. To simplify the specification of acoustic requirements in building regulations and to facilitate comparison of different building constructions, it is necessary to reduce all this frequency-specific information into a single-number quantity using a standardized rating procedure. For airborne sound insulation, one of the original solutions was to use a rating curve with a shape that corresponded to a solid, heavyweight separating wall that was known to provide reasonable sound insulation. This led to the definition of a weighted apparent sound reduction index R'_w and a weighted standardized level difference $D_{nT,w}$.[3] An alternative solution has been to define a single-number quantity such that subtraction of this quantity from the A-weighted SPL in the source room *with a known spectrum shape* equals the A-weighted SPL in the receiving room. This has led to an alternative single-number quantity that represents an A-weighted level difference that can either be calculated directly from the frequency band data, or by adding a spectrum adaptation term to the single-number quantity R'_{w} or $D_{nT,w}$.[3] The advantage of this alternative approach is that it makes a simple link to an A-weighted SPL in the receiving room that can be used to assess whether the sound insulation is reasonable. The disadvantage is that there is no single spectrum shape that adequately represents all airborne sound sources inside or outside a building. In practice, most sources can be described using one of two different spectra: a pink noise source that represents speech, radio/television, or fast-flowing road/rail traffic; and a noise source with an enhanced low-frequency component that represents urban road traffic, jet aircraft at long distances, or music with significant bass content.[3]

Once a suitable source spectrum has been identified, the choice of R' or D_{nT} as a descriptor depends on what is to be assessed or inferred from the field measurement. A potential advantage of R' in the field is that in the absence of flanking transmission, it is possible to identify poor workmanship by comparing it with R for the same type of separating element that was tested in the laboratory. However, unless the sound insulation is relatively low (e.g., single leaf lightweight wall, door) it is fairly rare to find buildings without significant flanking transmission, and it is also necessary for the boundary conditions of the test element in the laboratory to be representative of the field situation. An issue also occurs with the definition of R' when field sound insulation is measured between diagonally adjacent rooms, as these will not have a defined area for the separating element that is needed for its calculation. A particular disadvantage in setting requirements with R' is that the sound perceived by a listener in the receiving room is not related to the power ratio that defines the transmission coefficient; so for regulatory purposes, it is not suited to buildings where the rooms have very different dimensions. For this reason it is more logical to use D_{nT}, because it provides a direct link to the SPL in the receiving room. However, one must be cautious when presented with D_{nT} data from field sound insulation tests in different buildings if the intention is to demonstrate *pass* and *fail* statistics of a nominally identical type of construction for regulatory purposes. This is because the data set could be biased by the inclusion of very small or very large receiving rooms. Another issue with D_{nT} is that when testing between rooms of unequal volume, its value will depend on the choice of source and receiving room. If the receiving room is the larger room, L_{p2} will be lower than if the smaller room had been chosen because of higher absorption area in larger rooms. When carrying out testing to assess compliance with regulations, it is therefore common to require that the smaller room is used as the receiving room so that the lowest D_{nT} values are measured.

Building regulations typically specify the minimum requirement that must be met for the field sound insulation using a single descriptor. However, any requirement on the level of sound insulation should logically be linked to the typical background noise level in the receiving room. This is usually achieved through other regulations on the minimum façade sound insulation, or by assumptions on the use of open windows or other ventilation schemes.

7.4.1.2 Façade Sound Insulation

Façade measurements are used to determine the sound insulation of a single building element such as a window or door, or the entire façade.

The airborne sound insulation of a facade element can be determined using an outdoor loudspeaker pointing toward the center of the test element at an angle of 45° from the plane of the façade. This results in an apparent sound reduction index (because some sound will also be transmitted by the rest of the façade) that is specific to the 45° angle of incidence:

$$R'_{45°} = L_{p1,s} - L_{p2} + 10\lg\left(\frac{S}{A}\right) - 1.5 \text{ dB}, \qquad (7.32)$$

where $L_{p1,s}$ is the average SPL measured on the surface of the test element, and L_{p2} is the average SPL in the receiving room.

Alternatively, the existing environmental noise such as road traffic can be used as the sound source:

$$R'_{tr,s} = L_{p1,s} - L_{p2} + 10\lg\left(\frac{S}{A}\right) - 3 \text{ dB}. \qquad (7.33)$$

The airborne sound insulation of an entire façade can be measured using either a loudspeaker or the environmental noise source to determine a standardized level difference, $D_{2m,nT}$, given by:

$$D_{2m,nT} = L_{p1,2m} - L_{p2} + 10\lg\left(\frac{T}{T_0}\right), \qquad (7.34)$$

where $L_{p1,2m}$ is the SPL measured outside at a distance of 2 m in front of the center of the façade, and at a height of 1.5 m relative to the floor of the receiving room.

7.4.2 Flanking Transmission Between Rooms—Airborne Sound Insulation

This section looks at the effects of flanking transmission on the airborne sound insulation between adjacent rooms. In general, it is only under laboratory conditions in a transmission suite that we can consider flanking transmission to be suppressed to such a level that it is reasonable to assume all sound is transmitted via direct transmission across the separating element. In the field, it is common for there to be significant flanking transmission. Due to the wide variety of different flanking constructions that exist in practice, it is possible for flanking transmission to reduce the sound insulation in any single frequency band, any frequency range, or even all frequency bands across the building acoustics frequency range.

For airborne sound insulation between adjacent rooms, it is useful to have a form of notation to describe some of the different transmission paths; these are indicated in Figure 7.16. Direct transmission via the test element is denoted Dd and the twelve first-order flanking paths are denoted as either Df, Fd, or Ff, where the uppercase letter indicates a surface facing into the source room and a lowercase letter indicates a surface facing (and radiating) into the receiving room. The term first-order flanking path refers to structure-borne sound transmission paths that cross only one junction of walls/floors.

Based on work by Gerretsen [1979],[20] ISO 15712[21] describes a prediction model that can be used to estimate the sound insulation when there is both direct and flanking transmission by considering only the first-order flanking paths. This model is essentially the same as SEA path analysis but the equations can be written in terms of a flanking sound reduction index, R_{ij}, determined from the ratio of the sound power, W_1, incident on a reference surface area,

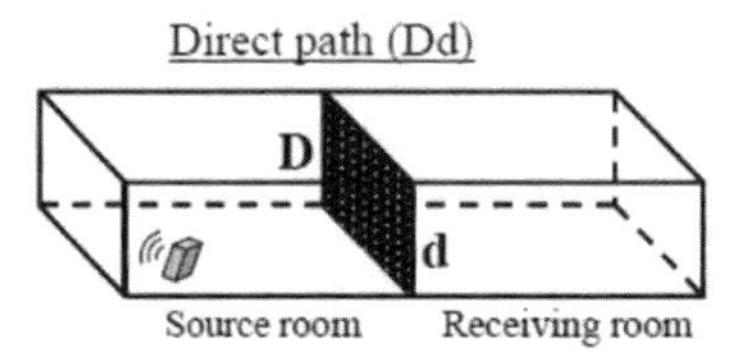

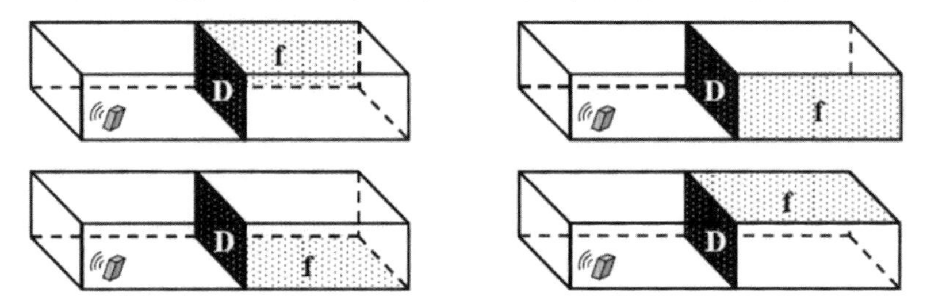

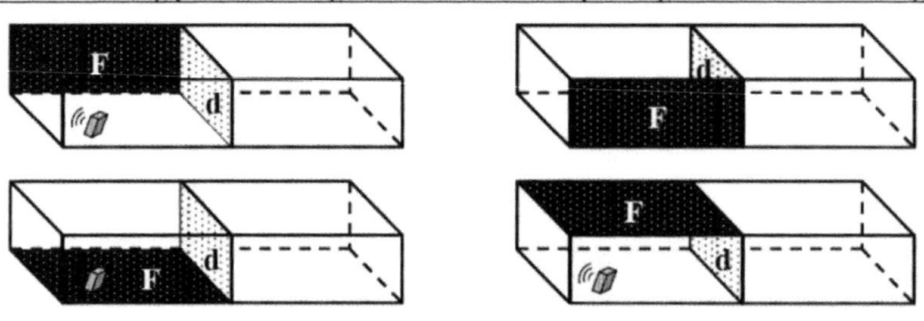

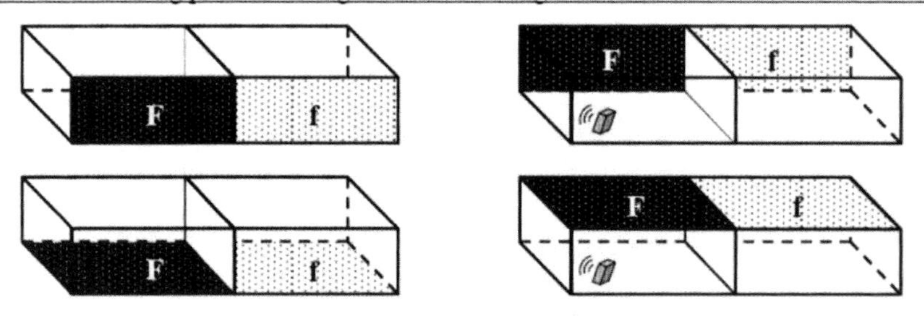

Figure 7.16 Direct and first-order flanking transmission paths for airborne sound insulation between adjacent rooms. For the first-order flanking paths, dark shading indicates the wall/floor excited in the source room, and light shading indicates the wall/floor that radiates sound into the receiving room.

S_S, in the source room to the sound power, W_{ij}, radiated by flanking plate j in the receiving room due to sound power incident on plate i in the source room, where:

$$R_{ij} = -10\lg\left(\frac{W_{ij}}{W_1}\right). \tag{7.35}$$

By setting the reference area, S_S, to equal the area of the separating wall/floor, the apparent sound reduction index for the combination of 12 flanking transmission paths is calculated from:

$$R' = -10\lg\left(10^{-R_{Dd}/10} + \sum_{\text{Paths}=1}^{12} 10^{-R_{ij}/10}\right), \tag{7.36}$$

where R_{Dd} is the laboratory measurement of the sound reduction index for the separating wall/floor that has been converted to the *in situ* total loss factor using Eq. (7.15).

In situ there will be many transmission paths involving structure-borne sound transmission across more than one junction, as well as transmission into and out of cavities contained within walls and floors. For this reason, the ISO 15712[21] prediction model that considers only direct transmission and 12 flanking paths for adjacent rooms is prone to overestimate the airborne sound insulation.[22,4] This limitation can be overcome by using SEA with the general matrix equation; this was introduced in Eq. (7.7) and can be expanded to any number of subsystems.

For many lightweight separating walls/floors it is possible to identify significant flanking transmission by comparing the sound reduction index for direct transmission from the laboratory, with the apparent sound reduction index in the field where there are additional flanking paths. Figure 7.17(a) and 7.17(b) show a construction that could occur in an office, school, or hospital where a continuous ceiling void runs between adjacent rooms. In Figure 7.17(a), the separating element is a studwork plasterboard wall. In the low-frequency range, the flanking construction has negligible effect on the overall sound insulation because the low-frequency performance of the lightweight separating wall is relatively low. However, in the mid-frequency range below the critical frequency of the plasterboard ceiling (3150 Hz one-third-octave band), there is a reduction in the sound insulation due to flanking transmission via paths across the ceiling void. A common solution to avoid this problem is to build the separating wall right up into the ceiling void to split the void into two separate cavities. Note that with this particular construction, there is no significant decrease in the sound insulation at the critical frequency of the plasterboard ceiling. This is because there is already significant radiation from the plasterboard that forms the separating wall at this critical frequency.

The same flanking construction is used in Figure 7.17(b) but with a dense aggregate masonry wall (plaster finish) for the separating wall. In this situation, there is not only a visible dip at the critical frequency of the plasterboard ceiling, but also a significant reduction in the airborne sound insulation in the low-frequency range. The latter occurs because the heavyweight separating wall has relatively high sound insulation in the low-frequency range due to its high mass per unit area, whereas the flanking paths involve single sheets of plasterboard that have low mass per unit areas.

Figure 7.17(c) shows a construction in a multi-story building where a floating floor is needed to meet the impact sound insulation requirements to the rooms below. In this situation, the screed has been laid continuously between horizontally adjacent rooms. At the junction of the lightweight separating wall and the screed, the wall has a negligible effect on vibration transmitted via the screed; hence this vibration is reradiated as sound into the receiving room. This structure-borne sound transmission path (Ff) via the screed results in a significant reduction in the airborne sound insulation. The solution to this problem is to lay

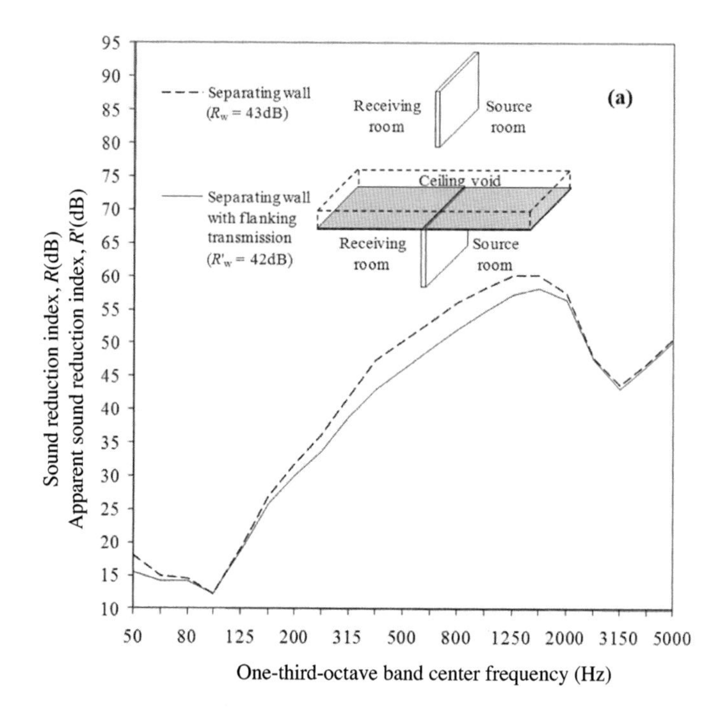

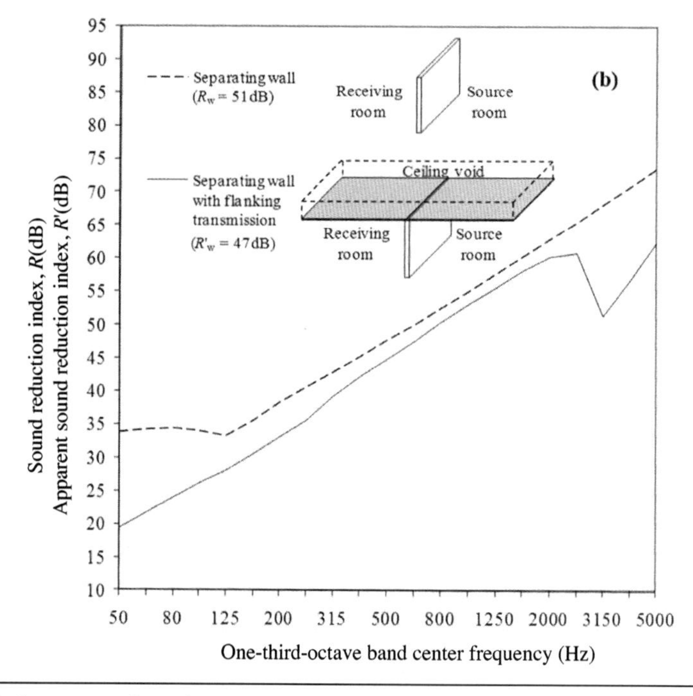

Figure 7.17 Airborne sound insulation across separating walls with different flanking constructions. (a) Separating wall: single sheet of plasterboard (≈ 9 kg/m^2) on each side of ≈ 70 mm steel C-studs spaced at ≈ 600 mm centers with low-density mineral fiber in the cavity. Flanking construction: ceiling comprises a single sheet of plasterboard with an empty ceiling void (0.5 m deep). (b) Separating wall: 100 mm masonry with plaster finish (200 kg/m^2). Flanking construction: ceiling comprises a single sheet of plasterboard with an empty ceiling void (0.5 m deep).

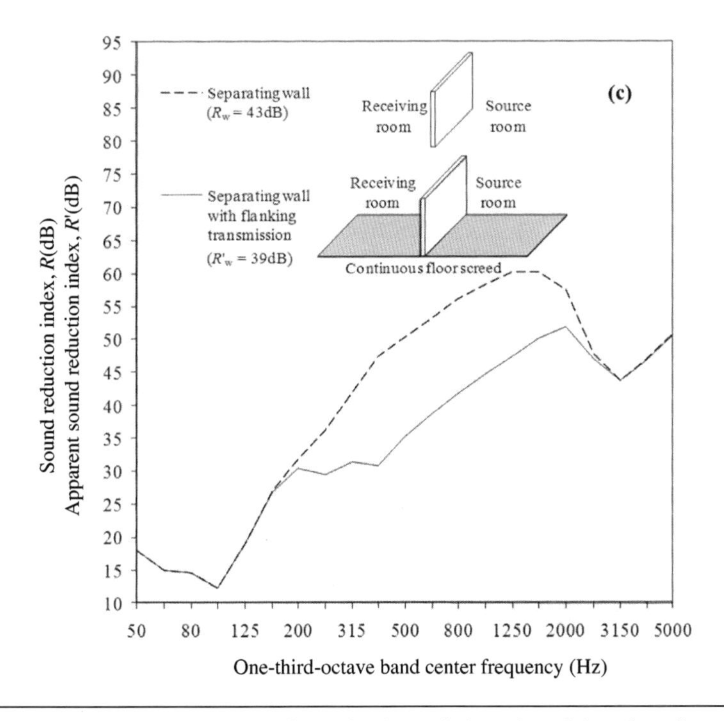

Figure 7.17 (continues) (c) Separating wall: single sheet of plasterboard (≈ 9 kg/m²) on each side of ≈ 70 mm steel C-studs spaced at ≈ 600 mm centers with low-density mineral fiber in the cavity. Flanking construction: 65 mm sand-cement floating floor screed on a resilient material and heavyweight concrete base floor.

individual screeds in each room with a break at the separating wall, so that the wall can be built off tracks fixed to the heavyweight base floor instead of the floating screed.

In the above examples for lightweight constructions, it was appropriate to consider only part of the flanking construction. However, for heavyweight masonry/concrete constructions, this approach is not usually possible. The main reason for this is that rigid connections are usually needed at the majority of wall and floor junctions for structural purposes. This allows structure-borne sound transmission to occur between all the walls and floors, which results in a large number of transmission paths that determine the overall sound insulation between rooms. Another factor to consider is that although bending waves are the main wave type of interest (because they are excited by sound fields and many other sources in buildings), when bending waves impinge upon a junction it is often possible for both bending and in-plane waves to be generated that propagate onto both the source and receiver plate. In order to account for all possible transmission paths and different wave types, the following examples use SEA models to illustrate the importance of flanking transmission in masonry/concrete constructions and the effect of in-plane waves.

Figure 7.18 shows the airborne sound insulation between two horizontally-adjacent rooms built from masonry/concrete elements. This allows comparison of direct transmission across the separating wall (using the *in situ* total loss factor) with *in situ* where there is both direct and flanking transmission. In this example the individual walls and floors support bending modes at frequencies below 100 Hz, but it is not until 630 Hz that they begin to support in-plane modes. Two SEA models are considered, the first model allows only bending waves to

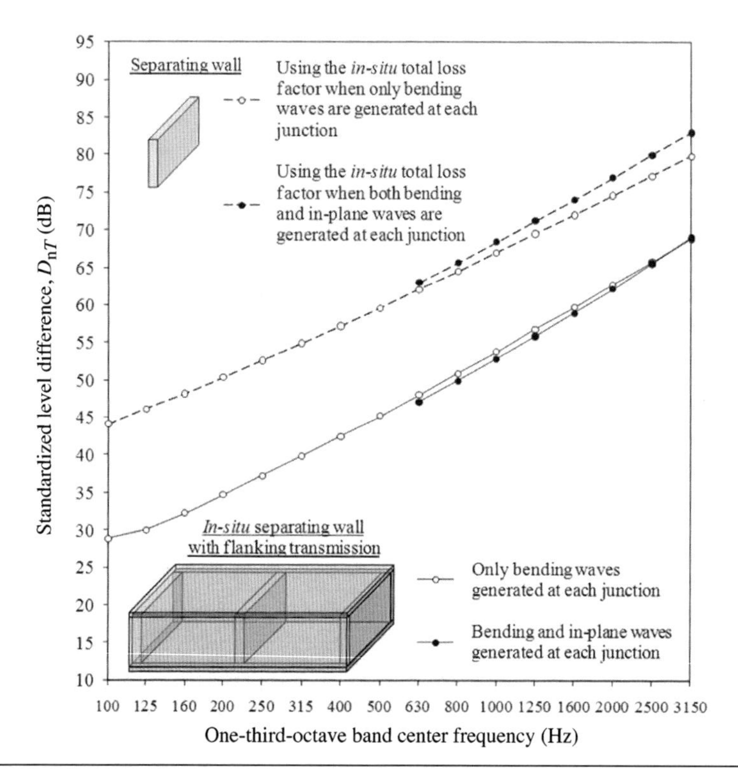

Figure 7.18 Airborne sound insulation across a separating wall in a heavyweight masonry/concrete construction. Separating wall: 215 mm masonry with a plaster finish (430 kg/m^2). Flanking construction: 100 mm masonry flanking walls with a plaster finish (200 kg/m^2), 150 mm concrete ground and upper floor (330 kg/m^2).

be generated at the junction, whereas the second model allows both bending and in-plane waves to be generated which is appropriate at and above 630 Hz. The results show that with both models there is significant flanking transmission across the entire building acoustics frequency range. In this example the two flanking walls radiate more sound power than the separating wall itself. For this reason it is rarely feasible to estimate the *in situ* performance by considering only the separating wall or floor.

Above 630 Hz the two models give similar values for the *in situ* sound insulation. However, structure-borne sound transmission across the four boundaries of the separating wall is significantly different when in-plane waves are generated at the junctions. The fact that the resultant sound insulation is similar does not necessarily mean that in-plane wave generation in the high-frequency range can be ignored for adjacent rooms. This is often done to simplify calculations where precise calculation in the high-frequency range is often unnecessary. However this simplification is not possible when predicting the sound insulation between non-adjacent rooms in large masonry/concrete buildings where the errors can be significant.[23]

7.4.3 Flanking Transmission Between Rooms—Impact Sound Insulation

This section looks at the effects of flanking transmission on the impact sound insulation between vertically adjacent rooms using SEA prediction models. The direct and first-order transmission paths for impact sound insulation between adjacent rooms are indicated in Figure 7.19.

The normalized impact SPL for the direct path is $L_{n,d}$, with a flanking normalized impact SPL, $L_{n,ij}$, defined as the normalized impact SPL due to the sound radiated by flanking plate *j* (wall/floor) in the receiving room with excitation of floor plate *i* by the ISO tapping machine. The ISO 15712 prediction model[21] gives the *in situ* impact sound insulation in terms of the normalized impact SPL using:

$$L'_n = 10\lg\left(10^{L_{n,d}/10} + \sum_{\text{Paths}=1}^{4} 10^{L_{n,ij}/10}\right), \quad (7.37)$$

where it is assumed that there is a box-shaped receiving room with four first-order flanking paths.

Figure 7.20 shows the impact sound insulation between two vertically-adjacent rooms built from a masonry/concrete construction. The results show a significant reduction in the impact sound insulation due to flanking over the entire frequency range, although this is not as large as the reduction that occurs for airborne sound insulation (refer back to Figure 7.18). In most buildings, a floor covering would be needed to improve the impact sound insulation; hence all the curves would be reduced by ΔL of the floor covering, but the reduction in the impact sound insulation due to flanking transmission would remain the same.

ACKNOWLEDGMENTS

Permission to publish data measured by the author at BRE is granted by the UK government under the terms of the Open Government Licence (www.nationalarchives.gov.uk/doc/open-government-licence/version/3) and BRE for Figures 7.4, 7.14, and 7.15, and from BRE Trust

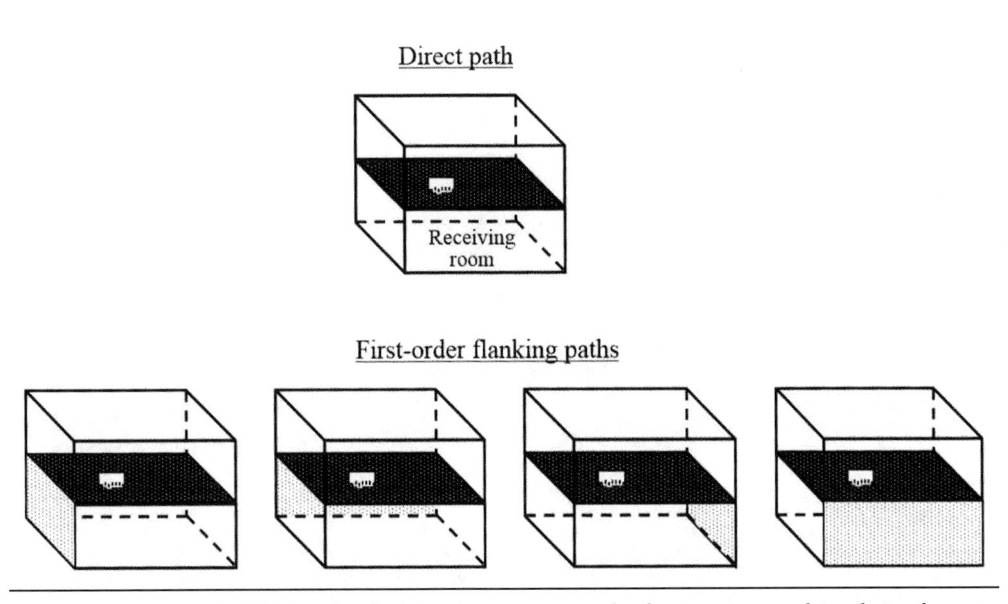

Figure 7.19 Direct and first-order flanking transmission paths for impact sound insulation between vertically-adjacent rooms. For the first-order flanking paths, dark shading indicates the floor excited by the ISO tapping machine, and light shading indicates the wall that radiates sound into the receiving room.

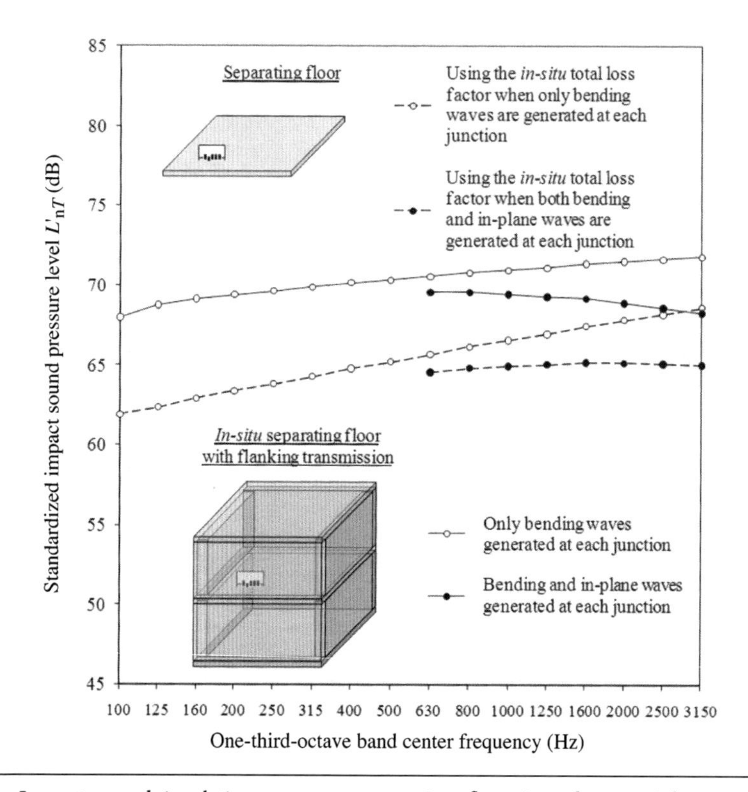

Figure 7.20 Impact sound insulation across a separating floor in a heavyweight masonry/concrete construction. Floors: 150 mm concrete slab (330 kg/m^2). Flanking walls in each room: Two walls are 215 mm masonry with a plaster finish (430 kg/m^2) and the other two walls are 100 mm masonry flanking walls with a plaster finish (200 kg/m^2).

(see Figure 7.7). Permission to use standard performance data from EN 12758[24] in Figure 7.9 is granted by BSI (www.bsigroup.com).

REFERENCES

1. Anon. ISO 10140 series. Acoustics—Measurement of sound insulation in buildings and of building elements. International Organization for Standardization.
2. Anon. ASTM E413-04 Classification for rating sound insulation. ASTM.
3. Anon. ISO 717. 1996, Acoustics—Rating of sound insulation in buildings and of building elements. Part 1: Airborne sound insulation. Part 2: Impact sound insulation, International Organization for Standardization.
4. Hopkins, C., 2007, Sound Insulation. Butterworth-Heinemann—An imprint of Elsevier. ISBN 978-0-7506-6526-1.
5. Lyon, R. H. and DeJong, R. G., 1995. Theory and application of statistical energy analysis, Butterworth-Heinemann, MA, USA. ISBN 0750691115.
6. Kihlman, T. and Nilsson, A. C., 1972, The effects of some laboratory designs and mounting conditions on reduction index measurements, *Journal of Sound and Vibration*, 24 (3), pp. 349–364.
7. Walker, K. W., 1993, 20+ years of sound rated partition design, Sound and Vibration Magazine, July, pp. 14–21.

8. Schultz, T. J., 1981, Impact noise testing and rating—1980, Report Number NBS-GCR-80-249 prepared for the National Bureau of Standards, Department of Commerce, Washington D.C., U.S.A by Bolt Beranek and Newman Inc.
9. Watters, B. G., 1965, Impact-noise characteristics of female hard-heeled foot traffic, *Journal of the Acoustical Society of America*, 37 (4), pp. 619–630.
10. Scholl, W., 2001, Impact sound insulation: The standard tapping machine shall learn to walk!, Building Acoustics, 8 (4), pp. 245–256.
11. Schultz, T. J., 1975, A proposed new method for impact noise tests, Proceedings of Internoise 75, Sendai, Japan, pp. 343–350.
12. Gerretsen, E., 1976, A new system for rating impact sound insulation, *Applied Acoustics*, 9, pp. 247–263.
13. Bodlund, K., 1985, Alternative reference curves for evaluation of the impact sound insulation between dwellings, *Journal of Sound and Vibration*, 102 (3), pp. 381–402.
14. Jeon, J. Y., Ryu, J. K., Jeong, J. H. and Tachibana, H., 2006, Review of the impact ball in evaluating floor impact sound. *Acta Acustica united with Acustica*, 92, pp. 777–786.
15. Anon. ASTM E989-06 Standard Classification for Determination of Impact Insulation Class (IIC). ASTM.
16. Warnock, A. C. C. (2004). Impact sound ratings: ASTM versus ISO. Proceedings of Internoise 2004, Prague, Czech Republic.
17. Anon. ISO 9052-1. 1989, Acoustics—Method for the determination of dynamic stiffness—Part 1.
18. Vorländer, M., 1995, Survey test methods for acoustic measurements in buildings, Building Acoustics, 2 (1), pp. 377–389.
19. van den Eijk, J., 1972, Sound insulation between dwellings: correction to 10.log S/A or to 10.log T/0.5? *Applied Acoustics*, 5, pp. 305–307.
20. Gerretsen, E., 1979, Calculation of the sound transmission between dwellings by partitions and flanking structures, *Applied Acoustics*, 12, pp. 413–433.
21. Anon. ISO 15712. 2005, Building acoustics—Estimation of acoustic performance of buildings from the performance of elements. Part 1: Airborne sound insulation between rooms. Part 2: Impact sound insulation between rooms. International Organization for Standardization.
22. Craik, R. J. M., 2001, The contribution of long flanking paths to sound transmission in buildings, *Applied Acoustics*, 62 (1), pp. 29–46.
23. Craik, R. J. M. and Thancanamootoo, A., 1992, The importance of in-plane waves in sound transmission through buildings, *Applied Acoustics*, 37, pp. 85–109.
24. Anon. EN 12758. 2002, Glass in building—Glazing and airborne sound insulation—Product descriptions and determination of properties. European Committee for Standardization.

8

Auditory Perception in Rooms

Jonas Braasch, Graduate Program in Architectural Acoustics, Rensselaer Polytechnic Institute

Jens Blauert, Institute of Communication Acoustics, Ruhr-University, Bochum, Germany

8.1 INTRODUCTION

As humans, we began building rooms that sounded *good* long before we understood how to influence a room's sonic qualities. One of the main reasons for this is that it used to be difficult to measure a room's influence on sounds in physical terms. In fact, this was not possible until the invention of the vacuum tube in the early 20th century, which enabled the development of sensitive electroacoustic measurement devices. Wallace Clement Sabine,[1] who is seen by many as the grandfather of modern architectural acoustics, had to rely on his ears to investigate the effect of absorbing acoustic materials in room reverberation. He did so by simply listening and using a stopwatch to measure a room's acoustic decay curve until he could no longer hear it. As a consequence, he is acknowledged as the first scientist to conduct a quantitative psychophysical study—as early as 1900—with the goal of understanding the perceptual consequences of room acoustics.

Listening in rooms is an illustrative example for emphasizing the relevance of binaural hearing. To clarify, this is a case where two ears are necessary to detect perceptual features, such as the positions and spatial extents of auditory events.* This is in contrast to monaural hearing, which relates to cases where one ear is sufficient to investigate the cues under examination. Pitch perception is a good example of a monaural task.

8.2 LOCALIZATION OF A SINGLE SOUND SOURCE

In order to understand how the auditory system processes sounds in a room, a brief introduction to binaural hearing is necessary. One of the first scientific discussions regarding binaural hearing started in the late 19th century and focused on the understanding of human sound localization. The problem concerned the questions of how the auditory system forms the position of the auditory event, given the position of a sound source. In 1882, S. P. Thompson (1882)[2] wrote a review of different theories with respect to this topic, where he listed the state of knowledge for his time in fundamental laws of binaural hearing. According to

* In this chapter the term *sound* denotes physical phenomena, that is, sound signals and waves. In contrast, auditory perceptual phenomena are denoted as an *auditory* event.

Thompson (1877),[3] he and A. G. Bell discovered the sensitivity of the auditory system to interaural phase differences. Their findings are based on the fact that the arrival times of a sound wave emitted from a single source arrive at different times at the left and right eardrums as a result of the different path-lengths from the source to the two ears. This arrival-time difference between the left and right ear is called the *interaural time difference* (ITD). A simple geometric model by Von Hornbostel and Wertheimer (1920)[4] predicts that the maximal ITD appears when a sound wave arrives from the side along the axis that intersects both eardrums (see Figure 8.1). For this case, the ITD is estimated as the distance between the eardrums ($\approx$ 18 cm) divided by the speed of sound in air ($\approx$ 340 m/s), which yields a maximum ITD of 529 μs. However, larger ITDs than this are observed in nature.

As a result of head shadowing and diffraction effects about the head, measured ITDs can be as large as 800 μs, depending on the size of the head used. Woodsworth and Schlosberg (1962)[5] proposed a model that estimated ITDs on the basis of the wave traveling around a sphere. Such a model was—and still is—a good predictor for the low-frequency range of hearing. The model later was modified to predict the ITD for all frequencies throughout the human hearing range, for instance, by Kuhn (1977),[6] Sayers (1964),[7] and Blauert (1974).[8] Toole and Sayers (1965)[9] demonstrated that a signal presented over headphones with electronically-generated ITDs induced a lateral shift of the auditory events along the *interaural axis* between both ears (see Figure 8.2—right panel). However, the existence of the skull between both ears not only determines the path the traveling sound wave has to follow but also results in the attenuation of sound waves at the contralateral eardrum. In this way, *interaural level differences* (ILDs)* are caused, in addition to ITDs. Artificially generated ILDs, when presented over headphones, also induce a lateral deviation of auditory events (shown in Figure 8.2—left panel).

By the end of the 19th century, a geometric model was established to estimate ILDs for various sound-source positions.[10] In contrast to ITDs, ILDs are strongly frequency-dependent. In the low-frequency range, the human head is small in comparison to the wavelength of sound and, therefore, diffraction has only a minor effect on the sound wavefront; this yields a smaller ILD. In the high-frequency range, the wavelength of sound is small as compared to the dimensions of the head. Consequently, much larger ILDs can be observed here rather than in the low-frequency range. In the high-frequency region, the ILDs are not only determined by the shape of the skull but are also greatly modified by the shape of the external ears—that is, the pinnae and the attached ear canals.

The frequency dependence of ILDs led to the idea of the so-called *duplex theory*, which claims that ITDs are the perceptually dominant cues used for localization in the low-frequency range, while ILDs are more dominant than ITDs at high frequencies. Lord Rayleigh (1907)[11] demonstrated, both theoretically and in a psychoacoustic experiment, that the head is very effective at attenuating the sound at the contra-lateral ear for high, but not for low frequencies. For this reason, in the low-frequency range, ILDs are considered too small to provide reliable localization cues. For high frequencies, it was concluded that the phase difference between both ear signals—and therefore the ITD—is ambiguous and, thus, auditory lateralization vanishes where the path-length difference between both ears exceeds the wavelength of signals from a sound source. Indeed, it was later shown by Mills (1958)[12] and others

* A *contralateral* sound source is a sound source that is located on the opposite lateral side of the head for the referenced ear. The same sound source becomes an *ipsilateral* source with regard to the other ear.

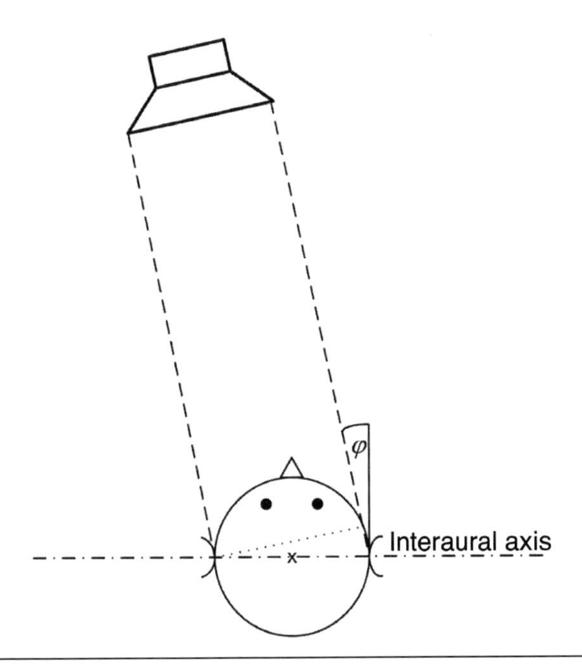

Figure 8.1 Model of Von Hornbostel and Wertheimer.

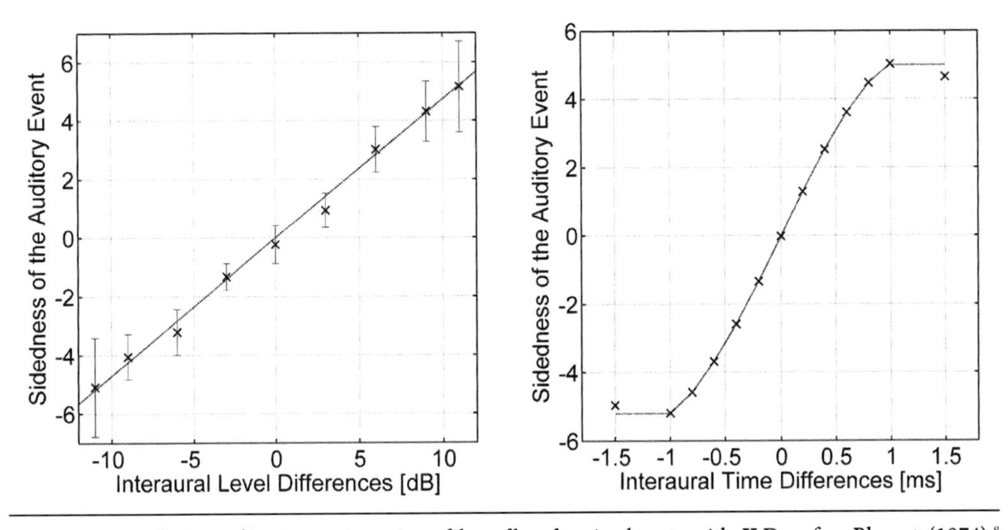

Figure 8.2 Left: Lateralization trajectories of broadband noise bursts with ILDs, after Blauert (1974).[8] Right: Lateralization trajectories curves of click signals with ITDs, based on data of Toole and Sayers (1965).[9]

for sinusoidal test signals that our auditory system is not able to resolve the fine structure of signal frequencies above approximately 1.5 kHz. Figure 8.3 shows the results of lateralization-blur experiments for various signals to demonstrate this effect. In a lateralization-blur experiment, the *just-noticeable difference* (JND) between two conditions (in this case by varying the ITDs) is measured. For high frequencies above 1.5 kHz, the auditory system is no

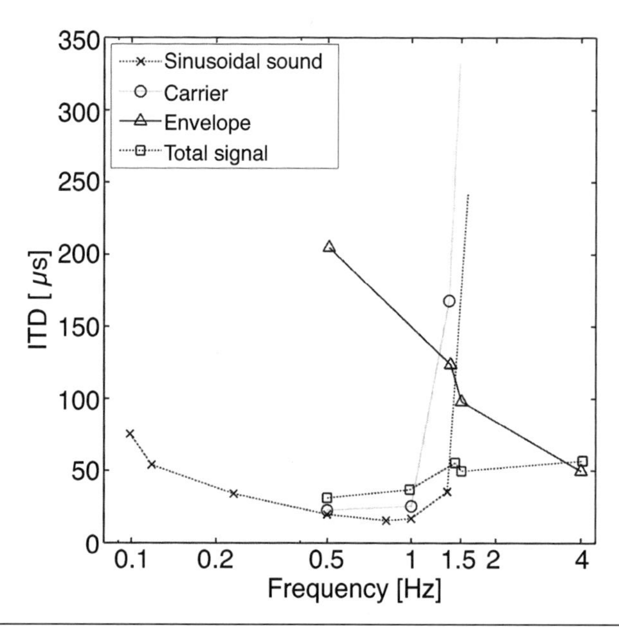

Figure 8.3 Lateralization blur for ITDs. The fine structure (carrier) of narrowband signals no longer serves as a lateralization cue above 1.5 kHz, while the envelope of the signal still carries useable information—based on data from Zwislocki and Feldman (1956),[13] Klumpp and Eady (1956)[14] for sinusoidal signals, and Boerger (1965)[15] for narrow-band bursts with a sinusoidal carrier and a Gaussian-shaped envelope.

longer able to extract ITD information from the fine structure of the signals, but it still can use ITD information obtained from the signal's envelopes.

For this reason, the duplex theory is no longer viewed as being accurate. It has become clear that ITDs are effective in the high-frequency range as well. This has been evaluated by means of signals with envelope fluctuations.[16] The analysis of the envelope, instead of the fine structure of the signals, also helps avoid phase ambiguities in the ITD-lateralization relationship. Furthermore, it has been revealed in detection experiments that our auditory system is equally sensitive to ITDs throughout the whole frequency range when *transposed* signals are employed at high frequencies.[17] In this study, a transposed signal was created by a half-wave rectifying a low-frequency sinusoidal tone, e.g., 125 Hz, and multiplying it with a sinusoidal tone of higher frequency, e.g., 2000 Hz. Assuming that the auditory system is analyzing the envelope of the composed signal to estimate ITDs, the signal analysis mechanism is more similar to that of a low-frequency signal than to that of amplitude-modulated signals. It was further shown in lateralization experiments[18] that the influence of ITDs in the high-frequency range can be improved by increasing the amplitude-modulation in the test signals.

So far, the question of how ITDs and ILDs are combined in the auditory system to form the lateral position of an auditory event has only partly been answered. For a long period of time, it was assumed that ITDs and ILDs are evaluated separately in the auditory system—namely, ITDs in the medial superior olive (MSO), as was first shown for dogs by Goldberg and Brown (1969)[19], and ILDs in the lateral superior olive (LSO). However, later neurophysiological findings revealed that ITDs in the envelopes of modulated sounds and even those in low-frequency carriers are processed in the LSO as well.[20, 21] In addition, David et al., 1959,[16] and Harris, 1960,[22] showed that the occurrence of the time-intensity-trading effect revealed a high complexity of processing in the auditory system. The trading effect describes the phenomenon that an auditory event often evolves midway between two expected positions in the case that ITD and ILD cues point to opposite directions. In other words, the cues compensate for each other. It has been suggested by Joris and Yin (1995)[20] that the combined sensitivity of single neurons in the LSO to ITDs and ILDs offers an easy explanation for the time-intensity trading effect. But keep in mind that the auditory event becomes spatially diffuse or even splits up into more than one auditory event when the ITDs and ILDs differ too much from natural combinations of ITDs and ILDs, such as observed in free-field listening—among others by Gaik (1990).[23]

Typically, the head-related polar coordinate system is used in binaural hearing-related studies (see Figure 8.4). In this system, the interaural axis intersects the upper margins of the entrances to the left and right ear canals. The origin of the coordinate system is positioned on the interaural axis, halfway between the entrances to the ear canals. The *horizontal plane* is defined by the interaural axis and the lower margins of the eye sockets, while the *frontal plane* lies orthogonal on the horizontal plane intersecting the interaural axis. The *median plane* is orthogonal to the horizontal plane, as well as to the frontal plane, and cuts the head in two symmetrical halves. Yet, it should be noted that the head is not perfectly symmetrical. For this reason, slight interaural time and level differences are also measured for sound sources in the median plane.

The position of sound sources and auditory events in three dimensions is described using the three polar coordinates azimuth, φ, elevation, δ, and distance, d. If δ is zero and d is positive, the sound source moves counterclockwise through the horizontal plane with increasing φ. At $\varphi = 0°$ and $\delta = 0°$, the sound source is directly in front of the listener, intersecting the horizontal and median plane. If φ is zero and d is positive, the sound source moves in front of the listener with increasing δ, upward along the median plane and downward behind the listener. Taking the model of Von Hornbostel and Wertheimer (1920),[4] ITDs of the same magnitude form hyperbolas in the horizontal plane, and at greater distances, the shell of cones in three-dimensional space is apparent—the so-called *cones of confusion*. Hence, there

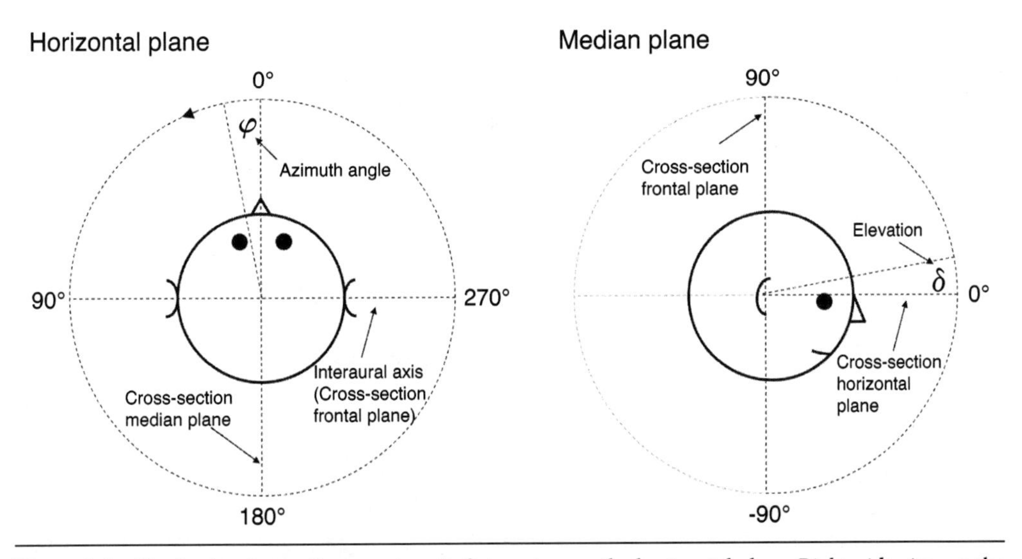

Figure 8.4 Head-related coordinate system. Left: top view on the horizontal plane. Right: side view on the median sagittal plane (median plane).

exist multiple positions with the same ITDs. Despite this, ITDs are still reliable cues for determining the left-right lateralization of a sound source.

In a detection experiment with sinusoidal sounds being masked by noise sounds, it was revealed that the auditory system analyzes sound in overlapping frequency bands,[24] called *critical bands*. The widths of these frequency bands correspond to a constant distance of approximately 2 mm on the basilar membrane. The relationship of frequency and distance on the basilar membrane between the place of maximum deflection and the helicotrema is approximately logarithmic, which means that the absolute frequency resolution is higher at low frequencies than for high frequencies. Therefore, the critical bands become broader with increasing frequency. Meanwhile, it is known that interaural cues are analyzed in such frequency bands as well. This fact is especially important for ILD evaluation due to their strong frequency dependence. By evaluating ILDs and ITDs across several critical bands, an unequivocal sound-source position can be determined for most directions. The median plane plays a special role in binaural psychoacoustics, as the interaural cues are very small here and cannot reliably be used to resolve positions. In this case, other cues become important.

It was found that for different elevation angles, the spectrum of the signals at the two ears is characteristically boosted or attenuated in different frequency bands as a result of diffraction and scattering at the head and external ears. It is assumed that the auditory system evaluated the frequency spectra of the incoming signals to determine the elevation of auditory events within the median plane and other sagittal planes. This fits the finding that for sinusoidal and other narrow-band signals, the vertical angle, δ, of the auditory event does not depend on the angle of sound incidence, but solely on frequency. For instance, for center frequencies of about 3 kHz, the auditory event appears in front, for 1 kHz behind, and for 8 kHz above—the so-called *directional bands* (Blauert, 1969/70).[25] For broadband signals, the following roughly holds. The auditory event is formed at an elevation given by the directional bands that receive the relatively highest amount of sound-signal power.

8.3 LISTENING TO MULTIPLE SOUND SOURCES

In many realistic situations humans have to resolve sounds from more than one source. In these cases, the desired source may be partly masked by concurrent sound sources. In general, there are two kinds of concurrent sources: *coherent* and *incoherent*, although any degree in between is possible, as well. Coherent sources (mirror sources) are generated by reflections of the desired primary-source on obstacles such as walls, ceiling, and furniture. Incoherent sounds are emitted statistically independent from the primary sound source, such as traffic noise or people talking in the background. Recognizing that interaural cues contribute significantly to our auditory assessment of rooms is essential for our understanding of how sounds in rooms are perceived.

In spaces with sound-reflecting walls, the sound that a source emits (e.g., an instrument) arrives at the receiver (e.g., human ears) via different paths (see Figure 8.5):

1. There is the direct path, which usually corresponds to the shortest distance between the source and receiver.
2. There are multiple reflective paths. The sounds along these paths change their directions whenever a reflective surface is hit. This means that they arrive at the receiver from different directions and with different delays, as compared to the direct sound.

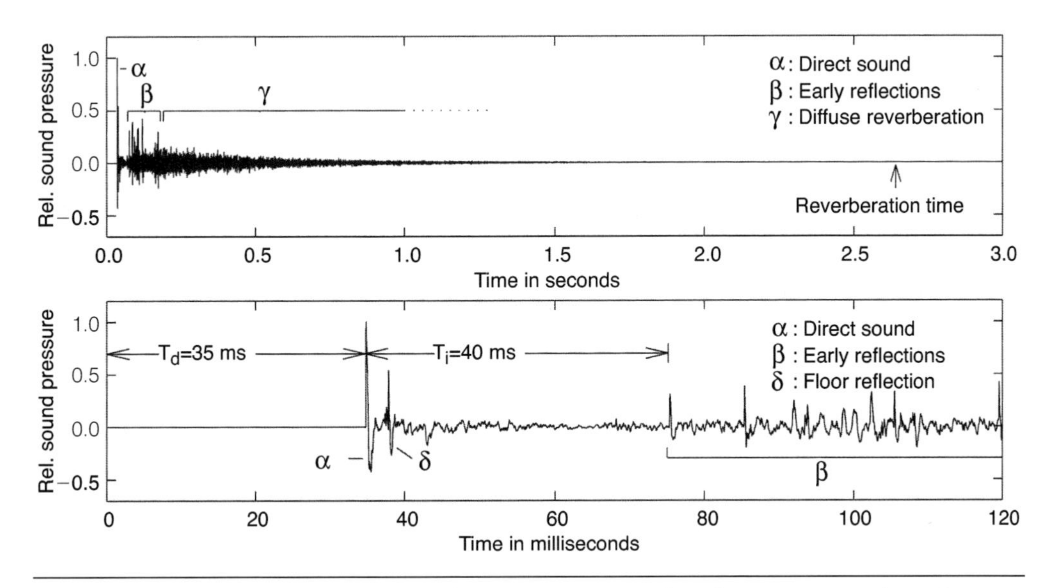

Figure 8.5 *Room impulse response* of the auditorium maximum at the Ruhr-University Bochum, Germany. Top: full response. Bottom: excerpt identifying direct sound, floor reflection, and further early reflections.

8.3.1 The Precedence Effect

The auditory system is capable of identifying the direct sound in these situations and, within a certain range of conditions, can form a homogeneous auditory event in a direction predominantly corresponding to the position of the primary sound source, thus, largely disregarding the directions of incidence of the reflective sounds. This complex and highly nonlinear phenomenon is called the *precedence effect*—or, historically, the *law of the first wavefront*.[26] This effect has been a central subject of scientific discussion for a long time.[27, 28, 29, 30]

To demonstrate the precedence effect in a laboratory, an often-used setup consists of two loudspeakers at different directions but from an equal distance with respect to a listener, as shown in Figure 8.6. The sound emitted by one loudspeaker, the *lag*, can be variably delayed with respect to the other one, the *lead*. For delays below about 1 ms, both loudspeaker signals contribute to the direction of the auditory event (called *summing localization*), an effect that is extensively exploited in stereo and surround-sound technology. For higher delays, there is still only one auditory event, but its position corresponds to the direction of the lead loudspeaker.

As noted previously, the directional information in the lag-loudspeaker signal is usually largely disregarded by the auditory system. However, it is important at this point to mention that this disregard only concerns directional information. A delayed sound can still be sensed by the auditory system and is, for instance, perceived as increased loudness, a colored timbre, and/or an increased spatial extent of the auditory event—as compared to what is heard when only the lead-speaker signal is presented. The spatial extent (width) of the auditory event is minimal for a delay of zero and increases monotonically with increasing delay. This holds true for any kind of broadband signal, such as impulses, noise, speech, and music; compare the discussion of *apparent source width* (ASW) later in the text.

For narrowband signals, the precedence effect is less distinct. In this case, the auditory event varies in its direction periodically with the inverse of the signal's center frequency in

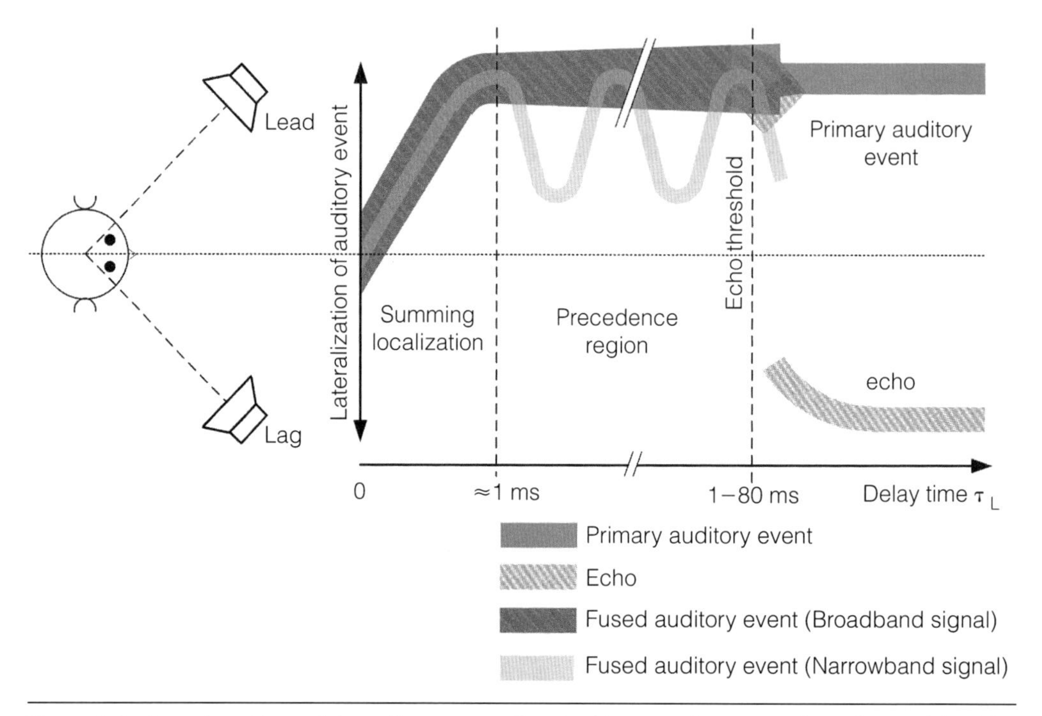

Figure 8.6 Lateral positions of the auditory events for a lead-lag stimulus pair in dependence of the delay time, τ_L, between lead and lag—after Blauert and Braasch (2005).[31]

the precedence-effect region. With decreasing bandwidth, this variation can extend to the midline between lead and lag, and even lead to failure of the precedence effect.[32] When the delay of the lag signal is further increased, the auditory event finally breaks up into a primary auditory event and an echo. The threshold where this happens, the *echo threshold,* varies depending on the source—from 1 ms for short impulses to upwards of 80 ms for organ music. It has been observed that the echo is perceptually shifted toward the direction of the lead when the delay between lag and lead is only slightly higher than the echo threshold—*summing localization of the second* kind.[8] With increasing delay time, this shift disappears (Djelani, 2002).[33] For delays slightly below the echo threshold, the auditory event may also shift slightly to the middle.

What holds for the case of only one simulated reflection (as described previously), is also true for multiple-reflection scenarios. Within a wide range of conditions, there is indeed only one homogeneous auditory event in the direction of the sound source—that is, the direction of the first incoming wavefront determines the direction of the auditory event. Obviously, the potency of our auditory system to deal with these situations in such a sophisticated way is the basis of our competence in localizing sound sources in enclosed spaces and other reflective environments.[31, 34, 35, 36]

Seraphim (1961)[37] took an alternative approach to measuring the influence of a room reflection. He was interested in the detectability of a single reflection or, said more precisely, the sound originating from a reflection. The data shown in Figure 8.7 were obtained using the *method of adjustment*, in which the listeners were instructed to adjust the level of the reflected sound until it was barely audible (detection threshold). For small delays, the listeners performed better in the diotic condition (lower threshold), but for high delays they did better

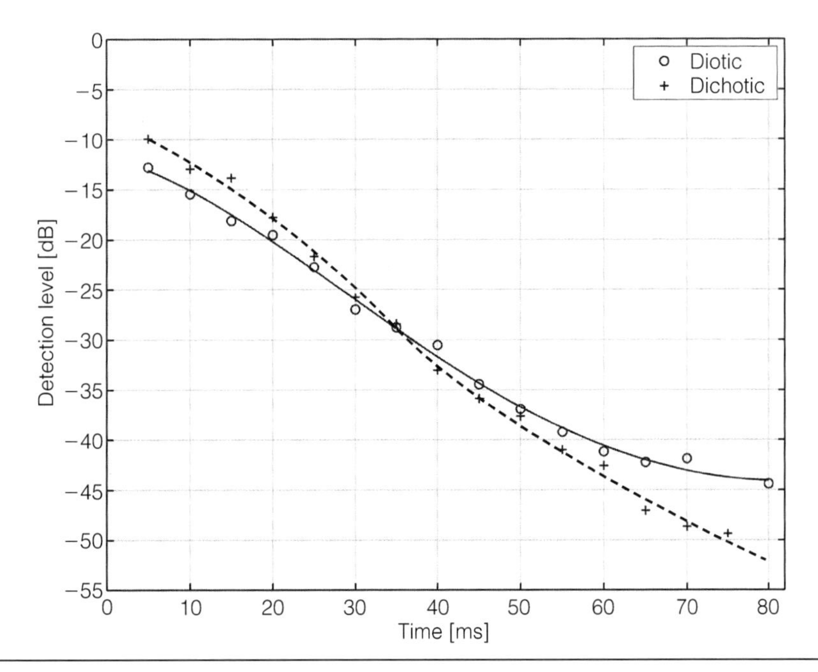

Figure 8.7 Detectability of a single reflection in a speech signal for a diotic* condition, where both the direct sound and the reflection are presented from 0° azimuth/0° elevation ('o') and a dichotic condition ('+'), in which the direct sound is presented from 0° azimuth and the reflection from 30° azimuth. The *solid* and *dashed lines* are 4th-order polynomial fits to the diotic and dichotic data points—based on data from Seraphim (1961).[37]

for the dichotic condition. It is assumed that temporal effects that lead to spectral distortions (the comb-filter effect) dominate for low delay times, while perceptual features, heard as changes in the spatial properties of the auditory event, are more important for larger delays.

8.3.2 Spatial Impression

The first physiology-related algorithm to estimate ITDs was proposed by Jeffress (1948).[38] The model consisted of two delay lines that were connected to several coincidence detectors. Cherry and Sayers (1956)[39] later substituted Jeffress' coincidence-cell network with an interaural cross-correlation algorithm, $\Psi(\tau)$. This is still the most commonly used approach in binaural modeling:

$$\psi(\tau)=\frac{\int_{t_1}^{t_2} p_l(t)\cdot p_r(t+\tau)dt}{\sqrt{\int_{t_1}^{t_2} p_l^2(t)dt\cdot\int_{t_1}^{t_2} p_r^2(t)dt}}, \tag{8.1}$$

with left and right ear signal pressures, $p_l(t)$ and $p_r(t)$, and integration times, t_1 and t_2.

Aside from taking advantage of an algorithm that is as commonly used in mathematics, another advantage of interaural cross-correlation is that the peak height of the normalized cross-correlation function, $\Psi(\tau)$, can be used to calculate interaural coherence, which is a good predictor of the spatial extent of the auditory event—that is, it can serve as an estimator

* *Diotic* ... identical sounds at both ears; *dichotic* ... the sounds differ interaurally.

of how spacious we perceive binaural acoustic signals. In this context, Blauert and Lindemann (1986)[40] reported on experiments where the spatial extent of auditory events for binaural sound signals with a varying degree of interaural cross-correlation has been measured. The setup that is typically used for this type of experiment is shown in Figure 8.8. Three independent noise generators produce signals of varying degrees of correlation. Although the stochastic character of each sound source is identical, meaning that their long-term spectra are identical, their fine structures do not correlate with each other. Two of the sound sources (Noise 1 and Noise 3) are directly sent to the ears via headphones, one to the left ear and the other one to the right ear. Since these two signals are uncorrelated, when played together they produce an interaurally-decorrelated signal.

The third signal (Noise 2) is sent diotically to both ears, meaning that both ear signals for this sound source are identical, producing a fully interaurally-correlated signal. Using a panning device as shown in Figure 8.8, the diotic third signal can be adjusted against the other two sound sources to produce signals with any degree of interaural correlation. Note that the power of two ear signals can be held constant while their correlation is modified.

In an earlier pilot study by Chernyak and Dubrovsky (1968)[41] that demonstrates the basic effect of binaural decorrelation, listeners were presented with signals as described previously and asked to draw the extent of their auditory events as intracranial images. Figure 8.9 shows

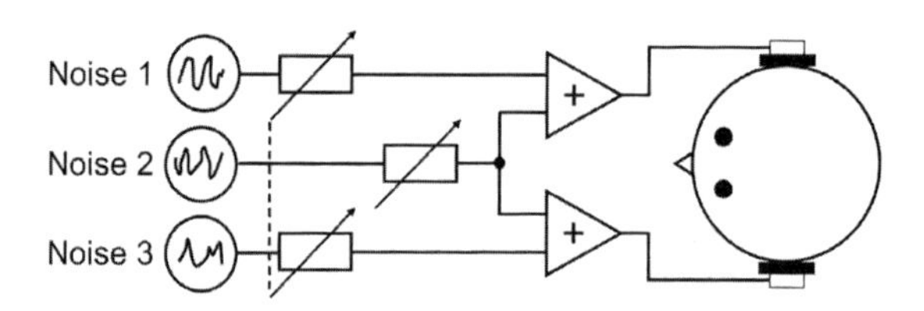

Figure 8.8 Experimental setup for generating two ear signals with an adjustable degree of interaural correlation. Note that the power of the two ear signals can be held constant while their correlation is modified.

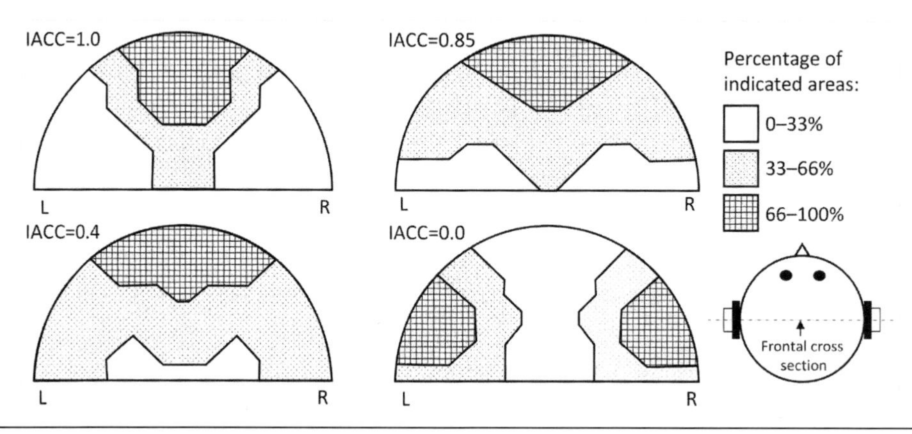

Figure 8.9 Position and extent of auditory events based on the degree of interaural correlation plotted in a frontal cross section of the head. The percentages indicate the relative number of occurrences in which the auditory event extended into the corresponding area—schematic after Chernyak and Dubrovsky (1968),[41] compare Blauert and Lindemann (1986).[40]

the culminating results. For a highly correlated signal, the widths of the auditory events were very narrow. The images became wider with decreasing interaural correlation, creating *auditory spaciousness*, until the auditory events split into two separate events for the left and right ears when the interaural correlation fell below a threshold.

The findings of this experiment are important when it comes to understanding perception in rooms, as reflected sound decorrelates both ear signals of a sound source in rooms. This holds for late reflections ($\tau >$ about 80 ms) and, in particular, for the diffuse late reverberation. In the older literature, this effect was called *auditory spaciousness*, but currently the term ASW is more frequently used for it.

Since the decorrelation effect of sideward reflections is much stronger than that of ceiling and floor reflections, Barron and Marshall (1981)[42] carried out experiments to study the influence of the early lateral reflections on human auditory perception.

ASW has been defined as the apparent auditory width that a performing music ensemble invokes for a listener seated in the auditorium of a concert hall. To study ASW, Barron and Marshall (1981)[42] used a setup as is frequently employed to study auditory perception in concert halls. In such a setup as shown in Figure 8.10, multiple loudspeakers are used to simulate:

1. The direct sound
2. Early reflections by means of delay lines
3. Diffuse reverberation, using a plate reverb or another device to artificially generate a stochastic decay curve

Barron and Marshall's experiment focused on the ASW for early reflections at different delay times, τ, compared to a reference pair of early reflections, τ_r. The delay time, τ, is also known as *the initial time-delay gap* (ITDG), which is defined as the arrival time of the first major early reflection measured from the arrival time of the direct sound source. In the room impulse response (RIR) shown in the lower graph of Figure 8.5, the ITDG, labeled T_i, is 40 ms. Note that the earlier floor reflection, γ, is not used to define the ITDG. In Barron and Marshall's study, the ITDGs for the left and right early reflections were slightly offset to avoid a coincidental arrival of both reflections, which would have resulted in a reduced ASW. Listeners were asked to match the ASW for both conditions by adjusting the levels of the reference reflections. As Figure 8.11 shows, the early reflections are less effective to produce a wide auditory event when the delay times are small compared to the reference delays. For example, at a delay time of 5 ms, the levels of the reflections in the reference condition had to

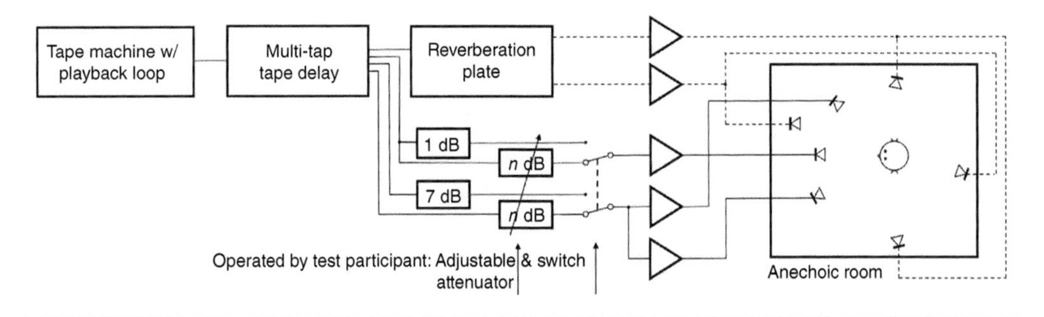

Figure 8.10 Early setup to simulate sound fields in concert halls—after Barron and Marshall (1981).[42]

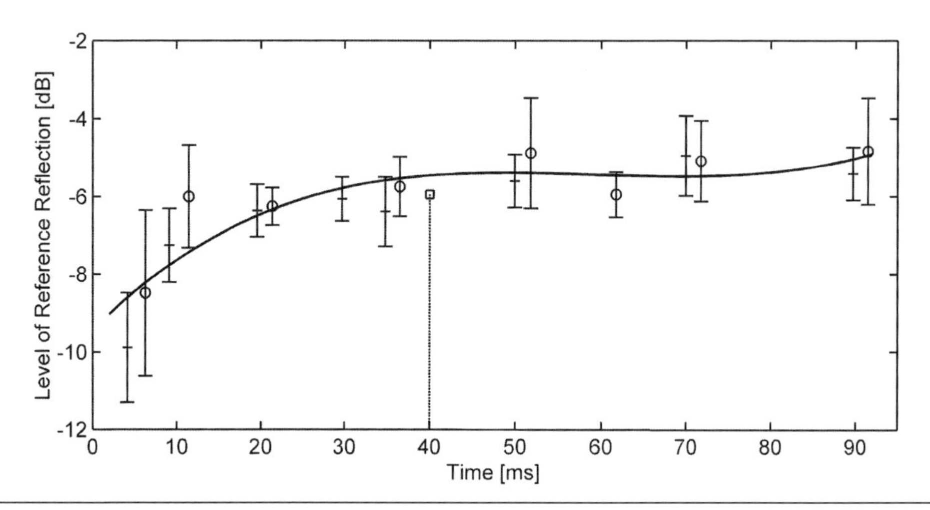

Figure 8.11 The ITDG of early reflections influences the ASW, ('-', Mozart motif; 'o', Wagner motif; '□', reference reflection). The y-axis shows the relative sound pressure level of the reference reflection compared to the sound pressure level of the direct sound. The x-axis shows the ITDG. For each ITDG, the test participants adjusted the level of a reference reflection pair at an ITDG of 40 ms to match the ASW of a test reflection pair at varying ITDGs—alternating between the test and reference conditions. The decline of the test-reflection level at low ITDGs indicates that ASW decreases with the ITDG. Data points and error bars show means and a 95% confidence interval (data from Barron and Marshall (1981),[42] 3rd–order polynomial curve fit by the current authors).

be reduced by 4 decibels (dB) for the Mozart motif as compared to the reflection levels in the test condition, in order to invoke the same ASW.

The results of this study emphasize the importance of paying attention to lateral early reflections in concert hall design. If the concert hall is too narrow, the path-length differences between the direct sound and the two side wall reflections is very low, resulting in small delay times that hardly contribute to a wide auditory event. As a result, orchestras in such a hall will be perceived as too narrow sound-wise. In contrast, if the hall is too wide, the path-length differences can lead to delay times beyond the echo threshold.

Another component which helps form the spatial impression of a room is called *listener envelopment* (LEV).[43, 44] LEV is defined as a listener's perception, for instance, in a concert hall, being enveloped by the sound field created by a performing music ensemble. LEV is primarily related to the reverberant sound field, which is generally said to begin about 80 ms after arrival of the direct sound.

A severe problem in room acoustics arises when sounds from a reflective surface arrive so late that they lead to perceptible *echoes,* that is, repetitions of primary auditory events. Echoes are among the biggest headaches an acoustical consultant typically has to deal with. Once the hall is built, there is often no straightforward way to eliminate the problem, since the geometry is fixed. Often, the best compromise is to put absorbing materials on the echo-producing walls, but this comes at the cost of reducing the levels of the side reflections which are essential for producing wide auditory images. Scattering surfaces may be an alternative in such cases.

8.3.3 Instrumental Indices for Perceptual Assessment of Rooms

It should be noted that in psychoacoustical experiments, such as the one described by Barron and Marshall (1981),[42] only one concert hall seat is typically simulated. However, a real concert hall needs to work for a large variety of seat positions. To achieve this in the design process of a concert hall is not trivial, especially when complex room geometries like those of vineyard-style concert halls are considered.

To help estimate the expected quality for performance spaces, a number of instrumental predictors (indices) have preferably already been defined in its design phase, which can be derived from physical data, measured or simulated (ISO 3382-1, 2009).[45] These indices are particularly useful to be applied in computer models[46] or physical scale models[47] of such spaces. Yet, it has to be stated at this point that the index values are often not well correlated with corresponding perceptual attributes. Further, among other deficiencies, they do not include information from other senses than hearing and, certainly, do not consider individual preferences.[48]

One of these indices concerns the ITDG. The ITDG index describes the arrival time of the first major reflection after the direct sound, but it does not take into account the level of the reflected sound, the direction of its incidence, nor information about whether or not it is accompanied by another reflection from the opposite side. Beranek (1962)[49] found that for the best-rated concert halls, the ITDG lies typically between 15 and 27 ms (see Figure 8.12).

Quite useful indices for ASW and LEV are based on measurements of interaural cross correlation of *binaural room impulse responses* (BRIRs). To measure these, typically a loudspeaker is placed on stage to simulate an acoustical source at the position of the orchestra's center and a binaural manikin is placed on a seat in the audience to measure the BRIR. For the ASW, the first 80 ms of a BRIR is analyzed as follows:

$$\psi(\tau)=\frac{\int_0^{80\,ms} p_l(t)\cdot p_r(t+\tau)dt}{\sqrt{\int_0^{80\,ms} p_l^2(t)dt\cdot\int_0^{80\,ms} p_r^2(t)dt}}, \tag{8.2}$$

with $\Psi(\tau)$ being the interaural cross-correlation function, and p_l and p_r being the left and right signals of a binaural impulse response. Usually, the maximum of this function within $-1\text{ ms} > \tau > 1\text{ ms}$ is considered for further evaluation.

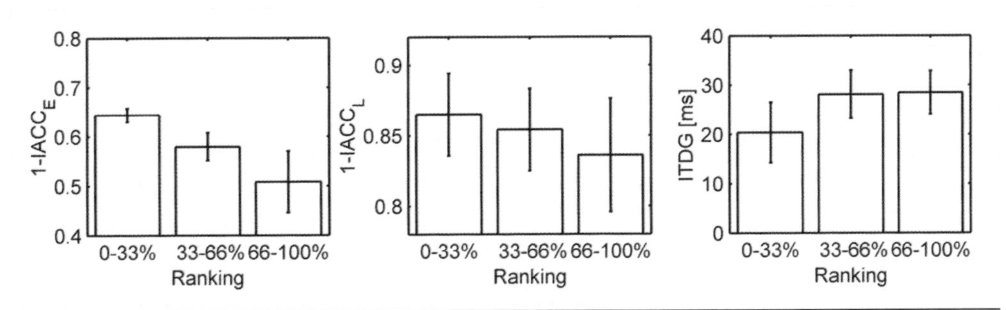

Figure 8.12 Room acoustical parameter values for differently ranked concert halls. Left: binaural quality index (BQI); center: [1-IACC$_{L3}$] corresponding to LEV; right: ITDG. Rankings: 0–33%...top-third ranked concert halls; 33–66%...medium-ranked concert halls; 66–100%...lowest-rated concert halls. The column height denotes the mean for each group, the error bar the standard deviation—based on data from Beranek (2003).[50] (For definitions of BQI, 1-IACC$_{L3}$, LEV, and ITDG, see text.)

LEV is predicted in a similar fashion, but this time, the remaining part of the BRIR after 80 ms is examined, that is:

$$\psi(\tau)=\frac{\int_{80\,ms}^{\infty} p_l(t)\cdot p_r(t+\tau)dt}{\sqrt{\int_{80\,ms}^{\infty} p_l^2(t)dt\cdot\int_{80\,ms}^{\infty} p_r^2(t)dt}}. \qquad (8.3)$$

Note that for both the ASW and LEV indices, the BRIRs are processed in octave-wide frequency bands. Typical center frequencies used for investigation are 500 Hz, 1000 Hz, and 2000 Hz—average values over a number of frequency bands are also usually given (e.g., Beranek, 2003).[50] This constitutes the interaural cross-correlation coefficient, $(IACC)_{E3}$, if the first 80 ms are used to measure $\Psi(\tau)$ [Eq. (8.2)], or $IACC_{L3}$, if $\Psi(\tau)$ is measured using the late part of the BRIR [80 ms – ∞, in Eq. (8.3)]. Since the acoustical quality of a concert hall improves with a decreasing interaural correlation, a *binaural quality index* (BQI) has been defined as [1-$IACC_{E3}$]. The BQI for favorable concert halls is typically around 0.65 (see Figure 8.12).

As to the IACC, it should be pointed out that the observable range of the interaural correlation no longer lies between 0 and 1 once the sound has been transformed with the ears of the binaural manikin and further band limited with the octave-band filter. Since processing with the binaural head—cross-talk between both ears in particular—and bandwidth reduction always partly correlate an incoming signal, the physically observable bottom range is substantially higher than zero with higher values at lower frequencies because the absolute filter bandwidth (which determines the increase in correlation) decreases with decreasing frequency for an octave filter. The auditory system processes sound in bands which originate at the basilar membrane—the critical bands as mentioned above. These bands are approximately one-third of an octave wide, and for this reason, graphic equalizers in audio engineering applications often use third-octave bands. Since auditory bands are narrower than the commonly used octave-filter banks for ASW and LEV measurements, the physiologically possible range of interaural correlation in the auditory bands is even smaller than is found in the octave band analysis.

A fixed threshold of 80 ms for discriminating between the ASW and LEV analyses in Eq. (8.2) and Eq. (8.3) has also been questioned. A better way of distinguishing both phenomena is to separate the components of BRIRs that belong to the early reflections and those that belong to diffuse (late) reverberation. The early reflections are usually of specular quality in concert-hall BRIRs as they result from the reflection off large plane walls. Consequently, they usually correlate highly with the direct sound. The diffuse reverberation is, as the name suggests, decorrelated from the direct sound. Practically speaking, the 80-ms criterion is an acceptable compromise in concert hall acoustics, as the first early reflections should arrive before this threshold (to avoid echoes) and most of the energy in diffuse reverberation is found beyond this threshold since it results from multiple reflections with long path lengths and delays.

What has been reported in the previous paragraph might lead to the impression that long impulse responses with much energy in the tail already make good concert halls. However, it has to be kept in mind that ASW and LEV are certainly not the only criteria for determining the quality of concert halls. Two further relevant criteria, among others, which can be instrumentally estimated, are features called *clarity*, *C*, and the loudness or *strength*, *G*, of a given music program perceived within the hall.

Clarity is an index for the ability of a hall to maintain the temporal transparency of music or speech in a given room. The late reverberation typically degrades the perceived clarity of a music program. The optimal quantity of reverberation is thus a compromise of achieving good clarity ratings while maintaining sufficient auditory spaciousness. Since clarity deals with the temporal aspects of impulse responses, a single-channel room impulse is used to determine a clarity index. It is obtained by replacing the binaural manikin with a single, omnidirectional microphone. Clarity is then computed by use of the following equation:

$$C = 10\log_{10}\left(\frac{\int_{0\,ms}^{t_e} p^2(t)\,dt}{\int_{t_e}^{\infty} p^2(t)\,dt}\right), \quad (8.4)$$

where p is the pressure signal of the measured impulse response and t_e is the upper integration threshold (traditionally 50 ms for speech and 80 ms for music). The threshold, t_e, is attached as a subscript (e.g., C_{80}) to indicate how the clarity-index values have been obtained.

The relationship between speech intelligibility and clarity is shown in Figure 8.13. *Clarity* (C_{50}) values above 0 dB typically result in an environment with good speech intelligibility. A similar threshold is found for the clarity of symphonic music in concert halls if the C_{80} measure is used. Reichardt et al. (1975),[51] proposed critical areas for clarity (C_{80}). If the C_{80} value is above 1.6 dB, a concert hall's clarity is typically very good, but if this value falls below −1.6 dB, the clarity is weak. If clarity is substantially below the critical area, the temporal details of symphonic music sound *washed out* but if it is too high above the critical area, a hall often lacks spatial impression.

While clarity is usually defined as a monaural parameter, the underlying mechanism has interaural components, as the effect of *binaural de-reverberation* demonstrates. With one ear in a reverberant room being occluded, the room will be perceived as more reverberant—as the experiment depicted in Figure 8.14 shows. An amplitude-modulated noise signal

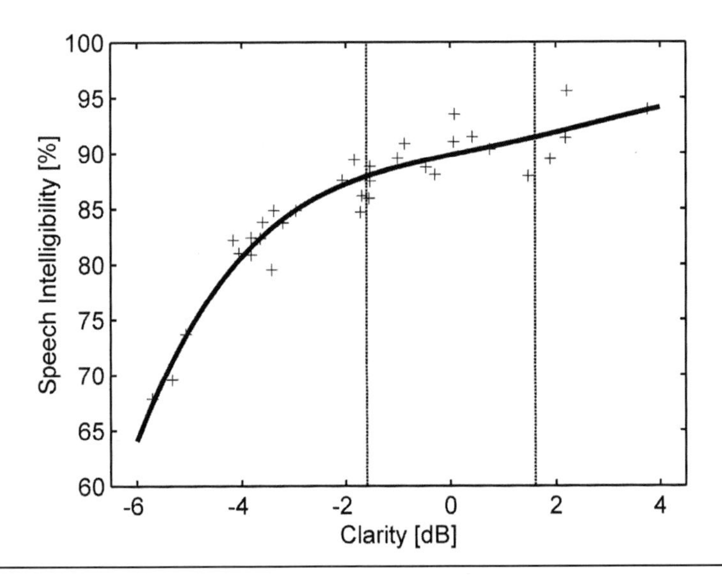

Figure 8.13 Speech intelligibility as a function of clarity (C_{50}). The solid line is a 4th-order polynomial fit. The vertical dashed lines indicate the critical area for clarity (C_{80}) of symphonic music in concert halls—based on data from Boré (1956)[52] and Reichardt et al. (1975).[51]

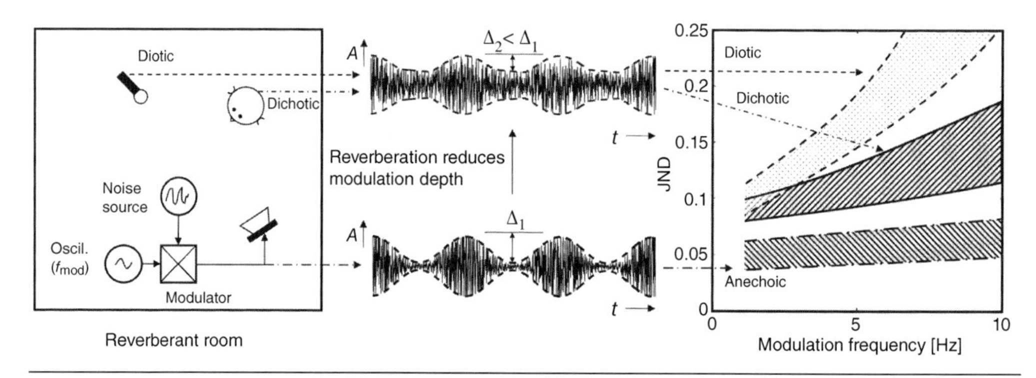

Figure 8.14 Measurement of binaural de-reverberation after Danilenko (1969).[53] *Left*: amplitude-modulated noise being emitted into a reverberant room with varying degrees of modulation and presented to the test participant diotically or dichotically. *Center*: waveform with reduced modulation depth in the reverberant condition (*top*) compared to the anechoic condition (*bottom*). *Right*: just-noticeable degrees of modulation for the reverberant diotic and dichotic conditions as well as the anechoic baseline condition.

is emitted from a loudspeaker in a reverberant room and recorded with either a single microphone or a binaural manikin. The impulse response of the room smears out the signal and thus reduces the amount of modulation. Depending on the modulation frequency, the modulation of the single-channel signal (the same signal is presented *diotically* to both ears) is detected at lower degrees of modulation than the binaural signal from the manikin. Consequently, the perceptual consequences of the smearing-out effect are less pronounced in the binaural condition.

As mentioned above, loudness is another important factor in judging the acoustics of a concert hall. A loudness-related index is called *strength*, *G*. It determines how well a concert hall carries the sound of a given ensemble in the hall. It is given as the ratio of the sound pressures measured in a hall and the sound pressure of the same source in a free field without room reflections. Strength has been defined as:

$$G = 10 \cdot \log_{10} \left(\frac{\int_{0\,ms}^{\infty} p^2(t)\,dt}{\int_{0\,ms}^{\infty} p_F^2(t)\,dt} \right), \tag{8.5}$$

with the sound pressure, p, measured in the concert hall, compared to the same value that would have been obtained at the same position in the free field, p_F. The same performance will be perceived louder in one hall than another if it has a higher strength rating. Control of strength becomes an increasing problem with hall size since the acoustic energy of an ensemble has to be distributed over a larger audience surface area. For large halls, it becomes critical to optimally guide the sound with hard surfaces rather than using absorptive materials, which reduce the strength.

At the beginning of the 19th century, the increasing size of spaces for musical performance concert halls led to an unavoidable reduction in strength. Thus orchestras had to become larger to reach similar sound pressure levels in the audience. Meyer (1978)[54] describes this trend for the music of Joseph Haydn, who started with an 18-member orchestra at Esterházy's court in 1766, but ended his career with 59 musicians at King's Theatre in 1795. Meyer's acoustical analysis reveals that the overall sound pressure level (90 dB in 1766 vs.

92 dB in 1795) was similar during large parts of Haydn's career, namely, because the venues in which his music was performed increased in size over time.

8.3.4 Limitations of the Room-Impulse-Response Concept

In this section, the limitations of the impulse-response measurement and analysis approach are discussed. In fact, this method is today's preferred method among acoustical consultants when assessing the acoustical characteristics of concert halls and to predict their acoustic quality. Yet, although the impulse-response method has its merits and has been the standard in the field for many years, it is a very simplified representation of how a human listener experiences sounds in these venues. The impulse response describes the acoustic pathway of a single stationary sound source to a stationary receiver. However, in a typical concert situation, we tend to listen to more than one musical instrument at a time. Even for the case of a solo concert, the directivity pattern of the sound radiated by the instrument into different directions changes over time unlike a loudspeaker, which has a constant directivity pattern over time. Traditional perceptual experiments, like the study by Barron and Marshall (1981)[42] among many others, reduced the size of the physical body of a large orchestra to a single point source, represented by one loudspeaker. This, of course, has implications when making judgments about the ASW of the virtual orchestra. Without early reflections, a full orchestra spread out on stage will be perceived as a wide sound object, whereas for the one-loudspeaker representation, early reflections are needed to provide the ASW.

Another limitation of the impulse-response method is that does not account for listener head and translational movements. It basically reduces a human to a passive system and not as an active being that actively explores his or her environment. It has been previously shown that a human listener can actively resolve front-back confusions of sound sources by turning the head a few degrees.[55]

It must also be questioned to what extent the auditory system possesses mechanisms to form a representation of the reflection pattern a sound source arrives with at both ears—so to say, an internal representation of a BRIR. Unlike the measurement process for impulse responses, the auditory system does not have access to the sound at the source position, and it needs to make all judgments based on sound analysis of the ear signals. A recent experiment has shown that human listeners have difficulties matching the reverberation time of diffuse reverberation tails for two different ongoing sound sources[56] (see Figure 8.15). In this experiment, listeners had no difficulties adjusting the reverberation time of a test signal to a reference signal, if both source signals were identical (conditions shown by arrows). For both the orchestra and click sound, the average mismatch was well below 50 milliseconds. However, the average mismatch grew to values as large as 250 milliseconds with large standard deviations if the test signal and reference signals differed in structure. It appears that different sound sources invoked different auditory percepts of reverberation for the same impulse response. The results suggest that the auditory system cannot form a detailed temporal representation of a diffuse reverberation tail from a running signal.

One form of displaying a BRIR in line with our current model of the auditory processes in the central nervous system is the so-called *binaural-activity map* (compare, e.g., Blauert, 2005; Blauert et al., 2013).[34, 57] The latter is a method to represent both the temporal and lateral positions of the reflected sound in a three-dimensional plot. One method of obtaining a binaural activity map is to segment a BRIR into temporal slices of 10–100 milliseconds

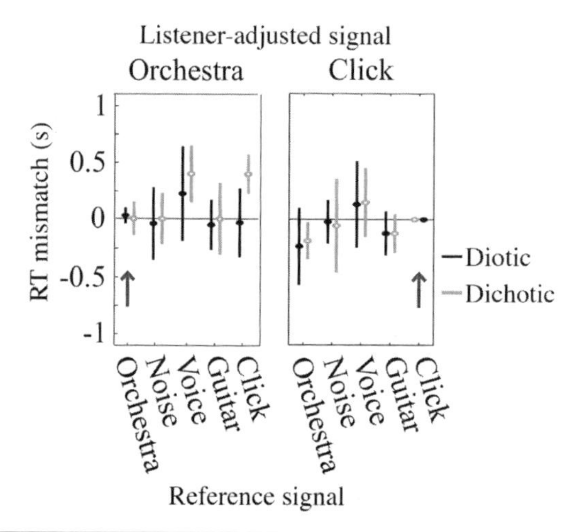

Figure 8.15 Reverberation time judgment of eight listeners, who adjusted the reverberation time of a diffuse, exponentially decaying reverberation curve for a test signal (orchestra, left graph or click, right graph) to the reverberation time of a reference signal as shown in the x-axis. The reverberation time of the reference signal was randomly selected between 0.8–2 seconds. The data shows the average mismatch in reverberation times (T_{60} mismatch) between the test signal and the reference signal for the diotic (black bars) and dichotic (gray bars) conditions. The arrows show the cases where the reference signal and the test signal are identical. The error bars show the standard deviations—after Teret et al. (2015).[56]

and then to perform a binaural analysis over each time slice. Figure 8.16 shows an advanced method[58] that can calculate a binaural activity map from two running ear signals without the need of a measured impulse response. The algorithm is able to predict the delays and lateral positions of early reflections from the running ear signals, but it cannot resolve the temporal pattern of diffuse reverberation like it appeared to in the case of Teret et al.'s experiment.

Possibilities for applying advanced models of binaural signal processing and perception in the context of prediction of and judgment on the quality of spaces for performances will most likely increase in the near future as the development of such a system is a current research focus.[57, 59, 60, 61]

8.4 THE *QUALITY OF THE ACOUSTICS*

So far, we have discussed how humans perceive sounds in rooms and how different characteristics of a room lead to perceptual differences, but we have not addressed the question of how these characteristics lead to preference judgments for a particular acoustical enclosure. Several studies have examined the question of how to rate the acoustical quality of concert halls including a field study by Beranek (1962, 2003),[49, 50] from which some results were already shown in Figure 8.10. In this study, conductors were asked in a letter to rate concert halls according to their past performance experience.* The ranking was then compared to the measurements of acoustical parameters for these halls. One can conclude from this study that a single parameter, such as reverberation time, does not necessarily lead to adequate

* Another method is to archive the sound of different concert halls, e.g., via binaural recordings and then have human subjects rate these recordings in a laboratory environment—see Schroeder et al., 1974.[62]

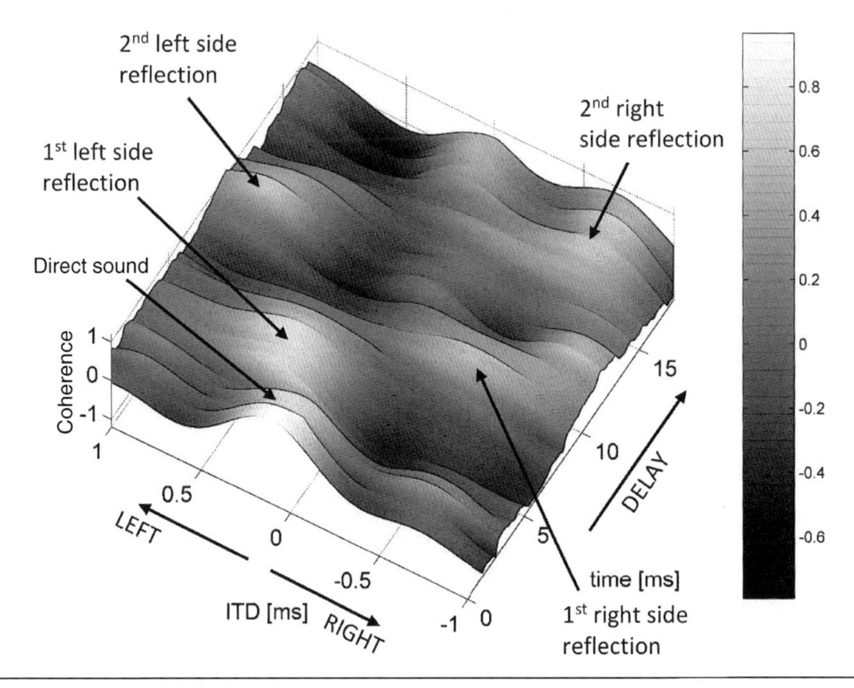

Figure 8.16 *Binaural-activity map* extracted from a 22-s speech clip (spatialized with head-related transfer functions) located in front (0-ms ITD, 0° Azimuth) and four early reflections arriving with delays of 6 ms (left side, 45° Azimuth corresponding to an ITD of +0.5 ms) and 9 ms (right side, 315° Azimuth, corresponding to an ITD of –0.5 ms), 12 ms (left side, 60° Azimuth, corresponding to an ITD of +0.7 ms) and 15 ms (right side, 300° Azimuth, corresponding to an ITD of –0.7 ms).

sound in a concert hall, even if the aforementioned parameter is within an optimum range. Instead, the parameters ranges can be seen as tolerance bands, and whenever an acoustical parameter of a concert hall falls well outside this range it is highly likely that this concert hall has acoustical problems. One has to keep in mind, though, that acoustical parameters are not fully independent of each other. Thus, in most cases, more than one acoustical parameter falls out of line in concert halls with problematic sounds, because one has to deal with a multilayered problem here.

For traditional concert-hall ratings, for instance, it is the compromise between clarity—which calls for a low reverberation time—and LEV—which rises with increasing reverberation time. This, of course, is very much task-dependent. In an opera, clarity becomes more important than for symphonic music because the singer needs to be intelligible, and therefore opera halls generally have a lower reverberation time than concert halls. To take care of interdependencies like this one, a revision of conventional quality judgment is needed. A historic case for the necessity of this was the introduction of public address (PA) systems in the 1920s. Until then, the only way to make a given sound source louder was by adding reverberation—see strength parameter G. After the introduction of the PA system, it became possible to achieve high sound pressure levels using dry signals. As a consequence, a number of architects began to favor electroacoustically enhanced venues like Radio City Music Hall in New York City's Rockefeller Center. Later, many of these aural designers reconsidered the positive aspects of reverberation.

A concept that goes beyond preference rating of concert halls for classical music is introduced in Blauert (2012),[63] following up previous work on product-sound quality.[64, 65] This novel *Quality-of-the-Acoustics* (QoA) approach does not aim at rating the acoustical quality of an acoustical enclosure per se. Instead, the adequacy of a performance space for a specific task is evaluated; such as its suitability to host, for instance, a symphony orchestra or a rock concert—compare also Blesser and Salter (2006)[66] in this context. Also, it is an important aspect of this approach that judgments on QoA are individually specific, that is, are subject to individual experience and taste, among other items.[67] In general, it is assumed that the listeners, when judging on QoA, perform a comparison of what they actually hear with internal references that they have developed beforehand, and that reflect their expectations and desires with regard to a high QoA.

Research is currently on its way to simulate this process with computer algorithms based on extended models of binaural signal processing.[57, 61] Figure 8.17 schematically depicts a possible architecture for such a QoA-assessment system. First, the *character* of the auditory scene concerned is evaluated by collecting information on all perceptual features that may be relevant for quality. Second, the internal references are determined that listeners possibly apply. Finally, a measure of the multidimensional distance (norm) between the character of the scene and character of the associated internal reference is established. The smaller this distance, the higher the QoA is rated.

It goes without saying that such an algorithm needs reliable data regarding the perceptual features concerned. To broaden the scope of applicability of the system, all task specific and individual aspects are dealt with in the character of the references, while the scene characters are kept largely neutral to enable applicability to a broad variety of tasks and listeners. Yet, one has, of course, to take care that the scene characters capture as many as possible quality relevant features. Here applies a rule that holds for all *automatic* recognition and assessment

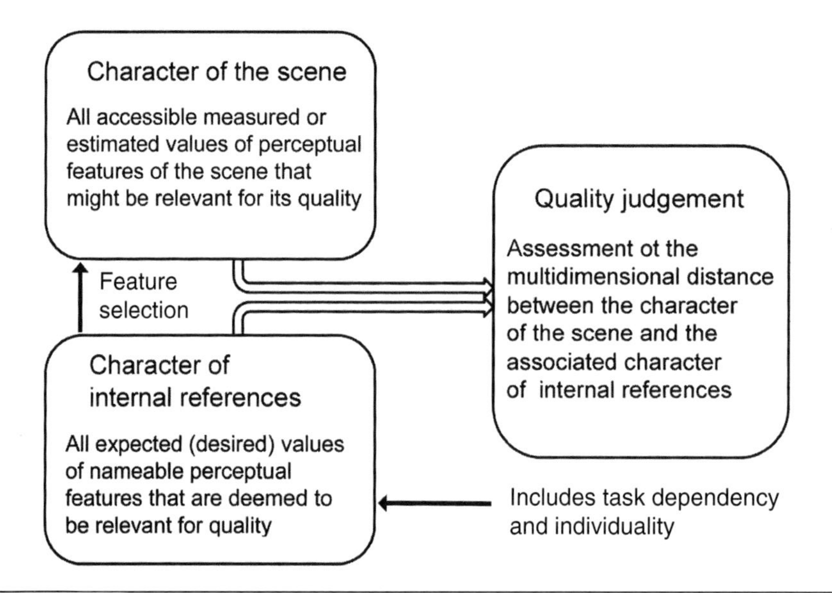

Figure 8.17 Architectural plot for a model of the quality-evaluation process according to Blauert and Jekosch (2007)[65] and Blauert et al. (2013).[57]

systems, namely, their performance depends on proper identification of suitable relevant features.

CONCLUSION

In this chapter, some of the remarkable capabilities are outlined that the human auditory system possesses to analyze the aural characteristics of rooms and sounds that are presented within these rooms. The auditory system's basic binaural mechanisms to localize sounds in rooms are described, based on interaural level and time differences. Thereafter an overview is given on how listeners perceive sounds in rooms, starting with the role of early reflections and late reverberation. The precedence effect is highlighted as a fundamental mechanism of the human auditory system, used to suppress information about the room reflections in order to analyze characteristics of the direct sound, for example, its direction of incidence. At the same time, the auditory system can extract properties of a room from the acoustic reflections. Subsequently, it is discussed how measured BRIRs can be used to instrumentally assess the characteristics of a room by the use of instrumentally derived indices—like, reverberation time, clarity, strength, ASW, and LEV. A section on the limitations of the BRIR analysis method follows suit. It is highlighted at this point that the method produces estimates that diverge from the actual practice of listening to music. This includes the restriction of the sound source to a point source and the fact that the method assumes a static, nonmoving listener. It is further briefly described how recent models can potentially overcome some of the limitations by extracting acoustic features directly from the ear signals to form a binaural-activity map without prior knowledge of the sound characteristics at close distance. The chapter concludes by introducing a novel QoA concept that can be used to determine the task-dependent perceptual quality of aural spaces by forming an expert panel to determine a reference set based on perceptual aural characteristics. This reference set can then be used by a second panel, preferably composed of an actual audience, for performing judgments on the QoA of a performance space.

ACKNOWLEDGMENTS

The authors thank Nikhil Deshpande for proofreading the manuscript. Figures 8.2, 8.3, 8.8, 8.9, and 8.14 have been adapted with permission from Blauert (1974).[8] The research results depicted in Figures 8.15 and 8.16 were accomplished with support from the National Science Foundation (IIS-1320059) and the European Commission (FP7-ICT-2013-C-#618075, www.twoears.eu).

REFERENCES

1. Sabine, W. C. (1922). *Collected Papers on Acoustics.* Harvard University Press. Cambridge, MA.
2. Thompson, S. P. (1882). On the function of the two ears in the perception of space. *Phil. Mag.* 13, pp. 406–416.
3. Thompson, S. P. (1877). On binaural audition. *Phil. Mag.* 4, 274–276.
4. Von Hornbostel, E. M. and Wertheimer, M. (1920). *Über die Wahrnehmung der Schallrichtung* [On the the perception of the sound direction]. In: Sitzungsber. Akad. Wiss., Berlin, pp. 388–396.

5. Woodworth, R. S. and Schlosberg, H. (1962). *Experimental Psychology*, pp. 349–361. Holt-Rinehart–Winston, New York, NY.
6. Kuhn, G. F., 1977. Model for the interaural time differences in the azimuthal plane. *J. Acoust. Soc. Am.* 62, pp. 157–167.
7. Sayers, B. McA., 1964. Acoustic-image lateralization judgement with binaural tones. *J. Acoust. Soc. Am.* 36, pp. 923–926.
8. Blauert, J., (1974 etc.). *Räumliches Hören* (1974), 1. Nachschrift (1985), 2. Nachschrift (1997). S. Hirzel Stuttgart. [English edition: *Spatial Hearing: The psychophysics of human sound localization*. Revised and extended edition, 1997], The MIT Press, Cambridge MA.
9. Toole, F. E. and Sayers, B. McA., 1965. Inferences of neural activity associated with binaural acoustic images. *J. Acoust. Soc. Am.* 37, pp. 769–779.
10. Steinhauser, A., 1877. The theory of binaural audition. *Phil. Mag.* 7, pp. 181–197 and pp. 261–274.
11. Lord Rayleigh, 1907. On our perception of sound direction. *Philos. Mag.* 13, pp. 214–232.
12. Mills, A. W., 1958. On the minimum audible angle. *J. Acoust. Soc. Am.* 30, pp. 237–246.
13. Zwislocki, J. and Feldman, R. S., 1956. Just noticeable differences in dichotic phase. *J Acoust. Soc. Am.* 28, pp. 860–864.
14. Klumpp, R. G. and Eady, H. R., 1956. Some measurements of interaural time difference thresholds. *J. Acoust. Soc Am.* 28, pp. 859–860.
15. Boerger, G. (1965). Über die Trägheit des Gehörs bei der Richtungsempfindung [On the persistence of the auditory system in direction perception]. In: *Proc. 5th Int. Congr. Acoustics*. paper B 27, Liège, Belgium.
16. David, E. E., Guttman, N. and Von Bergeijk, W. A., 1959. Binaural interaction of high-frequency complex stimuli. *J. Acoust. Soc. Am.* 31, pp. 774–782.
17. Van de Par, S. and Kohlrausch, A., 1997. A new approach to comparing binaural masking level differences at low and high frequencies. *J. Acoust. Soc. Am.* 101, pp. 1671–1680.
18. Macpherson, E. A. and Middlebrooks, J. C., 2002. Listener weighting of cues for lateral angle: The duplex theory of sound localization revisited. *J. Acoust. Soc. Am.* 111, pp. 2219–2236.
19. Goldberg, J. M. and Brown, P. B., 1969. Response of binaural neurons of dog superior olivary complex to dichotic tonal stimuli: Some physiological mechanism of sound localization. *J. Neurophysiol.* 32, pp. 613–636.
20. Joris, P. J. and Yin, T., 1995. Envelope coding in the lateral superior olive. I. Sensitivity to interaural time differences. *J. Neurophysiol.* 73, pp. 1043–1062.
21. Joris, P. J., 1996. Envelope coding in the lateral superior olive. II. Characteristic delays and comparison with responses in the medial superior olive. *J. Neurophysiol.* 76, pp. 2137–2156.
22. Harris, G. G., 1960. Binaural interaction of impulsive stimuli and pure tones. *J. Acoust. Soc. Am.* 32, pp. 685–692.
23. Gaik, W., 1990. *Untersuchungen zur binauralen Verarbeitung kopfbezogener Signale* [Investigations regarding the binaural processing of head-related signals]. Doct diss, Ruhr-Univ. Bochum, VDI-Verlag, Düsseldorf, Germany.
24. Fletcher, N. H., 1940. Auditory patterns. *Rev. Mod. Phys.* 12, pp. 47–65.
25. Blauert, J., 1969/70. Sound localization in the median plane, *Acustica* 22, pp. 205–213.
26. Cremer, L., 1948. *Die wissenschaftlichen Grundlagen der Raumakustik* [The scientific foundations of room acoustics]. Vol. I, S. Hirzel, Stuttgart, Germany.
27. Henry, J., 1849. Presentations before the American Association of Advanced Sciences on the 21st of August. In: *Scientific writings of Joseph Henry*, part II, pp. 295–296, Smithsonian Inst., Washington DC.
28. Wallach, H., Newman, E. B. and Rosenzweig, M. H., 1949. The Precedence Effect in sound localization. *Am. J. Psychol.* 57, pp. 315–336.
29. Haas, H., 1951. Über den Einfluß eines Einfachechos auf die Hörsamkeit von Sprache [On the effect of a single echo on the perceptual quality of speech], *Acustica* 1, pp. 49–58.
30. Haas, H., 1972. The influence of a single echo on the audibility of speech. *J. Audio Eng. Soc.* 20, pp. 146–159.

31. Blauert, J. and Braasch, J., 2005. Acoustical communication: The Precedence Effect. *Proc. Forum Acusticum Budapest 2005*, Budapest, Hungary.
32. Braasch, J., Blauert, J. and Djelani, Th., 2003. The Precedence Effect for noise bursts of different bandwidth. I. Psychoacoustical data. *Acoust. Sci. & Techn.* 24, pp. 233–241.
33. Djelani, T., 2002. Psychoakustische Untersuchungen und Modellierungsansätze zur Aufbauphase des auditiven Präzedenzeffektes [Psychoacoustical investigations and modelling approaches on the build-up phase of the auditory precedence effect]. Shaker Doct diss, Ruhr-Univ. Bochum, Aachen, Germany.
34. Blauert, J., 2005. Analysis and synthesis of auditory scenes. In: *Communication Acoustics*, J. Blauert, ed., pp. 1–26, Springer, Berlin–Heidelberg–New York.
35. Litovsky, R. Y., Colburn, H. S., Yost, W. A. and Guzman, S. J., 1999. The Precedence Effect. *J. Acoust. Soc. Am.* 106, pp. 1633–1654.
36. Zurek, P. M., 1980. The Precedence Effect and its possible role in the avoidance of interaural ambiguities. *J. Acoust. Soc. Am.* 67, pp. 952–964.
37. Seraphim, H.-P., 1961. Über die Wahrnehmbarkeit von mehrere Rückwürfe von Sprachschall [On the detectability of multiple reflections in speech signals]. *Acustica* 11, pp. 80–91.
38. Jeffress, L. A., 1948. A place theory of sound localization. *J. Comp. Physiol. Psychol.* 41, pp. 35–39.
39. Cherry, E. C. and Sayers, B. McA., 1956. Human 'cross-correlator'—a technique for measuring certain parameters of speech perception. *J. Acoust. Soc. Am.* 28, pp. 889–895.
40. Blauert, J. and Lindemann, W., 1986. Spatial mapping of intracranial auditory events for various degrees of interaural coherence, *J. Acoust. Soc. Amer.* 79, pp. 806–813.
41. Chernyak, R. I. and Dubrovsky, N. A., 1968. Pattern of the noise images and the binaural summation of loudness for the different interaural correlation of noise. In: *Proc. 6th Int. Congr. Acoustics*, Tokyo, paper A-3-12.
42. Barron, M. and Marshall, A. H., 1981. Spatial impression due to early lateral reflections in concert halls: The derivation of a physical measure. *J. Sound Vib.* 77, pp. 211–232.
43. Morimoto, M. and Maekawa, Z., 1989. Auditory spaciousness and envelopment. *In: Proc. 13th Int. Congr. on Acoustics*, Vol. 2, pp. 215–218.
44. Bradley, J. S. and Soulodre, G. A., 1995. Objective measures of listener envelopment, *J. Acoust. Soc. Am.* 98, pp. 2590–2597.
45. ISO 3382-1. (2009). Measurement of room acoustic parameters—Pt. 1. International Standards Organization, Geneva, Switzerland.
46. Kleiner, M., Dalenbäck, B. and Svensson, P., 1993. Auralization: An overview. *J. Audio Eng. Soc.* 41, pp. 861–875.
47. Xiang, N. and Blauert, J., 1993. Binaural scale modelling for auralization and prediction of acoustics in auditoria. *Applied Acoustics*, 38, pp. 267–290.
48. Blauert, J. and Raake, A., 2015. Can current room-acoustics indices specify the quality of aural experience in concert halls? *J. Psychomusicology, Music, Mind & Brain*, 25, pp. 253–255.
49. Beranek, L. L., 1962. *Music, Acoustics and Architecture*, Wiley, New York, 1962.
50. Beranek, L. L., 2003. Subjective rank-orderings and acoustical measurements for fifty-eight concert halls. *Acta Acustica united with Acustica* 89, pp. 494–508.
51. Reichardt, W., Abdel Alim, O. and Schmidt, W., 1975. Definition und Meßgrundlage eines objektiven Maßes zur Ermittlung der Grenze zwischen brauchbarer und unbrauchbarer Durchsichtigkeit bei Musikdarbietung [Definition and foundation of a measure to assess the border between usable and non-usable transparency of musical performances]. *Acustica* 32, pp. 126–137.
52. Boré, G., 1956. *Kurzton-Meßverfahren zur punktweisen Ermittlung der Sprachverständlichkeit in lautsprecherbeschallten Räumen.* [Short-tone measurement procedure to assess speech intelligibility in rooms excited by loudspeakers]. Doct. diss. Techn. Hochschule Aachen, Aachen, Germany.
53. Danilenko, L., 1969. Binaurales Hören im nichtstationären diffusen Schallfeld [Binaural hearing in a nonstationary diffuse sound field]. *Kybernetik* 6, pp. 50–57.
54. Meyer, J., 1978. Raumakustik und Orchesterklang in den Konzertsälen Joseph Haydns [Room acoustics and the sound of the orchestra in the Josph Haydn's concert halls]. *Acustica* 41, 145–162.
55. Perrett, S. and Noble, W. (1997). The contribution of head motion cues to localization of low-pass noise. *Perception and Psychophysics* 59, pp. 1018–1026.

56. Teret, E., Braasch, J. and Pastore, M. T. (2017). The influence of signal type on the perceived reverberance, *J. Acoust. Soc. Am.* 141, 1675–1682.
57. Blauert, J., Kolossa, D., Obermayer and Adiloğlu, K., 2013. Further challenges and the road ahead. In J. Blauert, ed., *The technology of binaural listening*, pp. 477–499. Springer, Berlin–Heidelberg–New York, NY, & ASA Press, New York, NY.
58. Braasch, J., 2016. Binaurally integrated cross-correlation auto-correlation mechanism (BICAM), *J. Acoust. Soc. Am.* 140, EL143–EL148.
59. Braasch, J., 2005. Modeling of binaural hearing. In: *Communication Acoustics*, J. Blauert, ed., pp. 75–108, Springer, Berlin–Heidelberg–New York.
60. Kohlrausch, A., Braasch, J., Kolossa, D. and Blauert, J., 2013. An introduction to binaural processing. In J. Blauert, ed., *The technology of binaural listening*, pp. 1–32. Springer, Berlin–Heidelberg–New York, NY & ASA Press, New York, NY.
61. Raake, A. and Blauert, J., 2013. Comprehensive modeling of the formation process of sound quality. *Proc. 5th Intl. Worksh. Quality & Multimedia Experience (QoMEX)*, pp. 67–81, Klagenfurt, Austria. (Accessible through IEEE Xplore.)
62. Schroeder, M. R., Gottlob, D. and Siebrasse, K. F., 1974. Comparative study of European concert halls: correlation of subjective preference with geometric and acoustic parameters, *J. Acoust. Soc. Am.* 56, pp. 1195–1201.
63. Blauert, J., 2012. Conceptual aspects regarding the qualification of spaces for aural performances, *Acta Acustica united with Acustica* 99, pp. 1–13.
64. Blauert, J. and Jekosch, U. (1997). Sound-quality evaluation: a multi-layered problem. *Acta Acustica united with Acustica*, 83, 747–753.
65. Blauert, J. and Jekosch, U., 2007. Auditory quality of concert halls—the problem of references. *Proc. 19th Int. Congr. Acoust.*, ICA 2007, Madrid, Spain, and Revista de Acústica 38, Paper RBA pp. 06–004.
66. Blesser, B. and Salter, L. R., 2006. *Spaces speak, are you listening? Experiencing aural architecture.* MIT Press, Cambridge MA.
67. Blauert, J. and Raake, A., 2015. Can current room-acoustics indices specify the quality of aural experience in concert halls? *J. Psychomusicology: Music, Mind, and Brain* 25, pp. 253–255.

9

Auralization

Michael Vorländer, Institute of Technical Acoustics, RWTH Aachen University, Germany

9.1 INTRODUCTION

Today, sophisticated tools for acoustic design of performance spaces such as concert halls and theaters are available. This includes the scale model technique, in situ measurements, and corresponding data analysis. Single-number metrics have been developed which are highly correlated with subjective impressions. The information obtained with these methods is well expressed in numbers such as T_{30} (reverberation time), C_{80} (clarity), G (strength), etc.—at least for the acoustician. The match or mismatch of such metrics with the design goal is relevant for decisions concerning the details in architectural design concepts. Numbers describing acoustic impressions, however, are not unambiguous in discussions with laymen. The next logical level of acoustic design methods is therefore the audio demonstration of *acoustics*, i.e., by listening to the data. The process of creating audible sounds from computer data is called *auralization.*[1] The result, the audio stream that we can listen to, is also called *the auralization* of the sound scene in the room or building.

Also, in industrialized countries and particularly in urban areas, noise control in buildings must be considered a matter of public interest and, thus, of the political discussion and participation of the population. The relevance of *good* acoustics being integrated in the architectural design of performance rooms is obvious anyway. Sophisticated tools are available as well, for obtaining information about sound insulation of building elements in the laboratory and, finally, in the building. The acoustic engineer is trained to evaluate and judge numerous temporal and spectral details which may lead to a well-grounded solution for the architectural design. Also this discussion on noise control in the built environment is simplified by reducing the description of the problem into single numbers; for instance TL, R_w, $D_{nT,w}$, etc. Single numbers are also important as a common basis for simple noise control measures for a harmonization of noise regulations and noise limits. But it is obvious that a more intuitive way of communication about acoustic phenomena, particularly for communication with acoustically untrained people is, again, the technique of auralization. This chapter focuses on the techniques for simulation of sound fields in architectural spaces and in buildings, the link between computer methods to audio signal processing tools, and the required technology for spatial sound reproduction.

9.2 DEFINITIONS AND STANDARDS IN ARCHITECTURAL ACOUSTICS

In this short introduction, the chapters dealing with prediction and computational models are referred to; see in particular—Chapters 1 and 2: Computational approaches to room-acoustic prediction and modeling; and Chapter 7: Sound insulation in buildings.

9.2.1 Impulse Responses in Rooms

If a source emits a short impulse, the sound propagates as a spherical wave into the room. It is reflected at the room boundaries and creates at the listener's ear a room response consisting of numerous repetitions of the original impulse. An alternative model of sound propagation is based on the concept of rays or particles that travel at the speed of sound along straight lines. This way, the emission and reflection can be modeled, the propagation amplitude delay determined, and the emission at the receiver calculated by recording the amplitudes and times in the *room impulse response* (RIR). The density of reflections on the temporal axis increases due to higher probability of similar ray paths at larger times, while the energies of the pulses decrease due to the spherical spread of sound energy and absorption (see Figure 9.1).

For the purpose of auralization, the impulse response (IR) must be transformed into a format suitable for *3-D audio signal processing* (see Section 9.3). The specific focus must be set on conservation of the perceptually relevant spectral, temporal, and directional features of sound incidence on the listener.

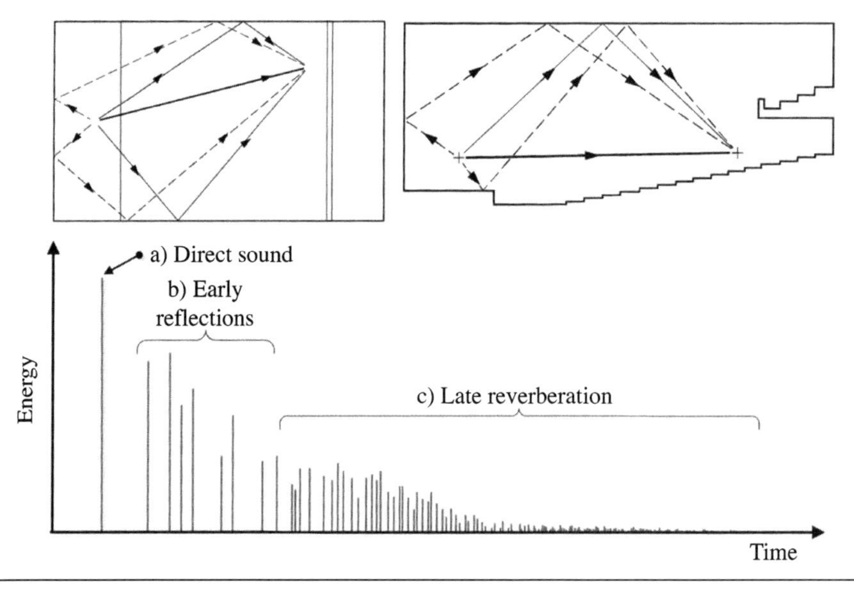

Figure 9.1 Top: room sketch in sheer plan and ground plan, bottom: RIR. a) direct sound, b) (t < 50 ms) early reflections, c) (t > 50 ms) reverberation.

9.2.2 Sound Transmission Between Rooms

Sound insulation data are given as sound level differences that are adjusted to certain standard situations of receiving room reverberation. The data are usually given in frequency bands between 50 Hz or 100 Hz and 3.15 kHz or 5 kHz (see Figure 9.2). Octave bands or one-third octave bands are in use.

For the purpose of auralization, this sound insulation data must be transformed into a format suitable for audio signal processing (see Section 9.4.2.2). Here, the particular focus is spectral envelope interpolation, equalizer settings, and level calibration of sound signals in the receiving room. Reverberation in the source room and the receiving room makes the auralized sound insulation effect more plausible. This can be added by using artificial reverberation.

9.2.3 Structure-Borne Sound in Buildings

Structure-borne sound data are given in sound levels adjusted to standard situations of receiving room reverberation. The data are given in frequency bands between 50 Hz or 100 Hz

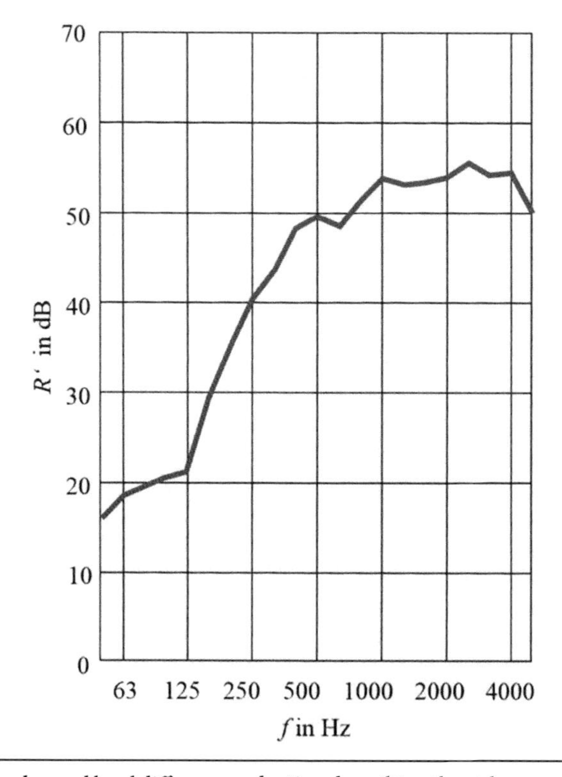

Figure 9.2 Standardized sound level difference reduction describing the airborne sound insulation of a wall (ISO 16283—Acoustics—Field measurement of sound insulation in buildings and of building elements—Part 1: Airborne sound insulation). Example: field test results in the frequency range between 50 Hz and 5 kHz.

and 3.15 kHz or 5 kHz (see Figure 9.3). Octave bands or one-third octave bands are in use. Problems in structure-borne sound are related to impact sound by walking, by housing equipment such as water installation, air conditioner systems, or elevators. All those sources are characterized by their force and their velocity injected into the building structure. One prominent example is the standard tapping machine with a standard force spectrum defined for a massive floor construction.

For the purpose of auralization, the impact sound level spectra must be transformed into a format suitable for signal processing (see Section 9.4.2.3). Also here, the particular focus is spectral envelope interpolation and level calibration of sound signals in the receiving room. Another important aspect is the consideration of the impedance contact between the source and the structure with complex coupling effects.

Before the aforementioned examples of auralization applied to rooms and buildings are discussed in more detail, some fundamentals of signal processing are introduced.

9.3 AUDIO SIGNAL PROCESSING FOR ARCHITECTURAL ACOUSTICS

Computational methods in architectural acoustics are based on geometrical room acoustics such as image sources,[2] ray tracing,[3] hybrid methods,[4, 5, 6] statistical methods of acoustic energy transfer such as the statistical energy analysis (SEA), and on numerical methods such

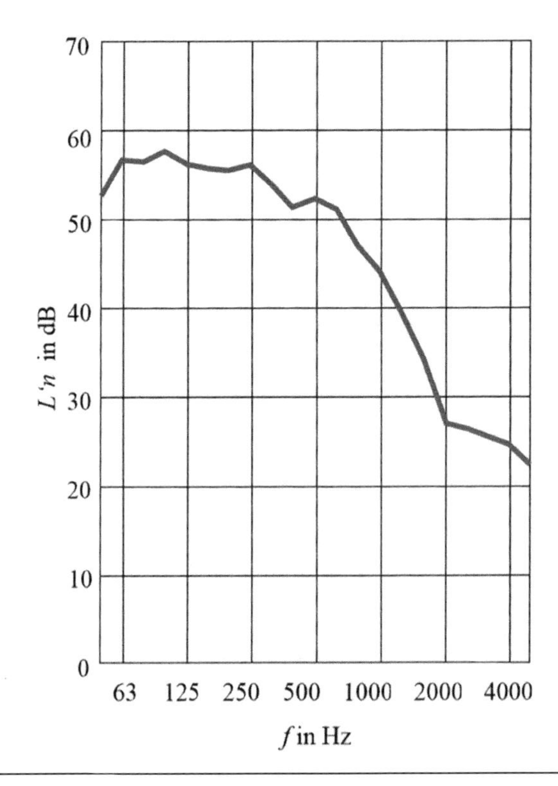

Figure 9.3 Normalized impact sound level describing the impact sound insulation of a reference floor (ISO 140-7). Example: Field test results in the frequency range between 50 Hz and 5 kHz.

as the boundary element method (BEM) or the finite element method (FEM). In rooms, as described previously, the primary results of these methods are IRs in time domain. The excitation considered in this description is a very short (Dirac) pulse. In the spectral domain, IRs correspond to stationary room transfer functions. The excitation signal considered in this description is a stationary sinusoidal signal (pure tone) at a given frequency. The whole transfer function is built up by a quasi-continuous ensemble of pure tones on the frequency axis. The two domains are connected with each other by the Fourier transformation.

With IRs at hand and with the concept of source-to-receiver signal processing, *digital filters* can be created for auralization of virtual spaces and buildings. Basically the data of the IRs *are* the filter coefficients in the finite impulse response (FIR) approach, if they are adapted to specific requirements of digital audio processing such as resolution in bits and audio sampling rates. The 3-D sound is implemented in the filters by using binaural technology or other systems for spatial sound reproduction. Hence, the filters include temporal, spectral, and spatial attributes of the space modeled.

For sound insulation problems the standard result is the sound transmission coefficient, STC, or the sound reduction index, *R*, in frequency bands.[7] These can be interpreted as filters, too, but in this case, rather as a kind of equalizer setting than as digital filter coefficients. The equalizer introduces spectral weights and decreases the transmitted sound pressure band by band in a straightforward manner, at least for the main effect on the sound transmitted, its reduction. For implementation of receiving room characteristics and 3-D sound effects, however, the equalizer approach is not sufficient.

The crucial part of auralization is the creation of appropriate filters based on the data that are available in daily practice: standard calculation and simulation data of architectural acoustics, such as sound transmission loss or energy decay curves. These results are a condensation or prediction of the *acoustics* in a rather simple standardized scheme. Typically, approximations are used at several points of such calculations. But we can take advantage of the fact that subtle details in the fine structure in amplitude and phases of room transfer functions, for example, is generally not audible. The question now is which features of the filters must be taken into account carefully and which features can be represented in rough approximations. An RIR, for example, consists of repeated impulses with time-dependent increasing density proportional to t^2. In frequency domain, the spectral density of room modes also increases quadratically, with f^2. Both findings have direct impact on the statistical modal overlap and the fine structure of the amplitude and phase response in the auralization filter. And the frequency responses in their details have only very little, or better to say, almost no audible effect, as long as a plausible critical band energy and a plausible general phase lag or group delay is chosen.

Accordingly, the common approach in all applications of auralization in architectural acoustics consists of filter construction with, to a certain extent, exact temporal envelopes of the exponential decay (basically corresponding to the decay curve) and spectral envelopes (corresponding to the coloration) over frequency bands, and of creation of a proper stochastic fine structure and phase response under those envelopes. More details about envelope versus fine structure will be discussed later.

9.3.1 Discrete and Fast Fourier Transformation

In order to allow processing in time and frequency domain, an efficient algorithm for discrete data is required. Discrete data are *samples* taken with an appropriate audio sampling rate or simulation results. Time-discrete data, however, have periodic spectra. One important prerequisite of sampling is that the *aliasing effect* must be avoided. The sampling rate must be sufficiently high to avoid such signal distortion. With a sampling rate, f_{sample}, larger than twice the largest frequency contained in the time signal and low-pass filtering, the corresponding analogue signal can be reconstructed without aliasing distortion and thus, replayed by using audio hardware without any losses of content or quality:

$$f_{sample} > 2 f_{\max,signal} \, . \tag{9.1}$$

The key to signal processing in time and frequency domain is the discrete Fourier transformation (DFT). In its discrete form for processing N samples of a time sequence $s(n)$ it reads:

$$\underline{S}(k) = \frac{1}{N} \sum_{n=0}^{N-1} s(n) \, e^{-j2\pi nk/N}; \quad k = 0,1,...,N-1 \, . \tag{9.2}$$

The result is the complex spectrum $\underline{S}(k)$. Here, n is the time index and k the frequency index. The spacing between the temporal data is:

$$\Delta t = 1/f_{sample} \, , \tag{9.3}$$

for example, 22 µs for f_{sample} = 44.1 kHz, or 0.1 ms for f_{sample} = 10 kHz. The frequency spacing between the spectral data is:

$$\Delta f = f_{sample}/N \, . \tag{9.4}$$

The inverse DFT reads:

$$s(n) = \sum_{k=0}^{N-1} \underline{S}(k) \, e^{j2\pi nk/N}; \quad n = 0,1,...,N-1 \, . \tag{9.5}$$

For solving the two Fourier series, N^2 complex multiplications are needed. A special solution, however, applicable only for certain block lengths, N, is the fast Fourier transform (FFT). It is not an approximate solution, but as exact as the DFT. The reason is circular symmetry in the complex exponential (harmonic) term and corresponding omission of summation terms of zeros. With FFT or inverse fast Fourier transform (IFFT), only $N \, \text{ld}(N/2)$ complex operations are required. The speed-up for the example of N = 4,096 is a factor of 372. FFT and IFFT, however, can only be applied for block sizes of $N = 2^m$ (= 4, 8, 16, 32, 64, 128, 256, etc.).

9.3.2 Convolution

The acoustic features of the interior spaces and of complete buildings are considered to be constant over time (time invariant) and linear. A linear and time invariant system is sufficiently described by its IR. If an IR $h(n)$ (the length of which is N') is fed with a (sound) signal $s(n)$ (the length of which is N), the resulting output signal, $g(n)$, is given by convolution:

$$g(n) = \sum_{n'=0}^{N-1} s(n) h(n-n') = s(n) * h(n) \, . \tag{9.6}$$

In the frequency domain the same can be expressed by:

$$\underline{G}(k) = \underline{S}(k) \cdot \underline{H}(k) \, . \tag{9.7}$$

After DFT, the convolution can be efficiently performed in the frequency domain because of a drastically increased efficiency of mathematical operations. The direct convolution sum

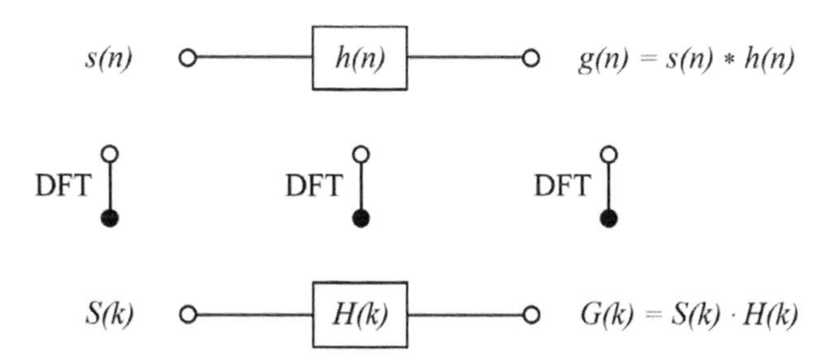

Figure 9.4 Discrete convolution in the time domain (top) and frequency domain (bottom).

of a shift-multiply process is then transformed into a multiplication scheme in the spectral domain (see Figure 9.4). Usually the signal audio stream is quasi-continuous whereas the filter IR is finite in length. Partitioned fast convolution methods are a standard class of algorithms used for such FIR filtering with long IRs. Here, a FIR is split into several smaller sub-filters. Short sub-filters are needed for a low latency, whereas long sub-filters allow for more computational efficiency. Thus, finding an optimal filter partition that minimizes the computational cost is an important task. Usually the FFT sizes are chosen to be powers of two, which has a direct effect on the partitioning of filters. Also, real-time low-latency convolution algorithms exist, which perform nonuniformly partitioned convolution with freely adaptable FFT sizes. Alongside, optimization techniques were developed that allow adjusting the FFT sizes in order to minimize the computational complexity for this new framework of nonuniform filter partitions.

9.4 THE CONCEPT OF AURALIZATION

One of the main tasks in auralization is simulation (or measurement) of an IR characterizing the room or the sound transmitting building element. This task can be solved by using simulation and prediction methods.[4–7] For the post processing, it is essential that the needs for high audio quality are fulfilled. One specific question in the beginning is the block length and resolution of temporal and spectral data. In order to achieve an undistorted time signal after filter processing with IR filters, the spectral density in the frequency domain (given by Δf) must be sufficiently fine to avoid time aliasing (see Figure 9.5). This is given if the total length of the filter IR is larger than:

$$1/\Delta f = T_{blocklength} > T_{\max,signal}\,. \tag{9.8}$$

The degree of exactness and the requirement of modeling of details for a sufficient auditory quality depend on the kind of application. In room acoustics the requirements are typically very high, and they include temporal, spectral, and spatial details in a best-possible authentic way. In auralization of sound insulation, the main task is conservation of the level difference between the source and the receiving room, the loudness and spectral features, while spatial effects and reverberation can be modeled in a less exact way since this is not the main effect introduced by the sound insulation building element.

Thus we can complete the signal chain as shown in Figure 9.6.

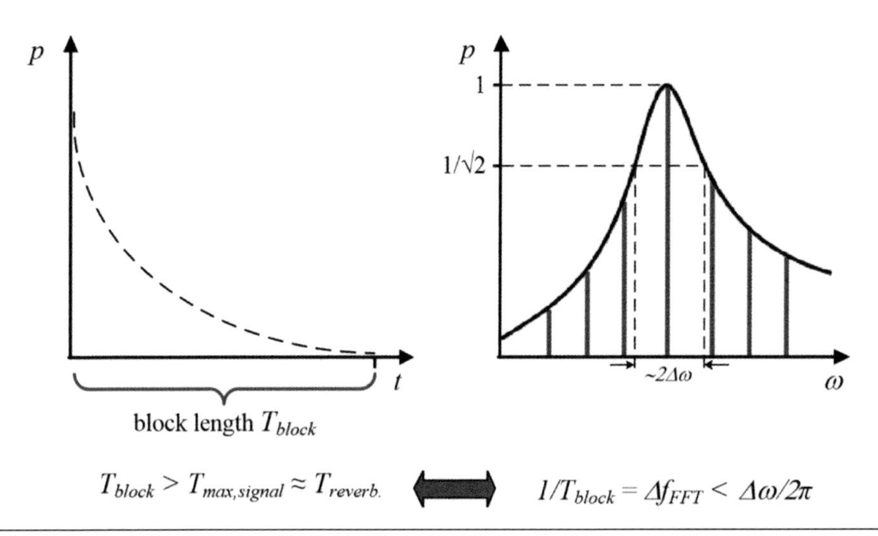

Figure 9.5 Equivalence of time domain block length and resolution of the discrete spectrum (spectral density) at the example of a resonance system.

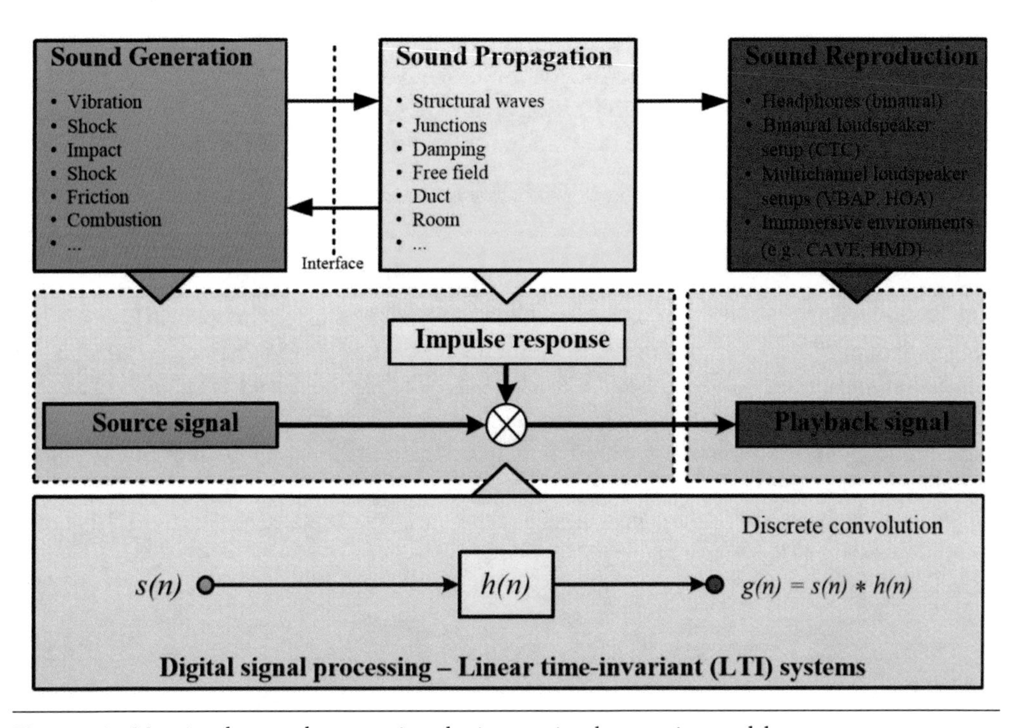

Figure 9.6 Mapping the sound propagation physics to a signal processing model.

Note that in the auralization approach the source signal and filter can be treated separately in the auralization, so that they can be freely chosen or changed (for instance by keeping the same RIR while changing source signals, or vice versa). In a pure recording in the specific building situation, which might well serve as a reference for comparison, the whole signal chain is captured in one piece, without changes possible. The appropriate playback arrangement is either by headphone presentation or free-field reproduction with loudspeaker systems (see the following text).

9.4.1 Source Characterization

Airborne sound sources for auralization in architectural acoustics can be recorded in free field or in anechoic chambers. This technique allows capturing of the time signal of the sound pressure radiated. This function can directly be used as an input signal for auralization. In the case of nearfield recording (<1 m distance), however, care must be taken to achieve a balanced frequency response with reference to the far-field situation. Basically, the same guidelines can be used as for sound power measurements in a free field.

Airborne sound sources have a radiation pattern with directivity (see Figure 9.7). For steady-state sound (running machines, devices, ventilation, stereo equipment, etc.) the directivity can be assumed to be constant. It is thus sufficient to capture the directivity with reference to a well-defined Null direction, and to use these data in the computer simulation.

Omnidirectional sources may serve as a reference for comparison of different rooms. For more natural auralization of instruments, however, the sources used should have a plausible directivity. In measurements with dodecahedron loudspeakers, an adjustable directivity can be implemented by driving the 12 units separately, one by one. The 12-channel room IR measured can later be combined by weighted adding in order to simulate specific source directivities. One way to establish a parameter set as a link between arbitrary directivities and the 12 channel loudspeaker is usage of spherical harmonics. This way, the source directivity is already included in the RIR.

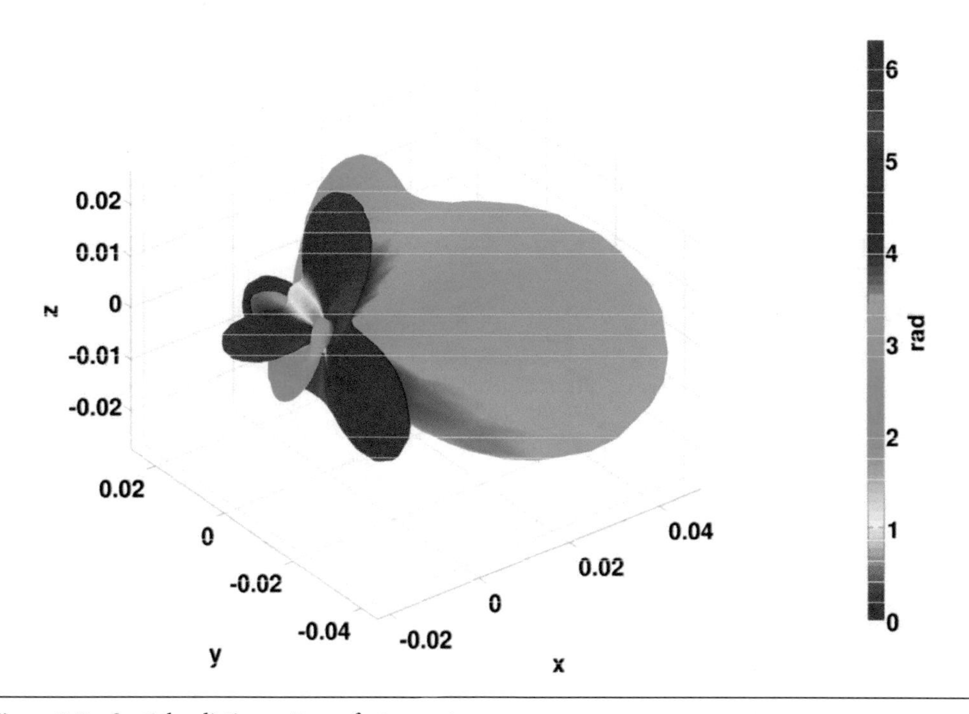

Figure 9.7 Spatial radiation pattern of a trumpet.

Some sources such as wind instruments, however, have a time-variant directivity depending not only on frequency but on the note played. This effect can only be integrated by multichannel spatial recording and auralization playback through an appropriate spatial mixing—either directly in dividing the binaural room simulation into spatial segments or again by using coding such as spherical harmonics with time-dependent data.

For structure-borne sources the situation is different and much more complex, as the coupling between source and transmission system (usually a construction of coupled beams and plates) or the mounting devices such as resilient layers play an important role. Therefore the characterization is more difficult in detail, but it can often be reduced to description of force and velocity of the free source, source impedance, and coupling impedance (see Figure 9.8).

Other crucial problems are crosstalk between various contact points of source and structure, the degrees of freedom of motion (six independent forces and moments), and in heavy sources placed on lightweight structures, most likely nonlinearities. For more information, refer to textbooks of structure-borne sound generation and propagation.

9.4.2 Filter Construction

IRs for auralization purposes must be generated with a temporal resolution adequate to the sampling rate of the test signals. It is well-known that 44.1 kHz is used as standard audio sampling rate. In room acoustics, music or speech may be sampled at half that rate, which still enables signals to be used with frequency components up to 10 kHz. This upper limit is also typically the limit of knowledge of absorption coefficients. Extensions above 10 kHz require some *guessing* about data extrapolation. Accordingly, the time resolution of the IR is very fine, less than 50 μs, which illustrates the high demands set for the simulation and the filter synthesis.

9.4.2.1 Filter Design from Room Impulse Response Data

Room acoustic simulation methods are the starting points. Typical simulation software uses hybrid methods, which are combinations of image source algorithms for the early specular reflections, ray tracing or radiosity-like methods for the scattered and late specular reflections, and *diffuse* reverberation tails for the late decay (see Figure 9.9). Note that there is no transition between the specular and scattered part because the scattered components begin at the first reflection. Pure specular reflections do not exist in room acoustics.

Image source algorithms for calculation of the specular reflection components are well qualified, at least up to a certain length of the IR until the computational effort gets too high, particularly in models with a large number of polygons. The results from image source

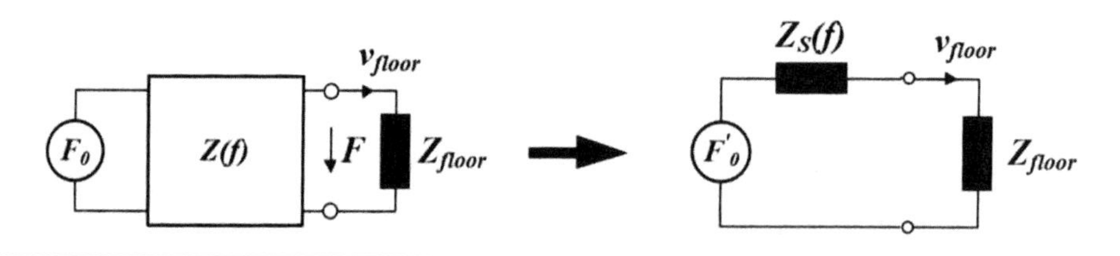

Figure 9.8 Impact—floor vibration source—two-port model.

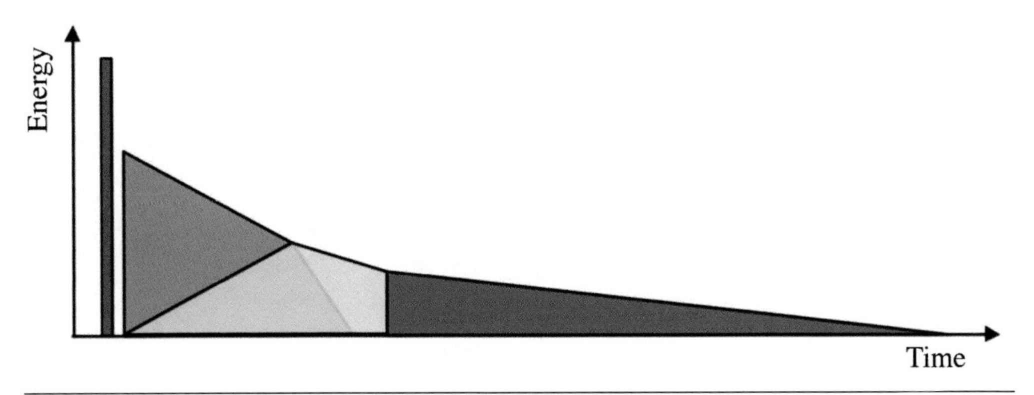

Figure 9.9 Division of the simulated RIR into early specular reflection components (green), scattered and later specular components (grey), and stochastic late diffuse decay components (violet).

algorithms can be resolved in time with any precision so that there is no problem concerning a sampling rate mismatch. It should be mentioned that image sources, their position in 3-D space, their propagation delay, and their amplitude (and phase response) can be also calculated by using beam tracing or ray tracing algorithms, but this should not be confused with the conventional stochastic ray tracing.[8]

The absorption coefficients are typically given in frequency bands, and this is the only part where the results from image sources (and from ray tracing) get frequency dependent. The corresponding complex reflection factors, $\underline{R}$, can be estimated from absorption coefficients by $|\underline{R}| = \sqrt{1-\alpha}$ and certain assumptions for the phase. Mostly, however, R is chosen to be real-valued with rather accurate results as concerns the perceptual cues of the total IR. Accordingly the simulation is performed representing sound absorption effects in the frequency bands separately and explicit bandpass filtering of each separate part. The broadband binaural RIR is finally obtained after summation (see Figure 9.10).

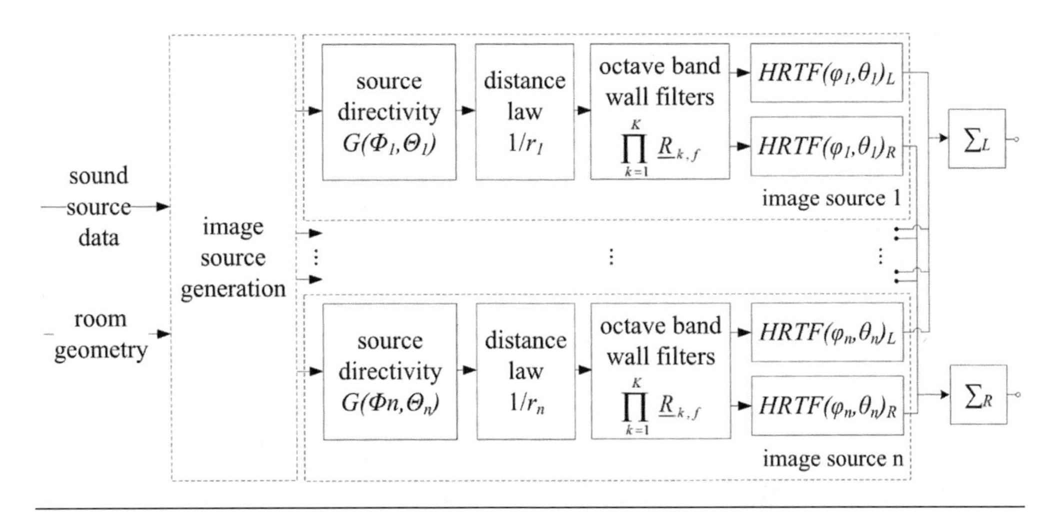

Figure 9.10 Schematic illustration of building binaural RIRs from image sources.

Results from stochastic ray tracing or radiosity algorithms cannot be transferred directly to signal processing tools because the statistical approach generates only temporal histograms in rough resolution, which is not adequate for audio sampling.

It is clear that auralization must involve binaural hearing. Mono reproduction gave some impression of the simulated sound, but a very important feature of hearing in rooms was then omitted. In binaural technology, interaural delay and diffraction around the head are basic parameters. Assuming linear sound propagation, both are included in *head-related transfer functions* (HRTFs) or in corresponding head-related impulse responses (see Figure 9.11). HRTF data were measured at different places and for different purposes. They can be determined with probe microphones for test subjects individually,[9] or they can be measured with dummy heads. This aspect is also important for comparison of simulated and measured binaural RIRs.

Overall, the binaural sound pressure IR is composed of numerous reflections. For each reflection the corresponding complex spectrum H_j can be calculated as follows (after [Dalenbäck, 1996][5]):

$$H_j = \frac{e^{-jkr_j}}{r_j} \cdot H_S(\vartheta,\varphi) \cdot HRTF(\theta,\phi) \cdot H_a \cdot \prod_{i=1}^{n_j} H_i, \tag{9.9}$$

with r_j denoting the distance between the image source and receiver, k, the wave number = $2\pi f/c$, H_S, the (directional) spectrum of the source in a source-related coordinate system, H_R, the HRTF (right or left ear, see upcoming text) in a receiver-related coordinate system, H_a, the air absorption, and H_i, the reflection factors of the walls involved, in frequency bands. The same equation can, of course, be expressed in the time domain by convolution of all relevant IRs. The total binaural IR (r, l = right, left ear) is then obtained by inverse Fourier Transformation:

$$h_{\text{total,r,l}}(t) = \boldsymbol{F}^{-1}\left\{\sum_{j=1}^{N} H_{j,\text{r,l}}\right\}. \tag{9.10}$$

Due to the fact that the latter part of the IR is not created with the same temporal resolution and accuracy as the early part, artificial reverberation or other data from hybrid methods

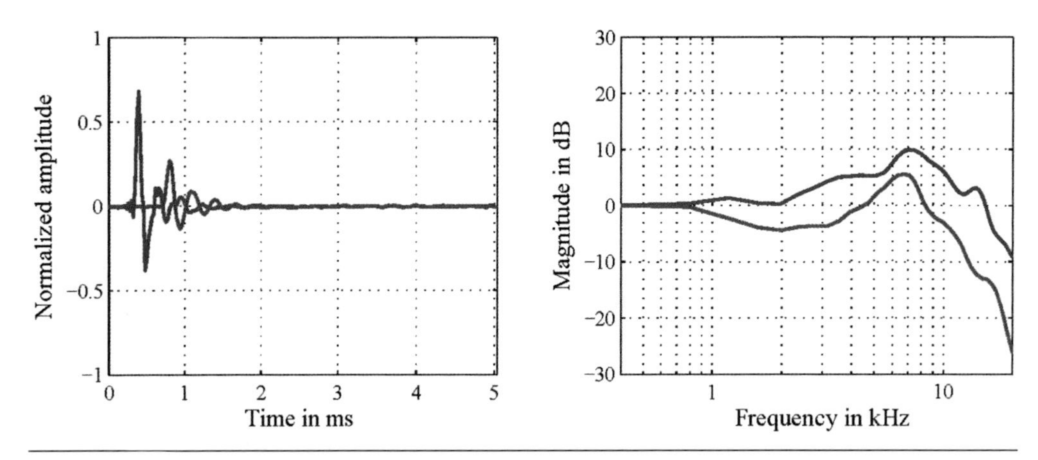

Figure 9.11 Head-RTF for the right ear (red) and the left ear (blue) at a specific incidence angle from the left (example) and corresponding head-related impulse response.

must be added by separating the simulation into two or more steps (see Figure 9.9). The acceptance of less accuracy can be justified by the low energy contained in the late part and by temporal masking. With stationary signals such as running speech or music, the latter part of the reverberation tail is masked and only noticeable in parts with impulsive or rhythmic signals. The early part is considered most important, up to a length of 400–500 ms. Typical *mixing times* published in studies on the perceptual effects of temporal/spatial masking are even shorter—around 100 ms or less.

One principle for creation of late reverberation tails is based on the assumption of a perfectly exponential decay envelope with a T_{30} estimated from Sabine's or Eyring's formula. The fine structure in this diffuse field approach is then purely stochastic and can be taken, for instance, from a Poisson process of Dirac pulses. Octave band filters and summation enable transition into narrowband or broadband sound pressure signals. This process is called noise shaping (see Figure 9.12). Another possibility replaces the exponential decay by the envelope found by a low resolved ray tracing or radiosity, which is both more accurate and more flexible than using Sabine's formula.

The criteria for binaural IRs can be summarized in Table 9.1.

9.4.2.2 Filter Design from Sound Transmission Data

In a typical room-to-room situation, the experience of listening to music or speech *through the wall* is a signal perceived to be more quiet and low-pass filtered. Auralization of this problem should therefore give an authentic representation of loudness and coloration. Spatial effects, like in room acoustics, have only minor importance. Instead, a *diffuse* reverberant field can mostly be assumed. The data available are usually expressed in band-filtered transfer functions.

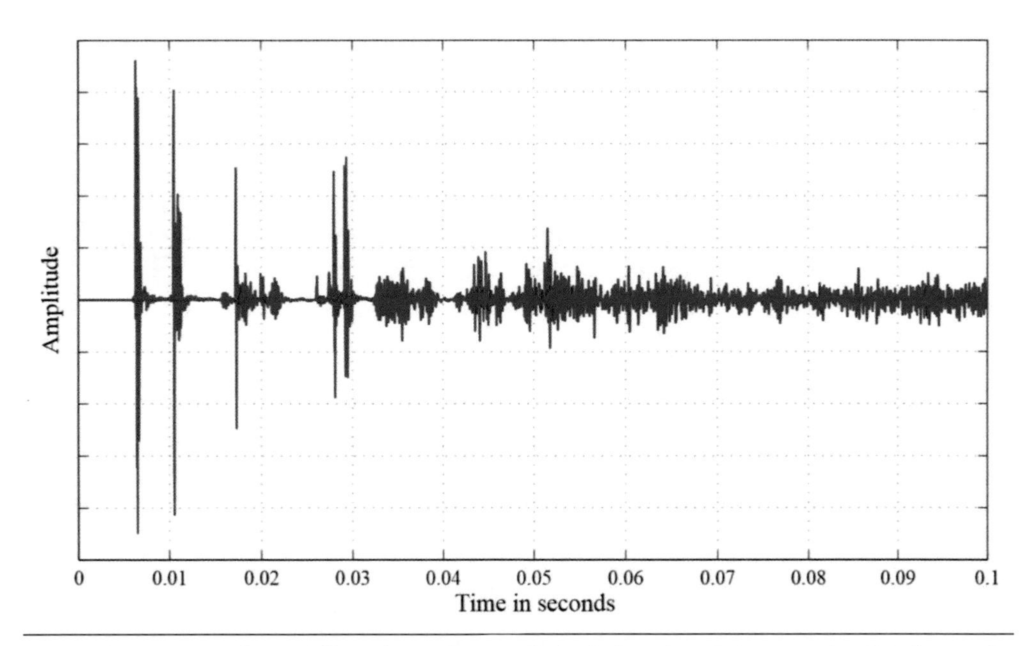

Figure 9.12 RIR auralization filter obtained using a hybrid algorithm. Direct sound and early specular reflections (red) in full precision; late reflections modeled as shaped noise (blue).

Table 9.1 Features of impulse responses for auralization

Parameter	Order of magnitude	Reason
Temporal resolution	50 μs	Sampling rate and Nyquist frequency
Upper frequency limit	>8 kHz octave	Highest components in music and speech
Lower frequency limit	<63 Hz octave	Lowest components in music and speech
Length	Half reverberation time: T/2	Minimum required at temporal masking of music or speech (not valid for percussion)
Accuracy of time structure (early)	50 μs	Early reflections (<100 ms) must have exact (binaural) delay
Accuracy of time structure (late)	5–10 ms	Statistical superposition of late reflections allows less resolution, as long as the envelope is correct

The method to determine the transfer function between source and receiving room must be adequate to cover these aspects. A physical model available for this task is a kind of first-order SEA approach.[7] This means that the sound energy is considered, its magnitude and its flow through the building elements, the energy exchange between adjacent building elements, and the energy losses (see Chapter 7).

Figure 9.13 illustrates the energy paths for a typical room-to-room situation and the corresponding transmission coefficients, τ (indicated by the arrows). The standardized sound level difference, D_{nT}, for instance, can be calculated by adding all energy contributions, provided the sound signals are incoherent:

$$\tau' = \sum_{i=1}^{N} \tau_i, \tag{9.11}$$

$$D_{nT} = -10\log\tau' + 10\log\frac{0{,}32\,V}{S} = -10\log\tau_{nT}, \tag{9.12}$$

with V denoting the receiving room volume in m^3 and S the separating wall surface in m^2. Basically, the same situation is present in a building acoustic measurement of the standardized sound level difference, D_{nT}:

$$D_{nT} = L_S - L_R + 10\log\frac{T}{0.5\text{ s}}, \tag{9.13}$$

with L_S denoting the average level in the source room, L_R the average level in the receiving room, and T the reverberation time in the receiving room. This equation can be written also in scales of squared sound pressures as:

$$p_R^2 = p_S^2\frac{\tau_{nT}\,T}{0.5\text{ s}}, \tag{9.14}$$

with p_S and p_R the sound pressure spectra in the source and the receiving room, respectively. τ_{nT} denotes the standardized transmission coefficient. It should be noted that τ_{nT}, like τ', is

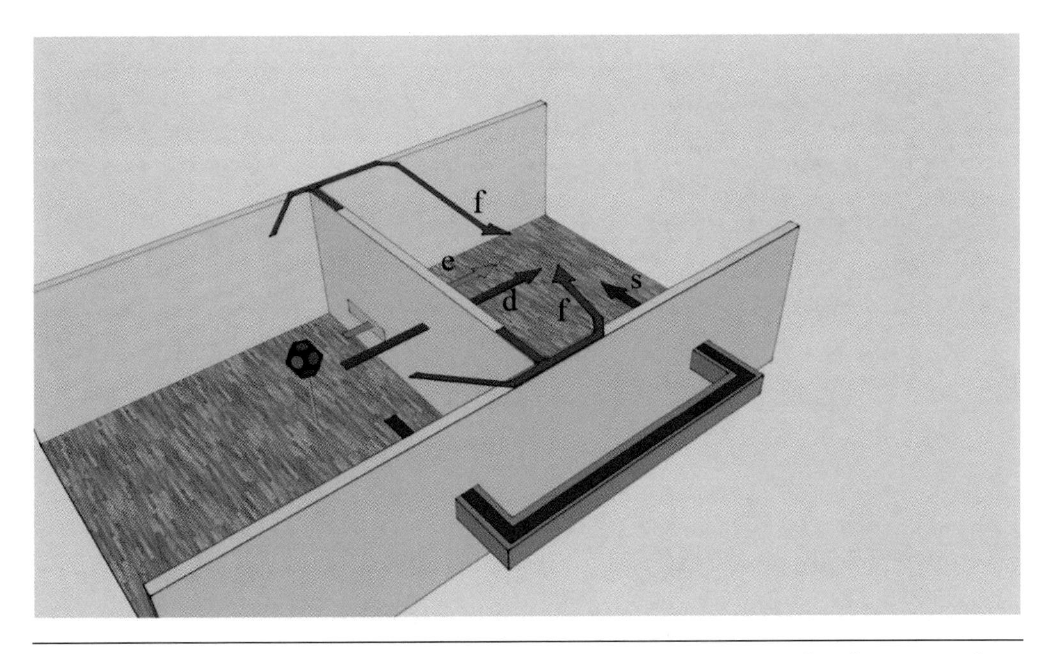

Figure 9.13 Room-to-room situation with sound transmission over various paths. The arrows indicate direct (d), flanking (f), small element (e), or duct (s) paths.

composed of the sum of all transmission paths (see Figure 9.13). In terms of sound pressure signals flowing through the building structure and rooms, the equation reads:

$$p_R(\omega) = p_S(\omega) \sum_{i=1}^{N} f_{\tau,i}(\omega) e^{-j\omega\Delta t_i} f_{\text{rev},i}(\omega), \tag{9.15}$$

with $f_{\tau,i}$ denoting interpolated filters related to the transfer functions between the source room and the radiating walls and Δt_i the relative delays in the receiving room. $f_{\text{rev,i}}$ is the binaural transfer function between the radiating wall and the receiver, including relative delay and direction of incidence of the paths (see Figure 9.14).

$f_{\tau,i}$ must have exactly the same one-third octave band spectrum as the corresponding path transmission coefficient, and $f_{\text{rev,i}}$ is the spectrum of the room IR between the wall and the receiver.

9.4.2.3 Filter Design from Impact Sound Data

For auralization of impact sound, it must be noted that all data of the impact noise levels of floors are defined on the basis of the tapping machine. If one attempts to auralize the noise of a person walking on the above floor on the basis of standardized impact sound levels, the tapping machine excitation must be extracted from the data. This could be achieved by dividing the impact sound spectra by the force excitation of the standard tapping machine:

$$p_{FS}^2(\omega) = \frac{F_{FS}^2(\omega)}{F_{TM}^2(\omega)} p_{TM}^2(\omega). \tag{9.16}$$

Thus, a magnitude transfer function can be defined by assuming the injected force to be invariant on various floor constructions (which is valid for heavy floors), with F_{FS} denoting the

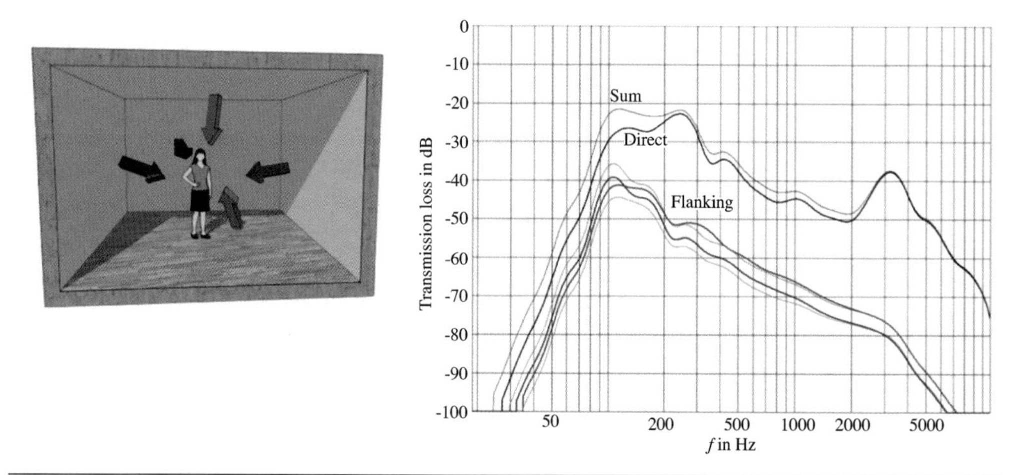

Figure 9.14 Auralization filter frequency curves (monaural part) of τ_{nT} for direct and flanking transmission (after [Vorländer and Thaden, 2000][10]).

spectrum of the force-time signal of the actual excitation, p_{TM} deduced from the normalized spectrum (L_n) of the tapping machine excitation, and F_{TM} the force spectrum of the tapping machine. The phase response is set to zero:

$$p_{FS}^2 = \frac{F_{FS}^2}{F_{TM}^2} \sum_i 10^{\frac{L_{n,i}}{10}} p_0^2 \frac{T}{T_0} \frac{V}{30}. \tag{9.17}$$

A filter dealing with the impact sound transmission can be created accordingly, also including reverberation in the same way as for airborne sound insulation.

$$\underline{p}_{FS}(\omega) = \underline{F}_{FS}(\omega) \cdot \frac{\text{"Filter"}}{|F_{TM}(\omega)|}. \tag{9.18}$$

Note that at this point the input signal is the complex excitation force (time signal and corresponding spectrum), and the sound insulation filter is equalized with the magnitude force spectrum of the tapping machine. The actual time sequence including the phase response of the impact is captured in the source recording of the actual force source.

This is, however, only a rough approximation since the injected force and the resulting velocity in the upper layer of the floor construction depends on the floor mobility. For lightweight floors (wooden floors) or lightweight covering such as floating floors this approach might produce significant prediction errors. There are no robust solutions to this problem yet.

9.4.3 Spatial Sound Reproduction

Spatial sound fields can be created with loudspeakers by using one of two general concepts. One can try to reproduce head-related signals, taking advantage of the fact that the hearing sensation only depends on the two input signals to the eardrums. The reproduction system may include headphones or loudspeakers (see the following text). Also, loudspeakers arranged around a listening point (*sweet spot*) may serve for a spatially distributed incident sound field. Furthermore, one can try to create a complete wave field incident on the listening area. The potential to involve more than one listener in the second approach illustrates the conceptual difference between the two methods.

Standard *surround sound technology* can be defined in various ways. One basic form of sound source imaging in the horizontal plane is the well-known stereo setup or a surround sound system. These approaches such as vector base amplitude panning (VBAP) make use of the psychoacoustic effect of phantom sources. Alternatively, the basis can be a multichannel microphone separating the incident field into an orthogonal coefficient domain, such as spherical harmonics. With proper reproduction setup, the 3-D sound field at the listener's point is reconstructed exactly. The accuracy of the spherical harmonics approach (*ambisonics*) depends on the frequency range and the corresponding order of the spherical harmonics functional basis.

Also in use for sound field reproduction is the method of wave field synthesis (WFS). Here, an approximation of the spatial sound field is created by using a microphone array rather than a specific listener point. The microphone arrangement is located on elementary geometric figures like straight lines or circles around the listening area. Using this approach, a large sweet spot can be created and more than one listener can be served with spatial sound.

Binaural sound reproduction over headphones is better the closer the characteristic of the HRTF matches the individual cues of the listener. For spatial hearing the binaural cues in the horizontal plane, interaural time differences, and interaural level differences are essential quantities. In the median plane these interaural differences do not contribute, but from experience we know that we can distinguish well between front and back sound incidence. Hence, monaural level differences (coloration of diotic signals) are also analyzed by the brain in order to localize in the median plane. Furthermore, head movements are crucial for enhancing the localization accuracy. It is generally agreed that the binaural cues are present in the eardrum sound pressure with only a negligible contribution of bone conduction or other effects. Research of human sound localization is highly developed and this information is available for binaural technology. The binaural cues are introduced into the eardrum sound pressure by diffraction of sound incident on the head and torso. Usually a plane wave is considered as reference. The amount of diffraction is described by the HRTF (see Figure 9.11). It is defined by the sound pressure measured at the eardrum or at the ear canal entrance and divided by the sound pressure measured with a microphone at the center of the head, but the head is absent. Accordingly, HRTFs are dependent on the direction of sound incidence.

The principal components of HRTF are above 200 Hz, where the (linear) sound field distortion due to diffraction becomes significant. Head and shoulders affect the sound transmission into the ear canal at mid frequencies, whereas pinna effects contribute to distortions in the higher frequency range (above 3 kHz). It should be mentioned that HRTFs are dependent on the individual anatomic features, but these individual HRTFs can be obtained within a few minutes.[9]

9.5 CHALLENGES AND LIMITATIONS

In order to draw conclusions on the benefit of computer simulation and auralization, the sources of uncertainties having a significant impact on the auditory impression are now discussed. In an architectural design task, it is not adequate to *calibrate* a computer model with an adjustment of input data in a way that, for instance, reverberation times or other damping effects are matched to measurement results. The objective for computer simulation should be to be independent of adjustment factors. It should be purely based on sound field physics and corresponding input data concerning the geometry and materials of the building.

It is reasonable to define a scale of psychoacoustic relevance of differences and, thus, comparing differences between results or quantitative uncertainties addressed to simulations with the just noticeable differences (JNDs) of human hearing. In a best case of the listening environment in the laboratory by using headphones, for instance, the JND for reverberation time is about 5%, for strength (level) 1 decible (dB), and for definition, 10% (ISO 3382). If uncertainties are smaller than these values, the simulation can be considered as sufficiently precise. For computer prediction and simulation including auralization, one could state the general rule of *don't compute what you can't hear.*

9.5.1 Level of Detail of the Room Model

A proper computer-aided design (CAD) model is essential for room acoustic simulations. The surface elements, usually polygons, must be large compared with wavelengths in three decades, in order to cover the audio frequency range. This is practically impossible. In engineering applications, therefore, compromises are used without any theoretical foundation. The results may be wrong due to a physically unfortunate choice of the level of detail. Also, a high level of detail will lead to unnecessarily long computation times. Accordingly, a large potential is identified in the acceleration of algorithms at low frequencies at low spatial resolution in the CAD model, and at late times in the IR where the late decay is built by scattering rather than by deterministic specular reflections in a detailed CAD model. It has been investigated which criteria can be used for choosing appropriate time and frequency-dependent levels of detail in CAD models. These findings will also be relevant for large volume simulations such as cathedrals, stadiums, airports, and train stations at reasonable computation times. As a rule of thumb, the maximum spatial resolution of details is in the order of a magnitude of 1 m, which is supported by results from listening experiments.

9.5.2 Diffraction and Seat-Dip Effect

Diffraction in room acoustics mainly happens for two reasons: There can be obstacles in the room space (e.g., stage reflectors), or there can be edges at the surroundings of finite room boundaries. In the latter case, either the boundary is forming an obstacle, such as columns or the edge of an orchestra pit, or the boundary is forming the edge between different materials with different impedances (and absorption). Since diffraction is a typical wave phenomenon, it is not accounted for by the basic simulation algorithms listed above. There are methods of including diffraction as a statistical feature into ray models presented by Svensson et al., 1999.[11]

The seat-dip effect is another example of an important wave effect with clearly audible consequences. The sound waves impinging on a seating or audience area at grazing incidence interacts with the rows of seats in a quarter-wavelength resonator effect. Thus a dip in the spectrum occurs in a frequency range between 100 Hz and 300 Hz and above—a quite important range for music and speech signals. Without a proper wave model, this effect cannot be integrated into simulation auralization programs.

9.5.3 Uncertain Absorption

Material data are usually taken from databases or textbooks, or they are integrated databases in software. These data are uncertain which is caused by operator influence and by uncertainties of material properties in the product specification from standard measurements or by manufacturing variations of the products. ISO 354 provides a standard method for measuring random-incidence absorption coefficients in reverberation rooms. The uncertainty inherent in the method can be expressed in Table 9.2.

Table 9.2 Uncertainty of absorption coefficients (ISO 354)

Low α ($\approx$ 0.1)	0.1
Mid α ($\approx$ 0.4)	0.1
High α ($\approx$ 0.9)	0.2

Source: Data extracted and condensed from (ISO 354, 2003)

Using error propagation theory, it can be shown that the resulting reverberation time uncertainty can exceed 10%, and the uncertainties of the strength, G, and clarity, C80, stay below 1 dB.[8]

9.5.4 Modes

For small- and medium-sized rooms and for studies of distinct wave effects such as modes, full wave-based models such as the FEM or the finite difference time domain method can be applied. Rapid calculations of IRs, even in large spaces, for frequencies up to and above the Schroeder frequency are possible. Above the Schroeder frequency, the *exact* sound field is stochastic, hardly time-invariant in a large room due to slow air movement and temperature effects. The fine structure of the stationary room transfer function depends on the specific geometric input data; on certain assumptions of medium properties such as density, temperature, and humidity; and on complex boundary conditions, which are hardly known to the required precision. Thus, the results are usually not as accurate as one might expect from discussing the numerical method as such. Another fact is that if small changes are introduced in a room (such as having more or less people present in the room during a measurement or a change in the air conditioning system), it will create a completely different fine structure of the room transfer function, particularly in the phase response. It is known, however, that the introduction of such changes are not audible as long as the mean transfer function (averaged in 1/3 octave bands or critical loudness bands) is constant. Hence, it is doubtful whether to use a computationally very expensive simulation approach to calculate a specific, but arbitrary complex transfer function which is more or less as accurate as the results from geometric models. This statement does not apply for the seat-dip effect, for example, or other specific diffraction effects dominating the direct sound or the early reflections.

9.6 REAL-TIME PROCESSING FOR VIRTUAL ROOM ACOUSTICS

The method of auralization can also be integrated into the technology of *Virtual Reality* (VR).[12,13] VR systems provide a multimodal (audio-visual) reproduction of computer-generated data in real time with the possibility to interact with a virtual scene. Accordingly, the three core features of VR are multimodality, interaction, and real-time performance. As a new challenge, the latency in the input-output auralization chain from tracking devices, audio hardware, signal convolution, and 3-D audio reproduction further reduce the maximum permissible computation time for acoustic simulation, filter construction, and 3-D audio reproduction. Real-time processing is only possible with a significant reduction of complexity. Physical and psychoacoustic evaluations usually help to find the space between simplifications and the period. In the following, data management and convolution problems are briefly discussed with respect to real-time processing.

The convolution engine processes the monaural audio signals of the virtual sources with these filters. As the sound propagation changes (e.g., movement or rotation of the listener), the room acoustic simulation is rerun and the filters, or parts of them, are updated and exchanged.

Direct-sound and early reflection filters are updated with high rates (>25–100 Hz). For the diffuse reverberation tail significantly lower rates (1–5 Hz) are mostly sufficient, as they do not diminish the perceived quality of the simulation. The diffuse sound field changes slowly with regard to the velocity of a walking listener in a scene. A smooth changeover of filters without any audible artifacts is achieved by crossfading in between the convolution results of the current and the next filter.[14]

It is not possible to explain all details of the very complex task of real-time auralization in architectural acoustics here. Instead, examples illustrate the possible benefit of virtual acoustics.

Concert hall acoustics is one of the classical examples where auralizations for research, planning, and teaching were first introduced. In the area of concert hall modeling, two new aspects are in the focus interest; both are related to input data of boundary conditions for the room surfaces. In numerical wave modeling, the question of complex material impedances and of locally versus nonlocally reacting surfaces is an interesting question (see Figure 9.15).

Another example is part of a project that aims at the virtual restoration of the sound of the Old Hispanic Rite, auralizing the Mozarabic Chant in pre-Romanesque churches of the Iberian Peninsula. For this purpose, an acoustic virtual model was created, according to archaeological documentation of the original building conditions. Anechoic recordings of several early Mozarabic Chant musical pieces were recorded and auralized corresponding to old Hispanic liturgical rites in multiple settings (see Figure 9.16).

DEFINING TERMS

Auralization: is twofold. First, the process of sound rendering, signal processing, and audio reproduction is called *auralization*. Second, the audible result, the sound file listened to, is called the *auralization* of the room or building situation.

Room impulse response: the temporal function of the direct and reflected sound after excitation with a short pulse.

3-D audio signal processing: a way of recording, simulating, processing, and reproducing spatial sound. The angles of sound incidence in 3-D space and their effect when exiting the

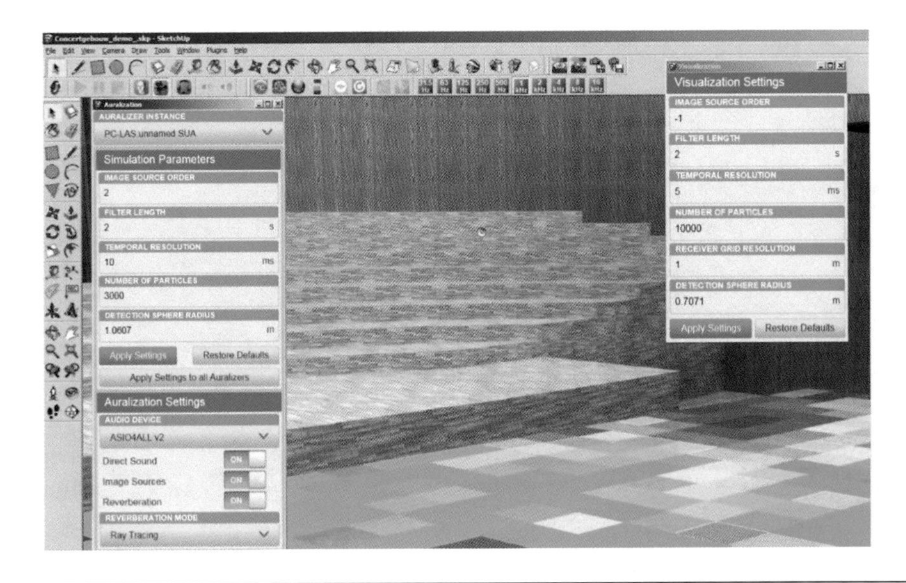

Figure 9.15 Real-time room acoustics simulation of a concert hall and calculation of room acoustical parameters using a plug-in for the CAD modeler SketchUp that triggers the real-time room acoustics simulation framework RAVEN.[15]

Figure 9.16 Simulation of a historic church in the aixCAVE environment of RWTH Aachen University (five-sided 3-D image projection and 3-D audio rendering).

human head and torso and the hearing system must be taken into account in various steps of the auralization process.

Digital filters: in auralization are used as signal processing components for forming the sound in a way that the effects introduced by the room impulse response or the sound transmission curve are included.

Surround sound technology: audio systems which provide a 3-D auditory impression. Examples are Ambisonics, vector base amplitude panning (VBAP), wave field synthesis (WFS), and binaural technology.

Computer-aided model: computer-aided design model of a room or building geometry created with usual architectural software.

Virtual Reality: computer and display technology for creating immersive, real-time, computer-generated 3-D virtual scenes with multimodal human interfaces to vision, hearing, and haptics.

REFERENCES

1. Kleiner, M., Dalenbäck, B.-I. and Svensson, U. P. 1993. Auralization—An Overview. *J. Audio Eng. Soc.* 41:861–875.
2. Allen, J. B. and Berkley, D. A. 1979. Image method for efficiently simulating small-room acoustics. *J. Acoust. Soc. Am.* 65:943–950.
3. Krokstad, A., Strøm, S. and Sørsdal, S. 1968. Calculating the acoustical room response by the use of a ray tracing technique. *J. Sound Vib.* 8:118–125.
4. Vorländer, M. 1989. Simulation of the transient and steady state sound propagation in rooms using a new combined sound particle—image source algorithm. *J. Acoust. Soc. Am.* 86:172–178.
5. Dalenbäck, B.-I. 1996. Room acoustic prediction based on a unified treatment of diffuse and specular reflection. *J. Acoust. Soc Am.* 100:899–909.
6. Funkhouser T. A., Carlbom I., Elko, G., Pingali, G., Sondhi, M. and West J. 1998. A beam tracing approach to acoustic modelling for interactive virtual environments. Computer Graphics, SIGGRAPH '98. Proc. 25th annual conference on computer graphics and interactive techniques. ACM New York, NY, USA: 21–32.
7. Gerretsen, E. 1986. Calculation of airborne and impact sound insulation between dwellings. *Applied Acoustics* 19:245–264.
8. Vorländer, M. 2013. Computer simulations in room acoustics: Concepts and uncertainties. *J. Acoust. Soc. Am.* 133:1203–1213.
9. Masiero, B. 2013. Individualized binaural technology measurement, equalization and perceptual evaluation. PhD Dissertation, RWTH Aachen University, Logos Publishers, Berlin, Germany.
10. Vorländer, M. and Thaden, R. 2000. Auralisation of airborne sound insulation in buildings. *Acustica united with Acta Acustica* 86:70–76.
11. Svensson, U. P., Fred, R. I. and Vanderkooy, J. 1999. An analytic secondary source model of edge diffraction impulse responses. *J. Acoust. Soc. Am.* 106:2331–2344.
12. Savioja, L., Huopaniemi, J., Lokki, T. and Väänänen, R. 1999. Creating interactive virtual acoustic environments. *J. Audio Eng. Soc.* 47:675–705.
13. Pelzer, S., Aspöck, L., Schröder, D. and Vorländer, M. 2014. Integrating real-time room acoustics simulation into a CAD modeling software to enhance the architectural design process. *Buildings* 2:113–138.
14. Wefers, F. 2015. Partitioned convolution algorithms for real-time auralization. PhD Dissertation, RWTH Aachen University, Logos Publishers, Berlin, Germany.

15. Vorländer, M., Schröder, D., Pelzer, S. and Wefers, F. 2014. Virtual Reality for architectural acoustics. J. Building Performance Simulation. doi:10.1080/19401493.2014.888594.

FOR FURTHER INFORMATION

Room acoustics by Kuttruff is the standard textbook and the most comprehensive introduction to theoretical and applied room acoustics.

Communication Acoustics: An Introduction to Speech, Audio and Psychoacoustics by Pulkki and Karjalainen is an excellent introduction to audio signal processing for recording, reproduction, simulation, and psychoacoustic evaluation.

Auralization by Vorländer gives an introduction into fundamentals in acoustics, signal processing, simulation, and auralization.

Current research results are published in the series of proceedings of the EAA Auralization Symposium and frequently by Acta Acustica, the journal of the European Acoustics Association, in the section *Virtual Acoustics.*

10

Room-Related Sound Representation Using Loudspeakers*

Jens Blauert, Institute of Communication Acoustics, Ruhr-University, Bochum, Germany

Rudolf Rabenstein, Chair of Multimedia Communication & Signal Processing, Friedrich-Alexander-Universität, Erlangen-Nürnberg, Germany

Available methods for room-related sound presentation are introduced and evaluated. A focus is put on the synthesis side rather than on complete transmission systems. Different methods are compared using common, though quite general criteria. The methods selected for comparison are: *intensity stereophony* after Blumlein (1931),[1] *vector base amplitude panning* (VBAP), *5.1-Surround* and its discrete-channel derivatives, synthesis with spherical harmonics (*Ambisonics and higher-order Ambisonics* [HOA]), synthesis based on the boundary method, namely, *wave-field synthesis* (WFS), and binaural-cue selection methods (e.g., directional audio coding [DirAC]). While VBAP, 5.1-Surround, and other discrete-channel-based methods show a number of practical advantages, they do not, in the end, aim for authentic sound-field reproduction. The *holophonic* methods that do so, particularly, HOA and WFS, have specific advantages and disadvantages that will be discussed. Yet, both methods are under continuous development, and a decision in favor of one of them should be taken from a strictly application-oriented point of view by considering relevant application-specific advantages and disadvantages in detail.

10.1 INTRODUCTION

Rendering of sound events and auditory environments to listeners has gained increasing importance both in scientific research as well as in the design practice in architectural acoustics. This is in agreement with the goals of audio technology, namely, to present sound fields to listeners in such a way that they experience an auditory perspective—that is, perceive auditory events in various directions and distances, which may then form complex auditory scenes. If some mobility of the listeners in the synthesized sound fields is desired, loudspeakers at fixed positions in space are usually employed. This kind of sound-field presentation is called *room-related*, in contrast to the *head-related* one as used in *binaural technology*.

*Reprinted with permission from an article published in *Archives of Acoustics* (Blauert, Rabenstein, 2012).

10.2 INTENSITY STEREOPHONY

In this long established two-channel method (more precisely, *amplitude-difference stereophony*) the horizontal angles of sound incidence are coded into amplitude differences of two loudspeaker signals. The auditory system then forms the direction of the auditory event from attributes of the two superposed sound fields of the two loudspeakers—a process called *summing localization*. This popular and surprisingly robust method is primarily restricted to 2-D presentation.

Figure 10.1 depicts a common coding scheme in this context.[1,2] Two spatially coincident figure-of-eight microphones, arranged under a mutual angle of 90°, may be excited by one sound source. They then produce two coherent microphone signals with a pure amplitude difference, the latter being unequivocally related to the angle of sound incidence. Instead of figure-of-eight microphones, cardioid or super-cardioid microphones are also in use. They are, for example, formed by means of a weighted superposition of an omni-directional characteristic upon the figure-of-eight microphone (see Figure 10.2). Figure-of-eight

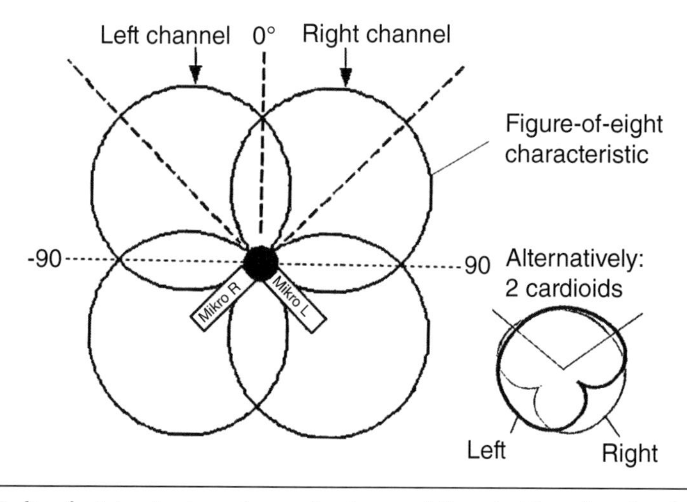

Figure 10.1 Coding for *intensity stereophony* using two spatially-coincident directional microphones.[1]

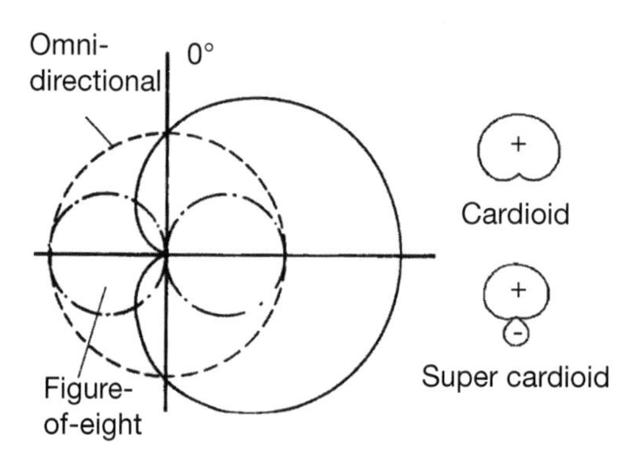

Figure 10.2 Typical directional characteristics of microphones as can be formed by weighted superposition of an omni-directional with a figure-of-eight characteristic.

characteristics can be realized by pressure-gradient microphones; omni-directional characteristics by pressure microphones.[3]

Intensity stereophony has been a wide-spread spatial-reproduction method for more than 50 years now. The fact that it works so well is due to the following acoustic effect: in the frequency range of up to 1.5 kHz, the two loudspeaker signals superpose in such a way that amplitude differences at the loudspeakers transform into arrival-time differences at the listener's ears (see Figure 10.3 and Wendt, 1963).[4] Interaural arrival-time differences, however, are the most robust attributes used by the auditory system to form the directions of auditory events.[2] Yet, this argument holds only up to about 1.5 kHz, as the human ear cannot detect the fine structure of ear-signal components of higher frequencies.

One advantage of intensity stereophony is, without a doubt, that loudspeaker signals stemming from the same source do not show any phase differences between them. Consequently, they can be electrically mixed without causing any coloration due to interferences, for instance, into a mono version. Further, proven microphone equipment is readily available, and mixing rules (panning rules) are simple. In fact, they follow the relationships given by cardioid and/or figure-of-eight characteristics.[5]

Besides the advantages of intensity stereophony, there are also significant disadvantages. One disadvantage is that good reproduction requires a listener to be positioned exactly as possible on the midline between the two loudspeakers (the so-called *sweet spot*), whereby the loudspeakers should be arranged under a horizontal angle of about 60 degrees (see Figure 10.3). Such a standardized play-back arrangement is, to be sure, somewhat restrictive in terms of listener mobility. Yet, it facilitates the production of stereophonic program material.

Another disadvantage is that intensity stereophony, unless special psycho-acoustical *tricks* are employed, renders auditory events only in the horizontal sector between the loudspeakers. This is often described by saying, "The orchestra comes into the living room." This saying also reflects the fact that this method can create spatial impression and ambience only in a very limited way. All auditory events appear predominantly at loudspeaker distance. Consequently, a convincing depth perspective is hard to achieve. Auditory events close to

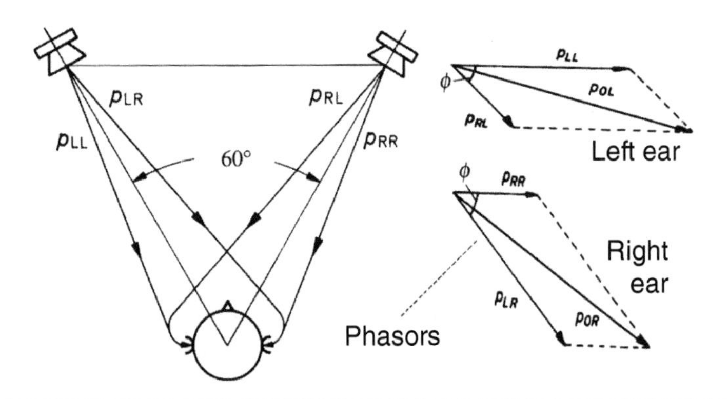

Figure 10.3 Formation of interaural arrival-time differences of the signals at the two ears due to amplitude differences of the loudspeaker signals—effective at frequencies below about 1.5 kHz. A phase difference of the ear signals, $\Delta\varphi$, is created due to the different delay times of each loudspeaker signal to the left or right ear, respectively. A larger amplitude of the left loudspeaker leads to a difference of the phase angles of the resulting phasors at the two ears, p_{OL}, and p_{OR}.

the listener are impossible. Auditory events at a larger distance than the loudspeaker can, to a certain amount, be simulated via the ratio of direct and reverberant sound, but the auditory perspective achieved in this way is not really perceptually convincing. Displacing the listener's head off the sweet spot gives rise to image shifts and coloration, due to interferences in the superposed sound field. Table 10.1 lists the advantages and disadvantages of intensity stereophony.

It should be noted at this point that there are many different methods for providing loudspeaker signals for stereophony, some of them employing inter-loudspeaker arrival-time differences solely or in addition to amplitude differences—for instance, using spaced microphones at the recording end. (For overviews and further discussion, see References 2, 5, 6, and 7.) In any case, stereophony, that is, reproduction with only two frontal loudspeakers, does not allow for the provision of surround sound. Nevertheless, stereophony has been mentioned here, taking intensity stereophony as an example, to introduce the psychoacoustic effect of summing localization, which is a basic effect for some surround-sound methods as well.

10.3 AMPLITUDE-DIFFERENCE PANNING

Intensity stereophony is based on the effect that the directions of auditory events are formed due to summing localization in a sound field that is superposed by more than one loudspeaker radiating coherent acoustic signals. More precisely, the loudspeaker signals are simultaneous in time but differ in terms of their amplitudes. In generalizing this basic idea, *domes* composed of triangles of loudspeakers have been built (see Figure 10.4 and Pulkki, 2001).[8]

In order to achieve high accuracy when predicting the perceived directions of the auditory events, it is of advantage to employ as few loudspeakers as possible for each direction to be synthesized. For this reason, a maximum of three loudspeakers is used, namely, those three that are positioned closest to the target direction.

There is no immediate support for direct recording, therefore this method is primarily used for sound-field synthesis from parametric auditory-scene representations, for instance, DirAC (see Section 10.7). 3-D presentation is possible, yet not in a precise way for all directions. Adaptability to specific loudspeaker arrangements is fairly easy. Yet, there are massive problems regarding the quite narrow sweet spot and the inadequate production of perceived

Table 10.1 Advantages (+) and disadvantages (–) of intensity stereophony

+	Simple panning rules (amplitude differences only)
+	Proven and readily available microphone equipment
+	Mono compatible
+/–	Standardized listening arrangement (restrictive)
–	Auditory events only in frontal sector
–	Limited listening area (sweet spot)—image shifts outside
–	Auditory events appear predominantly at loudspeaker distance
–	Limited possibility to create room impression and ambience
–	Coloration possible due to interference by sound-field superposition

distances. Further, summing localization is rather unstable for sideward auditory directions, thus stable auditory events cannot be provided laterally.[9] In Table 10.2, advantages and disadvantages are given in more detail.

In the 1960s, the method of *amplitude-difference panning* was known as *synthetic sound field* and was used intensively for scientific purposes; for example, at the Technical University of Dresden and the University of Göttingen,[10] both in Germany. The directions of sound incidence were visualized by so-called *hedgehog* plots (see Figure 10.5, left). Synthesis was performed with a dome of loudspeakers (see Figure 10.5, right). With a similar dome of loudspeakers, *Karl-Heinz Stockhausen* realized his famous performance of electronic music at the 1970 world's fair in Osaka, Japan.

Lately, the panning rules as applied in 2-D as well as 3-D mixing of directions are preferably notated in vector form.[30] Consequently, the method is currently often called VBAP (see Figure 10.4).

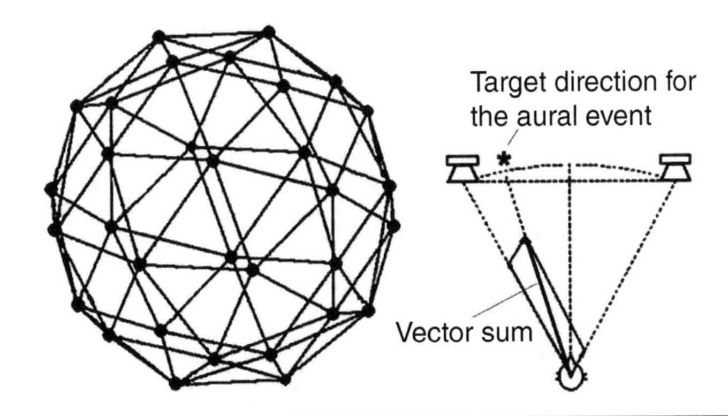

Figure 10.4 Triangulation for amplitude-difference panning. In *VBAP*, the vector sum is calculated with respect to a maximum of three loudspeakers that are adjacent in three dimensions. The simplified plot (right panel) only shows a two-dimensional case.

Table 10.2 Advantages (+) and disadvantages (–) of *amplitude-difference panning*

+	3-D presentation possible
+	Simple panning rules (amplitude differences only)
+	Easily adaptable to given loudspeaker arrangements
+	Low number of active loudspeakers per presented direction ($n < 3$)
–	No direct-recording technique available, thus for synthesis only
–	Limited listening area (sweet spot)—image shifts outside
–	Auditory events appear predominantly at loudspeaker distance
–	No precise localization in lateral directions
–	Serious problems with positioning auditory events at elevated directions (comb-filter effects)
–	Panning settings to be adapted to the specific loudspeaker arrangement used
–	Individual prediction of perceived direction hardly possible
–	Coloration possible due to interference by sound-field superposition

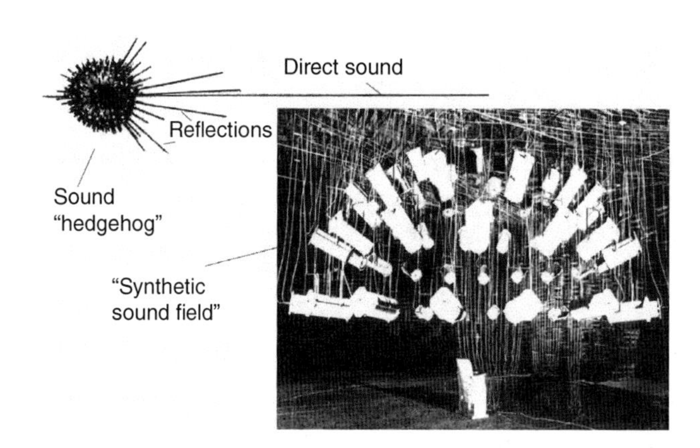

Figure 10.5 A historic *synthetic-sound-field* arrangement, used in the early 1970s at the Third Physical Institute in Göttingen.

The 2-D-panning rule as used by Pulkki (2001)[8] is based on the *tangent law.*[5] This law gives the panning functions, that is, normalized gain functions $g_l(\theta)$ and $g_r(\theta)$, for the left and the right loudspeaker positioned at azimuth angles of θ_1 and $-\theta_1$ with respect to the listener and for an auditory event at the angle θ as follows:

$$g_l(\theta)=\frac{\sin(\theta-\theta_1)}{\sin(2\theta_1)}, \quad g_r(\theta)=\frac{\sin(\theta+\theta_1)}{\sin(2\theta_1)}, \quad -\theta_1<\theta<\theta_1. \tag{10.1}$$

Figure 10.6 (left) shows these gain functions for a pair of loudspeakers with $\theta_1 = 30°$ and, $-\theta_1 = -30°$, that is, for intensity stereophony and/or VBAP with a desired auditory event midway between −30° and 30°. The contribution of one loudspeaker positioned at 30° azimuth in a circular arrangement with seven loudspeakers is shown in Figure 10.6 (right). The gain functions for the further loudspeakers are indicated by dashed lines.

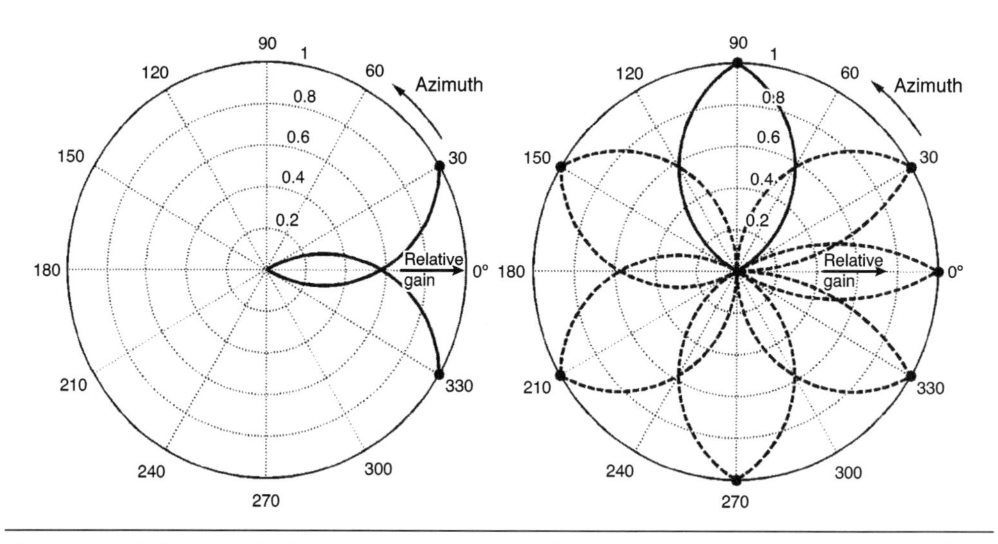

Figure 10.6 Left panel: gain functions for panning an auditory event between two loudspeakers with the tangent law. Right panel: gain functions of one out of seven loudspeakers (solid line) and of the other six loudspeakers (dashed lines) for VBAP.

10.4 SURROUND

A specific, very popular variant of the amplitude-difference-panning algorithm, is the standardized 5.1-Surround method. In this method, five loudspeakers are arranged according to Figure 10.7.[5] Formation of auditory-event directions is, again, based on summing localization. However, as has been mentioned before,[8, 9] the latter is rather unstable for lateral directions. In other words, precise synthesis of auditory events in lateral positions is hardly achievable. For this reason, in 5.1-Surround, the two rear loudspeakers are positioned at four and eight o'clock, respectively. This offers the possibility to create auditory events in a predictable way, at least in these singular directions. Further, there is a frontal center channel (*dialogue channel*) which is used to generate a stable position for dialogues in front—even when the listeners move their heads out of the sweet spot. This is of particular importance with the method being used in connection with a television image (*home-theater* set-up). An optional sixth channel may, for example, be used for a sub-woofer or for other effects.

The most important advantage of 5.1-Surround over conventional intensity stereophony is, without a doubt, that this method can provide a sense of *immersion*—that is, *room impression*—and a sense of *ambience*. Auditory events can be presented in all horizontal directions (surround!). Otherwise, all advantages and disadvantages as known from amplitude-difference panning remain valid. Program material that has been produced specifically for 5.1-Surround often exhibits a typical *cinema sound* that not everybody likes. The more important advantages and disadvantages of 5.1-Surround are presented in Table 10.3.

Following the idea of providing more loudspeakers than in stereophony to allow for a surround-sound impression, a number of further formats have been proposed—such as 7.1, 9.1, 10.1, and 11.1[11] up to 22.2.[12] Loudspeakers may not only be placed in the horizontal plane, but also above and below it. However, since all of these methods are based on pure amplitude-difference panning, their advantages and disadvantages are implicitly included in the statements contained in Tables 10.2 and 10.3. Yet, the more channels are employed, the more auditory-event directions can be presented without having to make use of summing localization—and thus coloration can be avoided as may appear due to interference in superposed sound fields.

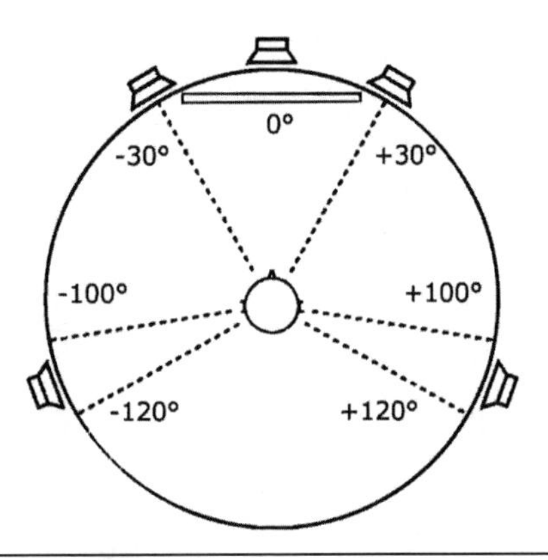

Figure 10.7 Loudspeaker arrangement for 5.1-Surround. There are one center, two front (left, right) and two rear (left, right) loudspeakers. A sixth channel is provided for, e.g., very low-frequency or effect signals.

Table 10.3 Advantages (+) and disadvantages (–) of 5.1-Surround

+	Less image shift for frontal direction due to center channel (dialog channel)
+	Simple panning rules (amplitude differences only)
+	Listening area broader than in intensity stereophony
+	Spatial impression and ambience can be experienced
+	Possibility to create special spatial effects
+/–	Standardized listening arrangement (restrictive)
–	Sweet spot limits possible listener positions
–	Auditory event predominantly in the horizontal plane
–	Auditory events appear predominantly at loudspeaker distance
–	No precise localization in lateral directions
–	Coloration possible due to interference by sound-field superposition
–	Often a characteristic "cinema sound"

10.5 SPHERICAL-HARMONICS SYNTHESIS

10.5.1 Classical Ambisonics

Looking back at Figure 10.1 opens a possibility for constructing a set of four super-cardioid microphones by applying a suitable combination of omnidirectional and figure-of-eight characteristics—each super cardioid accounting for one of the main horizontal directions. By adding one more figure-of-eight microphone with perpendicular orientation, two further super-cardioids can be generated, one of them directed upward and the other one downward. The idea for this arrangement originates from Gerzon (1973)[13] and was later dubbed Ambisonics. The set of four signals composed from the three figure-of-eight microphones plus an additional omnidirectional signal is called *B-format.* The B-format is considered to be a *portable* signal format, since it can be adapted to a given loudspeaker arrangement by purely real factors that cause appropriate shifts of the spatial characteristics. This kind of decoding can actually be performed by conventional mixing consoles.

Figure 10.8 depicts such a common decoding scheme, restricted here to the horizontal plane for simplicity. Each loudspeaker signal is weighted by a real factor that corresponds to the sensitivity of a super-cardioid receiver aiming at just that particular loudspeaker. It is apparent that all loudspeakers will be active in principle—though eventually with a 180 degree phase shift. By superposition of the sound fields of all active loudspeakers, a replica of the original sound field is achieved. Yet, this is only exactly true in the center of the synthesis area. Unfortunately, this is precisely the position of the listener's head—which thus disturbs the sound field. Please recall at this point that: Ambisonics is basically just another amplitude-difference panning method.

Ambisonics, in its classical form, never made it to wide application. Main reasons for this are listed in Table 10.4.

A closer analysis of Ambisonics reveals some interesting insights, for instance, that the method makes use of microphones with omnidirectional and figure-of-eight sensitivity characteristics which, as is well known, can be described by spherical harmonics of the 0th and

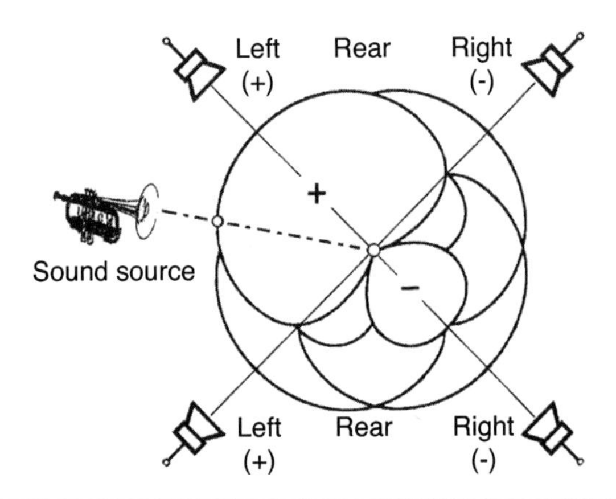

Figure 10.8 A decoding scheme for Ambisonics. All four loudspeakers are active to reproduce a single sound source, where two of them radiate 180° out of phase in the example as plotted here.

Table 10.4 Advantages (+) and disadvantages (–) of classical Ambisonics

+	3-D presentation possible
+	Proven microphone equipment available
+	Simple panning rules (amplitude differences only)
+	Easily adaptable to given loudspeaker arrangements
–	Very narrow sweet spot
–	Auditory events appear predominantly at loudspeaker distance
–	No precise localization in lateral directions
–	Panning settings to be adapted to specific loudspeaker arrangement used
–	Listener's head disturbs sound field (directional errors and coloration)
–	High localization blur in general, particularly in lateral and elevated directions

1st orders. Acousticians are usually acquainted with spherical harmonics since these are also used to mathematically describe the radiation by spherical sound sources.[3] Thus, a breathing sphere emits a spherical sound field of the 0th order, an oscillating, rigid sphere a 1st-order one. Higher orders are emitted when more complex radiation patterns on the surface of a sphere are given. Figure 10.9 shows examples up to the 2nd order. Spherical harmonics of the same order can be rotated by linear superposition, which is sometimes a useful feature. This is also how spatial characteristics are shifted into the actual loudspeaker directions for presentation—as is applied in Figure 10.8.

The spherical harmonics are *eigenfunctions* of the sound-field equation—namely, they are its solutions in spherical coordinates. Actually, spherical harmonics represent an orthogonal system of functions in which all practically relevant sound fields can be developed—similar to the harmonics in the conventional *Fourier* analysis of time functions.[14, 15] Classical Ambisonics makes use of this possibility by involving spherical harmonics up to the 1st order. As applied in this paper, the term Ambisonics usually denotes this classical form. Figure 10.10 schematically depicts all spherical harmonics up to the 2nd order.

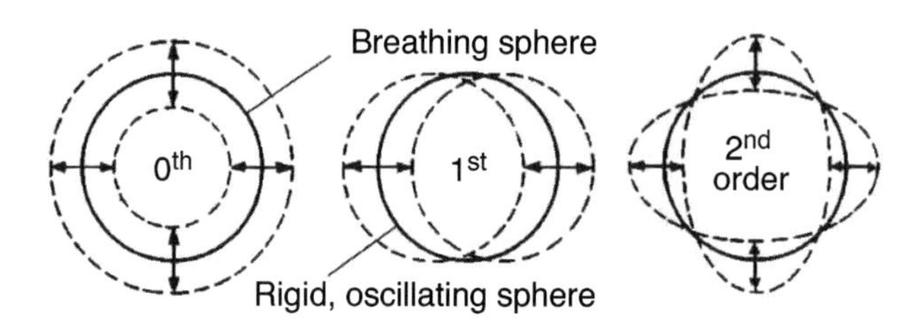

Figure 10.9 Spatial-radiation patterns of a *breathing* sphere, an *oscillating* (rigid) sphere, and a sphere with a *clover-leaf* pattern of motion—examples of spherical sound-fields of the orders 0 to 2.

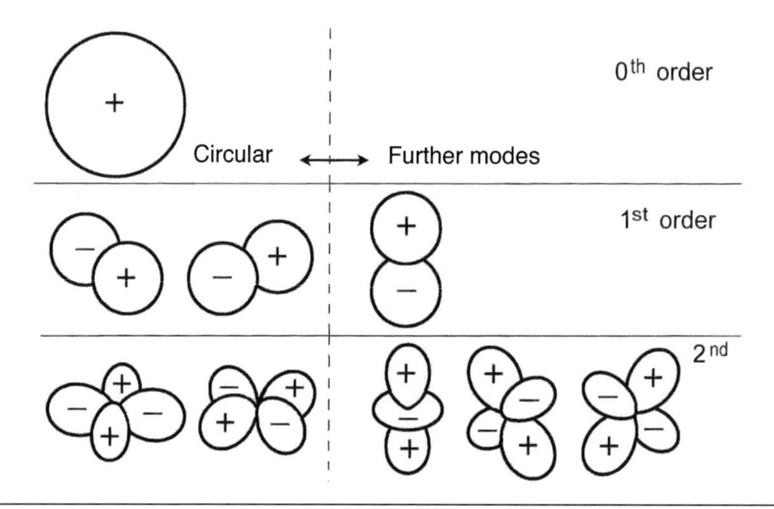

Figure 10.10 All spherical harmonics of the orders 0 to 2. In many current installations, the loudspeakers are arranged in a horizontal circle and, thus, only the circular modes are employed (left panel).

10.5.2 Higher-Order Ambisonics

HOA represents a further development of the classical approach by involving spherical harmonics of a higher order[16, 17, 18] as had already been suggested by (Gerzon, 1973).[13] The spatial selectivity of the method increases with the increasing order of the participating spherical harmonics. Accordingly, more loudspeakers are needed with the increasing order. With M being the highest order involved, the minimum number, N, of required loudspeakers corresponds to the sum of all linear independent spherical harmonics involved, namely:

– for spherical arrangements $N = (M + 1)^2$,

– for circular arrangements $N = (2M + 1)$.

This article is restricted to the discussion of planar arrangements—for example, a set of loudspeakers sitting on a planar circle (circular arrangement). Spherical, that is, 3-D loudspeaker arrangements for HOA have hardly been realized as of today, but are considered items of research.

All loudspeaker methods that have been dealt with so far in this article—namely, intensity stereophony, 5.1-Surround, Ambisonics, and consequently, HOA, are based on amplitude-difference panning and summing localization. It is thus to be expected that also here, the formation of the direction of an auditory event will be the more precise, the smaller the number of active loudspeakers is at a time. The number of active loudspeakers for a specific direction decreases with increasing spatial selectivity of the method. This relationship is illustrated in Figure 10.11—compiled from simulation data by Faller (2004).[19] It is evident that classical Ambisonics (orders 0–1) provides proper directional cues to the auditory system just in the very center of the synthesis area. Yet, with increasing participating order, the sweet spot becomes larger, until, for very high order, it finally extends over the complete synthesis area (Daniel et al., 2003).[16]

It is instructive to compare the gain functions of HOA to those of VBAP from Figure 10.6. The same loudspeaker arrangement can also be used for HOA. The gain functions for 2nd- and 3rd-order HOA are shown in Figure 10.12, left and right respectively. Similar to

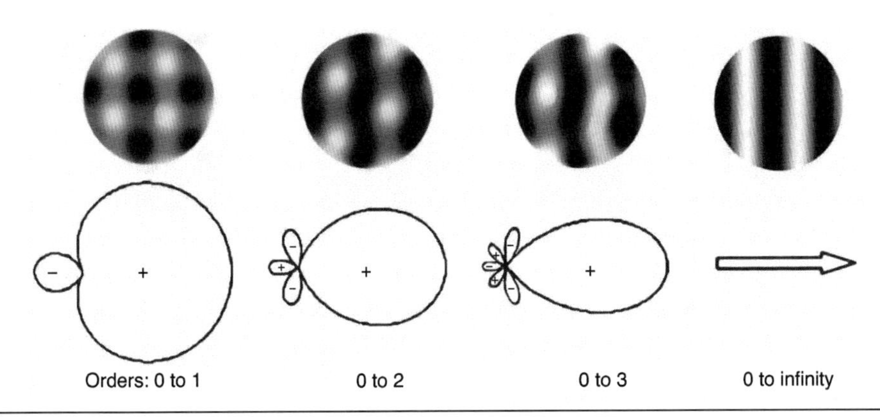

Figure 10.11 Circular sound-field synthesis of a plane wave with increasing order of participating spherical harmonics. The higher the order involved, the larger the area in which the sound field is correctly reproduced. The lower panel shows the directivity patterns with which the sound field is spatially sampled.

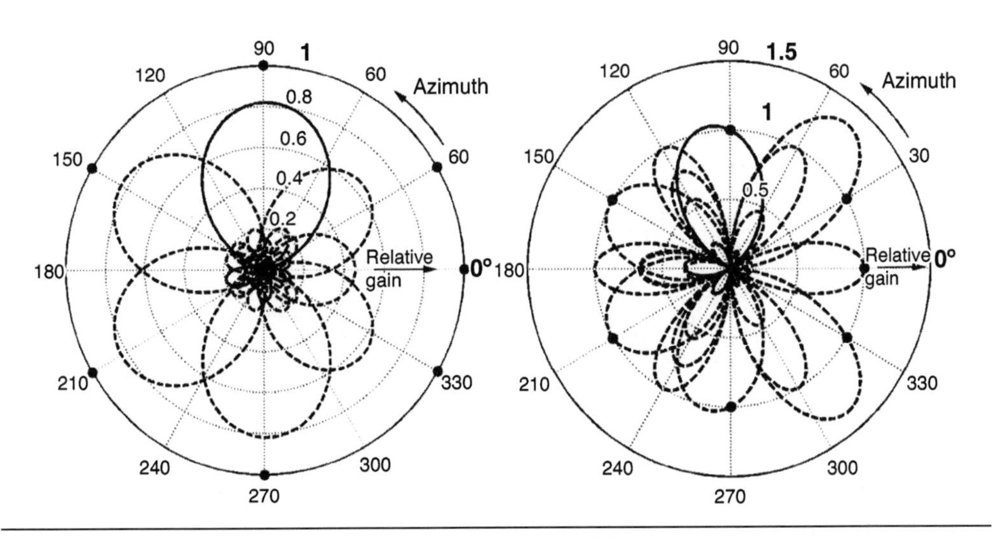

Figure 10.12 Gain functions for HOA. Left: 2nd order, right: 3rd order.

Figure 10.6, the gain functions for one of the loudspeakers is indicated by solid lines and for the other ones by dashed lines. It is obvious that, for HOA, all loudspeakers contribute to the reproduction and not only two, as for VBAP.

Microphone equipment for HOA is in the process of being developed in a number of laboratories.[20, 21] Commercially available models consist of a number of microphones in the surface of an as-small-as-possible rigid sphere. Other arrangements, such as directional microphones on an acoustically transparent spherical wire grid, are being tested. As of today, HOA recordings with up to the 4th order can be readily achieved. Regarding sound-field synthesis, there is no limitation in terms of the order. HOA is therefore a preferred format for many applications, for instance, for performances of electronic music. The portability of HOA-coded signals is considered beneficial in this context.

The original idea of spherical-harmonics synthesis was based on the assumption that plane waves (*sound rays*) from all possible directions aim concentrically at the center of the synthesis area. However, a sound source within the synthesis area cannot be rendered in this way. To enable this too, curved wave fronts must be synthesized, the amplitude of which naturally decreases with the assumed distance from the source. Actual loudspeakers emit such curved wave fronts anyway, that in fact, this leads to a well-known boost of the low-frequency contents of the signal spectrum.[3] By appropriate simulation of the low-end boost, the perceived distance of auditory events can be controlled. Suitable control enables virtual sound sources even within the synthesis area that are known as *focused* sources.[16, 18] However, for proper control of the perceived distances, the loudspeaker distances at the synthesis end must be known, thus restricting the portability of HOA-encoded signals. Table 10.5 sums up the most relevant advantages and disadvantages of HOA.

Table 10.5 Advantages (+) and disadvantages (–) of HOA

+	3-D presentation possible
+	Mathematically well defined by spherical-harmonics synthesis
+	Very broad sweet spot possible
+	Simple panning rules (amplitude differences only)
+	Easily adaptable to given loudspeaker arrangements
+	Localization blur decreases with increasing order of spherical harmonics
+	Graceful degradation at high frequencies (sweet spot becomes narrower but stays centered)
–	Higher-order microphones still under development
–	Auditory events appear predominantly at loud-speaker distance, unless compensated for
–	Panning settings to be adapted to specific loudspeaker arrangement used
–	Coloration possible due to interference by sound-field superposition
–	Head's shadow may cause localization errors and coloration—less with increasing participating order

10.6 WAVE-FIELD SYNTHESIS

WFS is a method that—very much like HOA—aims at synthesizing a sound field in a defined area such that it is actually a replica of a physically realistic sound field—be it real or conceptual. The theoretical approach to this problem, however, is significantly different. Namely, in WFS, arrival-time differences (that is, unwrapped phase differences) are applied in addition to pure amplitude differences of the loudspeaker signals. The theory of this method has been known for quite some time but can only be practically applied since the advent of micro-electronic signal processors.[22] An early figure by Steinberg and Snow (1934)[23] already illustrates the basic idea (see Figure 10.13). People sometimes talk of a transparent *acoustic curtain* in this context.[24]

The mathematical calculation of such virtual sound fields is performed with superposition methods as are also applied for the calculation of the directions of line arrays of monopoles—for instance, Rayleigh's integral equation, eventually complemented by Fraunhofer's approximation.[3] Figure 10.14 visualizes that various forms of superposed sound fields can be synthesized in this way. However, if the loudspeaker arrangement is not a line array—rather, for example, a rectangular (see Figure 10.15) or circular disposition—Rayleigh's integral equation, which is based on elementary monopoles only, does no longer suffice. Instead, the Kirchhoff-Helmholtz integral equation is employed. This equation states that the sound field within a closed boundary is completely determined by both the sound pressure, $P(\mathbf{x}, \omega)$, a scalar, plus the sound velocity or pressure gradient (both written in complex notation), respectively, a vector, everywhere on the boundary, namely:

$$P(\mathbf{x},\omega)=\int_{\partial V}\left[\frac{\partial}{\partial \mathbf{n}}G(\mathbf{x}\,|\,\boldsymbol{\xi},\omega)P(\boldsymbol{\xi},\omega)-G(\mathbf{x}\,|\,\boldsymbol{\xi},\omega)\frac{\partial}{\partial \mathbf{n}}P(\boldsymbol{\xi},\omega)\right]d\boldsymbol{\xi}, \tag{10.2}$$

where **X** denotes the position vector to any point within the closed boundary ∂V, ξ is an arbitrary point on this boundary with normal vector **n**, and ω is the angular-frequency variable.

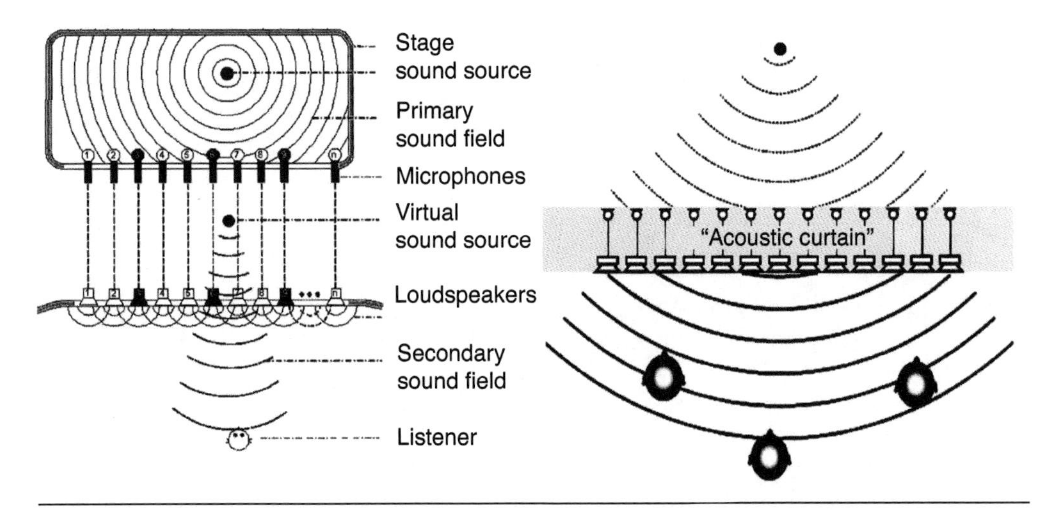

Figure 10.13 A basic idea in wave-field synthesis: the *acoustic curtain*—that is, synthesis with a line-array of loudspeakers. Left panel after Steinberg and Snow (1934),[23] right panel after Theile (2001).[24]

The Green's function from a surface point, ξ, to an interior point, **X**, is denoted by $G(\mathbf{x}|\xi, \omega)$. For a derivation of this equation see, for instance, Rabenstein and Blauert (2010).[14] Green's function describes the sound propagation from a source to a receiver and thus depends on the acoustic environment. In enclosures with low reverberation, it can be approximated by Green's function for the free field, with c being the speed of sound:

$$G(\mathbf{x}\,|\,\xi,\omega)=\frac{\exp\left(\dfrac{\omega}{c}|\mathbf{x}-\xi|\right)}{|\mathbf{x}-\xi|}. \qquad (10.3)$$

Real synthesis equipment always embodies a limited number of loudspeaker channels only. Thus, the sound field is sampled at a limited number of support positions. As known from quantizing time signals, at least two support positions are needed per period interval/length. If this condition is not fulfilled, distortions occur. In spatial sampling, the distortions are mirror images (*spatial aliasing*). In the mirror-image regions, a meaningful relationship of the directions of sound incidence and those of the auditory events is no longer given. Aliasing starts abruptly, right above a limiting frequency, f_{alias}, which is specific for the loudspeaker setting used. For line arrays of equidistant loudspeakers, it is, with s being the inter-loudspeaker distance:

$$f_{\text{alias}}=\frac{c}{\Delta\, s_{\text{loudspeaker}}}. \qquad (10.4)$$

Actually, the most realized loudspeaker arrangements for WFS are planar, that is, linear, rectangular, or circular—they may not even have a closed boundary. This leads to deviations from theory. Further, what is realized is hardly a 3-D boundary—not to mention a closed one. Restriction to planar disposition implies that cylindrical instead of spherical harmonics are engaged for the calculations.[31] The fact that it is hard to build dipole sources in reality is less critical, because dipole and monopole signals are highly correlated in most cases, such that employing common (monopole) loudspeakers is sufficient for most practical implementations.

Microphone arrays for WFS recording are being studied but are not yet readily available. To compose auditory scenes for WFS reproduction, special algorithms are necessary and available in the form of special WFS mixing consoles. An example is a freely available

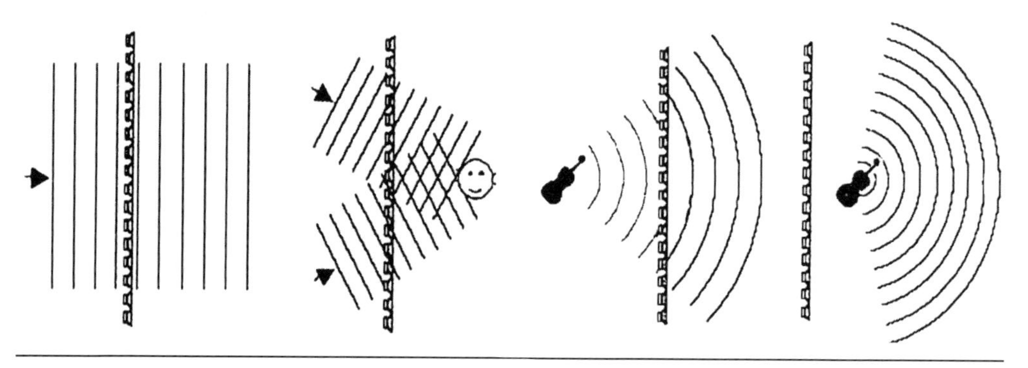

Figure 10.14 Examples of WFS with line arrays (plots courtesy of Sonic Emotion A.G.). Sound sources can also be reproduced to be placed in front of the array—that is, as *focused* sources.

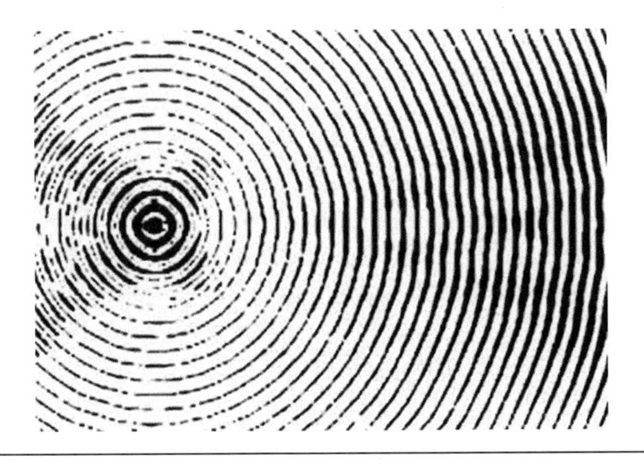

Figure 10.15 WFS employing a rectangular loudspeaker disposition creating a *focused* source (simulation data courtesy of ADA-AMC GmbH).

Table 6 Advantages (+) and disadvantages (–) of WFS

+	3-D presentation possible
+	Mathematically well defined (Kirchhoff-Helmholtz integral equation)
+	Listening area not restricted within synthesis area (no sweet spot)
+	Panning possible, but more complicated (amplitude- plus arrival-time differences)
+	Localization blur decreases with increasing number of channels
–	Coloration present above aliasing frequency
–	Proper directional information no longer available right above aliasing frequency
–	Suitable microphone equipment still rarely available
–	Panning settings to be adapted to specific loudspeaker arrangement used

rendering software called *Sound-Scape Renderer*, which handles WFS among other spatial-reproduction methods.[25, 26] Table 10.6 summarizes the most relevant advantages and disadvantages of WFS.

10.7 BINAURAL-CUE SELECTION

Application of all sound field synthesis methods dealt with in the previous paragraphs faces a severe problem, namely, the fact that data regarding actual auditory perception in synthetic fields are rare. In fact, even today, it is not understood in detail how summing localization really comes about. To avoid problems resulting from this lack of knowledge, synthesizing the sound fields in a physically as-authentic-as-possible way—the so-called holophonic approach—is often tried. This requires a lot of effort. Alternatively, one can try to first identify those attributes of the sound field that are perceptually relevant. Once these *cues* have been identified, they are then treated with preference—irrelevant ones being neglected in the further course of synthesis.

For the identification of perceptually relevant cues, it makes sense to start from the binaural ear-input signals—that is, the sound signals at the entrances to the ear canals. A common assumption is that the binaural auditory system forms a kind of interaural cross-correlation on these signals; the process being carried out in parallel frequency bands. Figure 10.16 schematically depicts the architecture of a common model of binaural processing.[2] A standardized output of such processing is called interaural coherence, k. In Figure 10.17, an example of a time function in a specific auditory frequency band, $k(t, f_n)$, is given.[19] By plotting the interaural cross-correlation function as a function of both the running time and the horizontal angle of sound incidence, one arrives at *binaural-activity maps.*

Figure 10.18 displays such a map for a case where a distinct binaural impulse response as recorded in a concert hall is used as input to the model of binaural processing.[27, 28] Experienced room acousticians can judge the acoustics quality of halls by interpreting such maps.

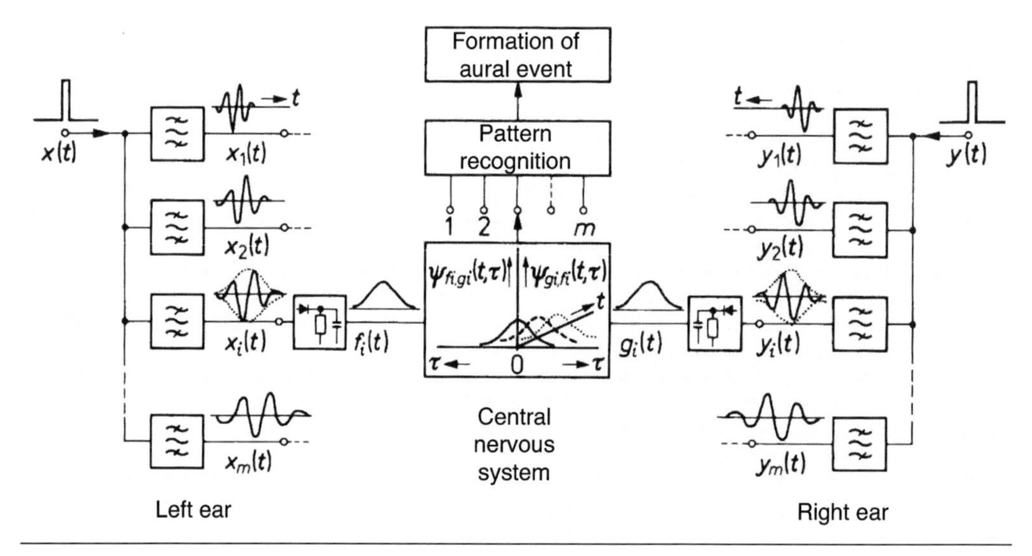

Figure 10.16 Architecture of a model of binaural processing (schematic). The sound signals as received at the ears are decomposed in spectral bands. For spectral components above about 1.5 kHz, the envelopes of the band-pass signals are extracted. Then, a set of interaural cross-correlation-functions is calculated (*binaural-activity map*), which forms the basis of further evaluation.

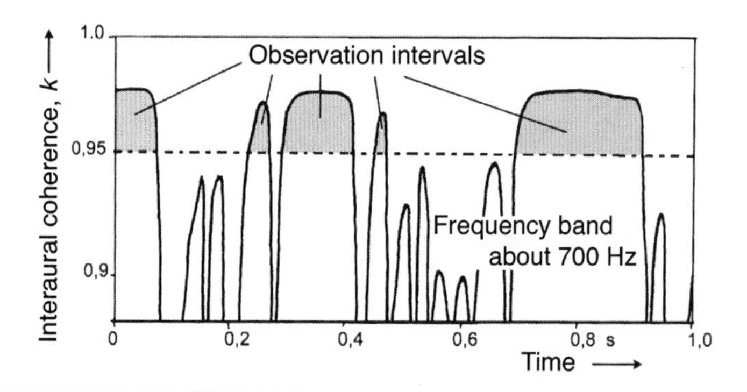

Figure 10.17 Interaural coherence, k—that is, a normalized interaural cross-correlation—as a function of time.[19] The amount of interaural coherence serves as a measure of confidence for detected interaural arrival-time differences (ITDs) and interaural level differences (ILDs), and, thus, for reliable spatial decomposition of auditory scenes.

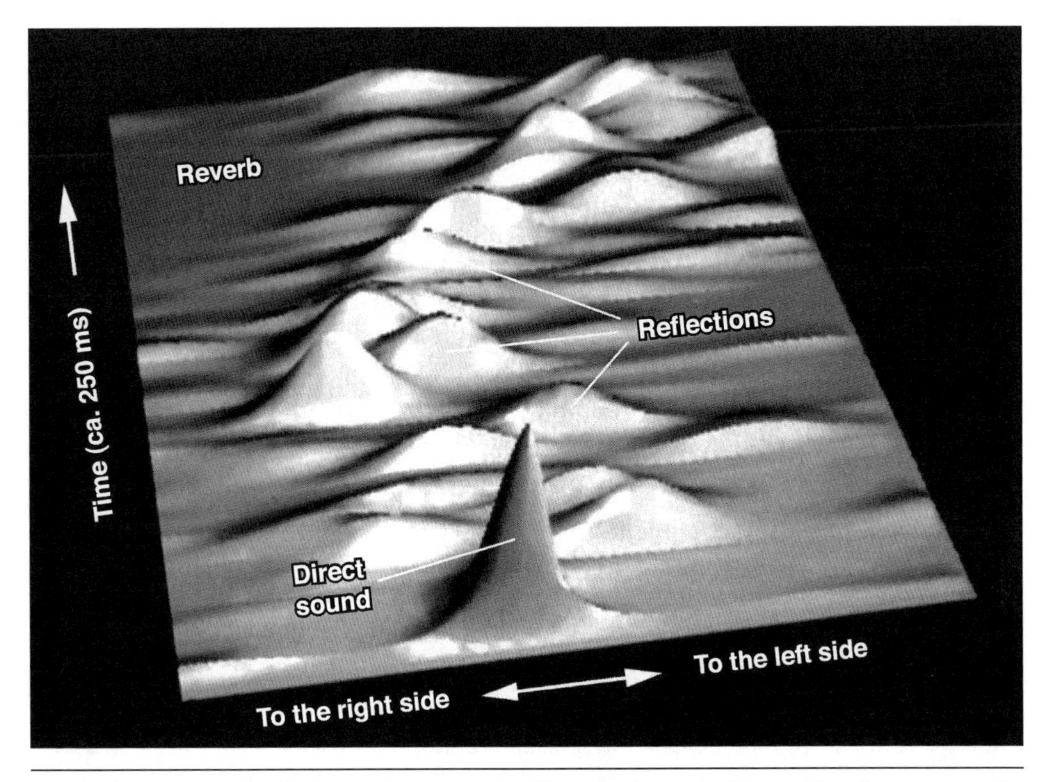

Figure 10.18 Example of a *binaural-activity map*. The map shows the binaural impulse response of a concert hall as seen through a model of binaural processing.

It is assumed that the auditory system collects interaural attributes from the ear-input signals at just those moments where they are highly correlated, namely, when k is relatively high (note the observation intervals in Figure 10.17), since at those instances the interaural attributes can be related with confidence to particular directions of sound incidence. In Figure 10.18, the instances of high coherence, and thus confidence, can be identified as peaks in the map. Assumingly, the auditory system determines the individual directions of sound incidence from the positions of these peaks.

Once the relevant perceptual attributes have been identified, a special focus can be put on them in the further process of auditory synthesis. This holds in particular when the primary goal is not a physically authentic, but rather a perceptually plausible synthesis.

Attributes of lower perceptual relevance may be added later with computationally less costly methods, such as artificial reverberation and back-ground noise (*ambience*). Methods for sound-field synthesis with prior binaural-cue selection have gained in importance recently.[19,29,30] A recent example in this context is the DirAC-technique.[31] The scientific foundations in this area are in the process of being intensively investigated.[32] They are, in fact, important for a number of further applications as well, for example, spatial coding of binaural signals such as in *mp3*, measurement of the *quality of experience* in speech-dialogue and multimedia systems, planning processes in architectural acoustics, enhancement of speech intelligibility, and *ease of communication* in hearing aids and public-address systems.

10.8 DISCUSSION AND CONCLUSIONS

There are different room-related methods available to generate spatial sound fields. In those cases where they are to be used for synthesis of auditory scenes only, problems regarding suitable recording techniques are completely avoided. When directional separation is the paramount issue, for instance, of speech and noise sources, the conventional intensity stereophony is fully sufficient—at least for directions in the frontal sector of the horizontal plane. If directions in further spatial directions are to be included, generalized amplitude-difference panning, such as VBAP, is adequate to create synthetic sound fields. However, if it is aimed at presenting a sound-field authentically in a spatially distinct synthesis area, methods like HOA or WFS must be employed. Both of these methods rest on exact solutions of the acoustical wave equation—this is why they are called *Holophony*. HOA and WFS can actually be transformed into one another in a mathematically conclusive way.[18]

In practice, arrangements with a limited number of loudspeakers are used. This leads to approximation errors that are specific for each of the two methods. Thus, which of the two methods is the best choice depends on the specific use case that it is dedicated for. Figure 10.19 illustrates some characteristic differences of the two methods—plotted for a circular 32-loudspeaker disposition with data from Daniel et al., (2003).[33] For further analysis of the approximation errors see Spors et al., (2008)[34] and Ahrens et al., (2010).[35]

For low frequencies with regard to f_{alias}, the sound field is rendered in a largely correct way, whereby, in the case of WFS, there are no spatial distortions, even in the direct vicinity of the loudspeakers. Both methods are capable of generating focused sources, that is, sources within the synthesis area; yet, HOA needs relatively high amplitudes for this purpose (not indicated in the plot). With increasing frequency the correct partition (sweet spot) in the synthesis area shrinks. In HOA, the sweet spot remains in the center, even above f_{alias}. In WFS, the sweet spot shifts rearward and dissolves abruptly with increasing frequency above f_{alias}, that is, when aliasing becomes effective.

There is the argument that the auditory system makes predominant use of directional cues in a frequency region below 1.5 kHz, which would imply that spatial distortions due to aliasing would not really matter perceptually. Yet, unfortunately, this assumption is not in accordance with common theories of auditory sound localization[2] and valid perceptual data regarding this issue, which is not yet well understood, are hardly available at this time.

Figure 10.19 Examples for a comparison of HOA and WFS. The plots depict the reproduced sound fields (plotted from data by Daniel et al., 2003).[33]

Regarding practical sound-field synthesis, the question is relevant of how to assign appropriate directions to each of the sound sources that finally comprise auditory scenes. Those methods that apply amplitude-difference panning—and all spherical-harmonics-synthesis variations belong to these—accomplish this by applying known amplitude-panning rules. In this context, it is a particularity of Ambisonics and HOA that synthesized sound sources can easily be rotated around the center. For WFS, this issue turns out to be more complicated since both amplitude- and arrival-time differences must be implemented. As a tool for determining suitable panning settings, special mixing consoles with intuitive graphical user interfaces are provided.[26]

Spatial sound fields as produced by adequate loudspeaker arrangement can be combined with natural sound fields, thus creating a particular kind of *augmented* auditory reality. An application of this idea, among other ones, is its use for enhancing the quality of auditory experiences in performance spaces.[36]

Some shortcomings of the holophonic methods can be compensated by exploiting psychoacoustics effects, such as *spatial masking* and the *precedence effect*.[2] Relevant research in this field is in progress.[32] In principle, sound-field synthesis can be restricted to really audible binaural cues by making use of binaural-cue selection methods as the DirAC method. Further suggestions include combinations with intensity stereophony and/or the use of head-related transfer functions (HRTFs)[36] to provide further support.

Those who consider the expenditure of multiloudspeaker systems to be too high, should consider *binaural technology* as an alternative. This technology is also capable of rendering auditory spatial scenes in a qualitatively excellent way. Instead of employing headphones to deliver the binaural signals directly to the listeners' ears, loudspeakers can also be used for this purpose. Figure 10.20 shows, as an example, a so-called *transaural* system, where the cross talk between the adjacent loudspeaker and the averted ear is compensated by a

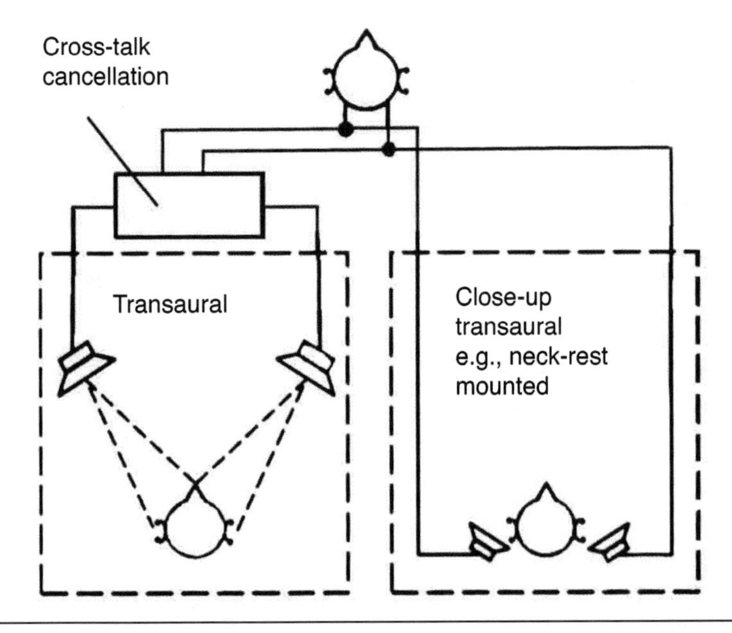

Figure 10.20 Loudspeaker methods for binaural reproduction. The cross-talk cancellation may be omitted when the two loudspeakers are positioned close to the head, such as in a neck-rest kind of arrangement.

dedicated electronic filter. In recent systems of this kind, the head position may even be tracked in order to continuously adjust the filter parameters. When the loudspeakers are positioned very close to the listeners' ears (neck-rest arrangement), cross-talk cancellation may even be skipped in less critical applications since interaural attenuation due to the head-shadow effect is substantial. Dispositions along this line of design are suitable for various applications, such as teleconferencing and video gaming.[37, 38]

FINAL REMARKS

When deciding on purchasing hardware for *Holophony*, it is certainly good advice to select the components in such a way that they can be utilized for different relevant reproduction methods—particularly for HOA and WFS. At this point in time, circular and spherical arrays offer the highest compatibility. Make sure to acquire loudspeakers and amplifiers of sufficiently high quality. The decision for a specific reproduction system will finally end up as being primarily an issue of software. In any case, the decision for a specific computational rendering algorithm should be taken from a strictly application-oriented point of view by considering the relevant application-specific advantages and disadvantages of the different methods in detail. The authors hope that this chapter provides some initial guidance to this end.

ACKNOWLEDGMENTS

Earlier versions of this material have been presented at the *OSA'2010* seminar in Gliwice, Poland (in English) and at the 2010 *ITG-Fachtagung Sprachkommunikation* in Bochum, Germany (in German). In the form of an archival-journal article, it was published in the Archives of Acoustics.[39] The authors are indebted to various anonymous reviewers for constructive remarks.

REFERENCES

1. Blumlein, A. D. (1931). *Improvements in and relating to sound transmission, sound recording and sound-reproducing systems*, British Patents #325 and #394.
2. Blauert, J. and Braasch, J. (2008). Räumliches Hören (Spatial Hearing), [in:] Handbuch der Audiotechnik, S. Weinzierl [ed.], Springer, Berlin-Heidelberg-New York.
3. Blauert, J. and Xiang N. (2009). *Acoustics for Engineers*, 2nd ed., Springer, Berlin-Heidelberg-New York.
4. Wendt, K. (1963). *Directional hearing in two superposed sound fields as in intensity- and arrival-time stereophony* [in German: *Das Richtungshören bei der Überlagerung zweier Schallfelder bei Intensitäts- und Laufzeitstereophonie*], Doct. diss., RWTH Aachen, Aachen, Germany.
5. Rumsey, F. (2001). *Spatial Audio*, Focal Press, GB-Oxford.
6. Theile, G. (2001). *Multi-channel natural music recording based on psychoacoustic principles*, 19th AES Int. Conf., Audio-Engr. Soc., New York NY.
7. Kamekawa, T., Marui A. and Irimajiri H. (2007). *Correspondence relationship between physical factors and psychological impressions of microphone arrays for orchestra recording*, 123rd AES Conv., Audio Eng. Soc, New York, NY.
8. Pulkki, V. (2001). *Spatial sound generation and perception by amplitude-panning techniques*, Doct. diss., Aalto Univ., Helsinki, Finland.
9. Plenge, G. and Theile G. (1977). *Localization of lateral auditory events*, J. Audio-Engr. Soc., 25, 196–200.

10. Meyer, E. and Thiele R. (1956). *Room-acoustical investigations in numerous concert halls and radio studios by means of novel measuring techniques* [in German: *Raumakustische Untersuchungen in zahlreichen Konzertsälen und Rundfunkstudios unter Anwendung neuerer Messverfahren*], Acustica, 6, 425–444.
11. Van Daele, B. and Van Baelen W. (2011). *Auro-3D: the advantage of channel-based sound in 3-D*, Proc. Int. Conf. Spatial Audio, ICSA, Detmold, Germany
12. Hamasaki, K., Hiyama, K. and Okumura R. (2005). *The 22.2 multi-channel sound system and its application*, 118th AES Conv., Audio Eng. Soc., New York, NY.
13. Gerzon, M. (1973). *Periphony: with-height sound reproduction*, J. Audio Eng. Soc., 21, 2–10.
14. Rabenstein, R. and Blauert J. (2010). *Sound-field synthesis with loudspeakers, part II—signal processing*, [in German: *Schallfeldsynthese mit Lautsprechern II—Signalverarbeitung*], ITG-Fachtg. Sprachkommunikation, Bochum, Germany.
15. Rabenstein, R. and Spors, S. (2008). *Sound-field reproduction*, [in:] Benesty, J., Sondhi, M.M. and Huang, Y. [eds.], Springer Handbook of Speech Processing, 1095–1114, Springer, Berlin-Heidelberg-New York.
16. Daniel, J. (2003). *Spatial encoding including near-field effect: introducing distance-coding filters and a viable, new Ambisonics format*, 23rd AES Int. Conf., Audio Eng. Soc, New York, NY.
17. Hollerweger, F. (2005). *An introduction to Higher-order Ambisonics*, www.create.ucsb.edu/wp/- FH HOA.pdf (last access, Febr. 2012).
18. Nicol, R. (2010). *Représentation et perception des espaces auditifs virtuels (representation and perception of virtual auditory spaces)*, Habilitation thesis, Univ. du Maine, Le Mans, France.
19. Faller, F. (2004). *Parametric coding of spatial audio*, Doct. diss., EPFL, Lausanne, Switzerland.
20. Moreau, S., Daniel J. and Bertet S. (2006), *3-D Sound field recording with higher-order Ambisonics —objective measurements and validation of spherical microphones*, 120th AES Conv., Audio Eng. Soc, New York, NY.
21. Meyer, J. and Elko G. (2010). *Analysis of the high-frequency extension for spherical eigenbeamforming microphone arrays*, J. Acoust. Soc. Am., 127, 1979.
22. Berkhout, A. J. (1988). *A holographic approach to acoustic control*, J. Audio-Engr. Soc. 36, 977–995.
23. Steinberg, J.C. and Snow, W.B. (1934). *Auditory perspective—physical factors*, Elect. Eng., 12–17.
24. Theile, G. (2005). *Spatial-audio presentation by means of wave-field synthesis* [in German: *Räumliche Tondarstellung mit Wellenfeldsynthese*], VDT-Magazin 2.
25. Geier, M., Ahrens, J., Spors, S. (2008). *The Soundscape Renderer: A unified spatial audio reproduction framework for arbitrary rendering methods*, 124th AES Conv., Audio Eng. Soc, New York, NY.
26. Geier, M., Ahrens, J. and Spors S. (2008). *The Sound-Scape Renderer.* http://spatialaudio.net/ssr/ (last access, Sept. 2016).
27. Lindemann, W. (1985). *Extension of the cross-correlation model of binaural signal processing by mechanisms of contra-lateral inhibition* [in German: *Die Erweiterung des Kreuzkorrelationsmodells der binauralen Signalverarbeitung durch kontralaterale Inhibitionsmechanismen*], Doct. diss., Ruhr-Univ. Bochum, Bochum, Germany.
28. Gaik, W. (1990). *Investigation regarding the binaural processing of head-related signals* [in German: *Untersuchungen zur binauralen Verarbeitung kopfbezogener Signale*], Doct. diss., Ruhr-Univ. Bochum, Bochum, Germany.
29. Merimaa, J. (2006). *Analysis, synthesis and perception of spatial sound—binaural localization modeling and multi-channel loudspeaker reproduction*, Doct. diss., Aalto Univ., Helsinki, Finland.
30. Merimaa, J. and Pulkki, V. (2005). Spatial impulse response rendering I: Analysis and synthesis. J. Audio Eng. Soc., 53, 1115–1127.
31. Pulkki, V. (2006). *Directional audio coding in spatial sound reproduction and stereo upmixing*, 28th AES Int. Conf., Audio-Engr. Soc., New York, NY.
32. Blauert, J., Braasch, J., Buchholz, J., Colburn, H. S., Jekosch, U., Kohlrausch, A., Mourjopoulos, J., Pulkki, V., Raake, A. (2009). *Auditory assessment by means of binaural algorithms—the* AABBA *project*, Int. Symp. Auditory Audiolog. Res., ISAAR'09, Danavox Jubilee Foundation, Ballerup, Denmark.

33. Daniel, J., Nicol, R. and Moreau S. (2003). *Further investigation of high-order Ambisonics and wave-field synthesis for holophonic sound imaging*, 114th AES Int. Conv., Audio Eng. Soc, New York, NY.
34. Spors, S., Rabenstein, R. and Ahrens, J. (2008). *The theory of Wave-field Synthesis revisited*, 124th AES Conv., Audio-Engr. Soc., New York, NY.
35. Ahrens, J., Wierstorf, H. and Spors, S. (2010). *Comparison of Higher-order Ambisonics and Wave-field Synthesis with respect to spatial-discretization artifacts in the time domain*, 40th AES Int. Conf., Audio-Eng. Soc, New York NY.
36. Woszczyk, W. (2011). *Active acoustics in concert halls—a new approach*, Archives of Acoustics, 36, 2, 379–393.
37. Kang, S.-K. and Kim S.-H. (1996). *Realistic audio teleconferencing using binaural and auralization techniques*, ETRI J., 18, 41–51.
38. Menzel, D., Wittek, H., Theile, G. and Fastl, H. (2005). *The Binaural Sky: A virtual headphone for binaural room synthesis*, Tonmeistersymposium, Hohenkammer, Germany.
39. Blauert, J. and Rabenstein, R. (2012). *Providing Surround Sound with Loudspeakers: A Synopsis of Current Methods*, Archives of Acoustics, 37, 1, 5–18.

11

Environmental Acoustics

Jian Kang, School of Architecture, University of Sheffield, United Kingdom

INTRODUCTION

With the increasing attention on our living environment, there have been considerable developments in the area of environmental acoustics in the last decade or so in various aspects, including sound propagation, regulation, standardization, perception, and creation. This chapter first analyzes the role of key factors in environmental sound propagation and noise mapping, followed by a section on sound propagation at micro-scale urban areas, such as in street canyons and urban squares, as well as the effectiveness of architectural changes and urban design options. Main environmental noise indicators and regulations/standards are then briefly summarized. This is followed by a section on another important facet of environmental acoustics, noise perception, where the influencing factors, both acoustic/physical and social/psychological/economic, are outlined. Finally, soundscape research, which considers the relationships between the ear, human beings, sound environments, and society, rather than simply reducing noise levels, is discussed, especially for urban open spaces.

11.1 ENVIRONMENTAL SOUND PROPAGATION AND NOISE MAPPING

Environmental sound propagation from a source to a receiver depends on the sound power level and directivity of the source, as well as the attenuation between the source and receiver due to various factors including geometrical divergence, atmospheric absorption, ground effects, barrier effects, and miscellaneous effects.[1] In this section, the role of key factors in environmental sound propagation is analyzed and noise mapping is briefly discussed.

11.1.1 Source Model

Much effort has been made recently, especially in Europe, on the prediction of the sound power level of traffic noise sources.[2] For road traffic, the sources on the vehicles are simplified into two point sources: the lower source at 0.01 m above the road surface, which is mainly due to tire/road noise; and the higher source, which is mainly propulsion noise with its height depending on the vehicle category. For each source the sound power level can be calculated based on vehicle speed and coefficients given in a one-third octave band for each vehicle category, with five main categories considered, including light vehicles, medium-heavy vehicles with two axles, heavy vehicles with more than two axles, other heavy vehicles, and two-wheelers. In each main category there are a number of subcategories. Some corrections are also made to the basic sound power levels, including for the road surface texture and condition, for directivity both in the horizontal and vertical plane, and for tires. Propulsion noise increases during acceleration and decreases during deceleration and correspondingly, a correction is considered.

Extended noise sources are divided into cells with each having their own characteristics. For example, a road segment can be composed of a series of point sources of different types depending on the percentage of vehicles of various categories and their speed.

11.1.2 Geometrical Divergence with Point, Line, and Plane Sources

In a free field, with a point source of sound power level, L_W, the sound pressure level (SPL) at distance d can be calculated by:

$$SPL = L_W - 10\log(4\pi d^2) = L_W - 20\log(d) - 11. \tag{11.1}$$

When the source is located close to a hard ground, Eq. (11.1) becomes:

$$SPL = L_W - 20\log(d) - 8. \tag{11.2}$$

For an ideal line source of infinite length in a free field, with L_W as the sound power level per meter, the SPL can be determined using purely cylindrical sound propagation:

$$SPL = L_W - 10\log(2\pi d) = L_W - 10\log(d) - 8. \tag{11.3}$$

Again, if a line source is located close to a hard ground, Eq. (11.3) becomes:

$$SPL = L_W - 10\log(d) - 5. \tag{11.4}$$

From Eq. (11.1) to (11.4), it can be seen that when the source-receiver distance is doubled, the SPL falls off at 6 decibels (dB) for a point source and 3 dB for a line source, respectively.

For a plane source, the sound radiation can be approximately calculated by considering the source as a number of evenly distributed individual point sources.

11.1.3 Ground Attenuation

Ground attenuation can be caused by the absorption of acoustic energy when a sound wave impinges on the ground. Effective ground absorption can be obtained from grass or other vegetation, ploughed fields, snow cover, or other kinds of sound absorber. Ground attenuation can also be caused by the ground effect, namely, interference between direct and reflected sound waves due to the change in phase of the reflection. The sound interaction with the ground depends on the geometry from the source to the receiver and the acoustic properties of the ground surface.

Accurate calculation of ground attenuation in practice is challenging, and some engineering methods have therefore been developed. For example, in ISO 9613, three separate regions—namely the source region, middle region, and receiver region—are used in determining ground attenuation for approximately flat ground, either horizontally or with a constant gradient [ISO, 1993].[1]

11.1.4 Atmospheric Absorption

There are two mechanisms by which acoustic energy is absorbed by the atmosphere, namely, molecular relaxation and viscosity effects. Air absorption is mainly dependent on temperature and relative humidity, and it is proportional to the source-receiver distance and is generally more significant at high frequencies.

In the case of wind, there is usually a velocity gradient where the wind speed increases with increasing height above the ground. Consequently, sound waves travel upwind at a greater speed near the ground and at progressively slower speeds with increasing height above the ground. The sound waves are therefore bent and less sound is received at a point upwind compared to windless conditions. Conversely, more sound is received downwind.

The effect of temperature gradient on sound propagation is similar to that of wind. An increase in temperature with altitude, which usually occurs at night when ground air temperature is considerably reduced, results in an increase in sound speed and consequently, the sound waves will be refracted downward in the absence of any wind. Conversely, daytime air temperature decreases with increasing altitude and thus, the sound rays will be continuously bent away from the ground and less sound will be received than if no gradient exists.

11.1.5 Vegetation

Tall vegetation—including corn, hemlock, brush, and pine—can cause significant sound reduction compared to open grassland. Wide belts of tall dense trees of a depth of 15–40 m appears to offer an extra noise attenuation of 6–8 dB at low (around 250 Hz) and high frequencies (>1 kHz).[3] Because attenuation from trees is mainly due to branches and leaves, sound energy near the ground will not be significantly reduced, and deciduous trees will provide almost no attenuation during the months when their leaves have fallen. It is also noted that the attenuation from dense plantings, say more than 30 m deep, will be limited by the flanking of sound energy over the top of the canopy of trees.

Tree and shrub arrangements are important. In random arrangements, the scattering contribution of trunks and branches is relatively minor and good sound attenuation requires high densities of foliage to ground level. Regular tree planting arrangements have been shown to offer useful *sonic crystal* effects, giving rise to more than a 15 dB reduction in transmitted sound in a particular frequency range.[4]

11.1.6 Noise Barriers: Basic Configurations

A noise barrier is a solid obstacle that impedes the line of sight between a source and receiver, as illustrated in Figure 11.1. For a single point source and an infinitely long barrier, a simplified equation can be used to calculate the insertion loss (IL), namely the SPL difference between them, with and without the barrier:[5]

$$IL = 20\log\frac{(2\pi N)^{\frac{1}{2}}}{\tanh(2\pi N)^{\frac{1}{2}}} + 5. \tag{11.5}$$

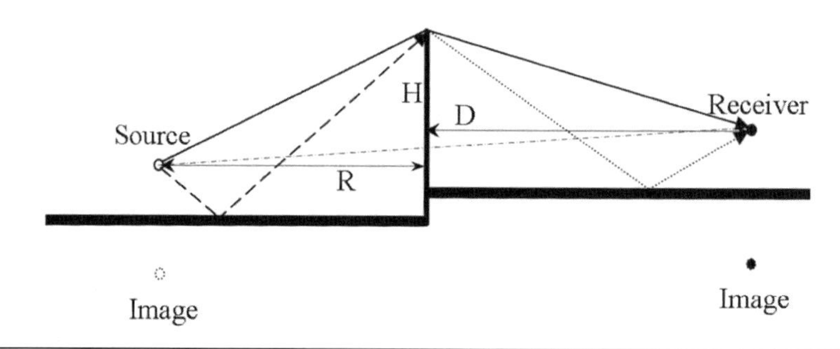

Figure 11.1 Diagram of calculating diffractions over a barrier, considering a point source and its image.

Equation (11.5) is applicable for $-0.2 < N < 12.5$, and for $N > 12.5$, there might be an upper limit. N is the Fresnel number, defined as:

$$N = 2\frac{\delta}{\lambda}, \tag{11.6}$$

where λ is the wavelength and δ is the path difference, namely the difference in distance between the direct path through a barrier from the source to the receiver and the indirect path over the barrier. If the source-receiver line is perpendicular to the barrier, as shown in Figure 11.1, δ can be calculated by:

$$\delta = R\left[\sqrt{1+\left(\frac{H}{R}\right)^2} - 1] + D[\sqrt{1+\left(\frac{H}{D}\right)^2} - 1\right], \tag{11.7}$$

where R is the distance between the source and barrier (m); D is the distance between the barrier and receiver (m); and H is the effective height of the barrier (m), namely the barrier height above the line from the source to the receiver.

The effect of ground reflection should be taken into account unless the receiver is much higher than the ground—say over 2 m. If the ground is perfectly reflective, the aforementioned method can be applied for the image of the source/receiver, as shown in Figure 11.1. Diffraction of sound also occurs around the ends of a barrier, and the algorithms described above can be similarly used. For relatively low frequencies, the interferences from various sound paths should be taken into account. It is also noted that the effectiveness of a barrier depends on the sound transmission loss through the barrier, which should normally be 10 dB higher than the diffracted sound.

The above theory assumes a knife edge at the barrier top and it is approximately valid when the barrier thickness is smaller than the wavelength. For thicker barriers such as buildings or banks, and for two parallel barriers, some engineering methods have been developed.[1,6]

11.1.7 Noise Barriers: Strategic Design

The beneficial effects of additional diffracting edges on the barrier top have been demonstrated with fir tree profile, T profile, Y profile, arrow profile, and branched barriers, with a typical insertion loss of about 3–5 dBA (i.e., a weighted decibel level).[7] The effectiveness of using absorptive treatments around diffracting edges on the barrier top is generally up to 3 dBA.

1. The possibility of reducing traffic noise using a series of parallel grooves in the ground has been studied, with an average insertion loss of around 4 dB.[8] With strategically

designed rib-like structures, the insertion loss could typically be 10–15 dB over a rather wide frequency range, where it seems that quarter-wavelength resonance and surface wave generation play a significant role in determining the performance at lower frequencies.[9] Fujiwara et al., [1998][10] found that a T profile barrier with a reactive surface on the top could produce an improvement of 8.3 dB in the frequency-averaged insertion loss.

Figure 11.2 shows a phase interference barrier[11] where the basic configuration is a three-sided barrier consisting of hollow passages at an angle to the ground. Sound waves are refracted when passing through this structural phase lag circuit, causing them to interfere destructively with top diffracted sound waves in some areas, resulting in noise reduction. With similar principles, Figure 11.2 also shows examples of wave guide[12] and phase reversal barriers.[13] Generally speaking, these devices would be effective for unique and dominant pure tones at low frequencies, whereas at higher frequencies the attenuation could be negative.

Picket barriers and vertically louvered barriers have also been explored, with which the dead weight wind loading on barrier foundations could be reduced. An improvement could be achieved due to destructive interference of low frequency sound between the sound transmitted through the gaps and that pass over the barrier top, and at high frequencies the performance could be improved by using sound absorbers in the gaps.

Along the barrier length, the wave front from a stationary point source meets the barrier top at different locations, causing coherent addition of the sound pressure. Barriers with randomly jagged edges along the longitudinal direction have been investigated,[14] and it has been shown that such barriers could give enhanced performance at high frequencies, but the low frequency performance could be even poorer than that of a straight-edged barrier.

2. Considerable noise barrier effects can also be achieved through creating shadow zones by strategic architectural/landscape designs, including earth mounds, roads in cuttings, roads elevated above the surrounding ground on embankments, or on other structures such as viaducts.

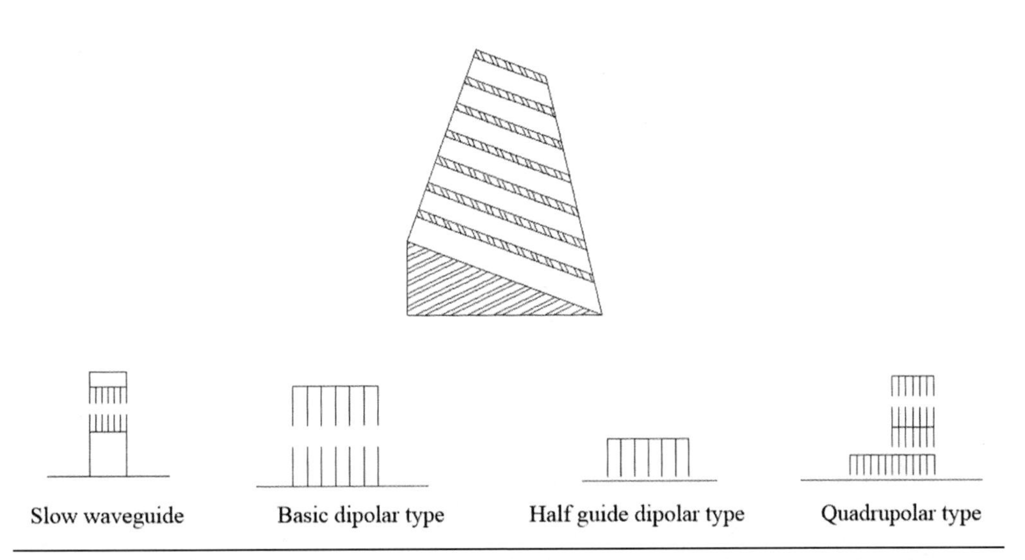

Figure 11.2 Cross-sectional view of phase interference (above) and phase reversal (below) barriers.

By angling the top section of a barrier toward the source, the diffracting edge of the barrier will become closer to the source, so that the barrier is more efficient. A galleried barrier is a substantial cantilevered barrier which covers the nearside traffic lane, forming a partial enclosure. Arrays of sound absorbing panels can be hung above a road to prevent any direct line of sight to the source when viewed at an oblique angle from receivers.

Trees and nonarboreal vegetation can be used conjunctively with noise barriers. Measurements have shown that the presence of trees produces an SPL decrease of 2–4 dBA downwind when the wind speed is 6–12 m/s.[15] Bio-barriers are also often used, which may be divided into four generic types, including A-frame and vertical corten steel, box wall, woven-willow, and stack and crib bio barriers.[3]

When there is another barrier or a building on the opposite side of a road, or when high-sided reflective vehicles run close to the barrier, the negative effects caused by multiple reflections should be considered. The single barrier insertion loss could be restored by relatively small angles of tilt, say 3–15 degrees, depending on the road width.[16] Dispersive barriers provide an alternative solution by scattering the sound waves, although such treatments are less effective for a line source parallel to the barrier, compared to point sources. Absorptive treatments, such as absorptive concrete, vegetation on perforated bricks, are also effective.

In addition to reduction of the noise level, the success of an environmental noise barrier depends on many other nonacoustic factors; for example, the consideration of structure, fixing, viewing at speed, pattern, texture, color, light and shade, material and design, visual neutrality and compatibility, safety, environmental impact, and cost.[3] Moreover, it has been demonstrated that a lack of effective public participation can negatively affect the perception of a barrier's effectiveness of mitigating noise.[17] Furthermore, a number of studies have shown that with the same insertion loss in dBA, people's perception on noise reduction could vary, depending on acoustic factors such as spectrum, as well as nonacoustic factors such as barrier materials.

11.1.8 Noise Mapping

Noise mapping, typically in a form of interpolated iso-contours to present the geographical distribution of noise exposure, has become an essential requirement, especially in Europe.[18] A number of noise-mapping software packages have been developed based on algorithms specified in various international and national standards. Noise maps normally predict long-term average A-weighted SPL, for which meteorological corrections are used to include the effect of varying weather conditions that occur over a time period of several months or a year.

Noise maps can be used for large urban areas, providing a useful tool for noise strategies and policies. They are also useful for relative comparisons in examining the effectiveness of certain noise mitigation measures. However, attention must be paid to its accuracy and strategic applications, since statistical methods and simplified algorithms are involved and 3-D models are also simplified, such as treating pitched roofs as flat roofs, and ignoring gaps between buildings.[19]

11.2 MICRO-SCALE SOUND PROPAGATION

While noise mapping is useful for strategic planning, to consider the noise level at specific receivers, accurate simulation of sound propagation at micro-scale, such as a street canyon or an urban square, is important. In this section some micro-scale models are summarized, most of which are based on simulation techniques originally developed for room acoustics (see Chapter 1 by Svensson et. al.).

11.2.1 Image Source Method

Consider an idealized rectangular square. By assuming the boundaries as geometrically reflective, especially in urban areas, a series of image sources can be created, as illustrated in Figure 11.3.[20] For the convenience of calculation, the image sources are divided into a number of groups. Consider an image source (j, k) $(j = 1\ldots\infty;\ k = 1\ldots\infty)$ in group I-iii, for example, the energy from an image source to a receiver R at (R_x, R_y, R_z) can be determined by:

$$E_{j,k}(t)=\frac{1}{4\pi d_{j,k}^{2}}(1-\alpha_A)^{k-1}(1-\alpha_B)^{k}(1-\alpha_U)^{j-1}(1-\alpha_V)^{j}e^{-Md_{j,k}}\left(t=\frac{d_{j,k}}{c}\right), \qquad (11.8)$$

where α_A, α_B, α_U and α_V are the absorption coefficient of façades A, B, U and V, respectively, M (Np/m) is the intensity-related attenuation constant in air, and $d_{j,k}$ is the distance from the image source (j, k) to the receiver:

$$d_{j,k}^{2}=(2jL-S_x-R_x)^2+(2kW-S_y-R_y)^2+(S_z-R_z)^2, \qquad (11.9)$$

where L and W are the square length and width respectively.

To consider the ground reflection, an image source plane similar to that in Figure 11.3 can be obtained. The energy from those image sources to receiver R can be determined in a similar manner as above, by taking the ground absorption α_G into account.

By summing the energy from all the image sources in all the source groups, and by taking direct sound transfer into account, the energy response at receiver R can be determined, with which a number of acoustic indices including reverberation time (T_{30}) and early decay time (EDT) and steady-state SPL can be determined. For urban street canyons, models can be developed using similar methods.[21]

For relatively narrow streets and/or low frequencies, interference effects due to multiple reflections from building façades and ground are important. A coherent model rather than energy-based and incoherent model should therefore be used. A comparison has shown that when the width of a street canyon is less than about 10 m, the differences between the coherent and incoherent models become more significant,[22] although with diffusely reflecting boundaries, energy-based and incoherent models can still be used for narrower streets, since the effects of phase cancellation and addition tend to be averaged.[23]

11.2.2 Ray-Tracing

A number of models based on ray-tracing, particle-tracing, or beam-tracing have been developed for urban areas, such as for interconnected streets, for single streets, for urban squares,[19] and for dynamic traffic noise distribution.[24] While a series of efficient ray-tracing techniques such as the QuickSort algorithm have been applied, some special methods relating to micro-scale urban environments have also been implemented. For example, since urban open spaces have a totally absorbent *ceiling*, the models would not generate any rays to that area. Similarly, any ray, after one or more reflections, once hitting the ceiling, will be stopped.

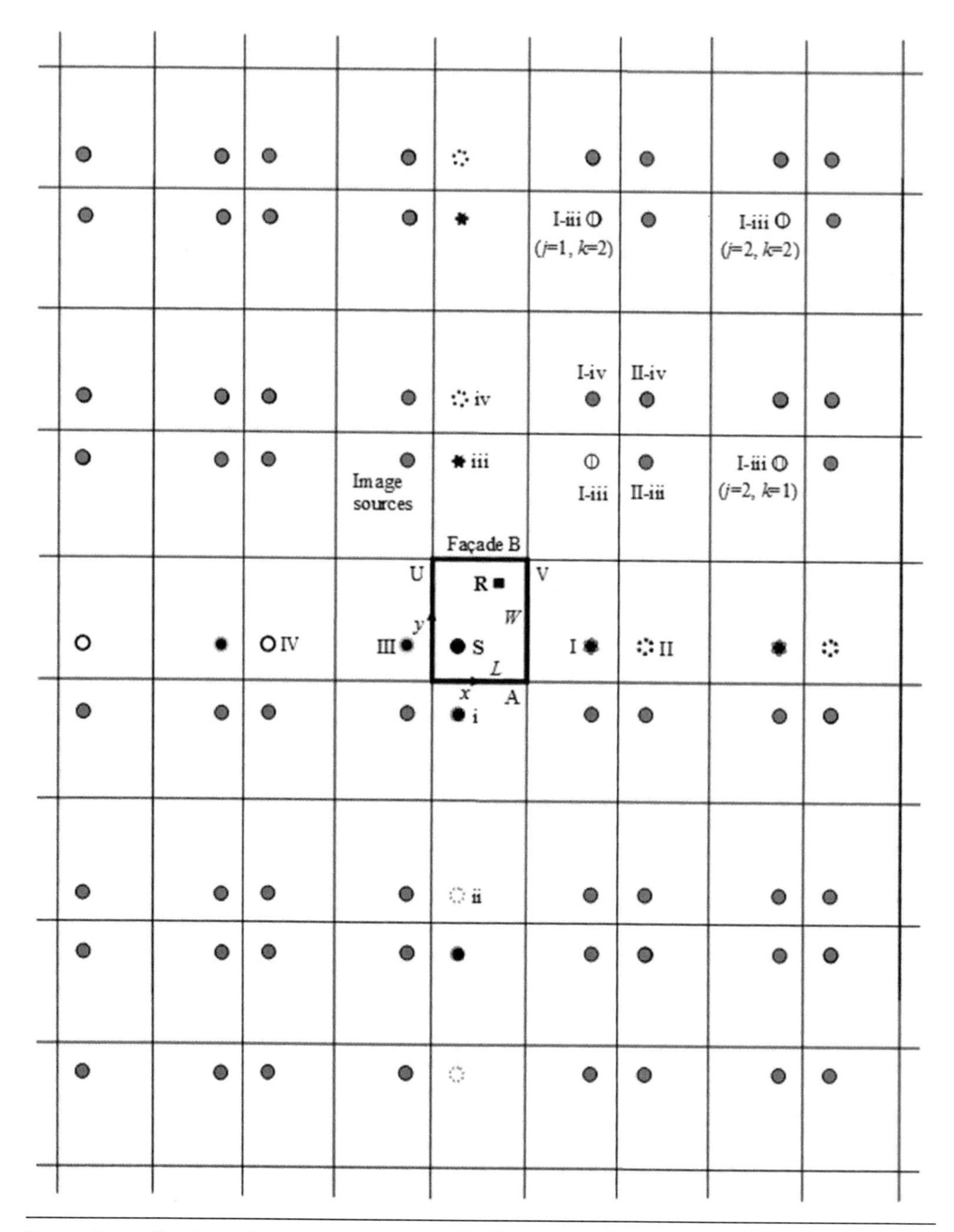

Figure 11.3 Plan view of image sources in an idealized square.

11.2.3 Radiosity Model

It is often necessary to consider diffuse reflections since there are always some irregularities on building or ground surfaces. The back-diffusion effect of reverberation has been demonstrated through measurements in a street canyon.[23] A study using physical scale models suggests that the scattering coefficient is about 0.09 to 0.13 for façades having surface irregularities similar to those found on real building façades, and the coefficient is not very sensitive to the degree of typical surface irregularities.[25] Although the coefficient appears to be small, the effect of multiple reflections is to make the diffuse reflection mechanism dominant at higher orders of reflection.

While several methods of considering diffuse reflections have been proposed, the radiosity method is an effective way.[19] With a similar approach to the radiosity techniques for long rooms, models have been developed for street canyons, cross-streets, and squares with diffusely reflecting boundaries according to the Lambert cosine law, and a series of validations have been made against measurements.[20, 26, 27, 28]

The model divides the boundaries of a street canyon or a square into a number of small patches, with each patch represented by a node in a network, and the sound propagation is simulated using the energy exchange between the patches. In the models the ground can be considered as either diffusely or geometrically reflective. A geometrically reflecting ground can be treated as a mirror and the sound source and the patch sources will have their images, as illustrated in Figure 11.4, using a street canyon as an example. The initial energy in the patch sources on the façades is from the sound source, S, as well as its image, S', for which the absorption coefficient of the ground is also taken into account. During the energy exchange process, the energy in a patch source on a façade, say A, is calculated by summing the contribution from all the patch sources on façade B and its image, B'. At receiver R, similarly,

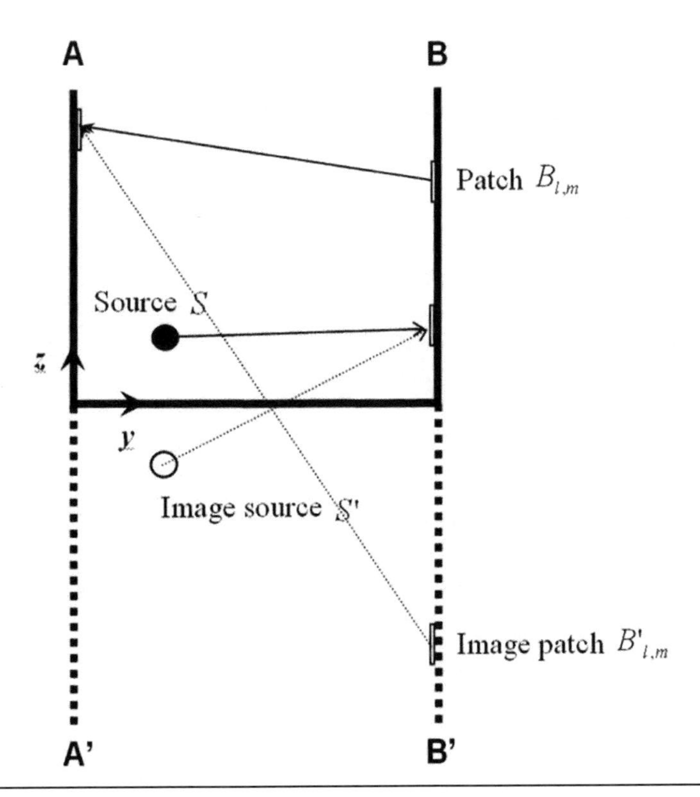

Figure 11.4 Cross-section view of radiosity modeling of sound propagation in an idealized rectangular street canyon with diffusely reflecting façades and geometrically reflecting ground.

for each order of energy exchange between patches, the sound energy contributed from each patch source and their images is summed.

11.2.4 Transport Theory

Based on the concept of sound particles and the application of the classic theory of particle transport, a model has been developed to predict the temporal and spatial sound distribution in urban areas.[29] The phonon obeys classic mechanics laws according to the Hamilton stationary action principle. In the model, a particle undergoes a straight line until it meets an obstacle, and it is assumed that the effects of phase cancellation and addition are averaged and the sound sources are not correlated. The model can consider partially diffusely reflecting building façades, scattering by urban objects, atmospheric attenuation, and wind effects. Under simplified conditions, it has been shown that the transport equation may be reduced to a diffusion equation. Jing et al. (2010) have been able to solve the transport equation without reducing to the diffusion equation for simulating long rooms (see Sec. 2.1.1 and Ref. 2 and Ref. 3 in Chapter 2 of this book).

11.2.5 Wave-Based Models

With the development of more powerful computers, a number of models based on numerically solving wave equations have been developed and applied in urban situations, including the acoustic finite element method (FEM) and boundary element method (BEM), the equivalent sources method (ESM), the finite-difference time-domain method (FDTD), and the parabolic equation (PE) method.

A 2-D BEM was used to study the sound field in the region of balconies in a tall building close to a roadway.[30] This is a typical situation where energy-based models are less appropriate since the wavelengths are not small compared to the dimensions of the balcony spaces and building elements.

The basic idea of the ESM is to reduce a problem to a simplified geometry with boundary conditions that are easy to handle. On boundaries with different conditions, virtual sources are placed. The method has been applied for the 2-D sound propagation in two parallel street canyons.[31] For the source canyon, the geometry is divided into two parts, the domain inside the canyon and the half space above. Compared with the standard BEM or FDTD, the ESM is relatively less computationally heavy since it only discretizes the opening of the canyon and the impedance patches.

The FDTD model is based on numerical integration of the linearized Euler equations in the time domain. An advantage of the FDTD compared with the ESM and BEM is its applicability for a moving, inhomogeneous, and turbulent atmosphere—namely the consideration of the effects of refraction. It can take into account the combined effect of multiple reflections, multiple diffractions, inhomogeneous absorbing, and partly diffusely reflecting surfaces.

The PE model is based on a one-way wave equation in the frequency domain. It is suitable for long-range sound propagation over a flat ground, but less suitable in situations with several reflecting obstacles and arbitrary wind fields.

A coupled FDTD-PE model has been explored, where the FDTD is applied in the complex source region and the PE is used for propagation over a flat ground to a distant receiver.[32]

11.2.6 Empirical Formulae

Based on both analytic theories and regressions of data obtained using computer simulation models, a series of formulae have been developed under various boundary conditions,[19] for example, in urban squares with diffusely reflecting boundaries:

$$RT = \frac{0.16V}{-S_0 Ln(1-\bar{\alpha}) + 4MV}\left(88.6 + 49\alpha_b + 2.7\frac{\sqrt{LW}}{H}\right) \quad (11.10)$$

$$L = L_W + 10\log\left(\frac{Q}{4\pi d_r^2} + \frac{3H}{W+L}\frac{4}{R}\right), \quad (11.11)$$

where S_0 is the total surface area and $\bar{\alpha}$ is the average absorption coefficient, both including an imaginary square ceiling. L, W, and H are the square length, width and height, $V = LWH$, α_b is the average absorption coefficient of boundaries, i.e., façades and ground only, $R = S_0\alpha_T/(1 - \alpha_T)$ and $\alpha_T = \bar{\alpha} + 4\ MV/S_0$. Q is the directivity factor of the source, and d_r is the source-receiver distance. Equations (11.10) and (11.11) are based on simulation using the radiosity method, with a range of urban square configurations: length, L = 20–200 m; width, W = 20–200 m, height, H = 5–100 m; and square area 400–40,000 m.[33] The length/width ratio is 1:1 to 4:1, and the side/height ratio $\sqrt{LW}/H$ = 0.5–40. Buildings are considered to be along two, three, or four sides of a square, with an absorption coefficient of 0.1–0.9.

11.2.7 Meso-Scale Models

A number of models have also been developed for sound propagation in urban areas at meso-scale, for example, the flat city model considering an urban area with homogeneous buildings (in terms of building height and absorption characteristics) as an almost flat plane with canyons containing streets and backyards criss-crossing the landscape;[34] the linear transport model where the sound propagation including multiple reflections and diffractions in a number of street blocks is treated as the flow of small packages of sound energy, namely phonons;[35] and the dynamic traffic model, which is different from conventional steady-state SPLs based on the average over a relatively long time period.[24]

11.2.8 Auralization

Techniques have also been developed to present the 3-D visual environment with an acoustic animation, which is useful for aiding urban soundscape design, including public participation. Different from auralization in room acoustics (see Chapter 9 by Vorländer), for urban areas considerations should be given to various urban sound sources, such as traffic, fountains, street music, construction, human voices, and birdsong; to dynamic characteristics of the sources, such as variation of traffic in a day; and to the movements of sources and receivers. On the other hand, compared to room acoustics, the requirements for urban auralization are relatively low.

A key issue of achieving fast urban auralization is to simplify the simulation algorithms, while retaining reasonable accuracy. A parametric study using beam-tracing shows that for urban squares, it is possible to reduce the reflection order and ray number compared to the parameters conventionally used in room acoustics.[19] In the meantime, since human sensitivity to a particular sound source might be reduced within a complex sound environment with multiple and moving sources, further simplification of algorithms has been suggested through a series of subjective experiments, although the degree of simplification depends on the source type.[36] For example, a low reflection order is generally acceptable in urban squares for music, fountains, and cars, but for human voices, a higher reflection order is needed. Based on these simplifications, a prototype of the auralization tool for the urban soundscape has been developed.

11.2.9 Physical Scale Modeling

In addition to computer simulation, acoustic physical scale modeling has been commonly used in simulating environmental sound propagation where some complex acoustic phenomena, such as diffraction around buildings, can be considered more accurately and thus used to validate computer models. Compared to real measurements, a significant advantage of scale modeling is that the geometry, source, and receiver condition, as well as background noise are relatively easy to control.

11.2.10 Noise Control in Street Canyons

This section explores the effects of architectural changes and urban design options in noise reduction, based on parametric studies using the radiosity model and the image source model, for diffusely and geometrically reflecting boundaries respectively.[19] Both single street canyons and an urban element consisting of a major street and two side streets are examined. For the latter only diffusely reflecting boundaries are considered. A point source is used except where indicated.

In single streets, the sound distribution in a cross section is generally even, unless the cross section is very close to the source. In terms of street configuration, when changing the width/height ratio, say from 0.3 to 3, the variation in the average SPL is typically 3–8 dB. A gap between buildings can provide extra sound attenuation along the street, and the effect is more significant in the vicinity of the gap.

For the urban element consisting of a major street and two side streets, with multiple sources or a moving source along a major street (as shown in Figure 11.5) the average SPL in a side street is typically 9 dB lower than that in the major street. In the meantime, the SPL attenuation along the side streets is significant, at about 15–17 dB. Conversely, despite the significant changes in the boundary condition in the side streets, the SPL variation in the major street is only about 1 dB, which suggests that with noise sources along a major street, the energy reflected from side streets to the major street is negligible.

An absorption treatment in a single street, such as increasing the boundary absorption coefficient from 0.1 to 0.5, or taking one side of buildings away, or treating the ground as totally absorbent, can typically bring an extra attenuation of 3–4 dB. With a very high absorption coefficient, an extra attenuation of 7–13 dB can be achieved. In the urban element consisting of a major street and two side streets, the effectiveness of boundary absorption on reducing noise typically varies between 6–18 dB, depending on the source position.

With a given amount of absorption, for both diffusely and geometrically reflecting boundaries, the sound attenuation along a single street is the highest if the absorbers are arranged on one façade, and the lowest if they are evenly distributed on all boundaries. For the urban element consisting of a major street and two side streets, with a given amount of absorption,

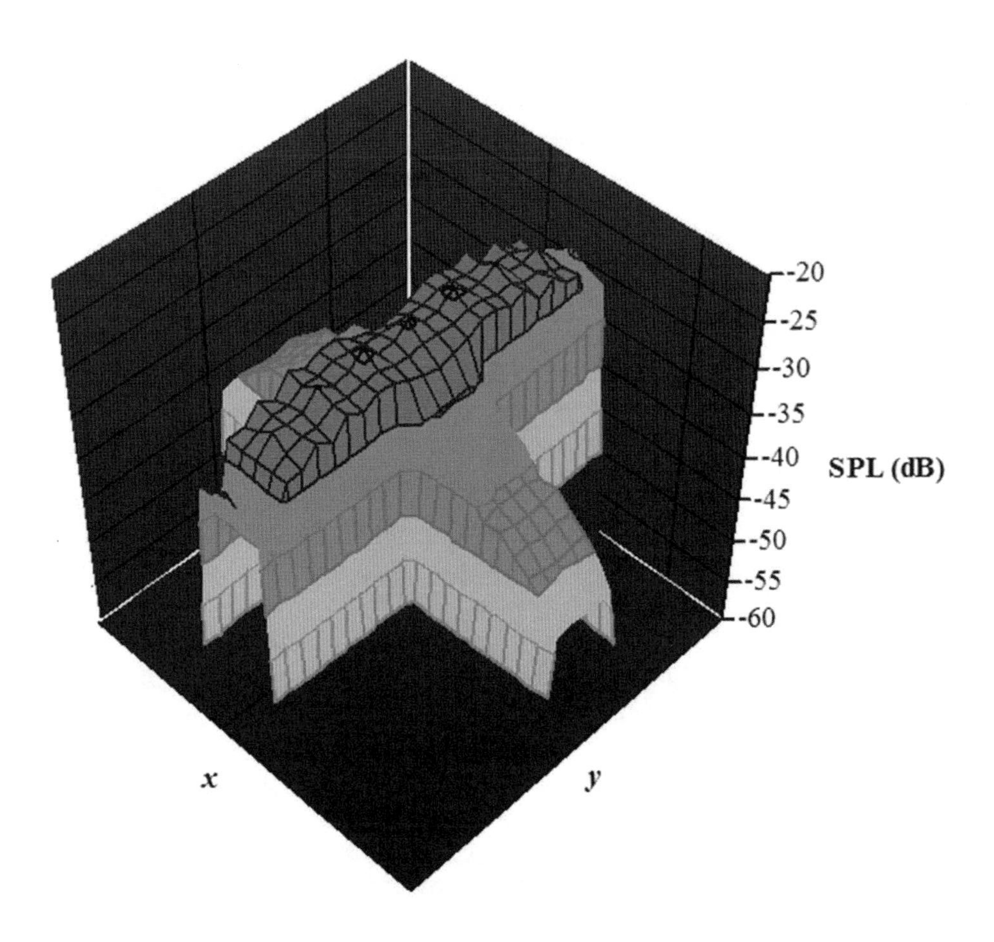

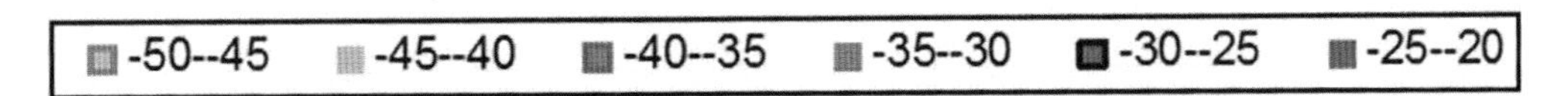

Figure 11.5 SPL distribution in an urban element consisting of a major street and two side streets, with multiple point sources in the major street.

if the absorbers are arranged near the source and on the boundaries with strong direct sound energy, the average SPL in the streets can be 4d B lower than that with other absorber arrangements. With diffusely reflecting boundaries, the extra attenuation caused by increasing boundary absorption is almost constant along the street length, whereas with geometrically reflecting boundaries, the extra attenuation increases with the increase of the source-receiver distance.

There are considerable differences between the sound fields resulting from diffusely and geometrically reflecting boundaries. By replacing diffusely reflecting boundaries with geometrically reflecting boundaries, the sound attenuation along the length becomes considerably less, typically by 4–8 dB with a source-receiver distance of 60 m; the T_{30} is significantly longer, typically by 100–200%; and the extra SPL attenuation caused by air or vegetation absorption is reduced. With moving traffic, about 2–4 dB extra attenuation can be obtained by using diffuse boundaries. Similarly, street furniture, such as lampposts, fences, barriers, benches, telephone boxes, and bus shelters, can also act as diffusers and thus be effective in reducing noise.

Reverberation is an important consideration in street canyons. With a boundary absorption coefficient of 0.1, T_{30} is typically 0.7–2 s in a single street, and 1–3 s in the urban element consisting of a major street and two side streets. With diffusely reflecting boundaries, the reverberation in a street canyon becomes longer with a greater street height and becomes shorter with a greater street width. For both diffusely and geometrically reflecting boundaries the reverberation increases systematically with increasing distance from the source. The T_{30} is rather even throughout all the cross sections along the length, whereas the EDT is only even in the cross sections beyond a certain distance from the source. With a sound source in the major street of the urban element, the reverberation in the side streets is systematically longer than that in the major street. This is particularly significant for the EDT.

The effectiveness of architectural changes and urban design options may become less if a receiver is very close to high-density traffic, where the direct sound is dominant. Nevertheless, if a barrier is inserted to reduce the direct sound, it is still important to consider these design strategies. With multiple sources, the extra SPL attenuation caused by a given acoustic treatment is generally less than that with a single source, but is still significant.

In the case of two parallel streets, where the diffraction over building roofs is important, a parametric study using the coupled FDTD-PE model has been carried out considering a two-dimensional idealized configuration with a coherent line source.[32] It is shown that the shielding is rather insensitive to the width/height ratio of the canyons, except for very narrow canyons. Rigid walls result in very poor shielding toward the receiver canyon, and the effectiveness of boundary absorption could be over 10 dB. In terms of diffusely reflecting boundaries, at 1k Hz an extra shielding of about 10 dB is gained with profiled façades.

11.2.11 Noise Control in Urban Squares

The sound fields in urban squares are also analyzed using the radiosity and image source models.[20] It has been shown that with either diffusely or geometrically reflecting boundaries, the SPL initially decreases significantly with increasing source-receiver distance and then becomes approximately stable. The T_{30} and EDT are rather long, about 2 s with diffusely reflecting boundaries and around 8–10 s with geometrically reflecting boundaries in a typical square of 50 × 50 m. The T_{30} is very even across a square, whereas the EDT is very low in the near field, and then becomes close to T_{30} after a rapid increase. Compared to diffusely reflecting boundaries, with geometrically reflecting boundaries the T_{30} and EDT are significantly longer, typically by 200–400%, and the SPL attenuation along a square is generally smaller unless the height/side ratio is high—say 1:1. It is interesting to note that when the boundary diffusion coefficient is increased from 0 to about 0.2, the decrease in the SPL, T_{30}, and EDT is significant, whereas when the boundary diffusion coefficient is further increased, the changes become much less.

If a relatively far field is considered, the SPL is typically 6–9 dB lower when the square side is doubled; 8 dB lower when the square height is decreased from 50 m to 6 m (diffusely reflecting boundaries); 5 dB (diffusely reflecting boundaries) or 2 dB (geometrically reflecting boundaries) lower if the length/width ratio is increased from 1 to 4, and 10–12 dB lower if the boundary absorption coefficient is increased from 0.1 to 0.9. When one façade is removed or made absorbent, the sound field near this façade is mostly affected, whereas when two or more façades are removed or made absorbent, the direct sound plays a much more important role.

11.2.12 Vegetation in Urban Context

In urban areas, multiple reflections play an important role and consequently, vegetation could be more effective through three mechanisms: sound absorption and sound diffusion, which occur when a sound wave impinges on the vegetation and is then reflected back; and sound level reduction, when a sound wave is transmitting through the vegetation. The analyses in the previous sections show that increasing boundary absorption can achieve a substantial extra SPL attenuation, and compared with geometrically reflecting boundaries, with diffusely reflecting boundaries there is a significant SPL reduction. When vegetation is used on building façades and the ground, the effectiveness of absorption can be greatly enhanced since there are multiple reflections. Similarly, due to multiple reflections, the diffusion effect of vegetation will be significant even when the diffusion coefficient is relatively low. The absorption and diffusion effects are also useful for reducing negative ground effects that often occur in outdoor sound propagation. While the transmission effect in an open field may not be significant unless the density and depth are considerable, the effectiveness could again be significant if multiple reflections are considered.

11.3 ENVIRONMENTAL NOISE INDICATORS AND STANDARDS

While mitigation efforts—such as developing quieter vehicles, improving traffic systems and urban texture, and designing better building envelopes including self-protection buildings—are effective for environmental noise reduction, a number of trends are expected to increase environmental noise pollution. This includes the expanding use of increasingly powerful sources of noise; the wider geographical dispersion of noise sources, together with greater individual mobility and the spread of leisure activities; the increase of roads, traffic, driving speed, and the distance driven; and increasing public expectations. Environmental noise indicators and standards are therefore of great importance.

11.3.1 Indicators

Along with various regulations, a large number of noise indices have been developed for different noise types and sound characteristics, in addition to the basic sound descriptors such as the SPL, weighted sound level, loudness and loudness level, noisiness and perceived noise level, and a series of psychoacoustic indices.

L_n is the level of noise exceeded for n% of the specified measurement period. In other words, if an N measured SPL are obtained in a time period, T, with a given time interval, and they are sorted in a descending order, then L_n is the $(100n/N)$th SPL in the order. By convention, L_1, L_{10}, L_{50}, and L_{90} are used to give approximate indications of the maximum, intrusive, median, and background sound levels, respectively.

The equivalent continuous sound level, $L_{eq,T}$, is a notional sound level. It is ten times the logarithm to the base ten of the ratio of the time-mean-square instantaneous sound pressure, during a stated time interval, T, to the square of the standard reference sound pressure:

$$L_{eq,T} = 10\log\left[\frac{1}{T}\int\frac{p^2(t)dt}{p_0^2}\right]. \tag{11.12}$$

If $p(t)$ is A-weighted before the $L_{eq,T}$ is calculated, then $L_{eq,T}$ will have units of dBA. With a series of measured shorter term L_{eq}, an overall $L_{eq,T}$ can be calculated by:

$$L_{eq,T} = 10\log\left[\frac{1}{T}\sum_{i=1}^{N}10^{0.1L_{eq,i}}t_i\right], \tag{11.13}$$

where N is the number of shorter term L_{eq}, and t_i is the time period of the ith L_{eq}. If the time period for all the shorter term is the same, then Eq. (11.13) becomes:

$$L_{eq,T} = 10\log\left[\frac{1}{N}\sum_{i=1}^{N}10^{0.1L_{eq,i}}\right]. \tag{11.14}$$

The day-night average sound level (DNL), or day-night equivalent sound level, L_{dn}, is the average over a 24-hour period but the noise level during the nighttime period, typically 22:00-07:00, is penalized by an addition of 10 dBA.

The day-evening-night level, L_{den}, currently being widely used in Europe (EU, 2002), is similar to DNL, but an evening period is considered, penalized by an addition of 5 dBA:

$$L_{den} = 10\log\frac{1}{24}\left(12\times10^{\frac{L_{day}}{10}} + 4\times10^{\frac{L_{evening}+5}{10}} + 8\times10^{\frac{L_{night}+10}{10}}\right)\ \text{(dBA)}, \tag{11.15}$$

where L_{day}, $L_{evening}$, and L_{night} are the A-weighted long-term average sound level, determined over all the day periods, evening periods, and night periods of a year, respectively. The time periods can be defined according to the national and regional situations.

Other noise indicators include the traffic noise index (TNI), the noise pollution level (NPL), the corrected noise level (CNL), the effective perceived noise level, and the sound exposure level. There are also a number of indices for evaluating aircraft noise.[19]

11.3.2 Standards and Regulations

Legislation on environmental noise is divided into two major categories: on noise emission of products such as cars, trucks, aircrafts, and industrial equipment; and on allowable noise levels in the domestic environment. The main factors in environmental noise legislations include adverse public health effects, annoyance of the residents in the neighborhood, and the risk management strategies of the legislatures.

There are two typical approaches to assess the environmental noise impact, based on absolute sound levels, or the increase of existing ambient sound levels due to a new or expanded development. The advantage of the former is that a noise ceiling is ultimately established, preventing a gradual increase of the noise level. The latter presumes that people are accustomed to the sound environment that presently exists, and if the change does not increase the existing sound level, they would not sense the change and thus not be significantly affected.

The World Health Organization's (WHO) health-based guidelines for community noise serve as the basis for deriving noise standards within a framework of noise management.[37] ISO 1996 is a major international standard concerning the description and measurement of environmental noise, including basic quantities and procedures, acquisition of data pertinent to land use, and application to noise limits.[38] No specific noise limits are given in ISO 1996, and it is assumed that these would be established by local authorities. In the EU, following the Green Paper[39] which aims to stimulate public discussion on the future approach to noise policy, a Directive of the European Parliament and of the Council relating to the assessment and management of environmental noise has been developed,[18] seeking to harmonize noise indicators and assessment methods for environmental noise, to gather noise exposure information in the form of noise maps, and to make such information available for the public.

Despite the development of various standards and regulations, environmental noise is still a serious concern. According to the 1999–2001 noise measurement,[40] the proportions of the UK population live in dwellings exposed to L_{den} <55, 55–60, 60-65, and 65–70 dBA were 33, 38, 16, and 13% respectively.

11.4 NOISE PERCEPTION

Noise perception is a complex system and it depends on both acoustic/physical factors and social/psychological/economic factors.[19] The main acoustic and physical factors include:

1. **Overall sound level**—Relationships between annoyance and noise exposure have been established for various noise types based on a considerable amount of existing data.
2. **Spectrum**—While an A-weighted sound level is commonly used in environmental noise regulations, there is increasing evidence that low frequency components play an important role in annoyance. More tonal components may increase annoyance too.
3. **Amplitude fluctuation or emergence of occasional events**—With a constant average sound level, noise annoyance may increase with a larger amplitude fluctuation or emergence of occasional events. The regularity of events, maximum sound level, rise time, duration of occasional events, spectral distribution of energy, and the number and duration of quiet periods could also affect noise perception.
4. **Situational variables**—Relatively long-term changes of several decibels in noise exposure may not cause any difference in noise annoyance. It is also interesting to note that the annoyance to a target environmental noise exposure is affected very little by the presence of another sound source qualified as ambient noise.
5. **Noise type**—In Europe it has been shown that aircraft noise is generally more annoying than road traffic noise, whereas road noise is more annoying than railway noise, but it seems that this is not always true in other cultures.
6. **Season and the time of day**—Case studies show that noise annoyance was greater in summer than in winter[41] and the effects of noise were greater in the evening and at the beginning of the night period.

Various studies suggest that the dependence of noise annoyance on acoustic/physical factors is only about 30% or less, whereas other aspects including socio/psychological/economic factors play an important role in annoyance evaluation:[19]

1. **Attitude**—There are generally six aspects that influence annoyance,[42] including fear, such as dangers related to a plane crash; cause of noise, relating to possible ways to control the noise or economic dependence on the activities generating the noise; sensitivity to noise, which could vary about 10 dB between different groups of people; activity, since noise could be more disturbing for certain activities such as oral communication, listening to a radio, and intellectual tasks; perception of the neighborhood; and the global perception of the environment, for example, due to the interactions between noise and light, noise and color, and noise and vibration.
2. **Demographic factors**—There are varied results regarding the age effect, whereas most studies seem to suggest that the effect of gender is not important. In terms of the effects of marital status, education level, house size/type, and family size on noise annoyance, the findings from different studies vary. Income and the economic status appear to be insignificant for noise annoyance, and so is the general state of health, measured by the frequency of visiting doctors.
3. **Noise experience, including exposure to noise at the place of work and over time**—The effect of length of residence seems to be insignificant for noise annoyance, whereas the time spent at home is important. Another influencing factor is the type of occupancy, namely owning or renting.
4. **Behavior and habit**—This includes opening and closing windows, using sleeping pills, using balconies or gardens, having a home sound insulated, and frequently leaving for weekends.
5. **Regional differences**—This includes cultural heritage, construction methods, lifestyle, and weather. A number of cross-cultural studies have demonstrated notable differences in noise annoyance. Moreover, for a given noise level, inhabitants of small towns may have different noise annoyance tolerances as compared to those of large urban communities, showing the effect of the environmental load.

To consider the combined effects of multiple sound sources, two kinds of methods can be used, namely mathematical summations of quantities of noise exposure, and models reflecting cognitive and perceptual mechanisms. Typical methods include: the energy summation model, which simply adds energy from all sources; the energy difference model, which describes the annoyance as the function of the total energy and of the difference between the energy of distinct sources; the dominant source model, where the total annoyance equals that of the most annoying source; the subjectively corrected model, which takes into account differences in the perceived annoyance due to each distinct source, using correction factors; and vector summation model, where the total annoyance is written as the square root of the sum of the squares of each noise source's perceptual variables.[33]

11.5 URBAN SOUNDSCAPE

Recent research has shown that reducing the sound level does not necessarily lead to better acoustic comfort in urban areas. In urban open spaces, for example, when the SPL is below a certain value, as high as 65–70 dBA, people's acoustic comfort evaluation is not related to the sound level, whereas the type of sound sources, the characteristics of users, and other factors play an important role.[43] Different from conventional noise control, soundscape research considers the ear, human beings, sound environments, and society, and regards environmental sounds as a *resource* rather than a *waste*. With the pioneering research in soundscape by Schafer in the 1960's,[44] the field has been mainly developed within the academic disciplines of anthropology, architecture, ecology, design, human geography, linguistics, medicine, noise control engineering, psychology, sociology, and more recently, computer simulation and artificial intelligence. As a global concept, it may also be fruitful to integrate insights from knowledge or values produced by every culture, therefore involving literature and musicology, and more generally, art, aesthetics, laws, and religious studies.

Urban open spaces are important components in a city, where the acoustic environment also plays an important role in the overall comfort.[45,46] In this section, soundscape evaluation, description and design are briefly examined in terms of four basic elements: sound, space, people, and environment.

11.5.1 Sound

Preference of individual sounds is important for the soundscape evaluation of urban open spaces. According to a field survey in Sheffield, as shown in Figure 11.6, people showed a very positive attitude toward the natural sounds, with over 75% of the interviewees being

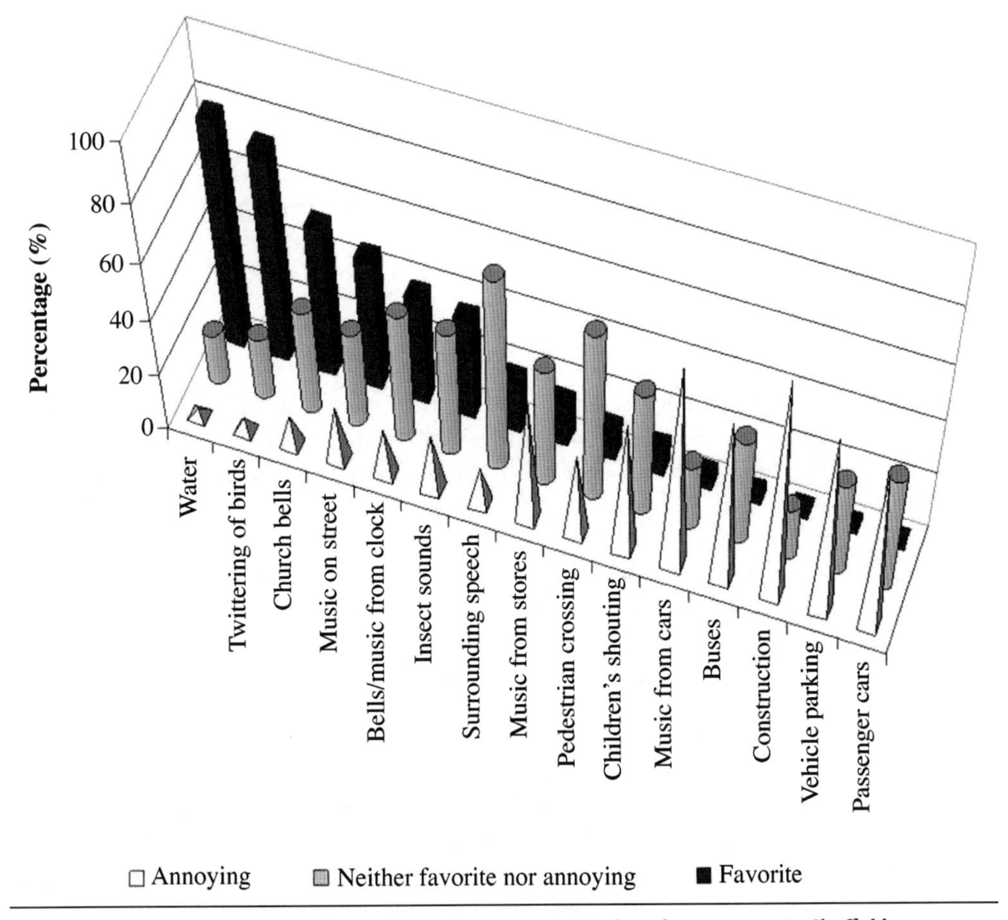

Figure 11.6 Preference of typical sounds in urban open spaces, based on a survey in Sheffield.

favorable to water sounds and birdsongs, for example. For culturally approved sounds, including church bells, music on the street, and bells/music from clocks, people also showed relatively high levels of preference. For human sounds such as surrounding speech, most people thought they were *neither favorite nor annoying*. The most unpopular sounds were mechanical sounds, such as construction sounds, music from cars, and vehicle sounds. It is interesting to note that for music from different sources, people's perceptions vary significantly: the rate of *favorite* was 46% for music on the street, 15% for music from stores, and only 2% for music from cars.[47]

Sounds can be classified as keynotes, which are in analogy to music where a keynote identifies the fundamental tonality of a composition around which the music modulates; foreground sounds or sound signals, which are intended to attract attention; and soundmarks, which are in analogy to landmarks.[44] From the viewpoints of designing soundscape, sounds can be classified as two types: those from human activities, defined as *active sounds*, and those from landscape elements, defined as *passive sounds*. Typical examples for the two types of sound are music and water sounds, respectively.

Loudness is another important consideration in addition to the type of sound. A study on the relation between loudness and pleasantness shows that the pleasantness of stimuli at intermediate loudness levels is not influenced by its loudness, but for sound at relatively high loudness levels, there is a good correlation between the two.[48]

Urban open spaces can be regarded as a product to a certain degree, so that the methodology developed in product sound quality is closely related to the soundscape evaluation, although in urban environments the complexity of multiple sound components should be taken into account. Moreover, the meaning of many environmental sounds may considerably influence the evaluation. Using semantic differential analysis, it has been demonstrated that for general environmental sounds, four essential factors include evaluation, timber, power, and temporal change.[49] For residential areas, the soundscape can be characterized in four dimensions, namely adverse, reposing, affective, and expressionless.[50]

11.5.2 Space

As mentioned previously, architectural changes and urban design options in urban open spaces could affect the sound field significantly. With a constant SPL, noise annoyance will vary with different reverberation. A suitable T_{30}, say 1–2 s, can make *street music* more enjoyable, although the requirement in urban open spaces is much less critical than that in room acoustics and also, the concept of reverberation in outdoor spaces may not be the same as that for enclosed spaces. Moreover, sound distribution and reflection patterns are important for soundscape evaluation. In urban open spaces there might be different sound zones, each with a different sound field as well as different sounds, associated with users' activities.

11.5.3 People

Based on large-scale field surveys the across EU and China, it has been shown that the effects of social/demographical factors including age, gender, occupation, education, and residential status on the sound level evaluation are generally insignificant, although occupation and education are two related factors and both correlate to the sound level evaluations more than other factors. The effects on the sound level evaluation by some behavioral factors including wearing earphones, reading/writing, and moving activities are also insignificant, but the watching behavior is highly related to the sound level evaluation. Compared to the social, demographical, and behavioral factors, the long-term sound experience, i.e., the acoustic environment at home, significantly affects the sound level evaluation in urban open spaces.[51]

In terms of acoustic comfort, however, the effects of social/demographical factors are significant. For instance, with the increase of age, people are more favorable to, or tolerant toward sounds relating to nature, culture, or human activities. Figure 11.7 shows the variation of sound preference among age groups for birdsong. By contrast, young people are more favorable to, or tolerant toward music and mechanical sounds, as also shown in Figure 11.7, using music from stores as an example.[47] It has also been shown that gender, occupation, and residence status would generally not influence the sound preference evaluation significantly,

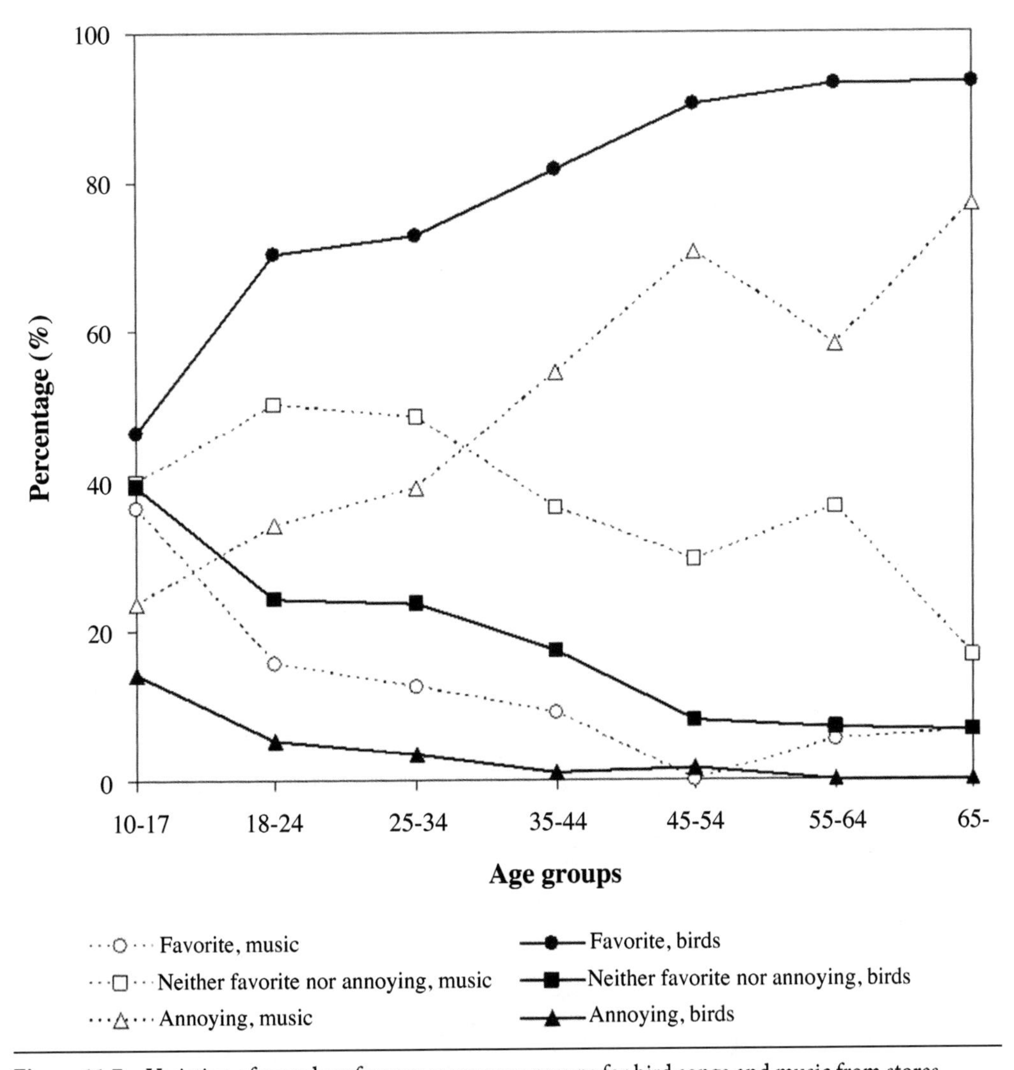

Figure 11.7 Variation of sound preference among age groups for bird songs and music from stores.

whereas with an increasing education level, people tend to prefer natural sounds and are more annoyed by mechanical sounds.

Culture differences could also lead to rather different acoustic comfort evaluations and sound preferences.[19] For example, for water sounds the preferences in Sheffield and Sesto San Giovanni are significantly different (p <0.01)—in the former, over 75–84% of the interviewees rated water sounds as *favorite*, whereas in the latter, this value was less than 28%.

11.5.4 Environment

The interaction between acoustic and other physical/environmental conditions is also a significant aspect in the soundscape evaluation of urban open spaces. For example, if an urban open space is very hot or very cold, the acoustic comfort could become less critical and less important in the overall comfort evaluation. To examine the relationships between the overall physical comfort evaluation of urban open spaces and the subjective evaluation of various physical/environmental indices (including temperature, sunshine, brightness, wind, view, and humidity), as well as sound level, the principal component analysis has been made based on a large-scale survey in Europe. Three factors have been determined. Factor 1 (22.8%), including temperature, sunshine, brightness, and wind, is the most important factor. Factor 2 (17.5%) is associated with visual and aural senses, showing that the acoustic environment is one of the main factors influencing the overall comfort in urban open spaces, and attention must be paid to the aural-visual interactions, for which considerable research has been carried out. Factor 3 (14.8%) is principally related to humidity, including humidity and wind.[43]

11.5.5 A Framework for Soundscape Description

A framework is proposed to describe the soundscape, as shown in Figure 11.8, in order to investigate the existing acoustic environment in an urban open space or to design a new soundscape. The description includes four facets, namely characteristics of each sound, acoustic effects of the space, social/demographic aspect of the users, and other aspects of the physical/environmental conditions.[19] Based on the framework, a neural network model has been developed, to predict the soundscape evaluation of potential users in urban open spaces at the design stage.[52]

Sound	**Space**	**People**	**Environment**
- Sound level - Spectrum - Temporal conditions - Location - Source movement	- Reverberation - Background sound - Surrounding sounds - Reflection pattern and/or echogram	- Social/demographic characteristics - Cultural background - Acoustic condition at home/work - Acoustic experience	- Temperature - Sunshine - Humidity - Lighting - Wind - Visual/architecture - Landscape

Figure 11.8 A framework for soundscape description in urban open spaces.

FOR FURTHER INFORMATION

More details on the results in this chapter can be found in the author's monograph Urban Sound Environment, published in 2006 by Taylor & Francis.

REFERENCES

1. ISO, 1993, *ISO 9613: Attenuation of Sound during Propagation Outdoors. Part 1 (1993): Calculation of the Absorption of Sound by the Atmosphere. Part 2 (1996): General Method of Calculation.* International Organization for Standardization.
2. Watts, G., 2005, "Harmonoise Models for Predicting Road Traffic Noise," *Acoustics Bulletin*, vol. 30, pp. 19–25.
3. Kotzen, B. and English, C., 1999, *Environmental Noise Barriers: A Guide to Their Acoustic and Visual Design*, E & FN SPON, London.
4. Umnova, O., Attenborough, K. and Linton, 2006, "Effects of Porous Covering on Sound Attenuation by Periodic Arrays of Cylinders," *Journal of the Acoustical Society of America*, vol. 119, pp. 278–284.
5. Kurze, U. J. and Anderson, G. S., 1971, "Sound Attenuation by Barriers," *Applied Acoustics*, vol. 4, pp. 35–53.
6. UK DfT, 1988, *Calculation of Road Traffic Noise (CRTN)*, UK Department for Transport.
7. Watts, G., 1996, "Acoustic Performance of a Multiple Edge Noise Barrier Profile at Motorway Sites," *Applied Acoustics*, vol. 47, pp. 47–66.
8. Van Der Haijden, L. A. M. and Martens, M. J. M., 1982, "Traffic Noise Reduction by Means of Surface Wave Exclusion above Parallel Grooves in the Roadside," *Applied Acoustics*, vol. 15, pp. 329–339.
9. Bougdah, H., Ekici, I. and Kang, J., 2006, "An Investigation into Rib-like Noise Reducing Devices," *Journal of the Acoustical Society of America*, vol. 120, pp. 3714–3722.
10. Fujiwara, K., Hothersall, D. C. and Kim, C., 1998, "Noise Barriers with Reactive Surfaces," *Applied Acoustics*, vol. 53, pp. 255–272.
11. Mizuno, K., Sekiguchi, H. and Iida, K., 1984, "Research on a Noise Control Device, 1st Report," *Bulletin of Japanese Society of Mechanical Engineers*, vol. 27, pp. 1499–1505.
12. Nicholas, J. and Daigle, G. A., 1986, "Experimental Study of a Slow-wave Guide Barrier on Finite Impedance Ground," *Journal of the Acoustical Society of America*, vol. 80, pp. 869–876.
13. Amran, M., Chvrojka, V. J. and Droin, L., 1987, "Phase Reversal Barriers for Better Noise Control at Low Frequencies: Laboratory Versus Field Measurements," *Noise Control Engineering Journal*, vol. 28, pp. 16–23.
14. Ho, S. S. T., Bush-Vishniac, I. J. and Blackstock, D. T., 1997, "Noise Reduction by a Barrier Having Random Edge Profile," *Journal of the Acoustical Society of America*, vol. 100, pp. 2669–2676.
15. Van Renterghem, T. and Botteldooren, D., 2002, "Effect of a Row of Trees behind Noise Barriers in Wind," *Acustica united with acta acustica*, vol. 88, pp. 869–878.
16. Slutsky, S. and Bertoni, H.L., 1988, "Analysis and Programs for Assessment of Absorptive and Tilted Parallel Barriers," *Transportation Research Record*, vol. 1176, pp. 13–22.
17. Joynt, J. L. R., 2005, *A Sustainable Approach to Environmental Noise Barrier Design*, PhD dissertation, School of Architecture, University of Sheffield, UK.
18. EU, 2002, *Directive (2002/49/EC) of the European Parliament and of the Council—Relating to the Assessment and Management of Environmental Noise*, Brussels.
19. Kang, J., 2006, *Urban Sound Environment*, Taylor & Francis incorporating Spon, London.
20. Kang, J., 2005, "Numerical Modelling of the Sound Fields in Urban Squares," *Journal of Acoustical Society of America*, vol. 117, pp. 3695–3706.
21. Kang, J., 2000, "Sound Propagation in Street Canyons: Comparison between Diffusely and Geometrically Reflecting Boundaries," *Journal of the Acoustical Society of America*, vol. 107, pp. 1394–1404.

22. Iu, K. K. and Li, K. M., 2002, "The Propagation of Sound in Narrow Street Canyons," *Journal of the Acoustical Society of America*, vol. 112, pp. 537–550.
23. Picaut, J., Le Pollès, T., LÕHermite, P. and Gary, V., 2005, "Experimental Study of Sound Propagation in a Street," *Applied Acoustics*, vol. 66, pp. 149–173.
24. De Coensel, B., De Muer, T., Yperman, I. and Botteldooren, D., 2005, "The Influence of Traffic Flow Dynamics on Urban Soundscapes," *Applied Acoustics*, vol. 66, pp. 175–194.
25. Ismail, M. R. and Oldham, D. J., 2005, "A Scale Model Investigation of Sound Reflection from Building Façades," *Applied Acoustics*, vol. 66, pp. 149–173.
26. Kang, J., 2001, "Sound Propagation in Interconnected Urban Streets: A Parametric Study," *Environment and Planning B: Planning and Design*, vol. 28, pp. 281–294.
27. Kang, J., 2002a, *Acoustics of Long Spaces—Theory and Design Guidance*, Thomas Telford, London.
28. Kang, J., 2002b, "Numerical Modelling of the Sound Field in Urban Streets with Diffusely Reflecting Boundaries," *Journal of Sound and Vibration*, vol. 258, pp. 793–813.
29. Le Pollès, T., Picaut, J. and Bérengier, M., 2004, "Sound Field Modeling in a Street Canyon with Partially Diffusely Reflecting Boundaries by the Transport Theory," *Journal of the Acoustical Society of America*, vol. 116, pp. 2969–2983.
30. Hothersall, D. C., Horoshenkov, K. V. and Mercy, S. E., 1996, "Numerical Modelling of the Sound Field near a Tall Building with Balconies near a Road," *Journal of Sound and Vibration*, vol. 198, pp. 507–515.
31. Ögren, M. and Kropp, W., 2004, "Road Traffic Noise Propagation between Two Dimensional City Canyons using an Equivalent Sources Approach," *Acustica united with acta acustica*, vol. 90, pp. 293–300.
32. Van Renterghem, T., Salomons, E. and Botteldooren, D., 2006, "Parameter Study of Sound Propagation between City Canyons with Coupled FDTD-PE Model," *Applied Acoustics*, vol. 67, pp. 487–510.
33. Berglund, B., Berglund, U., Goldstein, M. and Lindvall, T., 1981, "Loudness (or Annoyance) Summation of Combined Community Noises," *Journal of the Acoustical Society of America*, vol. 70, pp. 1628–1634.
34. Thorsson, P., Ögren, M. and Kropp, W., 2004, "Noise Levels on the Shielded Side in Cities using a Flat City Model," *Applied Acoustics*, vol. 65, pp. 313–323.
35. Thorsson, P., 2003, "Application of Linear Transport to Sound Propagation in Cities," *Proceedings of the 10th International Congress on Sound and Vibration (ICSV)*, Stockholm, Sweden.
36. Smyrnova, Y., Meng, Y. and Kang, J., 2008, "Objective and Subjective Evaluation of Urban Acoustic Modeling and Auralisation," *Proceedings of the Audio Engineering Society (AES) 124th Convention*, Amsterdam, Netherlands.
37. Berglund, B., Lindvall, T. and Schwela, D. H., 1999, *Guidelines for Community Noise*, World Health Organization.
38. ISO, 2003, *ISO 1996: Acoustics—Description, Measurement and Assessment of Environmental Noise. Part 1 (2003): Basic Quantities and Assessment Procedures. Part 2 (1998): Acquisition of Data Pertinent to Land Use. Part 3 (1987): Application to Noise Limits*. International Organization for Standardization.
39. EU, 1996, *Future Noise Policy*, European Commission Green Paper, Brussels.
40. Skinner, C. J. and Grimwood, C. J., 2005, "The UK Noise Climate 1990–2001: Population Exposure and Attitudes to Environmental Noise," *Applied Acoustics*, vol. 66, pp. 231–243.
41. Griffiths, I. D., Langdon, F. J. and Swan, M. A., 1980, "Subjective Effects of Traffic Noise Exposure: Reliability and Seasonal Effects," *Journal of Sound and Vibration*, vol. 71, pp. 227–240.
42. Nelson, P., 1987, *Transportation Noise Reference Book*, Butterworths, London.
43. Yang, W. and Kang, J., 2005a, "Acoustic Comfort Evaluation in Urban Open Public Spaces," *Applied Acoustics*, vol. 66, pp. 211–229.
44. Schafer, R. M., 1977, *The Tuning of the World*, Knopf, New York.
45. Raimbault, M., Bérengier, M. and Dubois, D., 2003, "Ambient Sound Assessment of Urban Environments: Field Studies in Two French Cities," *Applied Acoustics*, vol. 64, pp. 1241–1256.

46. Zhang, M. and Kang, J., 2007, "Towards the Evaluation, Description and Creation of Soundscape in Urban Open Spaces," *Environment and Planning B: Planning and Design*, vol. 34, pp. 68–86.
47. Yang, W. and Kang, J., 2005b, "Soundscape and Sound Preferences in Urban Squares," *Journal of Urban Design*, vol. 10, pp. 69–88.
48. Zeitler, A. and Hellbrück, J., 1999, "Sound Quality Assessment of Everyday-noises by Means of Psychophysical Scaling," *Proceedings of inter-noise*, Fort Lauderdale, USA.
49. Zeitler, A. and Hellbrück, J., 2001, "Semantic Attributes of Environmental Sounds and Their Correlations with Psychoacoustic Magnitudes," *Proceedings of the 17th International Congress on Acoustics (ICA)*, Rome, Italy.
50. Berglund, B., Eriksen, C. A. and Nilsson, M. E., 2001, "Perceptual Characterization of Soundscapes in Residential Areas," *Proceedings of the 17th International Conference on Acoustics (ICA)*, Rome, Italy.
51. Yu, L. and Kang, J., 2008, "Effects of Social, Demographic and Behavioral Factors on Sound Level Evaluation in Urban Open Spaces," *Journal of the Acoustical Society of America*, vol. 123, pp. 772–783.
52. Yu, L. and Kang, J., 2009, "Modeling Subjective Evaluation of Soundscape in Urban Open Spaces—An Artificial Neural Network Approach," *Journal of the Acoustical Society of America*, vol. 126, pp. 1163–1174.

12

Sound System Design and Room Acoustics

Wolfgang Ahnert, Acoustic Design Ahnert and Ahnert Feistel Media Group, Berlin, Germany

When designing a sound system, we must consider the following influences important for the success of the design:

- Purpose of the sound system
- Expected quality of the sound system
- Acoustic properties of the site where the sound system will work
- Place of the installation
- Available budget for the sound system

Independent of the purpose, the quality, the place to install it, and the available budget, the most important factor influencing the function of the system to be installed is the interaction with the existing environment in which the sound system will work. So, the first clarification a designer has to do is to figure out the natural acoustic properties of the facility for which the design has to be made.

In an existing room, a complete check of these properties in the form of extensive room acoustic measurements must be made (see Chapter 5). Here, mainly impulse response measurements will be executed and the needed corresponding acoustic measures (explained in the following text) will be derived.

In the case of a new building, computer simulation must be done with the same target to derive the room acoustic properties of the space in design.

Without an exact knowledge of these data, it can be expected that the sound system design will fail. Therefore we will, in this chapter, describe the room acoustic factors influencing the success of a sound system design in rooms or open spaces.

For a visitor of an event the overall acoustic impression will count. He will never differentiate between super room acoustics and a bad sound system or vice versa. He will speak about bad or good acoustics. So the overall quality is important, and we as sound system designers cannot make excuses for bad room acoustics. We have to adapt our design to the existing room acoustic properties, and if that is not possible, we have to push the client or the architect of the site to allow changes in room acoustics.

12.1 BASICS IN ROOM ACOUSTICS

12.1.1 Subjective Assessment of the Quality of Sound Events

The listener at a rock or pop concert or the visitor to a congress often pronounces judgment on the acoustic reproduction quality of a signal emitted by a natural source or via electro-acoustical devices. This judgment is often very imprecise, like *very good acoustics* or *poor intelligibility*. Such assessments combine objective causes with subjective experiences acquired through listening to broadcast and television transmissions, as well as CD, DVD, and other more advanced Hi-Fi reproductions.

For speech, one desires normally good intelligibility in an atmosphere unaffected by the room or by electroacoustical means. Far more sophisticated are the criteria for an assessment of music reproduction. Depending on genre, *good acoustics* means sufficient loudness, good sound clarity, and a spatial impression that is commensurate with the piece of music performed.

For determining the criteria governing subjective assessment of speech and music reproduction, definitions for the terms used were laid down in literature[1, 2] as well as in national and international standards.[3, 4] Originally these terms served mainly for the assessment of room acoustical circumstances and are therefore of importance not only for the understanding between the electro-acoustician and the room-acoustician, but also for assessing the electroacoustical reproduction itself. In the following, some of the important terms concerning reinforced sound reproduction are explained:

- **Acoustic overall impression:** Suitability of a room for the (scheduled) acoustic performances.
- **Reverberation and reverberation time (T_{60}):** Reverberation is the persistence of sound after a sound is produced. A reverberation, or *reverb*, is created when a sound is reflected causing a large number of reflections to build up and after the end of excitation to decay. The reverb time is the duration of perceptibility of the reverberation. The reverberation duration depends on the objective reverberation time (property of the room or manipulated with an enhancement system), the output level (sound signal), the noise level or the threshold of hearing, and the ratio between direct signal and room signal. The higher the absorption in the space, the shorter the reverberation time is. It is frequency-dependent.
- **Clarity:** Temporal and tonal differentiability of the individual sound sources within a complex sound event.
- **Spaciousness:** Perception of the acoustic amplification of a source as against its visual perception, especially in the lateral direction of the listener. The spaciousness depends on the one hand, on the sound level at the listener location, and on the other hand, on the direct sound-related intensity of the reflected original sound arriving up to 80 ms after this direct sound from lateral directions.
- **Intelligibility:** Clarity of the transmitted speech at the listener seat. Especially in the case of emergency call systems, the content of the messages has to be fully understandable.
- **Echo:** Reflected sound arriving with such an intensity and early-to-late difference that it can be discerned as a repetition of the direct sound.
- **Flutterecho:** Periodical sequence of echoes.

These are the most important subjective items a sound system designer has to be aware of. Only by considering these criteria and the derived acoustic measures will a successful design happen. Let us have a closer look at these and other room acoustic criteria.

12.1.2 Acoustic and Sound System Design Criteria and Quality Parameters in Rooms

12.1.2.1 Reverberation Time

Reverberation time (T_{60})[5] is the oldest room-acoustical measurable variable, and for many people the only known one, although it has been known for a long time that on its own it obtains no more than a limited statement on the acoustic properties of a hall. Nevertheless, it continues to be one of the main criteria and shall therefore, be also the first quantity to be explained here.

T_{60} is defined as the time during which the mean steady-state energy density of a sound field in an enclosure will decrease by 60 decibels (dB) after the discontinuanace of the energy supply. So after some corresponding transforms (see e.g., Kuttruff, p. 90)[6] we state that the T_{60} depends on the volume and the attenuation of the room:

$$T_{60} = \frac{0.163\,V}{4mV - S\ln(1-\alpha)} \approx 0.163\frac{V}{A}, \tag{12.1}$$

where V is the volume in m^3, A is the equivalent absorption area in m^2, α is the mean sound absorption coefficient (frequency-dependent), S is the total surface of the room in m^2, and m is the damping coefficient as a function of air absorption and frequency in m^{-1} (see Figure 12.1).

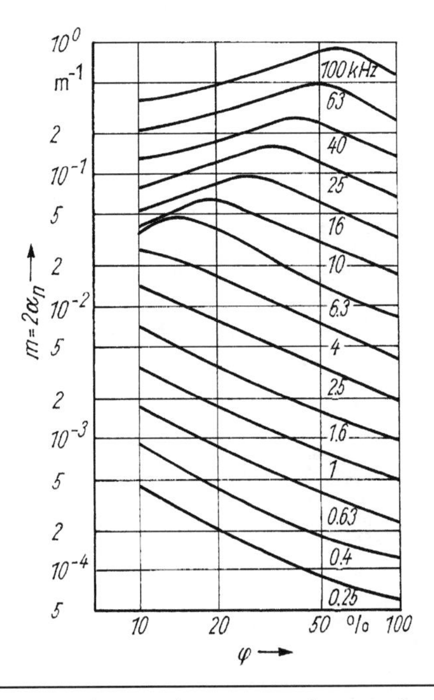

Figure 12.1 Damping coefficient, *m*, of energy decrement in the air in the direction of sound propagation, *n*, as a function of relative humidity, φ, at 20°C.

The equivalent sound absorption area, A, is calculated as:

$$A = \alpha S = \sum_i \alpha_i \, s_i + \sum_n A_n + 4\,m\,V \,, \tag{12.2}$$

where α_i is the sound absorption coefficient of the partial areas, S_i, and, A_n is the equivalent absorption area of objects and bodies.

The desired duration of reverberation is derived from the objective T_{60} recommended for determined applications (see Figures 12.2 and 12.3). One sees that the T_{60} should, in a speaking theater of usual dimensions, be between 1.0 and 1.3 sec. In a concert hall of equal dimensions the nominal values are even higher, i.e., between 1.6 and 2.1 sec. Figure 12.4 shows the recommended T_{60} as a function of frequency. In rooms for music performances an increase of T_{60} is recommended for the range below 250 Hz. This increase produces a warm, round sound pattern.

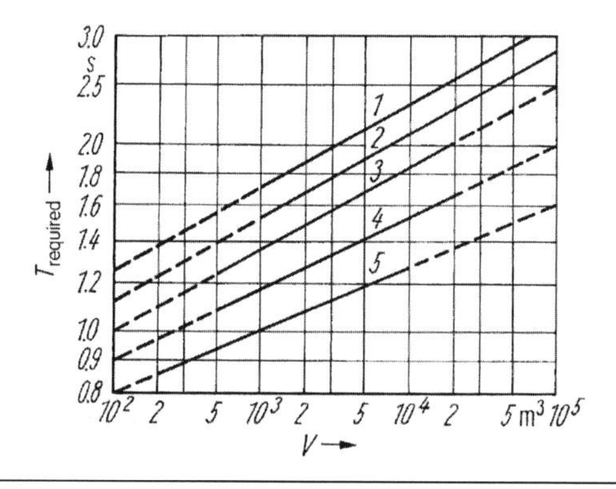

Figure 12.2 Recommended values of T_{60} at 500 Hz for different room types as a function of volume: (1) Rooms for oratorios and organ music, (2) Rooms for symphonic music, (3) Rooms for solo and chamber music, (4) Opera theaters, multipurpose halls for music and speech, (5) Speaking theaters, assembly rooms, and sports halls. T_{soll}

$RT_{required}$

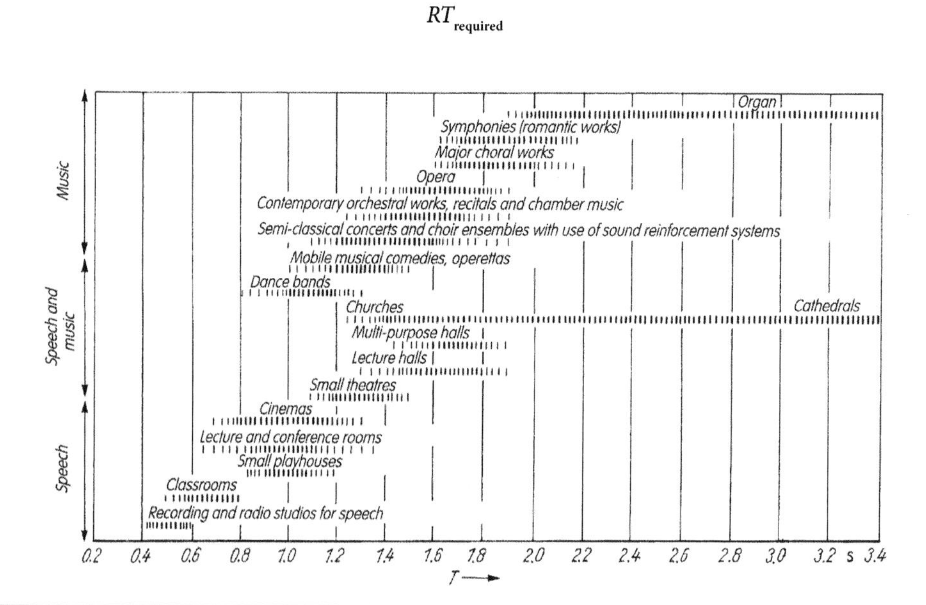

Figure 12.3 Optimum T_{60} at 500 to 1000 Hz (after Russell Johnson).

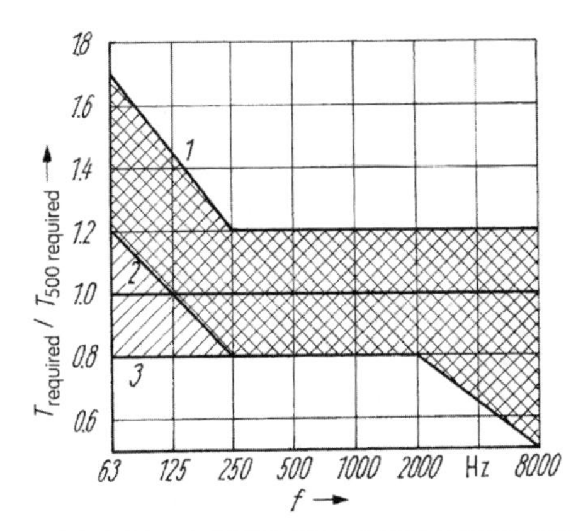

Figure 12.4 Tolerance ranges for the recommended T_{60} values according to Figure 12.2: Range 1–2 music Range *1–3* speech

In steady-state condition the absorbed power is equal to the sound power, *P*, fed into the room. Thus one obtains the *average sound energy* density in the diffuse sound field of the room as:

$$w_r = \frac{4P}{cA}, \tag{12.3}$$

c – Speed of sound in m per sec.

While the sound energy density, w_r, is approximately constant in the diffuse sound field, the direct sound energy and thus, also its density decrease in the near range of the source according to:

$$w_d = \frac{P}{c}\frac{1}{4\pi r^2}, \tag{12.4}$$

by the square of the distance r from the source.

In this range of dominating direct sound, the sound pressure decays as $p \sim 1/r$.

If the energy densities of direct sound and diffuse sound are equal, ($w_d = w_r$), Eqs. (12.3) and (12.4) can be equated—i.e., one can infer a special distance from the source, the *reverberation radius, r_H*. For a spherical source, it is in Meter (see Figure 12.5).

$$r_H = \sqrt{\frac{A}{16\pi}} \approx \sqrt{\frac{A}{50}} \approx 0.141\sqrt{A} \approx 0.057\sqrt{\frac{V}{T_{60}}}. \tag{12.5}$$

In the case of a directional source the reverberation radius, r_H, will be substituted by the so-called *critical distance, D_c*:

$$D_c = \sqrt{Q}^*\Gamma(\vartheta)^*r_H = \sqrt{g(\vartheta)}^*r_H \approx \sqrt{Q}^*r_H, \tag{12.5a}$$

Q – directivity factor (see Eq. 12.32), Γ – directional factor (see Eq. 12.30), (the product of $Q^*\Gamma^2(\vartheta)$ shall be called the angle-depending directivity deviation factor $g(\vartheta)$).

In Figure 12.6 the decay of the overall energy density level 10 lg *w* dB is plotted as a function of the distance, *r*, from the source, ($w = w_d + w_r$). In the direct field of the source one sees

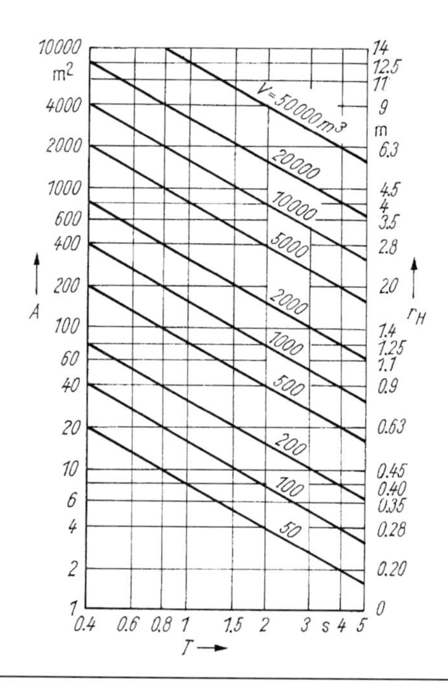

Figure 12.5 Reverberation radius, r_H, as a function of the equivalent absorption area, A, the volume, V, and the reverberation time, T_{60}, of a room.

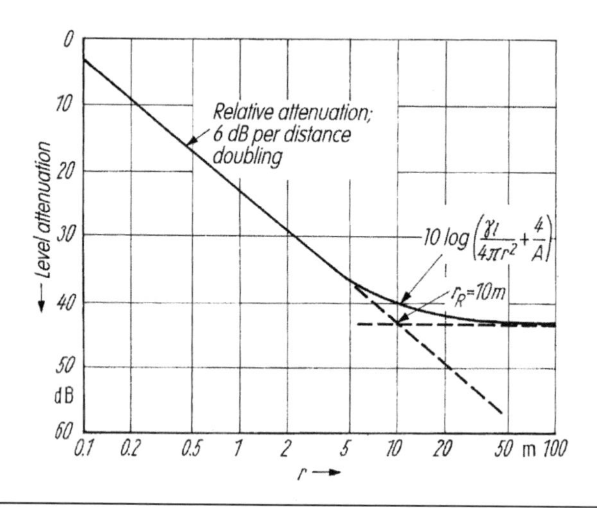

Figure 12.6 Sound level in enclosures as a function of the distance from the source.

a decrement of 6 dB per doubling of the distance. Beyond the *reverberation radius* (in the case of a spherical source) or beyond the D_c (shown in the figure as r_R), a range follows in which a constant diffuse-field level 10 lg w_r dB ~ −10 lg A dB prevails. In an absolute free field ($A \rightarrow \infty$) the free-field behavior (6 dB decrement per distance doubling) would continue (dashed line).

Figure 12.6 shows that the reverberation radius can also be derived graphically—owing to the directional effect of the source, in this case one must use the *critical distance*, D_c. The critical distance thus obtained is $D_c = r_R = 10$ m.

12.1.2.2 Energy Criteria

According to the laws of system theory, a room can be acoustically regarded as a linear transmission system that can be fully described through its impulse response, h(t), in the time domain.

Generally, the time responses of the following sound-field-proportionate factors (so-called reflectograms) are derived from measured or calculated room impulse responses, h(t):

Sound pressure: $p(t) \approx h(t)$

Sound energy density: $w(t) \approx h^2(t)$

Ear-inertia weighted

Sound intensity: $J_{\tau_0}(t) \approx \int_0^t h^2(t')\exp\left[(t'-t)/\tau_0\right]dt'$

with $\tau_0 = 35\ ms$

Sound energy: $W(t) \approx \int_0^t h^2(t')dt'$

Figure 12.7 shows two different impulse responses and some post-processed energy time curves of the same venue. After a sufficiently long time (depending on the T_{60}; in practice one usually assumes 500 ms) the reflected energy arriving at the listener's location becomes so weak that for $t \rightarrow \infty$, the final value, E_∞, of the energy sum no longer increases by a relevant margin.

The energy-time curve allows appreciating, in a simple way, the partial energy values arriving at the listener's location at, e.g., 50 ms, 80 ms after the direct sound. Of the overall energy, ($E\infty$), that arrives at the listener's location, only 5 to 15% can be attributed to the direct sound, (E_d), and 10% to the reverberation. About 80% of the energy is contained in the initial reflections. These are decisive for a large part of the subjective auditory impressions, which change from audience area to audience area along with the variation of these initial reflections.

Sound reinforcement systems provide the chance to improve unfavorable reflectograms by compensating electroacoustically for the missing energy in those time intervals in which room-acoustical reflections do not occur. All electroacoustically supplied energy shall normally increase the sound level in the region of initial reflections. The boundary between early and late energy portions depends on the genre of the performance and on the building-up time, and lies at about 50 ms for speech and at about 80 ms for symphonic music, after direct sound. Early energy enhances clarity, late energy the spatial impression. Laterally incident energy within a time range of 25 to 80 ms may even enhance clarity *and* spatial impression.[7] This is of crucial importance for the planning of sound reinforcement systems.

For the measuring or simulating of all speech-relevant room-acoustical criteria, an acoustic source with the frequency-dependent directivity pattern of a human speaker has to be used for exciting the sound field, while with musical performances, it is sufficient to use a nondirectional acoustic source for the first approximation.

The majority of the room-acoustical quality criteria is based on the monaural, directionally unweighted assessment of the impulse response. Head-related binaural criteria are still

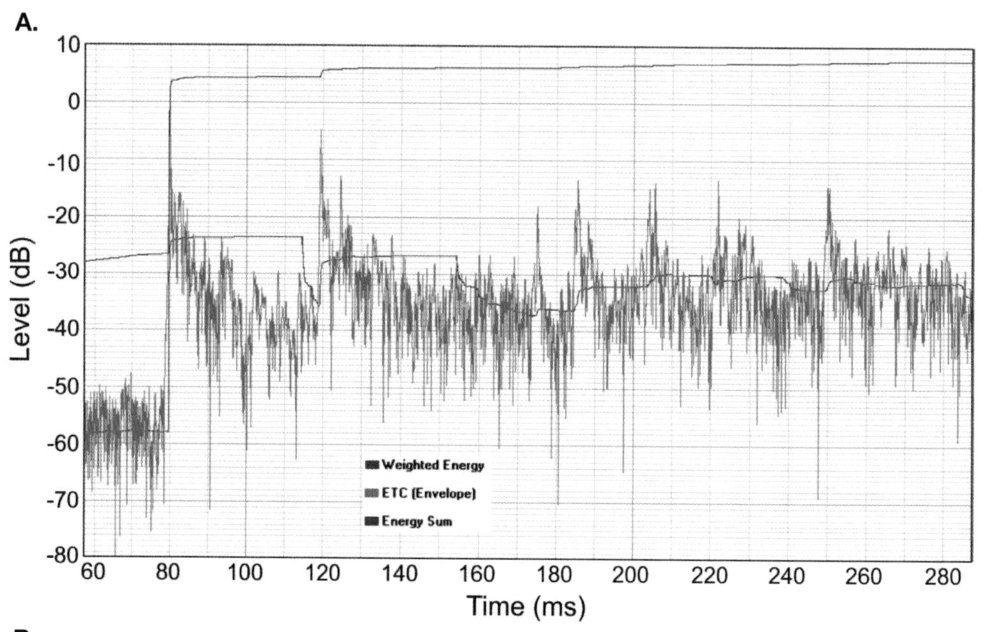

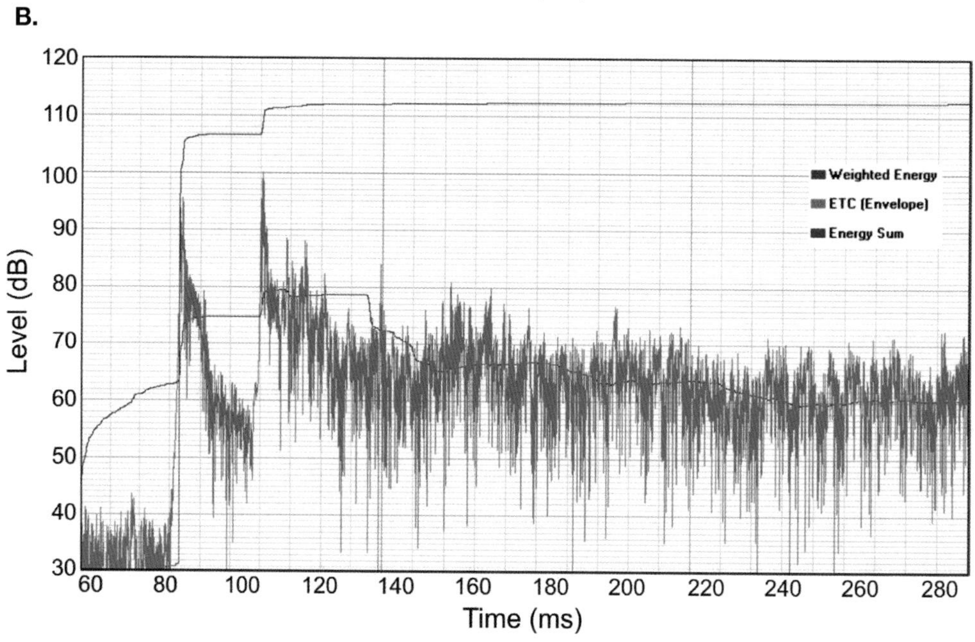

Figure 12.7 Time behavior of the energy time curve, ETC, of the weighted sound energy, $E_{\tau0}(t)$, integrated according to the inertia of the hearing system, and of the composite energy *sum*—(a) strong direct sound and (b) strong echo after 27 ms.

an exception (IACC, see Section 12.1.2.2.3). The influence of the sound-incidence direction of early initial reflections on room-acoustical quality criteria is principally known. Since subjective evaluation criteria are still missing to a large extent, however, this may also be generally disregarded when measuring or calculating room impulse responses.

12.1.2.2.1 Principal Measures for Speech Transmission

Definition Measure C_{50} for Speech

The *definition measure* C_{50} describes the intelligibility of speech and also of singing. C_{50} was derived by Ahnert[8] from speech clarity, *D*, as defined by Thiele.[9] It is generally calculated in a bandwidth of 4 octaves between 500 Hz and 4,000 Hz from the tenfold logarithm of the ratio between the sound energy arriving at a reception measuring position up to a delay time of 50 ms after the arrival of the direct sound and the following energy:

$$C_{50} = 10\lg\left(\frac{E_{50}}{E_{\infty} - E_{50}}\right) \text{ [dB]}. \qquad (12.6)$$

A good intelligibility of speech is mostly given when $C_{50} \geq 0$ dB.

The frequency-dependent definition measure, C_{50}, should increase by approximately 5 dB with octave center frequencies over 1,000 Hz (starting with the octave center frequencies 2,000 Hz, 4,000 Hz, and 8,000 Hz), and decrease by this value with the octave center frequencies below 1000 Hz (octave center frequencies 500 Hz, 250 Hz, and 125 Hz).

An equivalent, albeit less used criterion is the *degree of definition*, *D*, also called D_{50}, that results from the ratio between the sound energy arriving at the reception measuring position up to a delay time of 50 ms after the arrival of the direct sound and the entire energy (given in a percentage):

$$D = \frac{E_{50}}{E_{\infty}}. \qquad (12.7)$$

To reach an intelligibility of syllables of at least 85%, one should obtain $D = D_{50} \geq 0.5$, or 50%.

Speech Transmission Index (STI)[10]

The determination of the STI values is based on measuring the reduction of the signal modulation between the location of the sound source, e.g., on stage, and the receiving position with octave center frequencies of 125 Hz up to 8000 Hz. Here Steeneken and Houtgast[10] have proposed to excite the room or open space under investigation with a special modulated noise, and then determine the reduced modulation depth. They proceeded on the assumption that not only reverberation, noise, and masking reduce the intelligibility of speech, but generally all external signals or signal changes that occur on the path from source to listener. To derive this influence they used the *modulation transfer function* (MTF) for acoustical purposes. The available useful signal S (signal) is put into relation with the prevailing interfering signal N (noise). The so-determined *modulation reduction factor*, m(F), is a factor that characterizes the interference with speech intelligibility:

$$m(F) = \frac{1}{\sqrt{1+(2\pi F \cdot T_{60}/13.8)^2}} \cdot \frac{1}{1+10^{-(\frac{S/N}{10})}}, \qquad (12.8)$$

with F modulation frequency in Hz,
T_{60} in seconds,

S/N (signal/noise) ratio in dB.

To this effect one uses modulation frequencies from 0.63 Hz to 12.5 Hz in 1/3 octaves. In addition, the MTF is subjected to a frequency weighting known as the weighted modulation transmission function (WMTF), in order to achieve a complete correlation to speech intelligibility. In doing so, the MTF is divided into seven octave bands, which are each modulated with the modulation frequency, as shown in Eq. (12.14). This results in a matrix of 7 × 14 = 98 modulation reduction factors, m_i.

The influence of the second factor in Eq. (12.8) is shown in Figure 12.8. All the calculated m_i factors include an $m_{S/N}$ factor which reduces the calculated m_i values accordingly. Only in the case that the m_i factors are measured directly with modulated noise and the real noise floor has been active during the measurements, this second factor must not be considered. Otherwise, the signal-to-noise ratio is available in relative values in octave bands and has to be considered in all 98 modulation reduction factors m_i.

As another factor, a combined value for correcting the m_i values has to be introduced considering the threshold of hearing and the masking influence. Considered will be the octave band level of the speech signal, L_k, in relation to the threshold of hearing, $L_{rs,k}$, and to the level of lower and possible masked octave band, L_{k-1}. This additional reduction of the STI is shown for a certain speech level spectrum in Figure 12.9. For small speech signal levels below 60 dB(A), the first octave bands are below the threshold of hearing, and the STI values are decreasing. Between 70 and 100 dB(A), the maximal values are achieved and for higher speech levels the masking influence reduces the STI again. It starts a so-called self-masking. In Figure 12.9 the spectrum of the male speaker has been varied, too. Lower masking may be achieved by reducing the low-end level or shaping this part like white noise. But generally the masking effect cannot be avoided. This masking effect may prevent good intelligibility in large sport stadiums where needed high speech levels are needed. A compromise between noise influence and masking or the afore mentioned high pass filtering may help by reducing or shaping the low frequency range below 250 Hz. Here the remaining active bandwidth of the signal and the starting coloration of the speaker voice will set limits to do so.

After consideration of all these influences the (apparent) effective S/N ratio, X, can be calculated from the final modulation reduction factors m_i:

$$X_i = 10\lg\left(\frac{m_i}{1-m_i}\right) \; [dB]. \tag{12.9}$$

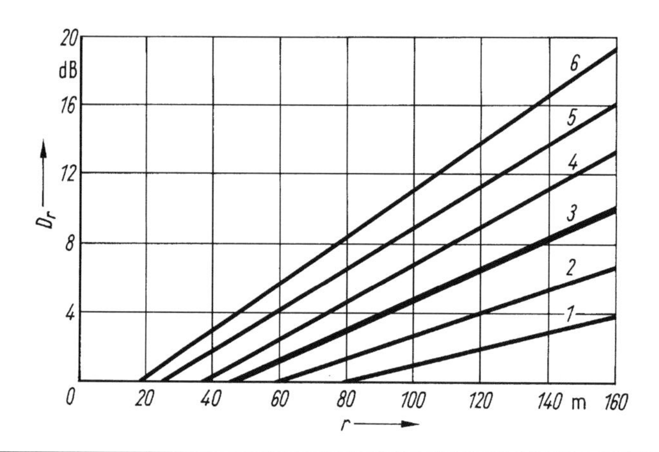

Figure 12.8 Modulation reduction factor m(S/N) vs. signal/noise level.

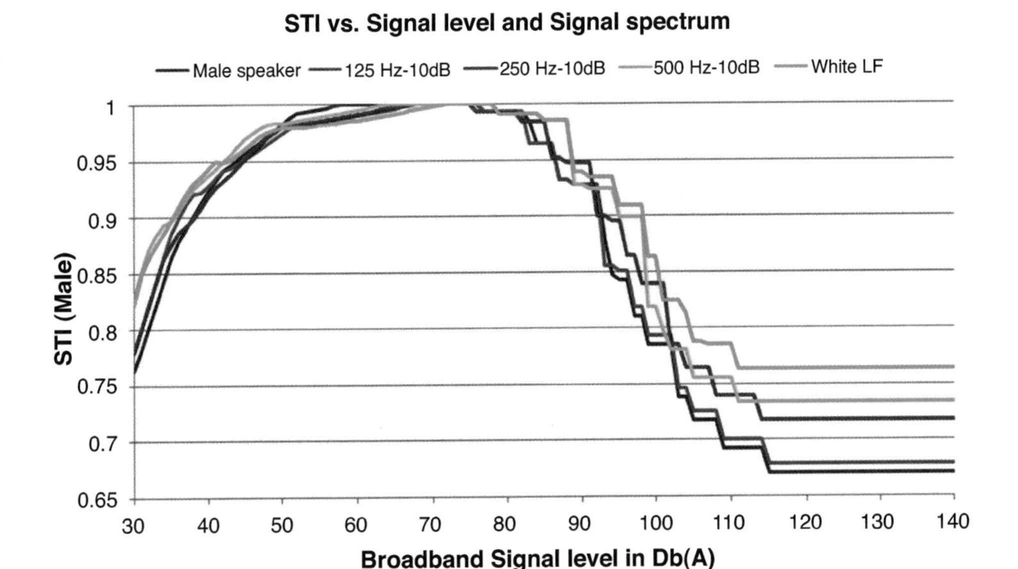

Figure 12.9 Reduction of STI as a function of speech level for a certain speech level spectrum.

These values will be averaged and for the seven octave bands, the Modulation Transfer Indices MTI = $(X_{average} + 15)/30$ are calculated. After a frequency weighting in the seven bands (partially separated for male or female speech's) you obtain the STI.

A sound source, having the directivity behavior of a human speaker's mouth, should be used to excite the sound field.

Note: Schroeder[11] could show that the 98 modulation reduction factors, m(F), may also be derived from a measured impulse response, h(t):

$$m(F)=\frac{\int_0^\infty h^2(t)\,e^{-j2\pi Ft}\,dt}{\int_0^\infty h^2(t)dt}. \tag{12.10}$$

This has been implemented with all modern computer-based measurement routines like MLSSA, EASERA, or Win-MLS.

A new method to estimate the speech intelligibility may be used instead to measure an impulse response and to derive the STI values against the excitation with a modulated noise. The frequency spectrum of this excitation noise is shown in Figure 12.10.

The ½ octave band noise may be recognized as it radiates through the sound system into the room. By means of a mobile receiver at any receiver location the speech transmission index for public address systems (STIPa) values can be determined.[12] Any laymen may use this method because no special knowledge is needed. It is increasingly used to verify the quality of emergency call systems (ISO 7240-19)[13], especially in airports, stations, or large malls.

According to definition, the STI value is calculated using the results of Eq. (12.9):

$$STI=\frac{X+15}{30}. \tag{12.11}$$

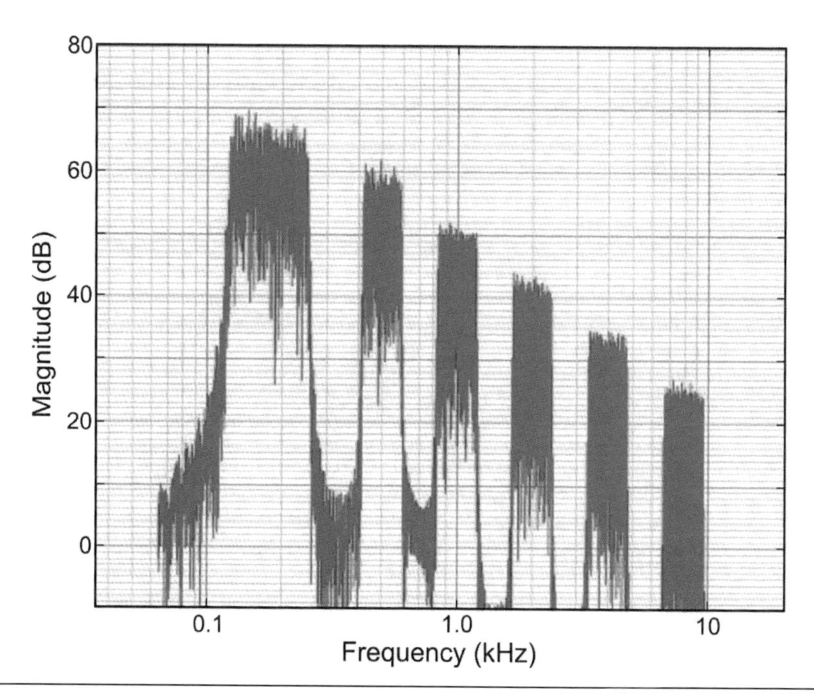

Figure 12.10 Signal in speech transmission index for public address systems (STIPa) in frequency presentation.

Based on the comparison of subjective examination results with a maximum possible intelligibility of syllables of 96%, the STI values are graded in subjective values for *syllable intelligibility* according to Table 12.1 (EN ISO 9921: Feb. 2004).

Because of the bad discrimination between good and bad STI values, a new scale is proposed in Edition 4 of the standard IEC 60268-16 (June 2011) with groups named with letters.[14]

By comparing Table 12.2 with Table 12.1, the so-called *satisfactory* STI values are after confirmation of the new scale now subdivided in four classes—E to H. This way a better evaluation becomes possible.

Table 12.1 Subjective weighting for STI

Subjective intelligibility	STI-value
Unsatisfactory	0.00 ...0.30
Poor	0.30 ...0.45
Satisfactory	0.45 ...0.60
Good	0.60 ...0.75
Excellent	0.75 ...1.00

Table 12.2 New proposal for subjective weighting for STI

1.0 0.72 0.64 0.56 0.48 0.40 0.0

A+ | A | B | C | D | E | F | G | H | I | J | U

0.76 0.68 0.60 0.52 0.44 0.36

Articulation Loss, Al_{cons}, with Speech

Peutz[15] and Klein[16] have found that the articulation loss of spoken consonants, Al_{cons}, is decisive for the evaluation of speech intelligibility in rooms. Starting from this discovery they developed a criterion for the determination of intelligibility:

$$Al_{cons} \approx 0.652\left(\frac{r_{QH}}{r_H}\right)^2 \cdot T_{60} \quad [\%], \tag{12.12}$$

r_{LH}: Distance sound source – listener in m
r_H: Reverberation radius or, in case of directional sound sources, critical distance D_c in m
T_{60}: Reverberation time in sec

According to Peutz[15] one can directly determine Al_{cons} from the measured room impulse response. Here we use the energy within about 25 ms to 40 ms (Default 35 ms) for the direct sound energy, and for the reverberation energy, the residual energy after 35 ms:

$$Al_{cons} \approx 0.652\left(\frac{E_{\infty} - E_{35}}{E_{35}}\right) \cdot T_{60} \quad [\%]. \tag{12.13}$$

Assigning the results to speech intelligibility yields (see Table 12.3).

Long T_{60} entail an increased articulation loss. With the corresponding duration, this reverberation acts like noise on the following signals and thus reduces the intelligibility.

Figure 12.11 illustrates the correlation between measured STI values and the articulation loss Al_{cons}. Acceptable articulation losses of Al_{cons} <15% require STI values in the range from 0.4 to 1 (meaning between satisfactory and excellent intelligibility). Via the equation:

$$STI = 0.9482 - 0.1845\ln(Al_{cons}), \tag{12.14}$$

it is also possible to establish an analytical correlation between the two quantities.

Subjective Intelligibility Tests

A subjective evaluation method for speech intelligibility consists in the ability to recognize clearly spoken pronounced words (so-called test words); these words are chosen on the basis

Table 12.3 Subjective weighting for Al_{cons}

Subjective intelligibility	Al_{cons}
Ideal intelligibility	≤ 3%
Good intelligibility	3 ... 8%
Satisfactory intelligibility	8 ... 11%
Poor intelligibility	> 11%
Worthless intelligibility	> 20% (limit value 15%)

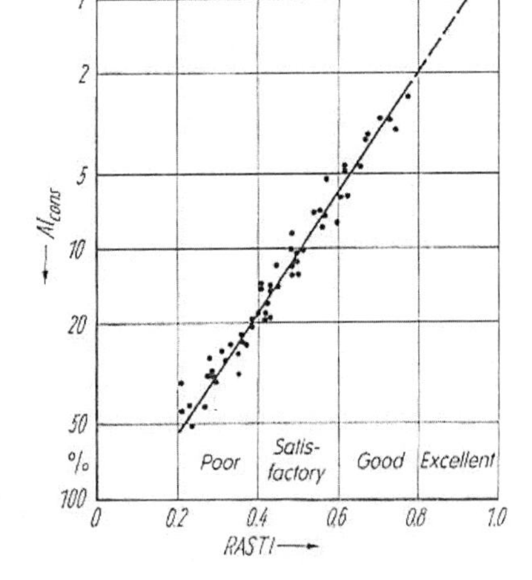

Figure 12.11 Relationship between Al_{cons} values and STI values.

of the word-frequency dictionary and a language-relevant phoneme distribution. In German intelligibility tests, *logatoms* are used for exciting the room (logatoms are monosyllable consonant-vowel-groups that do not readily make sense, so that a logical supplementation of logatoms that were not clearly understood during the test is not possible, e.g., *grirk*, *spres*). In English-speaking countries, however, consonant-vowel-consonant (CVC) test words as shown in Table 12.4 are used.[17]

Table 12.4 Examples of CVC words used in intelligibility tests

bad	cut	gap	jig	net	rob	stub
bag	dab	get	job	nip	rot	sub
ban	dad	glad	jog	not	rub	sum
bat	dam	glum	jug	nun	rug	sun
bed	dan	gob	kid	nut	run	swim
beg	den	grab	kim	pad	rut	tab
ben	did	gram	kit	pam	sad	tad
bet	dig	grid	lab	pan	sag	tag
bib	dim	grin	lad	pat	sam	ten
bid	dip	grip	lag	peg	tan	tim
big	dog	grit	lap	pen	tap	tin
bin	dot	grub	led	pet	sap	tip
bit	drip	gum	leg	pig	sat	top
blot	drop	gun	let	pin	set	tram
brat	dud	had	lid	pit	ship	trap
bud	dug	ham	lit	plan	shop	trim
bug	fan	hat	lob	plug	shot	trip
bun	fat	hen	log	plum	shut	tub
but	fib	hid	lot	pop	sip	tug
cab	fig	him	man	pot	sit	van
can	fin	hip	map	putt	skim	vet
cap	fit	hit	mat	rag	skin	wag
cat	flag	hog	men	ram	skip	wed
chat	flap	hop	met	ran	slab	wet
chin	flat	hot	mob	rap	slam	whip
chip	fled	hub	mop	rat	sled	wig
clam	flip	hug	mud	red	slid	win
cob	flop	hum	mug	rib	slip	wit
cog	fog	hut	mum	rid	slug	yet
cop	frog	jab	mutt	rig	snip	
cot	fun	jam	nag	rim	sob	
cub	gag	jet	nap	rip	spin	

For each test there are between 200 and 1000 words to be used. The ratio between correctly understood words (or logatoms or sentences) and the total number read yields the *word, syllable,* or *sentence intelligibility, V,* rated in percentages. Table 12.5 shows the correlation between the intelligibility values and the ratings.

The results of the subjective intelligibility test are greatly influenced by speech speed which includes the number of spoken syllables or words within the articulation time (articulation velocity) and the break time. *Predictor sentences,* therefore, are often used. These predictors are not part of the test and have to precede the words or logatoms. Such sentences consist of three to four syllables each, for example: "Mark the word..." "Please write down...," or "We're going to write..."

Strength Measure G

The strength measure, G, is the 10-fold logarithmic ratio between the sound energy components at the measuring location and those measured at 10 m distance from the same acoustic source in the free field. It characterizes the volume level:

$$G = 10\lg\left(\frac{E_\infty}{E_{\infty,10m}}\right) \quad [dB]. \tag{12.15}$$

Here, $E_{\infty,10\,m}$ is the reference sound energy component existing at 10 m (32.8 ft) distance from the free sound transmission of the acoustic source.

It was found[18] that the optimum values for musical and speech performance rooms are located between:

$$+1\text{ dB} \le G \le +10\text{ dB},$$

which means that the loudness at any given listener's seat in real rooms should be roughly equal or twice as high as in the open at 10 m (32.8 ft) distance from the sound source.

12.1.2.2.2 Measures for Music Reproduction

Clarity C_{80}

Extensive investigations were carried out for establishing a measure for clarity of traditional music. It was found that with symphonic and choir music it is not necessary to distinguish between temporal clarity and tonal clarity (the latter determines distinction between different timbres).[19] Both are equally well described by the *clarity measure* C_{80}:

$$C_{80} = 10\left(\lg\frac{E_{80}}{E_\infty - E_{80}}\right) \quad [dB]. \tag{12.16}$$

Table 12.5 Correlation between the intelligibility values and the ratings

Rating	Syllable intelligibility V_L in %	Sentence intelligibility V_S in %	Word intelligibility V_W in %
Excellent	90 ... 96	96	94 ... 96
Good	67 ... 90	95 ... 96	87 ... 94
Satisfactory	48 ... 67	92 ... 95	78 ... 87
Poor	34 ... 48	89 ... 92	67 ... 78
Unsatisfactory	0 ... 34	0 ... 89	0 ... 67

The value for a good clarity measure, C_{80}, depends strongly on the musical genre. For romantic music, a range of approximately:

$$-3 \text{ dB} \leq C_{80} \leq +4 \text{ dB},$$

is regarded as being good, whereas classic and modern music will allow values up to + 6 … 8 dB.

The *reverberance measure* is the spatial impression broken down into its two most essential components—spaciousness and reverberance.

Reverberance describes the subjective impression of the reverberation process. A distinction has to be made between the duration of reverberation and the reverberation level. The reverberation level characterizes the loudness of the reverberation process and thus, its perceptibility—i.e., the reverberation energy is related to the definition-enhancing early energy. One of the quantities characterizing this relation is, for instance, the reverberance measure:

$$H = 10\lg \frac{\int_{50ms}^{\infty} p^2(t)\,dt}{\int_{0}^{50ms} p^2(t)\,dt} \quad [dB]. \tag{12.17}$$

For symphonic music, reverberance measures of 3 to 8 dB are optimal according to Beranek.[20] With reverberance measures $H > 8$ dB, music is perceived as too reverberant, and with $H < 3$, as too dry. Also other measures, like the differently defined spatial impression measures, initial T_{60} algorithms and others, take the reverberation level into account. To measure the reverberance measure, omnidirectional sources solely must be used and only in this case yields $H = -C_{50}$.

Bass Ratio (BR)

Besides the T_{60} at medium frequencies, the frequency response of T_{60} is of great importance, especially at low frequencies, as compared to the medium ones. The bass ratio proposed by Beranek,[20] i.e., the ratio between the T_{60}s at octave center frequencies of 125 Hz and 250 Hz, and octave center frequencies of 500 Hz and 1,000 Hz (average T_{60}), is thus calculated basing on the following relation:

$$BR = \frac{T_{60}(125Hz) + T_{60}(250Hz)}{T_{60}(500Hz) + T_{60}(1khz)}. \tag{12.18}$$

For music, the desirable bass ratio is BR ≈ 1.0 ... 1.3, for speech, on the other hand, the bass ratio should, at most, have a value of BR ≈ 0.9 ... 1.0.

12.1.2.2.3 Measures for Music and Speech Reproduction and Binaural Measures

Echo Criterion (EC)

If we hear an echo in a room or in an open space there is no need to make measurements to verify the existence of echoes. But in many cases only trained musicians will hear the disturbing reflections, and this may become a disaster for a concert hall. So Dietsch[21] developed an algorithm to indicate the existence of echoes in a space. By applying:

$$t_s(\tau) = \frac{\int_0^{\tau} |p(t)|^n\, t \cdot dt}{\int_0^{\tau} |p(t)|^n \cdot dt}, \tag{12.19}$$

and using as exponents for the incoming sound reflection n = 0.67 with speech and n = 1 with music, we may calculate the following difference quotient:

$$EK(\tau) = \frac{\Delta t_s(\tau)}{\Delta \tau_E}. \tag{12.20}$$

We can discover echo distortions for music or speech when applying values of $\Delta\tau_E$ = 14 ms for music and $\Delta\tau_E$ = 9 ms for speech, derived by subjective tests.[21] The echo criterion depends on the motif. With fast and accentuated speech or music, the limit values are lower.

For 50% ($EC_{50\%}$) and 10% ($EC_{10\%}$), respectively, of the listeners perceiving this echo, the limit values of the echo criterion are (see also Figure 12.12):

echo perceptible with music for $EC_{50\%}$: ≥ 1.8; $EC_{10\%}$ > 1.5
determined for 2 octave bands 1 and 2 kHz mid frequencies
echo perceptible with speech for $EC_{50\%}$: ≥ 1.0; $EC_{10\%}$ > 0.9
determined for 1 octave band 1 kHz

Center Time t_s

For music and speech performances, the center time, t_s, is a reference value for spatial impression and clarity and results at a measuring position from the ratio between the summed-up products of the energy components of the arriving sound reflections and the corresponding delay times and the total energy component. It corresponds to the instant of the first moment in the squared impulse response and is thus determined according to the following ratio:[22]

$$t_s = \frac{\sum_i t_i E_i}{E_{ges}}. \tag{12.21}$$

The higher the center time, t_s, the more spatial is the acoustic impression at the listener's position is. The maximum achievable center time, t_s, is based on the optimum T_{60}.

For music, the desirable center time, t_s, is:

$$t_s \approx (70 \ldots 150) \text{ ms with a 1,000 Hz octave,}$$

and for speech:

$$t_s \approx (60 \ldots 80) \text{ ms}$$

with four octaves between 500 Hz to 4,000 Hz.

Interaural Cross-Correlation Coefficient (IACC)

The IACC is a binaural, head-related criterion and serves for describing the equality of the two ear signals between two freely selectable temporal limits, t_1 and t_2. In this respect, however, the selection of these temporal limits, the frequency evaluation as well as the subjective statement, are not clarified yet. In general, one can examine the signal identity for the initial reflections (t_1 = 0 ms, t_2 = 80 ms) or for the reverberation component ($t_1 \geq t_{st}$, $t_2 \geq T_{60}$).[23] The frequency filtration should generally take place in octave bandwidths of between 125 Hz and 4,000 Hz.

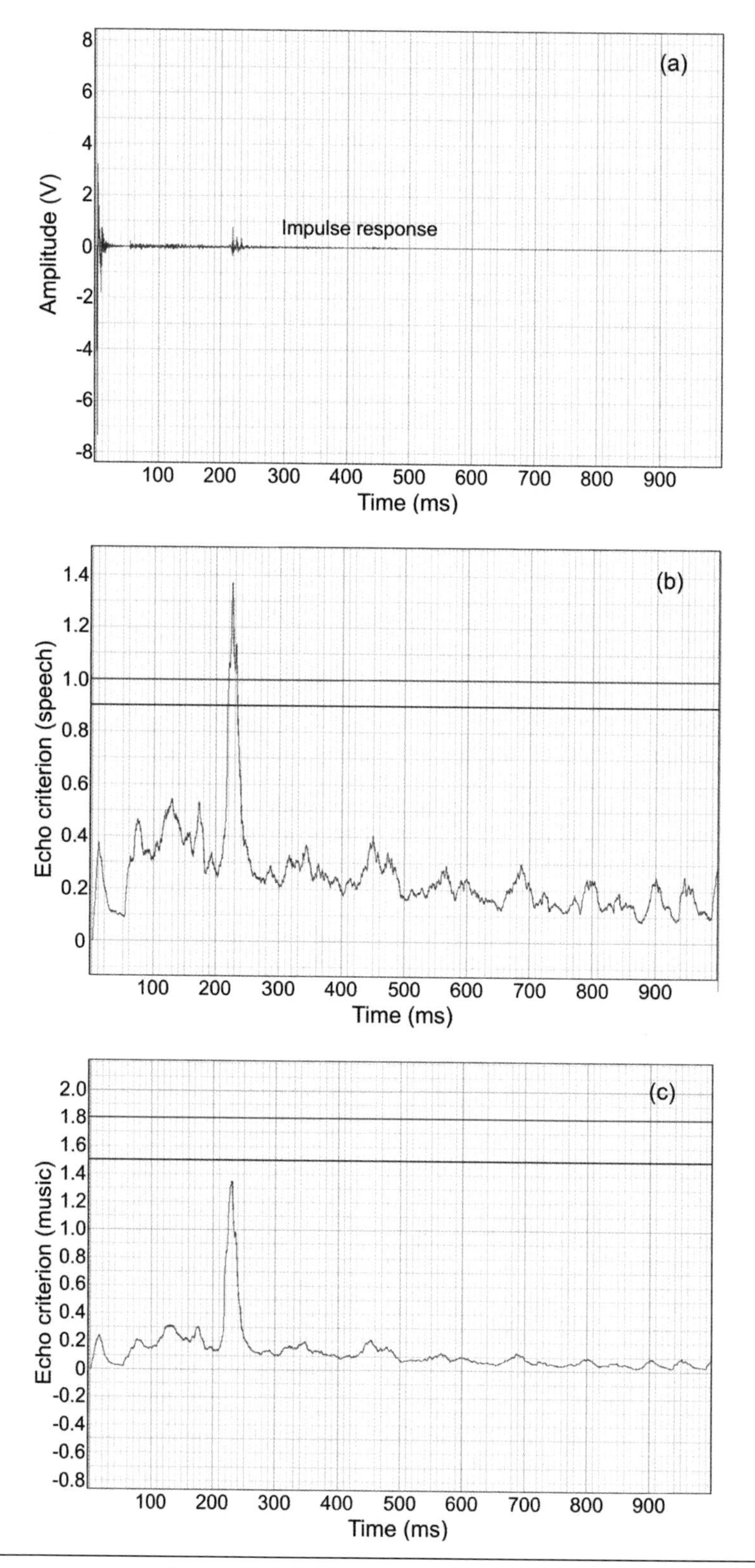

Figure 12.12 Echo detection by Dietsch: (a) impulse response with single echo, (b) strong echo detection for speech signals, and (c) weak echo detection for music signals.

The standard interaural cross-correlation function (IACF) is defined as:[23]

$$IACF_{t_1,t_2}(\tau) = \frac{\int_{t_1}^{t_2} p_L(t) \cdot p_R(t+\tau)\,dt}{\left[\int_{t_1}^{t_2} p_L^2(t)\,dt \cdot \int_{t_1}^{t_2} p_R^2(t)\,dt\right]^{1/2}}, \qquad (12.22)$$

with $p_L(t)$ impulse response at the entrance to the left auditory canal
$p_R(t)$ impulse response at the entrance to the right auditory canal

Then the IACC is:

$$IACC_{t_1,t_2} = \max\left|IACF_{t_1,t_2}(\tau)\right| \quad \text{for } -1 \text{ ms} < \tau < +1 \text{ ms}$$

t_1 and t_2 Integration time limits in ms:

for $IACC_{E(arly)}$: t_1 = 0 ms; t_2 = 80 ms

for $IACC_{L(ate)}$: t_1 = 80 ms; t_2 = 500 to 2,000 ms

for $IACC_{A(ll)}$: t_1 = 0 ms; t_2 = 500 to 2,000 ms

F Frequency range in Hz:

$IACC_E$ across 3-octave frequency ranges 500, 1,000 and 2,000 Hz, t_1 = 0 ms; t_2 = 80 ms

According to Beranek[1, 20] the value $\rho = (1 - IACC_E)$ correlates with the subjective perception of the *width* of the sound source (*apparent source width*) and the value $\varepsilon = (1 - IACC_L)$ correlates with the subjective perception of being *enveloped by the sound* (*listener envelopment*). For the values of $IACC_{E3B}$ or $\rho = (1 - IACC_{E;500,1,000,2,000Hz})$, Beranek specifies the following quality categories for concert halls:

Category: *Excellent to Superior* $IACC_{E;500,1,000,2,000Hz}$) 0.28 to 0.38

$\rho = (1 - IACC_{E;500,1,000,2,000Hz})$ 0.62 to 0.72

Category: *Good to Excellent* $IACC_{E;500,1,000,2,000Hz}$) 0.39 to 0.54

$\rho = (1 - IACC_{E;500,1,000,2,000Hz})$ 0.46 to 0.61

Category: *Fair to Good* $IACC_{E;500,1,000,2,000Hz}$) 0.55 to 0.59

$\rho = (1 - IACC_{E;500,1,000,2,000Hz})$ 0.41... 0.45

Lateral Efficiency for Music, Lateral Fraction, and Lateral Efficiency Coefficient

For the subjective assessment of the apparent extension of a musical sound source, e.g., on stage, the early sound reflections arriving at a listener's seat from the side are of eminent importance, as compared with all other directions. Therefore, the ratio between the laterally arriving sound energy components and those arriving from all sides, each within a time of up to 80 ms, is determined and its tenfold logarithm calculated therefrom.

If one multiplies the arriving sound reflections with $\cos^2\vartheta$, being ϑ the angle between the direction of the sound source and that of the arriving sound wave, one achieves the more important evaluation of the lateral reflections. With measurements this angle-dependent evaluation is achieved by employing a microphone with bidirectional characteristics:[24]

Lateral Efficiency (LE)

$$LE = \frac{E_{80\,Bi} - E_{25\,Bi}}{E_{80}} \tag{12.23}$$

E_{Bi}: Sound energy component, measured with a bidirectional microphone (gradient microphone)

The higher the lateral efficiency, the acoustically broader the sound source appears.
It is of advantage if the lateral efficiency is within the following range:

$$0.3 \leq \text{LE} \leq 0.8,$$

For obtaining a uniform representation of the energy measures in room acoustics, these can also be defined as lateral efficiency measure 10 log LE dB. Then the favorable range is between:

$$-5\text{ dB} \leq 10\log \text{LE} \leq -1\text{ dB}.$$

According to Barron,[25] it is the sound reflections arriving from the side at a listener's position within a time window from 5 ms to 80 ms that are responsible for the acoustically perceived extension of the musical sound source (contrary to Jordan[24] who considers a time window from 25 ms to 80 ms). This is caused by a different evaluation of the effect of the lateral reflections between 5 ms and 25 ms.

The ratio between these sound energy components is then a measure for the lateral fraction (LF):

$$LF = \frac{E_{80\,Bi} - E_{5\,Bi}}{E_{80}}, \tag{12.24}$$

where:

E_{Bi}: Sound energy component, measured with a bidirectional microphone (gradient microphone)

It is an advantage if the LF is within the following range:

$$0.10 \leq \text{LF} \leq 0.25,$$

or, with the logarithmic representation of the LF measure 10 log LF dB, within:

$$-10\text{ dB} \leq 10\log \text{LF} \leq -6\text{ dB}.$$

Both LEs and LFs have in common that, thanks to using a gradient microphone, the resulting contribution of a single sound reflection to the lateral sound energy behaves like the square of the cosine of the reflection incidence angle, referred to the axis of the highest microphone sensibility. Kleiner[26] defines, therefore, the lateral fraction coefficient (LFC) in better accordance with the subjective evaluation, whereby the contributions of the sound reflections vary like the cosine of the angle:

$$LFC = \frac{\int_{5}^{80} |p_{Bi}(t) \cdot p(t)| dt}{E_{80}}. \quad (12.25)$$

12.1.3 Basics in Sound Propagation for Sound System Design in Open Spaces

According to Eq. (12.26), the sound level decreases by 6 dB per distance doubling. The direct sound level of a spherical source is expressed by:

$$L_d = L_w - 20 \log r \text{ dB} - 11 \text{dB}. \quad (12.26a)$$

At a distance of r = 0.28 m from the source assumed to be punctual, the sound pressure level and sound power level are equal. At only 1 m distance (reference distance) both levels differ already by 11 dB, i.e., with a sound power of 1 W ($\Rightarrow L_w$ = 120 dB sound power level), the sound pressure level amounts to only L_d = 109 dB.

For open-air installations where the distance between loudspeaker and listener may be very large, an *additional propagation attenuation* D_r depending on temperature and relative humidity must be considered. In this case, the sound pressure level at the distance, *r*, is calculated for the assumed spherical source as:

$$L_d = L_w - 20 \log r \text{ dB} - 11 \text{dB} - D_r \text{dB}. \quad (12.26b)$$

Curve 3 in Figure 12.13 is the average value of the empirically derived curve family that should be used in practice. It reveals that up to a distance of 40 m, no additional attenuation is normally considered. This applies, e.g., to nearly all indoor rooms. The additional attenuation, D_r, increases along with the frequency (see Figure 12.14). An electroacoustical system is therefore required, especially for compensating high-frequency losses.

Figure 12.15 illustrates these frequency-dependent attenuation losses, D_r, in direct correlation with frequency and air humidity. With bad weather situations low frequencies are also affected by propagation losses.

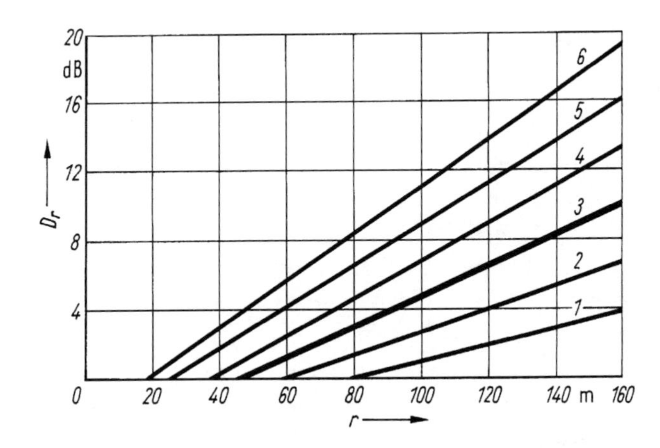

Figure 12.13 Additional propagation attenuation D_r with varying atmospheric situation as a function of distance, *r*. Range *1–2:* very good (dusk), Range *2–3:* good (overcast), Range *3–4:* mediocre (mean solar radiation), Range *4–5:* poor (heavy solar radiation), Range *5–6:* very poor (desert heat).

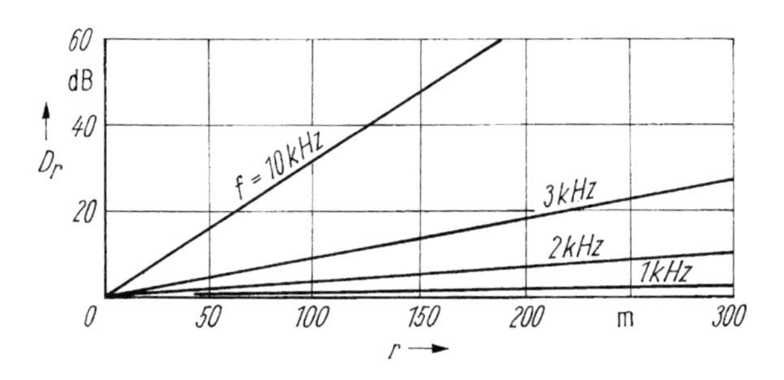

Figure 12.14 Atmospheric damping, D_r, as a function of distance, *r*, at 20°C, 20% relative moisture and very good weather situation—Parameter: frequency, *f*.

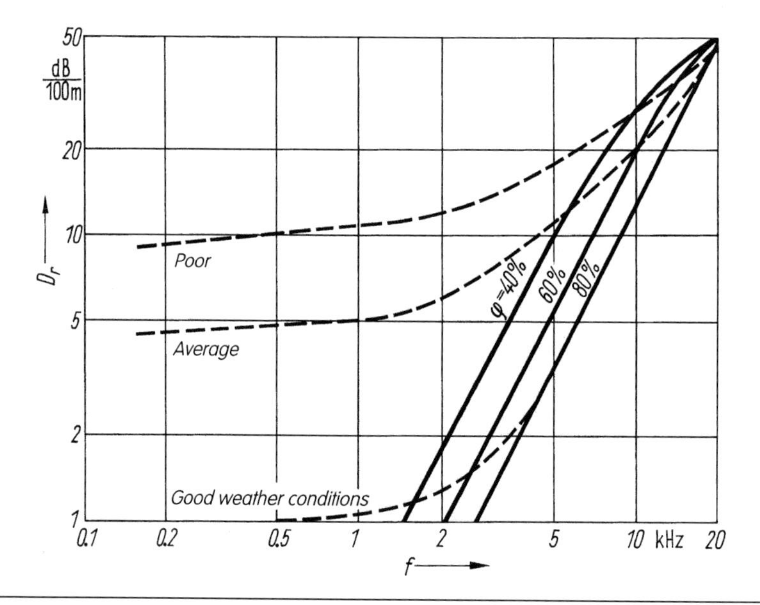

Figure 12.15 Additional propagation attenuation, D_r, with different weather conditions as a function of frequency with various weather conditions—φ, relative moisture

Owing to the heat expansion of the air, the speed of sound increases by about 0.6 ms per degree Kelvin. This implies that in a stratified atmosphere in which the individual air strata have different temperatures, sound propagation is influenced accordingly (see Figure 12.16).[27] If the air is warmer near the ground and colder in the upper layer, an upward diffraction of the sound takes place so that sound energy is withdrawn from the ground near the transmission path with ensuing deterioration of the propagating conditions (see Figure 12.16a). This case occurs, e.g., with strong sun irradiation on a plain terrain as well as in the evening over water surfaces that were warmed up during the day. Inverse conditions prevail with cool air at ground level and warmer air in the upper layer, as is the case over snow areas or in the morning over water surfaces. Under these conditions the sound energy is diffracted from the upper layers down to the lower layers where it boosts the energy, creating especially favorable propagating conditions (see Figure 12.16b).

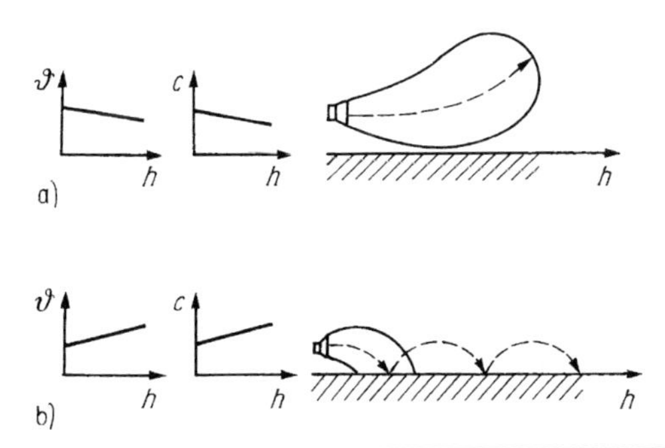

Figure 12.16 Sound propagation influenced by temperature (a) negative temperature gradient; sound speed decreases with increasing height, and (b) positive temperature gradient; sound speed increases with increasing height.

Given that the wind speed is relatively low by comparison with the sound speed (in a storm it is approximately 25 ms as against the average sound speed of c = 340 ms), sound speed is normally not significantly influenced by wind. Due to the roughness of the earth's surface, the wind speed is, however, lower at ground level than in higher layers, so that sound propagation may be influenced by the wind gradient in a similar manner as by the temperature gradient (see Figure 12.17). Thus, intelligibility may be heavily reduced by a very whirly and gusty wind.[27]

Two important psycho-acoustic items for sound system design would be loudness perception and masking effect. Loudness perception is limited downward by the threshold of hearing and upward by the threshold of pain.

The phenomenon of the threshold of hearing originates from the fact that a certain minimum sound pressure is required for producing a hearing impression. At 1,000 Hz, this minimum sound pressure, averaged over a large number of people, amounts to:

$$\tilde{p}_0 = 2 \bullet 10^{-5}\ \text{Pa} = 20\ \mu\text{Pa or a sound intensity of } J_0 = 10^{-12}\ \text{W/m}^2.$$

Following an international standard, these threshold values correspond to a sound pressure level of 0 dB. The hearing threshold depends heavily on frequency and rises significantly for lower frequencies, as shown in Figure 12.18.

The upper limit of sound perception is given by the pain caused by the reaction of the *clipping protection* of the auditory system (disengagement of the auditory ossicles). This limit lies at 10^6 times the sound pressure or 10^{12} times the sound intensity of the threshold value for 1,000 Hz. But also before this level is reached, nonlinear distortions are occurring that commence at a level of about 90 dB. Loudness perception widely follows a logarithmic law.[28]

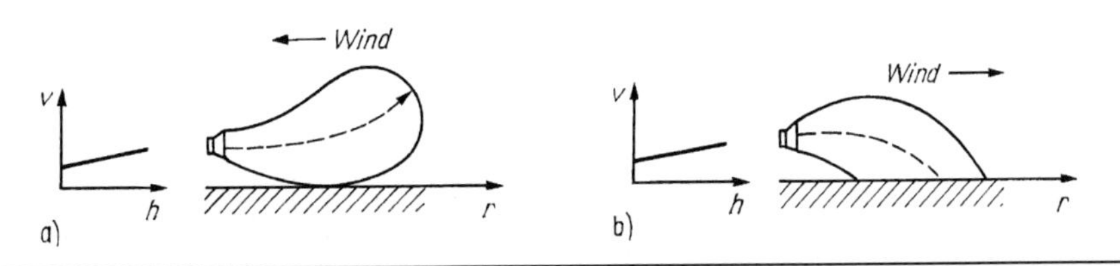

Figure 12.17 Sound propagation against the wind (a) or with the wind (b); wind speed increasing in both cases with height.

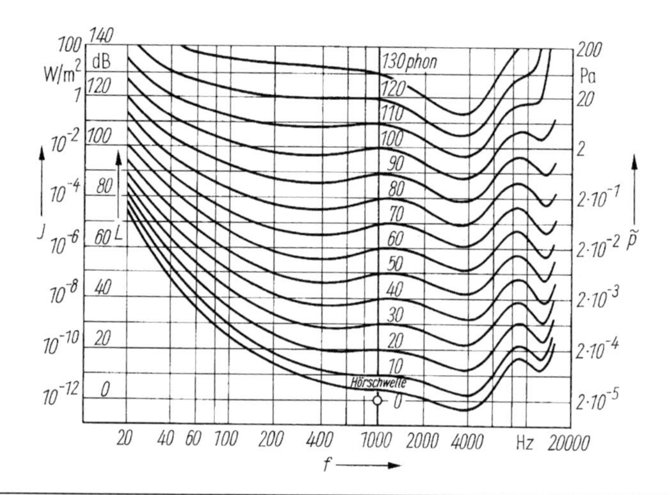

Figure 12.18 Curves of equal loudness level for pure sounds.

One scale for loudness perception is the subjective phon scale according to Barkhausen.[29] It is established by comparing a 1,000 Hz tone with a tone of another frequency and adjusting it to equal loudness. This way, one obtains curves of equal loudness levels that are similar to the curves shown in Figure 12.18 with some individual variations.

The reduced sensitivity of the auditory system for low and high frequencies at low sound levels is approximately simulated for determined loudness values by means of *weighting curves* in the sound level meter (see Figure 12.19). According to the IEC publication 651 of 1979, the A-weighted curve corresponds approximately to the sensitivity of the ear at 30 phons, whereas the B-weighted curve and the C-weighted curves correspond more or less to the sensitivity curves of the ear at 60 phons and 90 phons, respectively.[30] The A-curve, switchable in any sound level meter, however, is of high importance for sound reinforcement engineering. For measuring the sound level distribution of speech and information systems in a noisy environment, but also in the neighborhood of exit openings of ventilating and air-conditioning systems, it is recommended to using this weighting curve in order not to obtain incorrect measurement results caused by air turbulences or low-frequency sound.

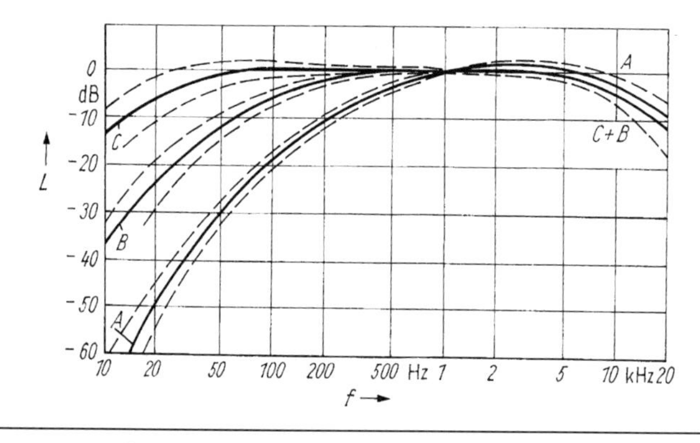

Figure 12.19 Frequency weighting curves recommended by the IEC for sound level meters.

If several tones or noises of different neighboring frequencies and different loudness levels sound simultaneously, the acoustical stimulus of low loudness may, under certain conditions, be inaudible, although its sound level lies above the threshold of hearing. In this case, the weaker noise is *masked* by the louder one.[31, 32] The audibility threshold of the weaker acoustical stimulus is determined by the *masked threshold of audibility*. Figures 12.20 and 12.21 show the masking effect under different excitation conditions. This effect is also based on the fact that on the basilar membrane, not only the narrow range corresponding to the stimulating frequency is excited, but the neighboring ranges are also, with greater loudness levels to an increasing degree. As can be seen from the figures, this affects the higher frequencies more than the lower ones. This effect is of great importance for sound reinforcement engineering. Its influence gives an explanation for the fact that narrow-band frequency response notches—which, e.g., occur frequently because of loudspeaker interferences—are normally

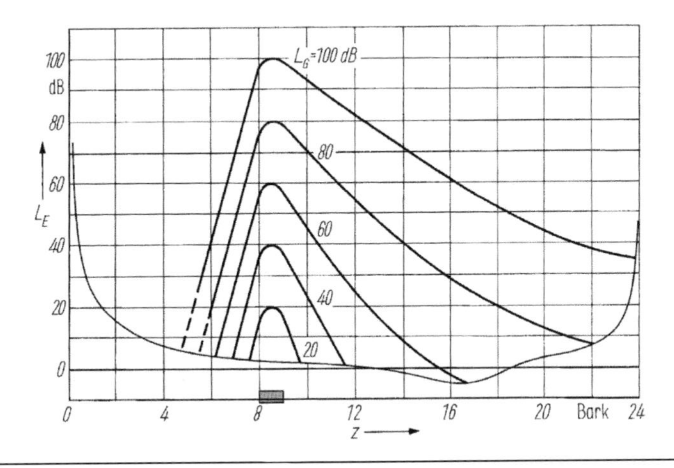

Figure 12.20 Excitation level, L_E, on the basilar membrane caused by masking narrow-band noise with a center frequency of 1 kHz and a noise level of L_G (indicated on the abscissa axis between 8 and 9 Bark)—z, subjective pitch.

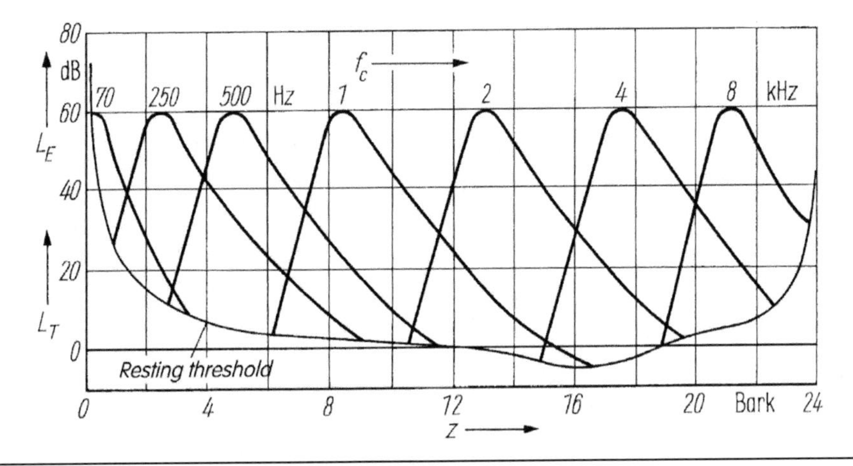

Figure 12.21 Excitation level, L_E, over the subjective pitch, z, (position on the basilar membrane), when excitation is done by narrow-band noise of L_G = 60 dB and a center frequency of f_c—L_T resting threshold.

inaudible, whereas formant-type presences may give rise to considerable timbre changes owing to the masking of the neighboring ranges. Also, speech intelligibility is influenced by masking, see Section 12.1.2.2.1.

12.1.3.1 Auditory Localization

Localization is achieved mainly thanks to the binaural structure of the auditory system. In the median plane (front, above, behind), the travel time differences between the two ears do not contribute to localization because both ears are equidistant from the median plane. *Monaural localization* of a sound event is nevertheless possible on account of so-called directional frequency bands. These result from the frequency-dependent shadowing by the pinna and allow assigning originating directions (behind, above, front) to narrow-band noises.[33] Small head movements will help here too.

Easier to explain is *binaural localization*. Lateral sound incidence from outside the median plane produces in the medium and upper frequency ranges a pressure increase in the directly impacted ear, whereas the shadowed ear is accordingly less impacted. The auditory system is thus more sensitive for lateral sound incidence than for frontal incidence. The direction of incidence is inferred from time and level differences at the ears. Figure 12.22 shows the travel time differences and Figure 12.23 shows the level differences at the ears.

Investigations have revealed that with frequencies below 300 Hz, direction is inferred mainly from travel time differences, and with frequencies above 1,000 Hz, from level differences.[34]

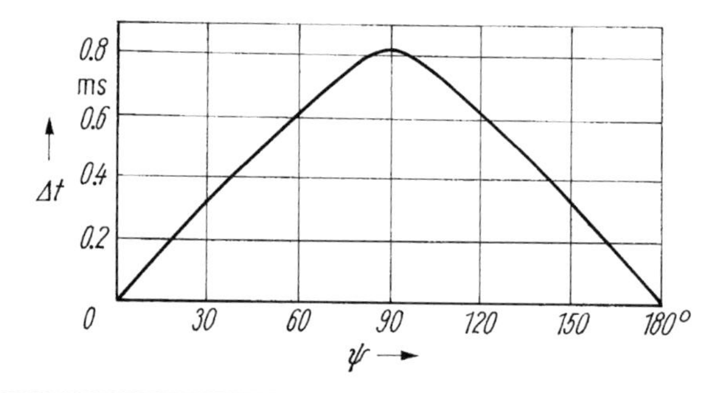

Figure 12.22 Early-to-late ratio of the sound reaching the ears, as a function of the incidence angle.

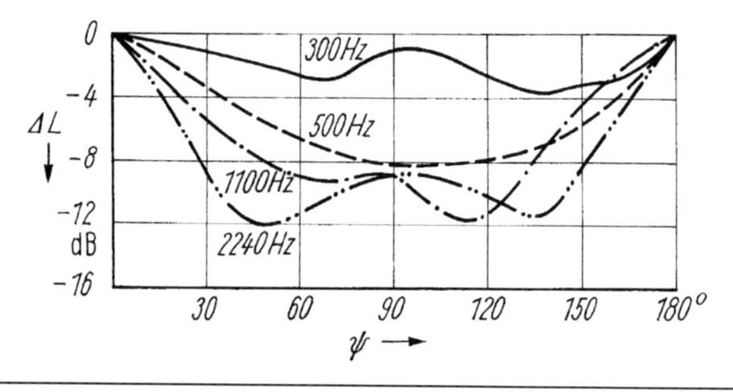

Figure 12.23 Variation of the sound level difference with discrete frequencies and horizontal motion of the sound source around the head.

In investigations on the influence of a simple echo on the intelligibility of speech, Haas established that the ear always localizes first the signal of that source from which the sound waves arrive first.[35] This holds true, however, only for finite time and level differences. The *precedence effect* derived from this and other investigations, thus establishes that the perception of the direction of incidence is determined by the first arriving signal. This is still the case even if the secondary signal (the repetition of the first one) has up to a 10 dB higher level and arrives within 30 ms (see Figure 12.24). Only with travel time differences of $t > 40$ ms, the ear slowly starts to notice existing reflections, but still continues to localize by the primary event. The blurring threshold lies here at about 50 ms. With longer delay times (from 30 ms on), one first notices timbre changes and, with times ≥ 50 ms, echoes (*chattering effect*).

The precedence effect is frequently used for the localization of sources in sound reinforcement systems. This will also be briefly explained in Section 12.5.2.2.

12.1.3.2 Effect of High Loudness Levels on the Auditory System

This section briefly discusses the detrimental effect of excessive loudness levels of transmitted music and, less frequently, speech. This aspect is gaining much more importance since—especially in rock and pop music, but also in other forms of entertainment music—high sound levels use to be presented over long periods and with only short interruptions. In order to overcome disturbing noise, the entertaining stimulus will have to be adjusted to high sound levels, causing hearing impairment. This negative impact has not yet been sufficiently recognized. Investigations show, however, that music of any kind may lead to hearing impairment whenever detrimental sound levels are produced at close range to the ear over long periods of time.

12.1.3.2.1 Mechanism of Hearing Impairment

For a better understanding of the mechanism causing hearing impairment, we would like to briefly touch on the implications of the impairment and its examination.

A negative influence on the auditory system is detected by a shifting of the hearing threshold, i.e., the deviation from a standardized threshold after exposing it to a certain acoustic

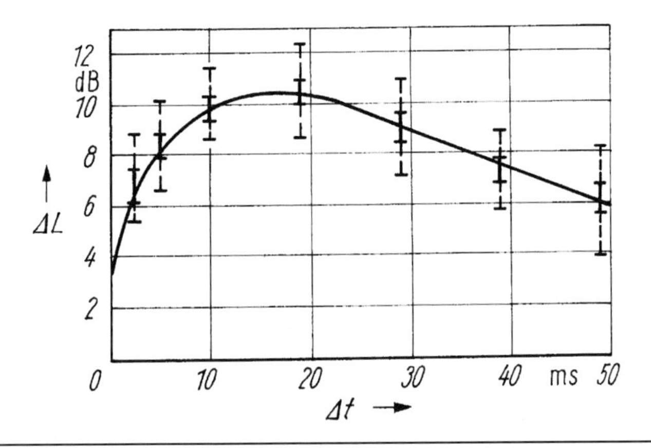

Figure 12.24 Critical level difference, ΔL, between reflection and undelayed sound producing an apparently equal loudness impression of both (speech) signals, as a function of the delay time, Δt.

excitation. First, a *temporal threshold shift* (TTS) normally occurs, which disappears after a period of recovery. If the ear is subjected to very long or very high and frequent acoustic excitations, however, it may come to a *permanent threshold shift* (PTS); the threshold of hearing has been shifted irreversibly by this value.

Figure 12.25 shows the PTS for 4 kHz as a function of the noise dose sound pressure in Pascal multiplied by the excitation time in seconds caused by music with very high loudness levels up to 120 ... 140 dB.[36]

12.1.3.2.2 Causes for High Sound Levels in Sound Reinforcement Systems

It is an undeniable fact that in performances of entertainment music and especially in juvenile rock music presentations, as well as discotheques, very high sound levels prevail, which in the course of the performance are increased for enhancing the emotional effect. At close range to the performance, there are produced permanent sound levels of over 90 dB(A) and up to a maximum of 115 dB(A). But permanently exceeding 85 dB(A) is not recommended.

The spectral analysis of such high levels reveals in most cases a very bass-accentuated signal which is intentionally produced in order to obtain the desired vibration of the abdominal wall. With regard to hearing impairment, this bass emphasis may even be considered as a positive factor, since most of the irreversible impairments occur in the range between 5,000 and 6,000 Hz.[36] According to the point of view of visitors to such events, this music has to be so loud in order to completely involve the listeners. Frequently, one has found that especially with bands of poor performance or low-quality reproduction, the music tends to be louder so as to make the ensuing distortions in the ear cover up the faults, whereas with better music performances, the levels produced are normally lower. Moreover, the musicians grow more and more hard of hearing in the course of their professional lives so that they are forced to play louder or must have a louder accompaniment for hearing each other. This leads in a cumulative process to higher loudness and thus, inevitably, to hearing impairment.

12.1.3.2.3 Possibilities for Reducing the Excessive Sound Levels

In the operation of sound reinforcement systems it is very difficult to avoid harmful sound levels of over 90 dB by technical means. A restriction of the sound level by limitation of

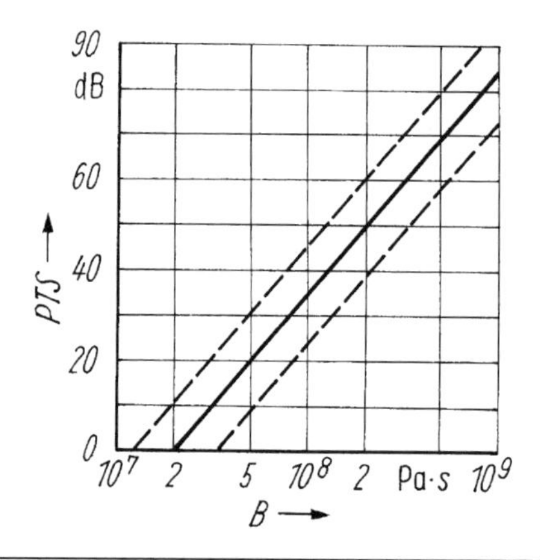

Figure 12.25 PTS at 4 kHz as a function of the noise dose, *B*, for a stationary noise from 79 to 105 dB(A); ——— regression line; - - - - - - - - limits of the ± 2 s range (*s* residual standard deviation),

the installed amplifier and loudspeaker power is not feasible, since in the interest of high overdrive safety, these components of a system should have a power handling reserve of at least 10 dB. Nevertheless, it is essential to adjust the levels of a system in such a way that the head room cannot be abused for excessive loudness enhancement. So-called *loudness indicators* have been installed in several countries in discotheques that show on an optical display that certain predetermined maximum sound levels have been exceeded, so as to advise the listener that he or she is exposed to a hearing impairment. Moreover, there exist automatic loudness limiters which are inserted in front of the power amplifiers and set a substantial preset attenuation when a certain maximum level is exceeded. The functioning of such devices should be constantly supervised by responsible authorities.

Apart from its harmful effect, excessively loud music contains an additional disturbing component for the environment not partaking in the sound event. For this reason, it is necessary for clubs and discotheques to take special building-acoustical noise control measures to avoid molesting the environment. In this respect the legislator has established exact limit values justifying intervention when being exceeded. This, however, is better done at the source of the noise pollution: that is, in the electroacoustical system.

12.2 LIMITS OF SOUND SYSTEMS IN ROOMS

12.2.1 Level Restrictions

Figure 12.26 shows an energy-time-curve (a squared impulse response), measured a in noisy environment, with an omnidirectional loudspeaker. So this response is a pure room acoustic one.

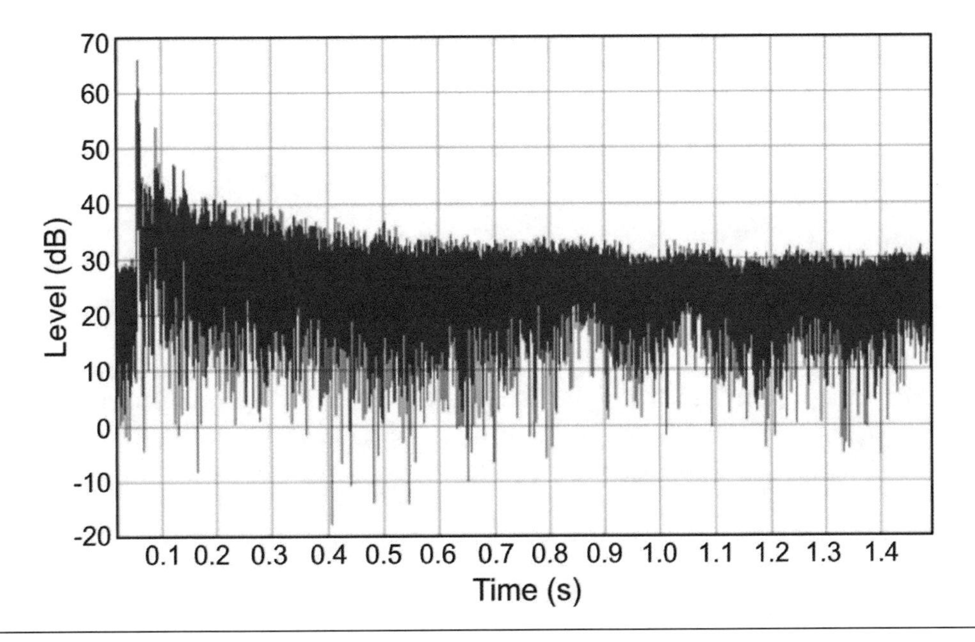

Figure 12.26 Energy time curve (ETC).

We may derive from this figure three kinds of information:

- Size of the direct sound
- Quantity of early reflections
- Influence of the noise floor

The possible conclusions in this specific case may be:

- Level too low, need for a sound system
- No supporting first reflections
- Too high noise floor, need of directed information to overcome noise

Let's assume that the use of a sound system is needed. For good speech intelligibility, a signal-to-noise ratio of a minimum of 10 dB should be achieved, with 15 dB being better.

The level of the sound system should be adapted to the possible noise floor, but cannot be higher than the threshold of pain. This is partially violated in rock and pop concerts and hearing damage is to be expected (see Chapter 5).

On the other hand, a minimum sound level must overcome the noise floor. In some areas this level may be controlled for fluctuating values of noise over time. This is quite often the case in railway stations and airport venues/concourses.

12.2.2 Primary and Secondary Structures of Spaces and Noise Floor Considerations

The primary structure of a space is determined by:

- The volume,
- The dimensions, and
- The shape of the space.

The secondary structure deals with:

- The absorption of wall and ceiling materials,
- The ratio between reflecting and scattering walls surrounding the space, and
- The shape and the size of audience areas.

By designing a sound system in a space, all of these issues interact and influence the quality of a sound transmission in that space. For better understanding of the interaction between room acoustics and sound design, we should illuminate the influence of the following items in more detail:

- Volume
- Reverberation time
- Location of speaker
- Distance to source and from source to source
- Width of stage

- Noise floor
- Pain threshold
- Listener threshold
- Stereo reproduction
- Echo effects
- Interaction with walls—binaural room impulse response (BRIR)
- Delay distortion

Only the main issues should be touched here. Let's start with the volume. In a closed room the size of the volume is important for the sound pressure level (SPL) produced by a single sound source. Figure 12.27 illustrates the relationship between the ratio of size of the volume to the T_{60} and the achievable SPL in the diffuse sound field. The parameter of the graphs is the radiated acoustic power in watts. The figure indicates that large volumes and low T_{60} values (similar to an open air situation) allow only very low achievable SPL values. In contrast, in a bathroom (small volume and higher reverberation) even a weak human voice may produce relatively high SPL values. These volume issues have to be considered when designing a sound system.

Already mentioned was the influence of the reverberation and, in particular, the duration of the decay process, the T_{60}. Depending on the user profile of the space, we have to expect T_{60} values between 0.3 sec (studio rooms) and 4 sec (organ spaces). Unwanted T_{60} values are even higher here, especially in modern airport venues/concourses or railway stations.

Besides the room acoustic criteria, sound systems are to be installed to supply a corresponding SPL above the noise floor and to assure needed speech intelligibility according to the corresponding standards (ISO 7240-19).[13] Figure 12.28 illustrates the relationship between the intelligibility measure STI and the ratio of D_{SL}/D_c (D_{SL} = distance source or speaker to the listener, D_c = critical distance). The parameter is the T_{60}.

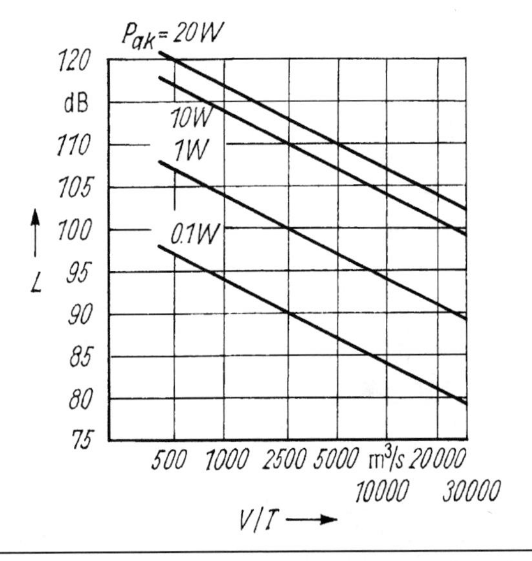

Figure 12.27 Sound pressure level, *L*, obtainable in the diffuse sound field as a function of the ratio between volume, *V*, and reverberation time, T_{60}—parameter: sound power P_{ak}.

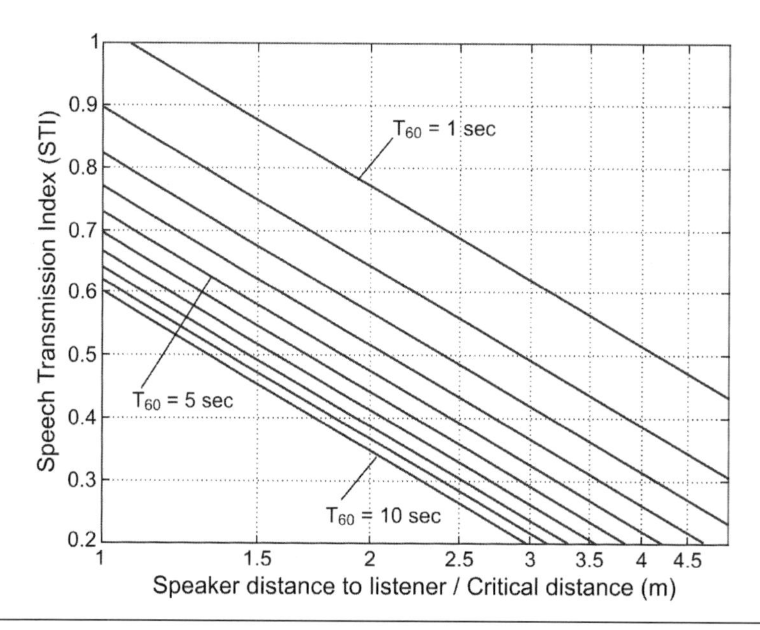

Figure 12.28 Speech transmission index (STI) as a function of source-receiver distance/critical distance (D_{SL}/D_c) with T_{60} = 1...10 s as a parameter.

For $D_{SL}/D_c \leq 5$ and $T_{60} \leq 1$ sec, the STI values are always higher than 0.45. In spaces with T_{60} = 10 sec and longer, you obtain only usable intelligibility for small ratios $D_{SL}/D_c \leq 1.5$, i.e., for the near field. Here in reverberant spaces (high T_{60}) the critical distance $D_c = \sqrt{Q^* r_H}$ (Q = directivity factor, r_H = reverberation radius = $\sqrt{(\text{amount of absorption}/16\pi)}$) is small. So we need small distance source-listeners D_{SL} to obtain STI values > 0.45. From the point of view of location of sources, you then need more speakers in the space, but unfortunately every distributed loudspeaker is the enemy of the other ones. A certain speaker supplies direct sound only to the area he is directed to, but supplies diffuse sound components to all the other areas. Therefore, a balance between the number of speakers in a reverberant space and the alternative solution to reduce the reverberation has to be achieved. Sometimes the last method is the only choice.

Another issue is the width of the stage and the distance of sources from each other to cover a space with sound. For even coverage by a sound system, we sometimes need more than one source location in case of large stage openings. This causes some problems which have to be considered:

- Uneven frequency response at the listener position caused by comb-filtering
- Interaction between speaker sound and close wall reflections
- Mislocalization of sound events on stage
- Weaker direct sound impact in center areas of the hall
- No stereo reproduction is possible (only in the center line of the hall)

With modern sound systems, most of the unfavorable hall sizes may be handled, but the larger and wider a hall is, the more effort is needed to diffuse the sound smoothly. Besides complicated delay lines, sophisticated sound tuning is important to avoid distortion, mislocalization effects, and echoes in the space. In order to design a hall for multipurpose use, close cooperation between the architect and the acoustician is needed, but should also include the sound system designer from the beginning. A separate design of the acoustician and the sound system designer must be avoided in these cases.

A final but important issue is the ratio between the source signal at the listener place and the noise floor existing there—both levels can be influenced. With acoustic treatment (mainly sound insulation) the noise floor may be reduced. But before doing this, the expected noise level must be evaluated. This can be done by measurements or by calculation (with or without computer simulation). Existing standards and norms help to focus on the right target values.

The noise floor should be known and has to be part of the sound system design. The signal level of the sound system should be 15 dB higher (with a minimum of 10 dB) than the noise floor. This ratio has, of course, limitations. In a railway station with a noise floor of 100 dB(A) (moving train in a station hall), we cannot increase the signal level to 115 dB or more. Here we are close to the threshold of pain and the ears of the passengers may be harmed. Such signal levels must be avoided. Thus, information of installed sound systems will not be understood if the noise floor is so high.

On the other hand, we know that signals may only be perceived by our ears if they are above the listener threshold. For a young person this threshold is, under normal conditions, always lower than any existing noise floor, i.e., any signal level must exceed the noise floor. But in the case of hearing impaired persons this may change. If the noise floor is not audible, the personal listener threshold becomes important in order to hear the signal or not. But this would mean, that a PTS (see Section 12.1.3.2.1) of 50dB, for instance, will demand a personal hearing aid in any case.

The sound system design will become successful only by considering the specialties of the primary and secondary structure of spaces.

12.2.3 Mono or Multipurpose Spaces

A concert hall or an opera house serves mainly one purpose—to be a space to perform concerts, operas, ballets, and other classical events. In all of these spaces the room acoustics are important, and the architect works together with an acoustician from the first moment to create a space of the highest acoustic quality. Other uses of these spaces have to be supported by a sound system. In a concert hall, conferences of well-known organizations are often held. The high reverberation of such a hall has a contra-productive effect to the speech intelligibility during a conference. But well-designed sound systems will provide communication using modern speaker set-ups including line-arrays. In theaters, sophisticated sound systems are needed anyway. All effect and playback signals are reproduced by a system of distributed speakers in the proscenium and stage area, and these speakers may also be used to carry out meetings and congresses.

That way a sound system design in so-called mono-purpose spaces is not trivial but follows known rules with audio equipment permanently in development. The situation is different in so-called multipurpose halls. Here it is important to know the main purpose of the hall. Shall it be:

- Case 1: a conference hall with some concerts or musical events during the year, or,
- Case 2: a hall for musical presentations with some conferences or congresses.

In coordination with the client and the architect, the acoustic layout will be different. In the first case we have a hall laid out for speech performances, so depending on the size of the hall, T_{60} varies between 1 and 1.4 sec. But what has to be done to adapt the acoustics for the musical events, where more reverberation is wanted? To enhance the reverberation we have two choices:

- Variable acoustics
- Electronic architecture

In the case of *variable acoustics*, the secondary and/or even the primary structure of the hall is changed. A well-known procedure is the exchange of wall materials—curtains or draperies along the wall surfaces are removed and reflecting surfaces become active. This may happen by vertical or horizontal movements of wall parts or even by turning wall and ceiling parts back and forth. Normally these movements are motor-driven and quite expensive, but they are widely accepted.

Another possibility to change the acoustic layout and to enhance the reverberation including first short-time reflections is the application of methods known as *electronic architecture*. This expression was first time used by C. Jaffe to describe such a system.[37] Around 10 different approaches are known to increase the spaciousness and reverberation, some have common features, and others are unique. But all of them pick up the source signal above the stage opening or in the diffuse sound field, post-process it in different ways, and radiate the signal over a multitude of loudspeakers hidden in wall and ceiling parts. By preprogrammed setups, different spaciousness and reverberation values may be produced in the hall. Because of the use of microphones and loudspeakers these systems are quite often named as sound systems, too. This leads to the rejection of such systems, especially among musicians; they don't want to become supported by sound systems. But the quality of modern enhancement systems is so high, that even world-known conductors recommend the use of such systems in halls with poor room acoustic conditions that are used for music performances. A detailed explanation of different enhancement systems can be found in the *Handbook for Sound Engineers.*[2]

After the selection is done to enhance the spaciousness in a multipurpose hall, we now have to make sure that spoken words will be understood under all acoustic conditions of the hall. Without the enhancement system or in a case with absorbing wall and ceiling parts, i.e., at low T_{60}, the sound system design is more simple. It becomes more complicated in the concert mode, i.e., at higher reverberation values. Here only modern sound systems like line-arrays or electronically steered sound columns will ensure that spoken words are perceived at a high quality and with corresponding intelligibility.

If a multipurpose hall would be designed according to Case 2, we have a hall suitable for the performance of concerts and musical events in high quality. Here again, depending on the size of the space, we have to expect T_{60}s in the range of 1.6 to 1.8 sec (higher values are important for pure classic concert halls or spaces mainly to perform organ concerts). Only variable acoustics may be used to decrease the spaciousness and reverberation for speech performances. Here, absorbing instead of reflecting and scattering wall and ceiling parts are becoming active and the T_{60} decreases to mid-frequency range values of 1.3...1.4 sec. An enhancement system may be used to increase the T_{60} additionally for special performances beyond 2 sec, to electronically decrease the T_{60} broadband is not (yet) possible.

The sophisticated sound system is similar to the system used in Case 1 (see previous text).

12.3 HOW TO DESIGN A SOUND SYSTEM

12.3.1 Introduction

A sound system consists of components to pick up a signal, process it, amplify it, and then play it back into the space. This section outlines some properties of these components where the main focus is on the sources, respectively the loudspeakers, and briefly on microphones. Their properties are interacting with the space where they are coming into operation. Only by knowing their behavior will we understand the interaction between a sound system and a room—we are only able to select a design which ensures high intelligibility on the one hand and desired spaciousness and reverberation on the other hand.

The use of signal processors, filters, and delay units is briefly explained. In context with prediction routines, the properties of wall materials—either absorbing, reflecting, or scattering the sound—are discussed as well.

12.3.2 Acoustic Sources and Loudspeaker Systems

We distinguish the following speaker types:

- Point sources
- Loudspeaker columns
- Line arrays
- Digitally controlled sound columns

For the use of these different sound radiators, the performance parameters of the same must be known. For this reason, we are going to mention the most important data to be specified in loudspeaker design. Let us start with the so-called point sources.

12.3.2.1 Point Sources

Point sources do not automatically show omnidirectional radiation behavior. Their directivity behavior is measured on a turntable and all directivity balloon data are referred to that point of rotation, which gives point sources their name.

The ratio between the sound pressure, $\tilde{p}$, and the voltage, $\tilde{u}$, is called sensitivity T_s:

$$T_S = \tilde{p} / \tilde{u}\ , \tag{12.27}$$

T_s is given in Pa/V.

The sensitivity level, G_S, is defined as the logarithmic quantity of sensitivity:

$$G_S = 20 \log \frac{T_S}{T_0} \ \ [\mathrm{dB}]. \tag{12.28}$$

The reference sensitivity, T_0, is preferably 1 Pa/V. If another value is chosen, it has to be indicated. The graphic representation of the sensitivity level as a function of frequency is called frequency response.

One of the quantities most frequently used in sound reinforcement engineering is the rated or characteristic sensitivity. According to standards like IEC 60268-5,[38] the logarithmic quantity of this expression is called characteristic sound level. L_K, but also sensitivity/dB. It is defined by the following:

$$L_K = 20\log\frac{E_K}{E_{K0}} = 20\log E_K + 94 \quad [\text{dB}]. \tag{12.29}$$

All loudspeakers used show a more or less pronounced directional dependence of radiation, which is frequency-dependent—just like beaming behaviors. This angular dependence of sound radiation is characterized by three quantities: the directional factor, Γ, the directivity factor, Q, and the directivity deviation factor, g(ϑ).

The *directional factor, Γ*, for a frequency or a frequency band, is the ratio between the sound pressure, $\tilde{\mathbf{p}}$, radiated at an angle ϑ, from the reference axis, and the sound pressure, $\tilde{\mathbf{p}}_o$, generated on the reference axis at an equal distance from the selected acoustic reference point (this reference point is selected by the speaker manufacturer and must be published in data sheets; generally it is the center of gravity of the speaker box):[38, 39]

$$\Gamma(\vartheta) = \frac{\tilde{p}(\vartheta)}{\tilde{p}_0}. \tag{12.30}$$

In general, $\Gamma(\vartheta) \leq 1$. Only if the maximum of directional characteristics does not occur at $\vartheta = 0°$, then may $\Gamma(\vartheta) > 1$.

The logarithmic quantity of the directional factor is the directional gain:

$$D(\vartheta) = 20\log\Gamma(\vartheta) \quad [\text{dB}]. \tag{12.31}$$

Figure 12.29 shows the directional characteristic of the horn loudspeaker in a polar plot of the directional gain. The main maximum is located at 0 degrees and several secondary maxima are at higher frequencies.

An important parameter for direct-sound coverage is the angle of radiation, Φ (beam width angle). It stands for the solid-angle margin within which the directional gain drops by a maximum of 3 dB or 6 dB (or another value to be specified) as against the reference value. The curves of equal directional gain are marked Φ_3, Φ_6, or generally Φ_n, the higher the directivity, the smaller the angle of radiation (see Figure 12.30).

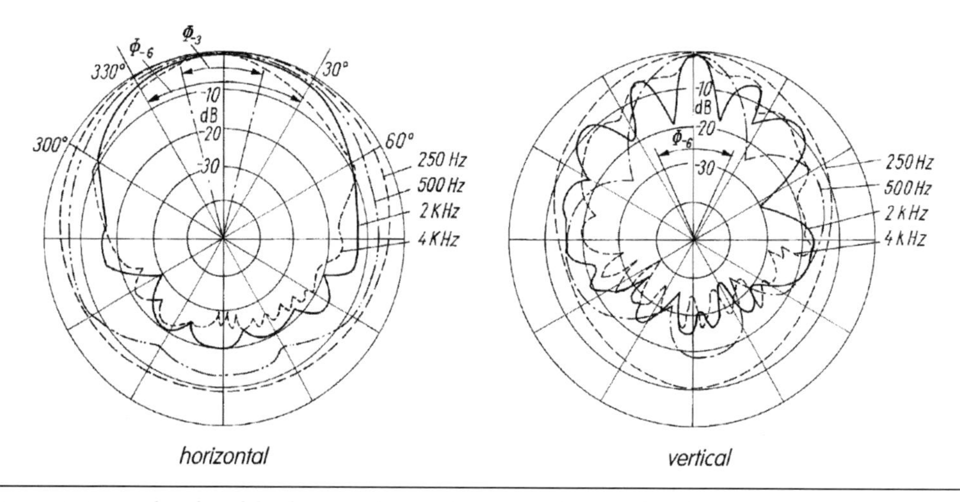

Figure 12.29 Polar plot of the directional gain of a sound column with indication of the radiation angles.

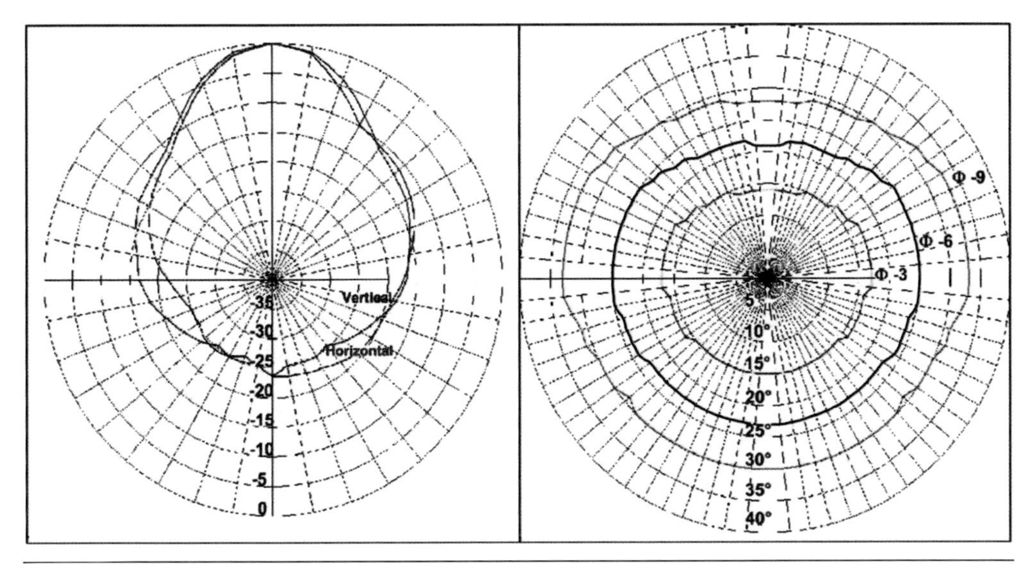

Figure 12.30 Plot of the directional gain and beam width curves of the radiator DML-1122 (Electro-Voice) frequency 2 kHz; front to random index 15 dB; max. sound level at 1 m distance: 125 dB—(a) directivity characteristic and (b) coverage area.

Because of the curves of equal directional gain and the sound distribution loss, the impact of the direct sound of a loudspeaker on a surface may produce elliptic curves that represent a calculated SPL isobar area of the direct sound coverage. In the planning of sound reinforcement systems, these isobar areas are important as coverage areas.

Another important parameter for approximating the sound-field conditions in rooms is the *directivity factor*, *Q*. It characterizes the relationship between the acoustic power that would be radiated into the room by an omnidirectional loudspeaker having the same free-field sensitivity as the real loudspeaker to be assessed, and the acoustic power of the real loudspeaker:

$$Q=\frac{\int_S \tilde{p}_0^2\, dS}{\int_S \tilde{p}^2(\vartheta)\, dS}=\frac{S}{\int_S \Gamma^2\, dS}, \tag{12.32}$$

where:

$\tilde{p}$ is sound pressure ($\tilde{p}_0$ is measured in the main front direction)

S is the globe surface around the speaker

ϑ is a room angle

The logarithmic quantity of the directivity factor is the *directivity index (DI):*

$$\mathrm{DI}=10\log \mathrm{Q}\ \ [\mathrm{dB}]. \tag{12.33}$$

For approximate calculations or also measurements of the directivity index, one has to cover the range between 500 and 1,000 Hz at least. Many manufacturers are indicating, in the data sheets of their products, the frequency dependence of directivity.

For combining the influence of the directional effect as well as that of the distribution between directional and omnidirectional energy, one uses the product of both quantities, and we obtain the *directivity deviation factor, g(ϑ)*:

$$g(\vartheta) = Q * \Gamma^2(\vartheta) . \tag{12.34}$$

The directivity deviation factor, g, is a function of the angle, ϑ, and sometimes one can find different Q values for different angles in some textbooks. That's unfortunately very common, but quite confusing. Nevertheless, this entity, g, is angle-dependent and therefore should always be referenced along with the corresponding angle. The logarithmic expression of the directivity deviation factor, g(ϑ), is the so-called directivity deviation index, G (also angle-dependent):

$$G(\vartheta) = 10 \log g(\vartheta) \quad [\text{dB}] . \tag{12.35}$$

The reader should be aware of the partially contradicting conventions, of which some are using Q and DI only for values of ϑ = 0°, and others employ Q and DI in an angle-dependent way, sometimes without clearly stating so.

By means of the DI and the nominal power rating P_n, it is also possible to describe the characteristic sound level of a loudspeaker system:

$$L_K = L_W + DI - 10 \log P_n - 11 \quad [\text{dB}] , \tag{12.36}$$

where:

L_W is the sound power level.

The efficiency, η, of a loudspeaker system is determined by the ratio between radiated acoustic power and supplied electric power:

$$\eta = \frac{P_{ak}}{P_{el}} = \frac{E_K^2}{\rho_0 c} \cdot \frac{4\pi r_0^2}{Q_L} 100\% , \tag{12.37}$$

where:

P_{ak} is the acoustic sound power
P_{el} is the electric power applied
E_K is the sensitivity of the speaker
r_0 is 1 m distance
Q_L is the directivity factor of the speaker
$\rho_0 c$ is the characteristic acoustic impedance of air = 408 Pa s/m³ at 20°C.

By combining all constants, one obtains the following approximation:

$$\eta = 3 \frac{E_K^2}{Q_L} \quad [\%] . \tag{12.38a}$$

Both Eqs. (12.36) and (12.37) may be combined, and after some rearrangements, we obtain (see also Figure 12.31):

$$L_K = 109 + 10 \log \eta + DI \quad [\text{dB}] . \tag{12.38b}$$

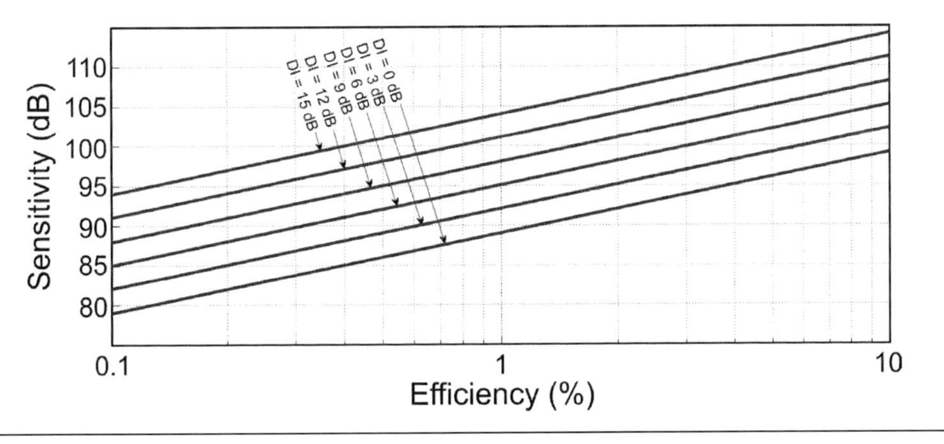

Figure 12.31 Sensitivity/dB as a function of efficiency with directivity index, DI, as a parameter.

The term, *transmission range of a loudspeaker*, is the frequency range usable or preferable for sound transmission. That region of the transmission curve in which the level measured on the reference axis in the free field, which does not drop below a reference level, generally characterizes the transmission range. The reference value is the average over the bandwidth of one octave in the region of highest sensitivity (or in a wider region as specified by the manufacturer). In the determination of the upper and lower limits of the transmission range, any peaks and dips whose interval is smaller than 1/8 octave are not considered. This definition implies that loudspeakers necessarily have to be checked as to their transmission range before being used in sound reinforcement systems.

Figure 12.32 shows an exemplary behavior of free-field sensitivity, frequency response, and DI of a loudspeaker.

Moreover, the transmission range is influenced, especially in the lower frequency range, by the installation conditions or the arrangement of the loudspeaker. This is due to the fact that arranging the loudspeaker in front of, below, or above a reflecting surface, causes interferences

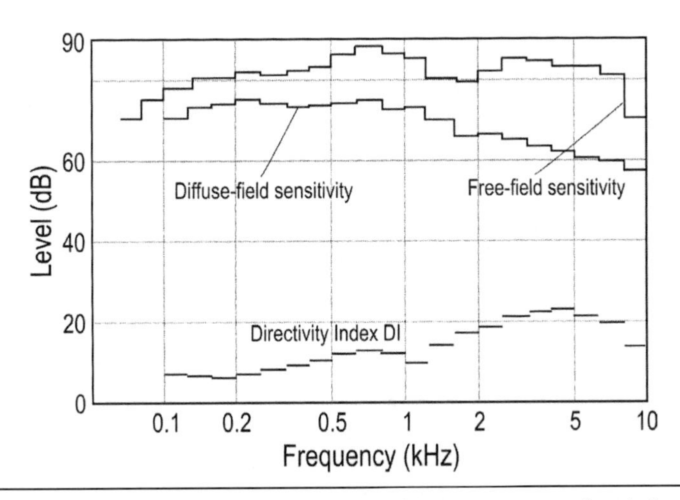

Figure 12.32 Frequency dependence of the directivity index, DI, as compared with the free-field and the diffuse-field sensitivities.

of the direct sound by the strong reflections that give rise to comb-filter-like cancellations which can be proven by a narrow-band analysis of the resulting signal.

As a rule, one can say that the ear does not normally perceive dips and peaks that are not measurable in a 1/3-octave band filter analysis (unless they show pronounced periodic structures).

A good bass radiation is produced if the radiating plane is embedded in a reflecting surface, for instance, a wall or a ceiling.

The different tasks of sound reinforcement engineering require different types of loudspeakers. These differ in size and shape of their enclosures, in the form of sound conduction, in the types of driving systems used, as well as arrangement and combinations of the same. In this way, one obtains different directional characteristics of sound radiation, sound concentrations, sensitivities, transmission ranges, and dimensions which facilitate solutions for diverse applications or even enables them at all.

The simplest speakers are single loudspeakers of smaller dimensions and ratings that are used in decentralized information systems—for instance, for covering large flat rooms or for producing room effects in multipurpose halls. Loudspeakers for such compact boxes are provided with an especially soft diaphragm suspension so that they cannot be easily used for other purposes.

Acoustically more favorable are the conditions with vented enclosures, the bass reflex boxes, or phase reversal boxes. Such box loudspeakers are used less as decentralized broadband radiators, but especially before the invention of modern line-arrays for high-power large-size loudspeaker clusters.

Another possibility for achieving a determined directional characteristic consists in the arrangement of sound-conducting surfaces in front of the driving loudspeaker system. Given that such arrangements are mostly of horn-like design, they are named *horn loudspeakers.* Because of the high characteristic sensitivity and the high directional characteristic, this loudspeaker design was very well suited for sound reinforcement in large auditoriums where the desirable frequency range and different target areas (coverage areas) require the use of different types of radiation patterns. Today line-arrays are used more in that situation. Different types of horns may be distinguished:

- **Bass horns.** Owing to the great dimensions involved, the design of bass horns requires extensive compromises. Practical models of bass horns receive a horn shape, as a rule, only in one dimension, whereas at a right angle to it, sound control is achieved by means of parallel surfaces. The power-handling capacity of such bass horns, which are mainly used in music or concert systems, is about 100 to 500 VA.
- **Medium-frequency horns.** The greatest variety of driver and horn designs is available for horn loudspeakers for the medium frequency range of about 300 Hz to 3 kHz. The drivers used are mostly dynamic pressure-chamber systems connected to the horn proper by means of a throat—the so-called throat-adapter.
- **Treble horns.** For the upper frequency range, two main types of horn loudspeakers are produced. These are the horn radiators showing similar design characteristics as the medium-frequency horns that function in the frequency range from 1 to 10 kHz and the special treble loudspeakers (calotte horns) used for the frequency range from 3 to 16 kHz.

12.3.2.2 Sound Columns

In the U.S. in the 1930s, horn speakers that were especially developed for the movie industry came into use. In contrast to this development, so-called sound columns had been designed first, in Europe. Sound columns consist of cone speakers (woofers) brought in a line arrangement and connected to the same signal. To have an effect, every single speaker must have a determined directivity. Each of the individual phase-identical loudspeakers radiates the sound spherically and the sound waves get favorably superposed in the far field, whereas the effect of the individual loudspeaker prevails in the near field. In this way you obtain columns that are capable of producing a high sound level at a large distance from their point of installation, while minimally affecting microphones located at close range to them. For the far field the following equation was given by Stenzel[40, 41] and Olson[42] for the directional factor, Γ, the so-called polars:

$$\Gamma = \frac{\sin\left[\frac{n\pi d}{\lambda}\sin\gamma\right]}{n\sin\left[\frac{\pi d}{\lambda}\sin\gamma\right]}, \tag{12.39}$$

where:

n = number of individual loudspeakers
d = spacing of the individual loudspeakers
γ = radiation angle
λ = wavelength of sound
l = (n − 1) d = length of the loudspeaker line

Figure 12.34 illustrates this directional effect of a loudspeaker line according to Figure 12.33 (Figure 12.34a—balloon 1k and Figure 12.34b—balloon 2k).[43] The line consists of nondirectional loudspeakers arranged with a spacing of 30 cm (12″ woofer).

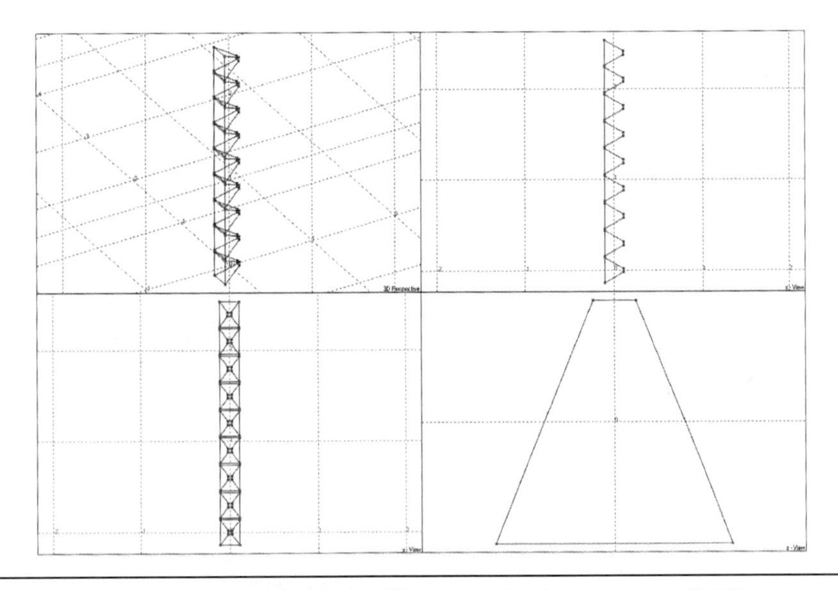

Figure 12.33 A line presentation with 9 Horns HP64 in the simulation program EASE.

A.

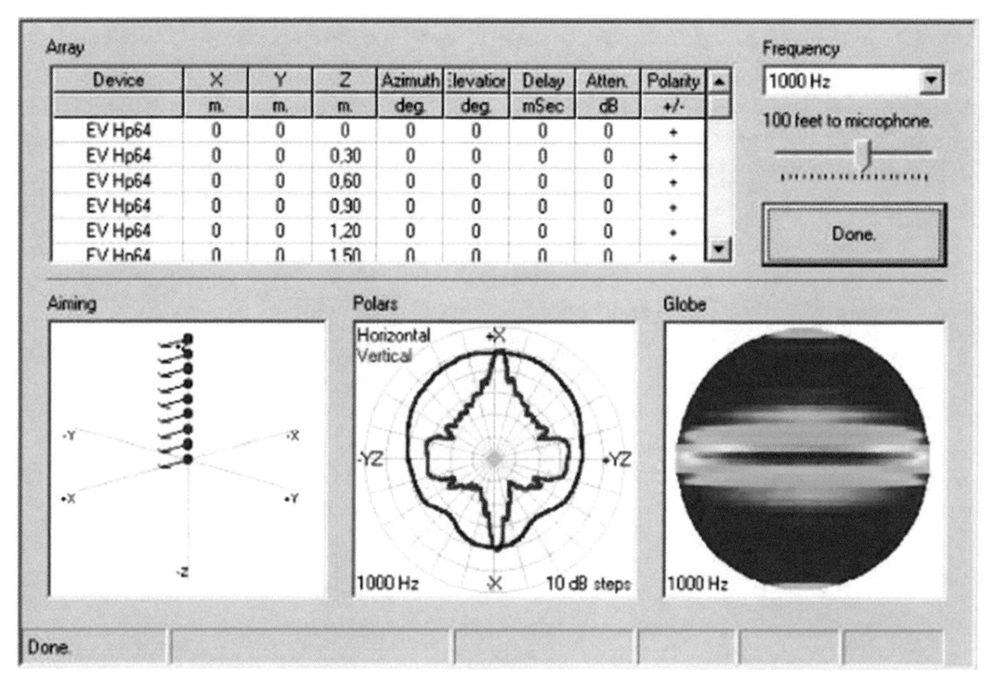

B.

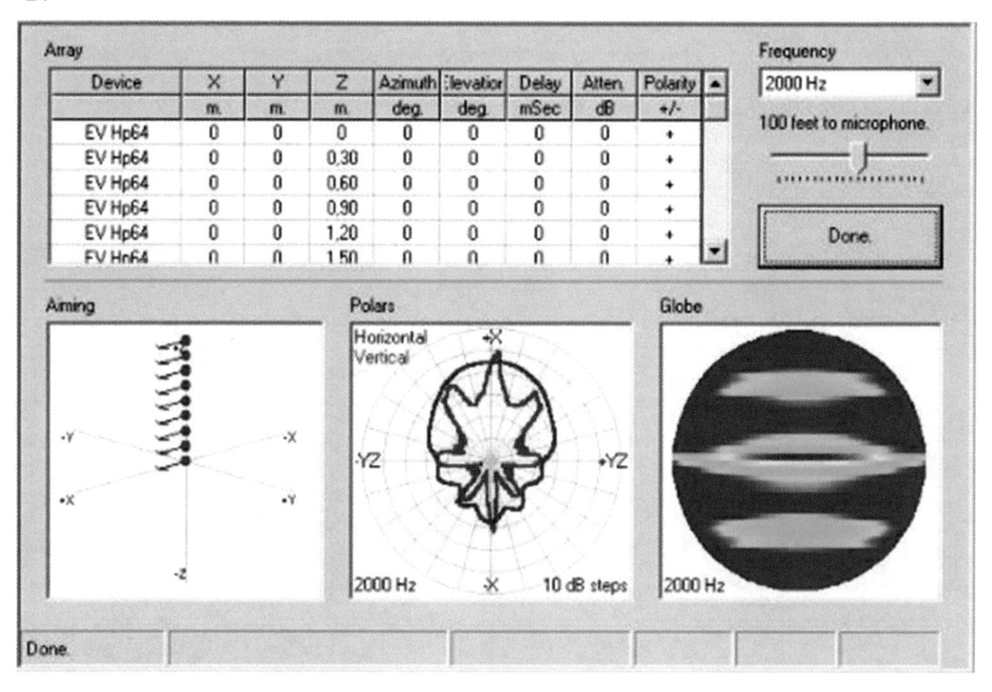

Figure 12.34 Balloon presentation of the line array according to Figure 12.33 in a simulation: (a) 1,000 Hz (b) 2,000 Hz.

The drawbacks of a sound column arrangement are as follows:

- The desired directional effect is given only in a range below a construction depending frequency, whereas above that frequency, additional secondary maxima occur.
- The directivity is frequency-dependent (directivity factor, Q, of the main maximum ≈ 5.8lf[44] (l is the length of the column in m and f the frequency in kHz).
- The directivity increase does not only occur in the *directivity domain* (main direction), but also, owing to the distances of the individual loudspeakers, in the *scattering domain* (radiation to areas not in the main direction), so the column is losing directivity at high frequencies.
- The effect of the nonomnidirectional radiation of the individual elements. This also causes lobing effects.

All of these frequency-dependent properties of the loudspeaker lines involve the possibility for timbre changes to occur over the width and depth of the covered auditory. Some manufacturers have developed sound columns with very tiny cone speakers. This way, the occurrence of side lobes is moved up to frequencies higher than 4 kHz (outside the speech range).[45]

12.3.2.3 Line Arrays

Modern line arrays do not consist of a lineup of individual cone speakers, but rather, of a linear arrangement of wave-guides of the length, l, which produce a so-called coherent wave front. In contrast to the traditional sound columns, these arrays radiate so-called cylindrical waves in their near field. This near range field is frequency-dependent and only valid up to the following distances, d:

$$d=\frac{l^2}{2\lambda}, \tag{12.40}$$

where both array length, l, and wavelength, λ, are in meters.

The characteristic feature of these systems is that the sound levels decrease in the near range by only 3 dB with distance doubling, and begin to decrease like those of spherical radiators only beyond the near range field.[46] This way it is possible to cover large distances with high sound levels and without having to use delay towers (see also Figures 12.35 and 12.36).

Figure 12.35 Photo of a Geo T series column by NEXO SA.

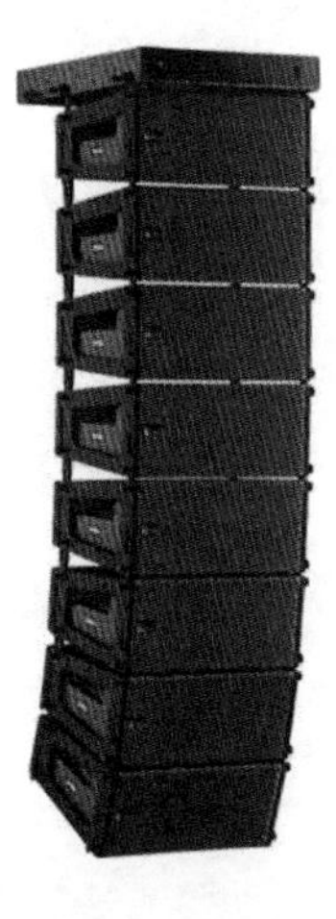

Figure 12.36 Photo of a MINA array by Meyer Sound Inc.

12.3.2.4 Digitally Controlled Line Arrays

A way of reducing the frequency dependence of the directivity of classic sound columns consists of supplying the sound signal with different phases and levels to the individual loudspeakers in an array. Practically, with increasing frequency the acoustic active length, l, of these line arrays is reduced by electronic means.[47, 48] Figure 12.37 illustrates different distances in the directional behavior in a three-dimensional representation. For the near field, the directivity effect will disappear (see Figure 12.37c).

Currently, a lot of manufacturers design such columns very successfully. The performance may be optimized with the following parameters of the columns:

- opening angle of the beams
- aiming angle of the beams
- focus distance

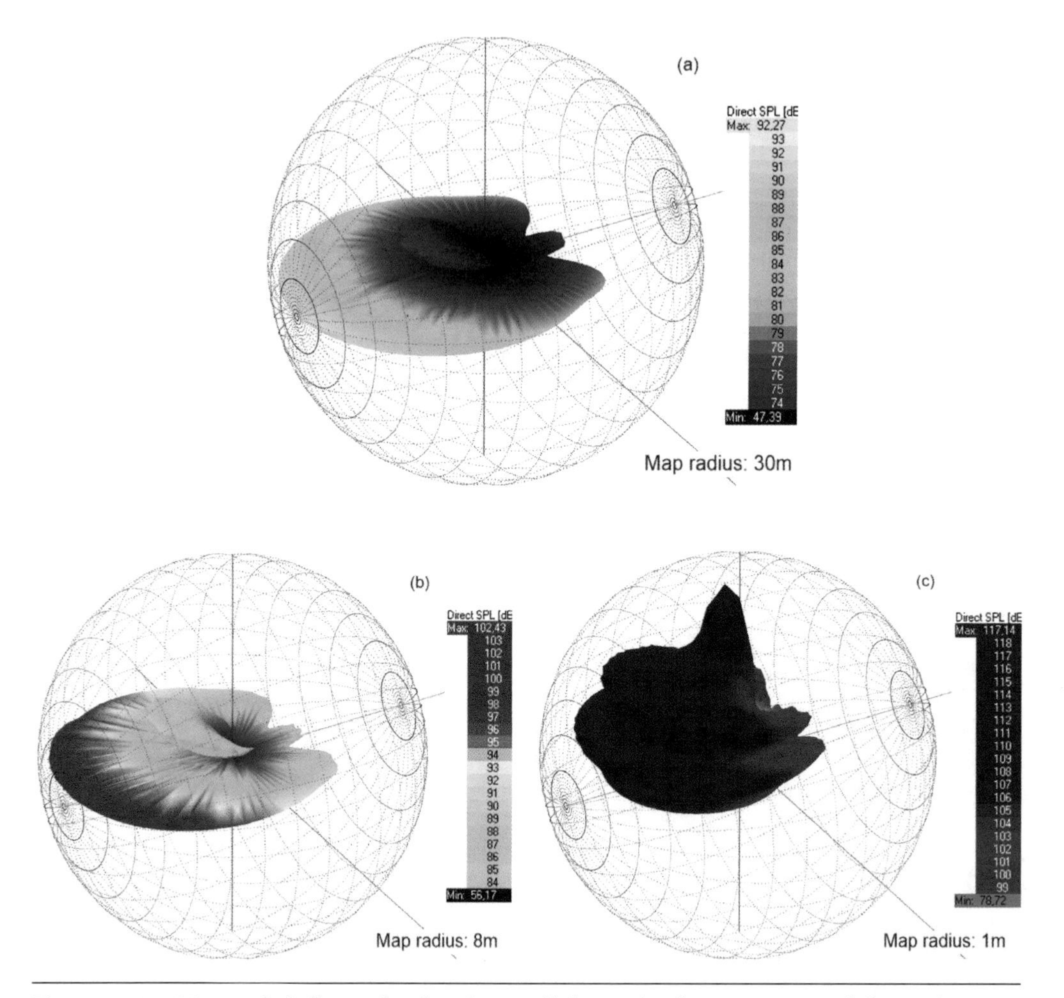

Figure 12.37 Magnitude balloon of a digital controlled sound column DC280 in different distances, analyzed at a 2 kHz 3rd octive band: (a) 30 m, (b) 8 m, and (c) 1 m.

- mounting height with respect to the audience area
- number of beams (for complex seating areas)

Further improvements in designing the directivity patterns of an array allow their adaption to existing audience areas; a uniform SPL distribution also becomes possible in complex-shaped audience areas.[49, 50]

12.3.3 Receivers and Microphone Systems

12.3.3.1 Acoustic Evaluation with Human Ears

The properties of human ears are explained in many books about psychoacoustics.[32, 33] In the real world we determine subjectively the acoustics of a space with our two ears. Over the years we've learned to distinguish between good and poor acoustic properties and evaluate correspondingly. On the one hand, the human being is also influenced by so-called master-evaluations. Some leading musicians may praise a new hall and the normal concert visitors will mainly do the same. On the other hand, a new hall may be talked down this way, although the hall does not have any poor acoustic properties—only because a conductor was unhappy with his own performance there.

Here, auxiliary measurements should be made to figure out the supposed deficits of the space. Acoustic measurements deliver impulse responses, and all the acoustic measures derived from these measurements should be compared with their own subjective evaluation. Also, subjective group tests during concert rehearsals in different seat areas may be conducted, so that afterwards, an acoustic judgment can be achieved. This will be the starting point for any serious considerations of a new sound system design. Without knowing the room acoustic properties of a space (hall or open area), a new sound system design may fail completely.

In the case of a new construction, acoustic evaluation or measurements are impossible. In all larger important spaces, acoustic simulations should be performed beforehand. In all such prediction programs the acoustic properties of a room or the free field environment are determined, not by experimental measurements, but by the numerical calculation of the impulse response. This response is calculated using different ray-tracing methods. For a single receiver point in space, the so-called monaural response is determined, and the result supplies not only the level at this receiver place, but also the frequency dependence, the angle of incidence for single reflections, and the run-time delay in comparison to the first incoming signal (direct sound). Using head-related transfer functions (HRTFs) measured with artificial heads or using in-ear microphones (compare Figure 12.38),[51] the binaural impulse response can be obtained. Acoustic measures may be derived, and they are used for the room acoustic evaluation of the future hall. Unfortunately, most of the room acoustic measures characterizing the acoustic properties of a space are monaural measures. For psychoacoustic phenomena like spaciousness or strength, binaural measures reflecting the subjective perception have not yet been introduced.

12.3.3.2 Microphones

The use of microphones in sound reinforcement systems requires observation of a number of conditions. To avoid positive acoustic feedback, it is frequently necessary to keep the microphone at a closer distance to the sound source, so that, often, considerably more microphones have to be used. Moreover, live conditions demand very robust microphones.

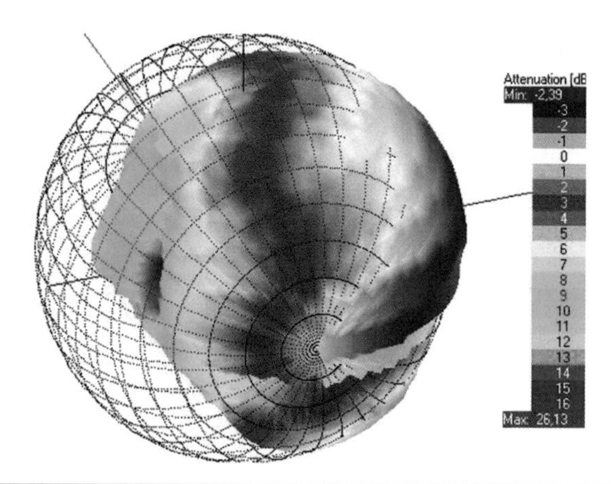

Figure 12.38 Head-related transfer function (HRTF) balloon of a left ear of an artificial head.

The microphone data are laid down in standards.[52] If sound systems must be designed, we have to know the acoustic properties of the space (see previous text). This way we do measurements with microphones having different properties as microphones for recordings. In Chapter 5 the different approaches and methods to perform room acoustic measurements are laid out and will not be repeated here. We understood that artificial heads, omnidirectional, and figure-eight microphones are used to acquire the required acoustic measures.

In this context, we will consider only those data which are important for computer simulation. To simulate the use of microphones, to precalculate the acoustic feedback threshold, or to simulate enhancement systems based on electronic processing requires an exact knowledge of the properties of the microphone types and their connection techniques. The following subsections explain only basic parameters.

12.3.3.2.1 Sensitivity

The magnitude of the output voltage of a microphone as a function of the incident sound pressure is described by the microphone sensitivity:

$$T_E = \frac{\tilde{u}}{\tilde{p}}, \tag{12.41}$$

in V/Pa or its 20-fold common logarithm, the sensitivity level:

$$G_E = 20\log\frac{T_E}{T_0}\,dB. \tag{12.42}$$

The reference sensitivity T_0 is normally specified for 1 V/Pa.

12.3.3.2.2 Directivity Behavior

The dependence of the microphone voltage on the direction of incidence of the exciting sound is called directional effect. The following quantities are used for describing this effect:

- The directional factor, $\Gamma(\vartheta)$, as the ratio between the (free-)field sensitivity, T_{Ed}, for a plane sound wave arriving under the angle, ϑ, to the main microphone axis and the value ascertained in the reference level (incidence angle 0°).

$$\Gamma(\vartheta)=\frac{T_{Ed}(\vartheta)}{T_{Ed}(0)}. \tag{12.43}$$

- The directional gain, D, as the 20-fold common logarithm of the directional factor, Γ
- The coverage angle as the angular range within which the directional gain does not drop by more than 3 dB (or 6 dB or 9 dB) against the reference axis

The relationship between the sensitivities by the reception of a plane wave and those with diffuse excitation characterizes the suppression of the room-sound components against the direct sound of a source. This energy ratio is described by the following parameters:

- The directivity factor, Q_M: If the sensitivity was measured in the direct field as T_{Ed} and in the diffuse field as T_{Er}, the directivity factor results as:

$$Q_M=\frac{T_{Ed}^2}{T_{Er}^2}. \tag{12.44}$$

- The directivity index as the 10-fold common logarithm of the directivity factor

While the directivity factor of an ideal omnidirectional microphone is one, that of an ideal cardioid microphone is three. This means that a cardioid microphone picks up only 1/3 of the sound power of a room compared to an omnidirectional microphone at the same distance from the source. This implies, for instance, that for an identical number of the sound power, the speaking distance for a cardioid microphone may be $\sqrt{3}$ times greater than that of an omnidirectional microphone (compare analogously Eq. (12.5a)).

12.3.4 Sound Processing Equipment

Various sound processing devices used in studio technique are also of special importance for sound reinforcement engineering. Here they are used for improving intelligibility, for reducing the risk of positive feedback, for adapting a reproduced sound pattern, and for other purposes.

12.3.4.1 Delay Equipment

Frequently, only delay devices enable a proper sound reinforcement by delaying electrical signals in such a way that echo disturbances do not occur even between widely-spaced loudspeakers. Some solutions for sound reinforcement problems realized by means of the delay technique will be explained in Section 12.5.2.2; here, however, we are going to present only the technical design of the devices.

Modern delay devices in studio quality rely extensively on computational processing of the digitized signal. These devices enable an automated control of the delay times, as well as a noise-free changeover during operation, thanks to the subdivision of the delay times into very small quantization steps. Figure 12.39 depicts the basic structure of such a delay device.[53]

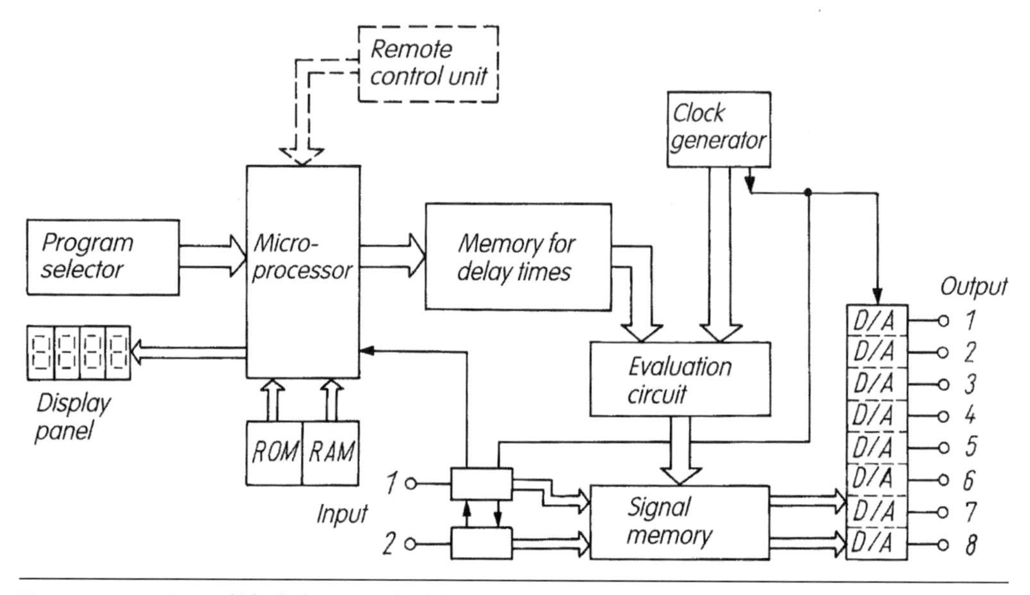

Figure 12.39 General block diagram of a delay unit with computerized signal processing.

12.3.4.2 Effect Devices

By bit completion or bit omission, and by changing the clock frequency, the procedures used in the digital delay technique may be employed for realizing diverse acoustical effects. Decreasing or increasing the clock frequency may produce a lower or a higher pitch, respectively. According to Webers[54] it is possible (in this way) to achieve *pitch variations* of up to 3 octaves (+1 to −2). Modern devices allow the possibility of a controlled dose mixing of the delayed and the undelayed signal. Moreover, it is possible to feed back a portion of the signal for signal iterations and echo effects.

Insignificant pitch shifts were used in the so-called *feedback suppressors* for reducing the positive feedback propensity in sound reinforcement systems. Because of the resulting timbre changes, however, such devices are hardly used currently.

The described processing devices are especially used in music electronics for realizing the phasing, flanging, and Leslie effects. All of the aforementioned effects can presently be realized by means of digital effect devices similar to those that have just been described.

12.3.4.3 Reverberation Equipment

Contrary to a delay device that merely repeats a signal with delay but without changing amplitude and phase, a reverberation device is expected to confer upon the original signal additional and delayed repetitions that, moreover, are to be incoherent with the original signal and grow denser as well as increasingly attenuated in the course of time. The duration of the decay process thus produced is, if possible, to be variable. The same applies to its frequency dependence, for which a faster decay is normally expected for the higher frequencies than for the lower ones.

The length of the audible decay process is the subjectively perceived *reverberation duration.* The time lapse required by a signal for decaying from an original level to a value 60 dB smaller is by definition called T_{60}, even if the decay process was previously swamped by the noise. Reverberation equipment should allow varying the T_{60} of a signal in a room from 1 to 10 sec.

To avoid timbre changes, a minimum density and uniformity of distribution of the eigenfrequencies are required. In this respect it must be considered that the human ear shows the highest sensitivity for timbre changes between 800 Hz and 1 kHz.

In sound reinforcement engineering, reverberation equipment is also needed for generating a so-called *reverberation tail.* Another application consists in directly reverberating picked-up signals—as required in sound reinforcement for avoiding positive feedbacks—in such a way that they can be perceived in an aesthetically satisfactory fashion.

Modern computerized reverberators[55] not only influence the frequency response of the T_{60}, but also delay the reverberation and insert various differently damped single reflections between the triggering sound signal and the reverberation process (see Figure 12.40). Some devices, e.g., TC electronic reverb 4000 or Lexicon 960, enable processor control not only of the decay process, but also of the overall room simulation, coupled in many cases with an effect device. They thus enable effects like, among others, extremely long and frequency-independent T_{60}s which cannot be realized by means of natural rooms or any older mechanical reverberation devices. So-called freeze effects allow constant maintenance of a sound with superposition of newly added sounds. Moreover, it is possible to realize decreasing echo loops, stereo phasing, etc.

By choosing the appropriate software, the modern computer-controlled equipment allows the achievement of very natural and adaptable settings which correspond to all requirements, including the naturalness of the reverberated sound pattern.

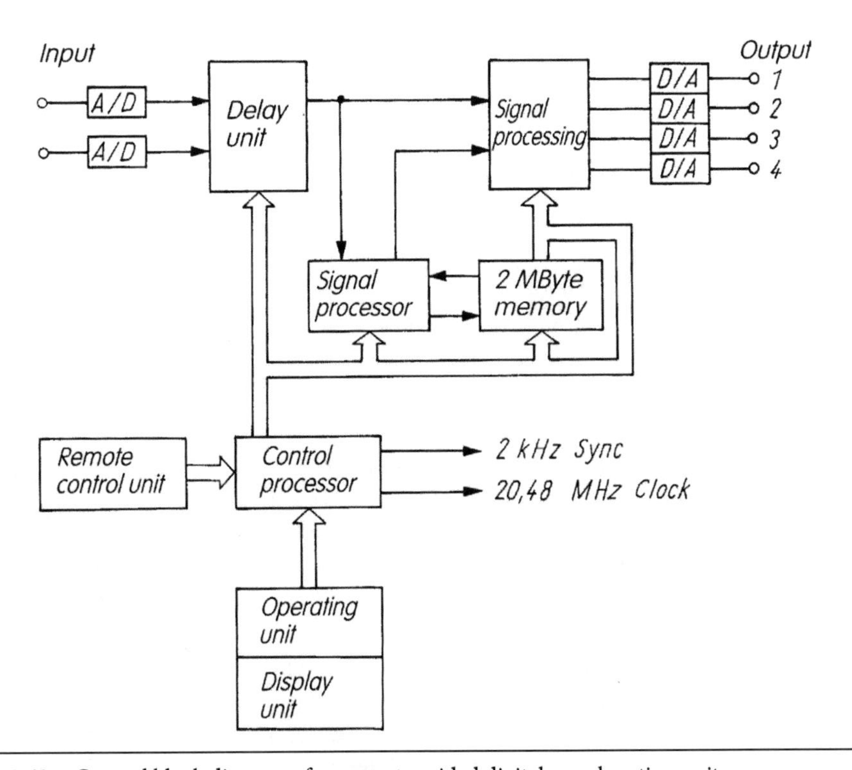

Figure 12.40 General block diagram of a computer-aided digital reverberation unit.

12.3.4.4 Feedback Suppressor

Every electric channel or system containing active elements may show feedback behavior. This is caused by the reaction of any output signals of a system to its input. The excited oscillation is sine wave-like and in case of an acoustic system or channel, we perceive an increasing sine signal in the form of howling or whistling.

In the case of the acoustic feedback we observe some specialties:

- The feedback loop of an electroacoustic amplifier channel consists not only of an electrical, but also of an acoustic part.
- Practically, it is impossible to split the feedback path in different parts (e.g., in the electroacoustic and the room-acoustic part).
- The feedback happens in a lot of different loops and paths; therefore, the nature of the acoustic feedback is more complicated as in pure electric networks.

To avoid acoustic feedback, the user or installer of sound systems needs to know the physical background of acoustic feedback. He has to know the basics of how to arrange microphones and loudspeakers in relation to each other. The loudness of monitor speakers on stage, often set very high, may produce problems that happen if a singer with a wireless microphone is performing in front of these monitor speakers.

Another possibility to reduce the probability of acoustic feedback is paying attention to the secondary structure of the space, including wall or ceiling parts that are close to microphones or loudspeakers that will be in use. Wall areas covered by absorbers reduce the occurrence of acoustic feedback caused by otherwise strong reflections there. Also, sound focusing effects in round spaces or concert shells must be avoided as a source for feedback support.

Normally a sound engineer does not have too much influence on the wall or ceiling design of a space, but certain knowledge of the character of acoustic feedback will help to reduce this unpleasant behavior. If such simple methods don't help to avoid acoustic feedback, some technical procedures like filters, frequency, and phase shifting, and other feedback suppressors may be used.

12.3.4.4.1 Use of Narrow Band Filters

The transmission curve of a space shows statistic irregularities with dips and peaks.[56] Investigations over the years have shown that such transmission curves display up to 70 eigen oscillations. Among these peaks up to 3 to 40, on average 12 … 25 oscillations may lead to acoustic feedback.

Automatic notch filters have been used to find the dangerous frequencies. Digital signal processing (DSP) allows flexibility in terms of frequency detection as well as frequency discrimination and the method of applying notches. Auto-notching is found more frequently among pro-audio users than any other method, because it is easier to manage the distortion. This way an escalating of frequency peaks will be prevented, the frequency response curve is smoothed, and the overall amplification of the sound system can be enhanced by the reduced difference between the smoothed peak and the average value of the curve.

The bandwidth of such notch filters is around 5 Hz, the attenuation is frequency-dependent—normally 3 … 30 dB. A professional feedback suppressor may enhance the loop gain until 5 … 8 dB. In this way, the level of a sound system may be increased on these values before feedback occurs (gain before feedback).

12.3.4.4.2 Frequency Shifter

Above a volume-depending frequency, the transmission curves in spaces show pure statistic properties.[57] Schroeder found that the average frequency distance between neighbored peaks and dips of a transmission curve is proportional to $4/T_{60}$. By means of a frequency shifter, the peaks and dips are shifted to overlapping positions and the response curve is smoothed this way. A loop gain of 6 … 8 dB can be reached practically. Frequency shifters are now used mainly for speech transmissions; pitch shifting in the case of music performances is critical and is not acceptable.

12.3.4.5 Filters

Filters for influencing the amplitude-frequency characteristic of the transmitted sound signals are among the classical means of sound processing used in sound reinforcement engineering. Two main fields of application have to be distinguished:

- Optimization of timbre within the reception area concerned, and
- Suppression of acoustic positive feedback frequencies.

The timbre optimization depends on the scheduled field of application of the system. A balanced frequency response over the entire frequency range may, for instance, be desirable for high-quality systems designed for music transmissions. For improving intelligibility in mere speech transmission systems, a reduction in the lower frequency range and an enhancement of certain formants in the range of about 2 kHz is, however, appropriate.[58] Figure 12.41 shows the recommended frequency response curves for different speech or music performances.[59]

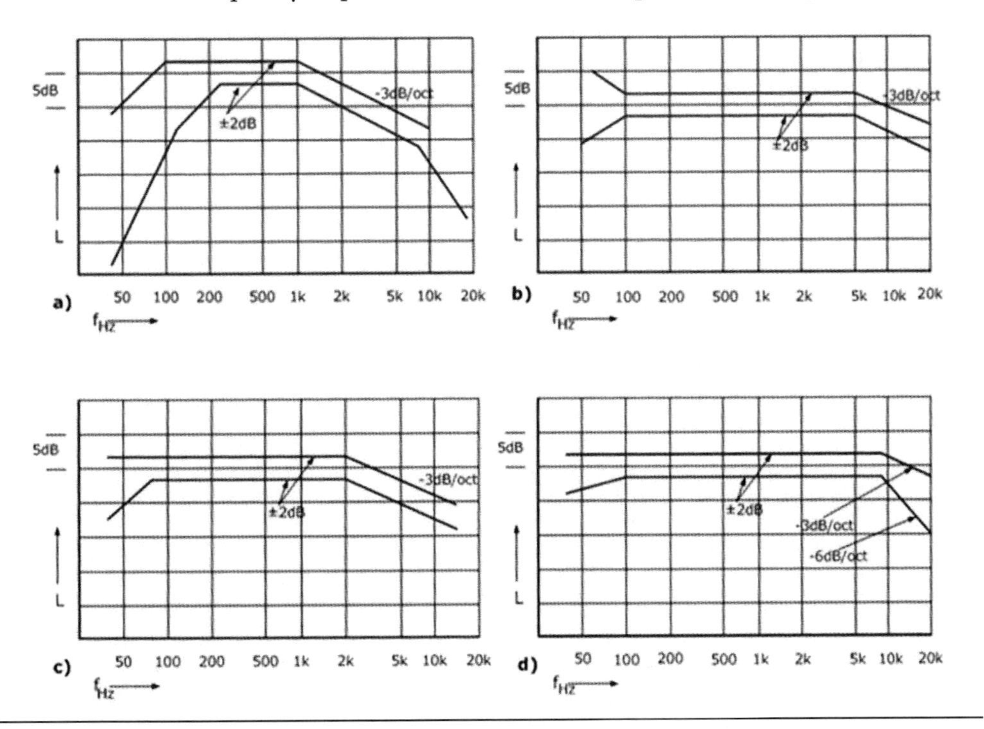

Figure 12.41 Tolerance curves for the reproduction frequency response in different applications: (a) recommended curve for reproduction of speech, (b) recommended curve for studios or monitoring, (c) international standard for cinemas, and (d) recommended curve for loud rock and pop music.

Quite different requirements may be decisive for optimizing the timbre of stage monitoring. In larger sound reinforcement systems there are filters used at various points. They are normally located in the input channel of the mixing console for influencing the microphone frequency response and, in most cases, also for suppressing the most essential positive feedback frequencies.

It has to be pointed out that the use of filters is nearly always at the expense of the maximum realizable sound level. For this reason, it is necessary to consider corresponding power reserves when designing the system.

One distinguishes *passive filters,* which operate without an additional power supply and thus do not offer any amplification possibility (level enhancement) nor any reduction of the signal-to-noise ratio, from *active filters,* which, thanks to their more universal applicability, their smaller dimensions, and their lower price, are presently used almost exclusively in studio equipment.

Another distinguishing characteristic is the influence of damping on the behavior of the filter curve. In this respect one distinguishes between filters of constant bandwidth and filters of constant quality, q (see Figure 12.42).

For reasons of expenditure, available equipment, and ease of operation, very different practical designs are used:

- Treble and bass correctors (shelf filters)
- Parametric filters (channel filters)
- Multi-bandpass filters (equalizers)
- Preset filters

Under the name *feedback controller,* microprocessor-controlled units have recently become available that trigger frequency-dependent attenuation in response to incipient positive feedback phenomena—such as timbre changes, reverberation effects, and fluctuations in level. These automatic filters are assigned to the individual microphone channels and are capable of enhancing the positive feedback limit of a system by up to 15 dB.[60]

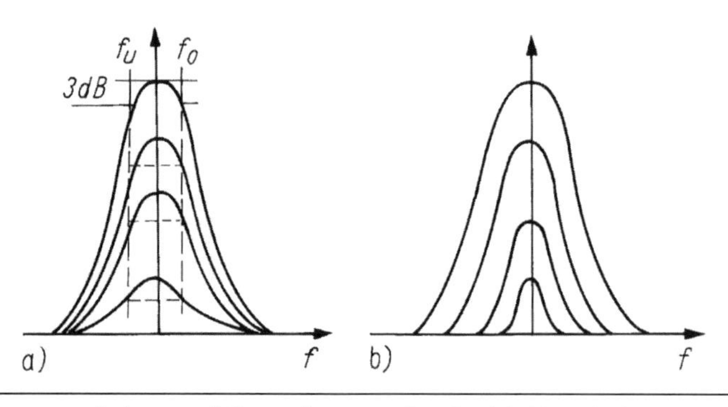

Figure 12.42 Attenuation behavior of filters of constant bandwidth (a) and of constant quality (b).

12.4 CALCULATION OF SOUND REINFORCEMENT SYSTEMS

Any sound reinforcement system is presently confronted by a great number of general demands; the focal points of which vary according to the planned use. Certain basic requirements, however, have to be complied with in all cases:

- The sound level produced by the system within the audience area of the hall or in the open air must be sufficient. The following quantities are related herewith:
  - Loudness (expectation value, adaptation to original sources),
  - Sound energy density (efficiency of the sound reinforcement system), and
  - Ratio between disturbing sound and useful sound, etc. (signal-to-noise ratio, dynamic range of reproduction).

The sound level distribution, which enables assessing the spatial distribution of the first-mentioned parameter, must be sufficiently uniform. It also determines the localization of the sound events and sources in the space. It depends on:

- The arrangement of the loudspeakers,
- The directivity characteristic of the loudspeaker, and
- The diffusivity of the room in which the reproduction takes place.

The clarity of reproduction must be up to the planned use. It depends on the following factors:

- Definition, clarity (R/D ratio),
- Masking of the reproduction signal, and
- Freedom from echoes.
- It is necessary to obtain a sufficient naturality of the transmission or in other words a perceptibility of a desired timbre change. Contributing factors in this respect are:
  - The timbre depending on the transmission range and the frequency response of the signal transmitted,
  - The frequency response, and
  - Freedom from distortions.

Sound reinforcement systems must be sufficiently insensitive to positive acoustic feedback. This implies demands concerning:

- The level of loop amplification,
- The level conditions around the microphone, and
- The directivity characteristic of the microphones and loudspeakers.

The mathematical background and technical conclusions to be drawn in this respect will be briefly discussed now.

12.4.1 Analytic Sound Level Calculation

12.4.1.1 Free Field (Direct Field of the Loudspeaker)

For calculating the sound level in the free field, one has to use the loudspeaker parameters valid for the free field. These include the characteristic sensitivity/dB, or in other words the characteristic sound level, L_K, according to Eq. (12.29), as well as the directional factor, Γ_L, (ϑ) according to Eq. (12.30), or the directional gain, D, according to Eq. (12.31).

The direct sound level in the direct field of a loudspeaker at a certain location at a distance of r_{LH} from the loudspeaker (with r_0 = 1 m) and at the angle, ϑ from the reference axis results as:

$$L_d = L_K + 10\lg P_{el} - 20\lg r_{LH} + 20\lg\Gamma_L\left(\vartheta_H\right);\ [dB]. \tag{12.45}$$

The 1 m-1W level is then $L_{d,1m,1W} = L_K$.

For larger distances (greater than 40 m), one still has to consider the meteorological propagation loss, $D_r = D_{LH}$, according to Figure 12.14. Eq. (12.45) is then transformed to:

$$L_d = L_K + 10\lg P_{el} - 20\lg r_{LH} + 20\lg\Gamma_L\left(\vartheta_H\right) - D_{LH};\ [dB]. \tag{12.45a}$$

The characteristic sensitivity as well as the directivity ratio and the propagation loss are frequency-dependent. To obtain the characteristic parameters of broadband loudspeakers one uses, therefore, an averaged value for the range between 200 Hz and 4 kHz. This range is generally recommended in most of the standards.

The aforementioned method for calculating the sound level to be expected in the free field is generally applicable to outdoor systems. The free-field propagation is also of interest for indoor sound systems. The sound level of the direct-field component of sound systems can be estimated using the aforementioned method as well. Moreover, the sound level of loudspeakers used in sound systems in theaters can also estimated this way (loudspeakers must be installed in the backstage area). In this case, one must assume that only the directly irradiated sound components passing the first stage portal (and in some cases also the second one) become effective in the audience, whereas all room reflections are absorbed in the stage house itself or become noticeable only as a long-delayed noise floor.

If in special cases one knows only the sound power level, L_W, of the loudspeaker and his directivity index, DI = 10 lg Q_L [dB], it is possible to calculate the required characteristic sound level, L_K, according to Eq. (12.36) and by means of $L_W = 10\lg(P_{ak}/P_0) = 10\lg(P_L/P_0)$ [dB], we obtain Eq. (12.38).

12.4.1.2 Diffuse Field

If the loudspeaker is located in a room, there exists a more or less homogeneous diffuse sound field. The homogeneity of this sound field depends on the size and shape of the excited room. For this reason, calculation of the sound level in this field is often carried out since the result provides an indication as to the value of the sound level minimally achievable in this room under given conditions.

Contrary to the calculation of the sound level in the direct field, the sound power, P_{ak}, of the loudspeaker(s) located in the room is of great importance. This power and the equivalent absorption area, A, of the room determine the sound pressure and thus, also the SPL:

$$\tilde{p}_r = \sqrt{\frac{\rho c P_{ak} \cdot 4}{A}}. \tag{12.46}$$

In the usual level notation, the diffuse sound level results from Eq. (12.46):

$$L_r = L_W - 10\lg A + 6 \quad [dB], \tag{12.47}$$

L_W is the sound-power level of the loudspeaker and can be derived from Eq. (12.36).

By means of Eqs. (12.47) and (12.36) it is possible to express the diffuse sound level in the following form:

$$L_r = L_K + 10\lg P_{el} - 10\lg A - DI + 17 \quad [dB]. \tag{12.48}$$

12.4.1.3 Real Rooms

In enclosures the sound field stems from the diffuse sound as well as the direct sound of the loudspeakers. By combining Eqs. (12.45) and (12.48) the sound level to be expected in the room at the distance r_{LH} is:

$$L = L_K + 10\lg P_{el} + 10\lg\left(\frac{\Gamma_L^2(\vartheta_H)}{r_{LH}^2} + \frac{16\pi}{Q_L A}\right) \quad [dB]. \tag{12.49}$$

This equation can be applied, however, only if there is one loudspeaker used or if the loudspeakers are arranged in a very concentrated form, as is the case, for instance, with a monocluster. Here the directional factor, Γ_{ges}, and the directivity factor, Q_{ges}, of the array must be known.

The transition from the spatially limited free-field behavior of the loudspeakers and the diffuse-field behavior of the same is characterized by the critical distance (see Section 12.1.2.1, Eq. [12.5a]). This critical distance will be reduced in the case of several loudspeakers arranged at a greater distance from each other.

12.4.1.4 Conclusions for the Practice

For rough calculations of the achievable sound reinforcement in enclosures and in open spaces, it is sufficient to consider only one reinforcement channel: that is, the one whose loop amplification is nearest to the feedback threshold. Below the feedback threshold of the critical channel, sound level values do not normally have higher results, neither with *n* reinforcement channels. Sound level distribution will, of course, mostly be better than with only one channel or with a centralized loudspeaker arrangement in the middle of the room or in the middle of an open-air auditorium.

But a more exact calculation of the sound level values can be realized only by means of a computer simulation program that excludes more or less approximations and considers exactly the interactions that exist between the different operating quantities. To get an overview, the above algorithm in this Section (12.4.1) will suffice.

Therefore, the use of prediction programs is indicated if more than one microphone or one loudspeaker is in use. Here a computer model with all the room-acoustic properties of the space must be built, the loudspeaker or line arrays placed, and in a very short time, calculated impulse responses or mapping figures will show the desired results. This will be explained in the following paragraphs.

12.4.2 Basic Tools and Parameters for Computer-Based Calculations

Nowadays the more sophisticated way to design a sound system is through verifying the intended design by prediction software. Any sound designer has a certain mental picture of the sound system. This is based on his experience and the knowledge he has to handle the job. He brings a preliminary design to paper and will discuss it with the client. After confirmation, the performance of the system must be proven, and this is done best with simulation programs. Such programs are, among others, Odeon,[61] CATT Acoustic,[62] EASE,[63] or Bose Modeler.[64] Common among these programs is the fact that it is necessary to have a full 3-D model and databases for wall materials and speaker systems. Let us start to discuss the issue of creating a model.

12.4.2.1 Computer Models

To create a model is time-consuming and requires experience. Different approaches are possible to build a model:

1. Using 2-D drawings and creating the model by entering faces based on x, y, z coordinates
2. Using prototypes or preinstalled sub modules to change the coordinates correspondingly
3. Imports from AutoCAD or SketchUp
4. Imports from other simulation programs

Most newcomers in sound-design simulation believe that importing from well-known CAD platforms will solve all of their problems. The architect can provide a 3-D model, and it should be possible to import it into a simulation program. But most of the time this doesn't work without additional effort.

Figure 12.43 a–c shows the same view of a computer model in AutoCAD, SketchUp, and in the simulation software EASE.

12.4.2.1.1 Wall Materials

With the creation of a computer model, we did realize the primary structure of the space of interest (shape, size, dimensions, etc). As the next step, we have to define the secondary structure, i.e., all boundary walls of the models need to have corresponding acoustic properties, such as absorption, scattering, and diffraction. Cox/D'Antonio[65] discusses these properties in detail. Some specialties that are important to know when doing computer modeling should be emphasized here.

12.4.2.1.1.1 Absorber Data

We distinguish between random-incident absorption coefficients measured in a reverberation chamber (Standard ISO 354)[66] or data that are angle-dependent. The latter absorption coefficient is very seldom measured and only available for special applications. For computer simulation, the absorption coefficient measured in a reverberation chamber will be used. They are often available between 63 Hz and 8 kHz in 1/3rd octave or octave bands. All these data are, meanwhile, published in table form and some simulation programs have more than 2,000 materials from different manufacturers on board.

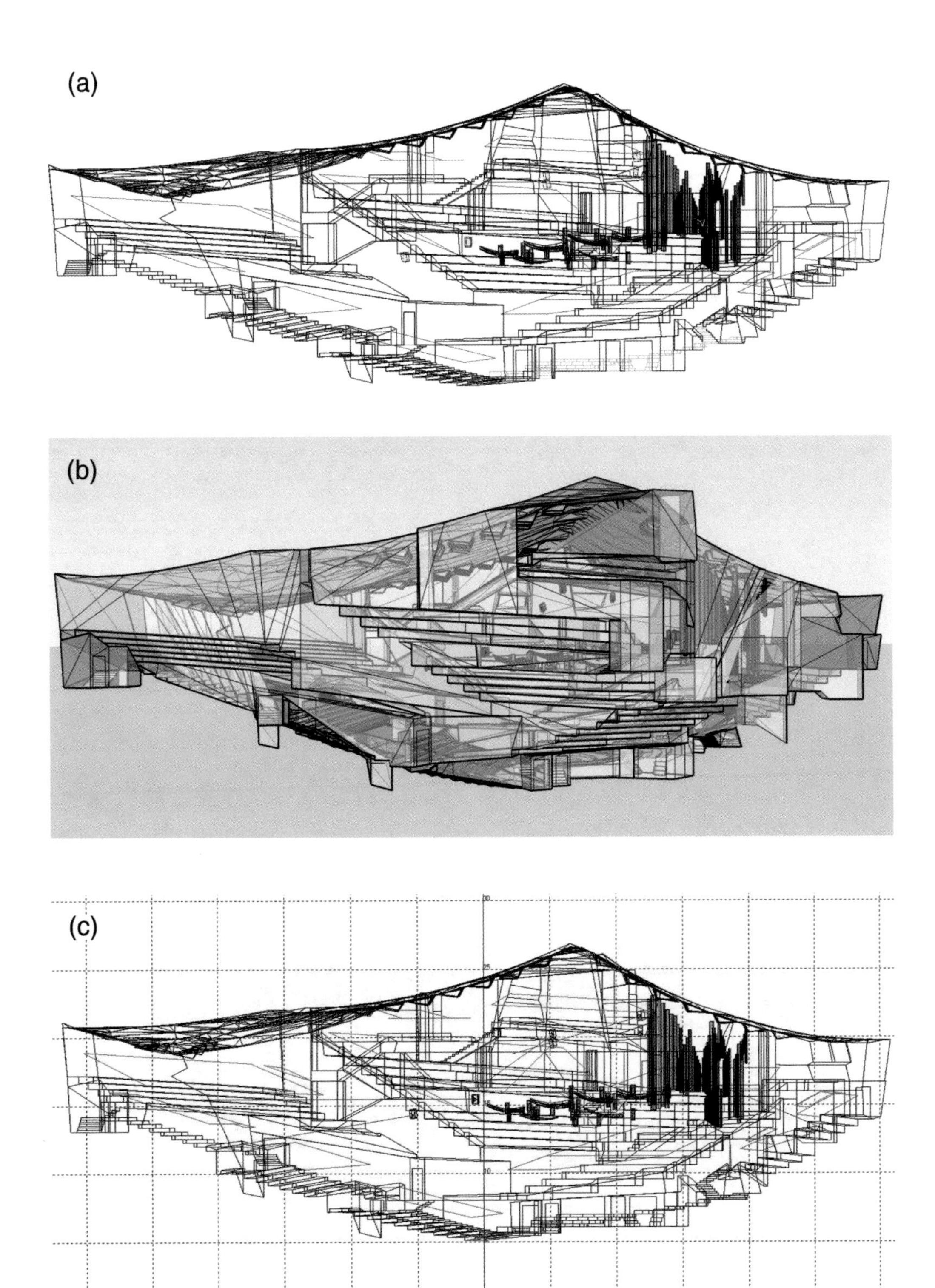

Figure 12.43 Room model of the Berlin Philharmonic concert hall in three different views: (a) AutoCad presentation, (b) SketchUp view, and (c) EASE Wireframe model.

12.4.2.1.1.2 Scattering Data

The practical values for the scattering coefficient, *s*, are between 0 and 1. So there are some rules of thumb to define the actual scattering coefficient in simulation software programs. Some programs give guidance to estimate the coefficients; other programs use special boundary element method (BEM) routines (see Figure 12.44) to derive the coefficient in a way as it should be measured according to the Standard ISO 17497-1.[67]

A scattering coefficient will never be generally available in tables (except for some special modules), because the way the interior architect uses the materials in a hall affects the scattering behavior as well. Therefore, the scattering behavior of wall parts in a computer model must be determined model-specific.

12.4.2.1.1.3 Diffraction, Low-Frequency Absorption

Computer simulation programs use different algorithms based on geometrical acoustics to calculate the impulse responses in model rooms. But these routines using particle radiation are only valid above the Schroeder frequency:

$$f_{\text{Schroeder}} = 2000\sqrt{\frac{T_{60}}{V}}, \tag{12.50}$$

T_{60} – reverberation time in sec and V – Volume in m^3.

For lower frequencies and especially in small rooms the assumption of particle radiation cannot be held anymore. Here, wave acoustics routines must be applied. An analytical solution is impossible so numeric routines have been developed. The finite element method (FEM) and the BEM are predominantly used. For applying an FEM routine, the computer model must be subdivided at first into small sub-volumes (meshes), where the dimensions of the mesh correspond with the upper frequency handled by the FEM. The higher the frequency, the smaller the dimensions and the longer is the calculation time. As an example, to

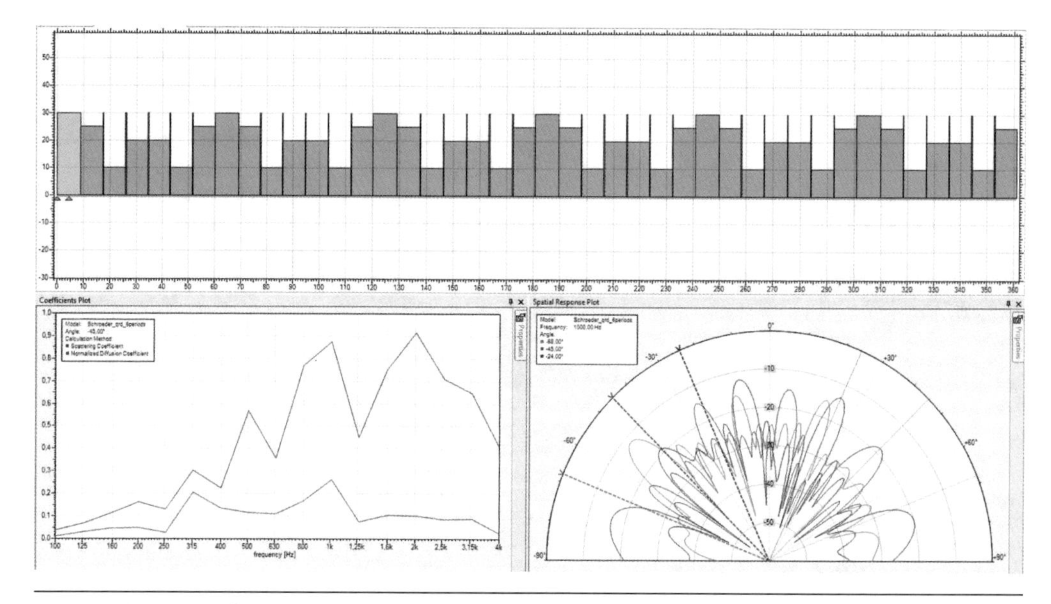

Figure 12.44 Boundary element method (BEM) based software tool for calculating scattering coefficients.

build a mesh in a hall of 10,000m^3 you need a mesh resolution of about 280,000 sub-volumes to apply the FEM up to 500 Hz. Figure 12.45 shows such a mesh grid for a church model. For the BEM, only the surface must be meshed accordingly.

After meshing, it is quite challenging in complex room structures to derive standing wave figures inside the model. So we may determine nodes, maxima, and minima of the sound field behavior in the room (see Figure 12.46).[68]

Figure 12.45 Meshed model in a plug-in module for EASE4.4.

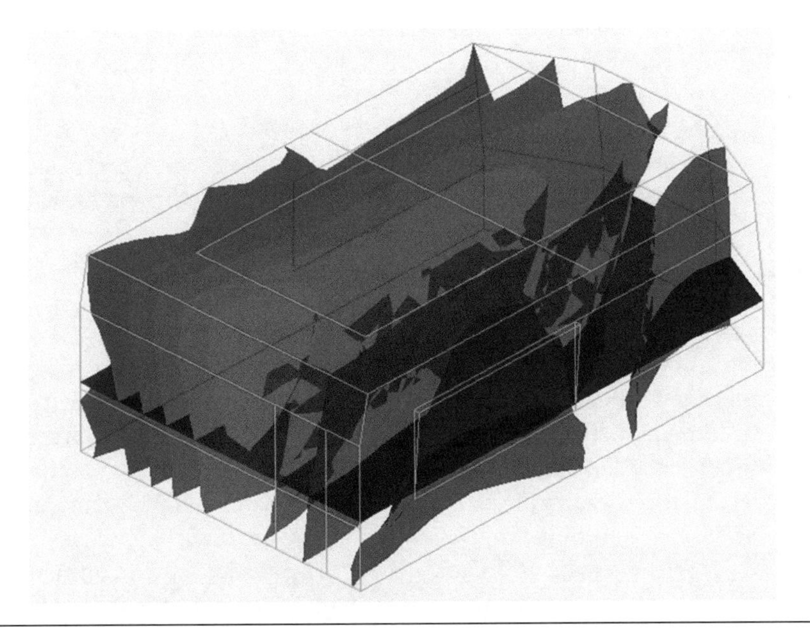

Figure 12.46 62 Hz contours in the finite element method (FEM) model according to Figure 12.45 (red–max, blue–min).

As a next step, the acoustic impedance of each individual wall part must be known. As a first approximation, the impedance of the wall material can be derived from the known absorption coefficient. Applying the well-known algorithm of the FEM, the transfer function at selected receiver positions may be calculated. By means of Fourier transform the impulse response is obtained in the time domain. By using this method, the transfer functions at receiver positions may be calculated also, even if the receiver is shadowed from the source position and the direct sound is only coming by diffraction to the receiver.

This method can be used effectively in small rooms below 300 Hz. A mesh of a control room of 135m^3 consists only of around 4000 sub-volumes (frequencies higher than 300 Hz are neglected with mesh size dimensions of around 1/3 of the smallest wavelength). In this way, efficient calculation results can be expected.

In some simulation programs, edge diffraction routines are introduced. So the diffracted sound from an orchestra in the pit to the visually shadowed stalls could be better calculated.[61, 62]

12.4.2.1.2 Transducer Data for Acoustic Simulation

Besides the primary and secondary structure of the room, loudspeakers and natural sound sources, as well as microphones and the human hearing system, have to be considered during the simulation. This section reviews existing practices and outlines advantages and disadvantages that the user of a software program should be aware of when applying performance data for a particular sound transducer—mainly for the purpose of the simulation of transducers with respect to the electroacoustic and room-acoustic prediction of the acoustic system as a whole.

Loudspeaker Data for Simulation

In computer-aided acoustic design and especially for sound reinforcement applications, the level of accuracy to which sound sources are modeled plays a crucial role. Accordingly, most simulation software packages have continuously developed their capabilities of describing loudspeakers by measurement data along with the complexity of the loudspeaker systems themselves.

In this sense, the measurement and simulation of loudspeaker systems can be roughly divided into two periods of time. The first period, until the late 1990s, was characterized by the use of simplified far-field data for almost any sort of loudspeaker and the assumption of a point source-like behavior. But with the advent of modern line array technology, for tour sound and speech transmission applications, new concepts had to be developed.

Simulation of Point Sources

For many years, the radiation behavior of sound sources, and loudspeakers in particular, was basically described by a three-dimensional matrix containing magnitude data in a fixed spectral and spatial resolution. Starting in the late 1980s, typical data files contained directivity data for the audible octave bands, such as from 63 Hz to 8 kHz, and for a spherical grid with an angular spacing of 15 degrees. Mostly, data was also assumed to be symmetric in one or two planes. With the need for higher data resolution and the limits of available PC memory and computing power changing at the same time, more advanced data formats developed eventually reaching a present-day typical resolution of 5 degrees in angular increments for 1 octave or 1/3 octave frequency bands (see the CLF also).[69]

This simulation model makes some significant assumptions:

1. The use of a spherical wave form assumes that both measurement and simulation happen in the far field of the device; that is, at a distance where the sound source can be considered as a point source.
2. It is assumed that the density of discrete data points is high enough and the frequency and angular dependency of the directivity characteristics smooth enough.
3. The use of magnitude-only data requires that the point of reference during the measurement can be reconstructed by the run-time phase in the model, and it requires that the source-inherent phase is negligible as well.
4. It is assumed that the concerned loudspeaker system is a fixed system that cannot be changed by the user or when its configuration is changed, its performance data is not affected.
5. For the use of such point sources in computations involving geometrical shadowing and ray-tracing calculations, the source is regarded as located at a single point and is thus either wholly visible (audible) for a receiver or not.

These assumptions have been made, especially in the early 1990s, in order to obtain and use loudspeaker directivity data in a practical manner. Important factors were the availability and accuracy of measurement platforms and methods, the storage size of the processed measurement data, and the PC performance with regard to processor speed available at that time.

However, these assumptions have a set of drawbacks that became most evident with the broad use of large-format touring line arrays and digitally controlled loudspeaker columns; but also with the increasing use of inexpensive DSP technology employed for multi-way loudspeaker systems. Some of the issues conflicting with the above five points are listed in the following:

- A large line array system of some meters in height cannot be measured adequately in its far field in addition to the fact that line array applications happen to take place mainly in the near field. Therefore, the simulation of a whole line array as a point source is invalid.
- Another problem often encountered is insufficient angular resolution. Loudspeaker columns—but also multi-way loudspeakers—exhibit significant lobing behavior in the frequency ranges where multiple acoustic sources interact at similar strength. Often too coarse angular measurements fail to capture these fine structures.
- While in many cases the phase of the sound pressure radiated by a simple loudspeaker is negligible—at least if it is considered on-axis and the run time phase is compensated for—the same is not true for most real-world systems.
- Loudspeaker systems become increasingly configurable, so that the user can adapt them to a particular application. Typical examples include almost all touring line arrays, where the directional behavior is defined mechanically by the splay angles between adjacent cabinets, and loudspeaker columns or multi-way loudspeakers, where the radiation characteristics can be changed electronically by manipulating the filter settings.
- In advanced computer simulations of sound reinforcement systems in venues, geometrical calculations must be performed. This is required to obtain exact knowledge as to which part of the audience might be shadowed by obstacles between the sound sources

and the receivers. Geometric considerations are also needed in ray-tracing calculation, in order to find reflections and echoes. For both processes, the reduction of a physically large loudspeaker system to a point source can lead to significant errors.

To resolve the problem of large-format loudspeaker systems, a subdivision into smaller elements is required to be able to measure them and use them for prediction purposes. To properly model the coherent interaction between these elements, complex measurement data—including both magnitude and phase data—is needed.

The most prominent solutions can be summarized as follows. Instead of measuring a whole system, so called far field cluster balloons were calculated, based on the far field measurement of individual cabinets or groups of loudspeakers.[70] To describe individual sound sources, phase data was introduced in addition to the magnitude-only balloon data or mathematical models providing phase information were applied implicitly—such as minimum phase or elementary wave approaches as well as two-dimensional sound sources.[71] However, these first approaches lacked generality and thus, their implementation into existing simulation software packages was specific, difficult, or even impossible.

Dynamic Link Library Concept for Modern Speaker Systems

In a first step to overcome the variety of issues related to the reduction of complex loudspeakers to simple point sources, the MS Windows dynamic link library (DLL) approach was developed.[62, 63] Technically speaking, the DLL is a program or a set of functions that can be executed and return results. It cannot be run *stand-alone*, but only as a plug-in of another software that accesses it through a predefined interface. The basic idea is to move the complexity of describing a sound source from the acoustic simulation program into a separate module that can be developed independently and that can contain proprietary contents. In this way, a clear cut is made between the creators for simulation software packages and the loudspeaker manufacturers who can develop product-specific DLL modules on their own.

Advanced Generic Loudspeaker Library Concept

Another concept, namely the generic loudspeaker library (GLL) developed by Feistel and Ahnert,[72] introduced a new loudspeaker data file format that is significantly more flexible than the conventional data formats and is designed to resolve most of their apparent contradictions. Based on the experience with many loudspeaker manufacturing companies and the implementation of simulation and measurement software packages, the GLL was developed as an object-oriented description language to define the acoustic, mechanical, and electronic properties of loudspeaker systems (see Table 12.6). Since for each physical entity the GLL language has a representation in the software domain, there is no need to make artificial assumptions in order to comply with rigid, reduced data structures. Basically, in the GLL philosophy, every sound radiating object should be modeled as such, and every interaction possible between engineer and loudspeaker in the real world should be imaged in the software domain. In this picture, transducers, filters, cabinets, rigging structures, and a whole array or cluster are present in the GLL with their essential properties and parameters (see Figure 12.47).

Typically, the GLL model of a loudspeaker consists of one or multiple sound sources, each with its own location, orientation, directivity, and sensitivity data. These sources can be simple point sources but also spatially extended sources, such as lines, pistons, etc. It can

Table 12.6 Comparison of speaker data formats

	EASE SPK	EASE XHN	ULYSSES UNF	CLF	EASE GLL
Data Type	Simple Data Table	Simple Data Table	Simple Data Table	Simple Data Table	Advanced Description Language
Balloon Symmetries	None, Half	None	None, Half, Quarter	None, Half, Quarter	None, Half, Quarter, Axial
Angular Resolution	5°	5°	5° or 10°	5° or 10°	1° to 90°
Frequency Resolution	1/3 Octave	1/3 Octave	1/1 or 1/3 Octave	1/1 or 1/3 Octave	Any
Phase Data	Limited	Limited	No	Limited	Yes
Individual Transducers	No	No	No	Limited	Yes
IIR and FIR Filters	No	No	No	No	Yes
Configurable	No	No	No	No	Yes
Line Arrays	No	No	No	No	Yes
Beam-Steering	No	No	No	No	Yes
Customizing	No	No	No	No	Yes

This comparison does not consider proprietary DLL solutions or data formats offered singularly by the respective software platforms

include multiple sets of filters, including infinite impulse response (IIR) and finite impulse response (FIR) filters, crossover, and equalization filters. The loudspeaker box is mechanically characterized by means of a case drawing and data for the center of mass calculations. Boxes can be combined into arrays and clusters. Available configurations are predetermined by the loudspeaker manufacturer, according to the functions available to the end user. Additional mechanical elements such as frames and connectors allow specifying exactly which configuration possibilities exist.

Practical Considerations

The acquisition of both magnitude and phase data requires more care than just the measurement of magnitude-only data. However, modern impulse response acquisition platforms deliver complex data in a sufficient frequency resolution. The representation of the loudspeaker directivity function based on impulse response wave files is included in a standard of the AES from 2009.[73]

In general, the computer model utilizing loudspeaker data can only be as good as the data of the lowest quality included. Presently, the accuracy of the loudspeaker data is often much higher than that of the material data of the model. Absorption and scattering coefficients are usually only known in 1/1 or 1/3 octave bands for random incidence. The user must be aware that although loudspeaker direct field predictions may be very precise, any modeling of the

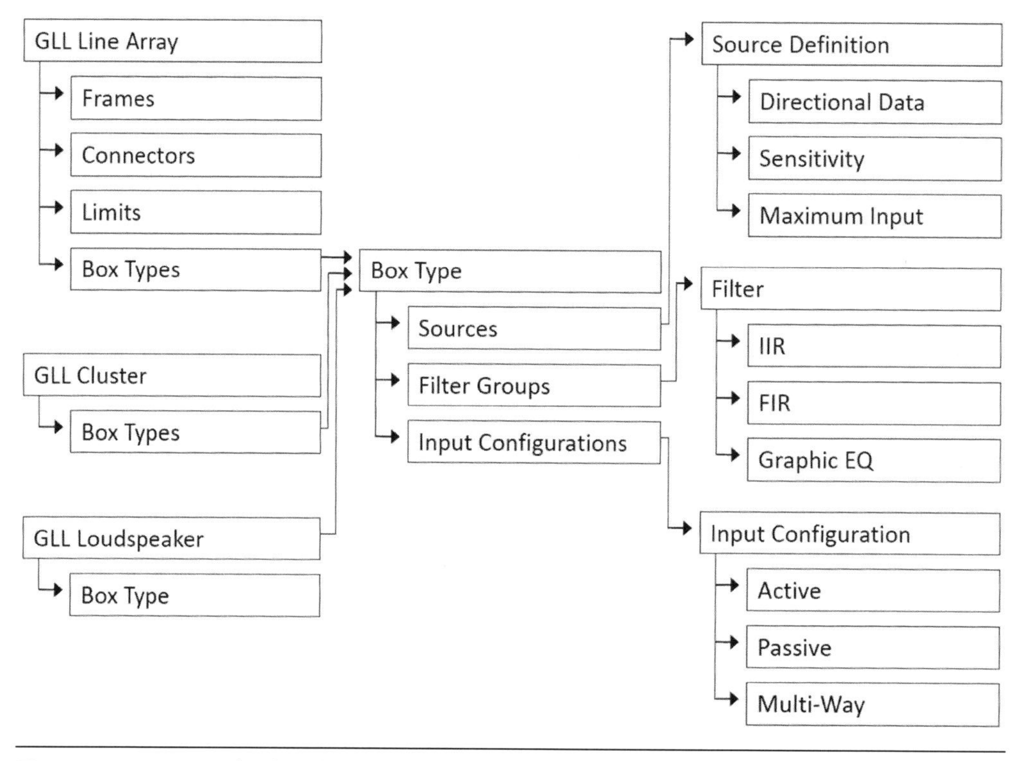

Figure 12.47 Generic loudspeaker library (GLL) object hierarchy.

reflections and the diffuse sound field in the room will be limited by the available material data. Furthermore, it is not very likely that there will ever be systematic, large-scale measurements of angle-dependent complex directivity data for the reflection and scattering of sound caused by wall materials.

12.4.2.1.2.2 Receiver Data for Simulation

For a complete acoustic model, the acoustic receivers must also be considered. Most important for auralization purposes are the characteristics of the human head and how this influences the sound that reaches the inner ear. Often, simulation software packages also allow utilizing microphone directivity data, in order to be able to image real-world measurement. However, the correct implementation of electroacoustic receivers has not yet received the same level of attention as for the sources.

Simulation of the Human Head

The HRTF is central to incorporating the characteristics of the human head into the simulation results and thus preparing them for final auralization purposes. Typically, this is a data set that consists of two directivity balloons, one for the left ear and a second one for the right ear. Each of them describes, usually by means of complex-valued data, how the human head and the external ear change the incoming sound waves as they arrive at the ear. It is critical for a satisfactory binaural auralization that the signal for each ear is weighted with an appropriate angle- and frequency-dependent directivity balloon.

The acquisition of measurement data for the human head is not a trivial matter. Since real human heads cannot be easily measured, an artificial head has to be built or in-ear microphones have to be used. Each ear of an artificial or a real head is equipped with two microphones. Balloon measurements are made similar to loudspeaker balloon measurements, except that the locations of source and receiver are inversed and a stereo set of data files is obtained.[51]

Newer research has shown that the inclusion of the human torso into the HRTF also has significant effect on the quality of the binaural reproduction.[74] Even more so, auralization results of the highest quality can be obtained utilizing a head-tracking system and a set of HRTF balloons, where each pair of balloons describes the transfer function for the left and right ear for a particular angular position of the human head relative to the human body. This data can then be employed to auralize sound samples of either a measured or simulated environment with speech and music contents.

Simulation of Microphones

The need for inclusion of microphones in acoustic simulation software has several reasons. On the one hand, to be able to compare measurements with computational results, the frequency response and the directivity characteristics of the microphone have to be taken into account. On the other hand, the possibility of simulating either recording or reinforcement of a talker or musician is of practical interest too. For example, by varying the location and orientation of the receiving microphones, the coverage can be optimized. Finally, by including microphones it becomes possible to simulate the entire chain of sound reinforcement, from the source, over the microphone to the loudspeaker, and back to the microphone. Only this enables the prediction of feedback and to estimate the potential gain before feedback.

However, the acquisition and distribution of microphone data must still be considered in its infancy. Available data consists largely of octave-based magnitude-only data that assumes axial symmetry. Measurement techniques vary significantly among microphone manufacturers, and measuring conditions (such as the measurement distance) are not standardized and often not even documented. Therefore, most users of simulation programs do not consider implementing microphone data into their models, or if so, they use generic data based on ideal directional behavior, like cardioid or omnidirectional patterns.

Previously, an advanced data model was proposed, suggesting use of an approach similar to the loudspeaker description language (GLL) introduced earlier, namely to describe receiver systems in a generalized, object-oriented way. This means especially that:

- Microphone data files should at least include far-field data (plane wave assumption), but can also contain proximity data for various near-field distances.
- A microphone model can consist of multiple receivers, that is, acoustic inputs, and can have multiple channels (electronic outputs).
- A switchable microphone should be represented by a set of corresponding data subsets.
- Impulse response or complex frequency response data should be utilized to describe the sensitivity and the directional properties of the microphone as appropriate.

12.5 COMPUTER-BASED CALCULATION OF SOUND LEVEL AND OTHER PARAMETERS

Today, an acoustic computer-aided drafting (CAD) program should be able to predict all needed acoustic measures precisely. A 100% forecast is certainly impossible, but the results of a computer simulation must come close to reality (maximum errors approximately or less than 30%). Then it also becomes possible that the acoustic behavior of a facility can be made audible by an auralization. We will listen to sound events just *performed* by means of the computer. The following pages will give some guidelines on how to use computer simulation today for sound field data acquisition.

Here we distinguish between room acoustic simulations to learn more about the acoustic properties of a space and the calculations used to study the acoustic properties of a sound system installed in this space.

12.5.1 Room Acoustic Simulation

To do a room acoustic simulation we need a corresponding computer model. This model should not be a copy of an AutoCAD or SketchUp model. If so, the resolution of the model would be too high, the results would be partially misleading, and the calculation time would be unnecessarily high. A rule of thumb should be to limit the number of wall areas to a maximum of 5,000, preferred are 1,000 … 1,500. Fine structures of surfaces should never be modeled in detail, here, properly selected scattering (and of course absorption) coefficients will reflect the acoustic behavior of these surfaces in a better way. Figure 12.43c illustrates an example of a model of the Berlin Philharmonic Hall (4,700 surfaces). All important fine structures are visible.

For sound sources, we select undirected room acoustic simulations, i.e., more or less omnidirectional sources.

12.5.1.1 Statistical Approach

12.5.1.1.1 Reverberation Time

The simplest way to obtain fast results is to use the direct sound of one or more sources (loudspeakers) and calculate the reverberation level of the room by means of the T_{60} equations, assuming the room follows a statistically even distributed sound decay (homogeneous, isotropic diffuse sound field—that is, the T_{60} is constant over the room). Based on known room data and the associated surface absorption coefficients, a computer program is able to quickly calculate the T_{60} according to the Sabine and Norris-Eyring equations (compare Eq. (12.1)). Even the volume of the closed space may be calculated—quite often an interesting number for architects, because the common design tools of the architects don't deliver such information.

Normally a set of frequency-dependent *target* T_{60}s is available for entering into the simulation program, so that the calculated T_{60} times can be compared with the target values. This way the program will draw (for each selected frequency band) the calculated T_{60} versus the target T_{60} and list the number of excess or deficient T_{60}s for each band relative to the target values, within a range of tolerance. Calculation of the early decay time should be possible, too.

With the simulation program directly, or by export of data to MS-Excel, it is possible to plot T_{60}s as multiple T_{60} values within a single graph, so as to show the impact of various audience sizes, proposed and/or alternative room treatments, etc., on the T_{60}. An option must show the desirable tolerance range of T_{60}s for a particular project. This way, measured or calculated T_{60} values can be checked as to whether they are inside the tolerance range or not.

12.5.1.1.2 Objective Room Acoustic Measures

Based on simple calculations (see Section 12.4.1), it is possible to derive the direct sound and the diffuse-sound levels and consequently a range of objective acoustic parameters. It goes without saying that this requires acoustical conditions of a space (room) having a statistically regular behavior (frequency response of the T_{60} that is independent of the location in the room). In practice, however, such behavior will hardly be found. For this reason, one tends to qualify such data as having only a preliminary guideline character and to have them confirmed by additional detailed investigations.

What is possible to calculate under these assumptions? As in all simple direct sound design programs,[75-77] the direct sound calculation always delivers correct results. So even for very sophisticated speaker or source arrangements, we obtain correct results if, of course, the radiation behavior of the source is modeled close to reality (see Section 12.5.2.3).

Based on these direct sound calculations and by assuming an exponential sound decay for the diffuse sound, we are able to calculate most of the basic parameters in acoustics. But, we should never forget that the exponential sound decay attached to a direct sound sequence is everywhere the same—what, in practice, will never happen.

So all objective measures—like clarity C_{80}, STI, or Al_{cons}—may be calculated, but these results have to be considered only as orientation values. A sound designer should not be surprised if his predicted values, calculated that way, are deviating from the real measured ones after the project is finished.

12.5.1.2 Ray-Tracing Approach

There are several ways to calculate the impulse response of a radiated sound event. In the image modeling method a receiving point is used instead of a counting balloon (in contrast to classical ray-tracing). Frequency response and interference effects (including phase investigations) are also easily calculated.

This method is very time-consuming and only used in combination with more advanced methods. So one gets usable results for models with face numbers <50 and maximum bounces i<6. For larger models and more complicated investigations, the next method is more advantageous.

In contrast to image modeling, in the ray-trace method the path of a single sound particle radiated under a random angle into the room along a ray is followed. All surfaces are checked to find the reflection points (with or without absorption or scattering). The tracing of the single ray or particle is terminated when the remaining sound energy has decreased to a certain level or when the particle hits an appropriately arranged counting balloon with a finite diameter, typically at the location of a listener in the room. The method runs significantly faster and the calculation time is only proportional to the number, N, of the model walls. This algorithm is rarely used alone, but more often in combination with other routines, called hybrid ray-tracing algorithms.

A newer algorithm, called AURA,[78] calculates the transfer function of a room for a given receiver point using the active sound sources. For this purpose, a hybrid model is employed that uses an exact image source model for early specular reflections and an energy-based ray-tracing model for late and scattered reflections. The transition between the two models is determined by a fixed reflection order.

The ray-tracing model utilizes a probabilistic particle approach and can therefore be considered as a Monte-Carlo model. At first, the sound source emits a particle in a randomly selected direction with a given energy. The particle is then traced through the room until it either hits a boundary or a receiver or its time of flight reaches the user-defined cut-off time. When the particle hits a boundary, it is attenuated according to the surface material, and its direction is adjusted according to the reflection law. An essential assumption of this Monte-Carlo approach is that attenuation due to air or surface reflections is taken into account as a reduction of particle energy while the propagation loss over distance is indirectly covered by the reduced detection probability for individual particles with increasing distance and fixed receiver sizes.

Per receiver and simulated frequency a so-called echogram is created that contains energy bins linearly spaced in time. When a receiver is hit, the energy of the detected particle is added to the bin that corresponds to the time of flight. Also, as a separate step, the contributions from the image source model are included. The particle model accounts for scattering in a probabilistic way. Whenever a particle hits a surface, the material's sound absorption part is subtracted from its energy. Then, a random number is generated and, depending on the scattering factor, the particle is either reflected geometrically or it is scattered under a random angle based on a Lambert distribution. After that, the particle is traced until it hits a receiver or a wall again.

Nowadays multithread and network calculations decrease the calculation time for complex situations from days to hours.

In the case of *cone tracing* (a special form is the so-called pyramid tracing), a directed ray radiation over the different room angles is applied. Because of this cone approach, fast ray calculations can proceed. The fact that the cones do not cover the source sphere surface turns out to be a complete disadvantage. It is necessary to overlap adjacent cones and an algorithm is required to avoid multiple detections or to weight the energy so that the multiple contributions produce (on average) the correct sound level. Some conical beam tracers are widely accepted.[79–82]

12.5.1.3 Results of All of These Calculations

Having impulse responses on listener positions in a room or in an open space, we may use them to derive all acoustic quantifiers and measures characterizing this environment from the acoustic point of view. By doing handclapping, firing a pistol, or by using modern measurement techniques, we obtain an impulse response of the space in seconds (see Figure 12.12a). In a simulation program this will take longer. A statistic approach may still take only minutes, but to calculate a full impulse response in more complicated rooms can take hours, even days.

The entire ray-tracing or image modeling methods that calculate impulse responses have to take into account the directivity of the sound sources and the absorptive and scattering characteristics of the surfaces encountered en route from the source to the receiving point.

Additionally, the dissipation of sound energy in air, i.e., the frequency-dependent air attenuation, must be considered too. As a result of all these calculations, impulse responses or energy-time functions can be obtained, as shown in the following figures.

The program CATT-Acoustic[62] shows the complete echogram with all input data (room, speaker, listener position, frequency) and presents all resulting room acoustic measures (see Figure 12.48). With EASE and AURA it looks different (see Figure 12.49).

Using the simulated impulse responses, the designer is able to calculate all room acoustic measures of the space, including:

- Reverberation time
- Clarity
- Intelligibility
- Echo behavior, etc.

12.5.2 Sound System Design

After the room acoustic properties of the corresponding space are known, the real sound system design may start. So we select the usable or recommended loudspeakers, insert them into the model, and bring them into the right direction to cover the existing audience areas—termed, *aiming* of the speaker.

12.5.2.1 Aiming

Aiming the individual loudspeakers is an important operation ensuring the proper spatial arrangement and orientation of the sound reinforcement systems. Once the corresponding room or open-air model is available and the mechanical and acoustical data of the speaker systems are exactly known, these systems are approximately positioned and then one may immediately begin the fine tuning the same. This also includes the determination of beam

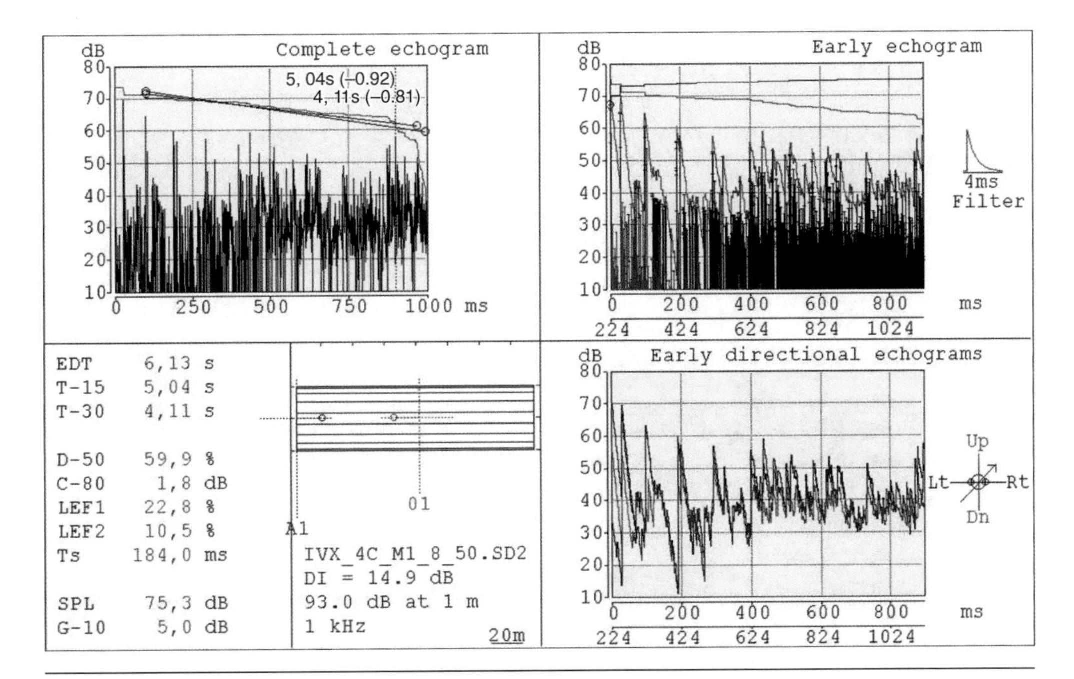

Figure 12.48 Echogram and data plot in CATT acoustics.

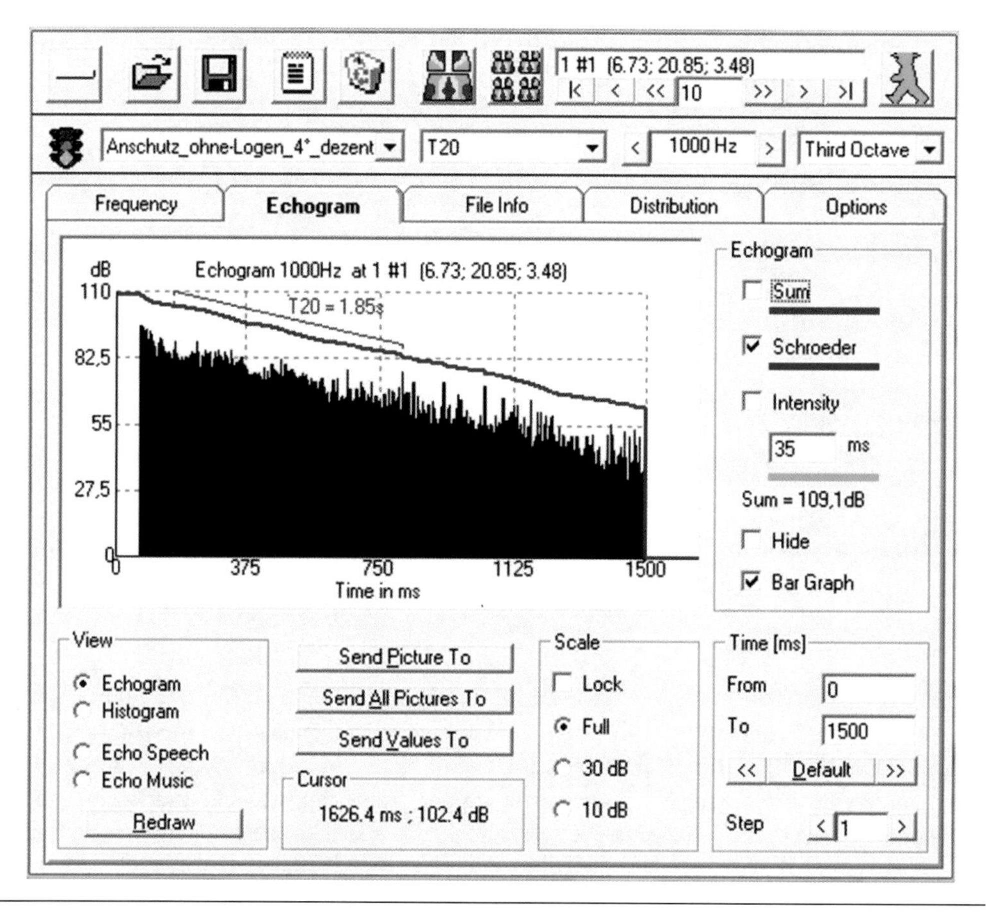

Figure 12.49 Echogram and data plot in EASE 4.4–AURA.

settings for digitally controlled arrays and/or their delays. A modern simulation program uses a kind of isobeam/isobar method to initially aim the loudspeakers, preferably utilizing the –3 dB, –6 dB or –9 dB contours.

Figure 12.50 shows various types of projection of the –3, –6, and –9 dB curves into the room. On audience areas, one can then also see superposed aiming curves for multiple speakers (see Figure 12.51).

12.5.2.2 Time-Arrivals, Delay, and Alignment

A graph of time-arrivals (direct, direct + reflected, reflected only) allows the user in all simulation programs to see the first energy arrival as required by the design, to adjust a single speaker delay, to bring the speakers into synchronicity, and to realize an acoustic localization of an amplified source—via distance and the HAAS or precedence effect (see Figure 12.52 (a) and (b)).

More complicated is the consideration of special requirements such as localization, stereo imaging, etc. Simulation programs allow determining the first wave front, as well as calculating initial time delay gaps or echo detections (see Figure 12.53 (a–c)).

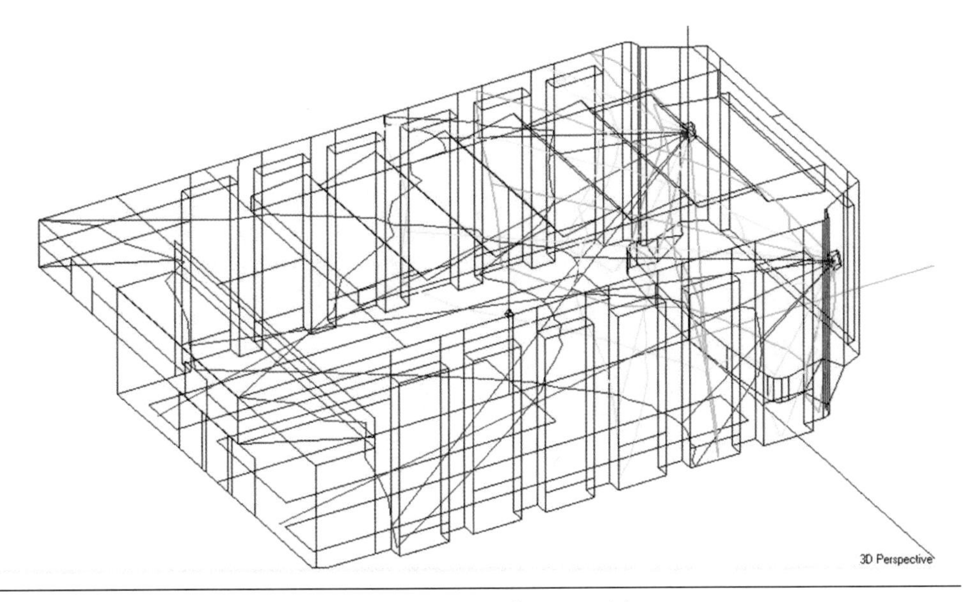

Figure 12.50 3-D aiming presentations in EASE 4.4 wireframe model.

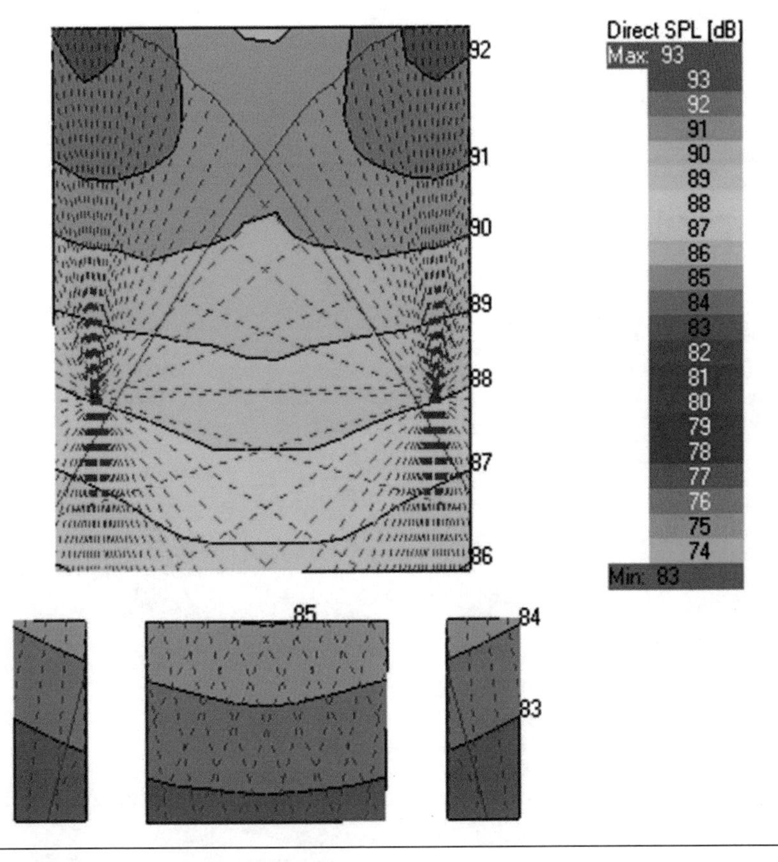

Figure 12.51 2-D aiming mapping in EASE 4.4.

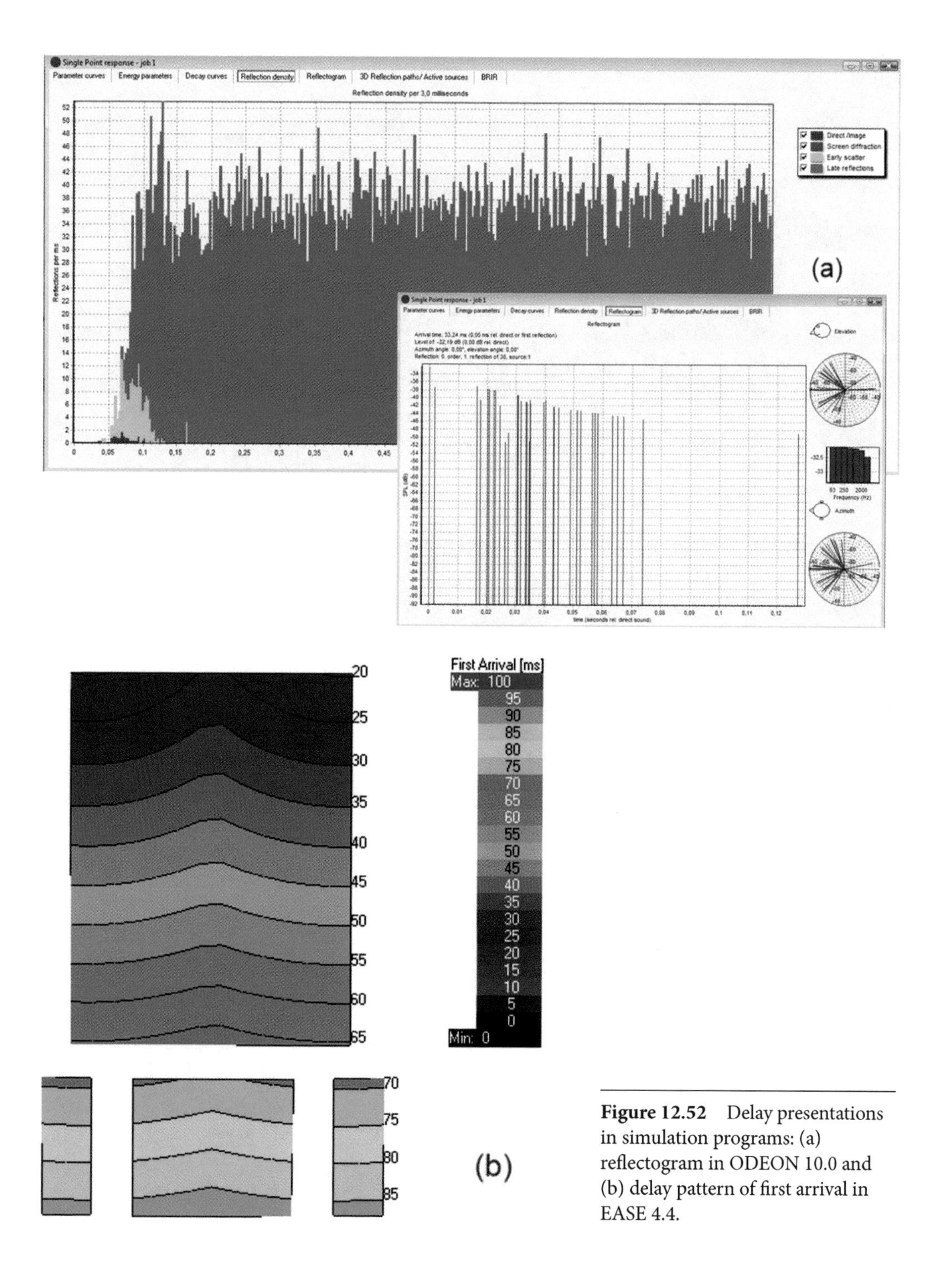

Figure 12.52 Delay presentations in simulation programs: (a) reflectogram in ODEON 10.0 and (b) delay pattern of first arrival in EASE 4.4.

Predicted array lobing patterns of *arrayable* loudspeakers are displayed by simulation programs, with the ability to provide signal delay and/or move the appropriate speakers, attempting to bring the array into acoustic alignment. A program today will have the ability to provide signal delay to the individual speakers to align them in time. The corresponding

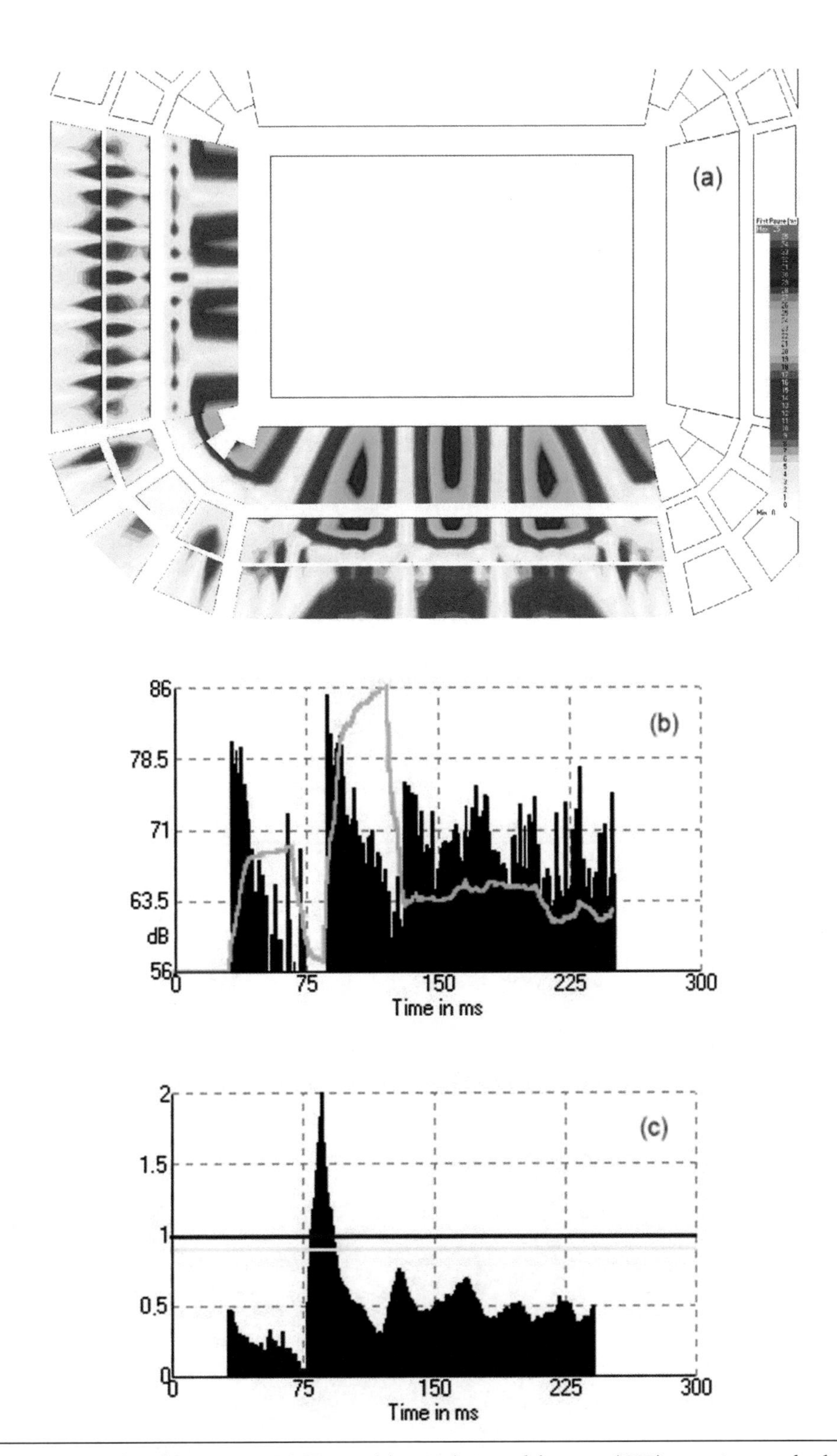

Figure 12.53 Echo detection in EASE 4.4: (a) initial time delay gap (ITD) mapping to check echo occurrence in a stadium, (b) echogram in weighted integration mode at 1 kHz, and (c) echo detection curve for speech at 1 kHz.

sound pressure calculations should take into account either measured phase data for the individual speakers or the run-time phase if phase differences among the components can be neglected.

12.5.2.3 SPL Calculations

After the loudspeakers have been correctly aimed and the delays are set correspondingly, one may begin to calculate the SPLs which may be achieved. The first results are given for the direct SPL, which is the same for statistic or ray-tracing-based calculation routines. As long as we predict a good direct sound coverage over the listener area, we also have to expect perfect intelligibility indices, of course, under the condition that the reverberation level is not too high.

A complex summation (phase conditions including travel-time differences should be included) has to be used as the standard method of calculating the DirectSPL. This method is exact for a planar wave, but only an approximation for the superposition of waves with different propagation directions. But anyway, the complex sound pressure components of different coherent sources must first be added and afterward squared to obtain SPL values. In so-called DLL or GLL approaches (see Section 12.4.2.1.2), one always calculates the complex sum of all sources in the array.

Today simulation programs are usually still only analyzing programs, capable of calculating which levels can be obtained by which loudspeakers and under which acoustical conditions. But questions are increasingly asked inversely. An advanced program feature should query the user for a desired average SPL of the system, and automatically adjust the power provided to each speaker (with a warning when the power required exceeds the capabilities for the speaker), based upon the desired SPL of the design, the sensitivity and directivity of the speaker, the distance of throw, and the number of speakers. Figure 12.54 illustrates the

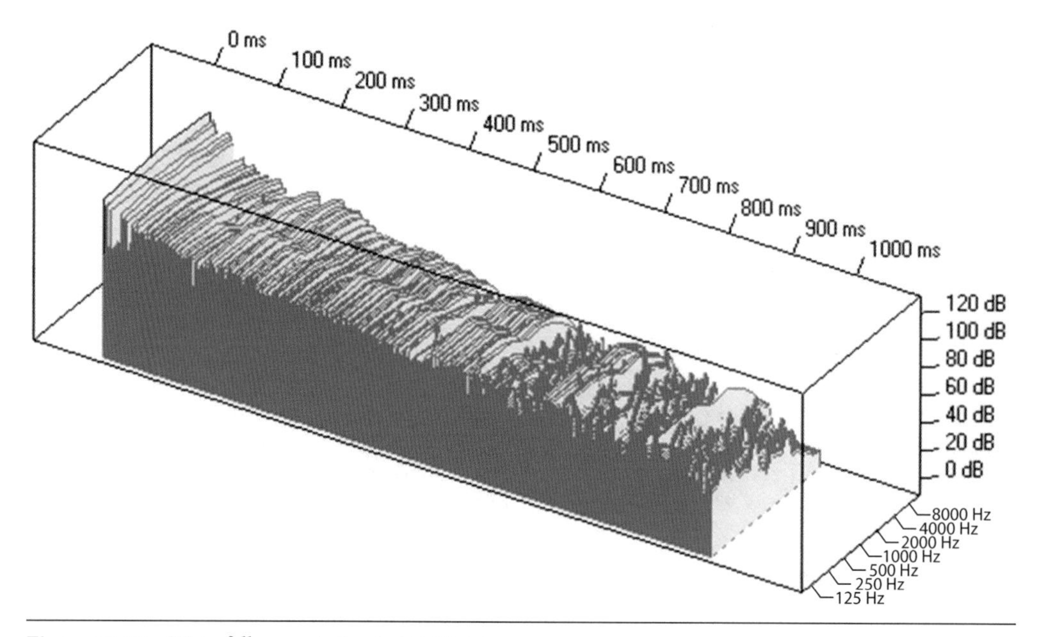

Figure 12.54 Waterfall presentation in EASE 4.4.

level-time-frequency-behavior of a loudspeaker cluster at a chosen listener seat in a room by a simulated waterfall diagram.

The target of all the efforts is to cover the whole audience area(s) evenly with musically pleasing and intelligible sound, while providing SPLs suitable for the intended purpose. All simulation programs today are widely lacking the algorithms required for computing the acoustical feedback. These, however, should soon become available.

12.5.2.4 Mapping, Single-Point Investigations

Once the aiming, power setting, and alignments are completed, all the programs provide a color-coded visual coverage map of the predicted sound system performance. This coverage map takes into account the properties of the speakers, as well as the impact of reflecting or shadowing planes or objects. Such maps provide, as a minimum, the following displays:

1. Predicted SPL, viewed at 1 octave or 1/3 octave band frequencies, and at an average of these frequencies (Figure 12.55 (a–c)).
2. Predicted intelligibility values, listed as STI values (see Figure 12.56 (a) and (b)).
3. Predicted acoustic measures (for octave or 1/3 octave band frequencies), listed in C_{80}, C_{50}, Center time, Strength, or other values according to the ISO standard 3382 (see Figure 12.57 (a) and (b)).

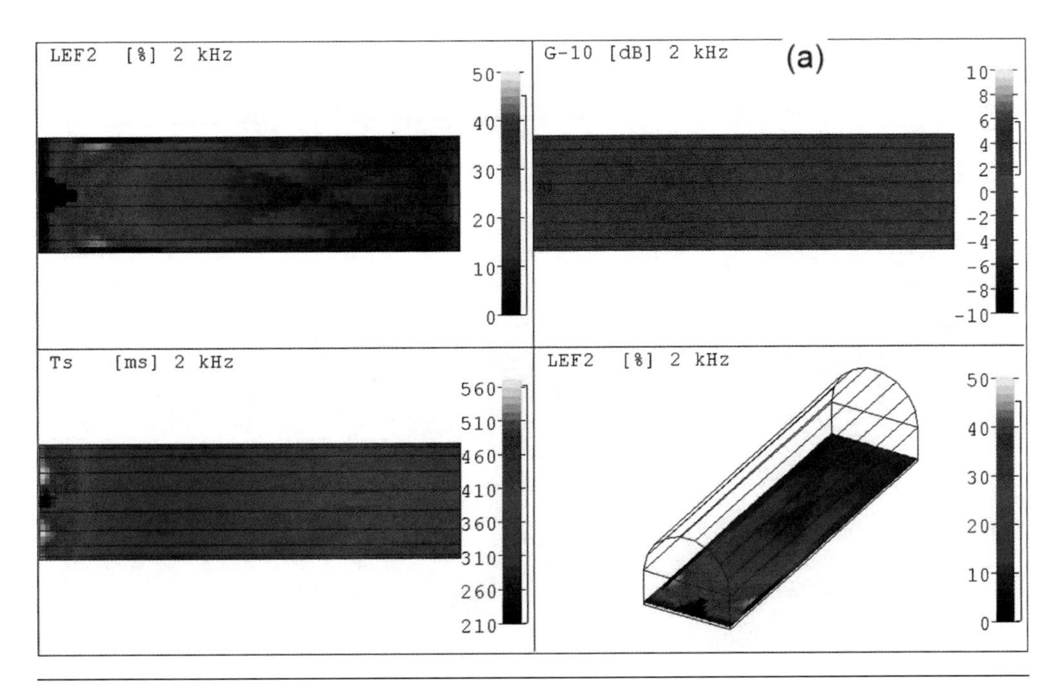

Figure 12.55 Sound pressure level (SPL) mapping in simulation programs (a) 2-D presentation in CATT acoustics, (b) narrow band presentation in EASE 4.4, and (c) broadband presentation in Odeon 10.0.

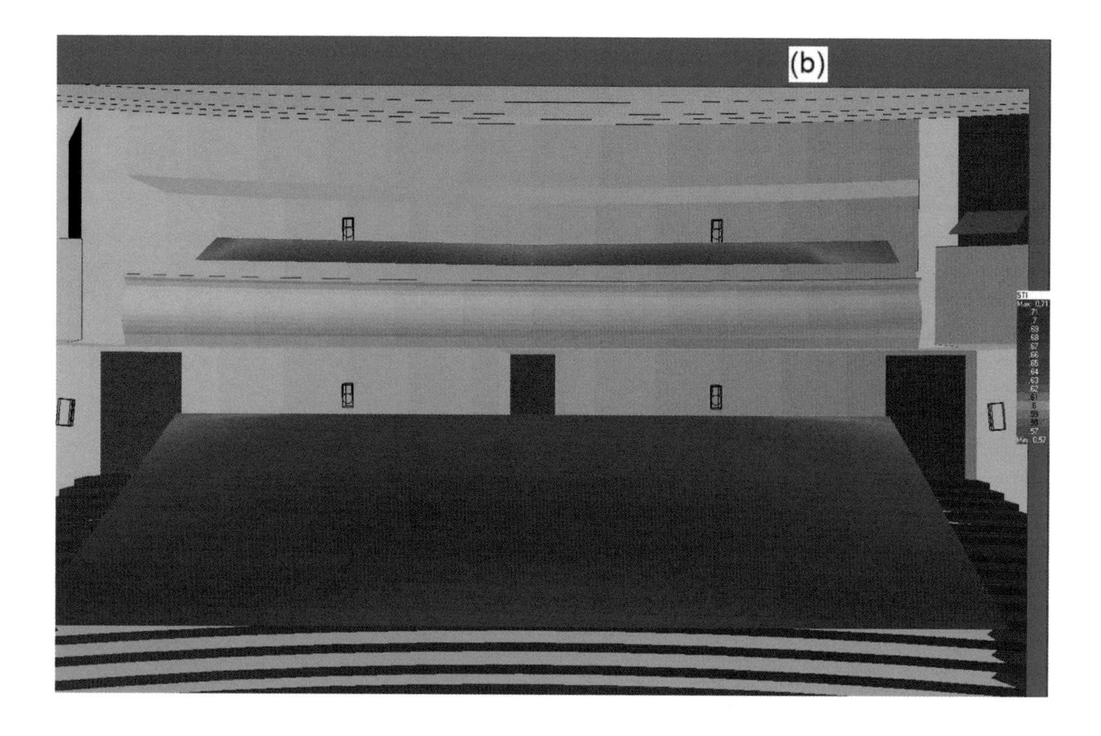

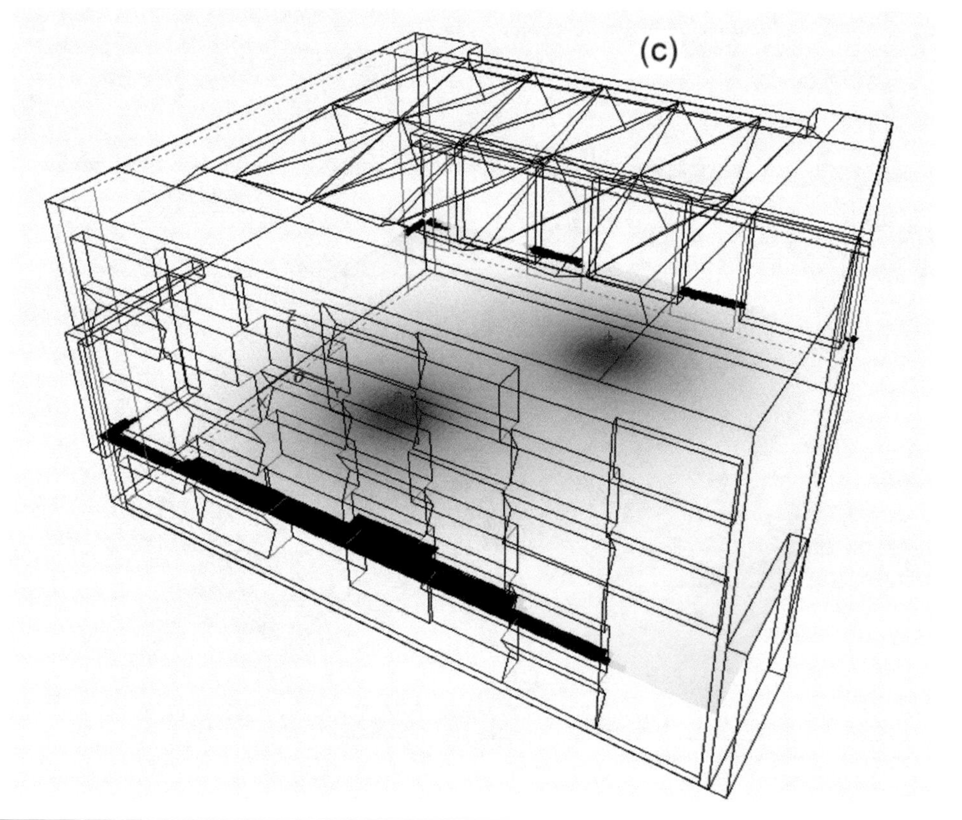

Figure 12.55 Continued

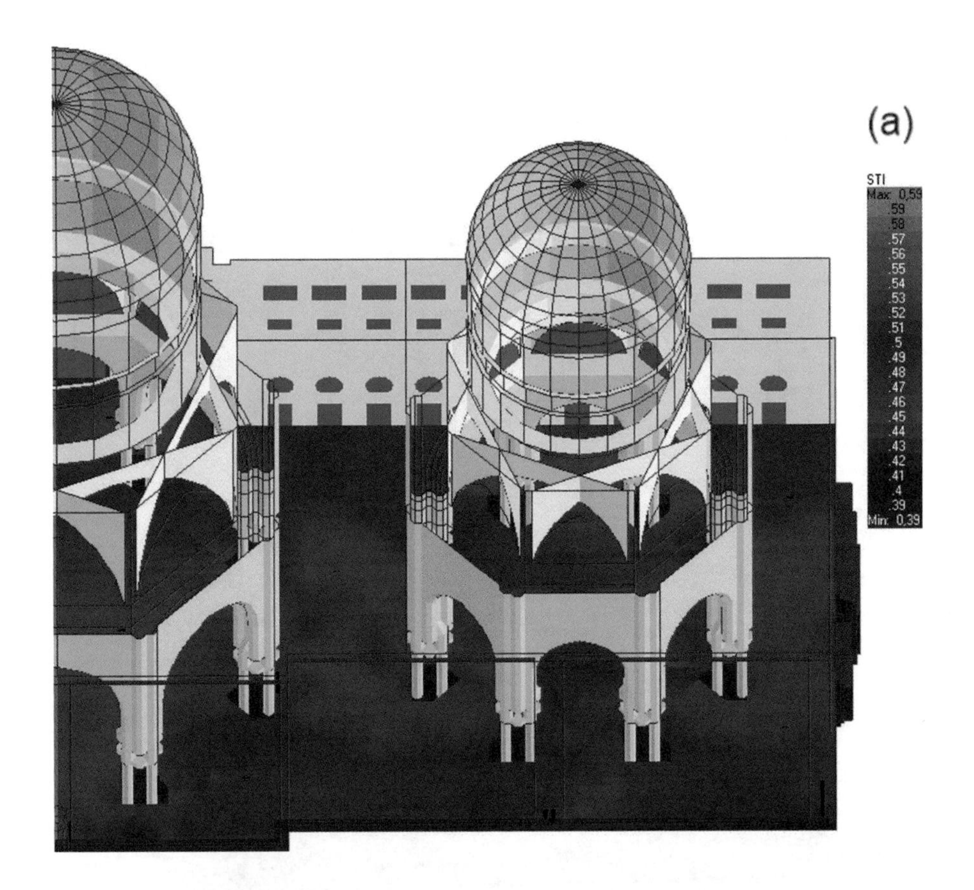

Figure 12.56 Speech transmission index (STI) presentations in the simulation program EASE 4.4: (a) Three-dimensional presentation in a mosque and (b) STI presentation in a theater hall.

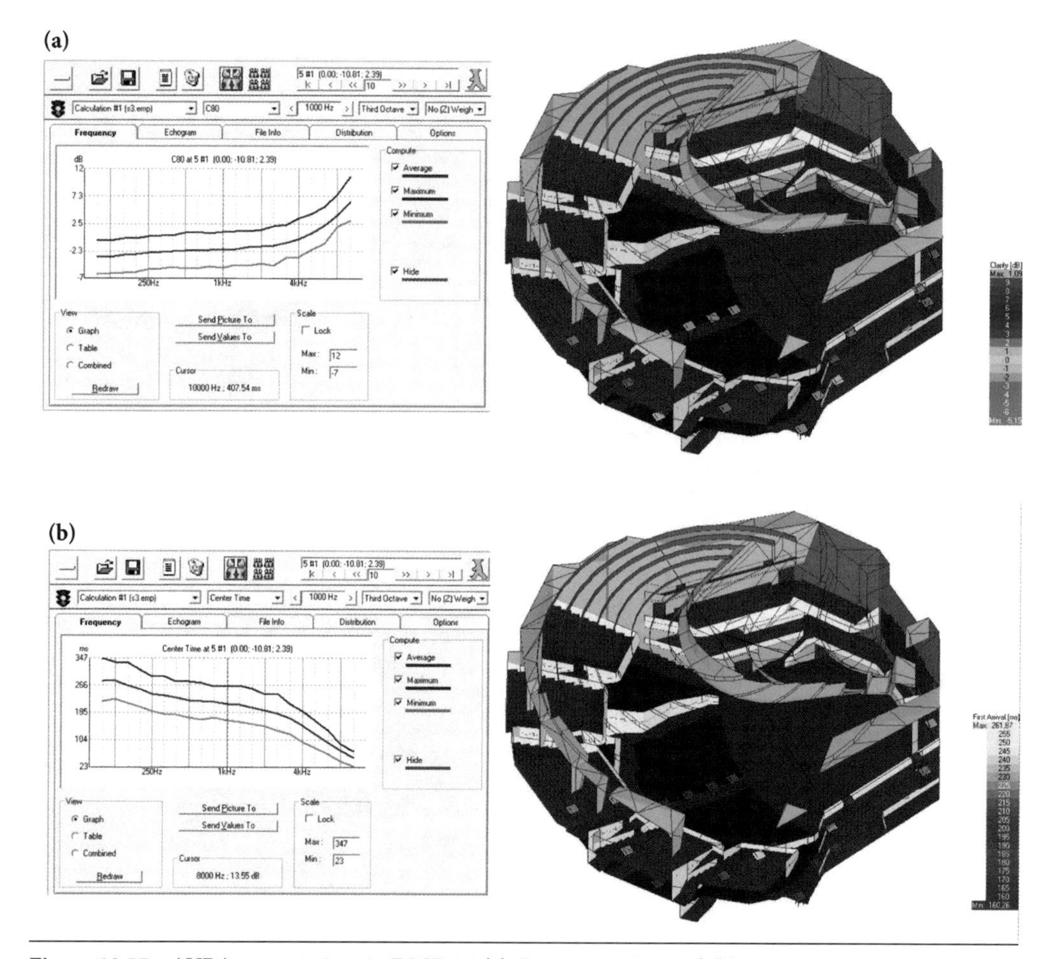

Figure 12.57 AURA presentations in EASE 4.4 (a) C_{80} presentation and (b) center time presentation

12.5.3 Auralization

Any simulation program should have the ability to transfer the calculated impulse responses to a post-processing routine, which will be used to auralize the simulated sound samples via BRIRs achieved with the sound system using music or speech material recorded in an anechoic environment. Of course the routine must generate a binaural data file in WAV or a similar format (see Figure 12.58) or other more complicated sound file formats like the known B-Format.

12.5.3.1 Useful Application of Auralization:

- Comparison of acoustical situation before and after application of room acoustic treatment
- *Audibilization* of the effect of a sound system or a natural sound source
- To make audible sound events or effects in rooms which don't exist anymore or which are not yet erected

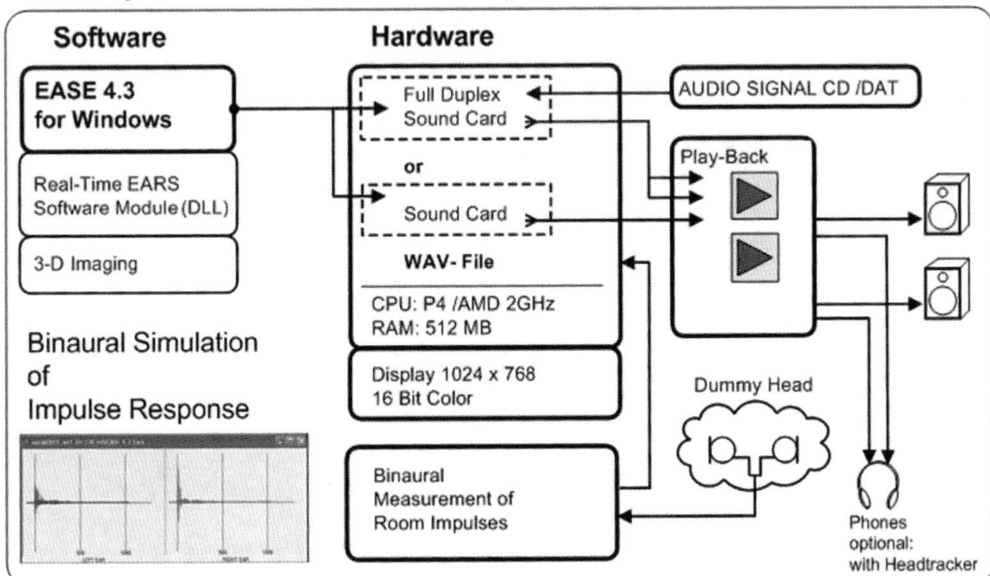

Figure 12.58 Block diagram of an auralization routine.

12.5.3.2 Limits and Abuse of Auralization:

- To make final decisions on the basis of just two auralizations of competing projects
- Use of auralization for acoustic design work alone

SUMMARY

This chapter describes the basic parameters in room acoustics that are important for the sound system designer. Some special acoustic measures for concert hall or opera house design are also found in the literature.

In this chapter, the data structure of transducer data that is important for correct prediction of room acoustic or sound reinforcement predictions is explained in detail. Here the use of not only magnitude but also phase data is explained. Finally, after a short introduction to analytical calculations of sound systems, the complex simulation of sound fields including the acoustic sources or loudspeakers is shown.

REFERENCES

1. Beranek, L. L., Concert halls and opera houses, music, acoustics, and architecture. New York: Springer, Second edition, 2004.
2. Ballou, G., (Ed.) Handbook for Sound Engineers, Chapter 9 (Ahnert/Tennhardt), 2015, Focal Press, 5th edition.
3. ISO 3382, Acoustics – Measurement of room acoustics parameters, 2008-2012.
4. EN60286-16, Objective Rating of Speech Intelligibility by Speech Transmission Index, 2004.

5. Sabine, W. C., Collected papers on acoustics. Cambridge: Harvard Univ. Press, 1923.
6. Kuttruff, H., Room acoustics. London: Spon Press, 2000, 4th edition.
7. Lehmann, U., Untersuchungen zur Bestimmung des Raumeindrucks bei Musikdarbietungen und Grundlagen der Optimierung. Diss. Tech. Univ. Dresden, 1974.
8. Ahnert, W., Einsatz elektroakustischer Hilfsmittel zur Räumlichkeitssteigerung, Schall-verstärkung und Vermeidung der akustischen Rückkopplung. Diss. Tech. Univ. Dresden, 1975.
9. Thiele, R., Richtungsverteilung und Zeitfolge der Schallrückwürfe in Räumen (Directional distribution and time sequence of sound reflections in rooms). *Acustica* (1953) Suppl. 2, p 291.
10. Houtgast, T. and Steeneken, H. J. M., A review of the MTF concept in room acoustics and its use for estimating speech intelligibility in auditoria. *J. Acoust. Soc. Amer.* 77 (1985), pp 1060–1077.
11. Schroeder, M. R., Modulation Transfer Functions: Definition and Measurement, *Acustica*, Vol. 49 (1981).
12. Jacob, K., McManus, S., Verhave, J. A. and Steeneken, H. J. M., "Development of an Accurate, Handheld, Simple-to-use Meter for the Prediction of Speech Intelligibility," Chapter 7 in Past, Present and Future of the Speech Transmission Index, TNO Human Factors, The Netherlands (2002).
13. ISO 7240-19.
14. IEC 60268-16, 2011.
15. Peutz, V. M. A., Articulation loss of consonants as a criterion for speech transmission in a room. *J. Audio Engng. Soc.* 19 (1971), H.11, pp 915–919.
16. Klein, W., Articulation loss of consonants as a basis for the design and judgement of sound reinforcement systems. *J. Audio Engng. Soc.* 19 (1971).
17. Technical report ISO TR 4870: Acoustics: The construction and calibration of speech intelligibility tests, 1991.
18. Lehmann, P. and Wilkens, H., Zusammenhang subjektiver Beurteilungen von Konzertsälen mit raumakustischen Kriterien (Relationship of subjective evaluation of concert halls with room-acoustic criteria), *Acustica* 45 (1980) 4, pp 256–268.
19. Reichardt, W., Abdel Alim, O. and Schmidt, W., Definitionen und Meßgrundlage eines objektiven Maßes zur Ermittlung der Grenze zwischen brauchbarer und unbrauchbarer Durchsichtigkeit bei Musikdarbietungen (Definitions and measuring base of an objective measure for ascertaining the boundary between useful and useless transparency). *Acustica* 32 (1975) 3, p 126.
20. Beranek, L. L., Music, acoustics and architecture. New York: Wiley, 1962.
21. Dietsch, L., Objektive raumakustische Kriterien zur Erfassung von Echostörungen und Lautstärken bei Sprach-und Musikdarbietungen (Objective room-acoustical criteria for registering echo disturbances and loudnesses in speech and music performances). Diss. Tech. Univ. Dresden, 1983.
22. Kürer, R., Einfaches Messverfahren zur Bestimmung der "Schwerpunktzeit" raumakustischer Impulsantworten (A simple measuring procedure for determining the "center time" of room acoustical impulse responses), 7th Intern. Congress on Acoustics, Budapest, 1971.
23. Damaske, P. and Ando, Y., Interaural Crosscorrelation for multichannel Loudspeaker reproduction. *Acustica* 27, pp 231–238.
24. Jordan, V. L., Acoustical design of concert halls and theaters, Applied Science Publishers Ltd. 1980.
25. Barron, M., Auditorium Acoustics and Architectural Design, Verlag E & FN SPON, London, 1993.
26. Kleiner, M., A new way of measuring lateral energy fractions. *App. Acoust.*, Vol. 27, 321 ff (1989).
27. Herrmann, U. F., Handbuch der Elektroakustik (Handbook of electroacoustics). Heidelberg: Hüthig, 1983.
28. Fletcher, H. and Munson, W., Loudness, its definition, measurement and calculation, *J. Acoust. Soc Am.*, vol. 5, pp 82–108, 1933.
29. Barkhausen, H., Ein neuer Schallmesser für die Praxis (A new sound meter for practice). Z. tech. Physik (1926). Z. VDI (1927) p 1471 ff.
30. Standard IEC 651, 1979—Sound Level Meters.
31. ISO-Standard 523 Ausg. 1975-07-15 Method for calculating loudness-level.

32. Zwicker, E. and Feldtkeller, R., Das Ohr als Nachrichtenempfänger (The ear as communication receiver). 2nd Ed. Stuttgart: Hirzel, 1967.
33. Blauert, J., Räumliches Hören (Stereophonic hearing). Stuttgart: Hirzel, 1974.
34. Jeffers, L. A. and McFadden, D., Differences of interaural phase and level detection and localization. JASA 49, (1971) pp 1169–1179.
35. Haas, H., Über den Einfluß eines Einfachechos auf die Hörsamkeit von Sprache (On the influence of a single echo on the audibility of speech). *Acustica* 1 (1951) 2, p 49 ff.
36. Kraak, W., Investigations on criteria for the risk of hearing loss due to noise. In: Hearing research and theory. New York: *Acad. Press*, 1981, Vol. 1, pp 187–303.
37. Sound system design for the Eugene Performing Arts Center, Oregon, Recording-Eng./Producer, S.86–93 (1982).
38. IEC 60268-5.
39. Ahnert, W. and Steffen, F., Sound Reinforcement Engineering, Fundamentals and Practice. London: E&FN Spon 2000, USA and Canada, New York: Routledge, 2000.
40. Stenzel, H., Über die Richtwirkung von Schallstrahlern, Elektrische Nachrichten-Technik, Band 4 (1927), Seite 239.
41. Stenzel, H., Leitfaden zur Berechnung von Schallvorgängen, Verlag von Julius Springer, Berlin, 1939.
42. Olson, H. F., Elements of Acoustical Engineering, D. van Nostrand Company, 2nd edition, New York, 1947.
43. Ureda, M. S., Wave Field Synthesis with Horn Arrays, 100th AES Convention Copenhagen/Denmark, May 1996, preprint No. 4144.
44. Benecke, H. and Sawade, S., Strahlergruppen in der Beschallungstechnik. Sonderdruck Funkpraxis, 1951.
45. European Patent Application EP 1 199 907 A2.
46. Heil, C., Sound Fields Radiated by Multiple Sound Sources Arrays, 92nd AES Convention, Vienna, 1992, March 24–27, preprint No. 3269.
47. Duran-Audio BV, White paper: Modelling the directivity of DSP controlled loudspeaker arrays, Zaltbommel/Holland, June 2000, under www.duran-audio.nl.
48. Heinz, R., "DSP-Driven Vertical Arrays," 2006, white paper, www.rh.com.
49. Van Beuningen, G. W. J. and Start, E. W., Digital directivity synthesis of DSP controlled loudspeaker arrays—A new concept, DAGA 2001, Hamburg-Harburg.
50. FIRMaker on http://firmaker.afmg.eu.
51. KEMAR-Head: http://www.gras.dk/00012/00330/.
52. EN 60268-4: Sound system equipment. Microphones.
53. Dynacord DDL 204 Digital Delay Line, www.dynacord.com.
54. Webers, J., Handbuch der Tonstudiotechnik (Handbook of Sound studio equipment), 9th ed. München: Franzis, 2003.
55. http://www.tcelectronic.com/Reverb4000.asp.
56. Schroeder, M. R., Frequency response in Rooms, *JASA* 34 (1962), pp 1819–1823.
57. Schroeder, M. R. and Kuttruff, H., On frequency response curves in Rooms, *JASA* 34 (1962), pp. 76 et seq.
58. Tool, F. E., Loudspeaker measurements and their relationship to listener preferences; JAES 34 (1986) 4, pp 227–235, 5, pp 323–348.
59. Mapp, P., Technical reference book "The Audio System Designer," edited by Klark Teknik in 1995.
60. http://www.behringer.com/EN/Products/DSP110.aspx.
61. ODEON software, version 10.1, www.odeon.dk.
62. CATT-Acoustic software, version 9, www.catt.se.
63. EASE software, version 4.4, www.afmg.eu.
64. Bose-Modeler, version 6.8, worldwide.bose.com.
65. Cox, T. J. and D'Antonio, P., Acoustic Absorbers and Diffusers: Theory, Design and Application, Taylor & Francis London and New York, 2nd edition, 2009.
66. ISO 354:2003: Acoustics—Measurement of sound absorption in a reverberation room.

67. ISO 17497-1: 2004 Acoustics—Sound-scattering properties of surfaces—Part 1: Measurement of the random-incidence scattering coefficient in a reverberation room.
68. Schmalle, H., Noy, D., Feistel, S., Hauser, G., Ahnert, W. and Storyk, J., "Accurate Acoustic Modeling of Small Rooms," presented at the 131st Convention of the Audio Engineering Society, New York, *J. Audio Eng. Soc.* (Abstracts), convention paper 8457 (2011 Oct.).
69. http://www.clfgroup.org/.
70. Ahnert, W. and Feistel, S., "Cluster Design with EASE for Windows," presented at the 106th Convention of the Audio Engineering Society, *J. Audio Eng. Soc.* (Abstracts), vol. 47, p 527 (1999 June), convention paper 4926.
71. Ahnert, W., Feistel, S., Baird, J. and Meyer, P., "Accurate Electroacoustic Prediction Utilizing the Complex Frequency Response of Far-Field Polar Measurements," presented at the 108th Convention of the Audio Engineering Society, *J. Audio Eng. Soc.* (Abstracts), vol. 48, p 357 (2000 Apr.), convention paper 5129.
72. Feistel, S., Ahnert, W. and Bock, S., "New Data Format to Describe Complex Sound Sources," presented at the 119th Convention of the Audio Engineering Society, *J. Audio Eng. Soc.* (Abstracts), vol. 53, pp. 1239, 1240 (2005 Dec.), convention paper 6631: GLL format specification, http://www.auralisation.de, http://www.sda.de.
73. AES56-2008, AES standard on acoustics—Sound source modeling—Loudspeaker polar radiation measurements.
74. Moldrzyk, C., Feistel, S. and Ahnert, W., "Verfahren zur binauralen Wiedergabe akustischer Signale," Patent No. DE 10 2006 018 490.4.
75. Software MAPP-XT, Meyer Sound Inc., www.meyersound.com.
76. Sound Vision Version 3.0, L-Acoustics, www.l-acoustics.com.
77. EASE Focus Software 2.0, SDA, www.easefocus.com.
78. Schmitz, O., Feistel, S., Ahnert, W. and Vorländer, M., Merging Software for Sound Reinforcement Systems and for Room Acoustics. Presented at the 110th AES Convention, 2001, May 12–15, Amsterdam, Preprint No. 5352.
79. Dalenbäck, B.-I., Verification of Prediction Based on Randomized Tail-Corrected Cone-Tracing and Array Modeling, 137th ASA/2nd EAA Berlin (March 1999).
80. Van Maercke, D. and Martin, J., The prediction of echograms and impulse responses within the Epidaure software, *Applied Acoustics* vol. 38 no. 2–4, p 93 (1993).
81. Naylor, G. M., Odeon—Another hybrid room acoustical model, *Applied Acoustics* vol. 38, no. 2–4, p 131 (1993).
82. Farina, A., Ramsete—A New Pyramid Tracer for Medium and Large Scale, Proceedings of EURONOISE 95 Conference, Lyon, 21–23 March, 1995.

This book has free material available for download from the Web Added Value™ resource center at *www.jrosspub.com*

13

Noise Control in Heating, Ventilation, and Air Conditioning Systems

Douglas H. Sturz, Acentech Inc., Cambridge, MA, USA

This chapter provides design guidelines, general design information, and references for detailed design data that are useful in assessing and controlling noise due to building mechanical systems. Information is provided for control of noise transmission via duct systems and for air flow velocities in ducts to achieve identified noise goals. There is also discussion regarding control of noise from equipment commonly found as part of building HVAC systems. Sound isolating constructions for building mechanical rooms are discussed in Chapter 7. A portion of this chapter is devoted to the control of vibration and structure-borne noise transmission from building mechanical systems. Further discussion and additional detailed design information can be found in the references listed at the end of this chapter; in fact, the presentation in this chapter relies on the detailed data found in those references.

13.1 NOISE CRITERIA

A variety of noise goal systems are available to use for design or to rate the acceptability of noise in occupied spaces in buildings, especially due to HVAC systems. Many commonly used criteria are presented in Chapter 7. The person responsible for noise control for a project will have to decide what criterion is appropriate for the various spaces.

Noise criteria (NC) goals are perhaps the most widely used, practical, and satisfactory criteria for a wide range of space types. For most sound spectrums, including many sound spectrums in buildings that result from HVAC systems, these provide a good evaluation of the suitability of a given noise environment for a variety of acoustical purposes. The largest body of knowledge regarding the application of noise criteria is probably for NC levels. The one potential drawback to the use of NC is that it does not evaluate well those sound spectrums with higher than nominal low frequency content. Many/most noise spectrums do not have excessive low frequency noise spectrums, but there are some systems and applications that routinely produce high levels of low frequency noise in occupied spaces and NC goals are often not the best for evaluating such spectrums.

To try to help improve the evaluation of sound spectrums with other than preferred spectrum characteristics and, particularly for the evaluation of low frequency noise, the room criteria (RC) and balanced noise criteria (NCB) schemes were developed. The NCB scheme is just as easy to apply as the NC, and is somewhat more stringent than NC at low frequencies so that it helps evaluate them better for a more preferred environment. The RC system is cumbersome to use and typically is applied for only the most refined of spaces or evaluations. RC seeks to achieve highly preferred spectrum shapes to ensure that the noise condition is *neutral* sounding and does not have excessive rumble, roar, or hiss. It is suggested that *neutral* sounding spectrums be achieved, but when evaluating HVAC systems for noise control, since not all the noise aspects of the systems can be evaluated or predicted, comparison with neutral spectrums is most often not possible for design. The RC process is better for evaluating measured conditions. Even at this, the criteria will tell one if the noise is neutral, rumbly, has roar, or is hissy, but when the condition is other than neutral, it does not give guidance as to how acceptable the noise condition might be for any particular application. You can't compare the perceived loudness or suitability of a non-neutral sound spectrum to the desired neutral spectrum standard.

A-weighted noise levels are another simple and commonly used noise criterion that easily assesses the subjective loudness of sounds with a wide range of spectral characteristics, but this works perhaps less well for sounds that are rumbly in character. To get the A-weighted sound level, the absolute sound pressures in the various frequency bands across the spectrum have weighting factors applied to them, corresponding to the perceived loudness of the sound at that frequency as compared to other frequencies. After the weighting factors are applied, the sound levels in the various frequency bands are summed as decibels (dB) to achieve the overall A-weighted sound level (dBA). There is no absolute and consistent translation between NC and A-weighted levels, but for many commonly occurring sound levels that are not substantially out of balance, A-weighted sound levels for a given case will be about four to seven points higher than the NC level rating of that sound. Most sound level meters that are available have built-in A-weighting functions so that these levels can be directly measured. Although evaluation of a noise condition/concern with A-weighting is perhaps not as refined as the other criteria mentioned above, this can be suitable for use in many cases. Note that when one only collects overall A-weighted sound level data for a noise condition, the spectral distribution of the sound is lost, and this can be an impediment for engineering of noise control.

13.2 DUCT-BORNE NOISE TRANSMISSION

The usual process for evaluating noise transmission via duct systems is to start with how much sound energy is emitted to the duct path from a source and then to evaluate how much attenuation each successive duct section provides along the path. In doing this, one can predict the sound level reaching the receiving space; and by comparing this to the desired noise goal, the magnitude of attenuation that needs to be inserted in the duct path, if any, can be determined. However, the process can also be worked in reverse—starting with the sound level that is desired in the occupied space, one can work back through the system to determine limitations on how loud the source can be to be consistent with the desired goal. With this information the allowable noise from the source can be specified. If the noise of the potential source can then be determined and this noise is greater than allowed with the natural system attenuation, the additional sound attenuation treatment for the duct system that needs to be added can be determined. Data is available in various references (see the end of this chapter) for the attenuation effect of many relatively standard duct elements, but it is also common for duct conditions not to exactly match the conditions for which data are available, and it is necessary to extrapolate between data points or to adjust the attenuation performance of some duct elements. The prediction procedure provides a methodical approach to evaluating sound transmission in duct systems, but because of the complexity of sound fields and the differences in sound field character that can enter each duct section, the procedure should only be considered approximate. Usually the precision with which attenuation can be calculated in the higher frequency bands is at least good, but with even modest attenuation treatment in the duct system there is often more high frequency attenuation than is needed, so precise prediction in this frequency range is usually not needed to resolve most noise issues, we typically just need to have the noise level be below a desired goal. Fortunately, at moderately low frequencies, where most noise concerns for HVAC systems exist, the prediction mythology is sufficiently precise to allow good decision making that results in noise levels sufficiently close to the desired target.

Rather than repeat design data that is available in common references, this presentation will deal with the general concepts; the reader will need to go to the references for the detailed prediction data. Alternatively, there are also computer modeling programs available from vendors.

13.2.1 Sound Attenuation in Straight Ducts

Even for simple, straight ducts, attenuation along the length of a duct section is complex. The attenuation is impacted by the character of the sound field entering the duct section, and the nature of the sound field will constantly vary along the length of the duct. The attenuation per given length of duct at any and every frequency band of interest will not be identical for each portion of a duct length. Nevertheless, we typically assume that the attenuation is the same per unit length of duct, and the data that is presented in references is usually given per unit length of a duct section that is moderately long.

Attenuation in ducts occurs when the sound waves interact with the walls of the ducts. Even hard, bare sheet metal ducts provide some attenuation. The attenuation in bare ducts is generally quite small at high frequencies, but at low frequencies, the sound field in the duct loses a noticeable amount of energy via transmission through the duct walls. When the duct is lined with sound absorptive material, there is still sound energy loss at low frequencies by the same mechanism described above, but there is also considerable attenuation at mid and higher frequencies due to dissipation in the lining. In straight ducts, very high frequency sound can *beam* down the duct and does not interact with the duct walls, so even with lining for longer duct lengths there can be less attenuation than anticipated. Fortunately, from the standpoint of achieving attenuation in the duct system, any change of direction of the duct will substantially attenuate the very high frequencies that might beam down a straight duct section because in the subsequent duct section, those sound waves become cross modes in the duct. Typically there will only be a concern about the potential lack of very high frequency attenuation if there is only a single straight duct between the source and receiving space. The attenuation values listed in the references are typical for ducts that meet the Sheet Metal and Air Conditioning National Association (SMACNA) standards for low pressure ductwork of the indicated sizes. Ductwork constructed to high pressure standards (that may be stiffer and heavier) will provide somewhat less attenuation. Because of the greater wall stiffness that is inherent with circular cross section ducts, they typically provide less attenuation than flat-panel rectangular ducts. The larger the duct cross-sectional size, and the more efficient the cross-sectional shape, the lower the attenuation will be per unit length because the sound energy interacts less with the duct walls.

13.2.2 Sound Attenuation by Duct Divisions

When acoustical energy propagating along a duct path comes to a branch, the sound energy is assumed to divide uniformly per unit area of the ducts leaving the junction. The effective attenuation effect in dB for the duct path that is being assessed is given by:

$$\text{Attenuation} = 10 \log (A_1/A_T), \text{ (dB)},$$

where, A_1, is the area of the duct of interest leaving the division and, A_T, is the total area of all branches leaving the division (not including the duct from which the sound is propagating).

At the low to moderately low frequencies that are of greatest concern for controlling noise in HVAC systems, this is probably a good prediction of the attenuation effect in the frequency range of greatest interest. However, at higher frequencies directivity effects in the ductwork make this estimation less accurate. Since most HVAC systems do not have concerns at very high frequencies due to the issues presented above, the imprecision of this aspect of the prediction model will seldom be of practical concern.

13.2.3 Sound Attenuation by Duct Cross Section Area Changes

When sound propagating along a duct encounters an abrupt change in cross section and there is nothing in the air stream to disturb the sound field, some sound is reflected back from the cross-sectional change, but a relatively large change in cross section is needed to produce significant attenuation for practical purposes. This effect is a strong function of frequency with the predominate effect at low frequencies. In ducts, this effect generally has little practical value because duct system designers rarely have large enough abrupt transitions (for aerodynamic reasons) to cause any significant effect. The most practical significance of this is where a duct terminates at a wall or ceiling surface and the area changes from the duct size to the room size. For this particular application of duct path cross-sectional change, the references refer to this as *end reflection*. However, in most cases there are diffusers or grilles at the duct terminations, and the presence of these devices reduces the attenuation effect. For most practical designs this effect can be ignored or should be substantially discounted.

13.2.4 Sound Attenuation by Elbows

When sound propagating along a duct path encounters a change of direction (elbow), the sound field after the turn has less energy because some sound energy is reflected back in the direction from which it was coming, and some is dissipated in extra duct cross modes just following the elbow. The larger the duct dimension in the plane of the turn, the greater the attenuation at low frequencies. High frequency attenuation does not vary a great deal with the width of the duct turn. The presence and type of vanes in the elbow will impact high frequency attenuation. Although mitered elbows without vanes provide the greatest high frequency attenuation, they will have greater flow resistance than vaned or radius elbows and the type of elbows to use should typically be determined by mechanical considerations. It may be that the increase in noise generated by the fan to overcome the extra flow resistance of a system without vanes in mitered elbows will be greater than the attenuation gained by not having elbow vanes, and this is not a good trade-off. Lining of elbows provides significant attenuation at high frequencies and this can be significant for short systems; for longer systems, there can be more elbows than needed to control noise transmission down the duct path at middle to higher frequencies. See the detailed attenuation performance data in the references for the various styles and sizes of elbows.

13.2.5 Prefabricated Silencers

Prefabricated duct silencers (also, attenuators or mufflers) are available from many manufacturers. These are available in a variety of standard constructions addressing a great number of common HVAC system applications. Custom size/shape silencers are also available from some suppliers. The sound absorptive media within silencers may be conventional dissipative materials such as glass fiber, mineral wool, or nylon wool covered by a perforated metal facing sheet to provide physical protection for the fibrous sound absorbing material from erosion by higher flow velocities and/or the turbulent flow through the narrow channels in the silencer. For very low velocity flows, the sound dissipating material may be protected by a thin surface layer of flow resistive facing such as that applied on the surface of duct liners. For special applications, to prevent any fibrous material from being shed into the air stream, the sound dissipating fill material can be sealed in thin plastic bagging. This typically will slightly increase the acoustical attenuation at specific low and mid frequencies and diminish attenuation more substantially at high frequencies compared to unfaced fill. Special internal construction details are typically used by suppliers to prevent the film (bagging) from sealing the holes in the perforated metal protective facing and to avoid chafing of the film on sharp edges formed on the back of the perforated metal in the punching process. Check the detailed characteristics of the bagging material for the application. There are also silencers without traditional dissipative fill, but with special acoustically reactive cavities lining the air channels to remove acoustical energy from the airstream. Such silencers are typically less effective and more expensive than conventional dissipative fill silencers.

Basic silencers have a straight-through air path with aerodynamic inlet and discharge geometry to help minimize pressure losses which are greatest at the flow transition points. Silencers are also available with special flow configurations for special applications. For HVAC system applications, elbow configuration silencers are particularly notable since these can help to avoid difficult flow conditions that could otherwise exist or even be caused by aerodynamically poor application of a straight silencer. Some manufacturers can also provide silencers in other complex geometries. The point of these alternate silencer geometries is to achieve the desired attenuation for the system without inducing excessive energy loss. Most often it is better in the long run to use a custom silencer with carefully conceived internal geometry and a known level of lower than nominal flow resistance than to use straight silencers in a poor aerodynamic configuration.

Silencers are typically tested for their sound attenuation, pressure drop, and flow-generated noise performance in accordance with ASTM E-477 or ISO 7235 standards—and the performance is catalogued. To create uniform conditions for all manufacturers, tests are made under rather ideal standard conditions. This is important for assessing the acoustical insertion loss and the self-generated noise of the device, but it is also important relative to the pressure loss that is expected. The system designer needs to take into account the manner in which the silencers are applied to understand how much pressure loss is expected for the actual application. Silencer manufacturers often provide guidance as to how much extra pressure loss may result from a variety of less than ideal flow conditions. For most typical HVAC system applications the pressure drop across a silencer that is applied near a fan and well away from occupied spaces should be limited to about 0.25″ to 0.30″ wg. The pressure loss can be higher, but the designer needs to be sure the extra pressure loss will not have undesirable impacts on the system. The energy consumed to overcome the pressure drop of the silencer should be considered in deciding what pressure loss is acceptable. Over the lifetime of the system there could be significant energy cost savings for having a silencer that has a particularly low pressure drop, which may be much more than the extra initial cost of the low resistance model. For a given application there are typically several silencers that might be able to provide the desired attenuation performance. The noise control and HVAC design engineers must select one that also yields a suitably low pressure drop and trade-offs will need to be considered between the length, cross section, and the resistance class of the silencer.

High velocity flow through the internal air passages within the silencer creates flow noise, and the designer needs to consider this by adding it to the sound field that propagates past the silencer. Manufacturers publish self-generated noise data for their silencers for a variety of face velocities and for a specific cross-sectional area. The designer needs to adjust the data for the area of the actual silencer used in the project, and this is done in accordance with the methodology presented by the manufacturer. The closer the silencer is placed to the receiving space and the quieter the receiving space, the greater the care that needs to be taken in controlling flow-generated noise, because there is less opportunity to attenuate it before it reaches the receiving space. For duct systems serving spaces where the noise goal is moderate (perhaps NC-35 to NC-40), silencers located near the occupied space (such as on the room-side of a terminal box) typically need to be sized for less than about 0.08″ to 0.10″ wg pressure drop, if there is no significant further attenuation between the silencer and the occupied space. This is just a guideline and the actual self noise of the silencer may need to be more rigorously assessed. Where the noise goal is more stringent than the example given here, the pressure loss through the silencer needs to be commensurately lower.

Poorly selected and positioned silencers, which produce excessive pressure drop, can sometimes cause more noise problems than they solve; so care needs to be taken in the application of silencers. To the extent that silencers are anticipated to be needed for a system, the original duct system planning would ideally create a good and logical location in which to include a silencer.

13.2.6 Sound Attenuation by Plenums

Plenums in duct systems can provide significant sound attenuation for the ducted path. The magnitude of the attenuation depends upon the size of the plenum compared to the connecting ducts, the orientation of the inlets and outlets, and on the absorptivity of the inner walls. The references provide design data for sound attenuating plenums of varying sizes and geometries. The accuracy of the attenuation predicted by the available methodologies is not considered to be especially good, and this is an area where research activity is being done.

13.2.7 Room Effect

Room effect is the translation of the sound power emitted into the room from the duct system to the sound pressure level that results in the space at a given location in the receiving space. The sound power emanating from the end of the duct system spreads out in increasingly larger *shells*, and as the energy is spread over the larger *shell* surfaces, the intensity level is reduced. The sound level that results at a given position in the receiving room is a function of the distance from the source, the directivity of the source, and the acoustical character of the receiving room. The references provide methodologies for determining the conversion from the sound power emission to the sound pressure level in the receiving room; and one simply needs to enter the tables, figures, or formulas that are available in the various references to determine the appropriate factors to use.

For cases where there is only one source of concern for design, using the available methods to determine the room effect is simple and sufficient. However, where there are multiple noise sources such as various diffusers or grilles with close to the same noise emission level and would produce close to the same noise level at the receiver position, the additive effect of multiple sources needs to be considered.

13.3 FLOW NOISE IN DUCTED SYSTEMS

13.3.1 Main Duct System Design

The noise generated by flow in duct systems is a concern for noise transmission along the duct path and for noise radiation from the duct surface. Noise generated at particular fittings or duct elements can be estimated in accordance with the methods presented in the references and based on this, noise, in the receiving spaces of interest, can be predicted. The noise generated by the flow in the duct needs to be added to the sound field propagating along the duct at each applicable duct section or component. This section only addresses general guidelines for controlling noise in ducted HVAC systems to be consistent with various noise goals.

The following guidelines are offered for duct sizing based on the location, type, and the class of ductwork for systems serving spaces with moderate to higher noise goals:

- <3,000 fpm (feet per minute, 1 ft = 0.305 m) velocity in round ducts in mechanical rooms and shafts well away from occupied spaces.
- <2,500 fpm velocity in rectangular ducts in mechanical rooms and shafts well away from occupied spaces.
- <2,000 fpm velocity in the ceiling of occupied spaces with mineral fiber ceiling for an NC-35 to NC-40 goal.
- <1,500 fpm velocity in the ceiling of occupied spaces with open or acoustically transparent ceiling for an NC-35 to NC-40 goal. (Acoustically transparent ceilings might be glass fiber panel, perforated metal, or open slat ceilings.) This would also apply for exposed ducts.
- <1,500 fpm velocity in larger final distribution ducts serving NC-35 to NC-40 spaces.
- A friction rate (pressure loss rate) of 0.10″ wg/100 ft of duct run in smaller final duct distribution serving spaces with an NC-40 goal.

- A friction rate (pressure loss rate) of 0.08″ wg/100 ft of duct run in smaller final duct distribution serving spaces with an NC-35 goal.

Air flow velocity considerations for the design of special low noise spaces (typically NC-30 and lower) are presented in the air flow velocity guideline for lined ducts in Table 13.1. Recommended air flow velocities for systems without a lining are about 20% lower in the ductwork near diffusers and grilles compared to the recommended lined duct velocities.

Duct systems serving particularly quiet spaces need to be designed to be as naturally balanced as possible. That is, if the system is turned on, the air flow naturally delivered from each outlet would ideally match the desired air flow with little need to throttle dampers to control the flow. Creating systematically branching duct arrangements to a regular array of diffusers or grilles is typically a good approach to achieve this goal. To achieve a naturally balanced design, it is generally best if the diffusers and grilles are within a reasonably cohesive grouping, rather than widely spaced-out. Avoid having a nice, tight cluster of diffusers serving a critical space and then have a long *tail* duct extending elsewhere, because it will be difficult to balance the system without excessive damper throttling, which will generate noise.

For systems serving especially quiet spaces, dampers should be provided in duct systems as necessary to control the flow, but they must be located sufficiently far from the terminal

Table 13.1 Recommended air flow velocities in lined duct systems[2,5]

Duct Element or Device	Air flow velocities (fpm) consistent with indicated noise criterion through net free area of duct section or device							
	NC 15		NC 20		NC 25		NC 30	
	Supply	Return	Supply	Return	Supply	Return	Supply	Return
Terminal device[1] (½″ min. slot width)	250	300	300	360	350	420	425	510
First 8-10 feet of duct	300	350	360	420	420	490	510	600
Next 15-20 feet of duct	400	450	480	540	560	630	680	765
Next 15-20 feet of duct	500	570	600	685	700	800	850	970
Next 15-20 feet of duct	640	700	765	840	900	980	1080	1180
Next 15-20 feet of duct	800	900	960	1080	1120	1260	1360	1540
Maximum within space[5]	1000	1100	1200	1320	1400	1540	1700	1870

Notes:

1. No dampers, straighteners, deflectors, equalizing grids, etc., behind terminal devices.
2. All ducts with 1″ thick internal sound absorptive lining.
3. Fan noise must be considered separately.
4. Reduce duct velocities (not diffuser/grille velocities) by 20% if ductwork is unlined.
5. Above mineral fiber panel ceiling. Lower velocities 20% if open or acoustically-transparent ceiling.

devices so that the noise they generate can be attenuated before it reaches the diffuser or grille at the space served; dampers at or near the face of the diffusers and grilles must be avoided for especially quiet systems. The duct system to serve the space should be crafted to create appropriate opportunities for locating what dampers are needed.

13.3.2 Diffuser and Grille Selection

The air flow velocity guidelines for diffusers and grills serving low noise spaces presented in Table 13.1 are guidelines based on a nominal diffuser face slot width of about ½ an inch (1.25 cm), the ductwork leading to the diffuser is configured to have nominal aerodynamic conditions, and other assumptions about room conditions. With larger nominal slot widths, somewhat higher velocities can cautiously be allowed. Even though these guidelines are rather general, experience with them has proven to be very good and appropriate in the design of particularly quiet spaces.

Manufacturers often provide data for the noise generated by their diffusers/grilles and typically, these are boiled down to a single NC rating for the diffuser/grille at a particular flow. These ratings are typically developed based on a standardized methodology. While the data are generally good, one has to be careful in application. Be careful in applying manufacturer's data because they are generated under ideal flow conditions, often include very favorable room effect factors, and they are typically for just one diffuser or grille serving the space. There is usually the assumption that the air flow is uniformly distributed over the active face of the grille, and this may not be the case in the field. Adjustments to the rated performance need to be made to account for the differences in getting to the conditions of the actual application. Especially note that the published data do not include any effects of dampers at or near the face of the diffuser; the presence of dampers at this location can have significant adverse results for the noise generated at the diffusers, especially if the duct system has characteristics that are particularly far from naturally balanced. Even though there are many variables in the noise conditions that may result due to flow through diffusers, it is often satisfactory to choose diffusers and grilles with manufacturer's noise ratings that are five to eight NC points below the desired goal for the space served, to account for nominal field conditions rather than the conditions of the test.

13.4 NOISE BREAK-OUT/BREAK-IN

When ducts containing high levels of noise pass though spaces with low or moderate noise goals, it is necessary to consider whether the noise that is in the duct may transmit out through the duct walls and cause a noise concern in the occupied space; this is typically called break-out noise. Similarly, when a duct in which the sound field inside has been quieted passes near a noisy piece of equipment prior to entering a receiving space, noise may break into the duct and travel down the duct to a receiving space of concern; this is typically called break-in noise. To minimize concern for noise break-out/in, it is usually best to position a high attenuation element in the duct system (a silencer) at or very near the penetration of the mechanical room boundary. When this is not possible, in order to address a noise break-in/out concern, it may be necessary to enclose the duct in soffit constructions (such as gypsum board on suitable framing) or to wrap them with special noise barrier wrap materials to reduce the unwanted noise transmission. Typically noise barrier wrap materials are less effective than constructions of gypsum board on stud framing and one needs to carefully consider the magnitude of noise control that is needed for the condition.

Sections in the various references present methods for predicting noise break-out/in conditions, and they also provide data on the noise control performance of some rather specific treatments that are commonly considered to control excess sound transmission when this is called for.

Note that walls of circular ducts are inherently stiffer than the walls of rectangular ducts, and for this reason, distribution systems using circular ducts are often less prone to break-out/in noise problems. There can be particular concern for break-out/in noise issues when ducts are exposed in occupied spaces, or highly acoustically transparent ceilings (glass fiber, perforated, or slotted ceilings) are used in occupied spaces in buildings. Where there are solid ceilings (like gypsum board), break-out/in noise is naturally less of a concern because of the higher transmission loss of the ceiling.

Ducts that drop out of the bottom of rooftop air handling units into the ceiling plenums above occupied spaces are a classic case where duct break-out noise is a problem, and this typically is worst when the supply fan discharges directly down into the ducts rather than first passing through a plenum within the unit above the roof. Elbow silencers with high transmission loss casings can be a convenient method of addressing this problem. If a straight silencer is used and is located well away from the elbow beneath the unit that turns in the ceiling space, there can be a significant amount of noisy duct between the roof penetrations and the silencers which can radiate noise to the occupied space and will likely need to be treated. For critical applications it is often best to avoid this condition by running the ducts across the roof so that suitable attenuation can be achieved before the ducts enter the building. In some cases, oversized curbs are provided, and the ducts run through the curb area where attenuation is achieved before penetrating the roof into the occupied space.

The most common case for a noise break-in concern is when the silencer for a duct system is located well inside the boundary of the mechanical space and then the duct runs over the top of a noisy piece of equipment (perhaps like a chiller or compressor) just before penetrating the mechanical room boundary construction into a noise sensitive space on the other side of the mechanical boundary construction. In this case, it may be necessary to enclose the ductwork in a noise barrier construction for some distance on one or both sides of the wall penetration, and it may be necessary to provide attenuation in a duct section at the wall penetration.

13.5 FANS

Fans are typically the major source of noise emission into duct systems that are of interest for analysis. The various fan types have differing noise spectrum and magnitude characteristics, and this is also a function of the duty. Certain fans are aerodynamically better suited for particular flow and pressure applications. This needs to be considered for the full operating range of the application. Selecting a fan that is not aerodynamically well matched to the duty can result in much more noise than is necessary for an application.

Manufacturers typically provide noise data for their fans based on standard test methods, and this is typically the preferred source of noise data for use in predicting sound propagation in duct systems or to the outdoors. Often there are several potential fan selections that will handle a particular duty, and it is wise to try to use the quietest of the reasonable fan selections for a given application. Including noise consideration in the fan selection process should help avoid being surprised by a particularly noisy fan, and this can help avoid the need to add excessive noise control treatments which will burden the system.

Particular care should be given to the inlet and outlet flow conditions at fans. Poor inlet flow can unbalance the flow over a fan wheel causing performance problems and unexpected excess noise generation. Air flow wakes (such as those due to an obstacle in the flow) that get sucked into some types of fans can cause the fan to be much more tonal than if the air flow into the fan is very smooth. The fan speed may need to be increased to compensate for unexpected system losses, causing even more noise than anticipated to deliver the necessary system flow. Inlet vanes or substantial flow distortion due to flexible connections protruding into the flow can easily cause an increase of noise on the order of 6 dB across the full frequency spectrum. Because airflow velocities on the discharge of the fan are typically high, system performance and fan noise generation are particularly susceptible to the aerodynamic condition that exists close to the discharge of the fan; designers should work to create favorable aerodynamic conditions here and must avoid poor flow conditions. Some fans and systems achieve much better noise conditions by having especially smooth flow conditions leading into or exiting the fans, and this can be highly desirable. However, be careful when using such systems that the necessary air flow geometries to achieve these results are maintained throughout the flow zone that influences these desired effects.

13.6 TERMINAL BOXES/VALVES

Many HVAC systems serving buildings include a variety of types of terminal boxes/valves to control the amount of air delivered to occupied spaces. What all of these boxes/valves have in common is that they produce a pressure drop from the higher pressure in the main duct system to a low pressure in the final distribution ductwork on the room side of the box. In throttling the pressure from the main to the final distribution ductwork, noise is generated—and this noise needs to be considered in the design to meet desirable noise levels in the occupied spaces. Typically the pressure loss across the box/valve is the primary determinant of how much noise is created, but the noise is also a direct function of the air flow quantity through the box. Boxes on a system that are near the fans typically see a higher pressure in the main duct than boxes that are near the ends of the system and so the boxes close to the beginning of the system will generate more noise than comparable box applications located toward the end of the system. Excess static pressure in the main ducts can be a significant factor in noisy systems as the terminal boxes in the system are throttling hard to control the excess pressure. To minimize noise, systems should be conceived in design to be able to operate at low pressures and, particularly, should have the smallest practical differential in pressure arriving at the various boxes. There are various duct design strategies to help minimize the system pressures, and these should be employed. In particular, keep flow velocities as low as reasonably possible. Also, use aerodynamic fittings, avoid poor aerodynamic fitting configurations, use ring ducts, and lower the velocity and flow resistance rate in progressing toward the end of the duct run rather than maintaining friction rate or velocity. Limit the length of runs and develop main distribution schemes with multiple short main runs. In the field, the systems need to be set to operate at the lowest pressure that meets the mechanical requirements. All too often, systems are left with excess pressure and this leads to excess noise. System control strategies that reset the system pressure to the lowest that can achieve the mechanical requirements are helpful to ensure that there is no more pressure drop than necessary taken by the boxes/valves and hence ensure the lowest noise operation all the time.

Manufacturers of boxes/valves generally provide noise data for their boxes rated at various flow rates and pressure losses, and these data can be used for noise analysis as presented above. Remember that the pressure loss that is of interest in determining box/valve noise is the actual pressure drop that the box is expected to produce in the field, not the minimum pressure the box/valve needs for control or some other arbitrarily selected pressure. You need to deliver the necessary pressure for the critical box on the system to operate, but then you need to consider the pressure that arrives at the closest boxes to the fan on the system that results in getting the necessary pressure to the ends of the runs.

There is also noise radiation from the casings of the boxes, which can impact the occupied spaces near them—to enable assessment of this, manufacturers publish casing radiated noise data. For fan-powered boxes, this data typically includes the noise that radiates from the fan out the plenum air inlet, and this can be a significant issue. In many cases it is necessary to provide attenuation for the plenum air inlet to the box. Most moderate size terminal boxes that are operating at nominal pressure drops can be located in the ceilings of occupied spaces where the noise goal is NC-35 or higher, when there is a conventional ceiling attenuation class (CAC) 35 mineral fiber ceiling. Where boxes are in the ceilings of spaces with noise goals more stringent that NC-35, where system pressures are anticipated to be high, and/or where ceilings with low transmission loss (low CAC ratings) are used, special consideration should be given to address casing radiated noise. Be cautious of exposed box applications or where glass fiber, perforated, or open slat ceilings are used. Also be careful where incomplete ceilings or ceiling clouds with openings around the edges are planned.

13.7 VIBRATION ISOLATION CONSIDERATIONS FOR BUILDING MECHANICAL SYSTEMS

Technical aspects of isolation of vibration sources are treated in detail in several of the referenced texts. Practical considerations and rather detailed general recommendations for vibration isolation of mechanical systems in buildings are presented in the ASHRAE *Applications Handbook*. Vibration isolation equipment vendors also provide guidance regarding the application of isolation devices for building mechanical equipment. This section is to provide a general overview of practical considerations regarding the application of vibration isolation systems and devices to building mechanical systems.

The isolation requirements for a particular application are based primarily on the type of equipment, the speed of the equipment, the location in the building, and the sensitivity of the application. For application to building equipment, the selection of vibration isolators is typically reduced to identifying the deflection of the isolation element under the static load, but underlying this, the deflection relates to the key dynamics considerations that affect isolation efficiency—the natural frequency of the isolation device. For typical isolators used for building mechanical equipment, the natural frequency of the isolators as a function of frequency is given by:

$$f_n = 3.13(1/d)^{0.5},$$

where:

f_n = isolator natural (resonance) frequency, Hz
d = isolator deflection, in. (1 inch = 2.54 cm)

To provide effective isolation, the natural frequency of the isolation system needs to be below the drive speed of the equipment by a factor greater than three and perhaps as high as about 10. At factors that are less than this, isolation efficiency is undesirably small and at a ratio of about 1.4, there is theoretically no isolation at all. You must avoid having the natural frequency of the isolation device be equal to (or even nearly equal to) the drive speed of the equipment because this will amplify vibration transmission, and movement of the isolated piece of equipment on the isolation mounts will be exacerbated. Typically there is little point in having an isolation system with a natural frequency more than a factor of 10 below the drive speed of the equipment.

Inertia bases are used with many pieces of equipment to lower the center of gravity of the isolated equipment and to reduce the vibratory movement of the equipment when it is resiliently supported. Potential movement due to starting torque, the static pressure a fan develops, or turbulent flow in large fan systems are also reasons that such bases are used. Some equipment is also designed in anticipation that its base will be intimately and uniformly in contact with a structural base. Inertia bases only reduce the motion of the equipment on the isolation system; they do not change the inherent unbalanced forces of the equipment, even though the magnitude of vibratory movement of the equipment on the base will be lower. The use of an inertia base does not change the vibration forces that are transmitted to the building, assuming that the deflection of the isolation system remains the same. The mass of an inertia base may be based on technical considerations, but more often there is no precisely correct weight of an inertia base, and the mass is determined by other considerations such as the requirement to create a suitably stiff base of a size to carry the piece of equipment. When this is done, for building mechanical equipment, the weight of the inertial mass is typically suitable relative to the weight of the equipment to accomplish the desired goals for use of the inertia base.

To achieve the desired isolation performance, the minimum static deflection of an isolator under load is specified and the contractors, together with the isolation product vendor, make the final selection of the isolators once the final equipment selections are known. In doing this, they need to consider the actual load distribution to each of the isolation mounts so that the correct selection can be made. Typically, because of economic factors, they will not provide a resilient element that substantially exceeds the specified requirements, but it is also important to check that they do not provide isolation devices that will not achieve the minimum actual deflection requirement because this will result in less isolation efficiency than is intended. The deflection of the isolation device under the actual applied load needs to be checked compared to the deflection performance criterion; it is not sufficient to simply look at the nominal (or rated) deflection series of the isolator.

Commonly presented, simple vibration isolation theory typically assumes that the support for the equipment looks into structures that are infinitely stiff and massive, such as the earth. When the supporting structure is substantially flexible or has little mass, isolation is less effective; you cannot isolate vibration from a structure that effectively looks like a trampoline and is soft compared to the typically applied isolation mounts. Building structures that are above grade can be, and typically are, suitably stiff for practical isolation purposes without undue adverse impact when there is at least a concrete slab involved in the construction and the span of the structure is not particularly long. If the structural span supporting the equipment is particularly long (and hence potentially flexible), it would often be wise to modestly increase the deflection of the isolators used for the application to account for the extra flexibility of the structure. When equipment is located at a roof condition where the roof is a simple metal deck and insulation system, this is a very light and flexible support condition, and care often needs to be taken regarding support of such equipment to avoid support to the metal deck between more substantial structural elements. Significant equipment items will likely need to be supported via dunnage frames to column extensions or to major structural members in the roof plane that are very stiff structural elements.

Seismic restraint of equipment, to the extent that is required, is an entirely separate issue from vibration isolation even though these issues are sometimes linked; seismic restraint is essentially a structural issue. Isolation and seismic restraint become related because many pieces of equipment that are resiliently supported need restraint, and sometimes isolation devices have integral features that can provide the desired restraint. In some cases it is convenient to incorporate seismic restraint into the isolators—products that do this are available from vibration isolation vendors, but it may not be necessary to use combination devices. It is essential that seismic restraints not compromise the performance of the vibration isolation system under normal operation, and this is not an easy task, considering the relatively small installation tolerances that are required for both seismic restraints and vibration isolators.

It is necessary to be sure that ducts, pipes, and electrical conduits, etc., that connect to vibrating equipment do not create paths for vibration transmission to the building structure that defeat the desired isolation of the basic equipment mounts. Often flexible couplings are used to control transmission by these paths, and sometimes it is necessary to go further and isolate the pipes, ducts, or conduits for some distance from the equipment. For electrical connections, flexible couplings or conduit sections usually provide sufficient control of vibratory energy via these connections. For ducts, a conventional four to six inch (10–15 cm) long flexible duct connection is common. For pipes, although flexible coupling devices are often used for mechanical purposes, these are only marginally effective for controlling vibration transmission along the pipes because significant vibratory energy is carried in the fluid within the pipe; resilient support of the pipe system is commonly required for some defined scope from the connected vibrating source.

13.8 OUTDOOR NOISE EMISSIONS

Many state and local authorities have noise regulations to control the noise impact within a community. Typically, either fixed noise limits for specific receivers are established or the allowable noise level is based on the existing ambient noise level in the impacted area. If the noise requirement is based on the existing ambient noise condition, to be rigorous about compliance, the existing ambient noise condition has to be measured and this may not be a trivial exercise. The details of the requirement need to be checked to be sure the correct assessment is made. Where residential neighbors are close by, where requirements are stringent, and/or where the receivers look down on the noise sources, meeting the noise requirements can be quite challenging.

Many pieces of mechanical equipment that are particularly economical (such as many air-cooled devices), can be quite noisy and may not easily conform to noise regulations. Some types of this equipment have available upgrades or accessories to lower noise emissions, but in some cases it may be necessary to change the mechanical schemes or equipment selections to reduce noise emissions to meet community noise requirements. Although such changes may entail greater initial expense, they can sometimes provide an energy-saving payback in the long run. Quieting the noise at the source is almost always the best approach to controlling outdoor noise emissions. Some equipment can be equipped with variable speed drives so that when the demand on the equipment is low, the equipment (typically fans) can operate more slowly and, therefore, with substantially lower noise emission levels. This is, of course, statistical noise reduction and will not produce noise reduction when full capacity/speed is required. However, this is particularly useful for equipment such as cooling towers or other fan systems for heat rejection because at night, when community noise limits are typically most stringent, it may not be necessary to move as much air to meet the demand. The noise reduction that results with variable speed drives often works naturally with the typical diurnal noise cycle and people's sensitivity to environmental noise. It is best to avoid equipment with a noise emission spectrum that is particularly tonal since many regulations prohibit, or more strictly limit, tonal sounds. Tonal sounds and other sounds that have an undesirable character are likely to be judged to be more annoying than an equivalent level of broad-band sound, so the likelihood of not generating a noise complaint is improved if this sort of equipment is avoided.

Sometimes, suitably quiet equipment cannot be purchased within practical or economical limits, and it is necessary to apply external noise control treatments. For many pieces of equipment and systems the desired noise reduction can be reasonably and economically accomplished by applying prefabricated silencers in the air flow paths or by building a noise barrier to block sound emissions in the direction of concern. Note, however, that there are often space, equipment performance, or air flow implications for the application of these treatments. The ability to adapt noise control treatments to equipment, if required, is a factor that should be considered in selecting the basic equipment and mechanical scheme. For small amounts of excess noise (typically less than about eight dBA), it may be possible to create a noise barrier with sound attenuating louvers which can allow some air flow to help mitigate adverse airflow obstruction due to the presence of the barrier. When the receiver of concern looks down on the source of the noise, a noise barrier is typically of little or no value for controlling noise reaching the neighbors of concern, because it is difficult to build a practical barrier to block the line of sight. Relocating a piece of equipment so that a noise barrier can be effective can be a good and cost-effective approach toward resolution of a noise concern. For some applications where the receiver is particularly low in elevation relative to the source, raising the equipment to the roof of a building can create very favorable geometry to help a noise barrier work better.

REFERENCES

1. ASHRAE, *Applications Handbook*, latest edition.
2. Cyril Harris, "Handbook of Acoustical Measurements and Noise Control, Third edition," McGraw Hill Book Company, New York, 1991.
3. Sound Research Laboratories Ltd., "Noise Control in Building Services," Pergamon Press, Oxford, 1988, Allan T. Fry, editor.
4. Mark E. Schaffer, "A Practical Guide to Noise and Vibration Control for HVAC Systems," ASHRAE, 1991.

5. Robert S. Johnson, "Noise and Vibration Control in Buildings," McGraw Hill Book Company, New York, 1984.
6. Istvan L. Ver and Leo L. Beranek, editors, "Noise and Vibration Control Engineering, Principals and Applications, Second Edition," John Wiley & Sons, Inc., 2006.
7. Trane Acoustics Program (TAP).

14

Acoustical Design of Worship Spaces

Ewart A. Wetherill, Acoustical Consultant, Alameda, CA, USA

14.1 INTRODUCTION

Even a cursory survey confirms that many worship spaces, whether new or ancient, present less than satisfactory conditions for hearing. Despite the wealth of information on acoustics of buildings, a substantial gap still remains between this understanding and how it can be translated successfully into design and construction. The intent of this chapter is to bridge the gap and thus help avoid perpetuation of poor hearing conditions in still more worship spaces. It describes how the fundamental principles of architectural acoustics—avoidance of acoustical conflicts by careful planning, control of unwanted sounds (*noise control*), and enhancement of wanted sounds (*room acoustics*)—can enable a designer with even an elementary understanding of acoustics to attain satisfactory hearing conditions in either a new or remodeled building.

Because of the diversity of religions, any of which may have several separate denominations, there are many different ideas on the form of worship, but all have a common need—the ability to hear well. The range of activities extends from silent prayer and meditation to understanding of speech and listening to sacred music of different cultures, all of which must be accommodated architecturally. Consideration should also be given to nonworship aspects of space use, such as meetings or concerts that may be appropriate as a way of providing operating funds.

To avoid reference to any specific religion, the neutral terms *worship space* or *sanctuary* are used to define the building or room within it where the formal act of worship takes place. Discussion of design and construction is based on contemporary North American building practice and usage of materials. However, although building customs and methods differ throughout the world, since the basic principles of acoustics are constant, the conclusions drawn here should be broadly applicable to all situations.

The requirements for good hearing within the sanctuary can be summarized as: a sufficiently quiet background noise level, adequate loudness of wanted sounds, suitable reverberation for the specific event, and avoidance of acoustical defects that would degrade the required hearing conditions.[2] How the building design incorporates these requirements may be summarized as:

Acoustics		*Architecture*
Quiet background	*requires*	Adequate enclosure
		Control of ventilation noise
Adequate loudness	*requires*	Interior shaping and finish materials
Control of reverberation		Sound amplification
Avoidance of defects		

14.2 FUNDAMENTALS AND PRINCIPLES

14.2.1 Requirements for Good Hearing

14.2.1.1 Quiet Background

The sanctuary should be quiet enough for softly-spoken words and quiet passages of music to be heard. This requires control of intruding noises from traffic, aircraft, and the ventilation system as well as avoidance of interference from other spaces.[1] Control of outdoor noise may begin with the selection of a quiet building site or shielding of the sanctuary from outside noise by other spaces. If this is not feasible, the walls and roof must be substantial enough to reduce indoor noise to an acceptable level. Obvious noise paths such as windows and doors must be designed so as not to reduce the effectiveness of a heavy enclosure

Within the enclosed volume, adequate mechanical ventilation is needed, which in turn requires control of noise and vibration from fans, air diffusers, and other equipment. Procedures for noise and vibration control of mechanical systems are well established and appropriate materials are readily available to meet specified design goals.

14.2.1.2 Adequate Loudness

Assuming that intruding noise has been controlled, loudness and projection of wanted sounds to all listeners can be enhanced by interior shaping and suitable surface materials. While the level of sound in free space will diminish as the distance from the source increases, sound-reflecting hard surfaces such as plaster or masonry will sustain and regenerate sounds throughout an enclosure. Within the sanctuary, orientation of sound reflecting wall and ceiling surfaces close to the choir and organ can enable the musicians to project and also hear each other well. By contrast, late-arriving reflections from rear walls or other distant surfaces (echoes) tend to obscure subsequent syllables and thus reduce clarity.

14.2.1.3 Suitable Reverberation

Reverberation is the persistence of sound in a space as a sequence of overlapping reflections after the source has ceased to produce sound. *Reverberation time* (T_{60})—the time required for a sound to decay to one-millionth of its initial intensity—provides a convenient indicator of acoustical character. The length of reverberation is determined by the volume of the space and the total amount of sound absorption within it.[2] Adding room volume will tend to increase T_{60}, while adding sound absorption will decrease it. For example, if the volume per occupant is less than 200 cubic feet (5.7 m^3), the T_{60} is probably reasonable for speech, while a volume of over 300 cubic feet (8.5 m^3) per occupant is more likely to be suitable for unamplified music.

14.2.1.4 Good Distribution of Sound

The interior shape of the space should allow strong and diffuse sound reflections to reach all listeners in the correct time sequence. Deep under-balcony spaces or focusing surfaces such as barrel vaults and domes do not distribute reflected sound evenly and are rarely desirable. Rear walls and ceilings should be shaped to provide diffusion and scattering, or they could be covered with a sound absorbing material to attenuate reflected sound. However, reflections from side walls can enhance singing in the congregation by helping the singers to hear each other clearly.

When designed to complement the natural acoustics of the space—and used with discretion—a speech amplification system is an important requirement for all but the smallest worship spaces. It can provide intelligibility for a wide range of speaking skills and an equally wide range of hearing acuity. In particular, the sound system must provide high intelligibility for people with impaired hearing.

14.2.2 Sound Propagation Outdoors

Sound outdoors radiates spherically from its source; its intensity is proportional to the square of the distance from the source in accordance with the inverse square law, $I_1/I_2 = D_2^2/D_1^2$. The sound level decreases by 6 *decibels* (dB) for each doubling of distance from the source (see Figure 14.1).

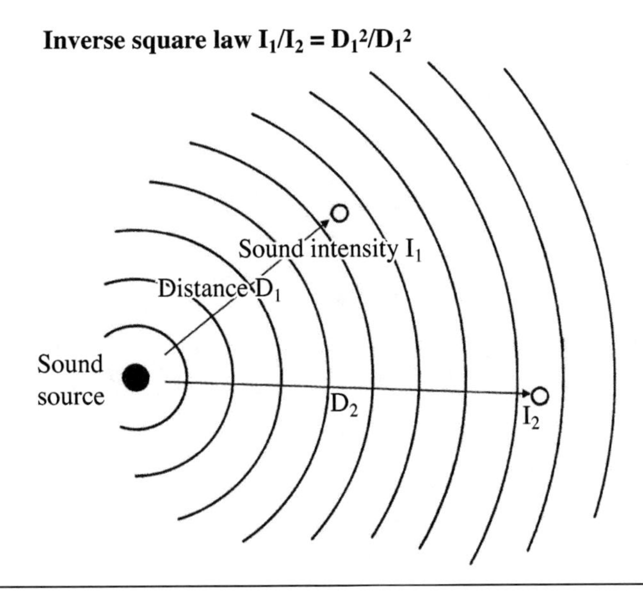

Figure 14.1 Sound propagation outdoors.

14.2.3 Sound Distribution in a Room

Sound indoors behaves in the same manner as outdoors until it encounters a boundary surface where it is either reflected, absorbed, or transmitted beyond the space. For surfaces that are large relative to the wavelength of the sound, reflections are specular, i.e., the angle of incidence equals the angle of reflection. Since at mid to high frequencies sound behaves approximately like light, it is convenient to represent a sector of a sound wave graphically by a *ray diagram*. This allows a first approximation of early reflections based on the interior shape and location of the source (see Figure 14.2).

A plot of the sound decay pattern for an impulsive sound (such as a balloon burst) shows each reflection as heard at the microphone location. Gaps in the time sequence indicate an absence of useful reflecting surfaces while strong peaks indicate noticeable *echoes* that could obscure later-arriving notes or syllables, as seen in Figure 14.3.

14.2.3.1 Acoustical Properties of Materials

Many aspects of building acoustics are common sense and in one form or another are familiar to most people from everyday experience—and yet they are often overlooked in the design process. The basic rules are simple: (a) an enclosure of heavy, impervious materials such as stone or concrete is effective for isolation from noise, whereas one of a lightweight, porous material, such as a glass fiber blanket, is not; (b) sound created within a masonry enclosure will excite reverberation, but if areas of carpet or heavy curtains are added, they will reduce reverberation by absorbing sound energy. In general, projection of sound to a congregation will be reinforced by a hard, sound-reflecting ceiling, whereas a sound-absorbing ceiling will tend to control excessive noise, such as in a dining hall where separate groups are conversing at the same time.

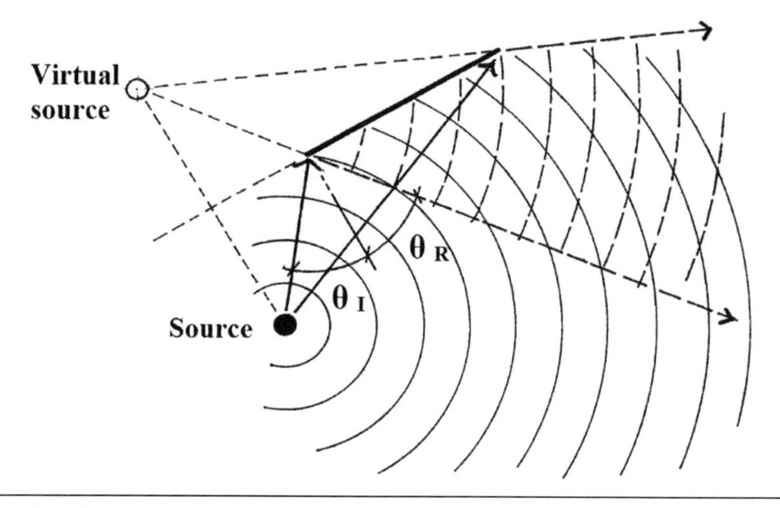

Figure 14.2 Sound propagation in a room.

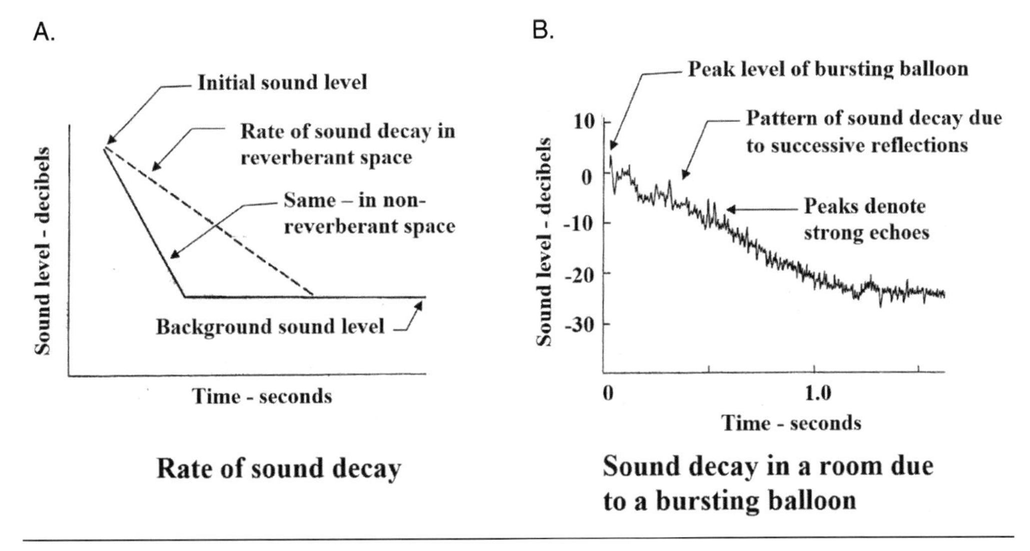

Figure 14.3 Decay of sound in air.

14.2.3.2 Sound Transmission Loss

Transmission loss (TL) of sound is a specific property of a building material that is the measure of its ability to resist transmission of sound through the material. For a built-up system comprised of several materials—possibly with sound *leaks* between them—loss of sound through the entire system is defined as *noise reduction* (NR).

14.2.4 Planning for a New Building

The design or remodeling of almost any building is a series of compromises in which benefits for any aspect of the design must be weighed against disadvantages for another, so the designer should know to what extent acoustical goals may be compromised without significantly affecting the usefulness of the building.

Since there may be no obvious connection between appearance or material selection and how the space will conduct the sound, acoustical quality of buildings is often overlooked in the initial design. In the normal course of events, the significance given to acoustical design will usually rank far below selection of the building site and the shape, structural form, and visual expression of the building. For example, in the process of designing and building or remodeling a worship space, many factors that are established by regulation and that may be completely inflexible—such as egress requirements—will always take precedence over hard-to-define acoustical properties that are intangible until the building is completed.

Consequently, the design must translate acoustic principles into architectural requirements and the details must be incorporated in the architectural drawings and specifications that form the basis for awarding a construction contract.[3] From the start, the designer must ensure close coordination and thorough cross-checking between disciplines. Possible changes that could be considered for the future, such as added seating or a T_{60} suitable for the eventual installation of a pipe organ, should be anticipated in the design from the outset. Figure 14.4 provides a model for tracking the simultaneous requirements to be included.

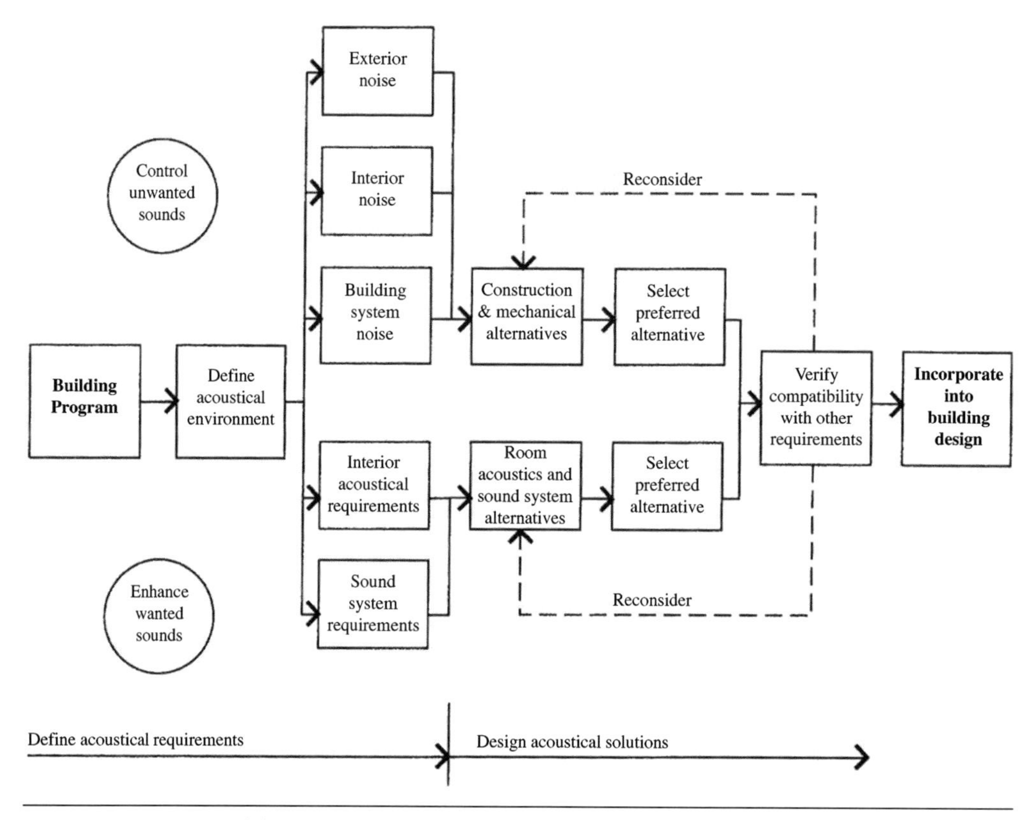

Figure 14.4 Acoustical design process [reprinted with permission from Crocker, M. J. ed., *Encyclopedia of Acoustics*, vol. 3, "Architectural Acoustics," 1997].

14.2.5 Selection of Acoustical Criteria

14.2.5.1 Use of Octave-Band Analysis

Sound level measurements are commonly analyzed in octaves, or one-third octaves, to define the distribution of sound energy and for ease of comparison with acoustical standards and other references, such as *noise criteria* (NC), as seen in Figure 14.5.

14.2.5.2 Criteria for Background Noise

Criteria for maximum background noise levels may be selected from tables of sound levels matched to specific uses. They are defined either as a single number or as a noise spectrum for which the level in each octave should not be exceeded. The latter is strongly recommended because most single-number rating systems (e.g., *A-weighted* sound level) do not take into account the low-frequency (bass) component of intruding noise that is often a source of distraction and annoyance in assembly spaces. However, if the difference—C-weighted level minus A-weighted level—does not exceed 20 dB, the balance between low and mid-to-high frequencies is likely to be considered acceptable.[4]

While any of several rating systems may be appropriate for this purpose, the important thing is to ensure that maximum allowed noise levels are clearly specified. For convenience,

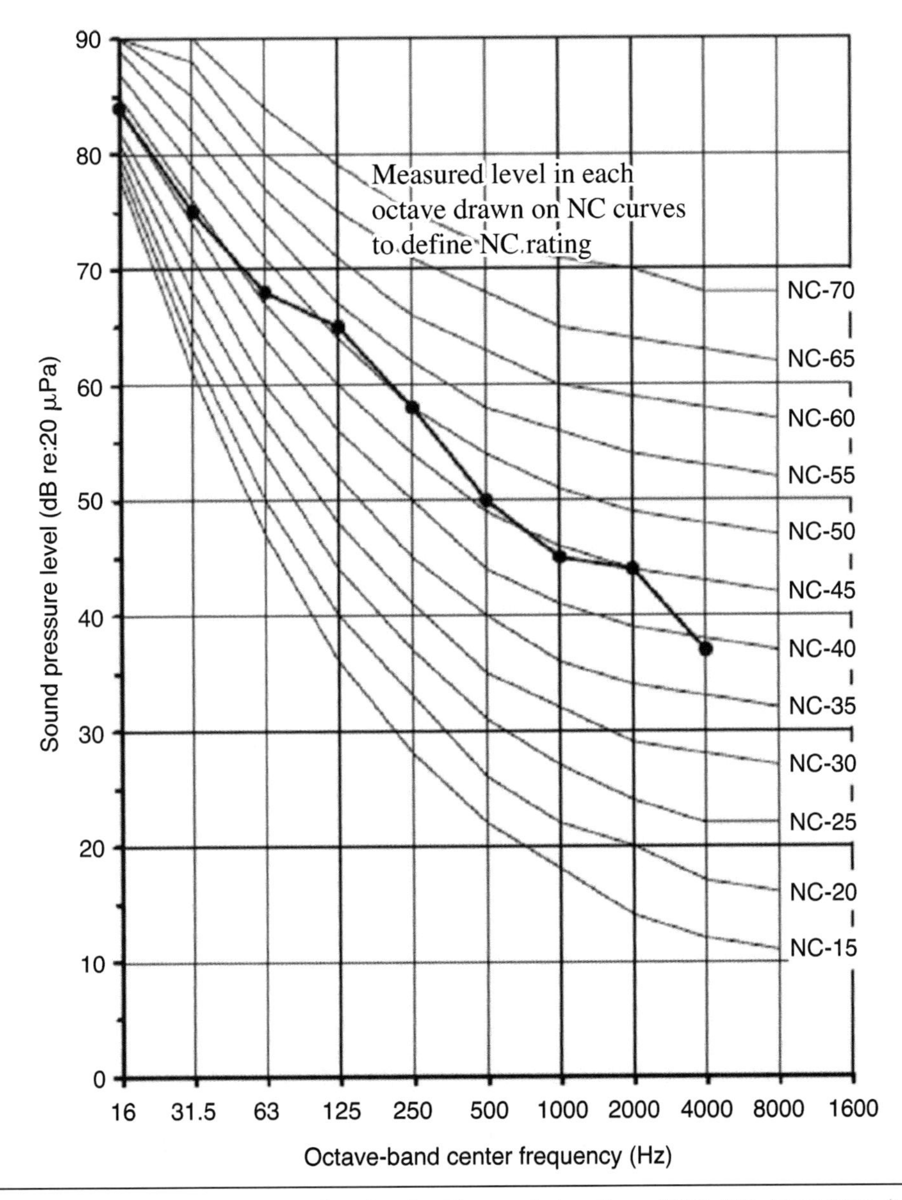

Figure 14.5 Noise criteria curves [reprinted with permission from ANSI-ASA S12.2-2008 (NC 51 spectrum shown)].

all references in this chapter are based on NC. Several alternatives and revisions to the original NC curves currently exist; the latest update (2008) is shown.

The maximum recommended background noise level for prayer or for performance of sacred music that may be broadcast is NC 25. Occasional slight variations from the specified values may not be significant, but large differences in level or strong peaks at certain frequencies should be avoided; and it should be free from noise intrusions such as fire sirens and aircraft. However, to enhance prayer or quiet music passages, or if digital recording is intended, the maximum level should never exceed NC 20. A cautionary note: although NC are simple to define and enable easy verification of compliance, careful analysis is needed to ensure that all building elements—and particularly the ventilation system—are actually designed and built to comply with the specified noise levels.

14.2.5.3 Criteria for Reverberation

Reverberation in a room is generally defined in terms of *mid-frequency* T_{60}, which can be measured in an existing space or calculated using the *Sabine equation.* There is general agreement on suitable T_{60} values for particular uses. A long T_{60} (e.g., more than 2.0 seconds) will enhance organ-choral music and congregational singing. A short T_{60} (e.g., around 1.0 second) is appropriate for speech clarity, but will not favor music. Figure 14.6 defines the spread of generally acceptable values with emphasis on the preferred range.

Recent studies have defined other important characteristics of sound decay that can be measured or derived from computer-based models. However, for an initial design study and a first estimate of building cost, calculations from an architectural plan and sections can provide a fast and reasonably effective way of approximating T_{60}, as well as defining the location and orientation of sound reflecting surfaces for sound projection and of avoiding potential echoes.

14.2.6 Control of Outdoor Noise

14.2.6.1 Site Evaluation

The minimum required NR of the building enclosure from outdoors to indoors is determined by subtracting the desired indoor level from the composite of maximum environmental noise levels due to existing and anticipated traffic, railcars, aircraft, etc., at the proposed site.

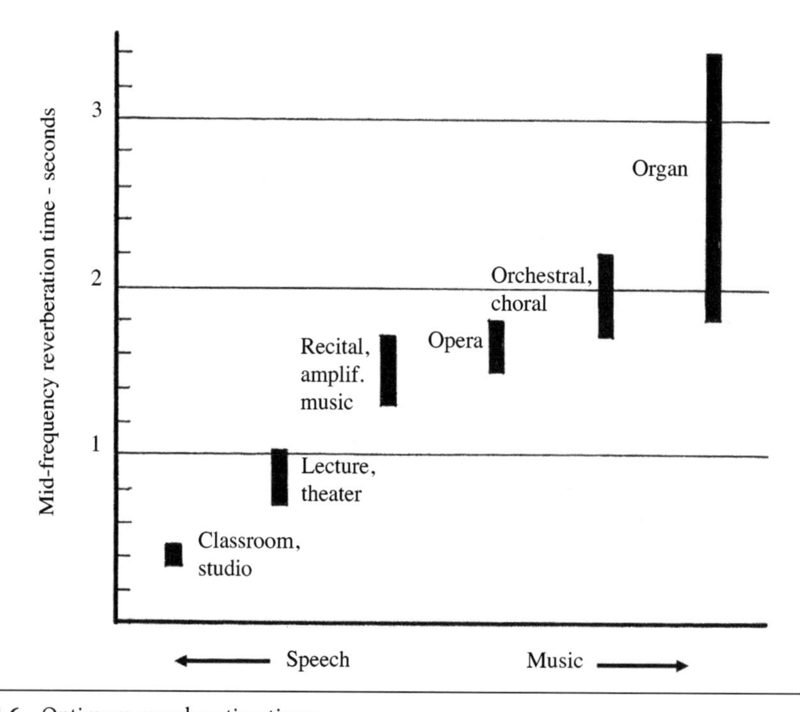

Figure 14.6 Optimum reverberation times.

The governing jurisdiction should be consulted to define allowed intruding noise levels for assembly and worship spaces based on U.S. federal highway standards [FHWA 23-772, 2011]. Since they typically specify single-number, time-averaged noise levels, these standards do not include either low-frequency levels or maximum overall levels. For this reason, octave-band measurements at the site should be made of the full audible frequency range, with a record of time of day and frequency of occurrence. Comparison with noise reduction capabilities of the desired building construction will indicate quickly whether a noisy site will be acceptable or not.

14.2.6.2 Outdoor Noise Level Minus *Indoor Level* Is Required Minimum *Noise Reduction*

A construction meeting this requirement should be selected. If it is too costly or not compatible with the proposed design, the building could either be relocated on the site or a more suitable site should be considered. If it is suitable except for noise from one direction such as a busy highway, noise exposure could be reduced by locating the sanctuary as far away as possible or shielding it with another building. Initial planning studies should also include alternatives such as facing the building away from the source of the noise, taking sound reflections from nearby buildings into account, and using heavy wall and roof construction on the noisier side. Each component of the building—including roof, windows, and doors—is a path for intruding noise and so requires evaluation.

14.2.6.3 Outdoor Equipment Noise

Transmission of noise may be *airborne* or *structure-borne*, each of which requires evaluation. Noisy outdoor equipment serving the new facility such as air-cooled chillers and transformers should be well isolated from both the worship space and adjoining residential areas by careful positioning and enclosures. Noise ordinances of many cities specify maximum noise limits at property lines, but in any case, it is always good practice to minimize annoyance for neighbors. Selection of outdoor equipment requires evaluation of manufacturers' published noise levels. Acoustical evaluation can confirm optimum site location or the need for enclosing equipment.

14.2.6.4 Noise Created by Building Enclosure

Some building types may require additional study because of their potential for creating distracting noise indoors and outdoors due to environmental changes. Typical examples include thermal expansion-contraction noise of metal cladding, noise from rainfall on metal roofs, and possible wind noise effects from nontraditional building forms. Since few of these conditions are well documented, any experimental design should be checked against existing buildings that have similar characteristics.

14.2.7 Control of Indoor Noise

14.2.7.1 Noise Control Between Spaces

The cost of effective noise control between spaces will depend largely on the ability of the designer to separate acoustically sensitive spaces, such as the sanctuary, from spaces with high noise levels. Exact requirements for noise control can be derived by comparing maximum noise levels of equipment or activities to criteria for quiet spaces. If noise control is not incorporated in the building plan, close attention to construction details and close inspection of the work will be needed.

When the required airborne NR between two spaces is defined, it is fairly easy to pick a wall construction that appears suitable from reference tables, but several pitfalls await the unwary designer. In the first place, the wall is only one element in a complex system comprising floor, walls, and ceiling. Each possible path by which sound can be transmitted must be blocked. Since many gaps and penetrations in construction are often hidden by finish surfaces, they are generally not of concern—except where noise control is required.

The second pitfall is to assume that the results of carefully controlled laboratory tests will in fact be emulated on site. It may be realistic to rate a supervised wall installation five dB below the test data published by a supplier, but without supervision, it could easily be 15 to 20 dB below the test values. Third, the single-number sound transmission class (STC) that is the common specification for noise isolation (see Figure 14.7) is often misapplied. While it may be a valuable tool in designing for speech privacy, it does not take into account low-frequency noise, such as the rumble of mechanical equipment or bass components of music, so a more detailed analysis may be required.

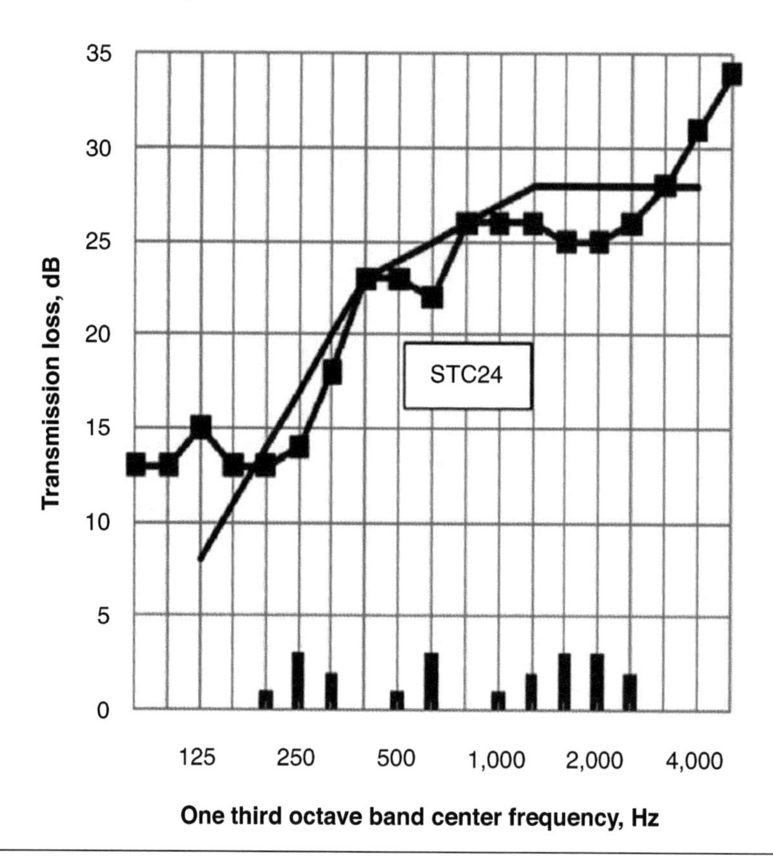

Figure 14.7 Sound transmission class (STC) contour [reprinted with permission from ASTM E413—Classification for Rating Sound Insulation].

14.2.7.2 Limitations of Divisible Spaces

Movable partitions can be effective for subdividing large spaces, but their advantages may be limited significantly where sound isolation is needed between the subdivided sections. Although movable partitions are a convenient option in some cases, other design possibilities, such as two partitions in tandem or the alternative of two simple nondivisible spaces, could add flexibility.

14.2.7.3 Control of Vibration and Structure-Borne Noise

Structure-borne noise isolation is especially difficult with wood frame construction because of its lightness and flexibility. Spaces such as corridors or classrooms, where footfall noise can always be expected, should never be built either above or connected structurally to the sanctuary. Single-story buildings can avoid such difficulties, but roof-mounted machinery such as air conditioning units and related piping should also be expected to transmit vibration and noise throughout a building if not located remotely and resiliently isolated by special mountings.

14.2.7.4 Impact Noise Control

Anyone who has lived in a wood-framed two-story house or apartment building is familiar with impact noise from footfalls on the floor above or on stairs. Control of impact noise is sometimes assumed to be simply a matter of installing carpet or other resilient flooring surfaces. However, avoiding impact excitation of structural resonances is usually difficult and is controlled better by complete structural separation. These resonances frequently occur at frequencies lower than the range defined by *impact insulation class* (IIC), thus, product advertising that claims to have high IIC values can be misleading and should always be verified by evaluation of existing installations (see Figure 14.8).

14.2.8 Ventilation and Air Conditioning Noise Control

14.2.8.1 Control of Equipment Noise

Until fairly recently, heating and ventilation systems for worship facilities usually consisted of fans and related equipment in a mechanical room kept remote from acoustically-sensitive spaces. However, cost-reduction strategies and changes in equipment design have led to the common use of factory-built HVAC units with built-in air conditioning. These package units, which tend to be noisy, are generally roof-mounted and are commonly installed with very short runs of ductwork between the unit and the space served. Most traditional noise and vibration safeguards could be neglected in such a system, resulting in background noise levels 10 dB or more above desirable levels. Consequently, control of noise and vibration—both indoors and outdoors—should be given priority as early as the schematic design.

14.2.8.2 Control of Fan Noise in Ductwork

Ducts transmit noise as well as air from fan to sanctuary. Both the supply and the return sides of the fan are noisy, and must be controlled to meet the required noise level by use of special acoustical duct lining or prefabricated duct silencers. Fan noise ratings, using standard

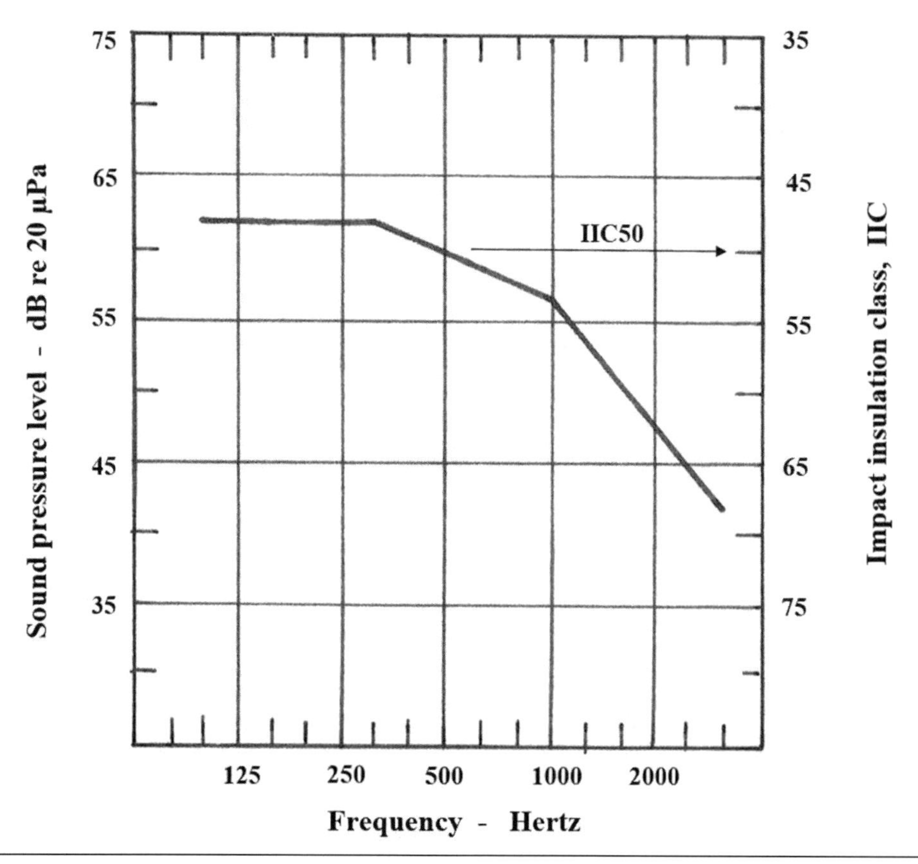

Figure 14.8 Typical impact insulation class (IIC) contour [reprinted with permission from ASTM E989—89 Determination of Impact Insulation Class].

evaluation methods based on the proposed system design, should be obtained from the fan supplier for calculation of required attenuation. A silencer should provide roughly the same attenuation as an acoustically-lined duct four to five times as long, so it may be economical to install, but it can add resistance to air flow and can also create noise itself if not sized or placed correctly.

14.2.8.3 Air Distribution—Supply and Return

For efficient and quiet air distribution, ductwork should split into branches arranged to deliver the required air quantity to each supply air diffuser without allowing cross-talk between spaces or creating excessive noise due to turbulence. Dampers which restrict air flow are commonly used for balancing air quantities between branches. However, they create noise and add airflow resistance, so careful layout and branching of ductwork is recommended to minimize dampers. For the sanctuary and other quiet spaces, dampers should never be installed close to supply air diffusers and actual air velocities should be checked against manufacturers' specifications. System controls should not be accessible for unauthorized casual adjustment.

14.2.8.4 Nonducted Air Return

To avoid the cost of return air ductwork, air distribution systems often rely on air returning to the fan room via above-ceiling plenums or through the corridors, requiring wall or ceiling transfer grilles or louvers between rooms. To maintain acoustical separation between spaces connected in this way, acoustically-lined transfer ducts should be used, rather than wall or door louvers. These ducts can usually be installed above a suspended ceiling for control of cross-talk or at the return air intake for control of fan noise.

14.2.9 Other Equipment Noise

Equipment that typically requires noise and vibration control includes elevators, transformers, water coolers, and exhaust fans. They should be kept at a distance from the sanctuary and may require resilient mountings. Dimmers, lighting control systems, and projectors generally must be close to the sanctuary and thus, may require special enclosures and mountings for control of noise and vibration from both equipment and cooling fans. Plumbing noise intrusion due to toilet flushing is best controlled by keeping such spaces and piping separated from quiet spaces. Resilient sleeves are recommended at all piping anchor points and penetrations for both air conditioning and domestic water systems.

14.2.10 Acoustics of Sanctuary

14.2.10.1 Background Noise Level

Assuming that intruding noise has been controlled, acoustical requirements for worship may vary from prayer, meditation, and possibly complete absence of music to choral, organ, and band music with bells, dancing, and processions. Selection of NC for the appropriate combination of uses should be a design priority together with any special liturgical constraints on plan layout, such as choir and organ location. Both speech and music can be enhanced by suitably oriented sound reflecting surfaces, which may also serve a dual purpose by forming part of the enclosure against noise intrusion.

14.2.10.2 Size and Shape of Space

The size, shape, and proportions of the space generally determine evenness of sound distribution and the *time delay* at the listener's ear between direct sound and reflected sounds. Concave and domed shapes, *transepts,*[5] or deep under-balcony spaces can create acoustical defects, but these could be mitigated by the shaping of interior surfaces to create diffusion of sound. For large places of worship it is desirable to consult someone experienced in room acoustics who will also take loudness, reverberation, and echoes into consideration.

The sanctuary plan may be rectangular or a shape that wraps the congregation around the central focus of worship. The latter is most suited to speech only, or to amplified speech and music. If a long T_{60} is not required, potential echoes or other acoustical deficiencies can be controlled by ceiling or wall-mounted sound absorbing material. In extreme cases such as a *mega church* accommodating several thousand people the most practical design option is to concentrate on the control of intruding noise and reverberation, relying primarily on a sophisticated sound amplification system for both distribution and quality of sound. However, for natural projection and support of sound from the choir, organ, and congregation, with amplification used only for speech, plan and section shaping and selection of wall and ceiling materials will require careful study early in the design.

14.2.10.3 Sound Reflecting and Absorbing Materials

The acoustical properties of a wall, floor, or ceiling surface depend on the frequency of sound and the nature of the construction material. Commonly used sound absorbing materials such as pew cushions and carpeting tend to reduce both reverberation and loudness, so while they may enhance speech clarity they should be limited or even avoided completely in spaces for music. Where room volume is limited, orientation of sound reflecting surfaces for projection of the choir and organ, with a minimum of sound absorption and a suitable speech amplification system, are recommended to maximize reverberation.

14.2.10.4 Reverberation Time

For the light frame construction common in North America, the T_{60} of most occupied worship spaces is likely to lie between 1.2 and 1.5 seconds. Since sound absorption of materials varies with frequency, T_{60} values should be calculated for the full audible frequency range. The low frequency component of organ sound is mostly absorbed by light wall and roof construction, while the mid to high frequencies are absorbed by clothing, carpet, and seat cushions. The T_{60} spectrum resulting from this combination will be highest in the mid-frequency range with lower values at high and low frequencies, so that both bass and treble sounds tend to be deficient. By contrast, the frequency spectrum of sound decay in a heavy masonry enclosure is typically strong in the bass range, favoring organ-choral music (see Figure 14.9).

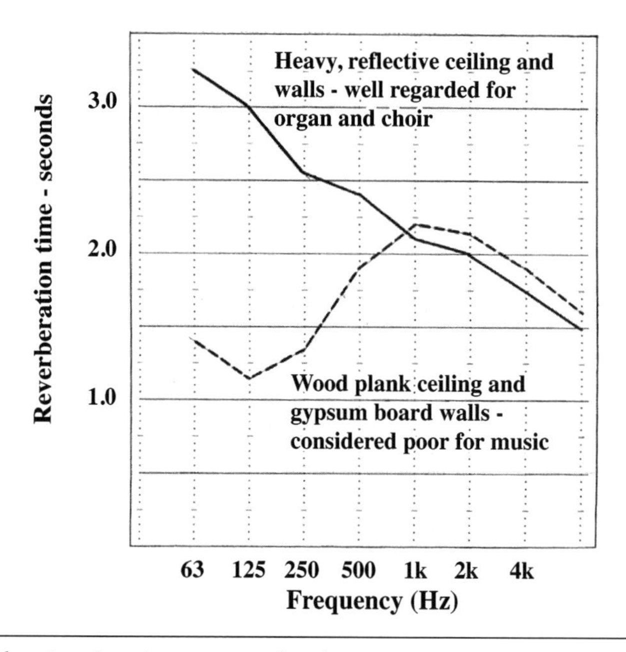

Figure 14.9 Reverberation times in two empty churches.

14.2.10.5 Requirements for Speech

While the space should be free of echoes, excessive reverberation and *flutter* effects that could obscure syllables, intelligibility of speech is governed largely by its level above the background noise level. For a listener with *normal* hearing, speech levels should be at least six dB above the background, but for people with impaired hearing, or for children and people unfamiliar with the language, speech levels should be at least 15 dB above the background.

To provide intelligibility without electronic amplification, several physical requirements must be fulfilled: a background noise level not exceeding NC 25 to 30 and good projection of sound from speaker to listener, with interior shaping and sound absorbing treatment to provide even sound distribution and low T_{60}, e.g., less than one second. Location and orientation of sound-reflecting walls and ceiling around the speaker for natural reinforcement of speech can provide loudness and clarity that would otherwise be dissipated.

For spaces seating more than around 60 to 100, room shaping alone may not ensure good sound distribution or speech intelligibility. These require a sound amplification system and trained speakers to provide clarity of speech for older listeners, the ones most likely to have impaired hearing. Since seniors are also the people whose contributions support most worship facilities, installation of a high-quality sound system to meet their needs should be seen as an investment rather than an expense. The requirements for music amplification can vary significantly so it may be advisable to keep them separate from speech requirements.

14.2.10.6 Requirements for Music

Where music requirements dictate a longer T_{60}, e.g., over 1.7 seconds, room proportions should minimize sound absorbing areas and maximize room volume. In general, a compact floor plan with a high ceiling tends to be more reverberant with better sound distribution than a wide, low space of similar volume. Choir-organ and other music groups should be located so that walls and ceiling provide strong sound reflections out to the *nave* or main volume of the worship space. A close review of readily-available literature from organ builders' associations defining desirable locations and space requirements for pipe organs may be helpful, but the guidance of an organ builder is strongly recommended.

14.2.10.7 Choir Rehearsal Room

A rehearsal room should be large enough for the choir to face the director around a piano, with a ceiling height of at least 12 to 14 feet (3.6 to 4.3 m) for good sound distribution and a T_{60} of 1.2 to 1.4 seconds. A combination of sound reflecting and sound absorbing surfaces on the ceiling and walls, possibly including retractable heavy curtains for adjusting reverberation, will enable singers to hear each other clearly. The room can also double as an intimate recital space, if the front walls and ceiling are shaped for good sound projection to the seating area.

14.2.10.8 Accommodating Nonworship Events

Nonworship events in the sanctuary for charitable or fund-raising purposes, etc., can often be accommodated by adding sound reflecting or absorbing surfaces and supplementary sound amplification. This requires easily-handled temporary elements that are stored nearby, although adjustable elements could also be incorporated in the architectural design.

14.3 APPLICATIONS TO DESIGN AND CONSTRUCTION

14.3.1 Transition from Design to Construction

In the transition from acoustical principles to construction, it is important to keep in mind the difference between conceptual design and what will actually be built. Whereas both architect and acoustical designers may think in terms of near-perfection, the builder signs an agreement with the owner, defined by contract drawings and specifications, for construction of the building with the intention of making a profit. If the drawings and specifications are not explicit, the building contractor could install at least cost whatever meets the stated requirements or seek an extra fee for each late revision.

14.3.2 Site Noise Control

14.3.2.1 Site Noise Measurements

Outdoor noise measurements should be made at times corresponding to the probable hours of use, with each location and the range of noise levels carefully documented. Maximum levels should then be compared with the desired maximum interior levels to define the required NR from outdoors. The NR can then be compared to the noise isolating capabilities of different enclosures. This evaluation, together with assessment of outdoor mechanical equipment noise, should preferably be part of a professional noise analysis. A schematic description of the site survey is shown in Figure 14.10, comparing outdoor and indoor noise levels and the required NR to be met by the selected construction.

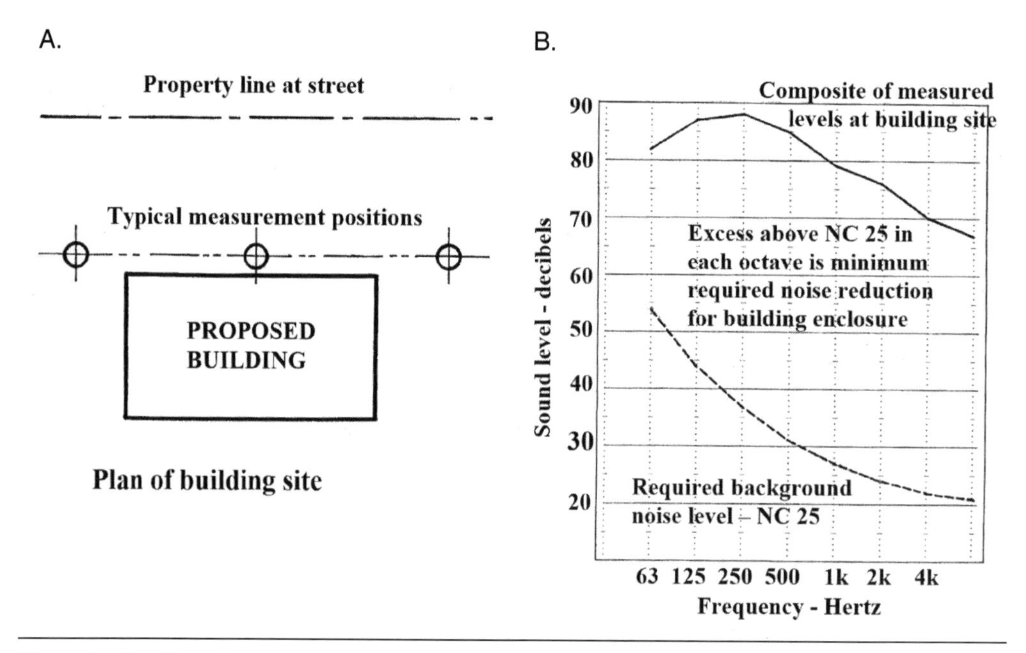

Figure 14.10 Site noise measurements.

14.3.2.2 Building Enclosure

Schematic plans and sections of the building enclosure should be checked for NR of each component. Doors and windows facing the source of noise, and whether windows are to be operable or fixed, will require evaluation. Large window areas or light roofs can offset the noise reduction of heavy walls and thus, should be avoided on severe noise exposures. Entrances should include a vestibule to provide two sets of doors between the outdoors and worshippers. Enclosures that may be suitable for different noise exposures include:

Little Noise Intrusion	*Distracting Noise Intrusion*	*Severe Noise Intrusion*
Lightweight construction with operable windows	Stucco or heavier enclosure, plaster ceiling, laminated or double glazing	Masonry enclosure, reduced window area, double glazing

Information on NR provided by different doors and windows may be found in reference sources, or preferably from manufacturers. Note that standard thermal double glazing may offer little acoustical advantage over a single layer of ¼″ (0.63 cm) thick plate glass, and that two layers of plate glass enclosing a sealed cavity at least four inches (10 cm) deep are recommended as a minimum for noise isolating windows. However, since adding a second glazing layer with a tightly sealed cavity over existing windows may cause deterioration of leaded glazing, details for venting the cavity should be reviewed with a glazing designer.

14.3.2.3 Outdoor Air Conditioning Equipment

The combined cost of the selected equipment and of controlling its noise should be considered from the start. Enclosures on grade should be as far as possible from occupied spaces, using a solid wall such as heavy stucco—not an open screen—with a solid door and no wall openings. It should be large enough to ensure good air circulation and high enough that neighbors cannot see the equipment. Other noisy equipment, such as transformers, may also be included in the same enclosure. In some situations, a weather-resistant sound-absorbing material may be needed on the inner surfaces of the enclosure.

To lessen the bulk of the enclosure around equipment with discharge fans, such as a cooling tower or chiller, an acoustically-lined inner screen could be mounted above the sides of the unit to provide added shielding of noise from propeller fans. It should be supported from the outer enclosure, with a resilient filler to close the gap between the equipment and inner screen, to minimize radiation of structure-borne noise from the equipment. The inner screen also provides the necessary separation between intake and discharge air (see Figure 14.11).

Roof-mounted equipment should be located above unoccupied spaces such as washrooms or storage, preferably on a concrete roof slab and with an enclosing wall if needed. It should be supported on short roof spans via external steel-spring vibration isolators and located far enough from the sanctuary to allow adequate length for duct lining or silencers for control of fan noise (see Figure 14.12).

Openings in the walls and roof must be sealed tightly around ductwork; the work should be closely supervised and no cost-cutting revisions should be permitted.

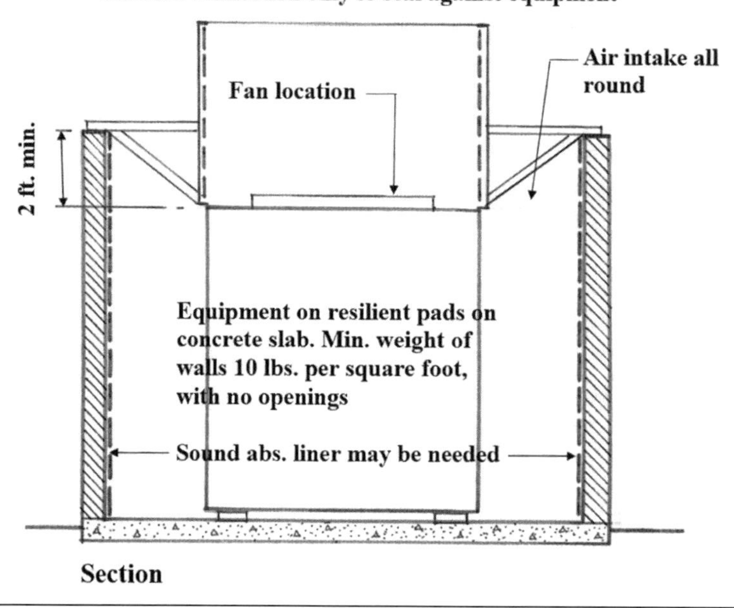

Figure 14.11 Chiller enclosure.

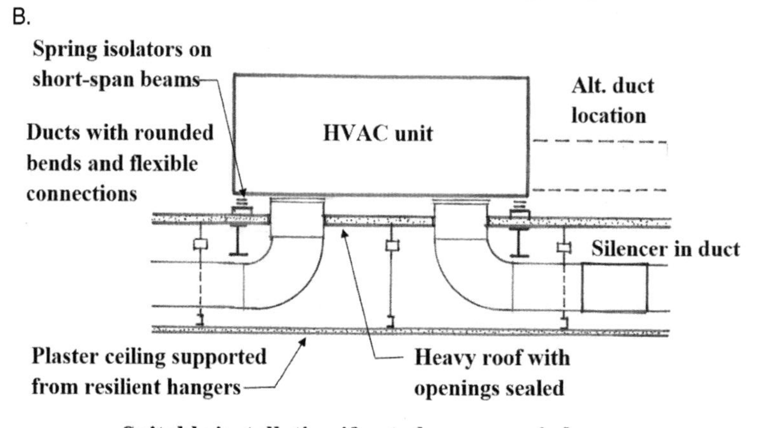

Figure 14.12 Roof-mounted equipment.

14.3.3 Control of Indoor Noise

14.3.3.1 Noise Isolation Between Spaces

The sanctuary should be isolated from activity spaces such as a lobby or corridor by construction having a sound transmission loss rating of at least STC 50 to 55 and if possible by vestibules with double sets of doors. In general, spaces such as for choir rehearsal or classrooms, washrooms, or mechanical spaces, should never be located above or next to the sanctuary, and balcony seating should never be left open to a narthex or entrance vestibule. If noisy spaces are on grade and well-separated from the sanctuary, the required NR may be attainable using standard gypsum board construction. Wall-ceiling details and duct layouts, etc., should be checked to avoid sound leaks (see Figure 14.13).

14.3.3.2 Movable Partition Details

Since effective sound isolation for a movable partition depends on a complete seal at all edges of each panel, the design must allow for the partition to be easily extended and retracted in a way that minimizes wear on edge seals. Suppliers should submit drawings defining the track layout, type of support and space required for storing the retracted partition. Top-hung, single-panel types with floor drop-seals are recommended. Compressible edge seals are easily damaged so occasional adjustment or repair will be needed.

Even expensive movable partitions provide limited sound isolation, and reducing their height to save on cost complicates the acoustical design of the undivided space by occluding sound projection from one side to the other. A related limitation is the acoustical effect on each divided space of the large partition area that can be neither contoured nor sound absorbing.

All associated construction costs should also be taken into account: for structural support of the overhead track, for ventilation, lighting, and sound systems that adapt to the subdivided sections, and for regular maintenance. The exits and ventilation system must be planned so that the divided spaces are usable independently, with no cross-talk. The basic acoustical limitations created by dividing a space with a movable partition are shown in Figure 14.14.

14.3.3.3 Impact Noise Control

Structural connections that allow impact noise to be transmitted to the sanctuary should be avoided if at all possible. Double framing for sanctuary isolation is possible with light-weight

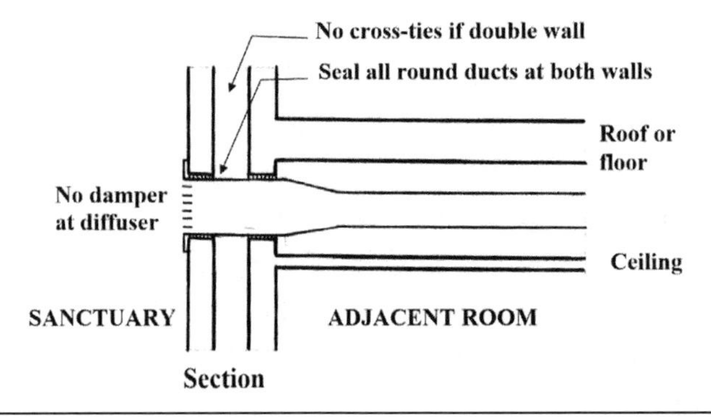

Figure 14.13 Typical duct penetration at wall.

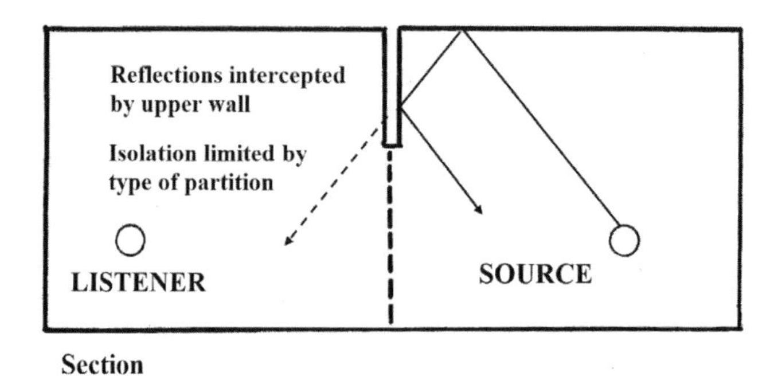

Figure 14.14 Limitations of movable partitions.

construction, but close inspection is needed to ensure that no rigid connections are overlooked.

14.3.3.4 Ventilation and Air Conditioning Noise Control

In low-budget designs, supply fans with heating-cooling coils are commonly installed in an attic space above either the *narthex* or sacristy, or package air conditioning units are mounted on a low roof adjacent to the sanctuary, creating the potential for both airborne and structure-borne noise intrusion. Supply air ducts are typically placed above side aisles, with wall-mounted supply diffusers and ceiling or wall intake grilles for return air. Adequate noise and vibration control of such a design may be impossible because of duct-borne fan noise, turbulence at air supply diffusers or return grilles, or inadequate vibration isolation (see Figure 14.15).

14.3.3.5 Air Distribution Noise Control

In many situations the ventilation design must adapt to liturgical and architectural conditions, requiring flexibility and the willingness to explore innovative layouts. However, if the

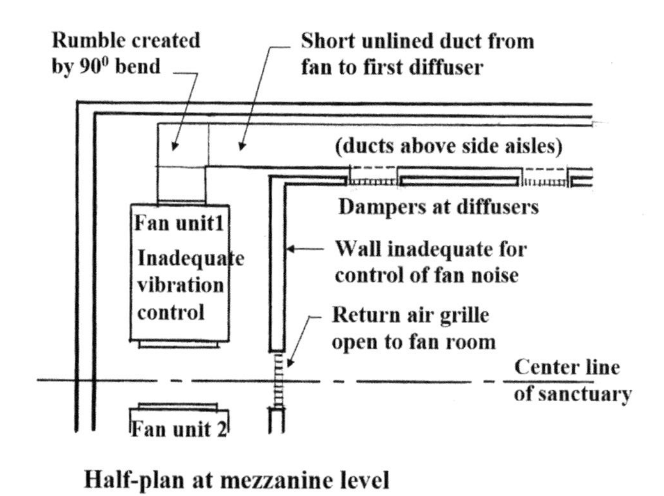

Figure 14.15 Example of unsuitable HVAC system.

initial design is unlikely to meet the selected NC, alternatives should be explored promptly. The required length of an acoustically lined duct to meet the background noise criterion can be calculated from the supplier's fan noise data following procedures defined in current ASHRAE publications.[6] Sheet metal ducts are recommended rather than flexible plastic for control of *breakout noise*. If standard ducts are still inadequate, convenient alternatives are heavier-gauge metal ducts or *lagging* the outside of ducts with cement plaster or layers of gypsum board. Since noise from a fan return air intake is roughly the same as on the discharge side, the same degree of noise control is required. This may be impossible for an installation with return grilles in the wall between the sanctuary and fan room unless there is space for adequate lined ductwork or a silencer.

Air supply by a single duct to a row of diffusers is commonly balanced by using a damper at each take-off, often creating noise due to turbulence and high air velocities. The combined noise level of several diffusers will also be higher than for a single diffuser. Installation of a lined branch duct at least three feet long between each damper and diffuser is recommended to attenuate damper noise and reduce air turbulence. While use of flexible plastic ducting may save time and cost for the installer, it is not recommended because it allows for misalignment that increases diffuser noise. A noisy duct configuration is compared here (see Figure 14.16) with one in which the dominant noise sources are controlled.

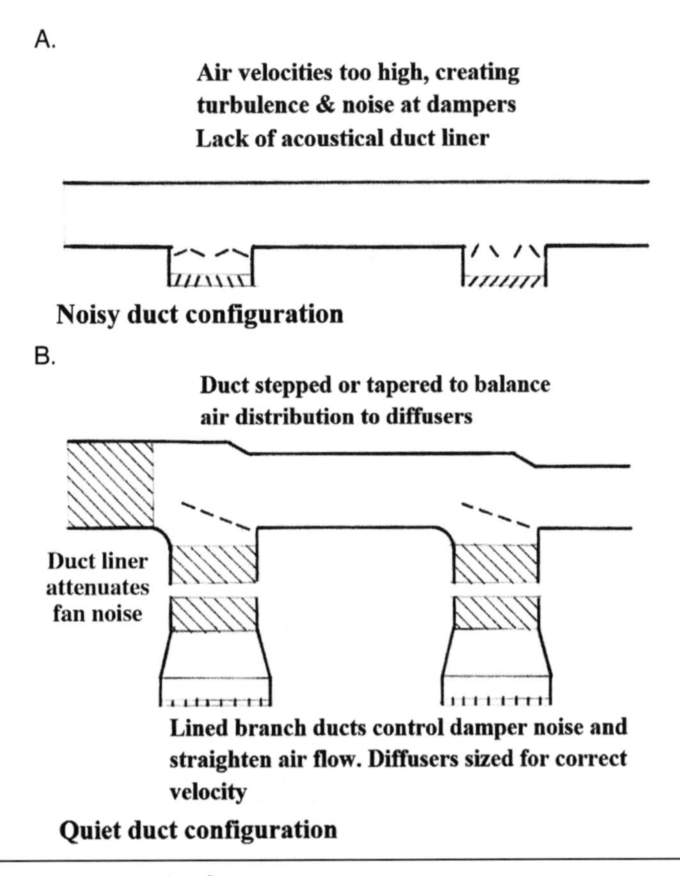

Figure 14.16 Control of fan and airflow noise.

14.3.4 Acoustics of Sanctuary

14.3.4.1 Size and Shape of Sanctuary

The minimum sanctuary area is typically dictated by regulations governing floor area per person and aisle and exit widths, while room volume determines the maximum possible T_{60}. Certain dimensions have been found to be the most suitable for specific uses. For example, in a narrow *shoe box* shaped room with a maximum width of around 65 feet (20 m), the time sequence of wall reflections from the sound source is generally considered optimum for loudness and clarity of orchestral and choral music, as seen in Figure 14.17.

By contrast, in wide sanctuaries that allow much of the congregation to be close to the celebrants, sound-reflecting side walls will be farther away from the sound source so that the loudness and clarity of a narrow space is replaced with indistinct and *muddy* sound. While a time delay at the listener's ear between *direct* and reflected sound of up to 30 milliseconds is considered excellent for reinforcement, reflections with a delay of over 50 milliseconds tend to obscure later syllables, reducing intelligibility.

Sound distribution and decay can be modified by surface shaping and selective location of either sound reflecting or absorbing materials. Effectiveness of side wall or ceiling sound reflections from choir to nave can be readily checked on a plan or longitudinal section by use of a ray diagram, e.g., drawing a line from the sound source to each ceiling area between cross beams and plotting its reflection to the congregation. If reflection paths are occluded by beams or deep *coffers,* the ceiling is ineffective as a sound reflector. A useful rule of thumb for coffering is that the depth of ribs should be no more than one fifth of the width of the coffer, with a maximum depth of less than two feet.

Scattering and diffusion of reflections from the walls and ceiling tend to envelop the listener, which is generally preferred to completely frontal sound. Architectural elements useful for sound diffusion include window recesses, protruding pilasters, and ceiling coffers, but only if their scale is appropriate. For example, if beams or their *haunches* are too deep and

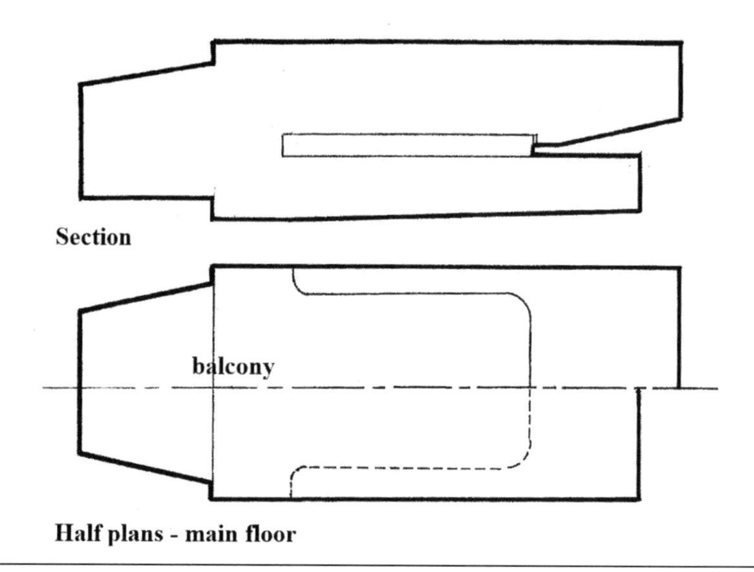

Figure 14.17 Typical *shoe-box* concert hall.

closely spaced they can occlude most ceiling and wall reflections, dividing the nave acoustically into separate compartments (see Figure 14.18). In such spaces, voice levels may be too high within the choir with little sound projection towards the rear of the congregation. Traditional architectural forms such as *vaulted* ceilings may create uneven sound distribution due to focusing of sound, although this could be mitigated by including coffers to modify continuous concave surfaces.

The addition of side aisles or side chapels can change the balance between early-arriving sound reflections and later-arriving reverberant sound. For certain configurations, such modifications could add diffusion and richness, but for others they could be detrimental. This effect is seen most dramatically in a cruciform plan where reflective side walls are replaced by transepts that generate long-delayed echoes. Where transepts are combined with a domed ceiling at the *crossing* few useful sound reflecting surfaces remain, resulting in a large seating area that is generally distinguished by less than satisfactory choir and organ sound. Where cruciform or fan-shaped plans are required, available sound reflecting surfaces should be oriented to maximize useful reflections and even distribution of sound (see Figure 14.19).

14.3.4.2 Organ and Choir

The choir, organ, and console (which is often separate from the organ itself) should be fairly close together so that the choir and organist can see and hear each other well, with the organ speaking over the choir so that it does not overpower the singers, while the overall shape of the organ-choir space should enhance projection of sound to the nave. Generally, this will entail a high space with side walls opening up to the congregation but with minor surfaces oriented to return just enough sound to the musicians for ensemble. Attention to such details of sound reflecting surfaces for a traditional organ-choir space can make the difference between one that projects sound poorly and one that adds loudness, clarity, and richness even for solo voices (see Figure 14.20). Since it may be difficult to attain the right balance in the initial design, the capability for later adjustment of wall or ceiling reflecting surfaces should be considered.

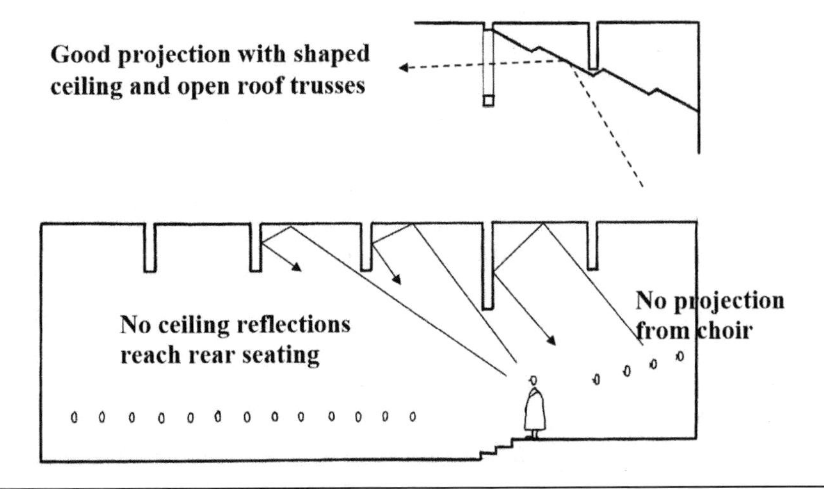

Figure 14.18 Effect of deep beams on sound projection.

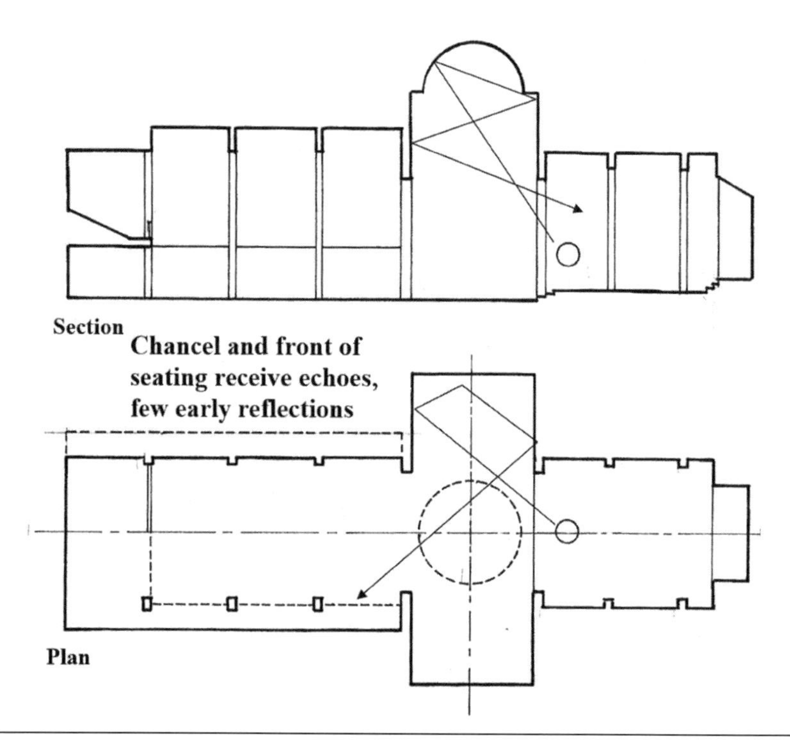

Figure 14.19 Effect of shape on sound projection.

14.3.4.3 Related Design Requirements

Other performance requirements such as screen locations, sight lines, and control of natural light for video presentations should also be established in conjunction with acoustical design. Close collaboration with lighting and audio visual designers is strongly recommended to avoid potential conflicts and noisy system components.

14.3.4.4 Sound Reflecting and Absorbing Materials

While a heavy, impervious surface such as thick masonry tends to reflect sound at all audible frequencies, thinner materials such as glass, metal, or plaster transmit sound at low frequencies but are reflective at higher frequencies. If a thin sheet covers an enclosed cavity it becomes a resonant absorber, absorbing sound in a narrow frequency range. Porous materials such as carpet or seat cushions absorb sound over a wider frequency range, depending on density and thickness. Absorption characteristics of various types of surfaces are compared, and sound absorption coefficients are defined for common building materials in Figure 14.21.

For support of organ music in light frame construction, low frequency sound absorption can be reduced a little by installing an extra-heavy plaster surface or by adding extra layers of gypsum board on walls and ceiling. The architecturally popular exposed wood plank roof is not recommended because it becomes increasingly sound absorbing as it dries out and joints widen between planks. An added layer of either plywood or heavy composition board nailed tightly to the plank surface can seal all joints, reducing sound loss to the outdoors

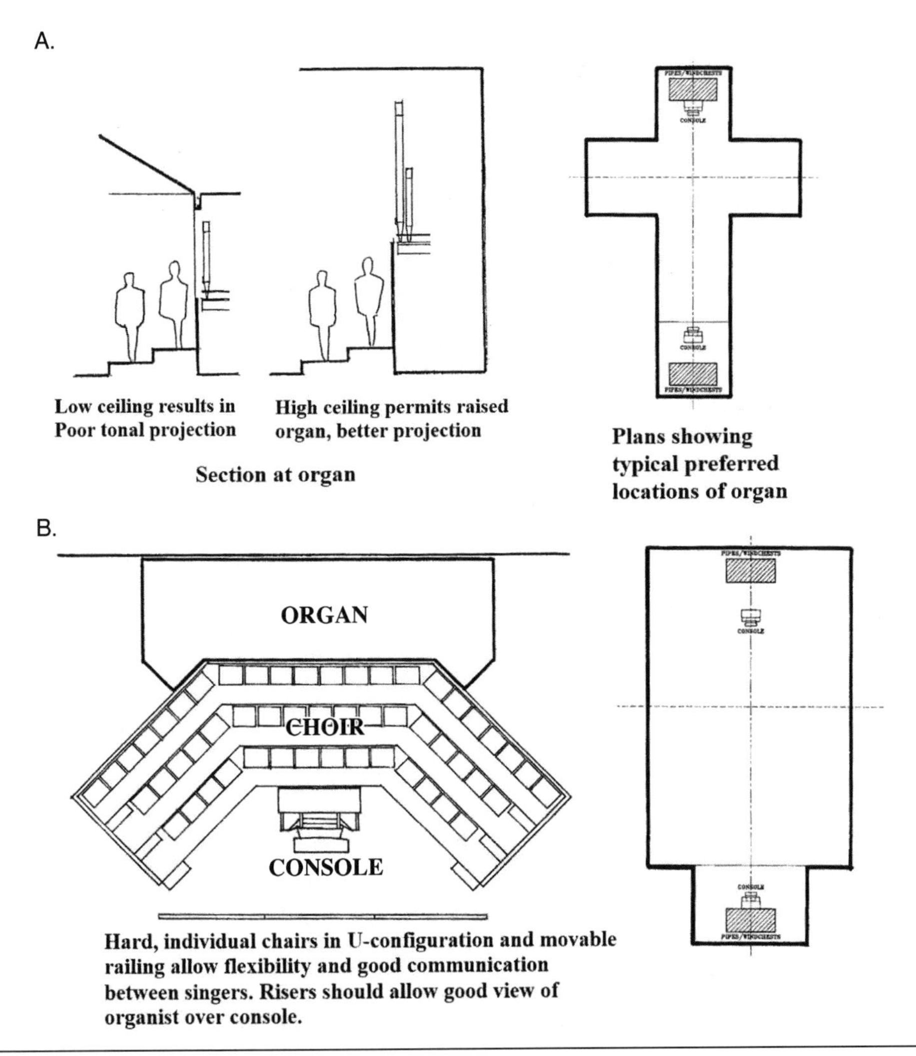

Figure 14.20 Planning for organ and choir [reprinted with permission of Associated Pipe Organ Builders of America].

and also improving NR from the outdoors. The wood plank appearance can be simulated by *kerf-sawing* a plywood surface.

14.3.4.5 Reverberation Time

Calculation of T_{60} using the Sabine equation requires compiling the total area of each interior surface multiplied by its *sound absorption coefficient.*[2] Published values of these coefficients may vary depending on the source and also on the method by which the areas are measured. Since the most significant sound absorbing surface in a worship space is generally the seating area, a check on the source of coefficients is recommended. Probably the most consistent information on sound absorption of seating areas is found in published studies on the design of concert halls where the *effective seating area* method is used, but further study is needed on sound absorption differences between wood pews and theater seating.[7]

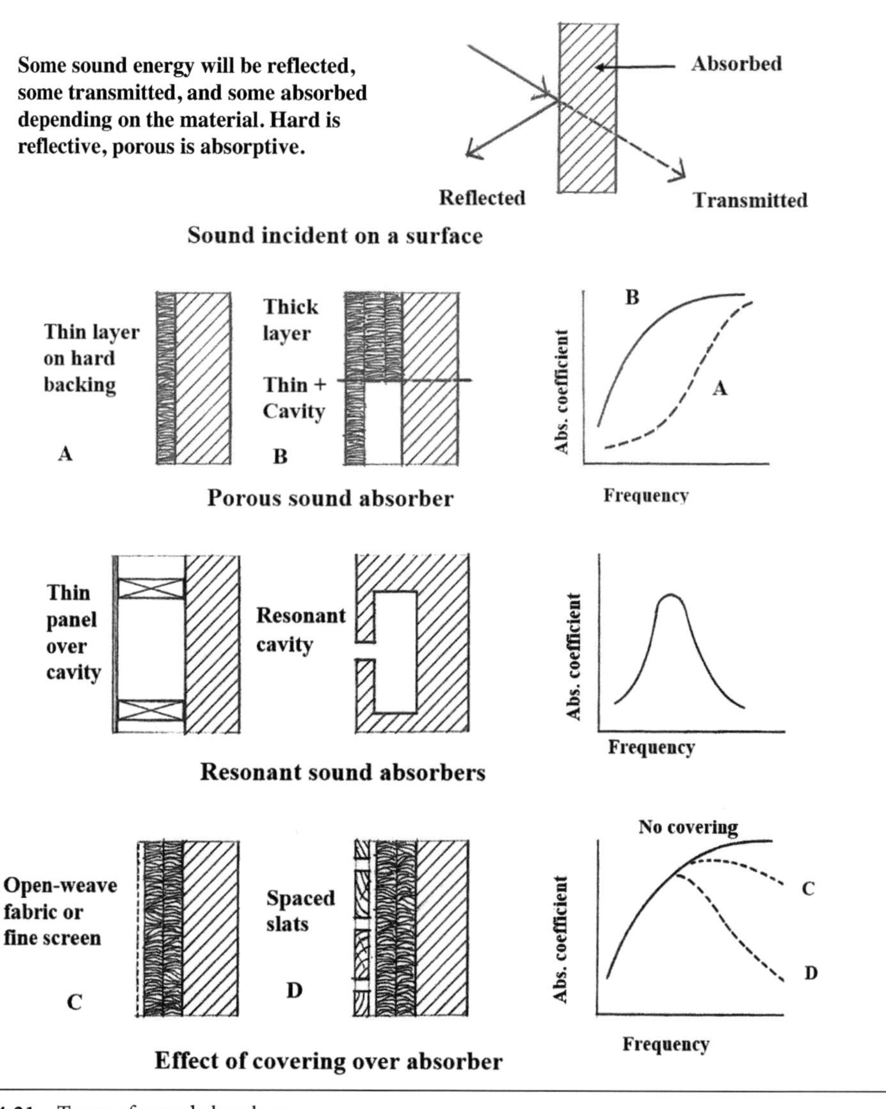

Figure 14.21 Types of sound absorber.

14.3.4.6 Estimation of Reverberation Time

The approximate T_{60} in seconds can be derived using the Sabine equation:

$$T_{60} = \frac{kV}{A_1 + A_2 + A_3 + \cdots}$$ where k = 0.16 in metric units or 0.049 in English units,

V is the volume in cubic meters or cubic feet, and A is the surface area of each interior material multiplied by its sound absorption coefficient, α.

For a preliminary estimate, absorption coefficients can be simplified by assuming one value for hard materials and one for soft materials:

Hard—e.g., wood, concrete—assume $\alpha = 0.1$
Soft—e.g., thick carpet—assume $\alpha = 0.8$

Example of calculation—worship space with a volume of 2,381.2 m³ (9.15 × 24.39 × 10.67):

Area 1: ceiling, walls, exposed floor = 979.66 m², $\alpha_1 = 0.1$, $A_1 = 97.97$

Area 2: empty seating (not upholstered) = 106.35 m², $\alpha_2 = 0.1$, $A_2 = 10.64$

Area 3: sound absorbing material = 49.63 m², $\alpha_3 = 0.8$, $A_3 = 39.70$

Total empty = 148.31 sabins

T_{60} empty = 0.16 × 2,381.2/148.31 = 2.57 seconds

Add a full congregation: 106.35 m², $\alpha = (0.8 - 0.1)$, Add 74.4 sabins

Then T_{60} full = 0.16 × 2,381.2/222.71 = 1.71 seconds

Reverberation at mid to high frequencies can be maximized by avoiding seat cushions and by limiting carpet only to thin glue-down aisle runners if needed for control of footfall noise. To sustain choir levels, particularly in the soprano range, flooring within or close to the choir area should always be sound reflecting. If still more reverberation is sought in an existing space after carpet and seat cushions have been removed, the perceived decay of sound can be prolonged by combining enhanced projection of sound from the choir and organ with the lowering of the background noise level (see Figure 14.22).

14.3.4.7 Choir Rehearsal Room

Sound diffusing lower wall surfaces can be built up to the door height from slightly convex plywood wall panels alternating with strips of 2″ (5.1 cm) thick, semi-rigid glass fiber board behind a protective covering of hardware cloth or other *acoustically-transparent* facing. The upper wall surfaces should be sound reflecting but capable of being covered with sections of medium to heavy velour curtains that stack in the corners when retracted. Recessed lighting fixtures with a hard plastic lens, selected for complete inaudibility and arrayed between

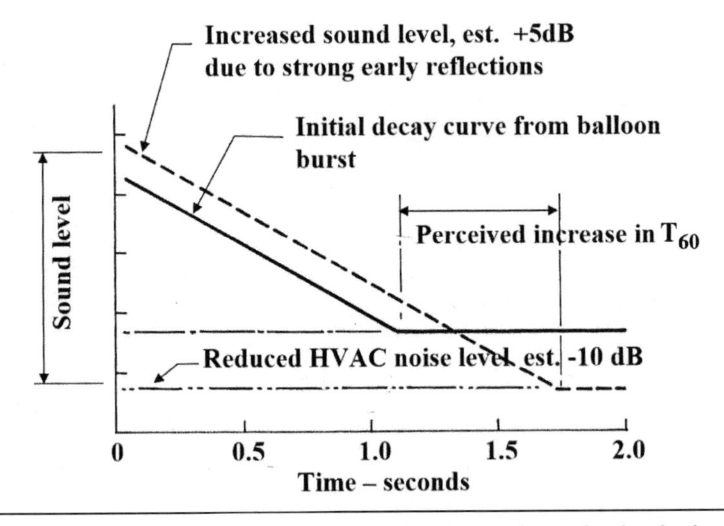

Figure 14.22 Perceived increase in T_{60} due to the reduced background noise level and enhanced projection from the choir.

suspended acoustic ceiling panels, can provide useful sound reflections between choir sections. For smaller spaces such as rehearsal rooms or recording studios, a check on the evenness of sound distribution at low frequencies is recommended.

14.3.4.8 Sound Amplification

The history of sound amplification for worship spaces is chiefly one of designing to maximize speech intelligibility, as distinct from loudness, by controlling excessive reverberation and providing good sound coverage for all listeners. This requires high-quality microphones serving high frequency directional loudspeakers to provide intelligibility, with supplementary low frequency loudspeakers to add bass for music and to make voices sound natural. Loudspeaker arrays are typically large, located above and slightly ahead of the person speaking and oriented so that they do not return strong echoes from the rear wall or ceiling.

For reverberant spaces, good speech intelligibility can be attained with pew-back loudspeakers using digital time-delays to match the arrival time of unamplified sound. These systems are expensive, requiring at least one loudspeaker for each three congregants, with concealed wiring and secure loudspeaker enclosures to prevent tampering. For low-ceilinged spaces, overhead loudspeaker arrays may be an appropriate alternative. Where low-budget systems are installed, reverberation is often deliberately suppressed to enhance intelligibility, although the additional cost of sound absorbing materials could make this a more expensive alternative than a well-designed high-quality sound system.

For speech intelligibility however, the still relatively new concept of digitally controlled loudspeakers that provide effective sound coverage without exciting room reverberation is superior to anything that has gone before. These new loudspeakers are also more compact than traditional loudspeaker clusters and can be integrated more easily into an architectural design.

The starting point for either a new or remodeled building should be an investigation of readily available speech amplification systems. A committee should be selected to listen throughout one or more worship services in other places of assembly and to review the merits of each system with its regular listeners. It is important to note, however, that sound system design is no longer the domain of the local audio dealer or an audiophile member of the congregation. Design and specifications should be prepared by someone experienced with current system design and the integration of sound amplification with natural room acoustics. The installer should be required to furnish references to comparable projects and should agree to prompt service if repairs are needed after installation. With normal maintenance a high-quality system should perform reliably for many years.

Close adherence to contract specifications and coordination of sound system installation with other trades are essential to ensure sound quality. While much of this work occurs late in the project, conduit and receptacles must be placed as required in floors and walls, free from possible interference from electrical and piping systems. When planning for revisions to an existing sanctuary, the cost saving from upgrading an existing sound amplification system should be weighed against the cost of architectural changes that may also be required for improved speech intelligibility. Common alternatives for the location of loudspeakers are shown in Figure 14.23.

14.3.4.9 Adapting for Other Events

For a performance by a small music group such as a string quartet, a sound reflecting *enclosure* close to the performers, consisting of lightweight removable panels will enhance

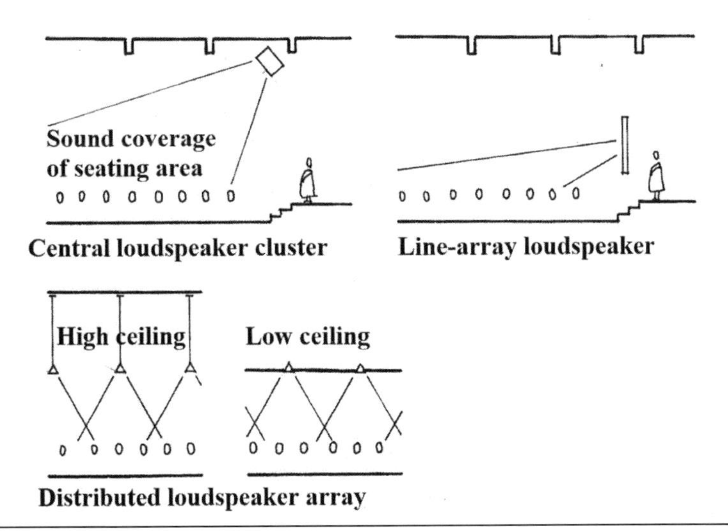

Figure 14.23 Typical loudspeaker configurations.

projection and enable them to hear each other well. A standard panel height of eight feet is generally adequate, although reflectors no more than two feet high have been used successfully in some situations. Panel design and construction should ensure stability when in place but should allow folding or stacking for compact storage.

For amplified events, wall reflections within the performance area can be controlled by temporary sound absorbing panels or heavy curtains that retract into wall pockets when not in use. Rear wall curtains, with possibly additional curtains on the upper side walls, can significantly reduce reverberation for the entire space. However, integrating large curtain areas so as not to be architecturally intrusive can be difficult and expensive.

14.3.5 Remodeling of Existing Facilities

If complete records of an existing building have not been kept, a search for architectural and engineering drawings and specifications should be made as soon as possible. While this will probably be difficult, it could be far less costly than hiring qualified professionals to prepare drawings and recalculate structural adequacy, heating and cooling requirements, and the like. If the original design firm no longer exists, it is quite possible that the required documents are in the archives of its successors. In preparation for future needs, contractual arrangements with the architect and engineers should include the cost of at least one extra reproducible set of all new documents, including all design calculations. These should be kept in a vault in the control of a responsible person who will allow only copies, but never the originals, to be taken away.

Remodeling should start with a clear assessment of what can be done with the funds available. First and foremost, items that require structural or planning changes must be checked against current standards because bringing older facilities in compliance with building codes always takes priority and could actually exceed the available budget. Initial cost estimates should be as thorough as possible but other hidden costs should be expected as the building is investigated. For example, building services such as heating, ventilating, and electrical equipment may require complete replacement. If existing ductwork and diffusers are to be retained, installing a new air conditioning system with increased air quantities could create unacceptably high noise levels.

14.3.6 Getting Things Built Properly

14.3.6.1 Contractual Arrangements

The agreement between building owner and designer should be clearly documented from the start on the basis of conceptual drawings and written performance requirements. As the design develops, design changes that may arise should be included formally as revisions to the original agreement, with the understanding that late requests for change will mean increased design fees. If the work is apportioned as separate contracts, the owner may save the additional charge for a general contractor to coordinate the work, but may also face the cost of last-minute revisions during construction when conflicts arise between building disciplines.

At the start of the contract a limited period should be allowed in which to clarify acceptability of any alternatives or possible substitutions for specified items. Even when products have been precisely specified and shop drawings reviewed, however, a contractor could substitute a less expensive product that is not seen until it is in place. At that point, it is often difficult to correct the situation without delay or further complications. This can usually be controlled by requiring samples of such products for approval, so firm enforcement of product selection should be established early in construction.

The owner-contractor agreement should specify demonstration by the contractor that acoustical requirements have been fully complied with in the completed building. Typically, part of each contract payment is withheld until the work is completed and the owner has confirmed that there are no outstanding *liens* against the project. A set period is also generally included, and may in fact be required by law, in which the contractor agrees to respond promptly to any problems that are discovered by the owner after occupying the building. This checkout period is the owner's most effective way of confirming that he is actually getting what was agreed to, so it should never be discarded lightly in contract negotiations.

14.3.6.2 Importance of Details

Unless special circumstances prevail, there is nothing in the competitive-bidding process that encourages a building contractor to take care with items that are not seen on the surface. For this reason, a building with requirements that may be unfamiliar to him, such as hidden details that are needed for noise control, will prompt a higher bid than for a better understood building type. Nevertheless, the interests of the owner are well served only if every effort is made to ensure completeness and accuracy of the work. Locations specifically defined for loudspeakers and rack-mounted equipment, etc., must be kept free of ductwork or other conflicting trades.

14.3.6.3 Ambiguities in Terminology

Two acoustical concerns that require careful specifications are the sealing of potential sound leaks and the selection of sound absorbing materials. Since it is difficult to convince the building industry that a wad of glass fiber is not a satisfactory way to seal a sound leak between adjacent spaces, careless terms such as *acoustical insulation* should never be used in place of a specific product. Similarly, naming the specific finish materials rather than saying *acoustical absorption* and specifying a review of samples for owner approval, will help to avoid last-minute *extras*. Other examples of ambiguity in the contractor's favor include terms such as *duct lagging* which legitimately mean different things to different trades.

14.3.6.4 Value Engineering

When cost estimates or bids are higher than expected, a cost-trimming process generally known as *value engineering* frequently takes place. Originally introduced as a way of checking that costs are kept under control, it can be effective in consideration of alternatives. On the other hand, an entire section of the specifications can be deleted by a pen-stroke, so the results could be devastating if a value engineering review takes place when the architect is not in a position to influence the outcome. The only way to avoid such unintended results is to ensure that proposed deletions or changes are fully examined by people with the necessary understanding of the work. The important consideration is to distinguish between simply reducing initial cost and spending money effectively, with long-term suitability and the cost of the end product as the final reference points.

14.3.6.5 Design Build

The design-build process, in which the building contractor is given responsibility for resolving construction details, effectively removes any control from the design team. In many cases this results in oversimplified or deleted acoustical controls that may cause occupant dissatisfaction. Having a summary of acoustical goals for each space may be sufficient to guide the contractor in avoiding serious omissions, but the need for corrective work after completion is probably more likely. In the last analysis control of details, along with inspections to verify that work is done as specified, forms the owner's only insurance that he is getting what he pays for, so eliminating these final steps is always a gamble.

14.3.6.6 Bidding Period

To ensure close liaison between owner and designer, the selection of a cognizant person to represent the owner is recommended. It will be this person's responsibility to understand the contract documents (i.e., drawings and specifications from the architect and consultants) and to keep a record of all decisions agreed upon. The same procedure will be required during bidding and construction, with the architect or other professional acting on behalf of the owner to see that products and construction methods are in accordance with the specifications. Where specific details are intended, they must be clearly defined in the bid documents to avoid disputes during construction.

14.3.6.7 Monitoring Construction

Coordination of building systems, allowing each its appointed space and separating vibrating from static systems, depends on alertness and tight control by the project superintendent. If he has planned well, each sub-trade will have a clearly defined time in which his work can be done with the least interference from other trades. If he has not, substantial conflicts may occur between subcontractors and with the general contractor, creating delays in the work schedule. It is in the hidden details that are essential for good noise isolation that shortcuts will most likely be taken, and any resulting acoustical problems could lead to litigation. However, none of this may be apparent when the list of bidders for the project is established, so references are important.

14.3.6.8 Acceptance Testing of Facilities

Final inspection and testing may be straightforward or very complicated, depending largely on the relationship established with the contractor throughout the work and the conditions written into the specifications. If the inspector is many miles away, confirmation in writing that the facility is completely ready for checkout is strongly recommended to avoid wasted trips. Assuming that all doors are in place, ventilating systems operating and at least roughly balanced, initial tests compare the measured background sound level in each space with specified values. Reasons for excessively noisy spaces should be tracked down and put on the punch-list for correction.

The second step is to shut down all building systems and make a preliminary assessment of noise isolation between spaces. This entails setting up a loud noise source in some rooms and listening in adjacent spaces for obvious sound leaks. There is no point in detailed acoustical measurements before correcting a hole through a wall behind a finish surface or a duct acting as a speaking tube between rooms. After such leaks are at least temporarily corrected, the measurement procedure is carried out according to existing standards, allowing comparison of the actual situation with the specified goal. Deficiencies should be summarized with as much information as possible, including photographs where useful, on how the required corrections are to be made.

Check-out requirements for the sound amplification system should include performance testing of individual components and of the entire system, typically including measurement of speech intelligibility. Microphones should be adjusted so that amplified sound levels are consistent for every location. Manuals and system diagrams should be supplied and the users should be fully instructed on microphone use.

Compilation of the punch-list of items to be corrected before the contractor's work is completed requires particular attention to detail. Where specific reasons for deficiencies can be documented, they perform the dual purpose of helping the contractor and simplifying the check on whether the corrections have been carried out. To cut his costs, a contractor may take shortcuts on remedial work so a careful recheck is needed. Ironically, at this stage the owner is generally trying to avoid further outlay, so, unless the consultant's contract includes such extra services, he may have difficulty getting paid for protecting the owner's interests.

SUMMARY

Providing satisfactory acoustics for any assembly space comprises many separate elements, all of which are easily understood but many of which require more than a *cookbook* of steps to be followed. It must be emphasized that if hearing conditions are not satisfactory on completion, the building will have failed to meet its primary requirement of good hearing.

Control of unwanted noise and enhancement of wanted sounds require careful coordination and attention to details. Techniques that still lie mostly outside the common parlance of the industry must be incorporated, requiring consistent and intelligent application from selection of the building site to acceptance testing of the finished building. This most definitely is not something that a building contractor should be expected to undertake correctly without specific direction. The design or remodeling of a worship facility that should be expected to perform well for a hundred years or more requires no less than the best.

ACKNOWLEDGMENTS

The advice and continuing support of the following colleagues is deeply appreciated: Leo Beranek, Jack Bethards, the late Warren Blazier, Bill Cavanaugh, Chris Papadimos, and Roman Wouk—together with a host of thoughtful suggestions contributed by musicians, builders, and others over the past fifty years.

REFERENCES

1. Federal Highway Administration, FHWA 23-772—guide to standards for traffic noise mitigation.
2. Sabine, W. C., *Collected Papers on Acoustics*, "Reverberation," Harvard 1921, reprinted by Peninsula Publishing 1994.
3. Wetherill, E. A., "Translating Acoustical Requirements into Architectural Details," *Sound & Vibration*, July 1997.
4. ASHRAE Handbook 2011, Chapter 48, Table 1—relationship of levels measured on A and C scales.
5. Harris, C. M., *Dictionary of Architecture & Construction*, New York, McGraw-Hill, 1993.
6. American Society of Air Conditioning and Refrigerating Engineers, ASHRAE—yearly handbook updates standards concerned with maintenance of indoor environments.
7. Beranek, L. L., *Concert Halls and Opera Houses; Music, Acoustics and Architecture*, New York, Springer-Verlag, 2004, p 575.

FURTHER READING

Apfel, R. F., *Deaf Architects and Blind Acousticians*, New Haven, Apple Enterprises, 1998.
Beranek, L. L., *Noise and Vibration Control*, Institute of Noise Control Engineering (INCE), 1988.
Cavanaugh, W. J., Tocci, G. C. and Wilkes, J. A., *Architectural Acoustics*, 2nd ed., New York, Wiley, 2010.
Crocker, M. J., ed., *Encyclopedia of Acoustics*, vol. 3, "Architectural Acoustics" (Chapters 90–98), New York, Wiley, 1997.
Egan, M. D., *Architectural Acoustics*, Fort Lauderdale, J. Ross Publishing, 2007.
Harris, C. M., *Handbook of Acoustical Measurements and Noise Control*, New York, McGraw-Hill, 1991.
Kleiner, M., Klepper, D. L., Torres, R. R., *Worship Space Acoustics*, Fort Lauderdale, J. Ross Publishing, 2010.
Long, M., *Architectural Acoustics*, San Diego, Elsevier Academic Press, 2006.
Lubman, D. and Wetherill, E. A., ed., *Acoustics of Worship Spaces*, Melville, N. Y., Acoustical Society of America, 1985.
Willimon, W. and Lischer, R., ed., Concise Encyclopedia of Preaching, Louisville, Westminster John Knox Press, 1995.

APPENDIX—DEFINING TERMS

Some definitions are abbreviated for reader convenience. Refer to an acoustics text for technical definitions.

A-weighted sound level, dBA: Sound level adjusted to de-emphasize low and high frequency components in a manner similar to the response of the human ear.

Acoustically transparent: With little resistance to sound transmission, such as an open-weave fabric.

Airborne: Sound transmitted in air.

Breakout noise: Noise radiated from a duct surface.

Coffer: Recessed section of a ceiling.

Crossing: Plan intersection of nave and transepts.

Decibel, dB: Compressed scale used to describe the full range of sound amplitudes.

Echo: Discrete reflection from a sound reflecting surface.

Effective seating area: Area corrected to include sound absorption of seating area edges.

Flutter echo: Repetitive sound reflections between hard parallel surfaces.

Haunch: Increased depth of a beam where it transitions to a supporting column or wall.

HVAC: Common abbreviation for heating, ventilating, and air conditioning.

Impact insulation class, IIC: Single-number rating of sound transmitted through a floor due to impacts.

Kerf-sawn: Partial-depth saw cut in a wood surface.

Lagging: Heavy covering layer on a duct to control breakout noise.

Lien: A claim on property as security for payment of a debt.

Mid-frequencies: Reference to the octaves centered on 500 and 1,000 Hertz (cycles per second).

Narthex: Enclosed vestibule at sanctuary entrance.

Nave: Main seating area for congregation.

Noise criteria, NC: Single-number ratings of background sound levels based on equal-loudness curves.

Noise reduction, NR: Measured reduction in level of sound transmitted between two spaces.

Ray diagram: Graphic representation of the direction of sound propagation.

Reverberation time, T_{60}: Time for sound in a room to diminish by 60 decibels after the source stops.

Room acoustics: Acoustical properties of an enclosed space.

Sabine equation: Derived by W. C. Sabine to calculate T_{60} of a room.

Shoe box: A long, narrow space such as a traditional concert hall.

Sound absorption coefficient: Fraction of incident sound not reflected by a surface.

Sound intensity: Rate of flow of sound energy through a unit area.

Sound transmission class, STC: Single-number rating of sound transmitted through a solid surface.

Structure-borne: Sound propagated in a solid.

Time delay: Interval between arrival of the direct sound and its first reflection at a listener's ear.

Transept: Transverse axis of a cruciform space.

Transmission loss, TL: Reduction in level of a sound passing through a solid material.

Vault, vaulting: Structural element of a ceiling, as in a medieval church.

15

Performing Arts Spaces

Ronald L. McKay, **David Conant**, and **K. Anthony Hoover** McKay Conant Hoover, Inc., Westlake Village, CA, USA

PROLOGUE

A. Types of Spaces

There are many types and sizes of spaces used for the performance and instruction of music and the dramatic arts. Included in this chapter are music and drama performance venues ranging in audience size from 100 to 2,500, plus music and drama instruction spaces ranging in size from a single-person music practice room to rehearsal or recording rooms suitable for an orchestra of 120. Such facilities are commonly found everywhere from some secondary schools and many universities to venues operated by professional companies. Not detailed in this chapter are outdoor amphitheaters, worship spaces, arenas, rock/pop venues, auditoria larger than 2,500 seats, and halls equipped with electronic music enhancement systems. All these spaces require careful acoustical design attention in the same manner as discussed throughout this chapter.

B. Chapter Sections

The following parts of this chapter are divided into three sections: Music Performance Spaces, Dramatic Arts Spaces, and Music Instruction Spaces. Some venues fit into more than one of these categories. An opera house is both a music performance space and a dramatic arts space, in which the music must sound rich and be enveloping, while sung words must be intelligible without listening strain. A college recital hall is both a music performance and a music instruction venue, which implies unique design provisions. A multipurpose civic auditorium typically must serve classical orchestral and choral music, opera and operetta, popular music of all varieties, Broadway musicals, classical drama such as Shakespeare, and speaking events of all types. In order to serve such wide-ranging music types plus the spoken word and serve them all well, a hall must have thoughtfully designed variable acoustics provisions, making it both a music performance and a dramatic arts space.

C. Obtaining Desired Results

The architectural acoustician always strives to master: (1) the critical music and drama listening experience and an understanding of which designs produce the best acoustical results; (2) the communication skills to elicit what the users wish to accomplish and, in the process, to help manage their expectations; (3) the architectural design and construction knowledge to guide the project toward the best acoustical results; (4) the understanding of and effectiveness in design team interactions with architects, engineers, theater consultants, code officials, and owners; (5) defending acoustical design provisions during value engineering critiques; (6) the knowledge of the standard design and construction process from programming through construction administration and commissioning; (7) the ability to test and listen to the completed project and to employ "tuning devices" to effect the best possible adjustments for excellent listening; and (8) the patience and enthusiasm to steadfastly guide acoustical aspects of design and construction over the long haul, often four or more years, from the initial concepts to a successful opening.

We would emphasize that nothing substitutes for the ability to draw upon abundant, critical listening experience and comparing those observations with other critical listeners. The committed acoustician attends performances regularly in different halls and associates closely with other practicing acoustical and theater consultants with extensive, proven experience in performing arts design. Such acousticians are highly active in the National Council of Acoustical Consultants (NCAC), the Acoustical Society of America (ASA), and the Institute of Noise Control Engineering (INCE), or comparable international organizations. Finally, time that the acoustician invests critically listening to halls of his/her own design always provides invaluable feedback. This also includes listening openly to observations by colleagues, the balance of the design team, stakeholders, reputable critics, and certainly the users.

15
Unit I
Music Performance Spaces

Ronald L. McKay, FASA

15.I.1 INTRODUCTION

The most prevalent types of music performance halls will be covered in this unit of Chapter 15—arranged in four groups, the members of each group having much in common. The groups are:

1. Concert, Recital, and Pipe Organ Halls
2. Multipurpose Halls with Variable Acoustics
3. Opera Houses
4. World, Country, Jazz, and Popular Music Halls

While the halls in Group 1 are largely homes to classical music from the 16th century through today, many also host Group 4 music ensembles because those performances are well attended. Group 2 halls serve all forms of music, plus drama, lecture, convocations, film, audiovisual displays, and more. Group 3 halls serve opera, plus ballet almost exclusively. Group 4 halls may specialize in one music form, such as "country" in the Grand Ole Opry, or they may accommodate all of their music family members, but are generally unattractive to Group 1 and 3 performers because their acoustics are normally too dry.

15.I.2 BASICS FOR ALL MUSIC PERFORMANCE HALLS

15.I.2.1 Design and Construction Processes

Most new and remodeling projects require about four years to design and build. The conventional project phases are: programming, schematic design, design development (often including value engineering), construction documents, bidding, construction administration, and commissioning. The owner, prime users, architects, and principal engineers will be engaged the whole time. And so must the acoustical consultant, who will serve vital roles: (1) in establishing a program responsive to the owner's and users' needs and budget; (2) in guiding design concepts along paths of high acoustic potential, and defending these designs during value engineering; (3) in continuing to refine designs, produce good details and construction documents, and assure that construction is done correctly; and (4) in testing and adjusting built rooms and systems, and advising owner/users on effective use. It is important for the acoustician to be very familiar with the design and construction process, knowledgeable regarding cost, concerned about visual aesthetics as well as sound, and reasonably competent in architecture and most building engineering disciplines. All this will contribute to a quality overall result.

15.I.2.2 Background Noise

Most music halls require background sound levels to be sufficiently quiet to ensure soft passages are clear in their entirety and noise is inaudible on recordings. Many noise and vibration sources may need to be considered, but building mechanical, plumbing, and electrical equipment always require acoustical planning and design. Exterior noise sources such as aircraft, railways, and roadways may require an above-average building exterior for noise isolation. Other interior spaces, such as a second auditorium, rehearsal rooms, lobby, scenery shop, and mechanical rooms will require remote location or above-average intervening construction.

The family of noise criteria (NC) curves is used to establish noise limits in all types of rooms. These curves establish maximum sound pressure level limits in each octave band from 31.5 Hz through 8,000 Hz. The curves are essentially equal-loudness contours, meaning that low-pitched sounds may have significantly greater sound pressure levels than high-pitched sounds, yet are perceived to be of similar loudness.

A variety of NC-like curve families exist—all similar but subtly different in order to best fit different applications. The 1971 preferred noise criteria (PNC) curves work very well for auditorium applications. These curves are shown in Figure 15.I.1. PNC-15 noise levels are virtually silent. PNC-25 noise levels are audible but quiet. PNC-35 noise levels are clearly audible and too loud for a music performance hall. PNC-15 to -20 is recommended for Groups 1, 2, and 3 halls in which amplification will be used for limited purposes. PNC-25 to -30 is recommended for Group 4 halls in which amplification will be used often, unless a recording program demands quieter noise levels.

Detailed noise and vibration control procedures are beyond the scope of this section. Several excellent texts on the subject are suggested at the end of this section.

15.I.2.3 Seating Capacities

Concert halls and flexible multipurpose halls between 800 and 2,500 seats are emphasized in this discussion. The best results for excellent classical music listening quality are typically halls with 1,500 to 2,000 seats, but very rarely more than 2,500 seats. Smaller halls fall into the recital hall category and are likely to be too loud for large instrumental ensembles. Larger concert halls for 3,000 or more seats as built during the last several decades of box-office pressure have proven to be of lesser music-listening quality.

Recital halls of high quality typically range between 200 and 400 seats and rarely more than 600 seats. These halls are for soloists, quartets, chamber music ensembles, and small vocal groups. They are visually and acoustically intimate.

Pipe organ halls of high quality typically range between 100 and 400 seats. They basically are recital halls for very expensive instruments of great pride to the owners. They often include a small platform for ancient or liturgical music, with instrumental or choral ensembles

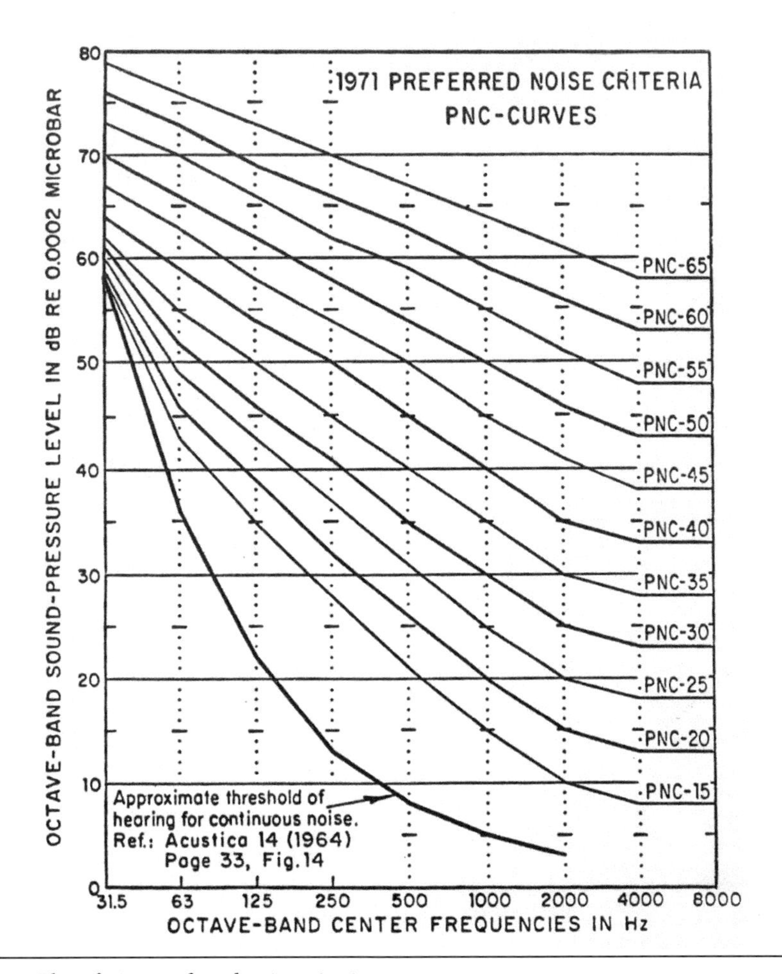

Figure 15.I.1 Plot of 1971 preferred noise criteria curves.

up to 20 persons. The platform typically is downstage, near the audience level and below the elevated organ.

Opera houses in the 1,200- to 2,500-seat range include the best sounding halls in the world. However, many of these halls are quite old and fall short on seating comfort and circulation safety requirements. More modern opera houses with excellent acoustics are typically in the 1,200 to 1,800 seat range. Note that the best operatic voices can fill only so much space.

World, country, jazz, and popular music halls can be acoustical successes for audiences of tens of thousands, but only with large, well-designed, and effectively operated sound amplification systems, and with controlled acoustics. In this section, such facilities of small (300 to 500) and moderate (500 to 1,500) seating capacity are emphasized. The small facilities may or may not use amplification; the moderate facilities most likely will.

15.I.2.4 Reverberation and Reverberation Times

Reverberation is the persistence of sound in a room after the source stops. This persistence is due to multiple reflections from the various room surfaces. By definition, reverberation time is the time it takes for a loud sound to decay by 60 decibels (dB) (virtually to inaudibility) once it is suddenly stopped. Reverberation time is often abbreviated as T, R_{60}, or T_{60} (or even T_n where decay is measured out only to n = 10 to 30 dB of decay, once decay is detected at a microphone to be 5 dB decayed). For simplicity, T_{60} will generally be used throughout this chapter.

Long reverberation times are popular for organ music and, to a somewhat shorter extent, for much classical orchestral and choral music. A long reverberation can envelop the audience from all directions and enrich listening to these unamplified music types. On the other hand, long reverberation times interfere with speech intelligibility and with clarity of popular music, especially under amplification.

Reverberation time may be measured and calculated using the following equation:

$$T_{60} = 0.049V/A + 4mV, \qquad (15.I.1)$$

where V is room volume in cubic feet, A is total room sound absorption in sabins in a particular frequency band, and m is the energy attenuation constant of air in inverse feet (1/ft)*, which varies with frequency, temperature, and pressure (as can be found in Number 7 on the next page and in standard texts and references at the end of this chapter).

A, the total sound absorption, can be derived from measurements, or calculated by the following equation:

$$A = \Sigma S\alpha = S_1\alpha_1 + S_2\alpha_2 \ldots + S_n\alpha_n, \qquad (15.I.2)$$

where S is the surface area in square feet and α is the corresponding material's Sabine absorption coefficient in a particular frequency band.

The previous equations were derived from experiments at Harvard University in the 1890s by Professor Wallace Clement Sabine. These equations and the unit of sound absorption, the sabin, are named after him.

The total sound absorption in a room may be calculated by determining the area of each of the individual surfaces (including audience and performance areas) in square feet, multiplying each area by its appropriate Sabine sound absorption coefficients, and then summing the results, as per Eq. 15.I.2. A sound absorption coefficient indicates the sound-absorbing efficiency, which may vary between 0 and 1, denoting 0 to 100% absorptive efficiency. Note that some materials are rated for more than 1.0 in some frequencies, but these are largely artifacts of the testing process, and in calculations, should be rounded down to 1.0, because a material, except for spurious edge effects, cannot absorb more than 100% of its direct-incident sound energy.

Absorption coefficients are determined by well-controlled laboratory or field measurements of the T_{60} in a room, with and then without a material sample present; the differences in T_{60}s, along with room volume and sample area, are used in the above equations to calculate the material's Sabine absorption coefficients. This procedure usually is done in the standard octave frequency bands centered at 125 through 4,000 Hz because the α's of most materials vary significantly with frequency. (See Number 7 in the following list for typical values of 4mV.)

Reliable T_{60} calculations can be complicated and subtle. A number of guidelines and precautions for T_{60} calculations are as follows:

1. The most common method of calculating T_{60}s at various frequencies uses the Sabine T_{60} formula. Most of the absorption coefficients measured and reported in the United

[a]An equivalent equation stated in metrics can be found in Chapter 12.

States are derived using the Sabine formula, in which volumes and areas are measured in feet and not in meters. Therefore, square feet for areas, the corresponding Sabine absorption coefficients, and cubic feet for volumes must be employed. Some absorption coefficients are based instead on metric measurements, so caution is advised.

2. Absorption coefficients and T_{60}s in rooms vary significantly with frequency. A range of 125 through 4,000 Hz must normally be considered, and occasionally 31 through 8,000 Hz.
3. The bulk of available sound absorption coefficients are derived from reverberation measurements made in large, reverberant laboratories with and without a large test sample present. Every effort is made to produce a diffuse and uniform sound field that is incident randomly from all angles upon the test sample. There are times when the absorption coefficient at normal (0° incidence) or grazing (90° incidence) or some other specific angle would be preferable, but such data may be difficult to locate. Normal-incidence α's are usually less than random-incidence α's, and grazing-incidence α's are usually higher.
4. When calculating a room's volume, include all of the room's coffers, niches, and other significant volumes that are openly connected with the room's primary volume. Include the volume "apparently" consumed by audience and performers plus their chairs. Include spaces above and below balconies where the minimum height of the space is equal to or greater than its depth—which is the strong design recommendation for new construction. If this balcony depth is larger than its opening height, exclude any such over-balcony or under-balcony volumes but substitute an appropriate absorption coefficient at the open "faces" of these volumes, such as approximately 0.5 for low- to mid-frequencies and 0.8 for higher frequencies.
5. The 0.049 constant in the Sabine formula is based on the (mean free path) distance of travel by sound between various surfaces in a large rectangular room, using feet as the measure. Therefore, square feet and cubic feet must be used for surface areas and room volume. A different constant is required for metric measures.
6. The audience and performers and their chairs are major sources of sound absorption in music halls. The areas of occupied or unoccupied audience and performer seating areas must be determined from the true area along the slope of the orchestra or balcony floor, and not the projected area shown on a plan drawing. Also, banks of audience in upholstered seats introduce an edge effect increasing their acoustically effective areas and making them larger than their physical plan area. Add a strip of audience area 1.67 feet (0.5 m) wide around all audience perimeters except for edges abutting a solid wall or balcony face.
7. Air, itself, absorbs significant high frequency sound of 2,000 Hz and higher—the drier the air, the greater the absorption. To obtain the air absorption in sabins at 68° F. and 50 percent relative humidity (typical of auditoriums), multiply the hall's volume in cubic feet by 4 m = 0.0029 for 2,000 Hz, by 0.0074 for 4,000 Hz, and by 0.026 for 8,000 Hz.

Middle-frequency reverberation times are the T_{60}s calculated or measured in the 500 and 1,000 Hz octave bands and then averaged. It is rather common to describe a room's T_{60} using the middle-frequency value only. Note that there is no "average" T_{60}, and T_{60}s should not normally be averaged between frequencies (with the important lone exception of this "mid-frequency T_{60}". The values at 500 Hz and 1,000 Hz are normally very close).

For classical instrumental, choral, and pipe organ music, low-frequency T_{60}s (31 through 250 Hz) should be greater than middle-frequency T_{60}s gradually rising with decreasing frequency until the T_{60} at 31 Hz is 1.5 to 2.0 times that of the middle-frequency T_{60}. The 1.5 multiplier applies to smaller halls, and the 2.0 to larger halls. This will help provide solid bass "warmth". For popular music, the above low-frequency guideline multipliers are reduced to 1.0 for smaller halls and to 1.5 for larger halls.

High-frequency (2,000 through 8,000 Hz) T_{60}s should equal middle-frequency T_{60}s less the unavoidable effects of air absorption. However, air absorption may be minimized via humidity control. A music facility should not get too dry (less than 50% relative humidity), which would be bad for high-frequency "sparkle" as well as for tuning stability of wood instruments.

Table 15.I.1 offers recommended middle-frequency T_{60}s for various music types performed by instrumental and/or vocal groups in halls with 800 to 2,500 seats and for pipe organ music performed in halls with 100 to 400 seats.

Another parallel quantity is the early decay time (EDT) or early reverberation, often referenced as T_{10}. Standard T_{60} involves determination of the time required for sound to decay by 60 dB, to near inaudibility. EDT or T_{10} concentrates on the early part of the decay, the time required for initial sound to decay by 10 dB (after its initial 5 dB of decay) and then multiplying the numeric result by 6 to make T_{10} comparable to T_{60}. The reason to consider T_{10} is that in most music, notes follow each other in short 50- to 80-millisecond (msec) intervals, so

Table 15.I.1 Recommended middle-frequency T_{60}s for music performance spaces

Instrumental and/or Vocal Music	Average T_{60} Value in the 500 and 1,000 Hz Octave Bands At Full Occupancy
Classical Music	
• Baroque period (1600 to 1750): Bach, Handel, Vivaldi	1.3 to 1.5 seconds
• Classical period (1750 to 1820): Beethoven, Hayden, Mozart	1.6 to 1.8 seconds Add 0.2 second for Beethoven
• Romantic Period (1820 to 1920): Berlioz, Brahms, Debussy, Strauss, Ravel, Tchaikovsky, Wagner	1.9 to 2.1 seconds Less for some composers
• Later 20th century period (1920 to 2000):	1.4 to 2.0 seconds depending on composer
• Southeast asian gamelan	As close to the outdoors as possible (0 to 0.5 second)
Current Music	
• Show tunes, soft rock, world country, jazz, popular music	0.8 to1.0 Second
• Hard rock	As close to the outdoors as possible (0 to 0.3 second)
Pipe Organ Music	
• Classical	3.0 to 4.0 seconds
• Current	2.0 to 3.0 seconds

the decay of a note can be heard for only 50 to 80 msec before a subsequent note makes the previous note's further decay inaudible; also, our ears and brain compares the strength of the subsequent note's early energy to the early decaying energy of the earlier note, and concludes that the music has either clarity or blend and richness. T_{10} (typically measured, not hand-calculated) can be a useful tool regarding music-listening quality, especially in conjunction with T_{60}.

15.I.2.5 Hall Volume

A music hall of 1,000 seats in which almost all performances will involve romantic classical music is best served by a 2.0-second mid-frequency T_{60} when fully occupied, whereas another 1,000-seat hall in which almost all performances will involve popular music is better served by a 1.0-second mid-frequency T_{60} when fully occupied.

Both halls will seat an audience of 1,000 and for simplicity, assume both halls employ one block of chairs on the main floor with continental seating (no internal aisles, only peripheral aisles). Seats are spaced 42 inches (1 m) back-to-back rather than the 36 inches (0.9 m) typical of conventional seating with internal aisles. Each 21 inch wide (53 cm) average seat plus the walking space before it will consume 6.125 square feet (0.57 m^2). So, 1,000 seats require 6,125 square feet (0.57 m^2). Proportion the seating area to fit within a 70 foot (21 m) wide hall with a 4 foot (1.2 m) wide aisle on each side, netting a 62 foot (19 m) wide seating area which must be just under 99 feet (30 m) front to back (to maintain some sense of visual intimacy) for the required area. A cross aisle will extend in front and in back of the seating. There will be a 62 × 99 foot (19 × 30 m) rectangle with a perimeter of 322 feet (101 m). The sound-absorbing edge effect of the seating block equals 322 feet × 1.67 feet = 538 square feet (101 m × 0.51 m = 30 m^2). Therefore, the effective acoustical area of the seating block is 6,125 + 538 = 6,663 square feet (569 + 50 = 619 m^2). See Figure 15.I.2; note that in this example, several sightline and circulation-planning refinements have been ignored.

The audience sound absorption in each of the halls is the same. Roughly, using a mid-frequency average absorption coefficient of 0.82 (0.82 × 6,663 square feet) we get 5,464 sabins ascribed to an audience in moderately upholstered chairs. This neglects performer absorption, but it will be greater in the classical hall because of the larger orchestra sizes. As a rule of thumb, allow 15 square feet (1.4 m^2) per person for instrumental performing groups and 6 square feet (0.56 m^2) per person for choral groups. Further, consider the musicians to be seated in lightly-upholstered chairs when determining their absorption coefficients, and add the 1.67 foot (510 mm) edge effect around the overall performance area.

Assume that both halls will be finished in appropriately heavy plaster on all exposed walls and ceilings (considerably heavier for classical music) and that only minor amounts of sound-absorbing materials will be needed to control echoes, etc. Because the classical hall will be of larger volume and with taller walls (hence more plaster area to absorb sound), the result is that we must add 842 sabins to the classical music hall (a 15% increase over audience absorption) and 626 sabins to the popular music hall (an 11% increase over audience absorption) to account for this greater exposed area in the classical music venue.

All of the above and the Sabine equation that $T_{60} = 0.049V/A$ shows that a doubling of reverberation time requires a doubling of hall volume, plus some extra volume to compensate for the added sound absorption of greater room surface areas. A more reverberant room design clearly involves increased cost, but this is justifiable if the need is demonstrated.

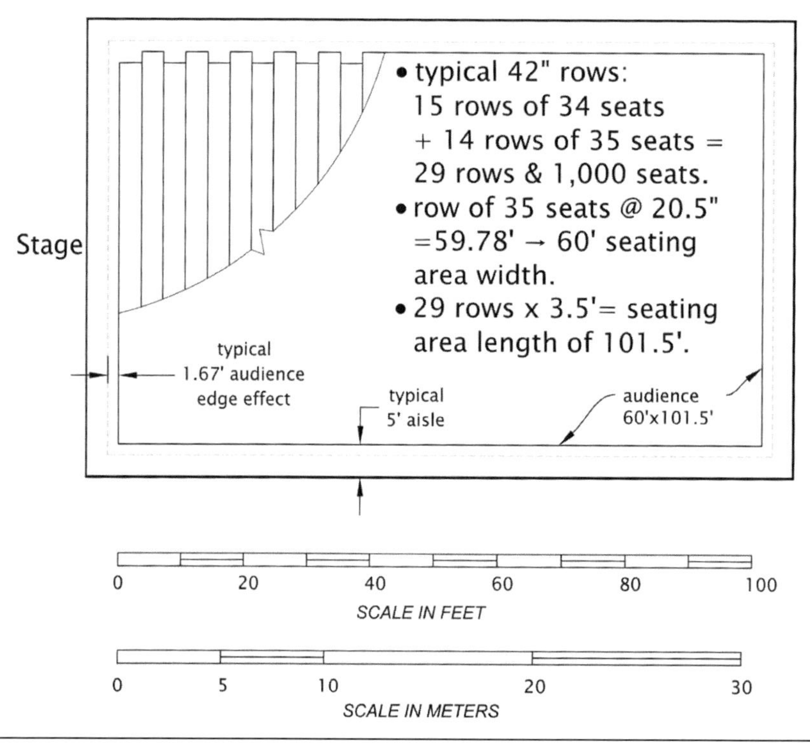

Figure 15.I.2 Seating layout for 1,000-seat music hall. Note, detailed seating area above = 6090 sq. feet (566 sq. m) or 1.9% more than the 5,978 sq. feet (555 sq. m), estimate based on a 20.5″ × 42″ (52 cm × 107 cm) per-seat area design, without parallel aisle and room design. This is a necessary and meaningful refinement.

15.I.2.6 Hall Finish Materials

15.I.2.6.1 Floors and Chairs

In classical music, multipurpose, and pipe organ halls, the floors beneath the audience should be hard in order to allow some vertical sound reflection with the ceiling. Painted concrete, vinyl flooring on concrete, and wood flooring glued directly to concrete are all appropriate. The aisles of organ halls should be hard surfaced. The aisles of classical music and multipurpose halls are best if hard surfaced, but thin (1/4 inch (6.4 mm) maximum thickness) carpet glued directly to concrete is acceptable in order to provide for appearance, safety, and especially footfall noise abatement. Wood floors are very popular throughout classical music and pipe organ halls because of their warmth of appearance, plus the notion that "wood is good" for music. In popular music halls, there is no need for hard floors below the audience or in aisles. In fact, carpet should be encouraged.

Music performance platforms and stages should be constructed of wood for sound reflection, and over an airspace so that they will resonate somewhat when driven by cellos and double bass violins, pianos, and percussion instruments. Common construction is hardwood or hardboard on plywood on wood sleepers on concrete, or the aforementioned without sleepers on wood joists and vertical framing. With sleepers, the airspace is only a few inches below the flooring. With joists and vertical framing, the airspace is typically a few feet deep, which is preferable for the best bass response for orchestral music.

Audience chairs can vary from essentially nonabsorptive wood or plastic, to somewhat absorptive leather covers over cushions, to significantly absorptive, sound-transparent fabric covers over open-cell foam cushions set in frames. The latter are the most useful because the unoccupied chairs approximate the sound absorption of a typical seated person, thus stabilizing a hall's reverberation regardless of occupancy, which is a highly desirable characteristic, especially during rehearsals.

Fabric-upholstered auditorium chairs come in three basic varieties: lightly, moderately, and heavily upholstered. The three varieties, in the same sequence, provide the following average mid-frequency sound absorption coefficients (per Beranek):

Unoccupied:	0.60, 0.69, 0.83
Occupied:	0.78, 0.82, 0.88

The heavily-upholstered chairs are exceptional for T_{60} stabilization with respect to occupancy in all auditoria, especially for drama, but note that they can diminish bass response in classical music and organ halls as a result of somewhat more low-frequency absorption.

Music performance halls are generally well served by moderately-upholstered chairs, detailed as follows: (1) chair backrest to be exposed and shaped plywood or steel (not plastic) without any cushion or fabric on the back side; (2) chair backrest front side to be upholstered with a "breathing", sound-transparent fabric over a 1-inch minimum thick cushion of open-cell sound-absorbing foam that is free of any sound-opaque skin covering; (3) chair seat bottom pan to be perforated metal with a minimum 30% open over at least half of the pan's area, with perforations open to the sound-absorbing seat-top cushion above; (4) chair seat top to be "breathing", sound-transparent fabric over a 2-inch (50 mm) minimum thick cushion of open-cell, sound-absorbing foam that is free of any sound-opaque skin covering.

Performers' chairs should be hard or leather covered or lightly upholstered.

Actual chair selection for a particular project is always a balancing exercise between acoustics, comfort, appearance, cost, code requirements, manufacturers' product lines, and availability.

15.I.2.6.2 Walls, Balcony Faces, Ceilings, and Soffits

All of the surfaces should be designed to return useful sound reflections to the audience and performers in concert halls, multipurpose halls, organ halls, and opera houses. Some of the surfaces should be used for the same purpose in popular music halls. To be effective sound reflectors, surfaces must be hard, impervious, stiff, and of medium-to-heavy density. The density variable relates to a surface's ability to reflect low-frequency sound, with a heavier density reflecting more low-frequency sound.

Halls for classical and/or pipe organ music should provide strong bass response in order to produce a strong sense of musical warmth. To sense low-frequency energy in one's chest cavity and/or to feel floor vibration in sync with the music adds to the experience.

The most bass-rich concert hall in this author's experience is Ham Concert Hall at the University of Nevada, Las Vegas, with the following measured, unoccupied T_{60}s (see Table 15.I.2). The 2,000-seat Ham Concert Hall has concrete floors, with thin carpet in aisles only, moderately upholstered audience chairs, painted/sealed stone-aggregate concrete block walls (with grout-filled cells to increase mass), an exposed steel roof deck consisting of roofing on rigid insulation on several inches of heavyweight concrete on steel decking on rather open

Table 15.I.2 Measured reverberation times in low-frequency range from 31.5 Hz to 1 kHz in Ham Concert Hall at the University of Nevada, Las Vegas

Frequency in hertz	31.5	63	125	250	500	1,000
T_{60} in seconds	4.85	4.25	3.90	2.85	2.40	2.30

and exposed steel trusses, a conventional wood stage floor over a 2.5 foot (0.76 m) airspace, and thick wood stage-surrounding walls 15 to 20 feet (4.6 to 6.1 m) high—all well-braced. This is heavy, basic construction and is the architectural prescription for strong bass.

The most bass-rich pipe organ hall in this author's experience is the Reyes Organ Hall of the DeBartolo Performing Arts Center at the University of Notre Dame. This 100-seat hall has an unoccupied middle-frequency T_{60} of 4.0 seconds and low-frequency T_{60}s that increase gradually from 500 Hz to near 6.0 seconds at 31 Hz. The 100-seat room has a nicely finished concrete floor below hardwood pews, a small choir performance platform of heavy wood construction over an airspace, stone on concrete walls, stone columns supporting a first and second balcony with concrete floors and open steel railing, a heavy, exposed steel deck roof (like Ham Hall) on heavy wood trusses, a glorious Fritts organ on the front wall at the first balcony level. This is another architectural prescription for warm bass.

There are other guidelines in the acoustic literature for appropriate warmth or bass richness such as bass ratio, relative bass strength, and others that have met with varying degrees of success in predicting or characterizing warm bass.

The "bass index", related to the loudness of the bass sound, and "listener envelopment" (LEV) are shown for most halls to be directly related to the mid-frequency value of G. Neither of these, however, seems to predict concert hall quality with good consistency. So, the above concert and organ hall examples are presented for some guidance.

The guidelines for strong bass support using plaster are 1-1/2 inch (38 mm) thick sand-aggregate gypsum plaster on metal lath [15 lbs/sq ft (73.2 kg/m^2)], perhaps with thin wood panels laminated directly to the plaster surface. Wood panels over airspaces and typical drywall construction are to be avoided, although multiple layers of gypsum board and/or plywood achieving 11.25 lbs/sq ft (55 kg/m^2) and curved to a radius of 15 feet (4.6 m) can be made to work acoustically. Note that curving further stiffens the paneling construction, and also assists with uniform and diffuse sound reflection within the hall. Of course, exposed concrete (precast or poured-in-place), stone, and brick plus thick precast plaster with fiber reinforcement also provides strong bass support.

Halls intended principally for popular music do not require strong bass response, but should have modest bass response with only moderate increase in T_{60}s below 500 Hz, from 1.0 to 1.5 times mid-frequency T_{60} at 31 Hz. The 1.0 multiplier applies to smaller halls and 1.5 to larger ones. This permits less massive room finishes, but "flimsy" ones (thin wood or one layer drywall over airspace) should be avoided.

In halls for classical and/or organ music and in opera houses, sound-absorbing finishes should be minimal, and should be employed only to treat unusual reflections and echoes, or other discreet problems such as footfall noise via thin carpet in aisles. In popular music halls, sound-absorbing finishes may be used as needed to control reverberation and to prevent echoes and other problematic sound reflections. Standard sound-absorbing finishes include carpet, heavy draperies, glass fiber and mineral fiber tiles and panels, and various glass fiber blankets or boards (often behind sound-transparent facings of fabric, grille cloth, perforated metal, wood or metal or plastic grillage, etc.).

15.I.2.7 Basic Hall Shaping

Wall, ceiling, and floor surfaces need to be properly shaped around the hall volume. Many auditorium forms and shapes have been built for performances of all types. Serious performing groups, performing arts sponsors, architects, theater designers, and acoustical consultants study these many halls worldwide; talk and deliberate with one another; and reach both agreements and differences. In the case of differences, they usually understand why and find the particular, governing preferences to be informative.

15.I.2.7.1 What Not to Do

There is a consensus among most design professionals about room and surface forms to avoid or to use with extreme care when designing auditoriums. Huge spaces such as arenas are not viable for unamplified music performances because of insufficient loudness and because the enclosing surfaces are too distant and cannot deliver timely sound reflections to the audience. That is one reason why this section has established a maximum of 2,500 seats, with a preference closer to 2,000.

Two- and three-dimensional concavely curved surfaces like vaults and domes have great negative acoustic influences. Reflections from these surfaces do not return uniformly, but focus in some areas creating excessive loudness and sometimes severe echo, while leaving other areas with weak or no reflections. There have been "acceptable" domed halls where sound reflectors of appropriate shape, size, and density were hung below the dome to provide uniform reflections and to break up the dome's focusing power. However, such geometry should be avoided when seeking excellent acoustics.

Fan-plan halls have a basic geometry that sends nearly all side-wall reflections to the rear corners of the room. Breaking the side walls into facets nearly parallel, or somewhat "recumbent", to the hall centerline and stepping them from front to rear can deliver important lateral reflections uniformly across the audience. However, music will be significantly quieter in back than in front because of energy spreading; again, a highly problematic (acoustically) choice of plan form.

A hard-surfaced vertical rear wall curved in plan to follow the curvature of seating rows will return focused and long-delayed reflections back to the stage and front audience. This is unacceptable in any hall and should not be built. If faced with such an existing condition, there are corrective measures by making the surface either diffusive (in the music dominant case) or absorptive (in the speech dominant case).

Placing hard, flat surfaces parallel to each other on opposite sides of a sound source leads to repeated inter-reflection between the surfaces. This condition is called "flutter echo", and its effect is to add unpleasant harshness and stridency to the sound source. The surfaces must be at least ten degrees out of parallel or be irregular and diffusive to avoid this flaw. Alternately, one of the surfaces can be covered with an efficient sound-absorbing material (α = 0.75 to 0.95 from 500 Hz upward).

15.I.2.7.2 Appropriate and Uniform Loudness

The loudness or strength of musical sound in a performance hall should be such that a pianissimo passage by a soloist or small ensemble is fully audible in all of its textures and subtleties, and a "fortissimo" passage by a large ensemble is exciting and enveloping, but not uncomfortable. Loudness is a well understood but subjective term. Strength is a quantitative term (G) that may be measured or predicted (via software) and expressed in dB.

The strength of musical sound usually is measured in each of the octave bands from 125 through 4,000 Hz. This often is done in two discrete time domain regions of a musical note arriving at the receiver: early sound, or the first 80 msec of the note, and subsequently, reverberant sound, or the remaining audible milliseconds of the note (1,920 msec in a hall with a 2.0 second T_{60}). This time division is explored so the very important early-to-late (or early-to-total) sound ratio may be determined. However, strength is determined by the total level (early plus reverberant) of the played note. A sound strength measurement normally is made without a musical source, but with analogous electronic equipment. Refer to this Chapter's Glossary for "Strength of Sound".

Leo Beranek reports G_{mid} (average of the 500 and 1,000 Hz octave bands) values measured in some 40 concert halls in Japan, Europe, and the United States in his excellent book *Concert Halls and Opera Houses: Music, Acoustics, and Architecture.* He concludes that a value of G_{mid} between 1.5 and 5.5 dB is preferred for concert halls over 1,400 seats, 9.0 to 13.0 dB for chamber music halls with less than 700 seats and −1.0 to 2.0 dB for opera houses with over 1,400 seats. Beranek's suggested G_{mid} values for chamber music halls seem applicable to the recital and pipe organ halls that are addressed here.

Before the strength measurement, G, was developed and employed on a wide scale, this author routinely placed a sound power reference noise source (see ILG in glossary) at the concert master's position on the platform of music performance halls that he helped design, and then tested at completion. Octave-band sound pressure level (SPL) measurements were then made in representative audience locations and averaged. For some ten halls in the 1,200–2,500-seat range, SPL_{mid} values of 55 dB ±2dB were found appropriate for classical music loudness. The typical variation over a given hall's seating area also was found to be ±2dB.

A guiding principle of music performance hall design, in fact any auditorium design, is to provide the highest possible uniformity of stage-produced sound over the entire audience. In the 1,000-seat concert hall discussed on page 409, the audience area, with front, rear and side aisles, will be about 70 feet (21 m) wide by 107 feet (33 m) long facing a stage that is about 50 feet (15 m) wide on average by 40 feet deep to accommodate an orchestra of 80, plus conductor and soloists. Imagine a woodwind player in an average mid-stage position near the hall's centerline. How can the hall be shaped to cause this musician to be heard uniformly across the audience?

Figure 15.I.3 shows a shoebox-shaped hall around the plan form just described. The shoebox will be 70 feet (21.3 m) wide by 147 feet (44.8 m) long by an average height above the stage and audience floor of about 35 feet (10.7 m), to achieve a 2.0 second mid-frequency T_{60}. The ceiling height is approximate; surfaces will be adjusted in order to achieve uniformity of source sound levels.

The 147 foot (45 m) long hall is divided into three near-equal subdivisions, front to back in the center diagram, and its plan is modified from a rectangle to an elongated octagon, and the ceiling's longitudinal section is adjusted to match the side walls. Sound reflections from our woodwind player to the audience are diagrammed, recalling that sound reflections, like light, obey the rule that angle of incidence equals angle of reflection. The result, as shown, is that the front ceiling and side walls provide reflections to the entire audience, the center ceiling and side walls to the rear two-thirds of the audience and the rear ceiling and side

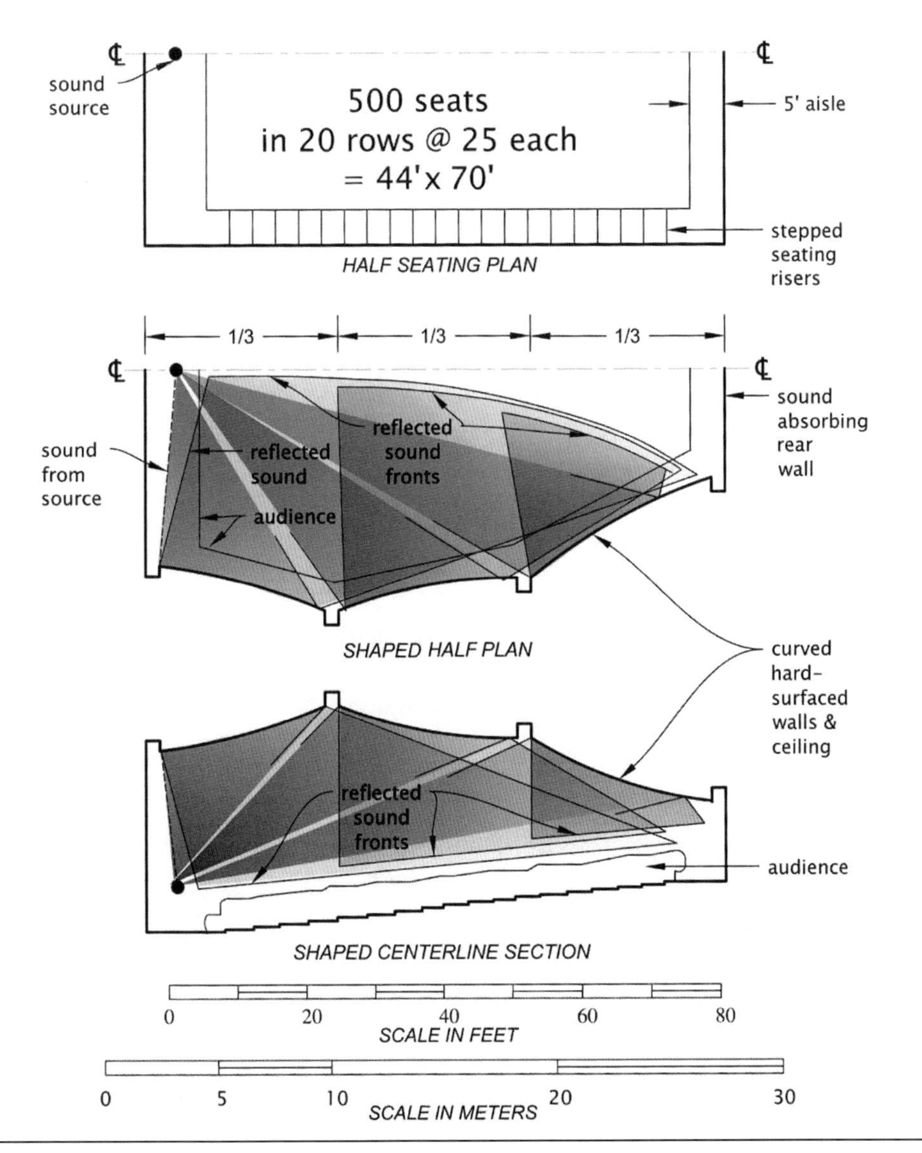

Figure 15.I.3 Room shaping for uniform loudness.

walls to the rear one-third of the audience. Note that the number of reflections delivered to the audience grows with increasing distance from the source to counteract inverse-square-law sound-spreading reductions. The resultant woodwind sound level variation over 1,000 seats is 2 dB, which is quite good. This is an excellent model for the provision of uniformity and should always be kept in mind. Understand, however, that other vital factors may force reasonable compromise.

15.I.2.7.3 Sound Arrival Times at Audience Members and Performers

Consider a violinist standing on stage before an audience, playing a single, short note that is radiated more or less spherically to the hall. The sound that travels the shortest possible path to any audience member is called the direct sound. Subsequently, the patron receives a series of sound reflections of the same note from various room surfaces until the note decays in sound level to inaudibility. The time between the arrival of direct sound and the arrival of the first hall reflection at an audience member is called the initial time-delay gap (ITDG). If the ITDG at mid-main-floor is 20 to 25 msec or less, the hall is likely to be considered acoustically intimate or to have acoustical "presence". The smaller the hall, the shorter the ITDG should be. This is a substantial music listening attribute. ITDGs of 35 or more msec exist in halls of unexceptional quality, at least in those seats exhibiting such values.

Because sound travels at 1,130 feet (344 m) per second at standard temperature and pressure, an ITDG of 20 msec indicates a travel-distance difference of about 23 feet (7 m). The clear design implication is that several sound-reflecting surfaces must be near the source and/or the listener, which is not easy in a large room with an orchestra-size stage, perhaps 2,000 seats and a ceiling as much as 50 feet (15 m) high. This is one of the major reasons why many quality concert halls are narrow, in order to provide quickly-arriving sound reflections from side walls and side balcony faces to the audience. These halls also may have a suspended canopy of sound reflectors 22 to 27 feet (6.7 to 8.2 m) above the performance platform and extending out and over front audience seats. If the platform's sidewalls, which are outwardly sloping (in plan) toward the audience, are a controlled 50- to 55-feet (15 to 17 m) wide at the platform's downstage edge and our standing violinist is downstage center, then the violinist is 25 to 27.5 feet (7.6 to 8.4 m) from each side surface and about 25 feet (7.6 m) less 4 feet (1.2 m) standing shoulder height = 21 feet (6.4 m) below overhead reflectors. This is a "balanced" architectural/theatrical/acoustical design.

Returning to the violinist's single, short, bowed note in our concert hall with a 2.0 second mid-frequency T_{60}, the bowed note will persist audibly for two seconds or 2,000 msec after the direct sound arrival at an average listening position. Extensive research shows that the arriving and then decaying sound sample may be divided meaningfully into "early sound" (first 80 msec after direct sound arrival) and "reverberant sound" (all remaining sound to the end of reverberation at, say, 2,000 msec). The selection of 80 msec as the divider between early and reverberant sound is related to the fact that many musical notes are played for 80 msec or less, so this is the time window we have to listen to note one before note two arrives. One can measure the sound level (in dB) of the early and, separately, the reverberant sound in the mid-frequencies and compare them to each other. This is an important measure of what the hall is doing to the violinist's bowed note. If the early sound exceeds the reverberant sound by 1 to 4 dB, listeners typically will say there is "clarity" or "good definition" and that one can perceive individual notes and musical detail with ease. This is especially appropriate for many piano works and for Baroque Period and small ensemble pieces. If the reverberant sound exceeds early sound by 1 to 4 dB, listeners typically will say there is "good blend" or "richness" or "envelopment". This is appropriate for Beethoven or Romantic Period symphony works.

An echo is an individual reflection or tightly-time-packed set of reflections with sufficient sound level to stand out clearly above other reflections arriving soon before and after. If one looks at the sound-level versus time plot of decaying reverberant sound, an echo will be seen as a short-duration spike that is well above its neighbors. Such echoes are unacceptable during either music or dramatic arts performances. They usually are caused by a distant hall surface with focusing capacity, such as a hard-surfaced rear wall curved concavely in plan. Platform sound will strike the hall's rear wall traveling a rising path, be reflected and focused upward to strike the ceiling, and then be redirected downward and to the platform. Performers and front-audience members listening to a greeting spoken from the platform will hear: good-good, morn-morn, ing-ing.

Performers, too, require early-arriving, clear sound reflections from the hall in which they are playing. The opera singer needs to hear the hall respond to his/her voice and return significant energy in order to hear that he/she is "filling" the hall. Instrumentalists also need to sense the hall's response as reverberant energy "washes" back to them in carefully-controlled amounts of reflection from rear-wall and balcony-face surfaces. Additionally, of course, a careful balance and blend of orchestral and vocal sounds are critical for all opera performers if the audience is to be captured by all the direct and reverberant sound correctly.

During an "ensemble", the ability of the members of an instrumental or choral group to hear each other well and perform in unison, is fundamental. The positioning of the musicians and the height of sound-reflecting surfaces above their heads are the major determinants of good ensemble. A choir should stand or be seated on risers so that the audience can see all mouths and so that each choir member can see all other members' mouths. Because the human voice is quite directional along the forward horizontal axis of the head, a choir should be seated in a shallow "U" plan—open to the audience but partially facing each other. Risers are most helpful for orchestra woodwinds so the downstage string sections and the audience have line-of-sight on their instruments. Risers are acoustically negative for trumpets and trombones, which frequently are too loud; it's better that they not be visible to downstage orchestra and audience—hidden by music stands or plastic shields or other orchestra members (who may complain).

Time after time, this author has encountered orchestra ensemble shortcomings when the group is seated under a high ceiling, even if stage rear and side walls are providing useful reflections between musicians. This is extremely common with secondary-school, college, and adult amateur orchestras. Professional orchestras can often overcome this high-ceiling problem, the Boston Symphony and Amsterdam Concertgebouw Orchestra being notable examples. In many experiments with professional and college orchestras seated under a movable sound-reflective canopy or orchestra shell ceiling, 27 feet (8.2 m) was found to be the maximum height above platform level before front-to-back and cross-stage timing plus sectional balance became problematic.

15.I.2.7.4 Lateral Sound Reflections to the Audience

Reverberation provides an audience listening to organ, classical orchestral, choral music, or to opera with a sense of "envelopment"—being surrounded by sound arriving from all directions. Lateral reflections to the audience from side walls, balcony faces, and ceiling irregularities provide a sense of "spaciousness" or of the musical-source seeming significantly wider than it appears. Spaciousness is a very important contributor to music-listening quality for the music performance types mentioned.

An orchestra playing outdoors will sound no wider than it looks and it will certainly not sound enveloping. However, in a good concert hall, the same group "will" sound significantly wider than it looks. It may seem to fill nearly the full 180 degrees in front of you. A sense of acoustical spaciousness is derived from multiple and relatively strong lateral reflections arriving from left and right, because we listen with two "laterally" spaced ears about six inches apart. One ear will receive a lateral reflection from its side just slightly sooner and louder than the other ear because of "shadowing" by the head. The brain will interpret this as wide sound or spaciousness if such lateral reflection differences occur frequently in the first 80 msec of arriving musical notes. The more times this happens during a listener's exposure to early sound, the more spacious the hall will seem, so multiple lateral reflections are important. This is one of the major reasons that shoebox halls sound better than fan plan halls (in which lateral reflections can exist only via intentional design and, even then, tend to be increasingly weak and ineffective toward the rear of the hall). Also, it is difficult to provide copious lateral reflections to the audience in a hall with seating that surrounds the stage in a large bowl as there are essentially no flanking walls to provide important, lateral reflections.

Here is an important point related to providing good and plentiful "early" lateral reflections to audiences seated in balconies. If one views a balcony seating plan *only* and shapes the hall's side walls to deliver platform sound laterally to the people there, ignoring what should happen to sound traveling in the vertical (3rd) dimension, the balcony audience will receive "no" early lateral reflections. Platform sound approaching side walls are rising vertically on a path to a balcony, and will continue rising after being reflected horizontally toward the balcony and will miss the balcony audience. Rather, it will strike the balcony ceiling, which may simply reflect it to the rear wall and thence back to the front of the room, possibly creating an echo. Side walls that can deliver lateral reflections to a balcony should be divided into panels that lean down and in toward the audience in order to deliver sound to them upon first reflection. Such panels should have minimum dimensions of 4 to 6 feet (1.2 m to 1.8 m), but if this is impossible, even a 9 to 12 inch (230 to 305 mm) horizontally projecting ledge that will turn sound downward will be of value. See Figure 15.I.4. Note the sloping side-wall panels in the first two photos and the horizontal ledges in the third.

Spaciousness is quantified in several ways. The method first developed involves the descriptor, lateral fraction (LF). One places an onmidirectional loudspeaker in several typical platform locations and delivers to it an 80 msec impulsive signal of 500-, 1,000-, and

Figure 15.I.4 Balcony and sidewall shaping.

2,000-Hz octave band pink noise. The impulses are received in several typical audience positions by two microphones—a figure-eight directional microphone with its null aimed at the source and an omnidirectional microphone. The total lateral energy received by the figure-eight microphone is divided by the total overall energy received by the microphone and then is averaged over the various source-receiver locations. This often results in fractions ranging from 0.24 to 0.12, which do not correlate well with perceived hall acoustic quality. Compared to other methods, LF is not recommended.

Another way of quantifying spaciousness is more complex, but indeed yields numerical results that are well-correlated with perceived hall acoustic quality. One first obtains the *interaural cross-correlation coefficient (IACC)* and then calculates the *binaural quality index (BQI)*.

The IACC is measured by placing an omnidirectional loudspeaker at several typical platform locations and delivering to it sound signals just as described previously for LF. The measurement difference is in the radiated sound reception and processing. The sound is received by two small (near) omnidirectional microphones placed in the ear canals of a realistic dummy human head positioned at several typical audience locations. Their signals feed two separate channels of an acoustical-analysis program—such as CATT *Acoustic* or *EASE/EARS*—that cross-correlates the left and right ear differences (in terms of IACC) during the early-arriving sound across the 500-, 1,000-, and 2,000-Hz octave bands. The $IAAC_{AVG}$ value then is converted to a BQI number via $BQI = 1 - IACC_{AVG}$. According to Beranek, good BQI values lie between 0.55 and 0.71 for symphony orchestra music in halls over 1,400 seats, between 0.70 and 0.76 for chamber orchestra music in halls less than 700 seats, and between 0.60 and 0.71 for opera houses over 1,200 seats.

Practical design guidelines to help achieve the aforementioned BQIs and the resultant important listening sense of spaciousness are: (1) keep the hall as narrow as possible, especially around an orchestral or choral ensemble and at the front of the audience chamber while still providing audience side aisles; (2) provide frontal side-wall balconies overhanging the main floor side aisles and provide faces that are closer to the hall centerline than the wall surfaces below; (3) design side-wall balcony soffits to interact with low side-wall surfaces to provide additional lateral reflections to an orchestra-level audience; (4) detail side-wall and balcony-front surfaces to be diffusive so they provide multiple lateral reflections (as in Figure 15.I.5); and (5) detail ceiling, cloud, and/or canopy surfaces to be diffusive so they provide both lateral and longitudinal reflections (as in Figure 15.I.6).

15.I.2.8 Platform and Stage Planning

A platform is a raised flat floor for performers that is placed so the audience can see and hear them well. Lighting must be provided so musicians can read sheet music easily and so that all performers can be seen clearly and pleasingly by the audience. Platforms are employed in concert, recital, and pipe organ halls and in many world, jazz, and popular music halls; they usually have no curtains or flyloft with rigging systems for scenery.

In strict theatrical parlance, a stage "does" include a full complement of curtains, a flyloft with a rigging system for both scenery and lighting, side stages for scenery and prop storage, plus performer entrances and exits. Opera houses employ stages (and orchestra pits) as do most multipurpose halls.

A problem arises when pipe organs, classical, or popular music ensembles perform on a stage with sound-absorbing curtains around them and more curtains plus scenery hanging above them.

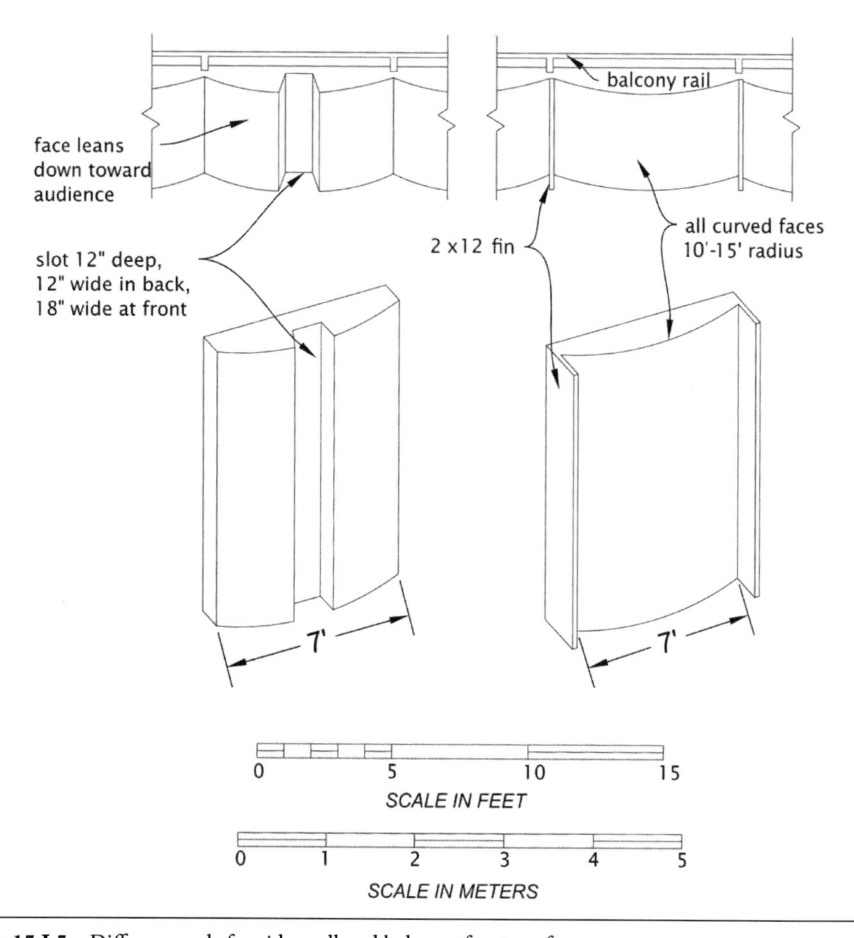

Figure 15.I.5 Diffuse panels for side-wall and balcony-front surfaces.

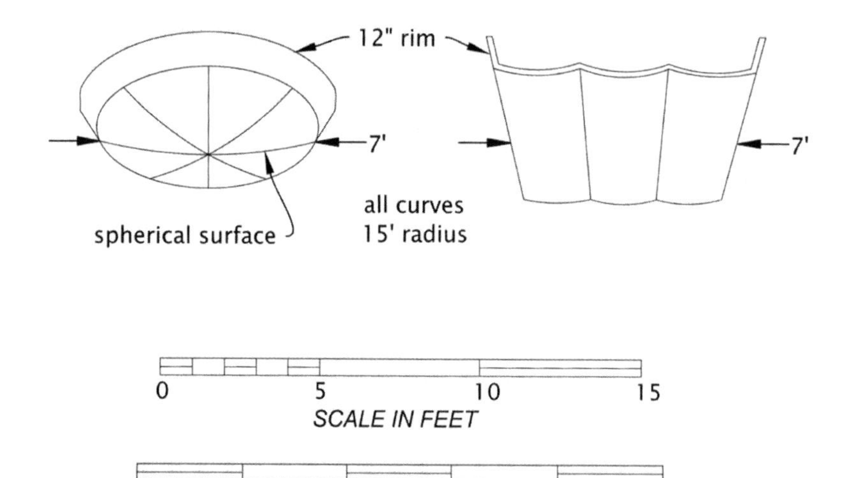

Figure 15.I.6 Diffuse surfaces for ceiling, cloud, and/or canopy.

For the performers, the acoustic environment is somewhat like being outdoors and ensemble among them suffers. For the audience, the performers sound weak, distant, and poorly blended.

The solution is to provide a demountable orchestra shell with a ceiling to close off the flyloft and vertical towers that form a surround on three sides and rise vertically to meet the ceiling. The ceiling is composed of nearly horizontal panels with built-in lighting (that can be rotated to the vertical and stored in the flyloft). The towers consist of panels mounted on multiple rolling metal frames so they can be moved to off-stage storage quickly by a few stagehands. The panels are predominantly sound reflective, best when shaped for diffusion and usually made of wood veneer or plastic laminate or hardboard on each side of a stiff honeycomb matrix or rigid foam.

Several music equipment companies sell modular orchestra shell components that can be configured in a variety of forms and also will build custom shells based on designs by others. Several theater equipment contractors also will build custom shells based on their designs or those of others.

Opera presents unique challenges because it must provide flexible and dramatic scenic display simultaneously with beautiful orchestral sound wafting from musicians who are largely hidden below the stage; all while creating a beautiful and articulate vocal sound for the entire audience. And this must be done in harmony with constantly changing performer movement, scenery, and lighting changes; no small task for anyone, including the acoustician.

To help themselves, many opera companies generally direct lead singers to position themselves downstage, as close to the audience as practical (on the stage apron just forward of the proscenium is best). Many companies also understand that stage scenery design should act like an orchestra shell as much as possible in order to boost projection of secondary singers and choruses, but this concept often bows to historic convention and/or other demands. When stage scenery is ill-conceived acoustically, the results are often highly problematic.

A well-considered orchestra pit design is largely left to the acoustician. It must make the musicians comfortable, able to see the conductor, able to hear each other, and able to play in controlled balance with singers as they are heard in the pit, on-stage, and certainly in the audience.

The principal elements of good and functional acoustic pit design (see Figure 15.I.7) are: (1) set the pit to be equal in width to the hall proscenium opening; (2) design the pit floor area to provide 10 square feet (0.93 m^2) for the conductor and each string and wind instrument plus 100 square feet (9.3 m^2) for percussion; (3) set the pit depth at 8 feet (2.4 m) below stage level; (4) plan the pit so that brass and percussion instruments are seated below the forestage overhang whose soffit is sloped up and out toward the audience at 10 degrees; (5) plan a fixed floor below the brass and percussion at 8 feet below stage level; (6) seat string and woodwinds and position conductor on a hydraulic lift downstage of the fixed pit floor with a play setting at only 7 feet (2.1 m) below stage level; (7) make the railing between the pit and audience solid and segmented with flat facets (not curved) on the pit side; (8) fit the pit's upstage wall and nearby overhead soffit with demountable, sound-absorbing panels from the baseboard upward and full-width panels of 2 inch (50 mm) thick, 7 lb/cu ft (112 kg/m^3) density glass fiber covered with perforated vinyl or polycarbonate; and (9) provide a sound reflector high over the open pit (e.g., an "eyebrow") to reflect string, woodwind, and soloist voices to main floor seats, at least.

Note that the loudest instruments are muted by being recessed and backed with absorptive panels and the softer instruments are open to the hall. Also note that the front main floor seats are shielded from the orchestra by the solid pit rail and that the back of this rail reflects

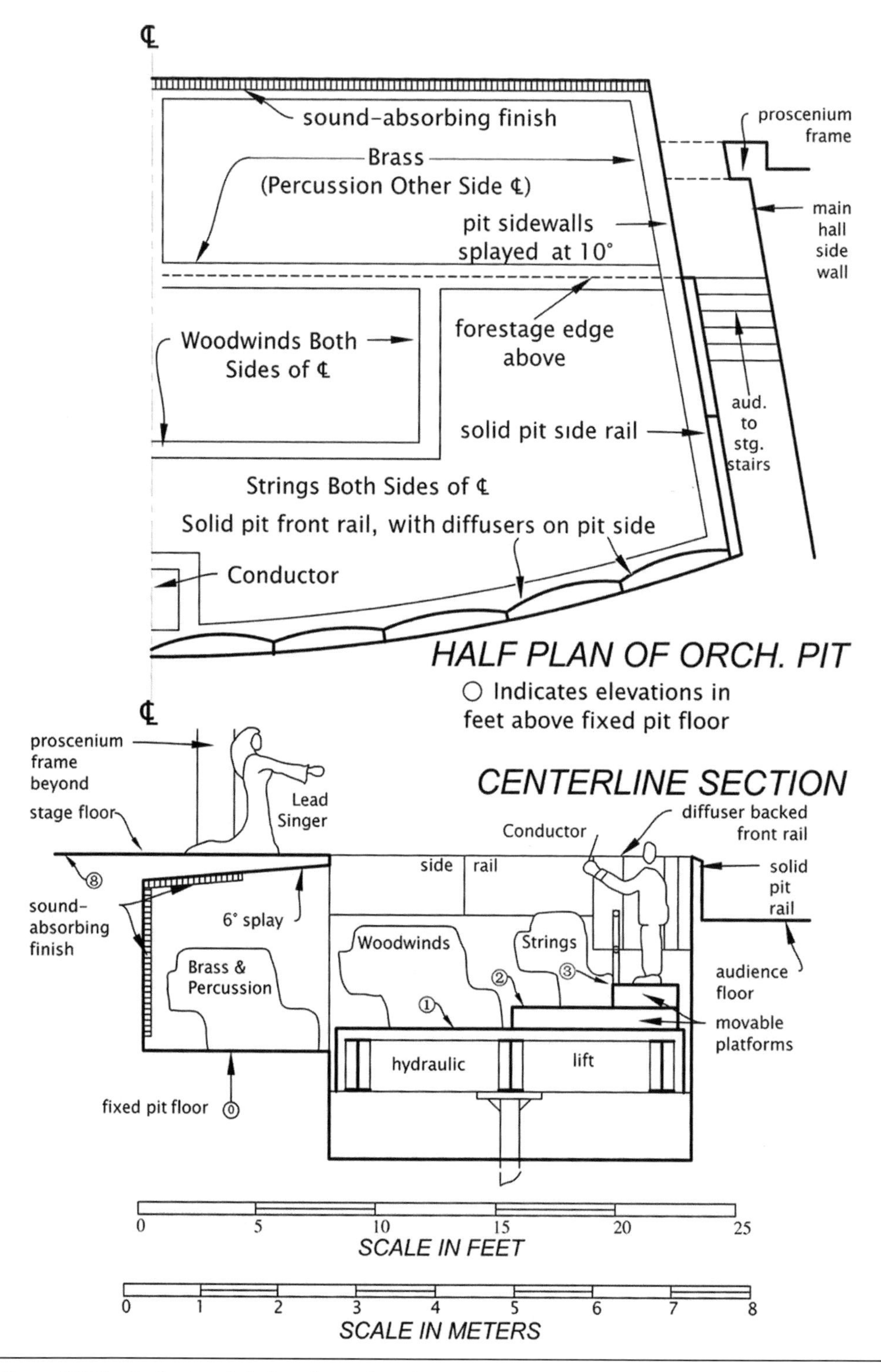

Figure 15.I.7 Orchestra pit design plan and section.

the orchestra to on-stage singers and vice versa—and do so in a way that minimizes strong reflections right back to pit musicians.

For an orchestra on a platform or stage, allow 15 square feet (1.4 m^2) per instrumentalist; plus a four foot (1.3 m) aisle at each side for musicians with large instruments to circulate; plus six feet downstage for the conductor, soloists, and a concert grand piano. Don't push the orchestra upstage against the stage back wall—leave two to three feet clear. Learn typical American and European orchestra seating plans. A typical orchestra will arrange itself in a 3 × 4 plan aspect ratio rectangle, the three being upstage-downstage and the four stage left-right.

Some orchestras prefer risers; others eschew them. Normally, woodwind sections are on risers, in two elevated rows—as they will be heard more clearly if they are not shielded from the audience by the string sections before them. On the other hand, shielding loud trumpets and trombones from other orchestra members simply by distance, music stands, and/or clear plastic barriers often is a helpful section-balancing and orchestra-pleasing measure. First violins almost always sit on the downstage-right side of the conductor (to conductor's left) with second violins *or* celli and double basses opposite, at downstage-left. Because there is a considerable distance from the back of the first violins to the back of the second violins or celli and basses (e.g., 40+ ft. [12m]), placing the rear parts of these sections on risers helps these opposing cross-stage musicians to hear each other more clearly and thereby benefit ensemble, if not balance.

Choirs larger than about 10 voices are best on risers, with or without chairs. With chairs, each chorister consumes a space three feet deep by 21 inches (580 mm) wide. Standing without chairs, the riser depth may be decreased. While choirs often are arranged in straight, parallel rows facing the audience, they are best arranged in parallel rows forming a shallow *U* shape opening toward the audience. This will help the several choir sections to hear each other better and improve their ensemble and balance.

Platform and orchestra shell side walls should not be parallel, but each splayed 12 to 15 degrees from the hall's centerline and opening toward the audience. This provides a good balance between the ensemble on the platform or stage and projection to the audience.

Surrounding platform walls may vary between full height to the room's ceiling and 10 to 15 feet (3 to 4.6 m) high to the tops of nearby balcony railings. These balconies may accommodate a pipe organ, choir, and/or audience. In all cases, the platform walls should be hard and sound diffusing—*not* smooth and flat. An excellent platform wall design consists of multiple vertical panels four to six feet wide and convexly curved in plan with a 15 foot (4.6 m) radius plus horizontal ribs 9 to 12 inches (230 to 305 mm) deep, applied every four to six feet vertically, starting about 10 feet (3.1 m) above the stage floor. Rather than projecting ribs, slots may be notched into the main panels to the same depth and on the same vertical centers as the ribs, although this is somewhat less effective acoustically (see Figure 15.I.5).

15.I.2.9 Audience Seating Configurations

In this section, we are dealing with concert hall audience capacities between 800 and 2,500 seats (see Figure 15.I.8). First, consider axial halls that have the stage and most audience seats arrayed along the room's centerline axis. An 800-seat concert hall normally has main floor seating only. A 1,000-seat hall might have main floor seating only or main floor plus a small rear balcony. 1,200- to 1,400-seat halls normally have main floor seating plus one rear

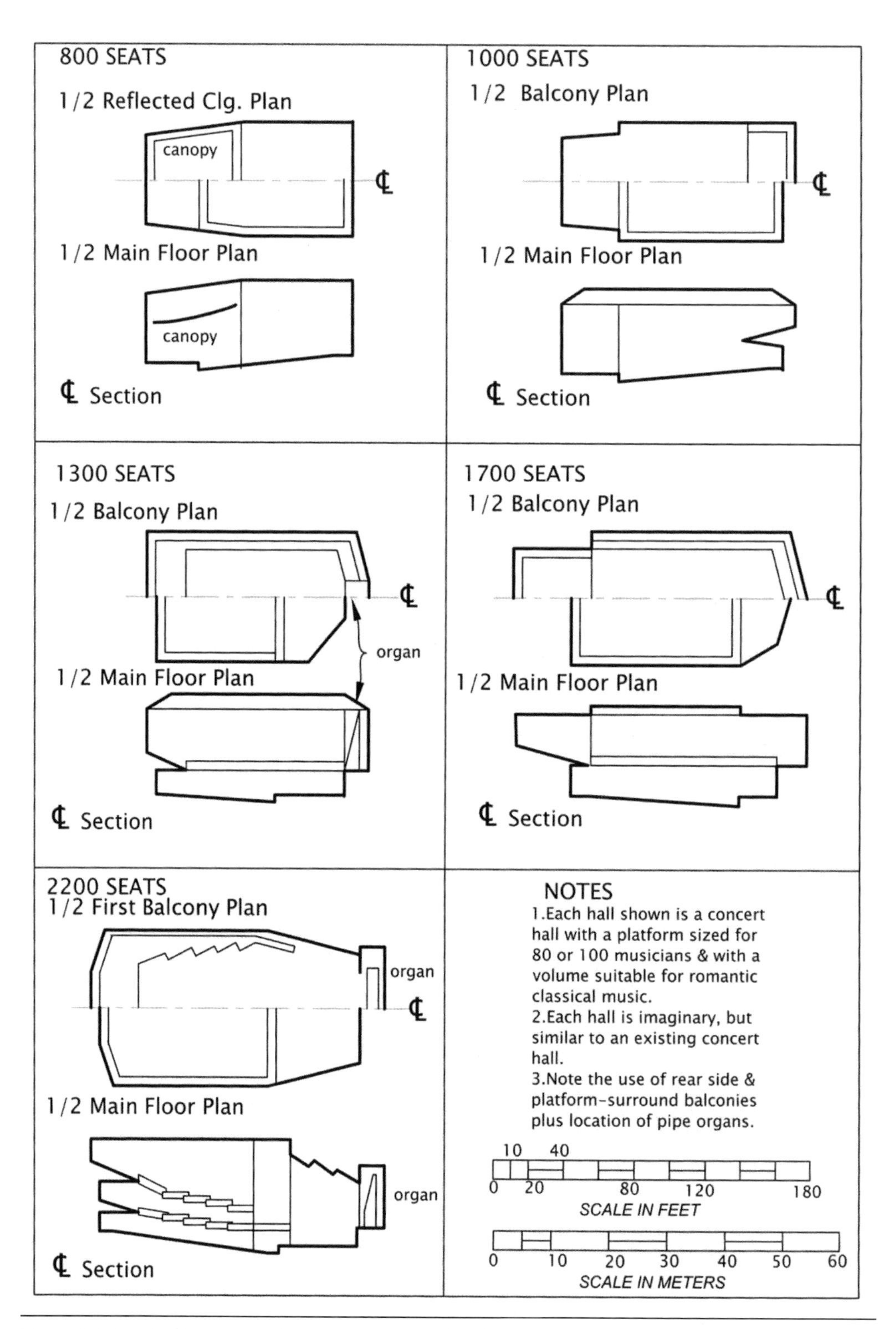

Figure 15.I.8 Audience capacities between 800 and 2,500 seats.

balcony and, possibly, a narrow balcony seating along side walls. 1,600- to 1,800-seat halls usually have main floor seating plus one or two rear balconies and, possibly, narrow side balconies. 1,900- to 2,500-seat halls usually have main floor seating plus two rear balconies and one or two narrow balconies on each side wall. Any of these halls may have narrow balconies behind or around the stage for organ, choir, and/or a small audience group.

All the above axial halls lend themselves readily to a shoebox or modified shoebox shape, but balcony configurations are critical to good overall results:

1. There should be no deep under balcony seating space at the rear or sides of a hall. The depth of such space, from balcony nose to rear or side wall, should be equal to or (preferably) less than the height of the opening to the space from the main hall volume. The opening height is measured from the bottom of the balcony nose to the floor directly below.
2. Side balconies, as a rule, should be shallow—one row of loose chairs or two or three stepped rows of fixed chairs. The balcony floors should be horizontal or gently sloped or stepped incrementally down toward the stage. They *must not* be steeply sloped because this can prevent sound-reflection-surface exposure by side walls for sound radiating from the stage and reduce or preclude valuable lateral reflections to the main floor seats.
3. All stage sound sources should be able to "see" continuous side-wall surfaces of generous height from front to back under all side balconies; the more side balcony soffits they can see, the better. Drawing sound-ray diagrams and/or constructing laser-pointer models are invaluable in "testing" how best to shape balcony undersides and their acoustical interaction with side walls.
4. There should be generous wall height exposed to all stage sound sources above the highest rear and side balconies. It is within this upper hall volume, with high and hard wall surfaces on four sides plus a hard ceiling that much of the room's reverberation will develop and "rain down" beneficially on the audience from all directions. This provides listeners with an important sense of "envelopment" by music, a truly positive attribute.

Surround seating has recently become a popular mode of new concert hall design worldwide. The platform is placed not at one end of a hall, but at the quarter or third point along the long plan axis (one example is seen in Figure 15.I.9). The audience is arranged 360 degrees around the platform in rising tiers and balconies that face the platform. This produces a strong sense of visual intimacy between the performers and audience, and of sharing among the audience. These visual attributes are most pleasant.

In the case of surround halls, we have an example of what was mentioned in the first paragraph of Section 15.I.2.7. It is a type of hall form which experts disagree about, but they understand why others have their particular preference. While this author thoroughly appreciates surround concert halls and believes many more will be built, he definitely prefers the classical music sound in shoebox or modified shoebox halls.

15.I.2.10 Clouds and Canopies

"Clouds" are sound reflectors suspended from a high ceiling or roof and usually are not intended to be adjustable. A canopy is an assemblage of clouds for a specific purpose and often is movable vertically and/or tiltable longitudinally (see Figure 15.I.10). Clouds and canopies

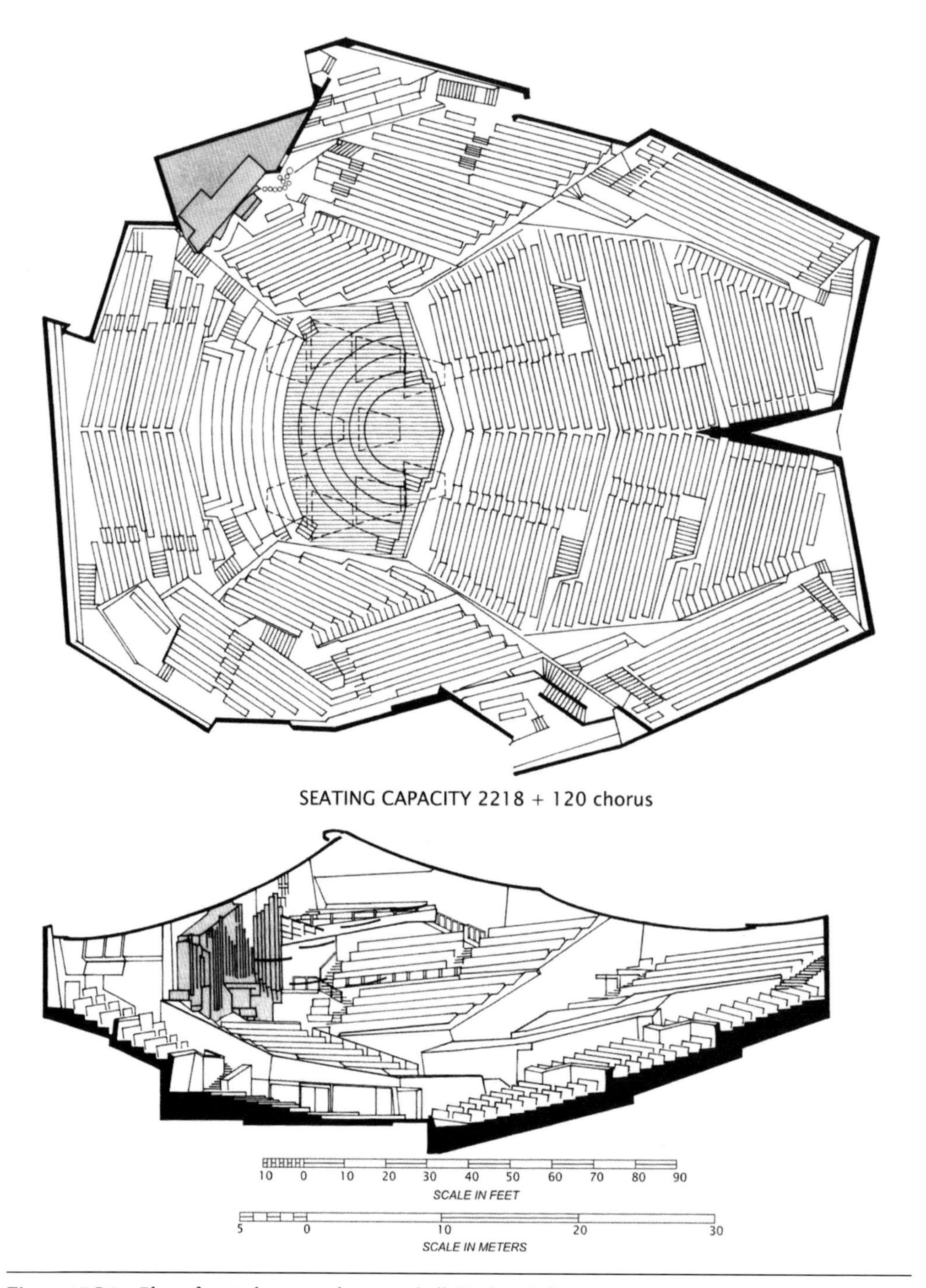

Figure 15.I.9 Plan of typical surround concert hall (Berlin Philharmonie, courtesy Leo Beranek.

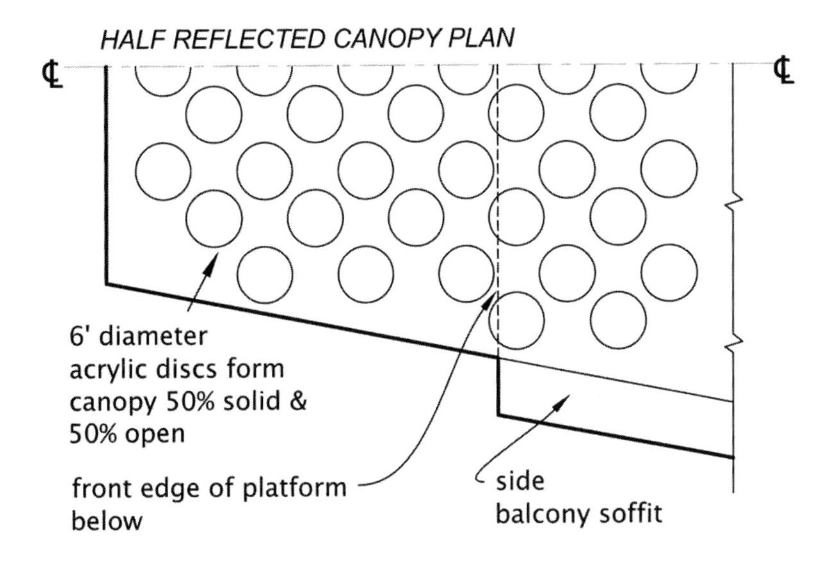

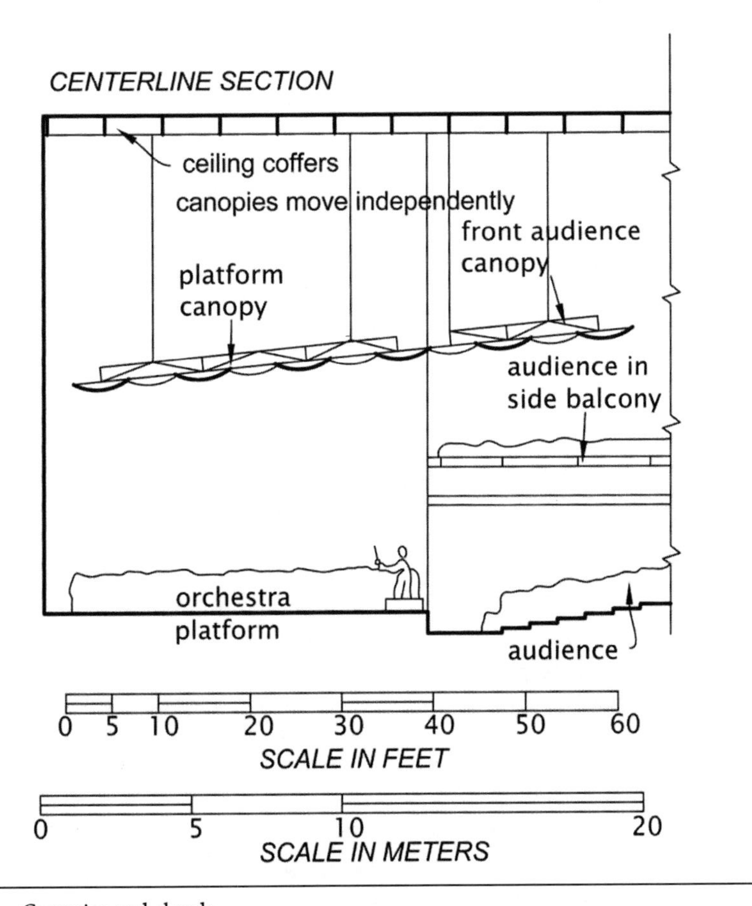

Figure 15.I.10 Canopies and clouds.

are employed when the ceiling or roof is too high to provide needed early reflections from performers to themselves and/or to the audience. Often the ceiling or roof is high for good reason, such as required volume.

Clouds usually are intended to provide early sound reflections in a frequency domain from about 350 Hz upward. The wavelength of 350 Hz is slightly under 3.25 feet (1 m). To be a specular reflector of 350-Hz sound, the cloud must have a minimum 6.5 foot (2 m) diameter or width, which may be shaved down to 6.0 feet (1.8 m) if essential. To serve principally speech, this dimension would drop to 4.5 feet (1.4 m), minimum, if necessary.

Clouds often are formed with a convex-curved, spherically shaped underside (i.e., compound-curved with its belly downward), using a radius of 15 feet (4.6 m). This shaping is applied most easily to a circular cloud plan form, but also may be imposed on squares, rectangles, etc. The curvature softens reflections from each cloud and spreads the listener area that is covered. This design facilitates a canopy of multiple clouds that is 50% solid and 50% open (a checkerboard) and that permits half the incident sound from below to be reflected quickly to listeners and the other half to rise into the upper hall to contribute to reverberation. The additional design advantage is that the canopy assembly delivers a series of early soft reflections spread over a short time span to any given listener. This is far better than one strong reflection from a large, flat reflecting plane. Such canopies are found over stages and often extend over front audience areas to serve both performers and front audience members.

Clouds frequently are made of translucent, cast acrylic or resin 3/8 inch (10 mm) thick and with a stiffening perimeter upturned edge or flange that also provides for a suspension connection. They also are formed of 5/8 inch (160 mm) thick moulded plywood, 3/4 inch (19 mm) thick plaster applied to metal lath and a metal frame, and of moulded glass-fiber-reinforced gypsum (GFRG) 1 to 2 inches (25 to 50 mm) thick.

Many professional orchestras do perform well on a stage under a 50 feet (15 m) high or higher ceiling that cannot provide early reflections from point to point within the orchestra to facilitate section-to-section hearing and overall orchestra cohesiveness (ensemble). While many pros can handle the situation, when provided with a canopy about 25 feet (7.6 m) above their floor, they profess pleasure and stress relief. Less-established professional orchestras, university and other nonprofessional orchestras do not have the experience of the long-standing major orchestras and often will lose their cohesiveness (ensemble) when seated under a high ceiling. Therefore, over-stage canopies are important in academic and other nonprofessional venues with high ceilings. Usually, the canopy is supported by a steel frame that is cable-suspended from winches above so that the canopy height and angle can be adjusted easily. This is necessary to find the best angle for the given hall, the best canopy height for the given orchestra, and to provide lower heights for smaller ensembles and soloists—plus the ability to raise the canopy up and out from occlusion of an upstage pipe organ.

15.I.2.11 Detailed Surface Shaping

Let's explore the important smaller-scale shaping of large individual room surfaces. Large ceiling, canopy, and wall surfaces, after being shaped to provide uniform sound level distributions and sufficient early and lateral reflections, should be further shaped via smaller-scale sound-diffusing elements. The purpose is to soften middle- and high-frequency reflections, spread them more widely in time, and to prevent high-frequency harshness or glare. We are dealing with the frequency range of about 3,000 to 8,000 Hz, so we must design irregularities with dimensions in the range of one inch to five inches (25 to 130 mm).

Imagine classical ceiling coffers in an approximate 4 × 4 foot (1.2 × 1.2 m) grid with applied plaster dentils, eggs, and darts about one to five inches (25 to 130 mm) in size. Further imagine walls with classical pilasters and cornices plus niches with statuary (like the Boston Symphony Hall). Imagine a contemporary hall with walls of convex cylindrical shapes about four feet (1.2 m) wide and interspersed with ribs or slots three to six inches (76 to 152 mm) high or deep. Design possibilities are extensive. See Figure 15.I.4 (contemporary) and Figure 15.I.11 (current classical).

Detailed surface shaping is most important in concert halls, recital halls, multipurpose halls, and opera houses. It is of lesser importance in pipe organ halls *if* the organ builder can be counted upon to voice his/her organ to prevent glare (high frequency stridency). If not, it becomes a most important acoustical-design factor. In halls serving world, jazz, and popular music, there are likely to be fewer sound-reflecting surfaces than in the halls just noted, but detailed surface shaping would serve to mellow live listening and recordings. This author recommends application of smaller-scale diffusive surfaces for high quality listening in these halls.

Figure 15.I.11 Side-view photo of Royce Concert Hall at the University of California in Los Angeles, showing classical ceiling coffers (see also Figure 15.I.19).

15.I.3 DESIGN HIGHLIGHTS: CONCERT, RECITAL, AND PIPE ORGAN HALLS

15.I.3.1 Introduction

Each of these three hall types is devoted to the performance of classical music and, in the case of organ halls, to liturgical music as well. Excellent sounding concert halls range in size from 800 to 2,500 seats, recital halls from 100 to 700 seats, and pipe organ halls from 100 to 400 seats. Recital halls and chamber music halls require essentially equal acoustical and architectural designs, while the chamber music halls will likely have a larger platform than recital halls and a higher average seat count.

Each hall type requires a performance platform and has no need for a stage house for scenery handling or orchestra pit. Each is appropriately designed by placing the performance platform and audience essentially in the same room where they share a common acoustic ambience. Each requires substantial reverberance much of the time.

15.I.3.2 Platform Designs

15.I.3.2.1 Concert Halls

The platform for a concert hall will typically provide for a maximum orchestra of 80 to 100 instrumentalists plus the conductor, a half-dozen soloists, and a concert grand piano positioned downstage of the orchestra. A choir of 100 to 150 voices and/or a pipe organ also may be required, usually above and behind or around the orchestra in low and shallow balconies. When a choir is not involved in a performance, their seats may be given over for audience use.

Figure 15.I.12 shows a design that accommodates an orchestra of 90, plus the conductor, soloists, and concert grand on the basic platform, plus a choir of 110 and a substantial pipe organ on surrounding low balconies. Also note that the low walls surrounding the orchestra are modular and movable so they can be moved downstage to support small ensembles and soloists both acoustically and visually.

15.I.3.2.2 Recital Halls

A recital hall's platform typically provides for a maximum of 30 to 40 instrumentalists and uses the technique of moving modular platform-surrounding wall panels downstage to support small ensembles and soloists. A driving force in such halls is acoustical and visual intimacy, so some recital hall platforms may provide space for a maximum of 20 instrumentalists or 50 choir members—or some combination thereof. There well may be an organ balcony above and behind the platform, but flanking choir balconies are unusual. However, shallow (one row of movable chairs) audience balconies around recital hall platforms are more and more common because they provide students with unique listening and viewing opportunities.

15.I.3.2.3 Concert Pipe Organ Halls

These are typically small-capacity halls of 100 to 400 seats for the display of an institution's proud and expensive pipe organ. The hall and its organ usually are for the benefit of the

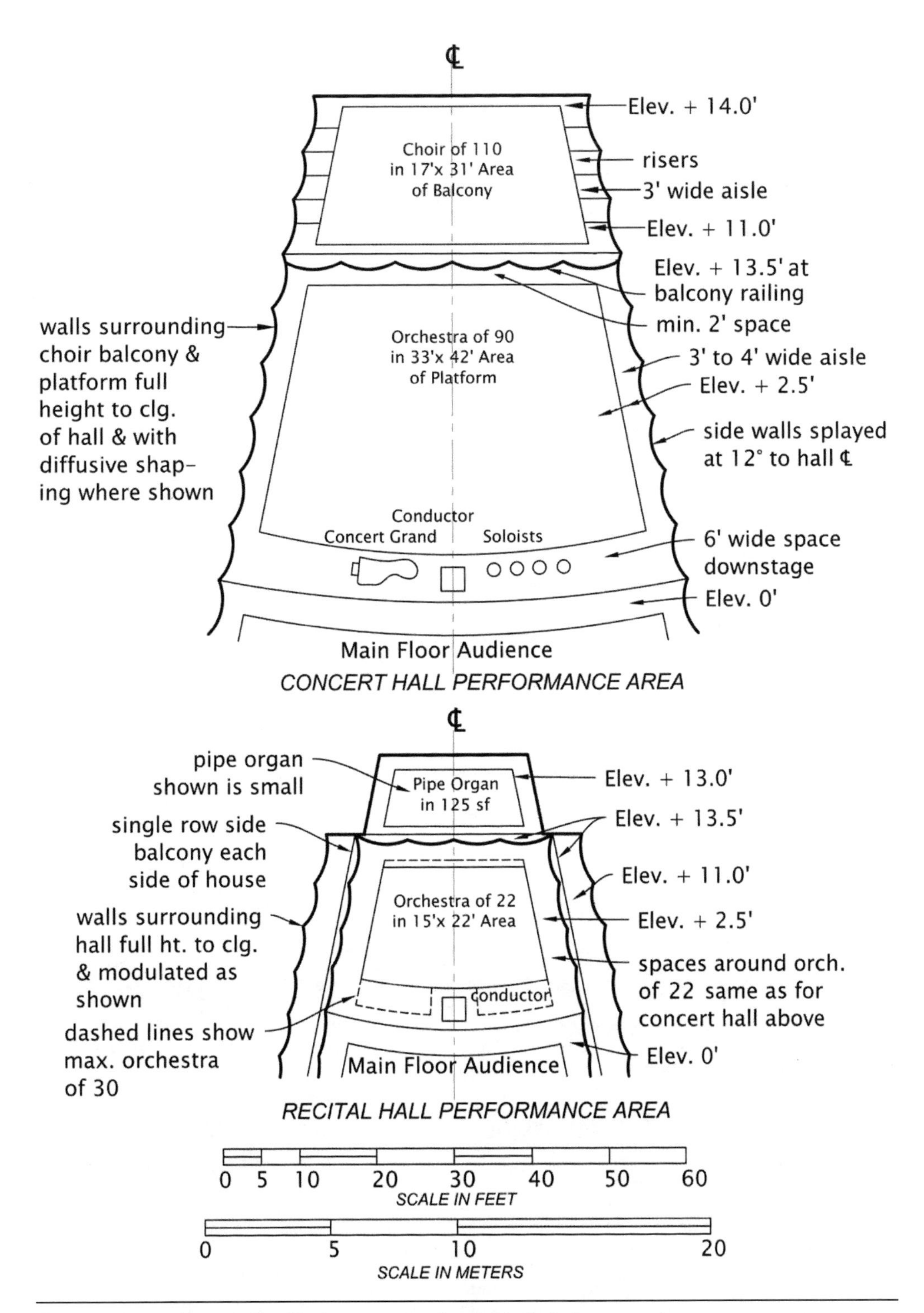

Figure 15.I.12 Acoustically effective concert and recital hall platforms.

institution's own—plus local aficionados, not to speak of its performance-grooming and student-learning benefits.

Pipe organ halls may have a small platform for only the organ console, organist, and several flanking students. The platform will be a few feet above the front of the audience and the organ pipes (with wind chests underneath) above and upstage.

Alternatively, organ halls may have the organ console and organist on an elevated balcony level at the front of the hall with pipes above and upstage of the console. In this case, a small platform may be provided a few feet above the front of the audience and downstage of the elevated organ console. The platform may be of a few hundred square feet for small instrumental groups of 10 or for choral groups of 20 performing early or liturgical music. See Figure 15.I.13 for a representative pipe organ hall, front-end view.

Figure 15.I.13 Front-end view of a representative organ hall (Reyes Hall at the University of Notre Dame).

15.I.3.3 Hall Shaping and Materials

The acoustics of each of the three hall types clearly will benefit from shoebox or modified-shoebox shaping in both plan and section. Modified-shoebox shaping (an elongated hexagon or octagon) will provide more sound pressure uniformity than conventional shoebox shaping (an elongated rectangle). Both forms of shaping, especially if the hall is kept narrow in width, will facilitate copious early lateral sound reflections to the audience. Both forms of shaping, with appropriate volume and considerable wall space provided above the heads of the seated upper audience, will facilitate long reverberation. The acoustical result will be a fine sense of musical "spaciousness" and "envelopment" for the audience—a rich listening reward.

Narrowing the width of each hall type, especially around the platform and the front audience, so that first sound reflections from musical sources arrive at mid-audience ears in 20 to 25 msec maximum after direct sound from the same sources, will ensure an important sense of "presence" or "immediacy" for listeners. The same uniform "quick" reflections from one performer to another on the platform will assure good "ensemble" (hearing each other easily and playing together) within the performing group. An operable canopy of sound reflectors over the platform can contribute to good ensemble among various group types and sizes and add to "presence" or "immediacy" in the audience.

One must be careful to keep the ratio of early sound (the first 80 msec) in proper balance with late or reverberant sound (that from 80 msec to the end of reverberation). This ratio is referred to as "clarity" and is denoted C_{80}. A positive number indicates "clarity" or "good definition" and a negative number indicates "richness" or "good blend". A large negative number indicates relative "muddiness" in music. While the most appropriate C_{80} value will vary with the music being performed, this author suggests the following ranges of values within the 500 Hz to 2,000 Hz bandwidth: for concert halls −4 dB to −1 dB, for recital halls −1 dB to +2 dB, for pipe organ halls −4 dB to −1 dB.

All three of the hall types discussed here require solid, rich bass sound, especially concert and pipe organ halls. This requires steadily increasing T_{60}s from 500 Hz down to 31 Hz—say from 2.0 up to 4.0 seconds. To accomplish this, "all" room materials must be hard, heavy, and necessarily stiff. In concert and recital halls, audience chairs should be moderately (not heavily) upholstered. In pipe organ halls, hardwood pews are the norm.

15.I.3.4 Examples

Drawings and/or photos of five halls are shown in the following pages. The first pair contrasts Boston's Symphony Hall, built in 1900 with 2,625 seats, with the University of Notre Dame's Leighton Concert Hall, built circa 2000 with 900 seats. The second pair compares two recital halls, both approximately 400 seats and built in the 1990s, the first at the University of California in Santa Cruz and the second at the Colburn School of Performing Arts in Los Angeles. The final hall is the Reyes Organ Hall at the University of Notre Dame, built circa 2000 with 100 seats.

In the Boston Symphony Hall (Figure 15.I.14), note the high volume with 60 feet (18.3 m) from the platform floor to the upper ceiling and the narrow width with only 60 feet (18.3 m) between balcony faces. Note the platform side walls are sloped about 15 degrees away from the hall's centerline. Note the diffusive shaping on the ceiling, upper walls, and balcony faces. Note the 35 feet (10.7 m) of side wall height above the upper balcony floor. Consider also that

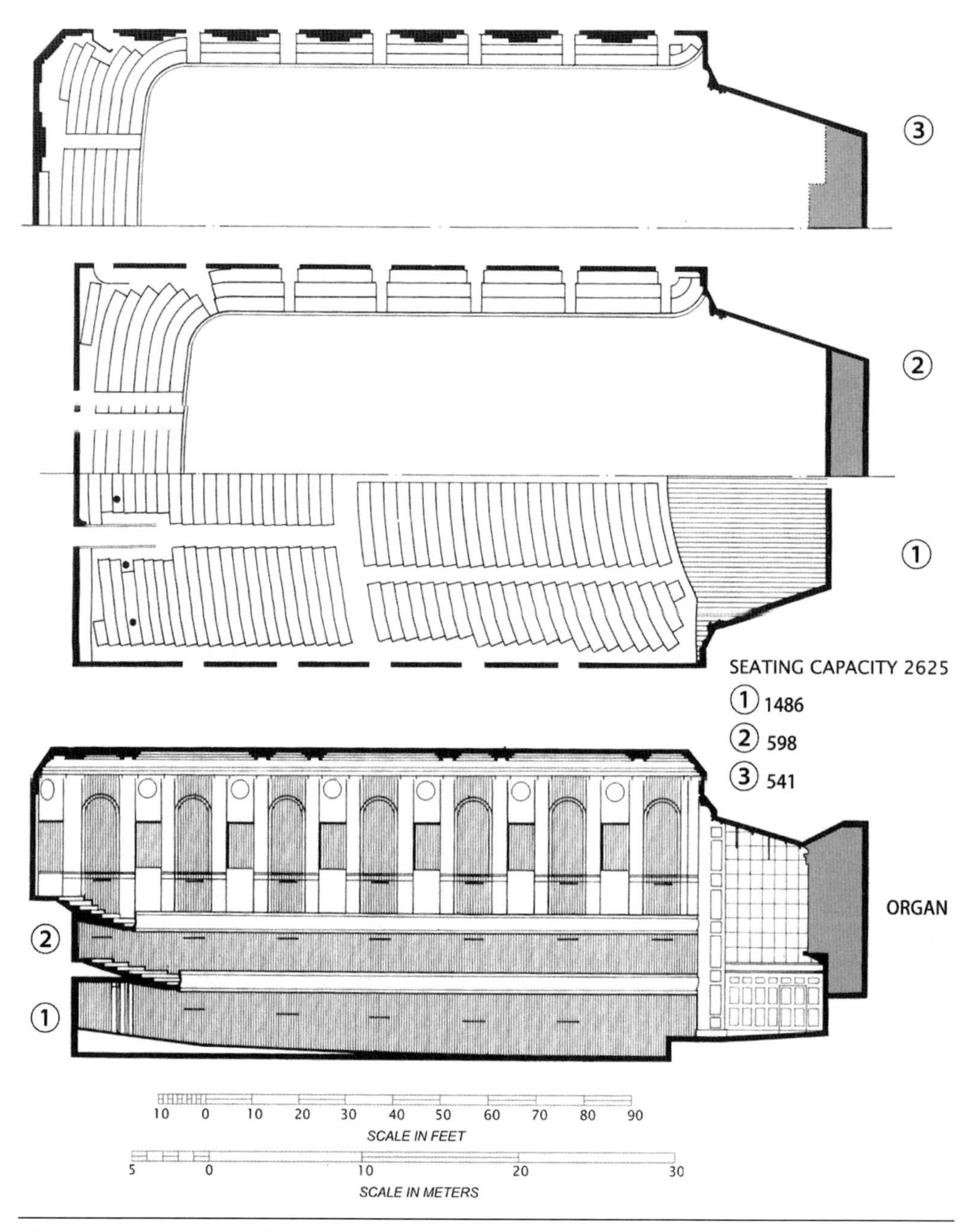

Figure 15.I.14 Drawing of a representative concert hall (Boston Symphony Hall, courtesy Leo Beranek).

temporary sound-reflecting canopies were suspended over the platform several times to the musicians' pleasure, but later removed for visual reasons.

In Leighton Hall (Figure 15.I.15), note the high volume, the narrowing of the hall by a surrounding balcony, the sloped platform side walls, the diffusive shaping of balcony faces, low stage walls, and overhead canopy sound reflectors. Note the high side surfaces (actually sloping roof surfaces) above the balcony to support reverberation and that these surfaces are

Figure 15.I.15 Photo of Leighton Concert Hall.

corrugated and provide high-frequency diffusion. Leighton has an operable canopy over the platform, which they value. Leighton is certainly smaller than Symphony Hall and *very* different architecturally—but the two halls both produce beautiful sound for symphonic classical music. Leighton, in addition, can be adjusted to serve a piano solo or a convocation by lowering the canopy and deploying fabric ceiling banners and sliding fiberglass panels that match the roof deck's color.

The two recital halls each seat about 400 and both were built in the 1990s. Both have steep seating slopes for excellent viewing of musicians, operating canopies over their platforms, splayed platform walls, diffusive lower walls and canopies, high upper volume, and variable acoustics. In Santa Cruz, hard-surfaced chambers surround the upper hall on all four sides. These chambers may be closed off by heavy velour curtains that store outside the hall, but can be moved into the hall (i.e., "exposed") to effectively reduce acoustical volume at 1,000 Hz and above and add sound absorption at 63 Hz and above. In Colburn's Zipper Concert Hall, a very large and high curtain storage "garage" is provided behind the upper rear wall. Eight vertical, dark-colored, heavy velour curtains move out of storage and longitudinally down the length of the hall above the arrays of sound-transparent but visually semi-opaque wood hoops. Both halls can vary their mid-frequency (occupied) T_{60} from just above one second (for jazz) to just above two seconds (for classical repertoire). Both halls are intimate, fine sounding for their programmed uses, and good teaching machines [see Figure 15.I.16 (a) and (b)].

The 100-seat Reyes Organ Hall (see Figure 15.I.13), built circa 2000 at the University of Notre Dame, is home to a handsome and splendid-sounding Fritts Organ. The hall is narrow and high, surrounded by colonnades and two balconies, and topped with a heavy timber

A

B

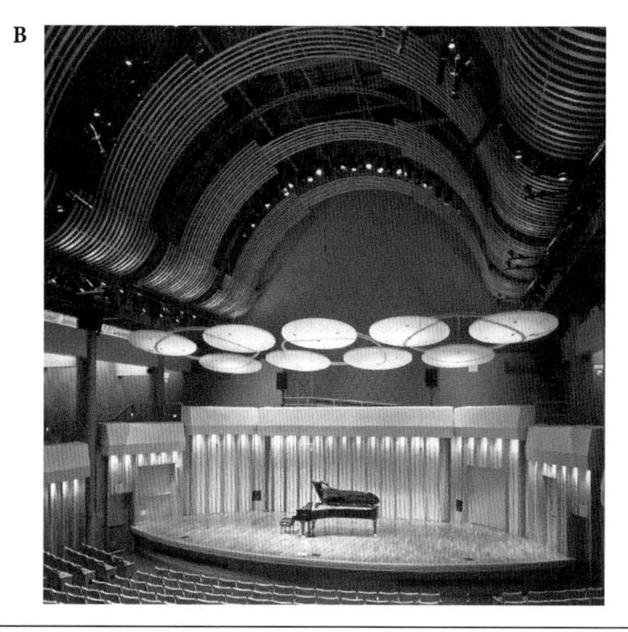

Figure 15.I.16 Recital halls: (a) Recital Hall at the Music Center of the University of California, Santa Cruz (b) Zipper Concert Hall at Colburn School of Performing Arts, Los Angeles.

roof structure. The organ console, wind chests and pipes are set against the front wall at the first-balcony level. A small platform for a chorus of about 20 is on the main floor just forward of and below the organ console and facing an audience seated in wooden pews. The balcony faces, surrounding round columns, and heavy wood trusses above are diffusive. All construction is variously of concrete, stone, heavy masonry or plaster, or wood. The first balcony is used as a platform for instrumentalists and/or singers to surround the audience with sound. The second balcony is used for variable acoustics. A curtain storage "garage" is located behind the upper-rear wall. Two heavy velour curtains, spaced about 2 feet (0.6 m) apart, track out of storage and extend along each upper side balcony to the front wall. When so deployed, the mid-frequency T_{60} drops from about 4.0 to 2.0 seconds. The organ, alone, generally is played and practiced in the 4-second environment. The 2-second environment is used for early and liturgical choral and instrumental music.

15.I.4 DESIGN HIGHLIGHTS: MULTIPURPOSE HALLS WITH VARIABLE ACOUSTICS

15.I.4.1 Introduction

Many municipalities and educational institutions want to present a wide variety of performing arts, but cannot fund a concert hall, opera house, Broadway theater, and drama theater; they can only afford one multipurpose hall. Many have been built in the last 50 years, and many have been unsatisfactory acoustically and in other ways. Since about 2001, however, a variety of multipurpose halls with variable acoustics have been built (or substantially renovated) and support many performance types with notable success.

Other recent experiences show that many "pure" classical music concert halls need variable acoustics because they necessarily augment their symphony orchestra's season with world, country, jazz, and popular music programs as well as optimizing for the various "classical" genres. Clearly, some of these do not fare well in a "fixed" classical music acoustic environment. Thus, variable acoustics systems are needed and the classical concert hall becomes at least semi-multipurpose (see Figure 15.I.17).

15.I.4.2. The Problems

The acoustical problems associated with presenting world, country, jazz, and popular music successfully in a concert hall designed for classical music do not require difficult solutions, provided they are developed during design. After a hall is built, solutions are more difficult and may be impossible to implement fully. There are two problem areas: (1) making the platform surround easily adjustable so it can be either sound reflective or absorptive and changing the performance area from large to small; (2) making the hall's upper volume larger, but reducible, and able to be sound reflective and/or absorptive by moving large areas of power-driven, metal-framed panels into and out of the space. The platform adjustments vary the local environment for performers. The upper hall adjustments vary the room's reverberance for all listeners.

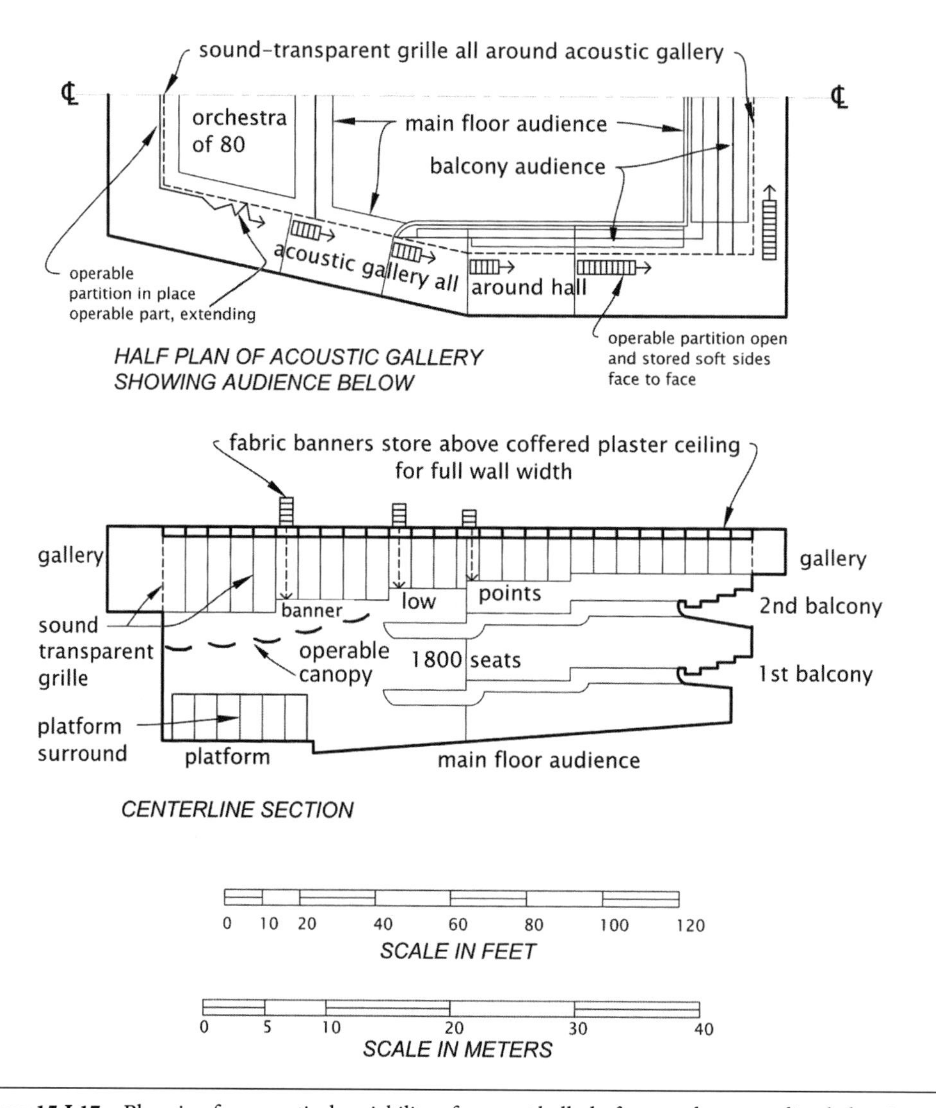

Figure 15.I.17 Planning for acoustical variability of concert hall platform and surrounding balconies.

The problems associated with presenting everything from Shakespeare to South Pacific to Puccini to Beethoven in, say, an 1,800-seat multipurpose hall *are substantial*, but certainly solvable with perception and determination. The various problems are:

1. Most drama, Broadway, operetta, and opera presenters require a stage house for scenery-handling capabilities. The classical music presenters will have to perform in whole or in part within the stage house, so a demountable acoustical shell is necessary to surround them, because stage houses are inhospitable to orchestras and choirs. The shell must be quick to assemble/disassemble with minimum manpower and requires substantial backstage or offstage storage space.
2. An orchestra pit forward of the proscenium opening to the stage house and about eight feet below stage level will be required for opera, operetta, and Broadway shows.

The pit either lengthens the hall or takes away what otherwise would be high-dollar-return seats at the front of the hall.

3. Provisions must be made to vary the hall's mid-frequency T_{60}s nominally between 1.0 and 2.0 seconds, while altering the room's sound-reflection patterns appropriately and simultaneously. Altering the room's sound-reflection patterns appropriately means leaving all low-ceiling or canopy, low-wall, balcony-face, and soffit surfaces that provide important early and lateral reflections unchanged, while introducing movable sound-absorbing materials to cover or shield high ceiling and wall surfaces that produce long-delayed reflections. Required, in addition, are upper-hall peripheral spaces that provide some of the volume necessary to produce a 2.0-second T_{60} "plus" substantial areas of solid-backed, sound-absorbing panels that are power driven, stored completely outside the hall, and can move into the hall to add large amounts of absorption and reduce hall volume by closing off the upper hall chambers (see Figure 15.I.17).

15.I.4.3 The Solutions and Key Details for a Concert Hall

First, solutions are needed *during design* for a classical music concert hall to enhance the room's acoustic environment for world, country, jazz, and popular music. At least, a stage house will not be required:

1. Provide an approximate two-foot-deep continuous recess around the stage's three walls to a height of about 12 feet. Design wheeled acoustical panels to fill the recess. Each panel should be just under 12 feet (3.7 m) high and 4 to 5 feet (1.2–1.5 m) wide, faced on one side with a diffusive sound-reflecting surface and on the opposite side with a sound-absorbing surface of durable, porous fabric or perforated vinyl over two- to three-inch (50–75 mm) thick glass fiber board. Rearranging the full stage surround only requires rotating the panels. Also, a group of panels may be rolled downstage to surround and support, both acoustically and visually, a soloist, quartet, or small ensemble (panel reflective surfaces forward for unamplified musicians).
2. At the hall's upper level, say the top 15 to 20 feet (4.6 to 6.1 m), construct a continuous open gallery around the room's entire perimeter above and surrounding the stage, control, and circulation spaces below. The gallery should be as wide (deep) as it is high. If the gallery can be 17.5 feet (5.3 m) high and wide around the full upper four walls, its volume will represent about 22% of the 1,800-seat hall's required total volume for a 2.0-second T_{60}. Closing off the gallery represents a T_{60} reduction of 0.45 second. In all, about 6,600 square feet (613 m^2) of glass-fiber-faced metal folding panels are needed to close off the gallery and add upper perimeter absorption.
3. The folding panels should be designed based on available folding steel partitions at least 2 inches thick, with damped 18-gauge steel facings and sound-absorbing glass fiber fill. The steel facing on the panel sides toward the gallery should be solid, while steel facing on the panel sides toward the upper hall should be perforated steel or polycarbonate at least 30% open.
4. Another 3,600 square feet (334 m^2) of heavy velour curtains will be needed in, say, four cross-hall ceiling-suspended curtains at 11.25 feet (3.4 m) high by 80 feet (24.4 m) long each to achieve the desired 1.0-second T_{60} at middle frequencies. These curtains

may be of several types. The author's preference is for theatrical velour weighing 32 ounces per linear yard (1 kg/m) of fabric at a standard 54 inches (1.37 m) wide. This is currently the heaviest available velour. When a curtain is extended, its fabric should hang in 100% folds [200 feet (61 m) of fabric to produce 100 feet (30.5 m) of extended curtain]. This version is for a horizontally-tracked curtain. Alternately, two layers of velour may be hung flat and spaced four inches apart and operated vertically by cables that pull up a bottom batten that gathers the fabric into folds as it rises to storage above. D-rings sewn to the inside fabric faces slide along the cables and keep the fabric-folding process under control. This version is called a banner.

15.I.4.4 The Solutions and Key Details for a Multipurpose Hall

Where multipurpose halls must handle a symphony orchestra in the presence of a stage house and orchestra pit as well as a host of other performance types, specific considerations apply.

1. Variable T_{60}s from 1.0 to 2.0 seconds at middle frequencies may be handled as described in Section 15.I.4.3.
2. The stage house's proscenium opening should be of a fairly common 50 foot width or slightly wider, depending on presenters' requirements. The proscenium height should be at least 30 feet (9.1 m), but 40 feet (12 m, with a vertically sliding teaser panel to adjust height as required) offers a significant acoustical opportunity and may save costs over time. A finish of 1 1/2 inch (38 mm) minimum thick black glass fiber duct liner board faced with protective hardware cloth up to at least 18 feet where not otherwise protected by counterweight arbor lines, should be placed on the surface of each stage side wall, full width from two feet six inches to at least 25 feet (7.6 m) above stage level. The purpose is to deaden a sparsely-set and curtained stage used for drama that otherwise would be too reverberant.
3. For details on platform layout for orchestras, see Section 15.I.3.2.1 and Figure 15.I.12.
4. Acoustically, it is best not to recess the entire orchestra fully within the stage house, even when an orchestra shell is employed, presuming that the hall's cheekwall region is well considered. Consider that within a pure concert hall the orchestra and audience share one room and one overall acoustic environment. To have an entire orchestra play through a possible 30 by 50 feet (9 by 15 m) aperture and hopefully couple its sound fully into an 1,800-seat, near 590,000 cubic foot (17,000 cubic meter) audience chamber is poor practice and will produce limited results.
5. Let us locate the front half of our orchestra (conductor, soloists, string sections, and front row of woodwinds) on a forestage in front of the proscenium opening and inside the audience chamber. The forestage must be nominally 50 feet (15.2 m) wide by an average 18 feet (5.5 m) deep, and the orchestra shell, for the back of the orchestra, approximately the same. The forestage will be composed of a fixed 4 foot (1.2 m) deep apron downstage of the curtain line plus two hydraulic lifts, each an average of seven feet front to back. A person may stand on the apron before a closed main curtain to address the audience when lifts are down. Each lift may be at the stage level or the audience-floor level supporting two audience seating rows or at the orchestra pit level 7 feet (2 m) below the stage. The pit will be about 50 feet (15.2 m) wide by 22 feet (6.7 m) front to back [8 feet (2.4 m) for brass and percussion recessed below the apron

and main stage floor and 14 feet (4.3 m) on the lowered lifts for woodwinds, strings, and the conductor]. The 50 by 22 feet (15.2 by 6.7 m) pit will accommodate an opera orchestra of about 60, plus conductor and concert grand piano. The pit's upstage wall and nearby ceiling should be fitted with demountable two inch (50 mm) thick, durable sound-absorbing panels to mute brass and percussion as necessary.

6. The on-stage orchestra shell will be composed of rolling towers to surround the musicians and overhead reflectors suspended from the stage rigging system to close off the large stage house "sound trap" above. Typical towers will each be about eight feet wide when their "wings" are unfolded and about 30 feet (9.1 m) high. Overhead reflectors also will be about eight feet wide and as long as the shell is wide. The tower and overhead reflector metal frames usually are faced with shaped plywood or laminated honeycomb panels that are hard and stiff and provide both wide frequency range and diffuse sound reflections. The towers store backstage or off stage in out-of-the-way locations. The overhead reflectors, which include lighting instruments, are rotated to the vertical and flown high in the stage house for storage. Several fine companies manufacture and install standard lines of orchestra shell components and also will construct custom shells conceived by concert hall designers.
7. The on-stage orchestra shell will be an average 45 feet (14 m) wide by 18 feet (5.5 m) deep and an average 25 feet (7.6 m) high. The side walls and ceiling will be sloped at 12 to 15 degrees, opening outward and upward toward the audience. The overhead panel or ceiling height is also in accord with the long experience that all but major professional orchestras find that hearing themselves easily in order to judge sectional cohesiveness begins to degrade significantly at a ceiling height in excess of 27 feet (8.2 m) above the stage. In our model design, the height of the ceiling's downstage edge is 27 feet (8.2 m).
8. A final improvement to the acoustic design of this hall's stage end involves two steps. First, raise the tower and overhead panel heights described previously so that the downstage ceiling edge just meets the desirable 40 feet (12 m) high proscenium, and increase tower heights to match this sloping ceiling at its edges. Second, along the ceiling profile (shown in Figure 15.I.18), provide an array of clouds as described in Section 15.I.2.10. Downstage of the proscenium opening, continue this array out over the forestage. This addition will provide good ensemble for the whole orchestra both within the stage house shell and on the forestage. It will permit the back half of the orchestra to couple much more readily with the audience chamber. It will deliver short-delay early reflections to the front audience and enhance their sense of musical "presence" and "intimacy". All of these are worthy additions.

15.I.4.5 Examples

Photos of two 1,800-seat halls are shown in Figures 15.I.19 and 15.I.20. Royce Concert Hall at the University of California in Los Angeles was built in 1929 and remodeled twice (in 1989 and 1999). Jackson Hall at the University of California in Davis (the keystone of the Mondavi Center for Performing Arts) was built circa 2002. The two halls are of sharply contrasting architecture but with equally successful acoustics. Each has a stage house and orchestra pit, and accommodates performances from a Beethoven symphony to a Shakespearean drama.

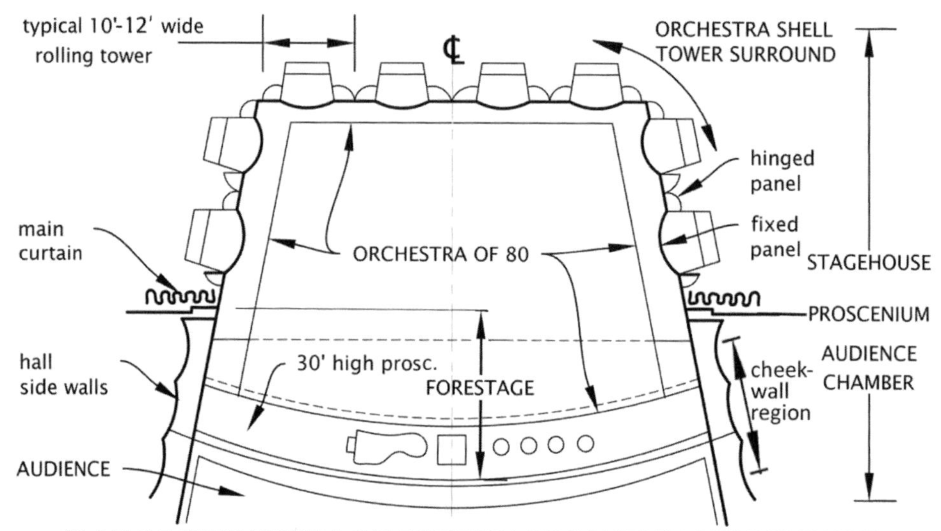

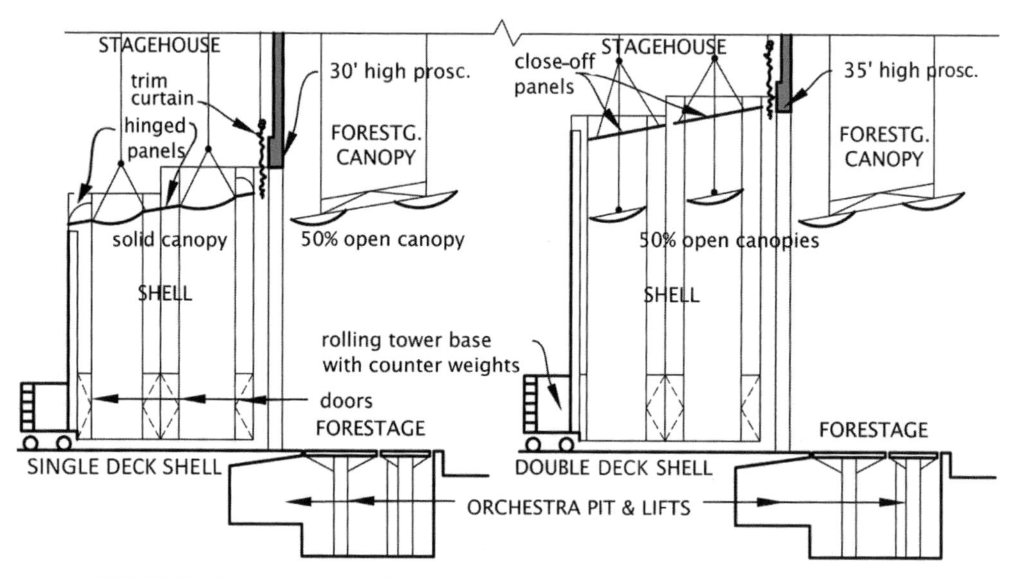

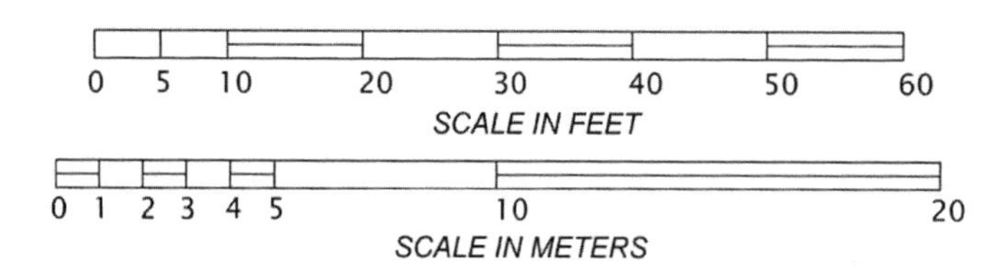

Figure 15.I.18 Orchestra shells and canopies.

A

B

C

D

Figure 15.I.19 Photos of Royce Hall at the University of California in Los Angeles (A thru F).

Figure 15.I.19 Continued.

Because Royce Concert Hall had excellent fundamental acoustical quality, even before the Northridge earthquake of 1994, a variety of renovation restrictions were minimally problematic. The restriction imposed by the old, small orchestra shell within the stage house was overcome by a new forestage with a carefully redesigned sound-reflecting surround (the proscenium "arch") so that much of the orchestra could move downstage into the hall.

Royce's height and volume are not great and were restrictive of reverberation. Note that large coves were cut into each side of the old ceiling up to the roof and that old upper-side-wall spaces that open to windows were reopened to increase the hall's volume and permit a 2.0 second mid-frequency T_{60}. Also note that Royce's 1929 highly-diffusive coffered ceiling and balcony faces were retained and its lower side walls made more diffusive to increase lateral-reflection density. In 1989, aisle carpet was replaced with wood or vinyl and in 1999 operable doors were added to close over-proscenium organ openings to add to reverberance.

Figure 15.I.20 Composited photo of Jackson Hall at the University of California, Davis.

New operable sound-absorbing banners were added to make the new side-ceiling coves act as reverberation reducers, when deployed. Doors were added to close off the reopened upper side-wall window spaces. The result was an appreciated mid-frequency T_{60} variability of 1.4 to 2.0 seconds. Finally, note the many upper side-wall horizontal ledges that deliver lateral high frequency reflections to the balcony.

The UC Davis Jackson Hall has the same seating capacity as UCLA's Royce Hall, and the same multiperformance-type requirements, but is of a highly contrasting architectural design. The two halls have beautiful, startlingly identical musical sound. This attests to the relevance of uniform acoustical design principles.

Figure 15.I.20 shows Jackson Hall set for an orchestra concert. The rear half of the orchestra is within a shell erected in the stage house volume. The front half of the orchestra is seated on lifts forming a forestage within the audience chamber. The canopy over the orchestra is solid within the stage house to close off the scenery and curtains above but about 50% open over the forestage to permit part of the orchestra's sound to reach the audience chamber's upper reverberant volume directly.

Note the balconies that narrow Jackson Hall's width, and the diffusiveness of the balcony faces and side walls. Also note the room's considerable height and the wood grillage that conceals operable curtains that surround the hall's upper space. The latter enable mid-frequency T_{60}s to range from 1.0 to 2.0 seconds.

15.I.5 DESIGN HIGHLIGHTS: OPERA HOUSES

15.I.5.1 Introduction

Opera is an old musical art form born in Italy probably in the late 1400s. Operatic form evolved and matured over the next 150 years or so until the first house specifically for opera was opened in Venice in the early 1600s. The two oldest opera houses in use today (after various reconstructions and remodelings) are Teatro di San Carlo, opened in Naples in 1737, and Teatro Alla Scala, opened in Milan in 1778. The former seats 1,400 and the latter 2,135. Bear in mind that these seat counts would be significantly less if current comfort and safety standards were enforced.

Both San Carlo and La Scala have horseshoe plan forms with the open end toward the stage, a sloped audience main floor within the horseshoe and five or six tiers of box seats arranged horizontally around the horseshoe and stacked up to the largely flat high ceiling. See Figure 15.I.21 for basic graphics. This design form has endured in most notable opera houses from the middle 1700s until recently. The 1997 highly-acclaimed Tokyo New National Theatre Opera House (1,810 seats) breaks from the horseshoe tradition, but maintains excellent opera acoustics. The world's most acclaimed opera house, Teatro Colón of Buenos Aires (2,487 seats), was opened in 1908. This hall adheres to the traditional horseshoe plan and tiered surrounding boxes, but it is larger than most (see Figure 15.I.22).

The driving acoustic challenge of opera house design is to deliver the perfect balance between orchestral music richness and sung music intelligibility with its own richness.

15.I.5.2 Stage and Orchestra Pit Design

Stage and flyloft design for opera is little different from that for a Broadway show or a light opera. The exceptions may be the need for a higher proscenium opening (perhaps 40 feet, 12 m) and more on- and off-stage storage space for the large and plentiful scenery to serve opera. For opera, as for most other stage-based performances, a fixed apron is required downstage of the closed main curtain. This is a 4 to 6 foot (1.2 to 1.8 m) wide fixed floor upon which performers may take bows or someone may address the audience. In fact, lead singers may position themselves on the apron during a performance.

Just downstage of the apron is the open orchestra pit, when the pit lift or lifts are depressed; or a forestage, when the lift or lifts are elevated; or additional audience seating, when the lift or lifts are at the front main floor level. See Section 15.I.2.8 and Number 5 in Section 15.I.4.4 on page 440 for pit design guidance.

15.I.5.3 Hall Shaping and Materials

15.I.5.3.1 Seating Capacity/House Size

It is best to hold the audience seating capacity to 1,800 or less to assist singers with "carrying the house". The house should be as narrow as possible to provide "intimacy" via early reflections with a short time-delay gap of 20 to 25 msec or less at the middle main floor and beyond. The narrowness will also assist the provision of "spaciousness" via profuse early lateral reflections in the 0 to 80 msec window.

15.I.5.3.2 Proscenium Size and Form

It also is best to limit the proscenium width to 50 feet (15.2 m) and height to 40 feet (12.2 m), again to assist with the provision of early sound reflections. These dimensions are not met by many opera houses. Some, usually with more than 1,800 seats, will have a width between 55 and 59 feet (16.8 and 18 m) and a height of 50 feet (15.2 m), which is an acoustic negative.

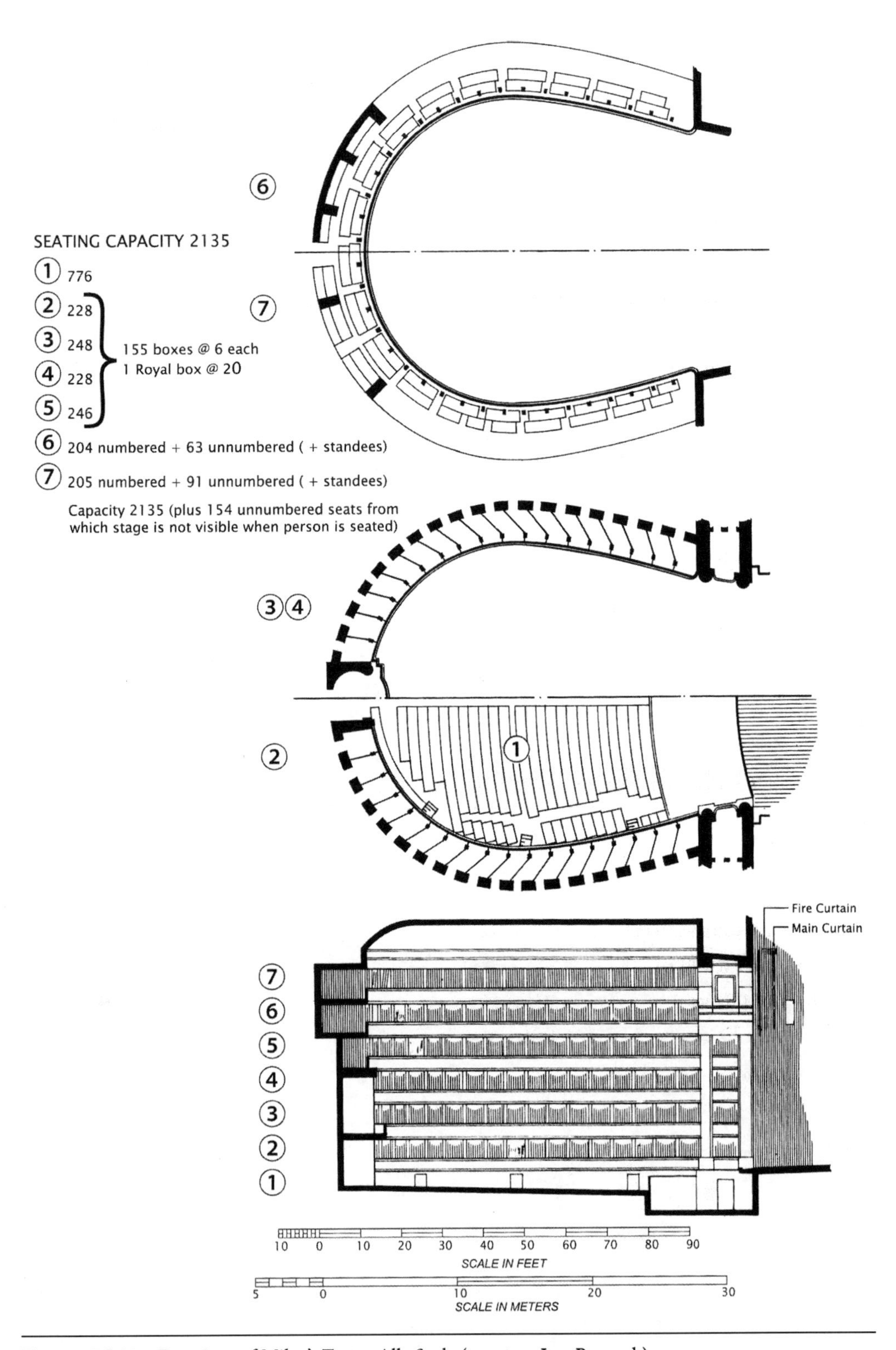

Figure 15.I.21 Drawings of Milan's Teatro Alla Scala (courtesy Leo Beranek).

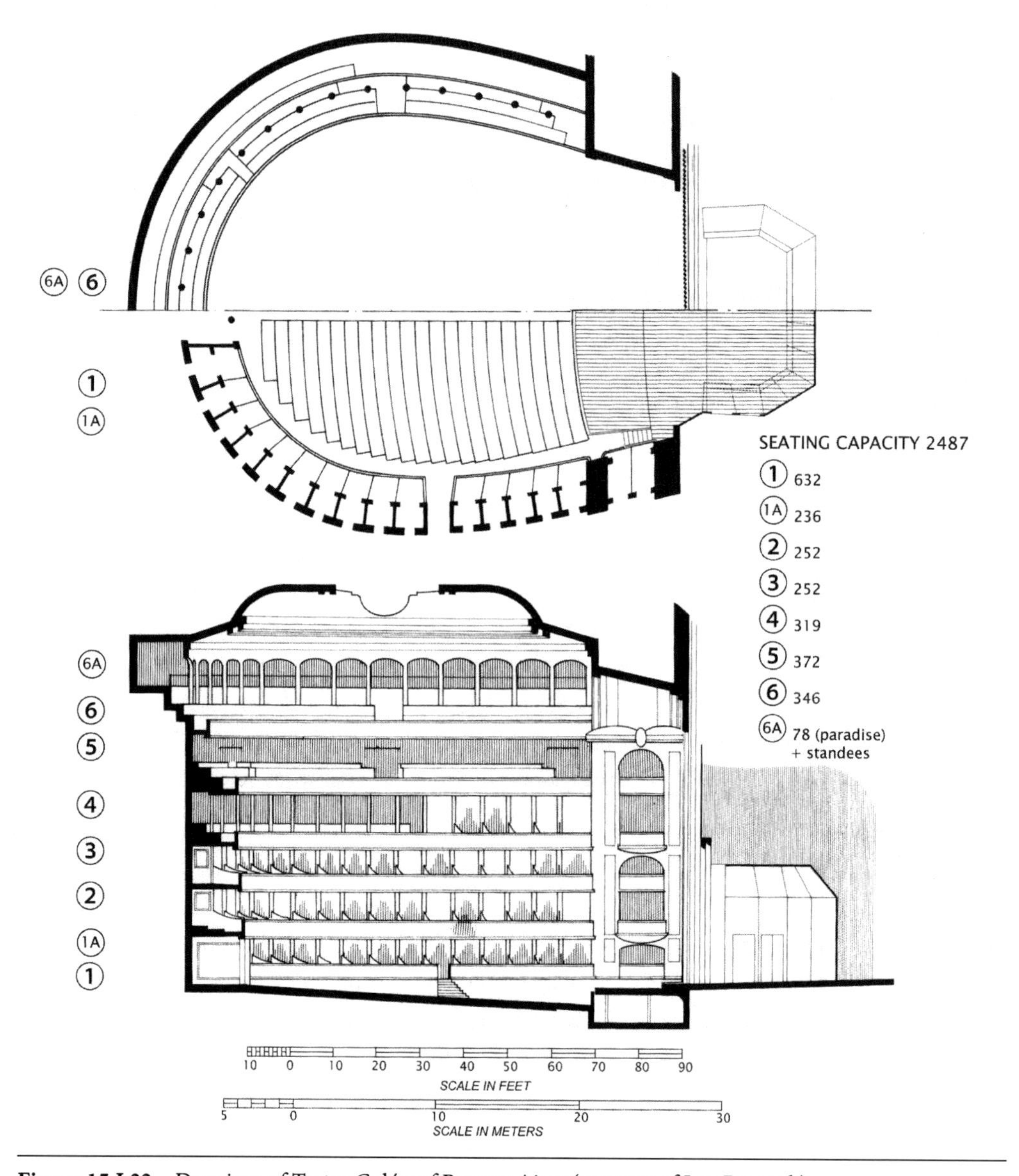

Figure 15.I.22 Drawings of Teatro Colón of Buenos Aires (courtest of Leo Beranek).

The proscenium frame usually matches the depth of the apron of 4 to 6 feet (1.2 to 1.8 m). Each side and the top of the frame should be splayed at 12 to 15 degrees outward toward the audience. Convex pillowed shapes surrounded by ribs or slots should sculpt the proscenium frame to spread and diffuse sound reflections (see Figure 15.I.23).

15.I.5.3.3 Basic Hall Shaping and Materials

The hall should have sufficient volume to support a 1.4 to 1.6 second mid-frequency T_{60} when fully occupied and a typical stage is set.

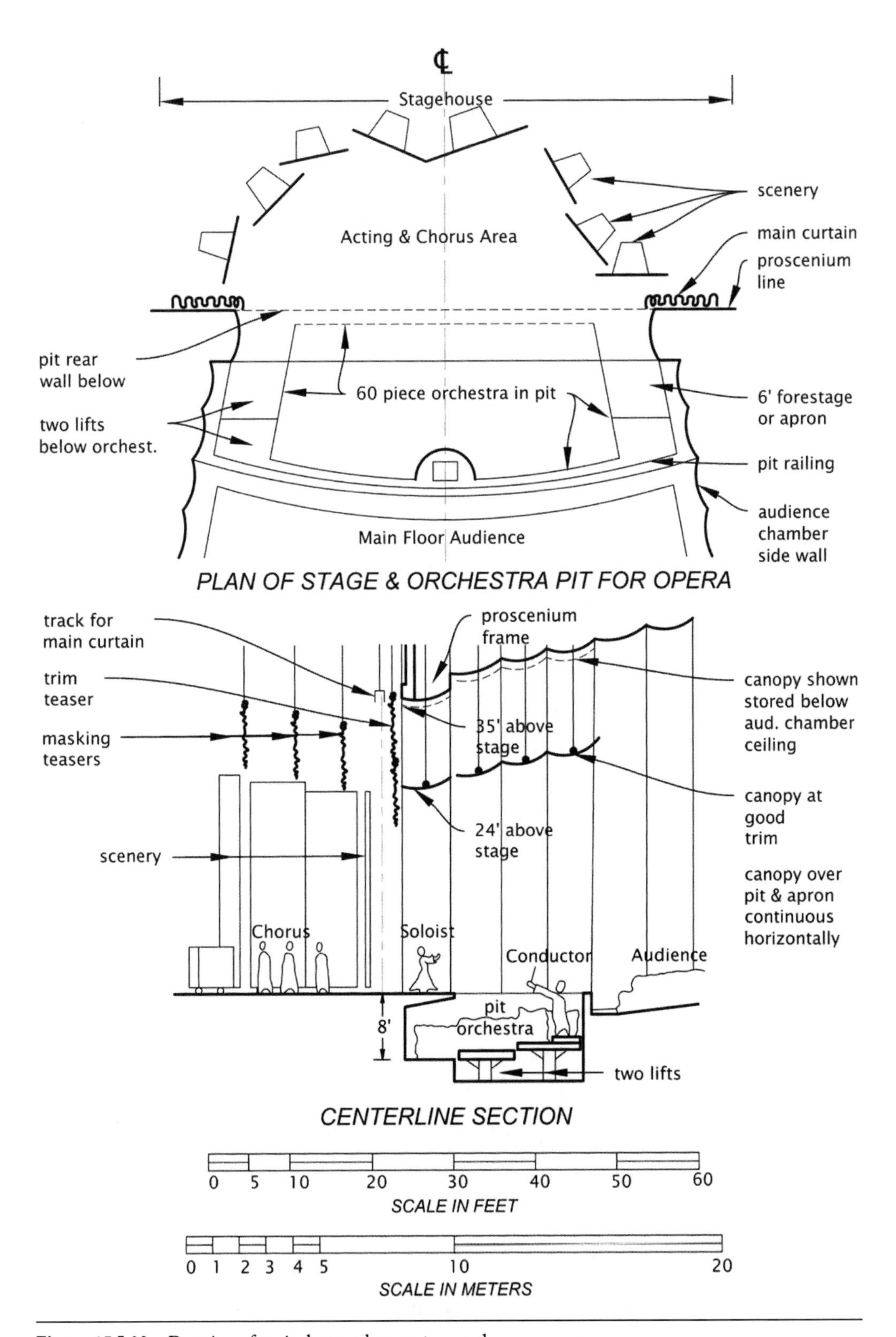

Figure 15.I.23 Drawing of typical opera house stage end.

The classical horseshoe plan form with multiple tiers of boxes and balconies stacked on top of one another is acceptable acoustically—with a few important admonitions:

1. The box and balcony faces must be sculpted to be quite diffusive of middle- and high-frequency sound to avoid serious focused reflections from the concave curve of each horseshoe tier. This diffusiveness also will aid the provision of copious lateral reflections to the audience.
2. If the boxes are outfitted with movable chairs set in several rows on a flat floor (typical of many classical halls), only those in the front row will be able to see the stage. Others will have to stand to see. Limiting the boxes to two rows with the second row elevated above the first is recommended.
3. In many classical halls, there is a full-height vertical divider between each box. The dividers limit horizontal sight lines and turn each box into an individual cell isolated to a significant extent from the hall's overall acoustic environment. It is recommended that such dividers be limited in height to approximately 2 1/2 feet (0.76 m) above the finished floor.
4. Box rear walls and doors should not follow the horseshoe's curved portion, but be shaped in a series of stepped, flat surfaces that approximate the curve but that will not focus sound reflections.
5. Carefully selected stage-facing surfaces should return "just enough" medium- and long-delayed sound reflections to singers so they clearly perceive the house's response to their voices. Rear wall surfaces may have to be adjustable to be reflective or absorptive and their final finish determined during pre-opening tests with singers.

The classical horseshoe opera house has a very high ceiling, often 60 to 65 feet (18 to 19 m) above the stage, even 90 feet (27 m) above the stage in the case of Teatro Colón in Buenos Aires. Such high ceilings are of no benefit to intelligibility of sung words. It is strongly recommended that a convexly curved sound reflector (called an eyebrow) be located over the front of the opera house extending from the proscenium wall out over the apron, orchestra pit, and the front audience rows as far as theatrical lighting will permit. The reflector should be at least as wide as the proscenium opening, should start at the proscenium's top at 40 feet (12 m) above the stage and should curve gently outward and upward. This provides reflective reinforcement of downstage lead singers and the pit's strings and woodwinds to the main floor audience. A valuable addition would be to make the reflector powered for vertical operation between about 25 feet (7.6 m) above the stage to near the hall's horizontal proscenium frame.

A successful opera house that is not shaped in the classical horseshoe plan with floor-to-ceiling surrounding boxes and balconies is quite conceivable. Indeed, one shaped in an operatic-friendly variation of a good concert hall opened in 1997 in Tokyo to high acclaim. The New National Theatre Opera House is illustrated in Figure 15.I.24 (on pages 452–453).

The room-finish materials of an opera house interior should be sound reflective as a rule, with diffusive detailing, which usually implies plaster on metal lath and/or fiberglass-reinforced cast plaster. Sand-aggregate gypsum plaster usually is appropriate and should be of sufficient thickness to assure that the hall's T_{60} increases gradually from 500 Hz to a value about 1.5 times as great at 31 Hz (say 1.5 seconds at 500 Hz to 2.3 seconds at 31 Hz). The goal is warm, but moderate, bass response from the hall. Hall audience chairs should be moderately upholstered. Dense carpet without underlayment is permissible in aisles. Any other sound-absorbing materials should be limited strictly to needed corrections and/or adjustments.

15.I.5.4 Examples

Shown previously were two acoustically excellent but contrasting opera houses. The first (Figure 15.I.21) shows the Teatro Alla Scala in Milan, opened in 1778, rebuilt and remodeled since, but always faithful to the original. This 2,135-seat hall is ranked among the world's best. The second (Figure 15.I.24) shows the New National Theatre Opera House in Tokyo, opened in 1997. This 1,180-seat hall is ranked equally with La Scala, the new hall's smaller size probably being a factor, along with its creative acoustical design.

La Scala fits the classical horseshoe-stacked-boxes model perfectly. The hall has a fine reputation with orchestra members, singers, and audiences; although it is said the sound is better in the boxes than on the main floor. This author suspects the sound is better at the front of the boxes than at the back. The hall was renovated recently; in part, by removing aisle carpet to expose wood flooring and by installing new seats with less upholstery. Before this renovation, the hall's middle-frequency, fully-occupied T_{60} was 1.24 seconds, lower than the 1.50 to 1.60 seconds of comparable houses. La Scala's new T_{60} has been measured as 1.35 seconds without an audience and 1.20 seconds with an audience.

The Tokyo Opera House makes many concessions important to opera, but also has the appearance of a concert hall. Visually, it contrasts strongly with the classical model, but acoustically it is rated in the top five—the other four being of the classical model. In the Tokyo hall, note the curved ceiling reflector at the front of the house and the three curved reflectors set in each front side wall, all intended to provide early reinforcing reflections from singers to the audience. Also note the stepped side walls and down-tilting balcony faces intended to provide lateral reflections across the audience.

15.I.6 DESIGN HIGHLIGHT: HALLS FOR WORLD, COUNTRY, JAZZ, AND POPULAR MUSIC

15.I.6.1 Introduction

Much of our planet is well populated with venues that present live and recorded world, country and jazz music, show tunes, soft and hard rock, and other popular music, often as heard daily on our media. Comedic, magic, and other speech programs often are intermixed. These prolific venues vary widely in size, shape, type, ambience, and sophistication. Included are Broadway-type theaters, Las Vegas-style showrooms, nightclubs, jazz venues, dance halls, ballrooms, etc.

Some of these venues were designed carefully and successfully regarding acoustics. Others were created with no thought to acoustics, but luckily provide good hearing conditions. Still others are marginal to poor for listening.

Because of the great number and types of venues and the widely-varying acoustic performance of existing venues, specific criticism and organized positive design recommendations are challenging. The following are some broadly-applicable acoustical design approaches.

15.I.6.2 Basic Concepts

1. Each hall should be designed to have a "neutral" acoustic environment—one that has a T_{60} of one second or less at all frequencies, including low frequencies, if at all possible.

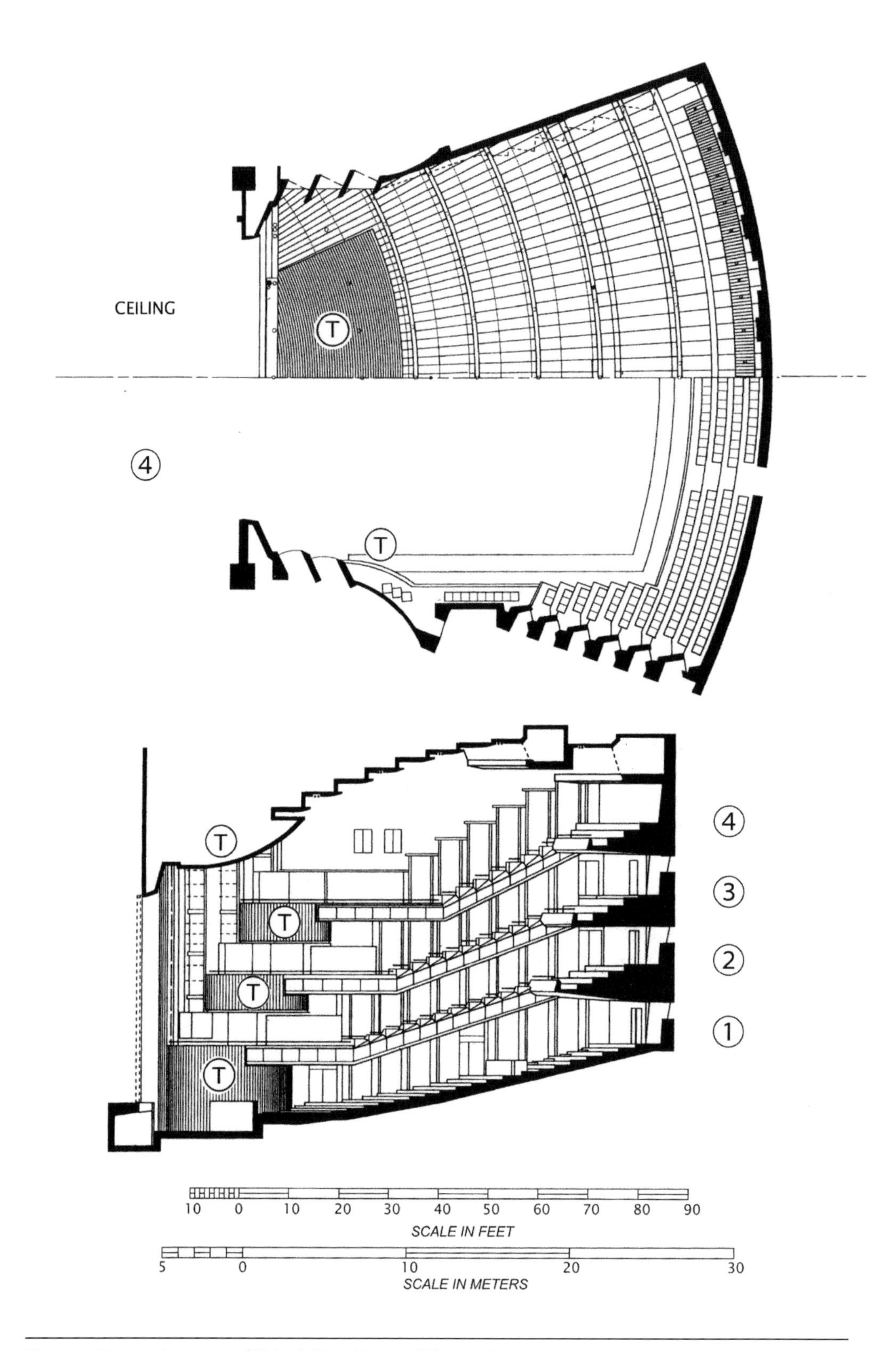

Figure 15.I.24 Drawings of Tokyo's New National Theatre Opera House.

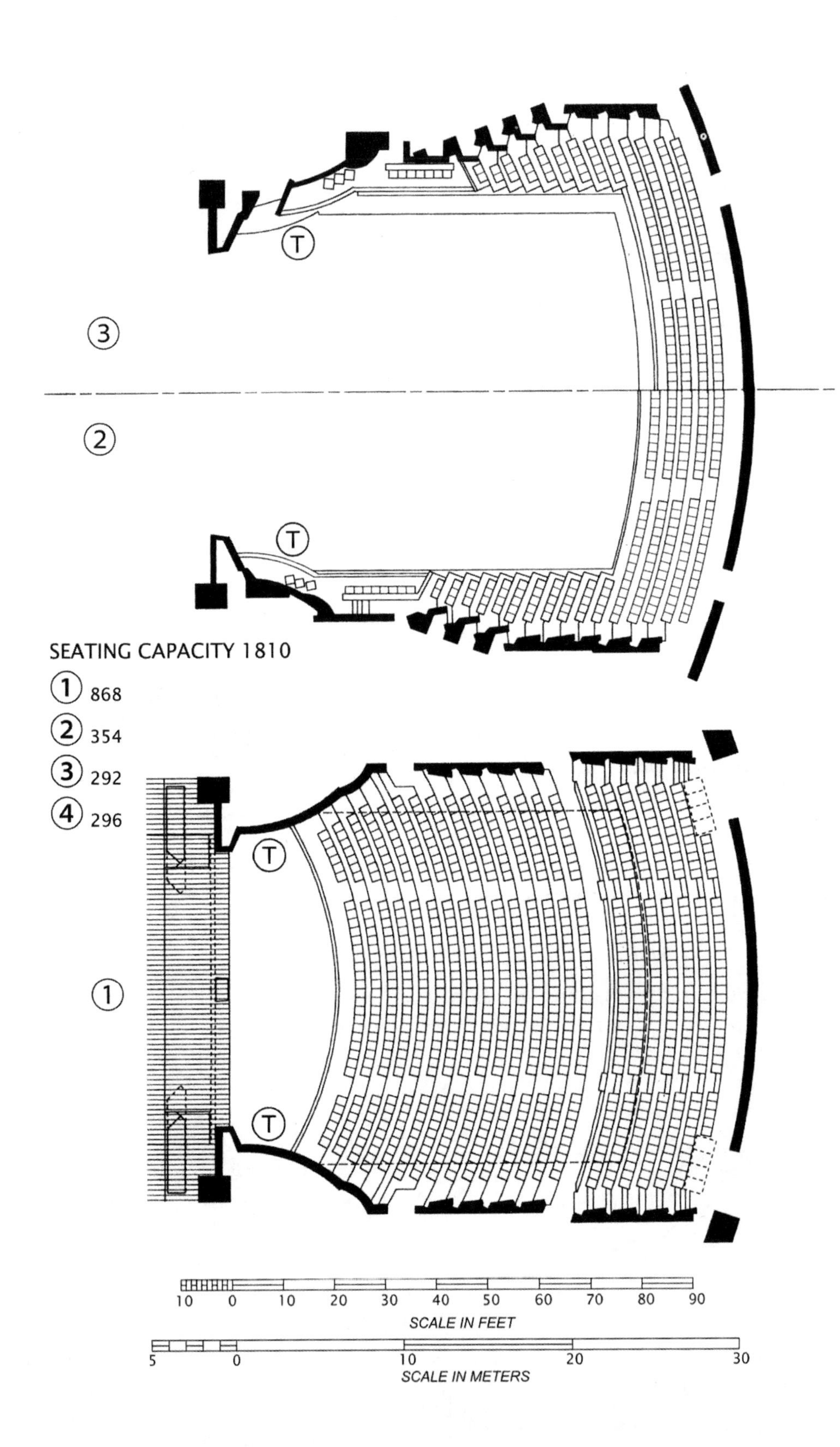
SEATING CAPACITY 1810
1 868
2 354
3 292
4 296
T
T
T
T
3
2
1
10 0 10 20 30 40 50 60 70 80 90
SCALE IN FEET
5 0 10 20 30
SCALE IN METERS

Each hall should have no negative, long-delayed reflections that garble speech or produce echoes, nor any hard, parallel surfaces that produce flutter and consequent harsh sounds.
2. Each hall should be provided with an appropriately quiet ambient sound environment (between PNC-25 and -35 values). The larger the hall and/or the softer the music usually performed, the quieter it should be.
3. Each hall should be shaped acoustically to provide the best possible uniformity of live sound. It is assumed, however, that an electronic sound system will be provided to contribute to combined (live and amplified) sound uniformity at appropriate loudness levels, *and* that the hall's acoustic design and sound system design will be integrated closely.
4. The performance area of each hall should be acoustically neutral (soft), or live (reflective), or a mixture of the two, or even variable. The choice depends on the performers and the music they play. The neutral approach best fits rock (especially hard rock). The live approach best fits show tunes and world music. The mixed approach best fits jazz, country, and other popular music. The variable approach applies when a hall presents ever-changing music types.
5. No matter the performance area design, each hall should include some nearby surfaces which return short-delay sound reflections to performers and more distant surfaces that return medium-delay reflections. If it is difficult to incorporate such reflectors into the hall's physical design and a sound system is included, system loudspeakers can provide "reflections". It is important that performers hear the hall responding to their music.

15.I.6.3 Author's Assorted Impressions

15.I.6.3.1 Broadway-Style Theaters

Most Broadway-style theaters (not necessarily located in New York) deliver fair to good sound via a combination of thoughtful room and audio system design. Some are excellent; some are poor. The typical shortcomings, when found, are a noisy ventilation system, excessive reverberation, and ineffective sound system design and/or operation.

15.I.6.3.2 Fortunate Halls

"Fortunate halls", as discussed here, are those which received little or no acoustical design attention but worked out well acoustically anyway. Once such is the Ryman Auditorium in Nashville (Figure 15.I.25). Built in 1892 before architectural acoustics arrived in America, this 2,362-seat hall originally was built as the Union Gospel Tabernacle and later became home for the Grand Ole Opry for 31 years. It continues today to serve a wide spectrum of top popular music performers. While one can observe the hall and see where some acoustical improvements could be made and where good acoustical luck served the owner/operator, one must concede that the Ryman sound is a long-standing success.

15.I.6.3.3 Las Vegas-Style Showrooms

Las Vegas-style showrooms are not exclusive to Nevada. Consider Branson, Missouri; like Broadway-style theaters, many of these showrooms deliver fair to good sound via thoughtful designs of rooms and audio systems—some are excellent and some are poor. The poor to fair

Figure 15.I.25 Photo courtesy of Ryman Auditorium in Nashville, TN.

group members frequently seem to suffer from excessive glitz. The typical shortcomings, like the Broadway-style theaters, are noisy ventilation systems, excessive reverberation, and ineffective sound system design and/or operation.

15.I.6.3.4 Jazz Venues and Nightclubs

These two types of facilities have much in common. Unfortunately, flat or near-flat floors which impair sightlines and hearing lines plus low ceilings which restrict uniformity of live sound distribution are all too common. On the plus side, many have well-designed and operated audio systems, and they usually are small in size.

15.I.6.3.5 Ballrooms and Dance Halls

These rooms typically have a significant floor area with ceilings 12 to 15 feet (3.6 to 4.6 m) high. Elevated bandstands are the norm. Sound-absorbing ceilings also are the norm and they control room reverberation, if an efficient material was selected. The most common acoustical shortcomings are noisy ventilation systems and hard, parallel walls that permit excessive flutter and a resultant harshness in the music. The best venues have nonparallel walls and/or sound-absorbing wall panels in key locations.

15.I.8 FURTHER READING

15.I.8.1 Books

1. Acentech and James Cowan, Senior Consultant (2000). *Architectural Acoustics Design.* (McGraw-Hill, New York, etc.).
2. Ando, Yoichi (1985). *Concert Hall Acoustics.* (Springer-Verlag, Berlin, etc.).

3. —— (1998). *Architectural Acoustics Blending Sound Sources, Sound Fields, and Listeners.* (Springer-Verlag, New York, etc.).
4. Baines, Anthony, Editor (1961). *Musical Instruments Through the Ages.* (Penguin Books, Baltimore, MD).
5. Beranek, Leo (1996). *Concert and Opera Halls: How They Sound.* (Acoustical Society of America, Woodbury, NY).
6. —— (2004). *Concert Halls and Opera Houses: Music, Acoustics, and Architecture.* (Springer-Verlag, New York, etc.).
7. Cavanaugh, William and Wilkes, Joseph, Editors (1999). *Architectural Acoustics Principles and Practice.* (John Wiley & Sons, New York, etc.).
8. Egan, David, Editor (1988). *Architectural Acoustics* (McGraw-Hill, New York etc.).
9. Hoffman, Ian; Storch, Christopher and Foulkes, Timothy (2003). *Halls for Music Performance.* (Acoustical Society of America, Woodbury, NY).
10. Jaffe, Christopher (2010). *The Acoustics of Performance Halls: Spaces for Music from Carnegie Hall to the Hollywood Bowl.* (W.W. Norton & Company, New York).
11. Sabine, Wallace Clement (1922, republished 1992). *Collected Papers on Acoustics.* (Peninsula Publishing, Los Altos, CA).

15.I.8.2 Papers in Technical Journals

1. Gulsrud, T. (2005). *Characteristics of Scattered Sound from Overhead Reflector Arrays: Measurement at 1:8 and Full Scale.* (Journal Acoustical Society of America, 107(1), Jan. 2000).
2. Hidaka, T. and Beranek, L (2000). *Objective and Subjective Evaluations of Twenty-three Opera Houses in Europe, Japan, and the Americas.* (Journal Acoustical Society of America, 107(1), Jan. 2000).
3. Hidaka, T. Nishihara, N. and Beranek L (2001). *Relation of Acoustical Parameters with and without Audiences in Concert Halls and a Simple Method for Simulating the Occupied State.* (Journal Acoustical Society of America, 109(3), Mar. 2001).
4. Hidaka, T. and Beranek, L (2001). *Mechanism of Sound Absorption by Seated Audience in Halls.* (Journal Acoustical Society of America, 110(5), Part 1, Nov 2001).

15 Unit II

Dramatic Arts Spaces

David A. Conant, FASA

15.II.1 INTRODUCTION

Whether serving professionals or aspiring young students, rooms supporting the unamplified voice are acoustically successful to the degree that vocal clarity, including the full range of emotional, vocal inflections is high throughout the audience. Equally important, but normally of much less concern than for dedicated music venues, is a need to achieve clarity without distortion or "coloration". Good design provides support for the voice via useful reflections while controlling noise interferences throughout the stage and audience areas. In an information theory sense, the unamplified voice "source" may represent the intended signal (S) material and the "noise" (N) represents everything else, including mechanical, electrical, and plumbing (M.E.P.); environmental noise; reverberation; etc. Often S and N are expressed in terms of the signal-to-noise ratio (expressed as S/N), but if each is expressed in decibels, then the ratio reduces conveniently to the arithmetic difference between the two levels, or S-N. At the audience, the level of direct plus early reflected vocal levels embodying the essence of the signal, summing to S, say, is influenced by the performers' gender, vocal effort, head orientation, distance from the audience, and the location, size, shape, and orientation of reflecting surfaces that direct clean (nonscattered) early sound reflections to each member of the audience. Because each theater's acoustic design is driven by factors unique to its intended programming of some mix of spoken word and music, a thorough knowledge of each of these influences materially informs design of halls serving musical theater, opera, and broad multipurpose uses. This section of the chapter addresses relevant planning and computational factors in the context of practical theater design presuming that electronic amplification is minimally employed, if at all. Certainly, although large audience sizes and pit orchestras may require audio amplification and mixing balance, to the extent that the parameters controlling the unamplified S/N ratio are addressed, then vocal "lift" by amplification can be minimized in most drama spaces of new design under about 1,200 seats. Here, we discuss rooms for what is termed dramatic arts—i.e., storytelling, drama, comedy, musical theater, opera, etc.—for spaces up to about 1200 seats. Similarly programmed spaces of larger size are commonly multipurpose venues and are treated in Unit I of this chapter.

For simplicity, the design principles described herein may be viewed as applicable to the unamplified singing voice to the same extent as for speech. That said, it is instructive to understand which acoustical design parameters are shared—and not shared—between drama venues and music concert halls. Clearly, instrumental-accompanied vocal works (especially musical theater and opera), require additional design considerations to account for the "competing noise of music" concurrent with the "voice signal". Indeed, although an entire chapter could be devoted to that matter alone, especially as it relates to orchestra pit design, the reader is directed to the references and glossary at the end of this unit.

In the following paragraphs regarding room design, and at the risk of some redundancy but for pedagogical benefit, we attempt to parse out the nearly inseparable: Room shaping versus finishes.

Finally, as for instrumental musicians, vocalists benefit from "support" or feedback to themselves, via either early reflections or foldback/monitor loudspeakers, in gauging how their vocal efforts are perceived by the audience. In other words, to the extent that singers hear a "return" from the hall and/or on-stage devices (e.g., early reflections, monitor loudspeakers, ear buds, etc.) their vocal production is informed and so the performer (and audience, consequently) benefit. Likewise, the stage set design can either materially augment the unamplified vocal work on stage or undermine it. While these issues are addressed here, treatment is necessarily brief.

15.II.2 WHAT YOU NEED TO KNOW: BROAD DESIGN PRINCIPLES

15.II.2.1 Source → Path → Receiver

Speech intelligibility is commonly assessed by parsing out for computation and consideration, its three principal components: Source, Path, and Receiver. Because each of these has unique characteristics driven not only by theater type and design, but varies among stage set designs, actor locations on stage and audience patron seat position, the acoustician must be mindful of these interplays during design. In an abbreviated view, these elements are discussed in the following paragraphs:

15.II.2.1.1 Source

Principal attributes of source spectrum level and directionality—*i.e., for a single performer: the age, gender, vocal level, head position, and orientation on stage.* One presumes at the outset, sufficiently uniform delivery, clear diction, competency in the language spoken or sung, and minimal dialectical issues. This sound source generally represents the prime "source signal" (S) material emanating from the actor/singer, although it could be viewed as that which comes from one or more loudspeakers. In the latter instance, each source is usually best examined separately with respect to source, path, and receiver issues—as the paths of sound travel will be rather different depending on the specifics of the loudspeaker location and its directivity pattern as a function of frequency. Additionally, some consideration in design calculations should be made to account for the likelihood of untrained voices (e.g., students) versus trained voices.

15.II.2.1.2 Travel Path(s)

Principal attributes are: room volume, direct sound path distance to audience patron, reflected energy (reflected distances to receiver, reflector locations, sizes, curvatures, spatial orientations, construction elements, and surface finishes)—*i.e., the room interior architecture that supports (or not) the sound waves propagating from actor to audience.* Note that discussions here presume fully, or nearly fully enclosed spaces, and not spaces that are principally outdoors, devoid of opportunities to provide at least a modicum of reflecting surfaces. Further, it should be recognized that the path(s) taken by the propagating sound have a material influence on the "source signal", yielding at the receiver the final "source" characteristics (including, especially level as a function of frequency and direction) that could be viewed in examining the S/N ratio as perceived by the receiver. Of course, unwanted noise, N, at a listener can include whatever is not a part of the intended signal, S, and thus not simply the conventional issues of M.E.P. noise contributions. For example, the "noise" within the path, such as excessive reverberation (which can muddy otherwise clear speech), may be viewed as a part of the overall N contribution and, in fact, is built into some of the instrument-measured and computationally intensive speech intelligibility descriptors.

15.II.2.1.3 Receiver

Principal attributes are: hearing acuity, ambient noise levels—*i.e., generally, this component is dominated by the M.E.P. and other environmental noise impacts near a specific receiver.* The audience is assumed to be of normal hearing acuity and substantially conversant in the language, dialect, etc., of the speaker and—although this can have a marked effect on intelligibility —not assumed to have a clear view of performers' lip movements.

15.II.2.2 Speech Intelligibility Descriptors

As planning of a space for dramatic arts unfolds into schematic drawing plans and sections, design decisions benefit by calculating any of the variety of speech intelligibility related measures. These include reverberation time (T_{60}), the articulation index (AI), the speech intelligibility index (SII) (or its predecessor the speech transmission index [STI, or RASTI]), or clarity (C_{50}). Because a considerable time investment is required to scour the drawings for acoustically salient parameters for computation, judiciousness in this exercise is in order for the practicing professional. In the end, significant differences among the more speech-related of these measures is generally not found in the practice of performing arts design, and such calculations are generally beneficial in design only for rooms over 300 seats. However, for venues serving over 500 seats that may also be pressed into significant use for music, deeper insight may be provided through software applications such as *EASE, CATT-Acoustic, Odeon,* and other computationally-intense packages (see Chapters 1–6). Among descriptors that are relatively simple to compute without necessarily reverting to proprietary software are T_{60}, AI, and C_{50}; and these should be examined periodically, often with increasing scrutiny, through the design process of larger drama venues. For AI and C_{50} hand calculations, the reader is directed to References 1, 2, and 8 at the end of this unit, and advised that while both AI and STI calculations incorporate ambient noise parameters, the C_{50} descriptor does not.

15.II.2.3 Aberrant Reflections

Throughout the design of venues larger than about 400 seats, it is prudent to examine sketches to identify, and design against, potential long-delayed reflections from the stage to individual seats. Thus, even if a speech intelligibility descriptor suggests good intelligibility, a strong sound reflection, or group of reflections, arriving too late after the arrival of the direct sound could be perceived as annoying—although not necessarily revealed as such by an STI value. This occurs under various combinations of reflected energy arrival time delay (after direct, unreflected sound) and level with respect to the direct energy. Such problematic anomalies are often identified by the hand calculation software indicated before, or as treated in other chapters in this handbook. The matter becomes more complex in the presence of a time-clustered group of low energy reflections arriving at, say, 10–15 milliseconds (msec) ahead of the paper-identified *problem* single reflection that may be identified. In such instances, the slightly earlier-arriving energy, if sufficiently strong, can forward-mask the slightly later-arriving problem reflection(s) to the point where it is not detected even by a trained listener. Packets of reflections arriving later still, more severely complicate the problem (backward masking) but in practice, these are rare, especially in drama venues. Similarly, spaces whose walls or ceilings are dominated by highly regular and fairly closely spaced linear shaping elements (e.g., picket fence-like) can yield in some seating sections noticeable "coloration" or "flutter" due to similarly regularly spaced (in time) reflection arrivals in those seating sections. We describe herein the more foundational design issues sufficient for setting an acoustically competent initial design, while also identifying subtler, albeit important points to avoid or perhaps investigate further and capitalize on for specific rooms. Finally, although Unit II is not intended to address instrumental music or audio amplification issues deeply, the matter of strong, long-delayed reflections that may be excited by percussive instruments or the unique source locations of loudspeakers (often excessively loud and at a too-severe elevation above actors, when used) should be examined at the same time as actor (source) locations when searching for annoying reflections and/or time delayed arrivals. This applies to both amplified and unamplified programming.

15.II.2.4 Guideline for Assessing and Designing for Individual Reflection Strength

In the course of sketching and performing calculations to ensure good early energy arriving at listeners, quick assessments of the efficacy of individual or an ensemble of reflections associated with speech intelligibility, including annoying echoes, can be drawn from the guidance in Figures 15.II.1, 15.II.2, and 15.II.3.

15.II.3 THREE BASIC THEATER FORMS

In the course of providing sound reflections that are important to aiding good speech intelligibility and minimizing aberrant acoustical effects, in addition to providing appropriate sound reflecting and absorbing surfaces, room shaping plays a key role, especially as the architect is concerned. If sufficient attention to detail of smaller-scale shaping can be provided, many acoustically problematic issues attendant to less-than-ideal, larger-scale plan forms and sections can be resolved without resorting to excessive absorptive treatment. Gross room shaping issues for new construction can unfold during the programming phase but should be formed at least broadly, by early schematic design. This is because drama stakeholders in such projects commonly engage a design team only after establishing among themselves a clear idea of the type (viz. form) of drama space they wish. Spaces for production of dramatic works tend to land in one of the three most commonly referenced and described in the

N O I S E P R E D I C T I O N S C E N A R I O
by McKay Conant Hoover Inc
Job #100000 for RPI in FILE-2.nps last saved on 3/24/10 @ 9:19AM

AI study for 450seat Rm w/ 2Reflct printed on 3/24/10 @ 9:25AM by CONANT

D E S C R I P T I O N	ft² or No.	63	125	250	500	1k	2k	4k
1. PeakPWL(77dBA) on-axis, avg. WOMAN'S NORMAL VOICE		28	59	74	77	72	67	66
EPA-600/1-77-025 May'77:K.Pearsons et al.								
2. Correction for 90deg off-axis from talker		0	0	-1	-2	-4	-5	-8
Manually entered miscelaneous data								
3. Wood Parquet on Concrete floor	**960**	-15	-16	-16	-18	-18	-18	-18
NRC6 1987 Data=-10log(Sabines)								
4. Plaster on lath	**7800**	-29	-30	-29	-27	-26	-25	-24
NRC6 1985 Data=-10log(Sabines)	...	for WALLS						
5. Occ.Med Uphol.Seat- (Hidaka, Nishihara, Beranek)	**3150**	-31	-32	-33	-34	-34	-34	-34
Manually entered: Data=-10log(Sabines)								
6. Glass, large panes of heavy plate	**60**	-9	-10	-6	-4	-3	-1	-1
NRC4 1987 Data=-10log(Sabines)	...	for CNTRLRM						
7. Plaster on lath	**3800**	-26	-27	-26	-24	-23	-22	-21
NRC6 1985 Data=-10log(Sabines)								
8. Decoustics 2" Fabric-Wrapped Panel	**900**	-23	-24	-29	-30	-30	-30	-30
NRC97 TYPE 1 Data=-10log(Sabines)								
9. Carpet, 0.3"thick on Concrete Floor	**1200**	-17	-20	-20	-24	-25	-25	-26
NRC21 Brtsh.Res.St Type-A Data=-10log(Sabines)	...	for CARPET						
10. Proscenium opening absorption[10log(1/Sα)]	**672**	-26	-27	-28	-28	-28	-28	-27
Manually entered: Data=-10*Log(Sabines)								
11.* RT60(sec.):135000cf.:Sα@3+4+5+6+7+8+9+10 =		2.1	1.7	1.4	1.3	1.3	1.2	1.1
Includes air absorption corr. @ 2kHz & 4kHz								
12. Q=2:r=40':n=2:Direct field only!:Lp-Lw=		-30	-30	-30	-30	-30	-30	-30
$Lp-Lw=10\log[(Q/4\pi r^n)+4/(\Sigma S\alpha+4mV)]+10$ dB:								
13. Cc corr.for 1, 4'wide, 20' radius convex reflector		0	0	0	-4	-5	-5	-5
Manually entered miscelaneous data								
14.* ARITHMETIC SUM of lines 1+2+12= 48dB		-2	29	43	45	38	32	28
Line #14 is SPL of direct energy only								
15. Q=1:r=47':n=2:Direct field only!:Lp-Lw=		-34	-34	-34	-34	-34	-34	-34
$Lp-Lw=10\log[(Q/4\pi r^n)+4/(\Sigma S\alpha+4mV)]+10$ dB:								
16.* ARITHMETIC SUM of lines 1+2+13+15= 42dB		-6	25	39	37	29	23	19
17.* ENERGY ADDITION of lines 14+16+16= 50dB		1	32	46	46	39	33	29
Line #17 is Energy SUM of direct + 2 reflections								
18. Curve of PNC20 line		46	39	32	26	20	15	13
19.* ARITHMETIC SUM of lines 17-18= 25dB		-45	-7	14	20	19	18	16
20.* AI @ #19=.47:Unaccep. Privacy /Fair Intellig		0	0	8	16	15	14	12
[uses #11 for T60correction (w/o aisle edge corr.)]								

Figure 15.II.1 Articulation Index;. simplified calculation example.

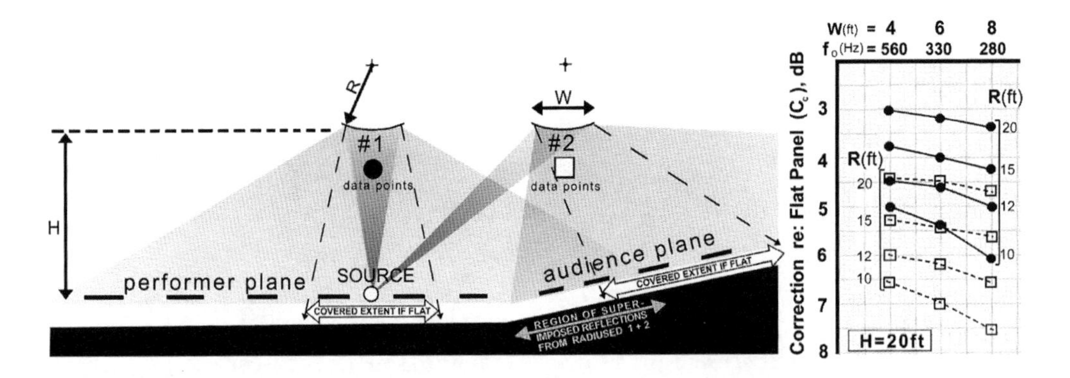

Figure 15.II.2 Overhead reflectors; example of energy adjustments associated with the radius of curvature, width, and orientation to source and receivers for singly curved reflectors.

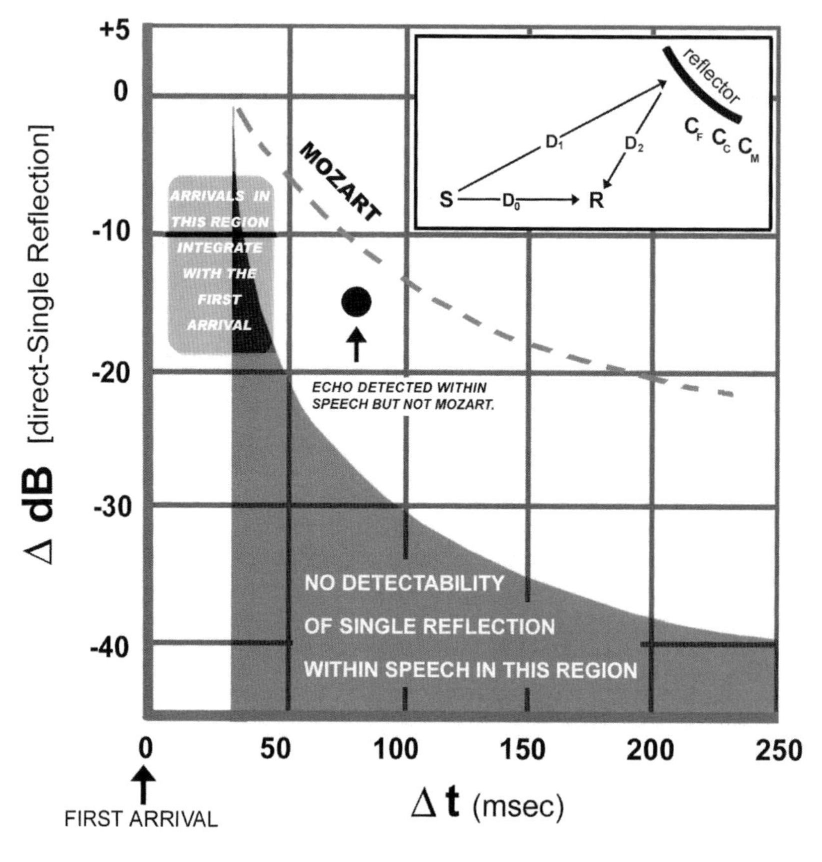

$$\Delta \mathbf{dB} = 20\log[D_0/(D_1 + D_2)] - C_F - C_C - C_M$$

D_0 = Direct, unbstructed distance Source to Receiver
$D_{1,2}$ = Distances: Source to reflector, reflector to Receiver, repectively
CF = Correction for energy losses due to reflector Finish construction
Cc = Correction for energy losses due to reflector Curvature
CM = Correction for energy losses due to Misc. (exposed size, etc.)
Δt = $(D_1 + D_2 - D_0)$/speed of sound in air

After Dietsch and Kraak (Acustica, Vol. 60, 1986)

Figure 15.II.3 Echo detectability for single echoes within speech.

following list. Within these, of special interest to the acoustician are the regions of the source end (sending or stage/platform) and the audience configuration(s)—as well, of course, as the form of surrounding surfaces (see Figure 15.II.4).

1. **The Proscenium Theater Form**: This can be described as having, at the stage end, a defined opening in the wall (the proscenium wall) that visually demarks separation between the audience and the principal acting region. The audience region assumes any of a variety of plan shapes, but commonly ranges between somewhat rectangular to fan-shaped. This form lends itself well to providing a recessed orchestra pit

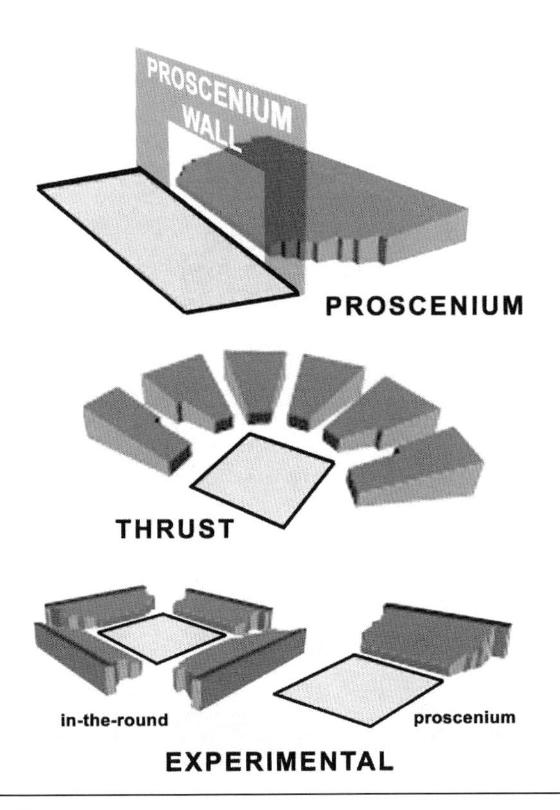

Figure 15.II.4 Theater form types.

between the stage and audience—more so than other forms. Seating in balconies, or at least galleries or parterres, tend to develop in this form when capacity approaches 450 seats, if there is an intention to promote visual intimacy of the space. In addition to the obvious, central performance opening in the proscenium wall, side stages flanking this opening (sometimes provided with operable, closure walls) may be provided as well.

2. **The Thrust Stage Theater Form**: The term "thrust stage" like "proscenium", focuses on the form of the region in which actors perform, wherein the acting platform thrusts noticeably into the audience area. Variations on the proscenium and thrust stage forms are many and extend from hybrid renditions, to theaters with exposed (or concealed) "caliper", or side stages often developed as opportunities for either additional staging or small pit orchestras. Where the thrust (or forestage extension) into the audience region is modest, a recessed orchestra pit is easily accommodated and such rooms acquire many of the attributes of proscenium theaters. For the deep (into the audience) thrust stage configuration as examined here, a proscenium opening, flytower, and balconies are effectively nonexistent and wraparound seating is developed between at least two audience entrance vomitories.
3. **Experimental Theater Form**: This form is characterized by much greater flexibility than other forms with respect to possible audience and actor configurations. The most flexible can offer theater-in-the-round as well as a myriad of actor entry points into the "audience chamber" itself. Between these extremes, both proscenium and thrust stage productions can be offered. Such spaces may be termed black box, flexible, or

studio theaters (or similar) with the designation influenced largely by predilections of the facility operators. This form normally accommodates less than 200 seats, although they can grow to over twice this size—with attendant intelligibility challenges, especially for unamplified speech.

15.II.3.1 The Proscenium Theater

15.II.3.1.1 Characteristic Elements

The term proscenium derives from early Greek forms as a theater's key architectural element—the wall demarking the plane separating the acting space from the audience space. Thus, the wall's opening through which the action is viewed is rather literally "in front of the scene". Certainly though, acting regularly occurs somewhat in front (downstage) of this wall as most proscenium theaters provide at least a modicum of performance area extending forward of the proscenium, or a forestage extension or modest thrust stage. Still, because the preponderance of actors' work occurs near, and somewhat upstage of the opening proper, the following key design elements require appreciation in terms of their influence on room acoustics.

15.II.3.1.1.1 The Proscenium Opening

The proscenium opening is generally rectangular and for drama works, is normally wider than it is tall. Its specific dimensions are established early by the theater consultant in consultation with the acoustician and other stakeholders and subsequently sometimes (visually, on the fly) further adjusted by the stage set designer as deemed best for each production. To the extent that the stage or performance platform may be used for acoustic music, its dimensions and the design of its sides and top requires acoustical scrutiny as these regions should provide important early sound reflections to the audience.

15.II.3.1.1.1.1 The Eyebrow

The proscenium opening top provides a critical opportunity to incorporate a sound-reflecting "eyebrow" (see Figure 15.II.5). This region can be made effective for overhead reflections which lend substantial clarity to unamplified sources, especially so, because the reflected sound normally won't suffer from attenuation losses via grazing propagation across the audience, as can occur with lower sidewall reflectors. Determining the number of such reflective eyebrow elements, their size, height, tilt angle, curvature, and material composition is developed largely by the acoustician. Not unexpectedly, this eyebrow region is of considerable interest to multiple disciplines and its final design parameters are necessarily the purview not only for the acoustician, but the audiovisual designer, the lighting designer, theater consultant, architect, and sometimes the mechanical engineer. Often the lighting designer wishes to provide downlighting from a catwalk here, the audio designer needs to situate loudspeakers along this region (serviced by the catwalk) often interfering with the eyebrow's sound reflections by scattering or blocking them. Perhaps more than any other region, a successful theater derives from successful collaboration of these several disciplines, and it regularly presents a visceral "test" of creativity and cooperation among theater designers!

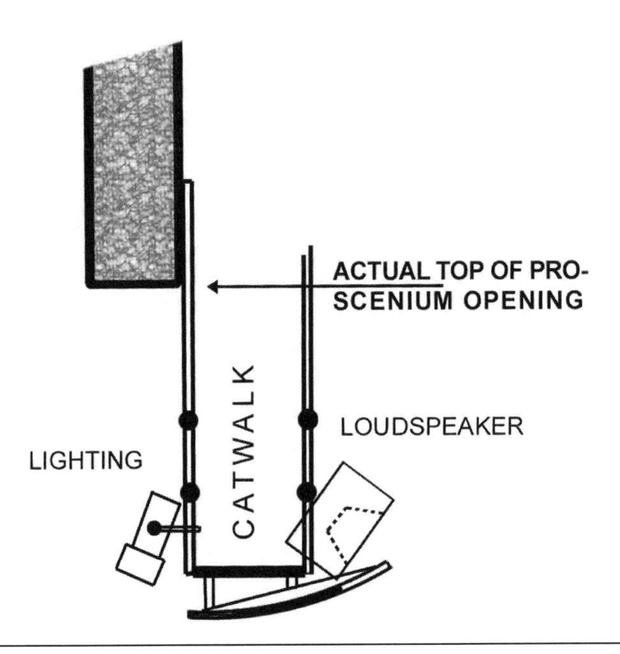

Figure 15.II.5 Proscenium eyebrow reflector with coupled opening (section).

15.II.3.1.1.1.2 The Cheekwalls

The sides of the proscenium opening also present sound-reflecting opportunities and for voice, can play a significant acoustical role in speech intelligibility especially for an audience seated within about 50 feet (15 m) of the proscenium (see Figure 15.II.6). Beyond this distance, without special additional acoustical attention, cheekwall reflections provide a diminishing benefit as these reflections are, unlike those from the eyebrow, increasingly prone to ever-higher attenuation due to grazing incidence propagation across the audience. To overcome the latter problem, creative shaping in the vertical section of this region to force additional downward reflections is often available, albeit within restrictions imposed by visual

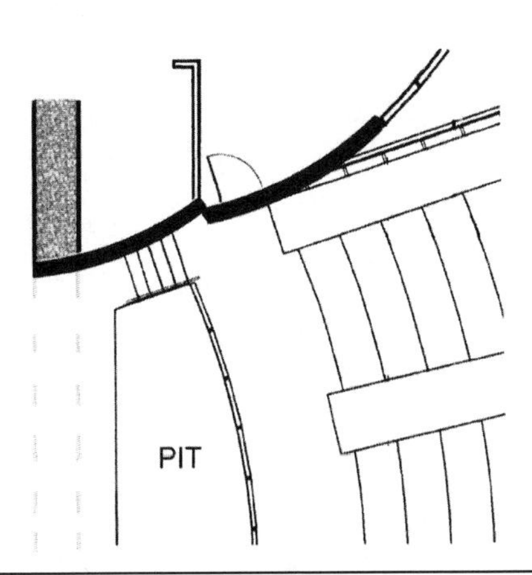

Figure 15.II.6 Proscenium cheekwall reflector (half-plan).

impact and accommodating theatrical lighting instruments (see Section 15.II.3.1.2.1.3 for a room shaping discussion). Because at above 12 feet (3.7 m) above stage level, cheekwalls become decreasingly beneficial for reflections to the orchestra-level audience, 10″–14″ (25–36 cm) deep, horizontal "ledges" are often provided on the cheekwalls to intercept and redirect sound back downward into audience areas.

15.II.3.1.1.1.3 The Set Design

Although clearly outside the realm of architectural design practice in the planning and design, the set designer, whose work arrives well after the opening ceremonies, can influence speech intelligibility within the audience chamber simply by material selections and set wall shaping. We mention this here because the hall's inherent acoustic design, if not fully optimized irrespective of set design, may be unable to overcome "path" challenges on stage especially with untrained "source" voices on stage. The author has been called upon more than a few times to assess speech intelligibility problems suddenly arising in theaters that had no such history, only to discover that the prime culprit was the acoustically challenged set. For example, to the extent that reflections from broad concave (to the audience) set surfaces, (especially with overhanging balconies) focus into one audience region, a much broader audience region may suffer without that energy. Thus, as actors move about certain stage areas, dialogue intelligibility may fluctuate markedly if not overcome by substantial reflections from non-set architectural surfaces.

15.II.3.1.1.2 Catwalks and Tension Grids

Catwalks and tension grids provide easy access to, and flexibility in, positioning lighting instruments, audio systems, variable acoustics, etc., while offering important sound-reflector opportunities. Most larger drama venues accommodate at least one of these. Theaters unable to accommodate either, require at least some array of lighting instruments on pipe grids and/or lighting trusses—neither of which provides any means of developing clean sound reflections to audience. Rather, the presence of lighting devices serves to scatter mid- to high-frequency sound thereby losing key articulation signals of the human voice.

15.II.3.1.1.2.1 Catwalks

These normally number at least two, parallel to the audience rows (i.e., transverse), plus at least one linking them, and are located by the theater/lighting consultant with acoustician input since their undersides easily accommodate reflectors. Where additional overhead reflections are desired, additional reflectors can normally be provided, suspended from the structure via rods. The most beneficial reflectors are positioned parallel to seat rows, are normally singly curved (i.e., cylindrical section), convex downward, at least the width of the catwalk (typically 4 feet [1.2 m] wide), and extend over most of the transverse catwalk breadth. They may form either continuous or segmented arrays and if segmented, depending on the seating plan and reflection needs, may be doubly-curved (i.e., spherical section, bowl-like, compound …).

15.II.3.1.1.2.2 Tension Grids

These devices, also termed Izenour Grids (after their inventor, theater consultant George Izenour), provide a greater acoustical design challenge than catwalks. Unlike reflectors residing just below catwalk gratings (which permit unobstructed reflection paths), accommodating reflectors with tension grids undermines the attraction such grids offer for lighting designers. Sound reflectors simply cannot perform their intended function if anything larger than a softball (including, especially, lighting instruments) interrupts the reflection path. This requires, then, that spaced reflectors (of an average 4–5 feet [1.2–1.4 m] plan form dimension) reside just below the plane of the tension grid, to the consternation of lighting designers who would prefer the entirely unfettered positioning of lighting instruments that tension grids provide. That said, carefully shaped and positioned reflectors below tension grids are installed and today provide major acoustical benefits with the full blessing of the project theater consultant and stakeholders.

15.II.3.1.1.3 Seating Above Orchestra Level

The seating above orchestra level where it exists (e.g., balconies, side boxes, parterres, and galleries), constitutes audience seating above the main or orchestra-level seating. The front-facing surfaces of these (e.g., balcony edges), as well as the undersides of balconies and side boxes, provide useful reflection opportunities to orchestra-level seating. Depending on their specific shape and location, they may, however, require sound diffusive and/or absorptive treatment to mitigate focusing or long-delayed reflections. Each instance deserves scrutiny in three dimensions, especially for rooms over about 300 seats. Balcony overhangs, since they can occlude important energy from ceiling reflectors to seats directly below them, require special attention. Care should be taken to provide that at least one, strong overhead reflection can reach all under-balcony seats and that the rearmost rows catch useful early (<30 msec) reflections from the rear wall wherever possible. Multiple sidewall reflections directly to under-balcony seats should be sought as well. In addition to the above prescripts, a useful rule of thumb is to provide balcony (and under-balcony) opening heights no less than their depths.

15.II.3.1.1.4 Orchestra Pit

An orchestra pit, if programmed, is best sunk at least 8 feet (2.4 m) below stage level (with a podium for the conductor's coordination of pit musicians with stage action) to help achieve appropriate levels and balance between stage singers and pit musicians. Considerable care should be taken in the design of orchestra pits intended for nonamplified musicians (especially percussion and brass) to provide sufficient sound absorption of low- to mid-frequency sound levels at these musicians' ears, as they are in a relatively small, partially enclosed pit volume. Glass fiber lay-in ceilings and 2″ (50 mm) thick material should be copiously applied where practical, especially on the upstage wall. Commonly, the upstage-facing wall behind the conductor is vertical and concave toward upstage in plan, and this presents both a substantial challenge and opportunity. We have found that if broad, convex (in plan) reflecting surfaces are developed along the structural wall and these are, in turn, tilted forward at their bottoms, then substantial benefit is derived for aural communication to/from the stage and pit while minimizing problematic back-reflections of musicians' sound to themselves from this surface. See Chapter 15: Unit I: Sections 15.I.2.8, 15.I.4.4.5, and 15.I.5.2 describing planning for orchestra pits in various related venues.

15.II.3.1.2 Design Principles for Proscenium Theaters

15.II.3.1.2.1 General Design Guidelines

The acoustical descriptor ranges shown in the upcoming text should be used with the following general understandings: as T_{60} approaches the upper limit, the ambient noise criterion should become correspondingly restrictive, and the converse is true. Similar comparative relationships with ambient noise can be ascribed to descriptors such as C_{50}. A cautionary note: aberrant long-delayed reflections, discussed elsewhere in Unit II, can become increasingly noticeable and problematic for speech as both T_{60} and ambient noise levels drop. Additionally, individual otherwise nonproblematic reflections can become annoying as strong, long-delayed reflections arriving about 20 msec prior to them are eliminated. This arises from a phenomenon known as forward masking and is established by reflections whose strength and timing effectively (but very briefly) lowers our hearing mechanism's sensitivity to detecting individual subsequent, and potentially problematic individual, strong arrivals. Variability in these descriptors will, of course, be found across the seating in any theater. With respect to developing shapes and finishes beneficial for useful, clean reflections bear in mind the dimensional and radius of curvature issues discussed in Section 15.II.2.4.

15.II.3.1.2.1.1 Design Goals

Reasonable starting point descriptor criteria ranges within the audience chamber follow:

- $0.6 < T_{60}$ mid < 1.3 sec., occupied (may be variable in some forms)
- C_{50}: ≥ 5.0 or AI ≥ 0.55
- NC/RC/PNC range: 22–25

15.II.3.1.2.1.2 Room Finishes

- Walls in the audience chamber should be generally sound-reflective near the stage or platform (e.g., gypsum board, wood, etc.) to at least 20 feet (6 m) from the proscenium opening, becoming absorptive at and near the rear wall (as required for the T_{60} goal) or shaped to provide early reflections, or a combination. Beyond this about 20 feet (6 m), sidewalls commonly remain reflective, taking on shaping to uniformly distribute reflected energy from actors' voices across the audience—at both orchestra and balcony (if any) levels. Audience in the parterres, sidewall galleries, or box seats normally derive somewhat less total benefit from such shaping due largely to architecture stemming from these features, as well as the fact that direct sound to such seats is usually unobstructed and unattenuated by not grazing across the audience, and often benefits as well by at least one good reflection off the stage floor. Note that rectangular plan forms lend themselves better than fan shapes to capitalizing on reflections at sidewalls, particularly for recumbent-angled rear-half sidewalls (i.e., the reverse of fan plans). The stage house left and/or right walls, particularly during rehearsals, can benefit by at least 1.5″ (38 mm) thick, direct-applied fibrous absorption such as coated duct liner board in order to control reverberance and flutter echoes that would normally be mitigated by full stage sets. Alternatively, curtain "legs" could be arranged to achieve comparable absorptive effect, albeit with somewhat more trouble.
- Ceilings are commonly shaped and finished to provide multiple clean, specular, early reflections to each patron, especially if walls cannot be sufficiently shaped, positioned,

or appropriately finished. Ceilings should be designed for some mix of sound scattering and absorption to achieve the T_{60} goal. Frequently, reflective elements (e.g., clouds or other reflective arrays perhaps extending from the eyebrow) are developed considerably lower than the underside of the roof, and these provide an upper cavity to accommodate catwalks, tension wire grids, ductwork, etc. Concomitant with growing desire in the industry for uniformity and control of dramatic lighting effects, visually "finished" ceilings for drama spaces often expose theater technology. Thus, except for visually apparent acoustical reflectors (generally under catwalks) and some ductwork, ceilings frequently disappear into the blackness at and above catwalks or tension grids.

- Flooring beyond about 20 feet (6 m) from the stage is often carpeted to mitigate footfall noise and reverberation but is generally hard-finished within the seating for maintenance purposes. Subfloor construction may be wood, concrete, etc. Rooms attempting to serve unamplified music are benefited especially by sound-reflective flooring within at least 20 feet (6 m) of the stage.
- Seating is normally upholstered at least on the bottoms and seat backs. Together with the seats proper, backrests are best selected to approximate the sound absorption of the seated audience in order to minimize apparent changes in room reverberance as seated occupancy fluctuates. Backs of seat backs and seat-bottom pans may be sound absorptive as well, in the interest of minimizing T_{60} changes—but perforated seat-bottom pans (to increase absorption) are a good choice in steeply raked audience areas where seat-bottom pans are more likely to be exposed to the sound field, especially when amplified. Issues concerning first costs and maintenance often override acoustical benefits of absorptive seat bottoms and back sides but should not be permitted to override value of appropriate absorption for finishes exposed to the stage and ceiling, even when occupied. These decisions should be balanced as well, with the programmatic expectations and desire for a good acoustic for unamplified music in multipurpose theaters.
- Occupancy influences on room acoustics can be marked, of course, even with seating acoustical specifications that attempts to approximate a seated patron when empty. Thus, T_{60} is best calculated to assess the anticipated range of audience occupancies. These may reasonably range from 15 percent occupancy for dress rehearsals to 85 percent for most performances. Best by far, to consider audience area in T_{60} calculations, then simply to count seats. This matter as well as others associated with T_{60} calculations are addressed thoroughly in Unit I of Chapter 15.

15.II.3.1.2.1.3 Room Shaping

- Reflective portions of ceilings and walls should provide a minimum of three clean, specular reflections from stage voice to each patron arriving within about 50 msec of direct sound, especially for an audience beyond 30 feet (9 m) from the stage. Design for at least two such reflections arriving via entirely unobstructed paths approaching from no less than 3 feet (0.9 m) above patrons' heads (i.e., via ceiling or upper sidewalls) because sound loses considerable energy traveling via a grazing path. In a renovation, to the extent that either a deep balcony overhang or a heavily coffered (historic) ceiling precludes useful reflections to the audience, considered sidewall and rear-wall shaping can become increasingly important for speech intelligibility—if it is possible at all. Similarly, to the extent that historically sensitive or ADA constraints on lower sidewall shaping conspire to preclude good lateral reflections, opportunities at the ceiling or upper sidewalls may be available. While it may seem that the first several rows of the audience would require few or no significant supportive reflections, due to their proximity to actors, this is often not the case. For these seats, at least one ceiling (via eyebrow) reflection is recommended in addition to a clean sidewall reflection (via cheekwall) where possible.

- Location and shaping of reflective elements to achieve requisite specular reflections are determined by ray-tracing studies on paper or computer. To be effective for speech wavelengths, surface dimensions of such elements exposed to the actors should generally be no less than 4 feet (1.2 m) across (they approach minimal effectiveness below 2 feet [0.6 m]). Additionally, the "subtended" dimension of any reflector, if it is to be effective, should be at least 2 feet (0.6 m) across in the "subtended angle" dimension as viewed from an actor on stage when considering angles of approach of a "sound ray" from the voice to the reflector. Sidewall shaping becomes acoustically meaningful for voice beginning about 30 inches (0.76 m) above the finished floor of a nearby audience.
- Rear walls provide opportunities, sometimes in conjunction with under-balcony ceilings—if these ceilings are not too low at the rear—to kick reinforcing reflections to the audience, generally from behind. It is best to capture these opportunities in lieu of simply adding absorption at the rear wall as long as sufficiently short time delay gaps between direct-arriving energy and these reflections do not exceed about 35 msec for any such seats. If these rear walls are not sufficiently effective in kicking their reflections downward, however, excessive low-frequency reflections can return to the stage or front seating rows. Care should be taken in this regard to avoid low frequency reflections of amplified music or percussion returning toward the stage and thereby generating a problematic slap-back echo; these same issues apply to rear walls of balconies, of course. Recall that the long wavelengths of low frequencies (10–20 feet [3–6 m] for 50–100 Hz) are not materially influenced by shaped surfaces whose dimensions are not at least 1.5 times their wavelengths—and this can lead to the alternative of thick absorption when shaping cannot solve long-delayed energy returning to the stage, for example.
- The previously described benefits of achieving early, clean (nondiffused) reflections to the audience from side and rear walls apply as well to capturing inter-reflections between side walls and the underside of side galleries, balcony parterres, and the like. Because sound could strike the underside of side galleries first (for example) and then the wall, prior to returning to the audience as well as first striking the wall and then the gallery underside, just one wall/ceiling intersection can provide two reflections. Further, the frontal balcony or side-gallery edge face provides another opportunity for downward reflections to the audience. Be careful, however, of concave-in-plan regions of such (otherwise) opportunities as they can produce problematic focusing toward the audience. Such issues can be mitigated by either directing reflected energy upward, diffusing it with convex or other diffusive shapes, or via absorption applied in these concave regions.
- Seating configuration (such as continental versus traditional or hybrid) is of somewhat less concern acoustically, for drama than music. That said, it is best in any case, to provide an aisle between the walls (either side or rear) and the nearest audience for as many patron seats as possible. This provides a greater likelihood of realizing another

useful, clean reflection that would otherwise be mitigated by the patron being too close to the reflecting surface.

15.II.3.1.2.1.4 Room Volume

Acoustically, this is dictated largely by T_{60} requirements and supporting calculations. The required acoustically free room air volume can normally be "backed into", given the need for generally hard-finished floors in seating areas for maintenance purposes as well as budgets and predilection for sound absorption applied on rear walls, at the roof underside, etc. Subfloor construction may be wood, concrete, etc. Initial, simple hand-calculated room volumes above catwalks may yield predicted T_{60} as much as 12–15% longer than actual due to the often inescapable plethora of ductwork and catwalks, supporting structure, exposed conduit, exposed stairs, etc., required for lighting and other such necessities. These elements not only increase sound absorption in such upper cavities, but they also reduce the mean free path of sound waves that might otherwise propagate freely in that realm. The net result is that absorption (sabins) grows while acoustically effective volume shrinks. This is not normally a problem for rooms used principally for speech, but it can markedly affect rooms requiring reverberant support for unamplified music, including opera.

15.II.3.1.3 Example: South Mountain Community College Theater

This 350-seat proscenium theater in Phoenix was programmed to serve storytelling, drama, and musical theater with full flytower, orchestra, chorus, and a strong jazz program. Thus, very unlike a deep thrust stage theater, this required a truly "multipurpose-on-purpose" design. Because such well-intentioned academic programming tends, during design, rather toward "multi-useless" in the presence of unenlightened planning and/or excessively Spartan budgets and project administrators, the acoustician should be aware of this natural tendency at the outset.

Despite its slim construction budget ($7M [c. 2002] included two classrooms, dance amenities, and black box/music recording facilities), this project was the substantial beneficiary of both a committed client and an enlightened design team. Especially gratifying was the fact that although this was the architect's very first performing arts venue, he was fully open to craft a design drawing directly and honestly from first principles. It was hoped that the plan form could evoke a sense that the patron was within a violin case! Conventional wisdom suggested that the attendant concave-inward surfaces would severely undermine uniform sound distribution, generating both regions of focusing and corresponding areas of insufficient loudness. Acoustician sketches followed, yielding diagrams that would capitalize on select convex (cheekwall) areas and place the areas of focused energy sufficiently beyond the realm of audience seating, thus permitting new, "virtual source" regions of lateral sound without excessive delay. The center half of the rear audience wall became broadly convex (single-curvature) double doors, permitting eager patrons to view the entire hall interior from the lobby through a 17 foot (5.2 m) wide, clear opening (see Figure 15.II.7).

Sidewalls were developed in two offset, articulated, lower and upper portions—each with its separate reflection-considered curvaceous shaping, appropriate to the needs of the audience region served. The ceiling was made open to above, albeit with an eyebrow in two parts, plus three transverse catwalks each with a long singly curved reflector.

Because of the building's simple box form, at and above the catwalk level the available room width grew by 12–16 feet (3.6–5 m) wider than the average lower sidewall separation,

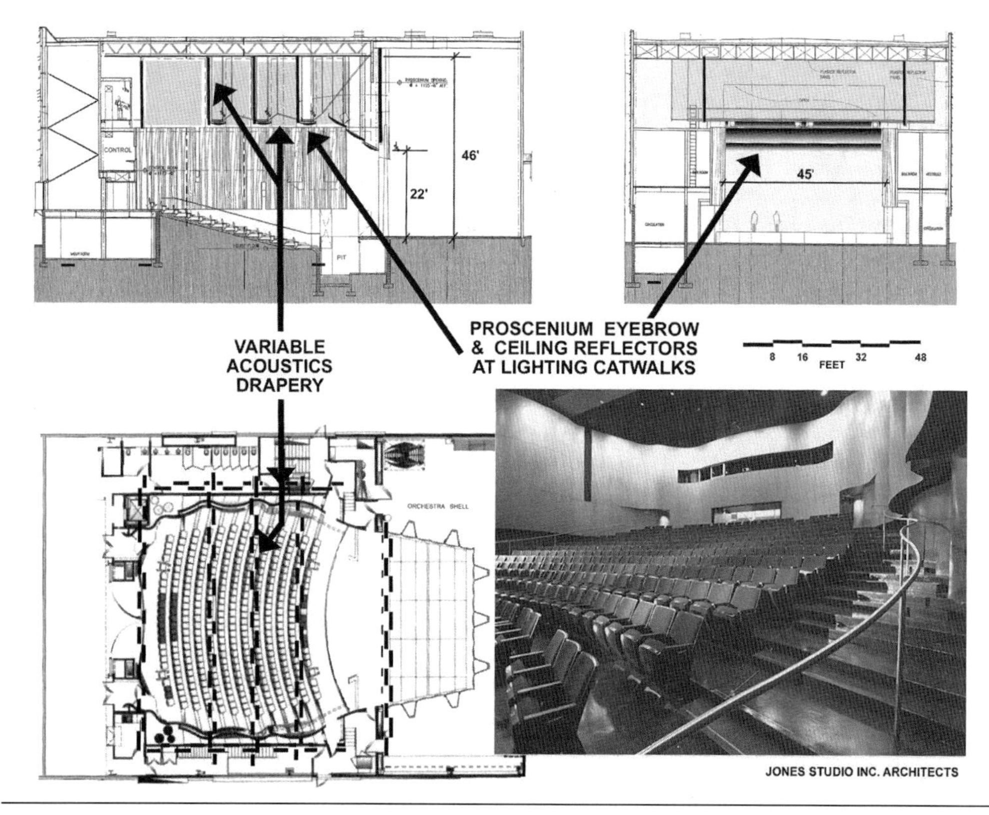

Figure 15.II.7 South Mountain Community College Proscenium Theater.

providing far more (desirable) acoustical volume than normal. To squelch the T_{60} to taste via motors, not only are four transversely tracked, acoustically absorptive, 100% fold curtains (variable acoustic, with storage boxes at 32 oz/yd [1kg/m^2]) provided, but six longitudinal sets (similar material, but double-sewn) running longitudinally between these in a lift-to-deploy configuration were designed to "chop off" the side volumes that arose as a consequence of the box form. The measured acoustical vitals are, unoccupied: $1.4 < T_{60}$ (sec.) < 1.7 (with complex set on stage, i.e., not empty orchestra shell), NC 23.

The room serves all its programmed functions beautifully and with its carefully designed reflectors augmenting the spoken word in a very quiet ambient, there is no need for audio "lift" of the live, spoken word. Again, precautionary notes are in order. Audio systems, even if "acoustically" unnecessary for most functions in such theaters, are to be anticipated and their loudspeakers coordinated with room surfaces. Accommodating full frequency range loudspeaker clusters, in eyebrows especially, can remove, if one is not cautious, excessive slices from these otherwise beneficial reflecting surfaces. Mechanized, articulating eyebrow reflectors are occasionally employed but usually at such expense and operational complexity as to beg justification on smaller (<500 seats) or inadequately funded projects. Finally, the acoustical consultant should be aware that it is the rare venue that, even with no or minimal need for audio lift, the sound board operator will frequently have a predilection to provide more gain on his system than ever seems warranted or desired by an audience. This may simply be human nature, since an apparent functioning, albeit excessive "audio presence" may be viewed as job security. Nonetheless, the project stakeholders, especially for high school and collegiate projects, should be carefully advised as to the real-versus-imagined need to provide voice amplification where the acoustician feels its use should be Spartan, only. Carefully controlled demonstrations of this are advised for technical staff and stakeholders alike at project completion.

15.II.3.2 Thrust Stage Theater Design

15.II.3.2.1 Characteristic Elements

Unlike theaters characterized by an obvious "frame" around a visual proscenium opening to the stage (commonly incorporating a sound-reflecting "eyebrow" across its top and/or acoustically shaped "cheekwalls" at the sides), dedicated thrust stage theaters with a significant thrust include no such opening to their stage or platform region. Rather, they feature an exaggerated forestage extension. These rooms are usually characterized by no stage house for flown sets but rather, because the audience wraps substantially around the acting region, a highly intimate connection develops between actors and their audience. Acoustically, these spaces, if seating more than about 350, tend toward poorer speech intelligibility than similar proscenium rooms, as the actors are as likely as not at any time, to be speaking 180 degrees away from much of the audience. Thus, as actors move about a thrust stage, speech intelligibility is likely to fluctuate substantially for any audience patron—sometimes clear, often very unclear—without special attention to engaging highly efficient sound-reflecting surfaces.

15.II.3.2.1.1 Stage Sets and Orchestra Pits

Among nonhybrid thrust stage theaters, of the several elements characteristic of proscenium theaters, only stage sets are recognizable and because they need to serve a nearly 160 degree audience wrap, even those are quite unlike proscenium sets. Acoustically, although these stage sets provide little-to-no reliable sound reflecting opportunities, audience seating rake is often sufficiently steep as to provide at least one good floor reflection, presuming there is no rug or carpet. Cheekwalls are so broadly splayed to accommodate the widely wrapping audience left and right walls that unless "plant-on" reflector devices are employed, their wide splay prohibits effective audience-supporting reflections. Accommodation for musicians can be provided offstage, because precious little opportunity lies in the usual region between the actors and audience for a conventional orchestra pit. However, because these venues are, by the nature of their thrust stage, so ill-suited to most musical theater or dance, the matter is essentially moot.

15.II.3.2.1.2 Catwalks and Tension Grids

Catwalks and tension grids, as for proscenium theaters, provide easy access to, and flexibility of, positioning lighting instruments, audio systems, variable acoustics, etc., and offer important sound-reflector opportunities near the ceiling. Most larger drama venues accommodate at least one type—occasionally both.

15.II.3.2.1.2.1 Lighting Catwalks

Lighting catwalks serving thrust stages require a tight, wrapping plan form closely tracing the seating—which, in turn, is informed by the thrust stage shape. Lighting catwalks normally number at least two and are located by the theater/lighting consultant with acoustical input as their undersides easily accommodate reflectors. Where additional overhead reflections are required, additional individually suspended reflectors can normally be provided. Because audience seating arcs rather severely and actors stage/audience orientation is so highly variable, a large number of doubly-curved, downward convex (i.e., bowl-like underside) reflectors is desirable as these can kick clean reflections in far more directions than a singly curved reflector.

15.II.3.2.1.2.2 Tension Grids

Tension grids offer a highly effective approach from a lighting design perspective, as their high degree of flexibility of lighting position and aiming direction accommodates the flexibility demanded by thrust stage work. Importantly, however, the reader is directed to the tension grid reflector discussion under Section 15.II.3.1.1.2.2 addressing Proscenium Theater applications.

15.II.3.2.2 Design Principles for Thrust Stage Theaters

15.II.3.2.2.1 General Design Guidelines

The thrust stage configuration treated in this section is that which fully capitalizes on the design implications of a deep thrust stage, and not simply a modest variant of the proscenium theater with its usual retinue of proscenium opening to a significant upstage area, an orchestra pit, etc. Although code-related distinctions arguably apply with respect to identifying thrust stages as "stages" versus "platforms", this treatment will simply refer to the performance area as "stage", proper. Ranges for descriptors shown in the following text should be used with the following understanding: as for proscenium theaters where speech intelligibility descriptors stray from optimal criteria, background noise levels should become correspondingly lower (see Section 15.II.3.1.2.1). This is especially true for thrust stages where actors will often be speaking 180 degrees away from both, other actors, and much of the audience. Further, stagecraft for lighting and set design for these spaces inevitably conspire against developing effective sound-reflecting surfaces as easily as for proscenium theaters; for seating capacities beyond about 300, scrutiny in all areas of acoustic design is required at the outset. Beware of significant variations in these descriptors across seating regions because of this "source directionality" sensitivity. With respect to developing shapes and finishes beneficial for useful, clean reflections, bear in mind the dimensional and radius of curvature issues discussed in Section 15.II.2.4.

15.II.3.2.2.1.1 Design Goals

Reasonable starting point descriptor criteria ranges within the audience chamber follow:

- $0.6 < T_{60} < 1.1$ seconds, occupied (rarely, if ever, variable)
- C_{50}: ≥ 5.0 or AI ≥ 0.55
- NC/RC/PNC range: 20–25

15.II.3.2.2.1.2 Room Finishes

- Walls are generally sound-reflective (commonly gypsum board, wood, etc., on studs) and require the utmost attention to shaping and angling to catch and redirect as much energy from the platform as possible to any and all regions of the audience. Walls within about 30 feet (9 m) of the sides of the performance platform are often necessarily so splayed (to accommodate a deep thrust stage proper), that their reflective value to the audience is minimal. Still, planar reflectors at tilt angles carefully assessed in either computer models or via laser pens and foil on foamcore physical models may be of some benefit in this region. Mostly rear walls plus a few sidewalls near the audience become candidates for catching voice energy and redirecting it back to an audience with arrival times between about 10 and 40 msec. At any one seat, unless additional reflections are determined to also arrive in this same time span, strong reflections extending beyond 50 msec are likely to degrade speech intelligibility. Wherever a wall area cannot be oriented to provide helpful reflections, they are candidates for absorptive treatment unless sufficient absorption (to meet the T_{60} goal) is already provided at the ceiling surface (described in the upcoming text). Again rear, audience-wrapping walls for deep thrust stage theaters are prime candidates for the most effective shaping—which commonly realizes itself in modest-to-severe downward tilting at the top. Of course, sufficient accommodation for circulation behind the audience at its uppermost level requires planning to ensure appropriate tilt angles and the plan form area consumed is available. Often, vomitories are provided for the audience (and some actors) entry/exit and it is wise to provide sound absorption on their walls.

- Ceilings are commonly visually unfinished (but mostly absorptive) and designed with suspended reflectors providing multiple clean, specular, early reflections. Thus, ceiling surfaces above reflectors become strong candidates to tune the room's T_{60} and as such may be simply "blacked out" with inexpensive direct-applied fibrous absorption such as glass fiber duct liner board or an equivalent. These are also often designed for some mix of sound scattering and absorption to achieve the T_{60} goal. To the extent that spaced absorptive treatment distributed across the (upper) ceiling surface is impractical, the top sides of the arrayed, suspended reflectors provide surfaces appropriate for resting (gravity) the same absorption. It would be unusual for successful, medium-to-large thrust stage theaters to have a hard, reflective exposed ceiling.
- Floors beyond about 20 feet (6 m) of the stage edge are often carpeted for footfall noise and reverberation control but generally are hard-finished within the seating area for maintenance purposes. Because these rooms normally have relatively steep rakes, carpeted aisles are indicated throughout. Subfloor construction may be wood, concrete, etc., but special care should be applied structurally to prevent creaking or booming noise upon footfall, especially in stepped aisles.
- Seating and occupancy issues are identical to proscenium theaters except that given the predilection to steeper rakes in thrust theaters, the argument to make seat bottoms absorptive becomes stronger to help minimize the T_{60} changes with versus without an audience, as these are somewhat more exposed.

15.II.3.2.2.1.3 Room Shaping

- Overall room shaping for thrust stage theaters with a deeply protruding forestage is largely dictated by the theater consultant who keeps a keen eye on the audience view and lighting angles, seating rake, aisles, and entry/exits. This requires the acoustician to establish early on, side and rear wall regions available for redirecting approaching direct sound energy from voices back to the audience with an arrival time within about

30 msec of the direct sound. Most all other reflecting surfaces can be expected to be sound absorptive to control T_{60}.
- Specific location and shaping of reflective elements on both walls and ceiling (typically with reflecting "clouds") to achieve requisite specular reflections are usually determined by ray-tracing studies on paper. To be effective for speech wavelengths, surface dimensions of such elements should generally be no less than 4 feet (1.2 m) (expect a negligible benefit under 2 feet [0.6 m]) and expose to the stage, this same minimum subtended dimension, when considering angles of approach of a "sound ray" from the voice to the reflector. Effective wall shaping nearest the stage is often unavailable below 8 feet (2.4 m) above the nearest floor. Considerably more effective wall reflections are available along the rear half of the room as downward-tilted, generally flat surfaces are available to kick energy back to the audience from behind the seating. Strive to achieve a time delay gap of between 15 and 35 msec between the direct arrival and these reflections.

15.II.3.2.2.1.4 Room Volume

- The same issues related to room volume as described for proscenium theaters apply to thrust stage theaters with the notable difference that since thrust stage theaters are far less likely to engage live music and far more likely to require a drier (less reverberant) acoustic, relatively lower volume and higher absorption than proscenium theaters is indicated.
- For the above reason, it's best to keep room volume from growing unnecessarily during design.

15.II.3.2.3 Example: Bistline Theater at Idaho State University

The 450-seat Bistline Theater at Idaho State University in Pocatello, as seen in Figure 15.II.8, epitomizes the benefits of multiple reinforcing reflections for such inherently problematic venues as thrust stage theaters. With its programmed large audience and unusually pronounced thrust stage, nearly all available ceiling and wall surfaces required acoustical engagement reflecting for good speech intelligibility to be realized throughout, irrespective of audience seat and actors' speaking direction.

15.II.3.2.3.1 Design Attributes and Features

Because of the usually wide fan shape required to accommodate audience seating around a deep thrust stage, the cheekwall areas offered minimal opportunity to develop early, useful reflections. To force would require aggressive and much-considered shaping on these surfaces were they to be engaged. For such a study, a rough foamcore model using light-reflecting foil applied to the sound-reflecting regions under examination might be developed at perhaps 3/16″–1/4″ scale and a laser pen engaged to check on the spatial distribution of early energy from the actors' head locations. A computer model (e.g., *CATT*-Acoustic, *EASE*, etc.) could be engaged as well—and provide more direct information on reflection time arrivals—but adjustments and studies using these, while helpful for refinement, are often excessively time-consuming for preliminary "quick testing" of early forms. If these widely splayed cheekwall surfaces are not engaged either as first-order reflectors to the audience, or in combination with ceiling reflectors for energy within about 50 msec of the direct-arrival, it can be best to absorb their upward-trending energy to minimize room reverberance. For this room and for a variety of reasons, a paper examination (only) in three dimensions was undertaken in lieu of such modeling.

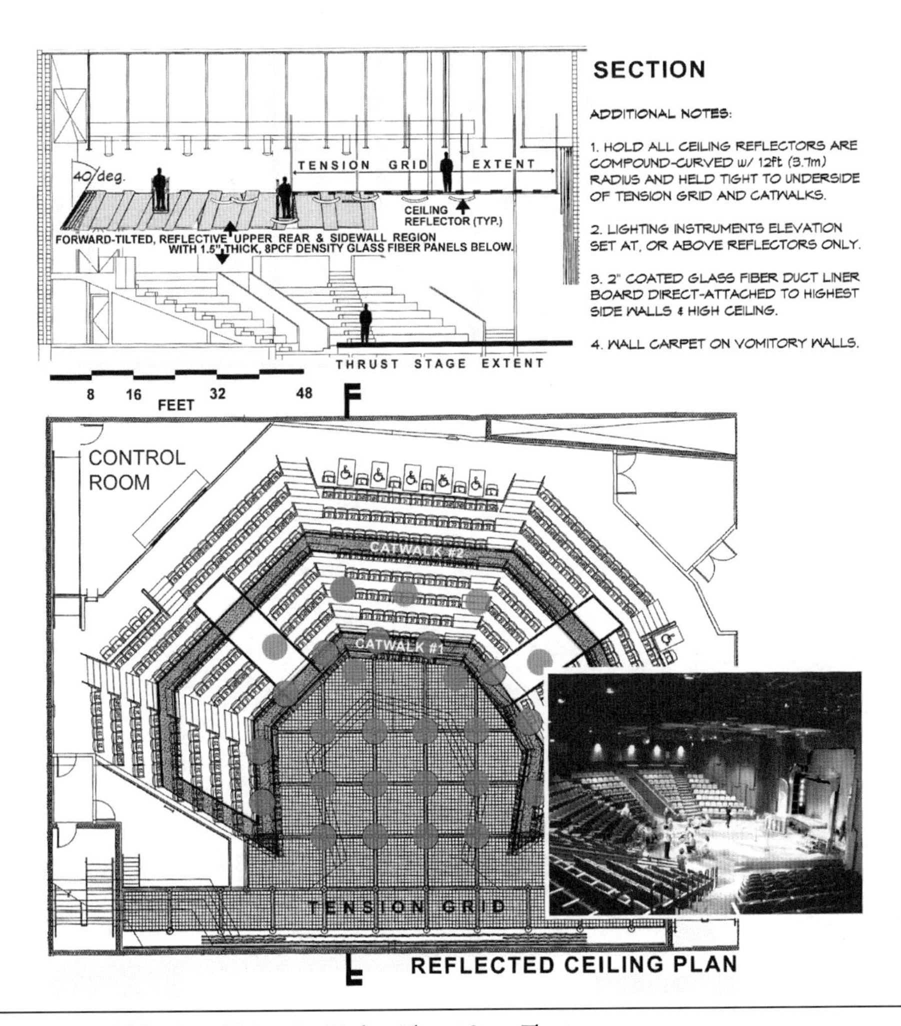

Figure 15.II.8 Idaho State University Bistline Thrust Stage Theater.

Upper portions of side and rear walls provide excellent opportunities to kick energy back (down from above) to the audience as long as the time delay difference between this energy and first arrivals does not exceed about 55 msec. Often, these wall regions require only some downward tilting to provide this useful energy and the specifics of where to tilt and at what angle is quickly assessed via section (2-D or model) studies. Here, the upper 6-foot segment of these walls was tilted at 45 degrees from vertical and thereby capture about 75% of the audience with specular reflections within 25 msec of the direct arrival. For a modest bit of added high-frequency reflection and diffusion, we panelized these into segments and offset adjacent panels by about 6″ (150 mm). Ceiling reflections here were achieved via twenty-eight compound curved (12 feet [3.7 m] radius), convex-downward reflectors of 5 feet (1.5 m). While some of these clearly interact a bit with the generally unshaped cheekwalls, most provide at least one good reflection (albeit each about 6 dB down with respect to a flat reflector [refer to Section 15.II.2.4]) to about 80% of the audience. Of course because of their number, most of the audience is getting at least 10 good reflections with slight time-delay offsets thus generating an effective "gain" approaching 10logN, where N represents the number of such time-clustered reflections to a patron. Additionally, the plethora of these, and especially their doubly-curved form, presents an optimal "reflective ceiling plane", effective for all possible actor positions and orientations with respect to the wide-wrapping audience.

Some cautionary notes are in order. Excessive packing of reflectors into a ceiling risks developing excessive reverberance for drama spaces. It is best to design toward a prudent number of them to achieve just the ceiling reflectivity desired, while permitting upward of 60% of the approaching energy to waft into the (usual) ductwork realm for subsequent broad scattering and absorption. Coordination of ceiling reflectors with ductwork, pipe grids, tension grids, and lighting instruments is key. To be effective, any properly sized reflector requires fully clear line-of-sight to both source and receiver. Avoid intervening, acoustically-opaque items of even a nominal 8″ (200 mm) dimension are to be avoided. Insufficient diligence on this matter results in an effective loss of reflection benefit. Energy will simply be randomly scattered. Thus, reflectors at or below the lighting plane of a tension or pipe grid are useful, and those above simply are not. Finally, the reflectors should be affixed so as to preclude staff from easily demounting them when they would prefer to substitute a lighting opportunity for a reflection opportunity. Permitting reflector relocation or removal carries the high probability of excessive loss of beneficial voice-augmenting reflections over time.

15.II.3.3 Experimental Theaters

15.II.3.3.1 Characteristic Elements

Within the stakeholders' framework of the wide-ranging sorts of anticipated productions, the major theater features are programmed and designed. When first developed, these spaces were principally spaces for the spoken word. Over time, ever more inventive "experimentation" arose involving increased use of lighting, sound and video technologies. In academic environments, with decreasing budgets for space allocation, these spaces are increasingly programmed to stray even beyond the highly inventive production available by audiovisual technologies into the realm of acoustic music rehearsal and performance. Such spaces, normally best treated as especially "dry" acoustically, to serve as suitable spaces for choral works! The design guidance provided here, however, will address the more conventionally programmed uses. In any case, given the usual highly flexible staging and layouts desired, the most common design features include a uniform "ceiling" plane on which to mount lighting instruments (catwalks, tension grid, or at least a pipe grid), a flat floor, multiple actor entry points, and one large door for loading sets, etc. When the theater is required to serve a few specific functions other than drama alone—such as music recording, rehearsal, dance, or receptions—a tendency will arise to ascribe at least one end as the most common area at which to establish bleacher or tiered seating. This, in turn, would inform optimal locations for audience entry and a control room. For flexibility in actor entry/exit, the space is commonly provided on two or three sides with doors to corridors or curtain openings linking them. Such perimeter curtains readily lend themselves to providing heavy acoustical drapery for additional "lower-wall" sound absorption near the audience.

15.II.3.3.2 Design Principles for Experimental Theaters

15.II.3.3.2.1 General Design Guidelines

Because these spaces can usually be configured in "arena" or "theater-in-the-round" forms, the very same principles and design issues arise as described for thrust stage theaters. Drama needs of such spaces are those that might be expected for a room seating between about 90 and 400 in the audience, in most any type of seating/staging aspect imaginable (proscenium, deep thrust, arena, etc.), although often (and this should be identified early in design), some specific seating arrangements are far more often to be set up than others. If drama alone fully described the room's programmatic needs, the following list of attributes could arguably be deemed sufficient:

- Generally sound absorptive on most, but not all walls; floor reflective
- Ceiling partially absorptive, partially reflective to avoid too dry of a sound
- Exposed supply air ductwork below an exposed roof/ceiling deck
- Sound and light locks at multiple perimeter points for both actor and audience entry/exit
- Theatrical velour drapery for audience/actor access and circulation from near floor to at least 10 feet above the floor along at least three walls. If this would serve a sound-absorption purpose, it's best in 100% folds and of a minimum 26 oz/yd (0.8kg/m) weight.
- Control room with windows or carved-out space, at least, at elevation above audience
- Some opportunities to support voice via reflection off of suitable elements around at least three sides of the room—sometimes at the ceiling, depending on room size and programming specifics.

The singular acoustical benefit that arises in designing the experimental theaters is that the vast majority of these seat less than 200 and any configuration of audience proximity to actors overcomes many of the challenges arising with larger audiences. Lighting design issues related to ceiling acoustical design issues are nearly the same as for thrust stage theaters only yet more flexibility is required.

15.II.3.3.2.1.1 Design Goals

Reasonable starting point descriptor criteria ranges follow:

- $0.5 < T_{60} < 1.6$ seconds, occupied, if variable to accommodate acoustic music
- $0.6 < T_{60} < 1.0$ seconds, occupied, if fixed
- C_{50}: ≥ 5.0 or AI ≥ 0.55
- NC/RC/PNC range: 22–27

15.II.3.3.2.1.2 Room Finishes

- Because these venues require ready access to lighting instruments in a fairly high "ceiling" plane (typically 18–24 feet [5.5–7.3 m] above floor level), there is commonly a defined lower volume (below lighting instruments) and upper volume (technical gallery, etc.). Walls up to at least 12 feet (3.7 m) require attention to abuse resistance, while attending to suitably high sound absorption (typically NRC 80, minimum) while above this, if the ceiling is made sufficiently absorptive, the walls can be treated less aggressively (say, NRC 55) or left untreated if budgets are tight or programmatic needs suggest. For example, to the extent the room has an important acoustic music program, the T_{60} should be able to climb higher than the range indicated above and wall shaping for diffuse reflections may be considered. It is best to assess early in the design, the several most likely, prospective seating arrangements and let this inform the acoustic design. Often, a great majority of performances will establish relatively steeply raked bleacher seating at one identified room end so this can inform optimal shaping for early wall reflections. Of course, this also establishes the fact that sound absorption along the wall behind the audience may not be an effective candidate for investing in permanent sound absorptive material. Among suitable fixed absorptive wall finishes are direct-applied 2″ (51 mm) thick duct liner board (3 pcf [48 kg/m³] density), protected at its face, where required, with Tectum™ or 23% (min.) open perforated metal or hardboard (1/8″ [3.2 mm] minimum holes) or shop-painted hardware cloth. Where curtains are indicated for dramatic purposes, these can be engaged with substantial acoustical absorption benefit as long as their track resides at least 6″ (150 mm) from the nearest reflective surface and the material provides sufficient "flow resistivity". That is, if the drapery material is 32 oz/yd (1 kg/m) theatrical velour and, when deployed is in 100% folds—or is double-sewn with a 2″ (5 mm), minimum, airspace. To realize the full range of T_{60} adjustment, all variable acoustic material should store out of the room's acoustic volume into pockets.

- Ceilings are either mostly, or entirely, sound absorptive and normally are essentially the roof assembly underside. Where the program deems acoustic music important, clean reflections from the ceiling are clearly important and this may be handled by some combination of scattered absorption (about 4′ × 4′ [1.2 m × 1.2 m] patches of absorptive material distributed among similarly-sized patches of reflective finish) and/or an array of compound-curved reflectors minimally 4.5′ × 4.5′ (1.4 m × 1.4 m) suspended just below the lighting instrument plane. Thus, like thrust stages, these ceilings (above reflectors) become strong candidates to tune the room T_{60}. A suitable absorptive finish would be the same as described for the walls. Apart from the obvious acoustical benefits of providing doubly-curved ceiling reflectors for acoustic music use of such spaces, the benefit for the spoken word cannot be discounted, especially when the room is used in anything other than a proscenium-like mode.
- Floors are inevitably hard and sound-reflective and nearly always sprung, on wood sleepers. Achieving resilience for movement via resilient mats alone is ill-advised if sound reflections off the floor would be deemed important—and they usually are.
- Seating is commonly simple folding chairs variously on the flat floor and risers. Expect padded seat bottoms and probably padded seat backs. Take some care to ensure, with the theater consultant, that these do not become noise-generators on their own, and that the risers are suitably quiet (minimal creaking, etc.) themselves.

15.II.3.3.2.1.3 Room Shaping

These rooms are normally highly rectilinear so in the interest of minimizing the likelihood of low-frequency standing wave resonances when music is played, it is always best to avoid square shapes in your plan and/or section—more rectangular than square is far preferred. The smaller the space (e.g., under a 99-seat audience), the more critical this can become. If this develops as a concern, a room modes calculation should be performed and assessed via the Bonello Criterion, (see Reference 11) or the like.

15.II.3.3.2.1.4 Room Volume

The same issues related to room volume as described for proscenium and thrust stage theaters apply to experimental theaters with the notable difference that if acoustic music is not an important programmatic function, the T_{60} should tend to the lower range, thus driving the volume to a minimum value, dictated only by ductwork and lighting instrument access requirements.

15.II.3.3.3 Example: South Mountain Community College Studio Theater

This 100-seat room in Phoenix is representative of a growing trend on college campuses toward highly multifunctional spaces. The program here was to accommodate the usual studio/experimental/black box theater functions while serving as both a fine music rehearsal and music recording space as well. Judicious application of variable acoustic technologies can permit a successful marriage of these fairly disparate program functions but not without a steadfast commitment to doing all that's required acoustically. Because drama functions are the focus of this chapter unit, this discussion will touch, but not elaborate on, the music acoustic aspects of this space (see Figure 15.II.9).

15.II.3.3.3.1 Design Attributes

In addition to the usual required attention to low background noise, sound isolation, appropriate reflections, and reverberation control for speech intelligibility, the following attributes for this space were necessarily considered as well, particularly because of their influence on music requirements:

1. Room volume (somewhat more important parameter for music than speech)
2. Room ceiling height (somewhat more important parameter for music than speech)
3. Audience seating count and configuration(s) (influence items 4–7)
4. Musician count and configuration(s) (influence items 6–7)
5. Control room design and location (closely related to items 3, 4, 6, and 7)
6. Variable acoustic systems (designed to be appropriate for the music genre and if required, for rehearsals, recordings, and performance)

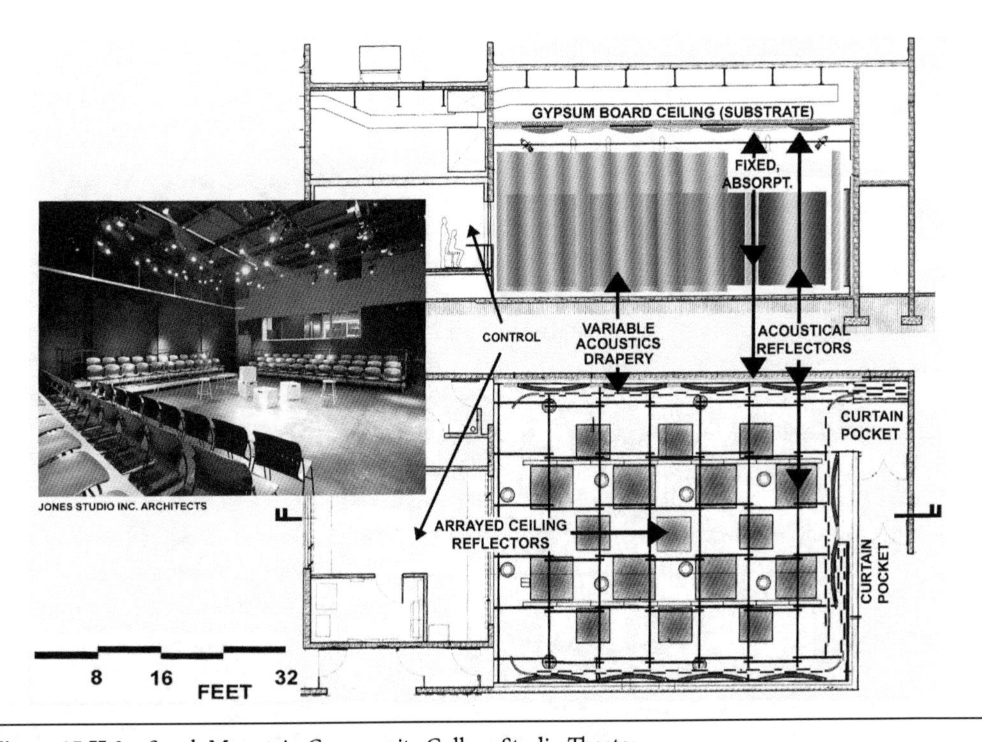

Figure 15.II.9 South Mountain Community College Studio Theater.

7. Coordination of acoustical with theatrical systems (especially lighting and rigging, if any)

15.II.3.3.3.2 Design Features

Being a rather multifunctioning space, its features stray from a traditional drama black box. Similarly, as a music rehearsal and recording room, it strays from traditional design solutions for that function alone.

Variable acoustics are provided by mechanized heavy draperies (32 oz/yd [1 kg/m] in 100% folds, drawn from storage pockets) across three walls. 2″ (5 cm) thick, 8 pcf (128 kg/m^3) fabric-wrapped fixed absorptive panels cover much of the fourth wall wherever the control room glazing permits. Fixed absorption via 2″ (5 cm) thick duct liner board is uniformly applied at the ceiling above arrayed, doubly-curved (compound) ceiling reflector panels. The three walls featuring the acoustical drapes each include large, spaced, singly-curved convex reflectors diffusing energy in the horizontal plane. Normally, we would engage at least one horizontal, 10″–12″ (250 mm–305 mm) wide "ledge" along their tops to kick some additional energy back downward from these walls. Between these are 2″ (5 cm) thick glass fiber panels as described for the wall common with the control room. The measured acoustical vitals are, unoccupied: $0.40 < T_{60}$ (sec.) < 0.83; NC 20. The T_{60} range and clarity (not measured) are somewhat less than originally anticipated and desired because the ceiling reflector elevation was raised at the very end of the construction documents phase—at the insistence of drama faculty—to reside above (in lieu of at the same plane as) the pipe grid and all its lighting instruments. This had the effect of substantially reducing important clean mid-high frequency reflections from the doubly curved ceiling reflectors.

15.II.4 CONSIDERATIONS FOR OPERA AND MUSICAL THEATER (Addressing Pits and Eyebrows)

15.II.4.1 Issues of Balance and Communications of Stage Voices with Pit Musicians

Among the challenges regularly facing directors of musical theater and opera regards achieving optimal timing and level balance between stage actors/singers and pit musicians. Often, there exist on both the stage and within the pit, multiple "hot spots" and "dead spots" of such communication, and it is left entirely up to the pit conductor to resolve such challenges. The problem often stems largely from a highly reflective and concave (to musicians) wall toward which the musicians are "aiming" from their realm beneath the stage edge. It is compounded by the fact that to the extent the pit musicians are tucked under the stage, there can arise uncomfortably high sound levels developed in their small, too-reflective volume, whose most-irradiated surface (pit's downstage wall face and/or pit rail) is shaped to focus much of this energy directly back upon the musicians. Consequently, the audience will suffer the aural challenges of this pit/stage level balance struggle, not only because ensemble among musicians' and singers' work is wounded but also because of challenges arising from getting the stage vocal work in a level balance with the orchestra as heard in the audience—from which the conductor is necessarily much removed! Often, the pit orchestra levels overwhelm both the musicians positioned there as well as stage voices, as heard within the audience.

15.II.4.1.1 The Pit Design

In addition to the following discussion, see Section 15.II.3.1.1.4. Next, we cover topics that help resolve issues that were previously described. Presuming the design program requirements establish a capacity along the order of 15 or more musicians for the pit, one must provide for the following:

- Approximately 1/3 of the musicians under the leading edge of the stage, 2/3 open to the hall ceiling, and preferably, a sound-reflecting proscenium eyebrow above.
- Approximately 11 square feet (1 m^2) per musician plus the conductor, as well as sufficient space for anticipated, larger instruments such as percussion, acoustic piano, etc.
- 2–4″ (5–8 mm) thick, abuse-resistant absorption on the pit's upstage wall—thinner for small pit orchestras—more efficient at low frequencies for large orchestras such as for Wagnerian opera and the like. Where applicable for prospective pit access at its upstage wall, heavy theatrical velour in 100% (minimum) folds may suffice.
- A pit ceiling comprised of uniformly varying sound absorption coefficients, at least in mid-to-low frequencies. A good choice could be an array of 2 foot × 2 foot (0.6 m × 0.6 m) lay-in glass fiber ceiling tiles alternating with a similar material with mineral fiber tiles or glass fiber with gypsum board backing.
- Attempt to shape 4 linear feet (1.2 m) of the ceiling region in the transition between the aforementioned treatment to the open hall ceiling with a broadly curved convex-downward reflector help distribute pit sound to the audience.
- A full width sound barrier (e.g., solid pit rail, equal to the stage height above the nearest audience floor) between the musicians and patrons sufficient to mitigate excessive energy to the nearest audience rows. Pit rail design features are reasonably influenced by the acoustician as well as the theater consultant and audio designer (to accommodate "front-fill" loudspeakers).
- The conductor's eyes must easily view not only all the musicians but the principal stage performers as well. This necessitates that his head be approximately at the stage floor elevation so a conductor's box and steps are in order.
- In order to minimize the "hot" and "dead" spots already described, and promote uniform distribution of meaningful acoustic communication signals between the pit and stage, the upstage-facing pit surface should be somewhat diffusive in the horizontal plane, while directing energy upward and backward to the stage. This may entail countering the usual concave surface form in the plane of the musicians with an array of convex surfaces tilted forward at their bottom. The upstage-facing surface of the pit rail barrier, being well above the musicians' plane, need not be so tilted, but will benefit from configuring either in multiple convex (toward upstage) elements or a reflective phase grating (e.g., QRD^{tm}).

15.II.4.1.2 The Eyebrow Design

In order to help "lift" singers' voices over a strong pit orchestra, incorporate a proscenium eyebrow as described for proscenium theaters. As a corollary to this application, the eyebrow design itself can be shaped to provide reflections from the orchestra back downward to the orchestra with no penalty in its service to singers. This latter effect can be especially appreciated by those pit musicians in the frontal 2/3, who have no other reasonably close overhead reflecting surface. Apart from the iterative design adjustments necessitated by co-ordinating the required acoustical attributes of a reflective eyebrow at the proscenium, one should strive to achieve the following attributes insofar as is possible.

- Width—overall, comparable to the proscenium opening width.
- Depth—front-to-back, minimum 6 feet (1.8 m) in the very smallest rooms, out to 20 feet (6 m).
- Height—curves upward toward audience as required, from the proscenium bottom edge opening.
- Composition material—any reflective material such as gypsum board, plywood, MDF, cardboard honeycomb sandwich (as for orch. shells), acrylic, resin, etc. For speech alone, heavy material such as plaster or glass fiber reinforced gypsum (GFRB), or the like, is unnecessary from an acoustical perspective. Unamplified music will benefit by reasonably stiff, dense material, however.
- Parts/pieces—may be monolithic or assembled as an array of elements of 3 feet × 4 feet (0.9 m × 1.2 m) minimum dimensions each. For music, these dimension should be at least 50% larger.
- Curvature—as required from ray-tracing studies, but normally between 15 feet (4.6 m) and 30 feet (9.1 m) radius.
- Tilt Angle—determined in conjunction with curvature, and as required from ray-tracing studies.
- Diffusion—for speech only, fine-scale articulation that could excessively scatter is generally to be avoided. More specular, rather than diffusion or scatter is preferred.
- Integral lighting and loudspeakers—downstage lighting and loudspeaker position opportunities abound for eyebrows and this feature characterizes much of the interest they garner from all design team members. Drama rooms of the sort discussed here, however, present only a modicum of interference with their intended acoustical function.
- Catwalk above—if either the audio designer or theater consultant require maintenance or adjustment access to lighting or loudspeakers, a catwalk directly above an eyebrow to serve this need rarely has an acoustical implication.
- Movable/demountable or not—unless the room were to serve small drama as well as grand opera (in which case the proscenium opening should be unusually tall for dramatic effect), only halls requiring both serious, unamplified music and road shows might dictate an eyebrow that mechanically articulates or is removable as the orchestra would likely require it, but the roadshow lighting and sound equipment would require it to be removed.

15.II.5 CONSIDERATIONS UNIQUE TO NONPROFESSIONAL VENUES (Addressing the Nonprofessional Vocal Effort in Training As Well As Professional Voices)

15.II.5.1 Acoustical Criteria

Contrary to "conventional wisdom", more restrictive acoustical criteria arguably applies to nonprofessional venues where untrained voices abound, than spaces where professional actors are more the norm. This applies to ambient noise as well as scrutiny of reflecting and absorbing surfaces. The descriptor values offered above for T_{60}, C_{50}, AI, and ambient noise remain valid but should not be arbitrarily relaxed under pressure.

15.II.5.2 Value Extraction Considerations

Certainly, construction cost issues arise but those should be clearly identified as drivers of "value engineering" exercises alone, and not tagged along with such arguments as; *"it's not a professional venue—only a school"*. The good acoustician is ever mindful of "value extraction" efforts exacted with well-intended but erroneous arguments within efforts to reduce costs. Physical acoustic and psychoacoustic matters stand alone as considerations in their own right and should be viewed and treated in design discussions as such. In any case, prudent acoustical design always bears with it the burden of being cognizant of relative cost-benefit relationships throughout all design phases. Often, we find in design team exercises a misguided perception that appropriate acoustical conditions must be more expensive to obtain than less good results. Often the opposite arises, and when it does it most commonly derives from architectural predilections of form and finish, rather contrary to the Bauhaus mantra of "form follows function".

15.II.5.3 Specific Additional Design Guidance

15.II.5.3.1 Construction and Finish Materials

Because of the relatively narrow bandwidth of the spoken word, a strong natural bass response of the room is of far less value than for important acoustic music spaces. Thus, if the walls or ceiling regions had acquired in early design unnecessarily heavy mass and stiff construction (CMU, plaster, concrete, etc.), these may be strong candidates for lighter and potentially less costly construction (stud construction with gypsum board, cement board, wood, etc.). Similarly, with respect to sound absorptive materials applied to control problematic long-delayed individual reflections, unless these materials contribute materially to controlling T_{60}, consider simply reshaping/reorienting such surfaces and keeping them hard (i.e., without the absorptive material applied).

15.II.5.3.2 Variable Acoustics

Optimizing T_{60} and on-stage acoustics by "variable acoustics" to meet specific goals may, or may not, be prudent in any one venue depending on multiple factors, not simply the associated construction cost and perceived acoustical value by the client. The decision to use any specific variable acoustic design approach must be balanced against the degree of complexity and time to deploy or store as well as the availability of suitable venue staff to effect the changes. This is a key topic to raise with the stakeholders during programming, as well as later in the design, to ensure expectations are correctly managed. That said, there are few secondary school and college performance venues that don't suffer from the absence of such a feature when we've recommended it. At the least, if such variability is deemed key to the eventual success of the space, accommodations can be made during design to help future implementation. The simplest such tasks include designing for, and possibly providing, drapery tracks and storage boxes during construction, as well as making early decisions about the weight of the drapery and whether it would be manually or mechanically deployed. Increased complexity and cost arises upon considering future use for serious music, in which case, the design should include the eventual room volume desired as well as considering the importance of strong bass response, along with the attendant mass and stiffness implications. One of our value engineering approaches provided for substantial upper volume—more than required for drama—as well as plaster and CMU wall constructions, enabling a fixed/variable acoustic option. At the apparent, but open grillage catwalk/ceiling plane, 2-inch (50 mm) thick, 4 feet (1.2 m) × 8 feet (2.4 m), 3 pcf (48 kg/m^3) glass fiber panels are rapidly deployed by hand, over a 10-minute period yielding a marked adjustment permitting dialing in appropriate response for drama, a string quartet, or an amplified pop show. Eventually, mechanized drapery can be installed.

15.II.5.3.3 Control Room

Ideally, a sound and lighting control room is at an elevation that provides operators a clean view of nearly the entire audience and acting area, as well as the loudspeakers over the proscenium. Practically, at least some portion of the audience will not be easily viewed by board operators due to geometric constraints associated with window sill heights and control board depths. An unusually wide operable window (openable to 12 feet [3.7 m] wide if possible) is best to facilitate easy communication during rehearsals and sound mixer's needs. Accommodation in design is best made to permit construction and outfitting of such a room but often equipment portability, costs, and practical needs argue for simply some space near the audience to be carved out for such operators—at least in the initial years of the venue's operation. This approach normally provides no significant operational challenges and can save considerable, initial construction dollars presuming vertical circulation to the future room is accommodated in the initial design.

15.II.5.3.4 More Audio

Even if "acoustically" unnecessary for most functions, anticipate audio reinforcement as an inevitability and coordinate this with room surfaces and sound control porch (or room) location. Accommodating full-frequency range loudspeaker clusters in eyebrow reflectors especially, can remove (if one is not cautious) excessive area from these beneficial surfaces. Mechanized, articulating eyebrow reflectors are occasionally employed, but usually at an unjustifiable expense and operational complexity on smaller (<500 seats) or inadequately funded projects. Finally, the acoustical consultant should be aware that it is the rare venue that, even with no or minimal need for audio "lift", the sound board operator will frequently have a predilection to provide more gain on his system than ever seems warranted or desired by an audience. This may simply be human nature since an apparent functioning, albeit excessive, "audio presence" may be viewed as job security. Nonetheless, the project stakeholders, especially for high school and collegiate projects, should be carefully advised as to the real-versus-imagined need to provide voice amplification where the acoustician feels its use should be Spartan, at most. Carefully controlled demonstrations of this are advised for technical staff and stakeholders alike at the project's completion.

REFERENCES

1. Marshall, J., "Speech Intelligibility Prediction from Calculated C_{50} values." JASA 98 (5), Nov. 1995.
2. ANSI/ASA S3.5-1997 (R2007). American National Standard Methods for Calculation of the Speech Intelligibility Index.
3. Pearsons, K. S., et al., "Speech Levels in Various Environments," BBN3281 became NTIS Stock#PB-270053, Ctrl: 323721638.
4. Dunn, H. K and Farnsworth, D. W., "Exploration of Pressure Field Around the Human Head During Speech." JASA 10 (2), Jan. 1939.
5. Hidaka, T. and Beranek, L., "Mechanism of sound absorption by seated audience in halls," JASA 110 (5), Pt. 1, Nov. 2001.
6. Hidaka, T., Nishihara, N. and Beranek, L., "Relation of acoustical parameters with and without audiences in concert halls and a simple method for simulating the occupied state," JASA 109 (3), Mar. 2001.
7. Gulsrud, T., "Characteristics of scattered sound from overhead reflector arrays: Measurements at 1:8 and full scale." JASA 117 (2) Pt. 2, 2005.
8. Kryter, K., "Methods of the Calculation and Use of the Articulation Index." JASA 34 (1), Nov. 1962.
9. Dietsch and Kraak, "Ein objecktives Kriterium zur Erfassung von Echostorungen bei Music—und Sprachdarbeietungen." Acustica Vol. 60 (1986).
10. von H. Niese, "Die Messung der Nutzschall—und echogradverteilung zur Beurteilung der Horsamkeit in Raumen." Acustica Vol. 11 (1961).
11. Bonello, O., "A New Criterion for Distribution of Normal Room Modes", J.A.E.S, Vol. 29, 9, Sep. 1981.
12. Hidaka, T. and Beranek, L., "Objective and subjective evaluations of twenty-three opera houses in Europe, Japan, and the Americas." JASA 107 (1), Jan. 2000.

FURTHER READING

1. *Theatres for Drama Performance: Recent Experience in Acoustical Design*, Richard H. Talaske and Richard E. Boner, eds., American Institute of Physics for the Acoustical Society of America, 1986 and similar, from the Acoustical Society of America.
2. *References for selected descriptors in this Section:*

   AI (Articulation Index), SII (Speech Intelligibility Index), STI, RASTI (Speech Transmission Index, Rapid Speech Transmission Index):

   a. Pavlovic, C., "The speech intelligibility index standard and its relationship to the articulation index, and the speech transmission index," JASA 119, (5), 2006.
   b. Also, the previous Reference #2.

   STC (Sound Transmission Class): "*Architectural Acoustics—Principles and Practice*," Cavanaugh, W. and Wilkes, J., Editors, J. Wiley & Sons, Inc., 1999.

   C_{50} (Clarity):

   a. Marshall, L. G., "An acoustics measurement program for evaluating auditorium based on the early-late sound energy ratio." JASA 96, (4) Oct. 1994.
   b. Also, the previous Reference #1 and ANSI Standard S1.11.

   NC (Noise Criterion)/RC (Room Criterion)/PNC (Preferred Noise Criterion):

   a. Tocci, G. C., "Room Noise Criteria—The State of the Art in the Year 2000." Noise/News International, Vol. 8, (3), Sep. 2000.

15 Unit III

Music Education Spaces

K. Anthony Hoover, FASA, INCE Bd. Cert.

15.III.1 INTRODUCTION

Facilities for music education are highly multipurpose, involving teaching and learning, practice and performance, music and speech—for young and old. These facilities must support and even encourage around-the-clock activity from persons who are deeply dedicated to improving their art. These are not ordinary buildings.

It can be difficult to fully clarify various subtleties, major components, and even overall concepts in a manner that all parties can easily understand. Among the goals of this chapter unit is to outline the fundamental acoustic concerns and to summarize the collective experience that has been found to be helpful in the design of successful music education facilities—in a manner that can provide a framework for reaching decisions and for managing expectations.

A successful design requires both science and art. The science of acoustics is instrumental for understanding and quantitative design decisions; much art is involved in the teamwork, expectations, explanations, and cooperation needed for the project to be completed to everyone's satisfaction. Expectations are always very high, and management of expectations can be one of the most important aspects of design of facilities for music education. Acoustics can often seem somewhat mysterious and confusing, due in large part to the complexity of acoustics and to the difficulties in communicating subjective responses and impressions.

A useful organizational principle to simplify and manage discussions is to generally sort acoustical concerns into the following four categories:

- Sound isolation,* both airborne and structure-borne
- Noise and vibration control of mechanical, electrical, and plumbing systems
- Surface finishes and room shaping
- Electroacoustics

These categories apply to all building types, but the ranges of frequencies, levels, variability, and expectations are typically increased for music education facilities. Additionally, occupiable spaces must be highly durable, flexible, and practical.

* "Isolation" is the term used in the USA, having essentially the same meaning as "insulation" in many other countries.

Although these categories can have aspects that occasionally overlap, they should not generally be confused with each other. For example, sound absorptive treatment on walls will not greatly affect the levels of sound transmitted through the walls. Also, a potential overlap is that the materials used for heating, ventilation, and air conditioning (HVAC) ductwork noise control can also be used to minimize the levels of sound that are transmitted through ductwork between rooms; however, HVAC noise control should be considered separately from sound transmission between rooms through ducts.

The fundamentals of these acoustical categories are covered elsewhere. This chapter unit will assume familiarity with the fundamentals, and intends to provide guidance for application for the various topics at hand.

There will be a variety of types of rooms in music education facilities, including larger recital halls, smaller practice rooms, classrooms, libraries, offices, and support spaces. There can be unique emphases related to different musical styles, performance methods, and technologies. There will almost certainly be differences in opinions, expectations, and understandings among the participants. Nevertheless, enthusiasm for overall cooperation can be cultivated by a common understanding of the acoustical concerns and goals, resulting in wonderful facilities that will be nurturing homes for teachers and students alike.

15.III.2 GENERAL ISOLATION CONCERNS

Music involves the entire range of audible frequencies. Furthermore, musical tones and discrete pitches are more easily identifiable and are more easily heard than broadband noise such as HVAC sound. Rhythm and tempo can further increase recognition and audibility of transmitted sound, and sometimes annoyance as well.

Large instruments, such as pianos and timpani, generate high levels of vibration that are transferred directly into the building structure, and propagate easily for considerable distances as structure-borne noise. After all, sound energy propagates more easily and quickly through solids than through air. As a rule of thumb, large instruments and loud levels of deep bass generate significant structure-borne noise transmission.

Structure-borne noise is most efficiently controlled by blocking at the source using vibration isolation techniques, such as floating floors, split wall constructions, and resiliently suspended ceilings. Most of the time, rooms designed for structure-borne noise isolation will inherently provide sufficient airborne sound isolation.

Some rooms, such as primary recital halls and recording studios, should be designed for virtually no audible sound transmission into them. Other rooms, such as critical listening rooms and music laboratories, can often tolerate faintly audible transmitted sounds.

Perhaps surprisingly, many types of spaces in music education facilities can function quite well with significant and audible levels of sound transmission. For example, audible transmission between practice rooms is often acceptable, provided that the levels and especially the tempo of the transmitted music are significantly quieter than the sound levels generated by one's own music rehearsal.

Overall layout of the facility is an important first step for proper design. Careful planning for proper usage can accommodate a layout that improves the potential for sound isolation. For example, grouping smaller practice rooms together makes sense for access and usability as well as for efficient sound isolation. Conversely, scattering practice rooms among classrooms or offices usually presents a complex isolation problem.

An efficient approach, which is generally best for ergonomics and for efficient sound isolation, is as follows:

- Identify those rooms which must be highly isolated and which must have quiet levels of background noise. Position those rooms as far as possible from mechanical rooms and other loud spaces. Design those rooms for isolated "box-in-box" constructions, including floors, walls, and ceilings.
- Position noisy service rooms, such as mechanical rooms, shops, and loading docks, away from the most noise sensitive rooms as much as possible. However, since some adjacencies cannot be avoided, proper isolated constructions and attention to entrances, duct penetrations, and other flanking paths are especially important.
- Identify rooms with similar usage, such as music practice rooms and teaching offices, and group them together. Often, rooms expecting similar usage can tolerate more transmitted sound, including clearly audible sounds, provided that most tasks are not unduly interrupted. Of course, such groupings are also ergonomically practical. Such suites can be additionally isolated from other areas with normal building provisions, such as corridors and expansion joints.
- Determine appropriate sound isolation goals within suites whose rooms will have similar usage. For example, administrative offices may have several rooms in which speech privacy is of high concern, but will typically require less sound isolation than music rooms. Additionally, if various music departments cannot be segregated, then louder rooms such as for percussion may require greater isolation, especially if they need to be located next to rooms with quieter instruments like woodwinds.

There will be exceptions and variations, but the overall approach can be summarized as separating the most sensitive from the loudest spaces, and grouping spaces that serve similar usage.

Several major cautions are as follows:

- Wood frame construction cannot be expected to provide the same levels of isolation as buildings with steel and concrete structures.
- New construction can almost always be designed for greater sound isolation at less cost than partial renovations.
- Sound-isolation constructions are usually heavier, thicker, and more expensive than more ordinary constructions.
- Mechanical systems and other services require attention to sound isolation integrity.
- Windows and doors should be carefully chosen for proper balance of isolation, aesthetics, and cost.

15.III.3 GENERAL MECHANICAL SYSTEM NOISE AND VIBRATION CONCERNS

The HVAC system is the primary focus for noise and vibration control, although plumbing, elevator, electrical, and lighting systems should be addressed.

Primary spaces, such as recital halls and recording studios, should be designed for very quiet levels of background noise. Typically, only those few rooms in which audio recording would regularly occur need to be designed for extreme quiet. Conversely, many other spaces greatly benefit from appropriate but perceptible levels of constant background sound, such as from HVAC operation. This is because constant broadband noise, such as from HVAC operation, will help to obscure or "mask" sound that is transmitted from other spaces.

In fact, most spaces throughout music facilities, other than those used for audio recording, perform quite well with levels of HVAC background noise such as might be expected in quality office buildings. However, the ductwork systems should be designed to prevent excessive ductborne transmission between rooms to a higher degree than in most office buildings, which in turn provides a quieter overall HVAC system.

Crosstalk between rooms should be addressed, so that sounds transmitted through the ductwork are quieter than the sounds transmitted via the typical airborne or even structure-borne paths. HVAC noise control measures, including both duct silencers and internal sound-absorptive duct lining, are effective at minimizing crosstalk.

Typically, duct silencers are needed to minimize and attenuate low-frequency sounds, especially from fans and air handling units. However, duct silencers also typically increase resistance to airflow, which can be problematic at the ends of the duct runs near rooms being served. As a result, internal duct lining is generally the most appropriate treatment for reducing crosstalk between rooms. Note that many fan coil units and terminal boxes can be provided with short "attenuators", which are actually 3 to 5 feet (0.9 to 1.5 m) long pieces of internally lined ductwork. Similarly, "acoustical flex ducts" are 5 to 7 feet long (1.5 to 2.5 m) round ducts with sound-absorptive lining that are typically attached to supply-air diffusers and return grilles. Such attenuators and acoustical flex duct are also effective at minimizing crosstalk, as would be expected.

A common method to address crosstalk is to run the main ductwork, both supply and return, through corridors, and with branches into individual rooms (see Figure 15.III.1). This is highly desirable, especially to minimize low-frequency break-in and break-out concerns that may compromise attempts to minimize crosstalk. Note that bare elbows without lining,

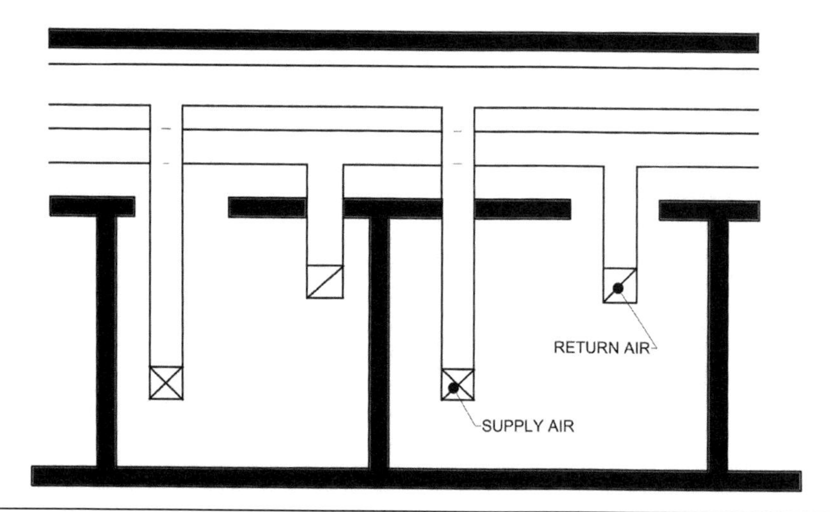

Figure 15.III.1 Ductwork layout to minimize crosstalk.

despite many misconceptions, do not provide significant attenuation. Although the plenum space above corridors is at a premium, this approach of running main ductwork outside of rooms served can often be accommodated with proper planning.

Integral control dampers located within supply diffusers and return grills are unnecessary in most of the rooms served, and with proper system design and proper balancing are generally unneeded. Furthermore, such integral control dampers have significant potential to generate airflow noise, and should be avoided.

Supply diffusers and return grilles are often provided with noise criteria (NC) ratings that are estimated under favorable operating conditions. Note that the assumptions used to convert from the operating sound power levels to the estimated NC ratings in rooms are typically "optimistic" by 5 to 10 NC points, especially in smaller rooms. Reputable manufacturers can sometimes provide sound power level data upon which their NC ratings have been estimated. Use of sound power level data is greatly preferred to the NC ratings, especially in sensitive rooms; on the other hand, selecting diffusers and grilles at 5 to 10 points quieter than the NC design goal will usually suffice.

It is not uncommon for major mechanical rooms to be positioned very near noise-sensitive spaces. For example, there are HVAC efficiencies in locating major mechanical rooms adjacent to recital halls or recording studios. Sometimes, judicious planning of corridors, service access, and less-sensitive rooms such as storage can minimize the need for partitions to provide high levels of sound isolation. Especially in situations where mechanical rooms must be adjacent to noise sensitive rooms, duct silencers should be positioned so that they butt into or even penetrate a high-isolation wall, which inherently helps to maintain the overall sound isolation integrity. As always, proper vibration isolation, sound isolation (both airborne and structure-borne), and ductwork noise control are highly important.

Thermal requirements and air quantities for music practice and performance rooms may be different from many other similarly sized rooms. Furthermore, some musical equipment such as amplifiers may represent significant heat loads, not to mention that many musical performances involve higher levels of performer and audience activity, all of which should be considered during the HVAC system design.

Windows and exterior views are of great desirability, as should be expected, especially in offices or other spaces that faculty members will occupy for major portions of the day. Occasionally, selected rooms may also require high sound isolation from exterior noise. In these situations, concerns for operability of windows (operable windows with high sound isolation can be expensive) and a wide range of heat load requirements (natural ventilation versus well-sealed rooms) must be considered. High sound isolation glazing systems are usually very effective thermal insulators.

Design criteria for sound isolation (both airborne and especially structure-borne) and for HVAC noise and vibration control must be carefully considered during the programming and planning of the facility. Post-construction renovations, fixes, and additions will certainly be more expensive and less effective. Details should be carefully scrutinized and checked during construction, in large part because most of these details will be concealed within the finished building fabric and will not be easily accessible for inspection, nor for any corrective measures when problems are identified.

Finish treatments should also be carefully considered during design, although they can more easily be addressed after the building has been occupied than can isolation and HVAC systems.

15.III.4 GENERAL SURFACE TREATMENTS AND SHAPING CONCERNS

Perhaps more than in any other type of building, most rooms in music education facilities serve a variety of purposes and functions. For example, faculty offices must serve as office space, sometimes requiring high degrees of speech privacy such as for counseling, but also must accommodate music rehearsal and music teaching. Proper sound isolation and HVAC noise control are important, and finish treatments need to accommodate all types and levels of music as well as speech.

Single-number ratings for sound absorptivity are most directly related to speech-like sounds. In particular, noise reduction coefficient (NRC) ratings do not address frequencies below 250 Hz, whereas many musical instruments extend into much lower frequencies. On the other hand, common furnishings and soft furniture can provide significant treatment for the lower frequency sounds. Furthermore, open bookcases and similar furnishings are very effective broadband treatments, including for musical sounds.

Very often, decisions for proper finish treatments are well-informed and greatly aided by input from the music faculty, perhaps to an even greater degree than regarding sound isolation and HVAC noise control. Furthermore, music faculty members have typically acquired great experience in other facilities, and are attuned to the qualities of sound that are influenced by finish treatments.

Most spaces are not well served with an abundance of sound absorptive treatment on all surfaces, which would tend toward an extremely "dry" response. On the other hand, there are a few exceptions that prefer a relatively short reverberation response, such as rooms devoted to music synthesis (because reverb, echo, and other responses are added artificially and electronically) or primarily-outdoor music (such as gamelan or many percussive music styles).

Smaller rooms, with occupancies fewer than roughly 12 persons, are often well served by an overall treatment scheme that is fixed and should accommodate a wide range of functions. Some large rehearsal rooms are well served with a variable acoustic treatment, where sound absorptive treatments can be easily and quickly deployed or retracted to satisfy a particular rehearsal requirement. For example, adding sound absorptive treatment helps to reduce overall loudness as well as reverberation, and can help instructors to distinguish those players who may be in need of greater assistance by reducing the overall blending of musical sounds. A common type of variable treatment is thick drapery that can be quickly deployed to cover one or two wall surfaces, or removed into curtain boxes. This is very similar, but in smaller scale, to variable acoustics in music and drama performance spaces.

A useful rule-of-thumb is that for each pair of parallel surfaces (i.e., floor/ceiling or opposing walls), sound-absorptive or sound-diffusive treatments should be applied to significant expanses of at least one surface of the pair of surfaces. The goal is to assure that sound is not reflected repeatedly between parallel surfaces, and treatment on at least one of the pair prevents many repetitions.

As spaces become larger, visual aesthetics tend to become increasingly important. Also, larger spaces lend themselves to small performances and audio recording. Thus, the overall environment becomes increasingly critical, and input from a variety of design team members should be expected, and is usually quite valuable. On the other hand, concerns for sound isolation and HVAC noise control should be deliberately addressed, and usually independently of the discussions concerning finish treatment.

15.III.5 ELECTROACOUSTICS

Most rooms used for any kind of instruction will have, at minimum, a basic sound system. These systems typically include a pair of speakers for stereo playback, and a wall-mounted rack of equipment that includes several formats for audio playback, basic signal processing such as equalization and sometimes reverberation, and amplifiers with a loudness control.

The equipment rack should be located for easy access by the instructor. The rack should be provided with security measures, such as a lockable front panel, and will need ventilation louvers. Cable runs, secure mounting of rack and loudspeakers, general wiring, power supply, and access must all be considered.

Each sound system should be designed for the intended usage of the particular room. For example, ensemble rooms and other rooms for live music require that the loudspeakers be sufficiently powerful for use as a sound reinforcement system. This would also imply that the equipment rack should include an appropriate mixing panel that can accommodate individual control of microphones and other inputs to the sound reinforcement system.

Smaller stereo loudspeakers are typically up to 2 cubic feet (0.06 m^3) in size, similar to home stereo systems. Sound reinforcement loudspeakers are significantly larger and heavier, in the range of up to 2 feet (0.6 m) wide, 2 feet (0.6 m) deep, and 3 feet (0.9 m) tall. Equipment racks are slightly over 2 feet (0.6 m) wide by 2 feet (0.6 m) deep, and height depends on the amount of equipment, typically ranging from 2 to 4 feet (0.6 to 1.2 m) tall.

Many schools will have a standard design for their basic systems, including equipment types. Personal preference will also play a large part in determining specific pieces of equipment, especially if variations on the school standard are possible. Larger systems for sound reinforcement may also have a standard design, but should accommodate the intended usages for the specific space.

Some schools also provide portable systems that float from room to room, as needed. These typically include a rolling cart, into which is built the equipment rack, and which carries the loudspeakers. These are very similar to home stereo systems, and so are typically very simple to operate, but can rarely accommodate sound reinforcement systems.

15.III.6 ACOUSTICAL CRITERIA

Single-number criteria are valuable for quickly comparing properties of various materials, constructions, and overall noise levels. There are many single-number values used in acoustical designs and evaluations. Of course, the single number values cannot address all of the complexities that may be of concern during design. The most common single-number ratings are well defined elsewhere, and are used throughout this chapter unit with particular notes for concerns that extend beyond the single number ratings, especially as they apply to music education facilities.

15.III.6.1 Sound Isolation: Sound Transmission Class (STC)

Proper isolation must carefully consider the extended frequency range of music, and concerns for structure-borne transmission.

Sound isolation considerations often begin with STC ratings, which are actually only appropriate for evaluating sound isolation in the range of speech frequencies. Musical sounds extend well beyond the range of speech, especially lower in frequency such as for bass sounds.

Low-frequency sounds are typically the most difficult to attenuate, and can involve greater mass, thicker constructions, and vibration isolation separations. Often, considerations for low-frequency sound isolation dominate the overall isolation design. Furthermore, with proper low-frequency isolation, higher frequencies will generally be addressed adequately, almost by default. Low-frequency isolation also minimizes the sensation of beat and tempo from other musicians, which can otherwise be especially distracting.

Recording studios, recital halls, and rooms that may often be used for audio recording require the highest levels of sound isolation, often including structure-borne isolation. Additionally, very loud rooms, such as those with drum kits, require high levels of sound isolation, including structure-borne isolation, especially those that are to be located within several structural bays of sensitive rooms.

Structure-borne isolation implies isolated box-in-box construction, including floating floors with a minimum three inch (7.6 cm) thick vibration isolators, walls whose inner and outer constructions either are not connected or are designed for a resilient connection, and ceilings whose upper and lower constructions are designed for a resilient connection. Service elements such as ductwork, wiring, and other penetrating elements require appropriate accoutrements.

15.III.6.2 HVAC Noise: NC (Noise Criteria)

Single-number ratings for HVAC and general background noise are covered extensively elsewhere. NC ratings suffice for most planning and initial design stages.

If the quality and characteristics of background noise itself are of concern, such as in recording studios, then the more-restrictive room criteria (RC) ratings can be substituted. RC ratings are similar in many ways to NC ratings, but RC curves reduce allowable levels of lower-frequency and higher-frequency noise. There may be some additional costs associated with achieving an RC rating compared to an NC rating.

Some devices may be provided with NC ratings at various operating points; a common example is supply-air diffusers. However, these NC ratings, which are based on the spectrum of octave band data, make simplifying assumptions that can be especially optimistic for smaller spaces with diffusers that will be close to listeners. A popular rule of thumb for such devices in smaller rooms is to add 5 to 10 points to the provided NC ratings, to account for more significant buildup of sound in smaller rooms and the inherent closer proximity to listeners.

15.III.6.3 Surface Treatments: NRC (Noise Reduction Coefficient)

For most music education facilities, NRC ratings as recommended by experienced acousticians may suffice for most of the planning and initial design stages; subsequently, detailed designs will be developed beyond the single number ratings. The most common enhancement is to increase the thickness of the absorptive material to several inches (7.6 cm or more) in order to improve low-frequency absorption.

Diffusion or scattering of sound is very desirable in most rooms for music. There are many advances in diffusion, and experienced acousticians will be best able to assist. An important characteristic is that significant size is needed for the "bumpiness" of diffusers. Open bookcases and open-faced instrument storage cabinets are of an excellent size and bumpiness, albeit difficult to quantify.

15.III.7 TYPES OF SPACES

15.III.7.1 Classrooms

Acoustical criteria for schools are covered in depth in the American National Standards Institute (ANSI), Inc. ANSI Standard S12.6 *(Acoustical Performance Criteria, Design Requirements, and Guidelines for Schools)*, which is a valuable resource, even though this standard is intended for typical schoolhouses with traditional lecture formats, and which may include a band room. However, music will heighten acoustical concerns, especially for sound isolation, and especially with loud or live music. Furthermore, most classrooms in music education facilities include music playback systems, and there may also be a desire for occasional live music presentation. The emphasis on music instruction and performance tend to go beyond the primary intent of the ANSI Standard.

Acoustical criteria for traditional lecture-format classrooms as covered in the ANSI standard suggest a minimum STC 50 for partitions, reverberation time of 0.7 secs, and background noise levels of up to NC 35.

For classrooms that would be adjacent to other occupied spaces and that can regularly expect moderate levels of music playback, sound isolation should be increased to STC 60. For classrooms that will be regularly used for live music presentation—including pianos, which can generate very significant levels of structure-borne vibration—consideration should be given to isolated room constructions.

A common approach is to improve sound isolation to STC 60 for classrooms with moderate music playback, and to select several primary classrooms for live music presentation, which would be provided with isolated room construction.

In addition, the primary classrooms for live music instruction may also increase detailing of some of the finish treatments, including slightly splayed walls and (partially sound-reflective) ceiling near the teaching/instrument areas to properly redirect sound, and increased ceiling height, especially if seating is tiered. Additionally, since some of the ceiling might have designed sound-reflective surfaces over the performance area, sound-absorbing wall treatments should be enhanced; a common approach is to place several 4 feet (1.2 m) wide by 8 feet (2.4 m) tall (very commonly available size) sound absorbing wall panels on sidewalls, installed individually with at least several feet of spacing of exposed wall between them (see Figure 15.III.2), and with somewhat greater density on the rear wall. Note that some of these classrooms take on some of the characteristics of small performance spaces, and may benefit from many of the concepts discussed in great detail in the music performance chapter.

Common criteria:

- Minimum STC 50, increased to STC 60 for regular music playback; selected classrooms may warrant isolated room construction
- Maximum NC 35, NC 25 to 30 for selected classrooms
- Acoustic ceiling tile system rated NRC 0.70; addition of scattered wall panels rated NRC 0.80 for larger classrooms; selected classrooms may warrant splaying of sound-reflective walls and ceilings in teaching/instrument areas

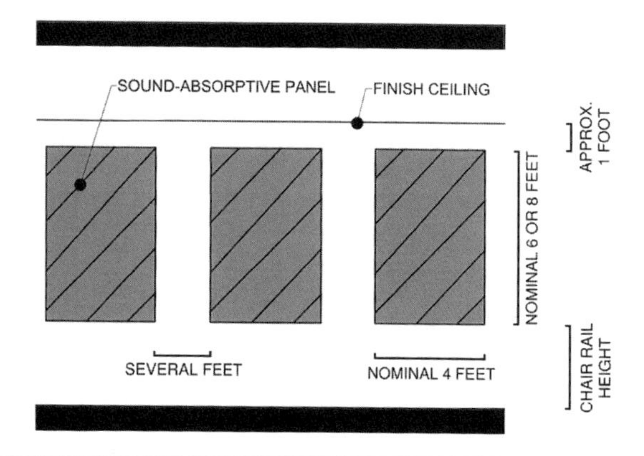

Figure 15.III.2 Sound-absorptive treatment using distributed wall panels.

15.III.7.2 Faculty Offices

Most faculty offices must serve all of the typical office functions, but must additionally serve as music instruction and rehearsal spaces. Most faculty offices are either single or double occupancy, and it is common to have associated open-plan areas that may serve as general work or meeting areas, or for part-time faculty. Of course, during music instruction or rehearsal, only one function can be served, and other occupants will relocate to other scheduled activities.

The faculty person's musical instrument is usually housed in the office, and the office must be sized accordingly. Furthermore, for instruments that are larger, louder, and extend to low frequencies, proper sound isolation should be provided accordingly. Whenever possible, it is best to group departments according to the type of musical instruments.

It is common, but with some exceptions, that faculty members of music schools are tolerant of transmitted sounds, especially when departments are grouped according to the type of musical instrument. For example, a suite for percussion faculty will tolerate transmitted percussion sounds more easily than an adjacent woodwind faculty member. However, many facilities do not have numerous faculty members teaching the same musical instrument type, which may raise the expectation for greater sound isolation.

The most common approach for sound isolation is to provide for high levels of airborne sound isolation, such as staggered-stud or separate-stud constructions, but not to provide for floating floor or split slab constructions. This allows for concentration of funds into highly isolated rooms that include floating floors for selected classrooms or other high-use ensemble rooms, especially those that are relatively near offices, libraries, or other quieter rooms. Music faculty may be somewhat tolerant of transmission of structure-borne sounds, especially if they have the opportunity to spend significant time in other areas of the facility.

Carpeting is highly desirable to minimize footfall and typical office impact noises, but will have almost no benefit for isolation of percussion instruments or large instruments with deep bass.

HVAC noise can usually approach typical office levels, because the steady sound helps to mask much of the transmitted sound.

Typical furnishings, especially open bookcases and upholstered office furniture, usually provide sufficient treatment. Some sound absorptive treatment should be added to at least one wall where there might be a pair of parallel and otherwise bare walls.

Common criteria:

- STC 55 minimum, STC 65 preferred; ductwork crosstalk should be controlled
- NC 35
- Typical furnishings and/or sound absorptive treatments on most or all of the wall surfaces
- Carpeting to minimize the need for a sound absorptive ceiling treatment and to minimize typical office floor impact noises, but not for percussion or deep bass isolation

15.III.7.3 Practice Rooms

Most music education facilities will have at least several, if not many dozens, of smaller practice rooms, for use by one or two persons at a time. These rooms can expect intensive around-the-clock usage, and are almost always in very high demand. These rooms will be used in one- or sometimes two-hour increments, with high rotation throughout the day. As a result, controlled scheduling and provisions for security are highly important. Highly durable finishes are required. Individual lock sets and door vision panels are very common.

Since access needs to be controlled, grouping small practice rooms into suites is highly desirable for control and security; this also enhances the potential for isolation from spaces of dissimilar usage.

Audible transmission of sound from adjacent practice rooms is usually acceptable, which is fortunate because high levels of sound can be generated in these rooms. The primary goal for sound isolation is that occupants can practice without repeated disruption, and to allow an occupant to clearly hear their own practice, even if sounds from other spaces are clearly audible during pauses in practice. Note that most small practice rooms have an upright piano, which is among the many types of instrument that can generate high levels of structure-borne sound, such that providing an isolated floor for the entire practice suite may benefit isolation to other areas of the facility. Moderate levels of background sound are actually desirable, providing some masking and thereby enhancing perceived privacy.

A common approach for treatment is to apply sound absorptive wall panels, roughly NRC 0.80, on two adjacent walls (see Figure 15.III.3). Of course, this places sound absorptive treatment on at least one of each pair of parallel surfaces, thereby eliminating the potential for a repeated reflection, such as would result in "flutter" echo. An added benefit is to provide one corner which is lively and reflective, and another corner which is dead and absorptive. Thus, someone who is rehearsing can take advantage of the natural properties of the room to suit their own needs. For example, some players may orient to hear the reflected sound from their instrument more clearly and more loudly, or vice versa, orient with the reflected sound from behind, which may simulate performance in a recital hall.

Another approach for treatment is to use a highly durable sound absorptive finish, with only moderate sound absorptivity, of roughly NRC 0.45, on all wall surfaces. The primary advantage is high durability. Note that high sound absorptivity of greater than NRC 0.50 on all surfaces should be avoided because this would result in an excessively dead room, which is challenging for music practice.

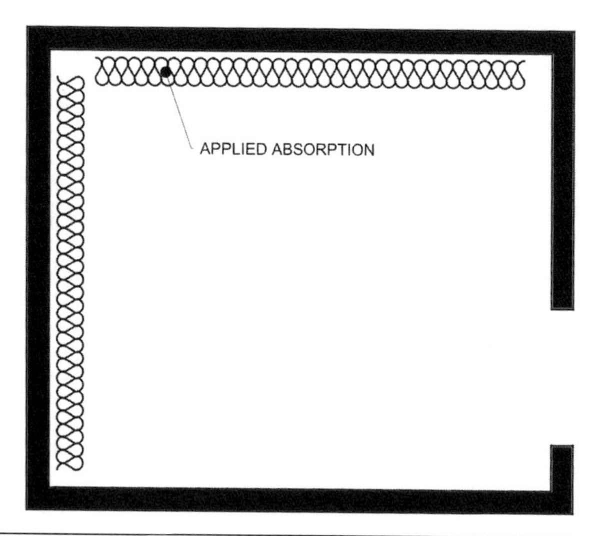

Figure 15.III.3 Sound-absorptive panels on two adjacent walls.

A third approach is to alternate sound absorptive panels with reflective wall surfaces of similar width all the way around the room (see Figure 15.III.4). This would result from applying 4 feet (1.2 m) wide sound absorptive panels of minimum NRC 0.80 spaced apart roughly 4 feet (1.2 m) from each other, thus revealing roughly 4 feet (1.2 m) of hard wall between the absorptive panels. Also, the absorptive panels typically can be applied from chair-rail height up to within a foot of the ceiling. Note that absorptive panels below chair-rail height are more prone to damage, and the added benefit of absorption below chair-rail height is usually minimal. It is best to stagger the treatment so that for parallel walls, an absorptive panel is opposite a hard surface, to minimize flutter echo above chair-rail height.

Prefabricated practice rooms are very common in many music education facilities. These rooms are typically of metal panel assemblies, complete with all provisions for doors, vision

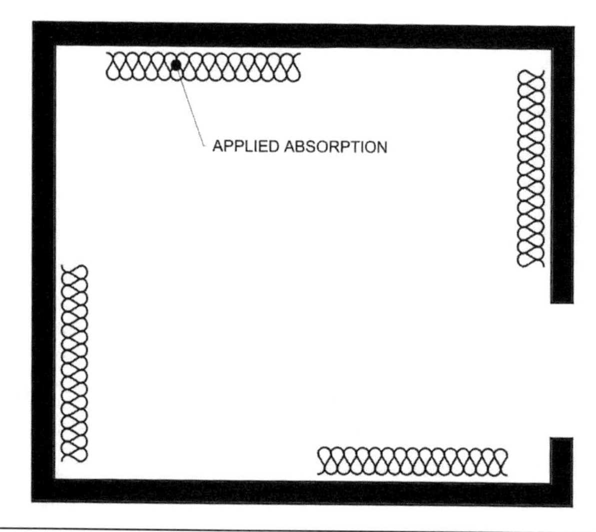

Figure 15.III.4 Sound-absorptive panels distributed on all walls.

panels, absorptive treatment, isolation, and even integral ventilation capabilities. An advantage is that these rooms can be relocated as may be desired as the music facility evolves. Some prefabricated practice rooms can also add electronic reverberation enhancement and simple recording capabilities. When implemented correctly, the reverberation enhancement systems are reported to greatly enhance satisfaction and efficiency of music rehearsal—by both an improved sense of musical quality as well as a sense of real concert performance. Space planning for prefabricated practice rooms is still important, including the need to purchase the proper components for the desired practice room sizes. Purchase costs should be compared to simplification of construction, scheduling, potential for relocation, options for electronic reverberation enhancement and simple recording, and connections to HVAC systems. Prefabricated rooms are usually treated with sound absorptive panels that alternate with hard panels similar to Figure 15.III.4, with desirable results.

Ceilings should be sound absorptive, minimum NRC 0.70, regardless of the wall treatments. Carpeting is desirable, but vinyl is acceptable for durability and minimum maintenance.

Common criteria:

- STC 60 to 65, with floated floors for more critical isolation
- STC 50 to 55 for moderate isolation
- NC 30 to 35
- Several treatments schemes; avoid parallel reflective surfaces, and avoid excessive treatment; ceilings minimum NRC 0.70; carpeting desirable but not essential

15.III.7.4 Ensemble Rooms

Ensemble rooms are rehearsal rooms for groups of musicians, typically three to eight, sometimes more. They are similar to small practice rooms regarding acoustics, but with some concerns related to the larger volume and floor area. Ensemble rooms are typically one-story tall, but a 12 foot (3.6 m) rather than an 8 foot (2.4 m) ceiling height is a desirable improvement.

Ensemble rooms nearly always permanently house basic "back line" equipment including a drum set; a piano; amplifiers for guitar, bass, and electronic keyboards; and a sound reinforcement system with mixing board, effects, and appropriate loudspeakers. These rooms will often generate higher sound levels than the small practice rooms; grouping ensemble rooms near other ensemble rooms and practice rooms is desirable for overall sound isolation efficiency.

Audible transmission of sound from other rooms at controlled levels may be acceptable, masking from HVAC noise is desirable, and floating floors are important. Security is essential, although ensemble nearly always includes the presence of a faculty member.

An ensemble room or two are often located next to a suite of small practice rooms, which is beneficial to overall sound isolation efficiency. Positioning ensemble rooms near noise sensitive rooms should be avoided, even with the best attempts at airborne and even structure-borne isolation.

Sound absorptive treatment nearly always involves distributing sound absorptive wall panels around all wall surfaces, at roughly 30 to 40 percent coverage of the wall surfaces more than 18 inches (46 cm) above the finish floor to within 1 foot (0.3 m) of the ceiling. Thickness of wall panels should be increased to 2″ or 3″ (5 cm or 7.6 cm) for improved low-frequency absorption. Ceilings should be sound absorptive, minimum NRC 0.70. Carpeting is desirable for acoustics and to minimize slipping of some instruments such as drum kits.

Common criteria:

- STC 60 to 65, with floated floors
- NC 30 to 35
- Distributed wall treatments; thicker wall treatments; ceilings minimum NRC 0.70; carpeting

15.III.7.5 Large Rehearsal Rooms

Large rehearsal rooms are usually differentiated into several main categories of usage, including orchestra rehearsal, choral rehearsal, and band rooms. All of these should accommodate at least 50 persons, and often 120 or even more, and should expect to generate loud sound levels. In addition to proper size to accommodate anticipated occupation, there are many factors to consider. Each usage may dictate different details of treatment, and often a single multipurpose rehearsal room must reasonably allow for a variety of usages, via variable acoustics or by a carefully considered overall permanent design. Details and variations are best addressed with an experienced acoustical consultant, but these large rehearsal rooms do share a number of similar requirements and characteristics.

There tend to be two directions for a large rehearsal room design. At one extreme, rooms can be designed to be extremely sound absorptive in order to minimize buildup of sound levels and to begin to simulate the lack of reflections as for an outdoor environment; this style is most applicable for elementary school bands, many high school bands, and most marching bands (since marching bands perform outdoors with virtually no reverberant support). Additionally, this emphasizes the ability to distinguish individual players from each other for more direct instruction by the conductor. In the other direction, large rooms may need to support a reverberant ambience and an overall blending of musical sounds; this style is more appropriate for colleges, professional level, orchestral rehearsal, and rehearsal in preparation for performance.

A very successful approach, especially for rooms that must accommodate a variety of usages, is to design large spaces that can support a reverberant ambience, but are provided with *variable* treatments, consisting of sound absorptive treatments such as heavy velour drapery that can be easily and quickly deployed to minimize buildup of sound energy and reverberation, or retracted from the space, revealing diffusive and reflective surfaces.

A popular scheme begins with a minimum two-story space, with the following treatments:

- **Ceiling**: The ceiling should be of an alternated treatment design, similar to a *checkerboard* layout, wherein panels of sound absorptive treatment alternate with panels of sound reflective and/or diffusive surfaces (see Figure 15.III.5). A budget-conscious solution would be a 2 × 2 ft (0.6 m × 0.6 m) or 2 ft × 4 ft (0.6 m × 1.2 m) grid system with ceiling tiles rated for minimum NRC 0.85 alternating with thin drywall or plywood (chosen to accommodate the grid system requirements). This treatment can be improved by covering the entire ceiling with a thick, highly absorptive treatment that provides an absorption coefficient of 0.9 from 125 Hz upward in frequency (such as 4″ thick glass fiber treatment), and suspending distributed reflector/diffusers of roughly

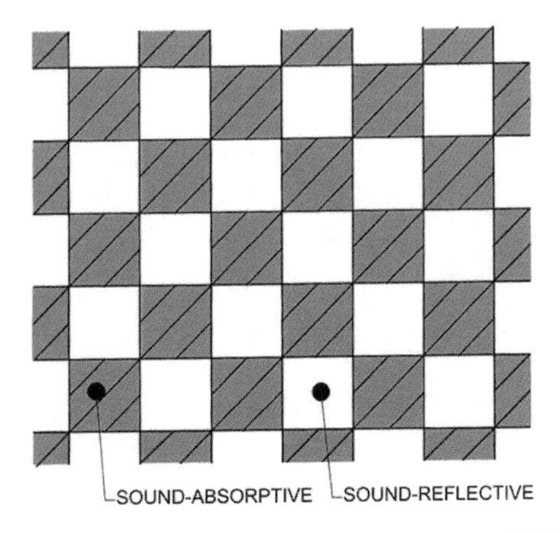

Figure 15.III.5 *Checkerboard* ceiling treatment.

4 ft × 4 ft (1.2 m × 1.2 m) at a foot or two below the treatment, with overall coverage of 35 to 50 percent of the ceiling area.

- **Walls**: The walls should have sound absorptive and/or sound diffusive elements *scattered* to at least 35 percent or more coverage of wall areas above the chair-rail height. Alternatively, at least two adjacent walls should be provided with thick curtains (minimum 24 ounce velour, 32 ounce preferred) that can be drawn across at least two adjacent wall surfaces with 100% fold (i.e., double the area of fabric needed to cover the wall section); these can be deployed or retracted for the desired ambience. Another budget-conscious scheme places 4 ft (1.2 m) wide × 8 ft (2.4 m) high sound absorptive panels (rated for minimum NRC 0.85), alternating with 4 ft (1.2 m) spaces between the panels, all the way around the room; note that 4 ft × 8 ft (1.2 m × 2.4 m) panels are the most common size, and are typically available with finished edges. A very desirable upgrade is to fur these panels two inches (5 cm) out from the wall, and with 2 inch (5 cm) batt insulation placed in the resultant cavity, which enhances low-frequency absorption (see Figure 15.III.6).
- **Floor**: Most floors will be hard, such as with a vinyl finish. Sturdy, portable risers are highly desirable.

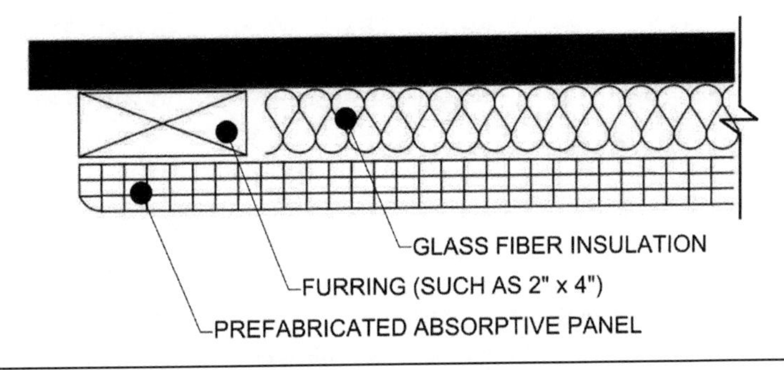

Figure 15.III.6 Panel furred out for enhanced low-frequency absorption.

A reverberant ambience requires at least a two-story tall band room space. Shallower ceiling heights restrict desirable reflections for a reverberant ambience, and prevent proper reflections for cueing between players across the room.

Sound transmission from other spaces into these large rooms is rarely of concern, unless there will be critical recordings. However, these rooms can generate very high levels of airborne and structure-borne sound, especially with percussion instruments or amplified instruments. It is highly desirable to provide an isolated construction, including a floated floor or separated slab construction. Sensitive rooms should not be located near band rooms, and should never be located directly under band rooms.

Moderately quiet background sound levels are typically acceptable, but very quiet levels will be necessary if the rooms are to be used for orchestral or chamber music, for musical styles with quiet and delicate passages, or for critical recordings.

Some large rooms need to also accommodate occasional performances, and should be designed accordingly, with considerations for lighting, seating, audience access, and audio recording.

Common criteria:

- STC 65, with isolated constructions; STC 80 between adjacent band and choral rooms
- NC 30; NC 20 to 25 for sensitive usages
- Common schemes include distributing treatments across wall and ceiling surfaces, as well as variable treatments; hard floors

15.III.7.6 Critical Listening Rooms

Critical listening rooms typically have high-quality sound systems, with at least one pair of stereo loudspeakers. An increasingly popular trend is toward *surround sound* systems with at least five loudspeakers plus a subwoofer. Loudspeakers are almost always freestanding, not installed flush into wall surfaces. Such rooms are often relatively small and with limited occupancy that is further hampered by an abundance of loudspeaker equipment.

Sound levels generated within critical listening rooms are only occasionally of high intensity; furthermore, there is considerable control over the sound levels being generated, since there are no live instruments. The primary concern for sound isolation is general freedom from intrusion from other spaces into critical listening rooms. A common criterion is to provide STC 65 constructions around the room, but provisions for structure-borne isolation are typically not implemented in favor of budgeting for isolating other spaces that may have even higher concerns for structure-borne isolation, such as recording studios.

There is a long-standing custom to splay sidewalls in critical listening rooms in order to lessen problems associated with low-frequency modes. However, such arrangements further reduce floor space, and they do not eliminate problems with modes. Instead, splayed walls will shift the frequencies of the modes somewhat. Thick absorptive treatments (3 inches [7.6 cm] thick or more) with additional *bass traps* can also address these low-frequency problems in a different manner, without the added complexity of splaying sidewalls.

It is very important to avoid room dimensions that are the same, or are integer multiples of each other—for example, avoid 9 ft (2.7 m) height with 18 ft (5.5 m) length (multiple of 2). Dimensions and their multiples should be different by a least a foot or two. There are various opinions about the optimum ratios of length to width to height, any of which will result in fine rooms.

A popular style for a critical listening room treatment is "live end, dead end" (LEDE), primarily for stereo mixing and playback. The live end is behind the listeners; the dead end is in front of the listeners. LEDE emphasizes a highly-absorptive wall (several inches [7.6 cm or more] thick to assure low-frequency absorption) behind the freestanding main stereo loudspeakers, and usually with this treatment extending several feet from that wall along the front sidewalls; the rear sidewalls and the rear wall behind the listeners are reflective, or more properly are diffusive. The general concept is that the loudspeakers do not have significant first reflections from the front of the room, thereby emphasizing the direct sound, and allowing some reverberant buildup from reflections from the other surfaces starting about 15 milliseconds after direct sound arrival at listeners.

Another popular scheme is to provide diffusive reflectors (three inches [7.6 cm] deep or more) alternating with absorptive panels (also three inches [7.6 cm] or more deep) every four feet (1.2 m) or so around all walls from chair rail to near the ceiling. Note that open bookcases and other open storage offer a highly diffusive treatment, although response is not as predictable as with commercially-available diffusers.

For the ceiling, a blend of absorption and diffusion is popular. The ceiling should be sound absorptive (minimum NRC 0.80) to roughly half of the surface area, and can be further optimized with several diffusive elements (nominally 2 ft × 2 ft [0.6 cm × 0.6 cm], two inches [5 cm] deep or more) distributed on the surface. The floor should be carpeted.

Very quiet background sound levels are essential in critical listening rooms. General experience suggests that freedom from constant background sound such as from HVAC systems is even more important than freedom from occasional intrusion of transmitted sounds from other spaces.

As with many other types of space in teaching facilities, personal preferences from faculty and teachers should be accommodated. This is especially true for the choice of loudspeakers and sound system equipment, and about the choice of a LEDE or other room design.

Common criteria:

- STC 65, structure-borne isolation optional
- NC 20
- Finish treatments as discussed above, generally providing diffusive/absorptive treatments on wall surfaces above the chair-rail height; absorptive and diffusive ceiling treatment; carpeting

15.III.7.7 Recording Studios

Goals for recording can vary greatly and will depend on the curriculum, some of which heavily promote recording engineering and recording arts.

Recording studios typically require a minimum of two rooms, one for the performance and microphoning/recording of the instruments (often called the "live" room or the tracking room) and one for the control, manipulation, processing, and engineering of the recorded signals (often called the control room or mixing room).

Most schools and professional studios with a recording program will have at least one live room and one control room. Additionally, there will often be microphone and cable connections to a variety of other spaces that can serve as ad hoc live rooms or performance rooms, such as auditoria, rehearsal rooms, ensemble rooms, or to other tracking rooms.

There will also be a need for several support spaces, including equipment rooms, administrative rooms, vestibules, and storage. Primary acoustical goals for most of the support spaces are relatively minimal, except as might affect sound isolation to live rooms; typically, security is of greater concern.

Tracking rooms share many characteristics with rehearsal rooms because many factors will influence the design, including intended styles of music, budget, space allocation, and ceiling height. A fundamental decision involves the height of the tracking room. In order to develop a proper reverberant ambience for most recording styles, ceiling height must be at least two stories tall, and sometimes higher. This high ceiling allows for a variety of treatments, such as were discussed for rehearsal rooms, including reflection, diffusion, and variable treatments. If the ceiling height is limited to one story, initial time delays will be too short and it is unlikely that satisfactory reverberant ambience can be provided, and recording will tend toward close/direct microphoning, with reverberation, delay, and the other effects added by signal processing. Similarly, smaller ad hoc tracking rooms, such as ensemble rooms, will probably rely on added signal processing for desired ambience.

Tracking rooms require extremely high levels of sound isolation, both airborne and structure-borne; even faintly audible intrusive sounds can easily be picked up by sensitive microphones. Tracking rooms also require extremely quiet levels of background sound from HVAC systems for the same reason. Moreover, microphones can often record unwanted sounds that are not initially heard by the artists or the recording engineers during a session, but instead reveal themselves later on the resultant recording, and can become highly problematic as a result.

Control or mixing rooms are typically much smaller, with many characteristics in common with critical listening rooms, including concerns for different loudspeaker systems such as surround sound. Control rooms are typically 1 to 1½ stories tall, house a great deal of recording equipment, with seating for two or three persons (occasionally larger control rooms add seating for additional involved persons such as producers, observers, students, and additional musicians), and must have sufficient space for easy access to the front and rear because of the variety of recording equipment.

Control rooms often have a large window for direct sight into an adjacent tracking room. The window usually consists of two large panes of glass with several inches (3 cm or more) of air space. One or both panes are tilted to minimize a double visual reflection, to redirect potential acoustic reflections, and to direct light reflections from ceiling fixtures away from people's eyes. Interestingly, the tilted glass has little to do with sound isolation, contrary to popular myth.

LEDE treatment, often with modifications to taste, is very common in control rooms. However, LEDE is made difficult when the large window occupies much of the wall that would otherwise be treated with sound absorptive material. At the very least, this control-room side pane should be tilted to minimize anomalous effects of reflections, and thick drapery should be considered to provide the option to deaden that wall, especially during critical mixing when there is no need for visual connection to the tracking room.

Recording studios are notorious for using "found" spaces for added variety; for example, a guitar amplifier might be placed in a vestibule, thereby providing some isolation and allowing the amplifier to be operated more loudly for certain styles. Live rooms may also house *isolation booths*, which are smaller rooms that allow for greater separation between musicians. Isolation booths are very popular for recording solo vocalists (reducing the level of musical sound heard by the singer and the microphone from other instruments) and also for drummers (who for many styles will be the loudest of the instruments). Isolation booths are typically provided with large windows for maximal visual intimacy and connection; small to moderate levels of sound isolation are usually adequate and acceptable, being limited by the isolation of the glazing system.

Common criteria:

- STC 65 minimum, with isolated constructions
- NC 20 to 25 or RC 20 to 25 maximum, NC 15 or RC 15 for critical rooms
- Final designs of all aspects of recording studios are highly dependent on the style, intention, and goals of the particular curriculum or market, especially regarding finish treatments and surface shaping

15.III.7.8 World Music Rooms

World music refers to a great number of non-Western styles of music and associated instrumentation. Many of these styles were originally performed outdoors. Outdoor performance suggests that any indoor spaces for world music should be quite dead, thus simulating an outdoor environment.

The finish treatment in world music rooms would tend toward highly sound absorptive treatments throughout, including sound absorptive panels on most wall surfaces, and a highly sound absorptive ceiling. On the other hand, some performers have come to prefer a somewhat more responsive or live ambience; such matters of taste should be carefully deliberated.

Many world music rooms are of limited height, typically one story. Significant reverberation is unlikely to develop in shallow rooms, further suggesting treatment toward a relatively dead ambience.

Walls and ceiling should be liberally treated with sound absorptive finishes, including NRC 0.80 panels and ceilings; additionally, treatments should be spaced away from the walls to allow for 2 or 3 inch (5 or 7.6 cm) thick glass fiber insulation backing to further enhance low-frequency sound absorption.

One popular world music style is Gamelan, featuring primarily percussion instruments, most of which rest on the floor and are struck by mallets or hammers. Some Gamelan styles can be quite loud with intensive percussion. Isolated floor constructions and high sound isolation capabilities should be considered. It is common for Gamelan instruments to be permanently set up, ready to play, with instruments covering much of the floor area; as a result, security is of great concern, in part because Gamelan instruments can be extremely delicate and intricate, and easily damaged.

Some styles of world music tend toward a very quiet and delicate music, requiring a very quiet HVAC system. On the other hand, the louder and highly intensive percussive styles may not require the quiet levels of background noise. Since world music implies variety, and since there will certainly be new influences and interests, world music rooms should be designed to accommodate all styles, including both loud and quiet. Also, variable acoustics should be considered.

World music rooms can benefit greatly if they are located on exterior walls; an operable wall could open such world music rooms to the outdoor environment, increasing the audience and the performance size, as well as providing a clear reference to the original outdoor and traditional performances.

Common criteria:

- STC 65, and isolated constructions, especially the floor
- NC 25 to 30
- Minimum NRC 0.80 with added thickness on walls and ceiling; less treatment for added reverberation

15.III.7.9 Libraries

Most music education facilities will include a library, which must house a variety of materials (including sheet music, audio recordings, and books), and which usually have music and AV playback stations. Libraries should avoid significant levels of transmitted sound. On the other hand, steady background sound such as from an HVAC system is usually quite desirable.

Music education libraries should be designed like any other library, but with the added concerns for sound isolation from other spaces, and with concerns for the music and AV playback stations which may require specialized design assistance. There may also be unique requirements for storage of certain types of media and historical documents. Some libraries save floor space using rolling storage; such systems should operate slowly and safely, and should be checked to ensure a minimum of vibration during operation that could generate structure-borne noise, especially if located above or near sound-sensitive spaces.

Consideration should be given to locating libraries on upper floors, since they will generally generate the least sound energy; furthermore, music rooms that are located above libraries will require a floating floor and isolated constructions, and even then, may transmit unacceptable levels of sound into the library.

Common criteria:

- Controlled sound isolation from other spaces
- NC 35 to 40
- Acoustic ceiling tile system, minimum NRC 0.80

15.III.7.10 Lobbies/Atriums

Grand lobbies and large atriums can serve as wonderful ad hoc or pre-function music performance spaces, especially in conjunction with concerts, celebrations, and convocations. Of course, these public spaces must serve their primary functions for ticketing, information, security, access, and traffic.

Background noise should be reasonably quiet for good listening conditions and to promote a comfortable environment. A design goal of NC 30 should be considered, even though this is quieter than typical criteria for public spaces.

Large flat or concave surfaces that might cause harsh reflections should be avoided. Sound absorptive treatments should be selected and located with care to minimize general noisiness without unduly impairing musical performance, and should be concentrated near ticketing and information booths where speech intelligibility is important.

COMMON GROUND FOR DISCUSSION

The various concepts in this chapter are intended to provide general guidance and to stimulate questions about details and specific issues.

Perhaps the most important objective is to offer some common ground for discussion with an experienced acoustical consultant, who can assist with questions, ideas, concerns, and variations, and who can help to navigate through the ocean of acoustical information and misinformation. An experienced acoustical consultant will inherently improve efficiency and value, resulting in the savings of time and money, with managed expectations and predictable outcomes.

ACKNOWLEDGMENT

Thanks to Brandon Cudequest for assistance with figures and final editing.

FURTHER READING

1. Cavanaugh, W. J. and Wilkes, J. A., Editors (Second Edition 2010) *Architectural Acoustics, Principles and Practice*, John Wiley and Sons, Inc.
2. McCue E. R. and Talaske, R. H., Editors (1990) *Acoustical Design of Music Education Facilities*, The Acoustical Society of America.
3. Benson, K. B., Editor (1988) *Audio Engineering Handbook*, McGraw-Hill Book Company.
4. Egan, M. D., Editor (1988) *Architectural Acoustics*, McGraw-Hill Book Company.
5. Hoover, K. A., (1991) *An Appreciation of Acoustics*, KAH Books.

15

Glossary

The reader is directed to the American National Standards Institute (ANSI) Standard S1.1 and the International Standards Organization (ISO) Standard 3382 for greater detail.

1. Basic sound quantities

a) **Frequency** (f), **tone, pitch, and timbre**. Each of these terms is closely associated to the others but in music, the term "frequency" identifies the aspect of a note to its apparent position on a musical scale. Most sounds comprise many frequencies with a few carrying substantially greater sound energy than others. Frequency units are hertz (Hz). Middle 'C' on a piano, for example, is dominated by sound energy at about 262 Hz.
b) **Sound pressurel level** (Lp or SPL). Units are decibels (dB) referenced to 0.0002 microbar. Level differences are also expressed in dB, without reference to 0.0002 microbar.
c) **Sound power level** (Lw or PWL). Units are decibels (dB) referenced to 10^{-12} watt.

2. Noise criteria

a) **PNC criteria** refer to a family of curves plotting Lp versus octave band center frequency (31.5 Hz through 8,000 Hz) developed in 1971. These are similar in application to the more common NC and RC (and curves) described elsewhere in this book. As referenced in this chapter, PNC curves differ from noise criteria (NC) largely by requiring tighter restriction on noise levels in both low and high frequencies than the corresponding NC curves.

3. Sound loudness and strength

a) **Loudness**, as used in this chapter, is a subjective term without abbreviation or technical definition. Technical use is related to the terms "phons" and "sones", not discussed here.
b) **Strength of sound (G)**, as used in this chapter, is determined by measurements in accordance with ISO Standard 3382: 1997 (E). Its value is the difference in dB between Lp measured in a hall at an identified location from a reference sound source on stage and that which would be measured in an anechoic (free field) environment at 10 m distance from the same source. G normally is measured in octave bands over a full frequency range, although often simply reported at middle and low frequencies. It is closely related to "loudness". The sense of feeling enveloped by a sound field appears directly related to the mid-frequency value of G.
c) **ILG fan** (or similar, calibrated, somewhat omnidirectional) reference sound source comprising a small centrifugal fan with multiple forward-curved blades arrayed around its wheel. Such fans are designed and calibrated to serve as a consistent and known sound power level source for comparison with other sources to determine their

sound power levels and also to assess the sound absorption (sabins) in some spaces. They can be used in assessing sound strength as well.

d) **Warmth** is a subjective term associated with apparent strength of bass and a sense of richness in halls for unamplified music.

e) **Bass ratio** (BR) attempts to quantify "warmth" by measuring the reverberation times of a hall in the 125, 250, 500, and 1000 Hz octave bands and then calculating the following: $(T60_{125} + T60_{250})/(T60_{500} + T60_{1000})$. This metric proves not to be an especially reliable indicator of "warmth". Excessive reverberant energy between 125 Hz and 250 Hz can lead to undesirable "boominess".

g) **Relative bass strength** attempts to quantify "warmth" by measuring a hall's strength in the 125, 250, 500, and 1000 Hz octave bands and then calculating the following: $(G_{125} + G_{250}) - (G_{500} + G_{1000})$. This metric also proves not to be a reliable indicator or warmth.

h) **Bass index** is an apparently successful attempt by Beranek (Ref: J. Acoust. Soc. Am. 129, 3020 (2011)), at relating bass strength to subjective assessments of a hall's ability to positively influence a sense of immersion in (unamplified) music sound fields. It is defined simply as $G_{125} - G_{mid}$ where G_{mid} is the average of G_{500} and G_{1000}.

4. Sound arrival times at listeners

a) **Direct sound** is that sound energy that travels directly from an on-stage source to an audience member or performer. It is presumed to travel via the shortest possible path and without reflection(s).

b) **Initial time-delay gap** (ITDG) is the time difference in seconds (milliseconds) between the arrival of direct sound and the first sound reflection from a hall surface at an audience member or performer.

c) **Early sound** is the total of direct and all reflected sound energy of a musical note, say, played by a performer and arriving at a listener within an 80 millisecond time window after direct sound arrival. For speech, the more appropriate timespan is 50 milliseconds.

d) **Late or reverberant sound**, is the total of all reflected sound energy of a musical note, say, played by a performer and arriving at a listener between the conclusion of the note's early sound and the end of the note's audible reverberation (typically between 80 and 2,000 milliseconds in a 2-sec T_{60} hall). The above assume a single played note, not running music.

e) **Echo** is a readily-detected, and long-delayed sound reflection from a distant surface, often one with focusing concave curvature, arriving at a listener at a sound level well above all other sound reflections that may be arriving at about the same time. The listener's perception may correspond to acoustical "stutter".

5. Sound reverberation and absorption

a) **Reverberation time** is abbreviated as T_{60} (often T60, RT, or RT_{60}) and is the time in seconds required for a loud sound to decay by 60 dB after the source is suddenly stopped. It is of great importance as a sound-quality characteristic of halls serving either music or speech and is directly proportional to room volume and inversely proportional to sound absorption within it. Thus, because presence of an audience so often dominates absorption, it is best to identify if a T_{60} value applies to an occupied or unoccupied condition. A T_{60} of 1.1 seconds may be appropriate to a fully occupied drama theater while

2.2 seconds may be appropriate to an occupied hall featuring a Tchaikovsky symphony.

b) **Early decay time** is abbreviated as EDT or T_{10} and is the time in seconds required for a loud sound when suddenly stopped to decay by 10 dB during a typical portion of its early decay (say between −5 and −15 dB). The result is multiplied by 6 to make the metric comparable to standard T_{60} with a 60 dB decay. The argument in favor of EDT is that the early decay may be heard between successive notes of running music, but the long decay of T_{60} may be heard only at the end of a music passage. The authors prefer to evaluate both metrics in parallel, either subjectively or computationally.

c) **Sound absorption coefficients** represent the fraction of sound energy within an identified frequency range that is not reflected to that which arrives from a specific, and well-defined surface. In the USA, it is commonly measured in a reverberation chamber and calculated using the USCS system of units (English feet and inches). The measured coefficients, therefore, are applicable only in calculations of random-incidence (reverberant-field) effects using English units of measure.

d) **Sabin units of sound absorption** typically are used in the USA for audiences and room finish materials using the formula $a = \Sigma S\alpha$, where a = total room absorption in English sabins and $\Sigma S\alpha = S_1\alpha_1 + S_2\alpha_2 + S_n\alpha_n$. S is an audience or material area in square feet and α is the English-unit sound absorption coefficient in the frequency band of interest.

e) **Air absorption of sound** affects only the higher frequencies—i.e., from about 2000 Hz upward. The drier the air and the higher the frequency, the greater is the sound absorption. Air absorption is quantified by 4mV in the Sabine T_{60} equation, where m is the appropriate correction coefficient as a function of frequency and relative humidity and V is the room volume. This can materially influence T_{60} measurements and assessment of a hall's apparent (subjective) "brightness".

6. Music-specific acoustic descriptors (formal descriptions of some of these terms are found in the International Standards Organization [ISO] Standard 3382)

a) **Clarity** (C_{50} or C_{80}) expressed in dB, expresses the ratio of early-arriving to late-arriving sound energy. Where the interest is in speech or singing, the time division between "early" and "late" energy is commonly 50 msec, while for instrumental music it is more often set at 80 msec. Later energy is then all that which arrives after the time division and would extend out to between one and two or three seconds.

b) **Early-to-late sound ratio** or **early-to-reverberant sound ratio** is the Lp of the first 80 milliseconds from the onset, say, of a played musical note less the Lp of the continuing reverberant sound of the same note. A positive value is often associated with good "clarity" or "good definition" of sound; a negative result more likely indicates less clarity, but better "blend" or "richness" plus, often, good "envelopment".

c) **Spaciousness** regards the subjective sense of the degree to which a musical ensemble, for example seems "horizontally-wide" (even up to 180 degrees wide). Several measureable quantities, apparent source width (ASW), lateral fraction (LF) and binaural quality index (BQI) calculated from measured interaural cross-correlation coefficient, have been used to quantify spaciousness. LF does not correlate well with listener perception of spaciousness; BQI does correlate well. Refer to *J. Acoust. Soc. Am.* 104, 255 (1998), and subsequent papers.

d) **Lateral fraction** is abbreviated as LF and is the fraction of the energy arriving within the first 80 milliseconds horizontally approaching a figure-8 microphone (excepting that arriving directly from the source) as measured or modeled from impulse responses. It attempts to provide a measure of spaciousness, but does not correlate as well with listener perceptions as other descriptors.

e) **Binaural quality index** is denoted by BQI and is calculated from measurements of interaural cross-correlation coefficient (IACC). $BQI = 1 - IACC_{avg}$. BQI correlates well with listener perception of spaciousness. Know that $IACC_{avg}$ is that derived from multiple, averaged 80-ms pulses of combined 500, 1000, and 2000 Hz octave-band sound from selected stage to several representative audience positions. Refer to J. Acoust. Soc. Am. 104, 255 (1998).

f) **Ensemble** is a musical term, not an acoustical term. Ensemble may refer to a specific group of musicians or, more pertinent here, to the "togetherness" (with respect to timing) of a group of musicians attempting to play in unison. The acoustician must understand and design the architectural conditions that facilitate good ensemble.

7. Architectural elements

a) **Platform.** A raised performance area set before audience seating that is so configured that all members easily can view all performers and scenic elements. Platforms usually are flat, elevated about 2½ feet (0.77 m) above the nearest audience, provided with easy egress and effective theatrical lighting. Curtains and scenery handling the usually not provided.

b) **Stage.** Is much like a platform but larger in plan area for audience viewing and with peripheral stages for scenery handling and actor-group egress. Usually has multiple sets of curtains and a high stagehouse for scenery handling and storage.

c) **Cheekwall.** That region of wall in a proscenium theater or concert hall that is between the audience and the performers who are closest to them. It performs the acoustically critical function of directing unamplified early-reflected sound (early energy) from the platform or stage to patrons. If correctly designed (angled, curved, and provided with down-kicking protruding ledges), it can do much to provide clarity of sound for patrons throughout much of the audience.

d) **Cloud.** A sound reflector suspended from an auditorium roof, ceiling, or catwalk, or from a stagehouse overhead grid to reflect performer sound in designed patterns to themselves and/or the audience. Usually 6 feet (1.8 m) or more in diameter or width and convex-curved (downward) or otherwise carefully modulated on its underside.

e) **Canopy.** A group of clouds, often supported by a framework suspended via cables from winches above, that provides height and angle adjustability.